The Handbook
of Homogeneous
Hydrogenation

Edited by
Johannes G. de Vries
and Cornelis J. Elsevier

1807–2007 Knowledge for Generations

Each generation has its unique needs and aspirations. When Charles Wiley first opened his small printing shop in lower Manhattan in 1807, it was a generation of boundless potential searching for an identity. And we were there, helping to define a new American literary tradition. Over half a century later, in the midst of the Second Industrial Revolution, it was a generation focused on building the future. Once again, we were there, supplying the critical scientific, technical, and engineering knowledge that helped frame the world. Throughout the 20th Century, and into the new millennium, nations began to reach out beyond their own borders and a new international community was born. Wiley was there, expanding its operations around the world to enable a global exchange of ideas, opinions, and know-how.

For 200 years, Wiley has been an integral part of each generation's journey, enabling the flow of information and understanding necessary to meet their needs and fulfill their aspirations. Today, bold new technologies are changing the way we live and learn. Wiley will be there, providing you the must-have knowledge you need to imagine new worlds, new possibilities, and new opportunities.

Generations come and go, but you can always count on Wiley to provide you the knowledge you need, when and where you need it!

William J. Pesce
President and Chief Executive Officer

Peter Booth Wiley
Chairman of the Board

The Handbook of Homogeneous Hydrogenation

Edited by
Johannes G. de Vries and Cornelis J. Elsevier

Volume 1

WILEY-VCH Verlag GmbH & Co. KGaA

The Editors

Prof. Dr. Johannes G. De Vries
DSM Pharmaceutical Products
Advanced Synthesis, Catalysis, and Development
P.O. Box 18
6160 MD Geleen
The Netherlands

Prof. Dr. Cornelis J. Elsevier
Universiteit van Amsterdam
HIMS
Nieuwe Achtergracht 166
1018 WV Amsterdam
The Netherlands

Library of Congress Card No.: applied for

British Library Cataloguing-in-Publication Data
A catalogue record for this book is available from the British Library.

Bibliographic information published by the Deutsche Nationalbibliothek
The Deutsche Nationalbibliothek lists this publication in the Deutsche Nationalbibliografie; detailed bibliographic data are available in the Internet at http://dnb.d-nb.de.

© 2007 WILEY-VCH Verlag GmbH & Co. KGaA, Weinheim, Germany

Printed in the Federal Republic of Germany
Printed on acid-free paper

Composition   K+V Fotosatz GmbH, Beerfelden
Printing   Betz-Druck GmbH, Darmstadt
Bookbinding   Litges & Dopf Buchbinderei GmbH, Heppenheim

ISBN: 978-3-527-31161-3

Contents

The Handbook of Homogeneous Hydrogenation.
Edited by J.G. de Vries and C.J. Elsevier
Copyright © 2007 WILEY-VCH Verlag GmbH & Co. KGaA, Weinheim
ISBN: 978-3-527-31161-3

Foreword

Homogeneous hydrogenation of organic compounds catalyzed by metal complexes is undoubtedly the most studied of the entire class of homogenously catalyzed reactions. Indeed, advances in hydrogenation systems have contributed significantly to progress in homogeneous catalysis more generally, mainly because of the involvement of intermediate metal hydrides in a wider range of catalytic processes. The historical development of homogeneous hydrogenation is documented in my 1973 text on this topic, which was intended to represent an exhaustive treatise on the subject (the process, prior to the computer era, was certainly *exhausting* as over 1900 multi-language references were compiled).

Before outlining the content of *The Handbook of Homogeneous Hydrogenation*, I will briefly note here a few early facts chronologically for the appropriate context. Melvin Calvin first used the term "homogeneous hydrogenation" in 1938 for some non-aqueous, Cu-based systems, and a year later an M. Iguchi was the first to record the use of Rh species for hydrogenations in aqueous media. Jack Halpern was the first to study (in the mid-1950s) the kinetics and detailed mechanisms of such hydrogenations, while notably R. J. P. (Bob) Williams was the first (in 1960) to suggest in an equation the possibility of an $M(H_2)$ species, long before their true characterization in the early 1980s! The majority of the systems studied up to the early 1960s (for homogeneous catalysis generally, as well as hydrogenations) were in aqueous media – a fact frequently overlooked by current researchers – but developments at that time in the isolation and characterization of transition metal hydrides (pioneered especially by Joseph Chatt's group), including their stabilization by tertiary phosphines, led to increased studies in non-aqueous systems. Cleaner and "greener" aqueous systems are preferred for industrial processes, and intense interest remains in the incorporation of, for example, water-solubility enhancing, polar groups (sulfonate, carboxylate, hydroxide, etc.) into phosphine-containing ligands, protonation of N-atoms in P-N donor ligand, and more general use of cationic or anionic species for catalysis. The completion of the cycle back to aqueous systems is now in progress.

The classic 1961 paper by Halpern *et al.* (the '*al.*' being John Harrod and myself) on the catalytic hydrogenation of unsaturated acids using chlororuthenium(II) species in aqueous acid solutions certainly motivated the work of Geoffrey Wilkinson's group on Ru- and Rh-triphenylphosphine hydrogenation catalysts; these

The Handbook of Homogeneous Hydrogenation.
Edited by J.G. de Vries and C.J. Elsevier
Copyright © 2007 WILEY-VCH Verlag GmbH & Co. KGaA, Weinheim
ISBN: 978-3-527-31161-3

findings, published in the mid-1960s, are now legendary. The next highly significant step was the use of chiral phosphines with Rh precursors, first reported in 1968 by the groups of Knowles and Horner, the work providing the first examples of catalytic enantioselective hydrogenation (of unsaturated acids and a-substituted styrenes). Thousands of subsequent publications have recorded the development of catalyst systems containing chiral ligands (such as phosphines, sulfoxides, oxazolines, nitrogen-ligands, combinations of P/N/O/S-donors, carbenes, etc.) for hydrogenation of a wide range of prochiral substrates including alkenes, ketones, and ketimines. Processes reaching close to 100% enantiomeric excess (e.e.) are no longer uncommon, and, in a few cases, a remarkable degree of understanding of the mechanistic pathways has been achieved. About one dozen industrial, catalytic enantioselective homogeneous hydrogenation processes are now operating, and the potential for chiral catalyst systems within fine chemical industries (particularly for pharmaceuticals and agrochemicals) remains enormous. The first such process, that went on-line in 1970, was for the synthesis of L-dopa, a drug for treatment of Parkinson's disease; the system involved hydrogenation of a prochiral enamide using a Rh-chiral phosphine catalyst. In the production of the herbicide Metolachlor, a process that went on-line in 1996, a chiral amine is generated by hydrogenation of a ketimine using an Ir catalyst, and this currently provides the largest scale industrial process for an enantioselective synthesis of any type. The discovery and explanation of non-linear effects in enantioselective reactions, first reported by Kagan's group in 1986, are also noteworthy, and should lead to improved applications in enantioselective synthesis and, more importantly, a better understanding of the origins of enantioselectivity.

Advances thus far in experimental enantioselective hydrogenation have stemmed largely from empirical studies. More trendy and certainly more effective is high-throughput experimentation using ligand libraries, a methodology that is being increasingly promoted by researchers in the fine-chemical industries. Novel attempts to find the best catalyst by purely theoretical work that involves screening virtual catalyst libraries are also being published.

Hydrogenation catalysis in the petrochemical and related industries remains in the domain of heterogeneous systems, because of the practicality of separating and recycling the catalyst, although advances in the use of multiphase systems might find eventually use in relatively small-scale systems where a requirement is high selectivity, a vital property that can be engineered with a homogeneous catalyst. The design of supported metal complexes, including dendrimers, the use of size-exclusion filtration methodology, and the use of biphasic systems with all their ramifications (fluorous solvents, ionic liquids, and supercritical fluids), continue to be areas of intense current interest, and the findings should lead to further industrial uses of homogeneous catalysts, particularly in the small-scale synthesis of high value products.

The classic division between heterogeneous and homogeneous catalysts appears to becoming increasingly blurred and, in some cases involving colloidal/nanoparticle and metal cluster catalysts, the difference is difficult to determine experimentally. The large majority of reported homogeneous hydrogenation catalysts for aromatic residues now appear to be colloidal systems; in terms of activity within a particular reaction, the true nature of a catalyst may be considered somewhat irrelevant, but this is key when catalyst separation/removal for the purposes of recycling and residual toxicity levels is considered.

The exponential increase in the homogeneous hydrogenation literature over the last three decades shows no sign of abatement, and indeed, with the "replacement" of phosphine ligands by the increasingly popular carbenes, and the use of various two- and multi-phase systems, a further endemic literature expansion in homogeneous catalysis, and especially in the most understood area of hydrogenation, is guaranteed.

There is no question that much general knowledge on homogeneous catalysis has stemmed from studies on homogeneous hydrogenation, and it is fitting that *The Handbook of Homogeneous Hydrogenation* is published about 50 years after Halpern's first reports on the mechanistic aspects of such reactions. The editors have assembled an impressive list of eighty-one international experts that review the field from several aspects noted above. The first six chapters are categorized according to the catalyst metal used (most often the more traditional group 8–10 noble metals, although data on the early transition metals are presented), and there are chapters on the use of metal clusters and nanoparticles. A separate chapter appears on the kinetics commonly observed in hydrogenation systems, and there is one chapter on ionic hydrogenations. Three well-known techniques for studying homogeneous hydrogenation are then each presented in a chapter that discuss: NMR methods in general, the PHIP (parahydrogen induced polarization) NMR method, and the application of mass spectrometry. There are chapters on hydrogenation of organic substrates that are generally assembled according to the nature of the unsaturated function present in the organic, while separate chapters describe hydrogen transfer processes, CO_2 hydrogenation, and Rh-catalyzed, hydrogen-mediated, carbon-carbon bond formation. A large number of the chapters appropriately cover the many aspects of enantioselective hydrogenation, including a synopsis of current industrial applications. The final chapters deal with the fundamental problem associated with applications of homogeneous catalysis: deactivation, separation and recovery of the catalyst, and related engineering aspects.

The editors and authors are to be congratulated on assembling what is destined to become a classic in the area of Homogeneous Hydrogenation, which over the years has earned its title in capital letters.

Brian R. James
(University of British Columbia)

Preface

It is truly astonishing that such a simple reaction as the addition of one molecule of hydrogen to a double or triple bond can have so many facets.

When we had chosen the title of Handbook of Homogeneous Hydrogenation for our book we meant it to be a comprehensive work of reference. In this respect we are quite satisfied. We only had to skip the chapter on dehydrogenations, for which we could not find an author. We are extremely grateful to the other 88 authors for dedicating so much of their valuable time to writing the 81 marvellous chapters included in these volumes. We had envisaged an average of 30 pages per chapter but in the end this was not enough, necessitating an expansion of the two projected volumes to three.

One may wonder how long this handbook remains up-to-date. Indeed many areas are continuously undergoing new developments. In addition, new topics that were hardly emerging five years ago seem to develop at a very fast pace. Colloidal hydrogenation catalysts, for instance, which until recently were seen as the Cinderella of both homogeneous and heterogeneous catalysis – too soluble to be heterogeneous and too ill-defined to be homogeneous – have become quite respectable since they were recognized to be part of nanotechnology. Reductive coupling reactions can be considered as a green method to construct carbon-carbon bonds without taking resort to leaving groups. Indeed, not only this class of reactions, but all hydrogenations are of course extremely environmentally benign. Also the number of substrates is continuously expanding; carboxylic acid derivatives and heteroarenes are good examples of substrates recently added to the existing pool.

Just when everyone thought the chiral-ligand-boom was coming to an end, extremely simple monodentate ligands turned out to be quite effective, also allowing a combinatorial approach using mixtures of ligands. There is now a bewildering choice of chiral ligands available, increasing the chances of application. Indeed, the number of industrial applications is steadily increasing; an important breakthrough in this area was the advent of high throughput experimentation, which allowed for the first time to find a chiral ligand with good performance within a matter of weeks.

Our insight into the mechanisms of hydrogenation reactions has grown tremendously, thanks to advances in spectroscopic techniques, but also thanks to the hard work of many organometallic chemists. Many authors now also recog-

The Handbook of Homogeneous Hydrogenation.
Edited by J. G. de Vries and C. J. Elsevier
Copyright © 2007 WILEY-VCH Verlag GmbH & Co. KGaA, Weinheim
ISBN: 978-3-527-31161-3

nise the importance of the rate of these reactions. Turnover frequencies of hydrogenations are listed throughout the book.

One aspect that remains underdeveloped is the insight in deactivation pathways. Our knowledge in this area is growing, but the pace is slow. We have devoted an entire chapter to this topic, since the economics of many processes could benefit a lot from more insight in ways to reduce catalyst deactivation.

So far none of the industrial processes recycle the catalyst. Yet the number of ways to do this has grown far beyond simple immobilization. Two-phase catalysis now comes in many flavours.

Looking into the future, we expect that hydrogenation reactions will also be tremendously important for the conversion of renewable resources. Going from carbohydrates to valuable chemicals will require deoxygenating reactions. Thus, hydrogenation of alcohols, aldehydes and carboxylic acids will become very important topics.

We hope the readers will appreciate as well as enjoy the contents of this book. Any comments you may have are of course very welcome.

Hans de Vries   September 2006
Kees Elsevier

List of Contributors

David Ager
DSM Pharma Chemicals
9650 Strickland Road, Suite 103
Raleigh NC 27615-1937
USA

Terry T.-L. Au-Yeung
Open Laboratory of Chirotechnology
Institute of Molecular Technology
for Drug Discovery and Synthesis
Department of Applied Biology
and Chemical Technology
The Hong Kong Polytechnic
University
Hong Kong

Joachim Bargon
Institute of Physical and Theoretical
Chemistry
University of Bonn
Wegelerstraße 12
53115 Bonn
Germany

Sharon Bell
University of Basel
Department of Chemistry
St.-Johanns-Ring 19
4056 Basel
Switzerland

Claude de Bellefon
Laboratoire de Génie des Procédés
Catalytiques
CNRS-ESCPE Lyon
69616 Villeurbanne
France

Steven H. Bergens
Department of Chemistry
University of Alberta
Edmonton
Alberta
Canada T6G 2G2

Claudio Bianchini
ICCOM-CNR
Area della Ricerca CNR
Via Madonna del Piano snc
500019 Sesto Fiorentina/Firenze
Italy

A. John Blacker
Avecia Pharma (UK) Ltd.
Research and Development
Leeds Road
Huddersfield
HD1 9GA
UK

The Handbook of Homogeneous Hydrogenation.
Edited by J.G. de Vries and C.J. Elsevier
Copyright © 2007 WILEY-VCH Verlag GmbH & Co. KGaA, Weinheim
ISBN: 978-3-527-31161-3

Hans-Ulrich Blaser
Solvias AG
WRO-1055.6.28
Klybeckstrasse 191
4002 Basel
Switzerland

Elisabeth Bouwman
Leiden Institute of Chemistry
Leiden University
P.O. Box 9502
2300 RA Leiden
The Netherlands

John M. Brown
Chemical Research Laboratory
Oxford University
Oxford OX1 3TA
UK

R. Morris Bullock
Brookahven National Laboratory
Chemistry Department
Upton
New York 11973-5000
USA

Daniel Carmona
Departamento de Química Inorgánica
Instituto Universitario de Catálisis
Homogénea
Universidad de Zaragoza
Instituto de Ciencia de Materiales
de Aragón
C.S.I.C.-Universidad de Zaragoza
Zaragoza 50009
Spain

Shu Sun Chan
Open Laboratory of Chirotechnology
Institute of Molecular Technology
for Drug Discovery and Synthesis
Department of Applied Biology
and Chemical Technology
The Hong Kong Polytechnic
University
Hong Kong

Albert S. C. Chan
Open Laboratory of Chirotechnology
Institute of Molecular Technology
for Drug Discovery and Synthesis
Department of Applied Biology
and Chemical Technology
The Hong Kong Polytechnic
University
Hong Kong

Yongxiang Chi
Department of Chemistry
104 Chemistry Building
The Pennsylvania State University
University Park
PA 16802
USA

Chang-Woo Cho
University of Texas at Austin
Department of Chemistry
and Biochemistry
Austin
Texas 78712
USA

Hong Yee Cheung
Open Laboratory of Chirotechnology
Institute of Molecular Technology
for Drug Discovery and Synthesis
Department of Applied Biology
and Chemical Technology
The Hong Kong Polytechnic
University
Hong Kong

Matthew L. Clarke
School of Chemistry,
University of St. Andrews
St. Andrews
Fife KY16 9ST
UK

Christoper J. Cobley
Dowpharma
Chirotech Technology Limited
The Dow Chemical Company
321 Cambridge Science Park
Milton Road,
Cambridge CB4 0WG
UK

Christophe Copéret
Laboratoire de Chimie
Organométallique de Surface
UMR 9986 CNRS – CPE Lyon
CPE Lyon – Bât. 308
69616 Villeurbanne Cedex
France

Robert H. Crabtree
Yale University
Chemistry Department
350 Edwards Street
New Haven
CT 06520-8107
USA

Berth-Jan Deelman
Arkema Vlissingen B.V.
P.O. Box 70
4380 AB Vlissingen
The Netherlands

André H.M. de Vries
DSM Pharmaceutical Products
Advanced Synthesis, Catalysis,
and Development
P.O. 18
6160 MD
Geleen
The Netherlands

Johannes G. de Vries
DSM Pharmaceutical Products
Advanced Synthesis, Catalysis,
and Development
P.O. Box 18
6160 MD Geleen
The Netherlands

N. Koen de Vries
DSM Research Geleen
P.O. Box 18
6160 MD Geleen
The Netherlands

Elwin de Wolf
Debye Institute
Dept. of Metal-Mediated Synthesis
Utrecht University
Padualaan 8
3584 CH Utrecht
The Netherlands

Hans-Joachim Drexler
Leibniz-Institut für Organische
Katalyse an der Universität Rostock e.V.
Albert-Einstein-Str. 29a
18059 Rostock
Germany

Cornelis J. Elsevier
Institute of Molecular Chemistry
Univeristy of Amsterdam
Niewe Achtergracht 166
1018 WV Amsterdam
The Netherlands

Ben L. Feringa
Stratingh Institute
University of Groningen
Nijenborgh 4
9747 AG Groningen
The Netherlands

Robin J. Hamilton
Department of Chemistry
University of Alberta
Edmonton
Alberta
Canada T6G 2G2

Ulf Hanefeld
Organic Chemistry
Delft University of Technology
Julianalaan 136
2628 BL Delft
The Netherlands

Detlef Heller
Leibniz-Institut für Katalyse
an der Universität Rostock
Albert-Einstein-Str. 29 a
18059 Rostock
Germany

Philip G. Jessop
Department of Chemistry
Queen's University
90 Bader Lane
Kingston
Ontario
Canada K7L 3N6

Ferenc Joó
Institute of Physical Chemistry
University of Debrecen
1, Egyetem tér
4010 Debrecen
Hungary

Abdallah Karim
Laboratoire de Chimie
de Coordination
Faculté des Sciences Semlalia
Université Cadi Ayyad
BP 2390
Marrakech
Morocco

Ágnes Kathó
Institute of Physical Chemistry
University of Debrecen
1, Egyetem tér
Debrecen
4010 Debreden
Hungary

Dirk Klomp
Organic Chemistry
Delft University of Technology
Julianalaan 136
2628 BL Delft
The Netherlands

Alexander M. Kluwer
Van't Hoff Institute for Molecular
Sciences
Nieuwe Achtergracht 166
1018 WV Amsterdam
The Netherlands

Stanton H. L. Kok
Open Laboratory of Chirotechnology
Institute of Molecular Technology
for Drug Discovery and Synthesis
Department of Applied Biology
and Chemical Technology
The Hong Kong Polytechnic
University
Hong Kong

Michael J. Krische
University of Texas
Depart. of Chemistry
and Biochemistry
1 University Station A5300
Austin TX 78712-0165
USA

Wing Sze Lam
Open Laboratory of Chirotechnology
Institute of Molecular Technology
for Drug Discovery and Synthesis
Department of Applied Biology
and Chemical Technology
The Hong Kong Polytechnic
University
Hong Kong

Laurent Lefort
DSM Pharmaceutical Products
Advanced Synthesis
Catalysis & Development
P.O. Box 16
6160 MD Geleen
The Netherlands

Walter Leitner
Lehrstuhl für Technische Chemie
und Petrolchemie
Institut für Technische
und Makromolekulare Chemie
RWTH Aachen
Worringer Weg 1
52070 Aachen
Germany

Matthias Lotz
Solvias AG
P.O. Box
4002 Basel
Switzerland

Stephan Lütz
Institute of Biotechnology 2
Research Centre Jülich
52425 Jülich
Germany

Andrea Meli
ICCOM-CNR
Area della Ricerca CNR
Via Madonna del Piano snc
500019 Sesto Fiorentina/Firenze
Italy

Adriaan J. Minnaard
Stratingh Institute
University of Groningen
Nijenborgh 4
9747 AG Groningen
The Netherlands

Paul H. Moran
Dowpharma
Chirotech Technology Limited
The Dow Chemical Company
321 Cambridge Science Park
Milton Road
Cambridge CB4 0WG
UK

Robert H. Morris
Department of Chemistry
University of Toronto
80 St. George St.
Toronto M5S 3H6
Canada

André Mortreux
Laboratoire de Catalyse de Lille
UMR 8010 CNRS, USTL, ENSCL
BP 90108
59652, Villeneuve d'Ascq Cedex
France

Ryoji Noyori
Department of Chemistry
and Research Center
for Materials Science
Nagoya University Chikusa
Nagoya 464-8602
Japan

Taheshi Ohkuma
Graduate School of Engineering
Hokkaido University
Laboratory of Organic Synthesis
Division of Molecular Chemistry
Sapporo 060-8628
Japan

Luis A. Oro
Departamento de Química Inorgánica
Instituto Universitario de Catálisis
Homogénea
Universidad de Zaragoza
Instituto de Ciencia de Materiales
de Aragón
C.S.I.C.-Universidad de Zaragoza
Zaragoza 50009
Spain

Qinmin Pan
University of Waterloo
Department of Chemical Engineering
Waterloo
Ontario
Canada N2L 3G1

Paolo Pelagatti
Dipartimento di Chimica Generale ed
Inorganica
Chimica Analitica, Chimica Fisica
Università degli Studi di Parma
Parco Area Scienze 17/A
43100 Parma
Italy

Nathalie Pestre
Laboratoire de Génie des Procédés
Catalytiques
CNRS-ESCPE Lyon
69616 Villeurbanne
France

Joop A. Peters
Organic Chemistry
Delft University of Technology
Julianalaan 136
2628 BL Delft
The Netherlands

Andreas Pfaltz
University of Basel
Department of Chemistry
St.-Johanns-Ring 19
4056 Basel
Switzerland

Karine Philippot
Laboratoire de Chimie
de Coordination du CNRS
UPR 8241
205, route de Narbonne
31077 Toulouse Cedex 04
France

Angelika Preetz
Leibniz-Institut für
Katalyse an der Universität Rostock e.V.
Albert-Einstein-Str. 29 a
18059 Rostock
Germany

Corbin K. Ralph
Department of Chemistry
University of Alberta
Edmonton
Alberta
Canada T6G 2G2

Garry L. Rempel
University of Waterloo
Department of Chemical Engineering
Waterloo
Ontario
Canada N2L 3G1

Geoffrey J. Roff
Clarke Research Group
School of Chemistry
University of St. Andrews
St. Andrews
Fife KY16 9ST
UK

Alain Roucoux
UMR CNRS 6052
Synthèses et Activations
de Biomolécules
Ecole Nationale Supérieure de Chimie
de Rennes
Institut de Chimie de Rennes
Ave. du Général Leclerc
35700 Rennes
France

Roberto A. Sánchez-Delgado
Brooklyn College and the Graduate
Center of the City University
of New York
New York
USA

Thomas Schmidt
Leibniz-Institut für Organische
Katalyse an der Universität Rostock e.V.
Albert-Einstein-Str. 29a
18059 Rostock
Germany

P. Schulz
Lehrstuhl für Chemische
Reaktionstechnik
Universität Erlangen
Egerlandstraße 3
91058 Erlangen
Germany

Ottó Balázs Simon
Department of Organic Chemistry,
University of Veszprém
P.O. Box 158
Veszprém
Hungary

Attila Sisak
Research Group for Petrochemistry of
the Hungarian Academy of Sciences
P.O. Box 158
Veszprém
Hungary

Felix Spindler
Solvias AG
P.O. Box
4002 Basel
Switzerland

Wenjun Tang
Department of Chemistry
104 Chemistry Building
The Pennsylvania State University
University Park
PA 16802
USA

Marc Thommen
Solvias AG
P.O. Box
4002 Basel
Switzerland

Imre Toth
DSM Research
Department of Industrial Chemicals-
Chemistry and Technology
Urmonderbaan 22
6167 RD Geleen
The Netherlands

Michel van den Berg
Stratingh Institute
University of Groningen
Nijenborgh 4
9747 AG Groningen
The Netherlands

Paul C. van Geem
DSM Research
Department of Industrial Chemicals-
Chemistry and Technology
Urmonderbaan 22
6167 RD Geleen
The Netherlands

Francesco Vizza
ICCOM-CNR
Area della Ricerca CNR
Via Madonna del Piano snc
500019 Sesto Fiorentina/Firenze
Italy

Peter Wasserscheid
RWTH Aachen
Institut für Technische
und Makromolekulare Chemie
Worringer Weg 1
52056 Aachen
Germany

Jialong Wu
University of Waterloo
Department of Chemical Engineering
Waterloo
Ontario
Canada N2L 3G1

Takamichi Yamagishi
Department of Applied Chemistry
Graduate School of Engineering
Tokyo Metropolitan University
Tokyo
Japan

Xumu Zhang
Department of Chemistry
152 Davey Laboratory
University Park
Pennsylvania, PA 16802-6300
USA

Part I
Introduction, Organometallic Aspects
and Mechanism of Homogeneous Hydrogenation

The Handbook of Homogeneous Hydrogenation.
Edited by J.G. de Vries and C.J. Elsevier
Copyright © 2007 WILEY-VCH Verlag GmbH & Co. KGaA. Weinheim
ISBN: 978-3-527-31161-3

1
Rhodium

Luis A. Oro and Daniel Carmona

1.1
Introduction

Homogeneous hydrogenation constitutes an important synthetic procedure and is one of the most extensively studied reactions of homogeneous catalysis. The homogeneous hydrogenation of organic unsaturated substrates is usually performed with molecular hydrogen, but it is also possible to derive hydrogen from other molecules acting as hydrogen donors, such as alcohols; these are termed hydrogen transfer reactions. The impressive developments of coordination and organometallic chemistry have allowed the preparation of a wide variety of soluble metal complexes active as homogeneous hydrogenation catalysts under very mild conditions. These complexes are usually derived from transition metals, especially late transition metals with tertiary phosphine ligands, having partially filled d electron shells. These transition metal complexes present the ability to stabilize a large variety of ligands with a variability of the coordination number and oxidation state. Among those ligands are unsaturated substrates containing π-electron systems, such as alkenes or alkynes, as well as key σ-bonded ligands such as hydride or alkyl groups.

Although the reaction of hydrogen with alkenes is exothermic, the high activation energy required constrains the reaction. Thus, the concerted addition of hydrogen to alkenes is a forbidden process according to orbital symmetry rules. In contrast, transition metals have the appropriate orbitals to interact readily with molecular hydrogen, forming metal hydride species, allowing the transfer of the hydride to the coordinated alkene. Rhodium is a good example of such transition metals, and has played a pivotal role in the understanding of homogeneous hydrogenation, providing clear examples of the two types of homogeneous hydrogenation catalysts that formally can be considered, namely monohydride and dihydride catalysts.

The Handbook of Homogeneous Hydrogenation.
Edited by J. G. de Vries and C. J. Elsevier
Copyright © 2007 WILEY-VCH Verlag GmbH & Co. KGaA, Weinheim
ISBN: 978-3-527-31161-3

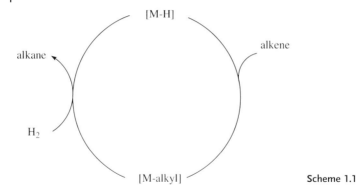

Scheme 1.1

1.1.1
Monohydride Hydrogenation Catalysts

Monohydride (MH) catalysts, such as [RhH(CO)(PPh$_3$)$_3$], react with substrates such as alkenes, according to Scheme 1.1, yielding rhodium–alkyl intermediates which, by subsequent reaction with hydrogen, regenerate the initial monohydride catalyst. This mechanism is usually adopted by hydrogenation catalysts which contain an M–H bond.

1.1.2
Dihydride Hydrogenation Catalysts

Many of these catalysts are derived from metal complexes which, initially, do not contain metal hydride bonds, but can give rise to intermediate MH$_2$ (alkene) species. These species, after migratory insertion of the hydride to the coordinated alkene and subsequent hydrogenolysis of the metal alkyl species, yield the saturated alkane. At first glance there are two possibilities to reach MH$_2$ (alkene) intermediates which are related to the order of entry of the two reaction partners in the coordination sphere of the metal (Scheme 1.2).

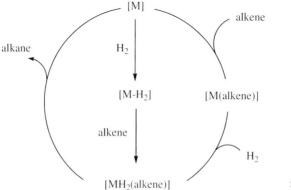

Scheme 1.2

The hydride route involves the initial reaction with hydrogen followed by coordination of the substrate; the well-known Wilkinson catalyst [RhCl(PPh$_3$)$_3$] is a representative example. A second possible route is the alkene (or unsaturated) route which involves an initial coordination of the substrate followed by reaction with hydrogen. The cationic catalyst derived from [Rh(NBD)(DIPHOS)]$^+$ (NBD = 2,5-norbornadiene; DIPHOS = 1,2-bis(diphenyl)phosphinoethane) is a well-known example. The above-mentioned rhodium catalysts will be discussed, in the detail, in the following sections.

Homogeneous hydrogenation reactions by metal complexes have been investigated extensively during previous years. The first authoritative book on this subject, containing interesting and detailed historical considerations, was written by James, and appeared in 1973 [1]. Following this, other monographs [2, 3] were published, as well as several reviews and book chapters [4], reporting on the rapid and impressive development of this field.

1.2
The Early Years (1939–1970)

Calvin made the first documented example of homogeneous hydrogenation by metal compounds in 1938, reporting that quinoline solutions of copper acetate, at 100 °C, were capable of hydrogenating unsaturated substrates such as *p*-benzoquinone [5]. One year later, Iguchi reported the first example of homogeneous hydrogenation by rhodium complexes. A variety of organic and inorganic substrates were hydrogenated, at 25 °C, by aqueous acetate solutions of RhCl$_3$, [Rh(NH$_3$)$_5$(H$_2$O)]Cl$_3$, or [Rh(NH$_3$)$_4$Cl$_2$]Cl, whilst more inert complexes [Rh(NH$_3$)$_6$]Cl$_3$ and [Rh(en)$_3$]Cl$_3$, were inactive [6]. The initial hydrogen reduction of these and other related rhodium(III) complexes with nitrogen donor ligands seems to proceed via hydride rhodium intermediates formed by the heterolysis of hydrogen [1, 7]. In the case of the Iguchi rhodium(III) catalysts, however, after some time, partial autoreduction to rhodium metal occurred. Another seminal contribution of Iguchi, which appeared few years later in 1942, was the observation of the ready absorption of hydrogen by cobalt cyanide aqueous solutions; the formed cobalt hydride, [CoH(CN)$_5$]$^-$, was seen to be highly selective for the hydrogenation of conjugated dienes to monoenes [8].

During this period, the most significant advances in homogeneous hydrogenation catalysis have been the discovery of rhodium phosphine complexes. The hydridocarbonyltris(triphenylphosphine)rhodium(I) complex, [RhH(CO)(PPh$_3$)$_3$], was reported in 1963 [9], and its catalytic activity was studied in detail by Wilkinson and coworkers a few years later [10–13]. The most important rhodium catalyst, the chlorotris(triphenylphosphine)rhodium(I) complex, was prepared during the period 1964–1965 by several groups [14]. It is easily synthesized by treating RhCl$_3$·3H$_2$O with triphenylphosphine in ethanol. It can also be prepared by displacement of the coordinated diene 1,5-cyclooctadiene from [RhCl(diene)]$_2$ complexes [15]. Wilkinson and coworkers extensively studied the remarkable catalytic properties of this complex, which is usually known as Wilkinson's catalyst. It was the first practical hydrogenation system to be used routinely under very mild conditions – usually room temperature and atmospheric pressure of hydrogen. This compound catalyzes the chemospecific hydrogenation of alkenes in the presence of other easily reduced groups such as NO_2 or CHO, and terminal alkenes in the presence of internal alkenes [14 b, 16]. The remarkable success of this catalyst inspired the incorporation of chiral phosphines, which in turn prompted the birth of the catalytic enantioselective hydrogenation with pioneering contribution from the groups of Knowles and Horner (see [38, 39, 45 b] in Section 1.1.6).

The synthesis of cationic rhodium complexes constitutes another important contribution of the late 1960s. The preparation of cationic complexes of formula $[Rh(diene)(PR_3)_2]^+$ was reported by several laboratories in the period 1968–1970 [17, 18]. Osborn and coworkers made the important discovery that these complexes, when treated with molecular hydrogen, yield $[RhH_2(PR_3)_2(S)_2]^+$ (S = solvent). These rhodium(III) complexes function as homogeneous hydrogenation catalysts under mild conditions for the reduction of alkenes, dienes, alkynes, and ketones [17, 19]. Related complexes with chiral diphosphines have been very important in modern enantioselective catalytic hydrogenations (see Section 1.1.6).

Rhodium(II) acetate complexes of formula $[Rh_2(OAc)_4]$ have been used as hydrogenation catalysts [20, 21]. The reaction seems to proceed only at one of the rhodium atoms of the dimeric species [20]. Protonated solutions of the dimeric acetate complex in the presence of stabilizing ligands have been reported as effective catalysts for the reduction of alkenes and alkynes [21].

It is noteworthy to comment that the remarkable advances of experimental techniques made during the past few decades has allowed, in many cases, information to be obtained about key intermediates. As a consequence, detailed mechanistic studies have now firmly established reaction pathways.

1.3
The $[RhH(CO)(PPh_3)_3]$ Catalyst

This complex was first prepared by Bath and Vaska in 1963 [9], and studied in detail by Wilkinson and coworkers some years later [10] as an active catalyst for hydrogenation [11], isomerization [12], and hydroformylation [13] reactions.

The crystal structure of the $[RhH(CO)(PPh_3)_3]$ complex [22] shows a bipyramidal structure with equatorial phosphines. This coordinatively saturated 18-electron complex should be considered strictly as a precatalyst (or catalyst precursor) which, after dissociation of PPh_3 and creation of a vacant coordination site, yields the real catalytic species, $[RhH(CO)(PPh_3)_2]$. It should be remembered at this point that homogeneous mechanisms involve multistep processes, characteristic of coordination and organometallic chemistry (e.g., ligand dissociation or substitution, oxidative addition, reductive elimination), and therefore several types of intermediates are involved. The intermediates proposed by Wilkinson and coworkers in 1968 [11], based on kinetic studies, are shown in Scheme 1.3. Thus, the first step is the *ligand dissociation* of PPh_3 (step a), yielding

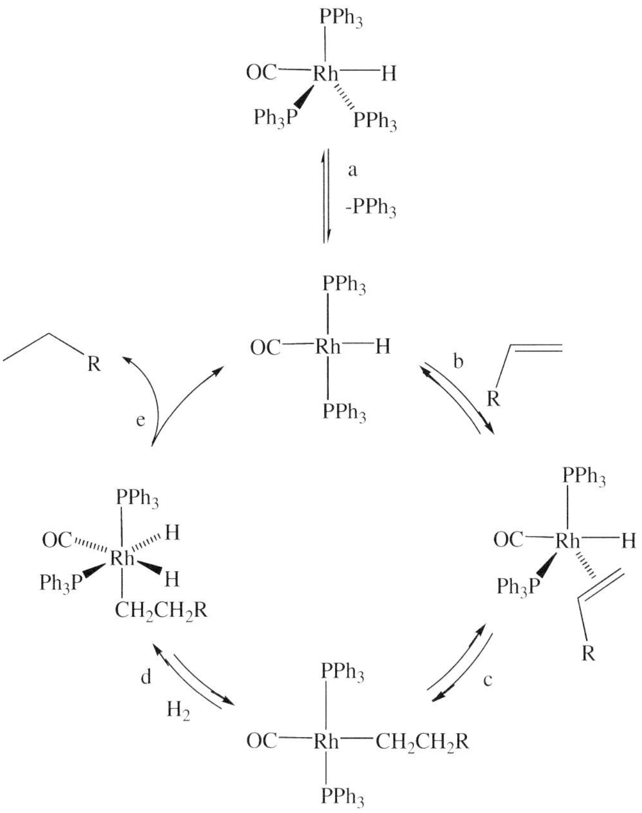

Scheme 1.3

[RhH(CO)(PPh$_3$)$_2$]. A vacant coordination site has been created and *coordination* of the 1-alkene substrate to the [RhH(CO)(PPh$_3$)$_2$] catalyst is feasible (step b). This is followed by the *migratory insertion* of the hydride to the coordinated alkene (step c), and then *hydrogenolysis* of the metal alkyl species to yield the saturated alkane. This hydrogenolysis proceeds, in this case, through two steps – *oxidative addition* of H$_2$ (step d), followed by *reductive elimination* to regenerate the [RhH(CO)(PPh$_3$)$_2$] catalyst (step e). Further support for steps c and d has been obtained by studying the reaction of [RhH(CO)(PPh$_3$)$_2$] with tetrafluoroethylene that yields the stable *trans*-[Rh(C$_2$F$_2$H)(CO)(PPh$_3$)$_2$] that, under reaction with hydrogen and in the presence of phosphine, gives C$_2$F$_2$H$_2$ and [RhH(CO)(PPh$_3$)$_3$] [13].

Rhodium species in oxidation states I and III are involved in the process. Rhodium-catalyzed hydrogenations generally involve oxidative addition reactions, followed by the reverse process of reductive elimination in the final step. Another common elimination process is the so-called *β-elimination*, which accounts for the frequent side reaction of isomerization of alkenes, according to Eq. (1):

$$RCH_2CH_2CH_2 \quad\ RCH_2CH=CH_2 \quad\ RCH_2CHMe \quad\ RCH=CHMe$$
$$| \rightleftharpoons \quad | \rightleftharpoons \quad | \rightleftharpoons \quad | \quad (1)$$
$$M \qquad\qquad MH \qquad\qquad M \qquad\qquad MH$$

This reversibility explains the frequent isomerization of alkenes by metal hydride complexes, even in the absence of hydrogen. The formation of the secondary alkyl should be unfavored when bulky ligands are present. This steric argument explains why internal alkenes are isomerized by $[RhH(CO)(PPh_3)_2]$ but are not competitively hydrogenated, as well as the high selectivity of this catalyst for the hydrogenation of terminal alkenes compared to 2-alkenes due to the lowered stability of the secondary alkyl species. The formation of the real catalyst $[RhH(CO)(PPh_3)_2]$ requires the dissociation of PPh_3 from the rhodium precatalyst. Thus, an addition of excess PPh_3 prevents the dissociation and inhibits alkene hydrogenation. On the other hand, at very low concentrations, the $RhH(CO)(PPh_3)$ intermediate, which is still more active, is formed by further dissociation of $RhH(CO)(PPh_3)_2$ according to Eq. (2).

$$[RhH(CO)(PPh_3)_3] \xrightleftharpoons{-PPh_3} [RhH(CO)(PPh_3)_2] \xrightleftharpoons{-PPh_3} [RhH(CO)(PPh_3)] \quad (2)$$

1.4
The [RhCl(PPh₃)₃] Complex and Related Catalysts

The widely studied $[RhCl(PPh_3)_3]$ complex, usually known as Wilkinson's catalyst, was discovered independently in 1965 by Wilkinson (a recipient of the Nobel Prize in 1973) and other groups [14]. This compound catalyzes the chemospecific hydrogenation of alkenes in the presence of other easily reduced groups such as NO_2 or CHO, and terminal alkenes in the presence of internal alkenes [16]. The rate of hydrogenation parallels their coordination ability (Scheme 1.4), but tetrasubstituted alkenes are not reduced.

This burgundy-red compound can be easily prepared by reacting $RhCl_3 \cdot 3H_2O$ with triphenylphosphine in refluxing ethanol. In this reaction, rhodium(III) is reduced to the rhodium(I) complex $RhCl(PPh_3)_3$, whilst the phosphine is oxidized to phosphine oxide according to Eq. (3).

$$[RhCl_3(H_2O)_x] + y\,PPh_3 \xrightarrow{EtOH/78°C} [RhCl(PPh_3)_3] + OPPh_3 \quad (3)$$
$$y > 4$$

The most accepted mechanism for alkene hydrogenation is mainly due to Halpern [23], and is supported by careful kinetic and spectroscopic studies of cyclo-

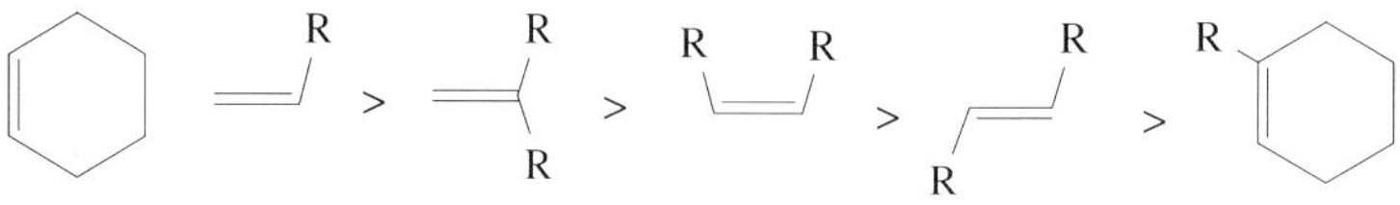

Scheme 1.4

hexene hydrogenation. This mechanism is shown in Scheme 1.5, where the route dominating the catalytic cycle is surrounded by the dotted line. According to this scheme, the predominant hydride route consists of oxidative addition of a hydrogen molecule prior to alkene coordination. Both the associative pathway *via* a 16-electron complex [RhCl(PPh₃)₃] and the dissociative pathway *via* a 14-electron species RhCl(PPh₃)₂ could function for hydrogenation, depending on the concentration of free PPh₃. However RhCl(PPh₃)₂ reacts with H₂ at least 10 000-fold faster than [RhCl(PPh₃)₃], and is therefore the active intermediate. Thus, the hydride path is much more efficient than the alkene path. However, in the absence of hydrogen, RhCl(PPh₃)₂ showed a remarkable tendency to dimerize, yielding [(PPh₃)₂Rh(μ-Cl)Rh(PPh₃)₂] species with bridging chloro ligands; the formation of alkene complexes of general formula [RhCl(alkene)(PPh₃)₂] can also be observed. The addition of small amounts of free phosphine inhibits the formation of these dimers, thereby enhancing the catalytic activity of the complex.

The rapid oxidative addition of hydrogen to RhCl(PPh₃)₂, followed by alkene coordination, affords the 18-electron octahedral dihydride alkene complex [RhH₂Cl(alkene)(PPh₃)₂]. The unsaturated [RhH₂Cl(PPh₃)₂] intermediate is also capable of ligand association with PPh₃, being in equilibrium with [RhH₂Cl(PPh₃)₃]. The rate-determining step for the whole process is the intramolecular alkene insertion into the rhodium–hydride bond of [RhH₂Cl(alkene)(PPh₃)₂], to produce the alkyl hydride intermediate, [RhH(alkyl)Cl(PPh₃)₂]. The next step, the reductive elimination of alkane from this alkyl hydride intermediate to regenerate RhCl(PPh₃)₂, occurs rapidly. The proposed cycle implies changes in the oxidation state (I and III) in the oxidative addition and reductive elimination

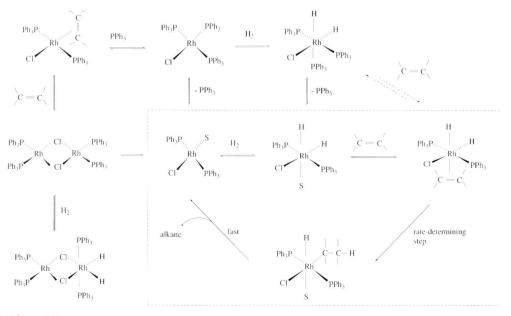

Scheme 1.5

steps, as well as changes in the electronic environment (14-, 16-, and 18-electron species) and coordination numbers, from 3 to 6. Nevertheless, in some cases, some coordination vacancies could be occupied by polar solvent molecules, such as ethanol, which are frequently added to the usual aromatic organic solvents.

This cycle was pieced together from the results of a variety of experiments. The complexes $[RhCl(PPh_3)_3]$, $[RhCl(alkene)(PPh_3)_2]$, $[RhH_2Cl(PPh_3)_3]$, $[(PPh_3)_2Rh(\mu\text{-}Cl)_2Rh(PPh_3)_2]$ and $[(PPh_3)_2Rh(\mu\text{-}Cl)_2RhH_2(PPh_3)_2]$ were directly observed, but the species of Scheme 1.5 which are inside the dashed line, and are thus responsible for the hydrogenation process, were not observable. Furthermore, when styrene was used as substrate, a further path involving the coordination of two styrene molecules was found to be operating. It seems important to remember that, in some catalytic reactions, those compounds that are readily isolable might not be the true intermediates in the catalytic cycle.

Recently [24], using parahydrogen-induced polarization (PHIP) NMR techniques, it has been possible to observe the dihydride $[Rh(H)_2Cl(styrene)(PPh_3)_2]$, which is involved in the hydrogenation process. Although traditionally it has been depicted with *trans* phosphines, this technique shows that the two phosphines are arranged in a *cis* fashion. When PHIP NMR was used to examine the reaction of $[Rh(\mu\text{-}Cl)(PPh_3)_2]_2$ with parahydrogen, the dinuclear hydride complexes $[Rh(H)_2(PPh_3)_2(\mu\text{-}Cl)_2Rh(PPh_3)_2]$ and $[Rh(H)_2(PPh_3)_2(\mu\text{-}Cl)]_2$ were detected and characterized [25]. The same reaction, when carried out in the presence of an alkene, revealed signals corresponding to new dinuclear dihydrides of the type $[Rh(H)_2(PPh_3)_2(\mu\text{-}Cl)_2Rh(PPh_3)(alkene)]$. Kinetic data showed that hydrogenation *via* this intermediate may be significant [26].

Related triarylphosphine complexes of formula $[RhCl(P(C_6H_4\text{-}4\text{-}X)_3)_3]$ are also active hydrogenation catalysts. Studies on cyclohexene hydrogenation show an increase in the relative rates when the basicity of the triarylphosphine increases: $P(C_4H_4\text{-}4\text{-}Cl)_3$ (1.8) $< P(C_6H_5)_3$ (41) $< P(C_6H_4\text{-}4\text{-}Me)_3$ (86) $< P(C_6H_4\text{-}4\text{-}OMe)_3$ (100). However, complexes with more basic tertiary phosphines, such as PEt_3 or $PPhEt_2$, are practically inactive.

A simple method for the *in-situ* preparation of Wilkinson-type catalysts consists of the addition of the appropriate amount of the triarylphosphine to the rhodium dimers, $[Rh(\mu\text{-}Cl)(diene)]_2$ or $[Rh(\mu\text{-}Cl)(cyclooctene)_2]_2$, according to Eqs. (4) and (5). The best results are usually obtained for a rhodium/phosphine ratio of 1:2.

$$[Rh(\mu\text{-}Cl)(COD)]_2 + 2\,n\,PPh_3 \longrightarrow 2[RhCl(PPh_3)_n] + 2\,COD \qquad (4)$$
COD = 1,5-cyclooctadiene

$$[Rh(\mu\text{-}Cl)(COE)_2]_2 + 2\,n\,PPh_3 \longrightarrow 2[RhCl(PPh_3)_n] + 4\,COE \qquad (5)$$
COE = cyclooctene, n = 3 or 2

The water-soluble analogue of Wilkinson's catalyst, $[RhCl(TPPMS)_3]$ [TPPMS = $PPh_2(C_6H_4SO_3Na)$], prepared *in situ* from $[Rh(\mu\text{-}Cl)(diene)]_2$ and TPPMS, reacts with hydrogen in aqueous solution to yield $[RhH(TPPMS)_3]$, instead of $[RhH_2(TPPMS)_3]$, according to Eq. (6):

$$[\text{RhCl(TPPMS)}_3] + \text{H}_2 \longrightarrow [\text{RhH(TPPMS)}_3] + \text{Cl}^- + \text{H}^+ \tag{6}$$
$$\text{TPPMS} = \text{PPh}_2(\text{C}_6\text{H}_4\text{SO}_3\text{Na})$$

The presence of $[\text{RhH(TPPMS)}_3]$ causes substantial changes in the mechanism of hydrogenation, that most probably follows a conventional monohydride mechanism as shown in Scheme 1.1. This is also reflected in the rates and the hydrogenation selectivities [27].

1.5
The Cationic[Rh(diene)(PR₃)ₓ]⁺ Catalysts

The catalytic potential of $[\text{Rh(diene)L}_n]^+$ ($n=2$ or 3; L=phosphine, phosphite or arsine) complexes as hydrogenation catalysts was discovered by Osborn and co-workers during the period 1969 to 1976 [19, 28]. Under a hydrogen atmosphere the diene is hydrogenated, generating the reactive $[\text{RhH}_2\text{S}_x\text{L}_n]^+$ species which, in some cases such as in $[\text{RhH}_2(\text{solvent})_2\text{L}_2]^+$ intermediates, can be isolated relatively easily from coordinating solvents such as acetone or ethanol. In contrast to Wilkinson's catalyst, a large number of donor ligands can be used and several easy preparative routes are available. NBD was the preferred diene for the $[\text{Rh(diene)L}_n]^+$ catalyst precursors, but other dienes such as 1,5-cyclooctadiene (COD) or tetrafluorobenzobarrelene (TFB) have also been used [28, 29]. The dihydride complexes $[\text{RhH}_2\text{S}_2\text{L}_2]^+$ (S=Me₂CO, EtOH) have two solvent molecules bound through their oxygen lone pairs in the coordination sphere. They are excellent hydrogenation catalysts, and some unusual properties seem to depend on its ionic character.

The dihydride complexes $[\text{RhH}_2\text{S}_x\text{L}_n]^+$ are in equilibrium with monohydride species according to Eq. (7).

$$[\text{RhH}_2\text{S}_x\text{L}_n]^+ \rightleftharpoons [\text{RhHS}_y\text{L}_n] + \text{H}^+ \tag{7}$$

The equilibrium can be shifted by the addition of acid or base, and is also sensitive to the nature of the ligands and solvents. Scheme 1.6 qualitatively accounts for the experimental observations involving three possible pathways by which an unsaturated substrate can be hydrogenated.

Path A involves a neutral monohydride species which both extensively isomerizes and also hydrogenates alkenes, and is favored in the presence of NEt₃ or when the catalyst contains basic phosphine ligands. In path B, which is favored in the presence of acids, cationic dihydride species are the active catalyst; they hydrogenate alkenes less efficiently, but with limited isomerization. Path C involves formation of $[\text{RhH}_2\text{S}_x(\text{alkene})\text{L}_n]^+$ from $[\text{RhS}_x\text{L}_n]^+$ by intermediacy of $[\text{RhS}_x(\text{alkene})\text{L}_n]^+$. In the proposed cycles the substrate enters the coordination sphere by displacing a solvent molecule. The experimental observations suggest that the neutral monohydride species RhHS_yL_n are considerably more efficient hydrogenation

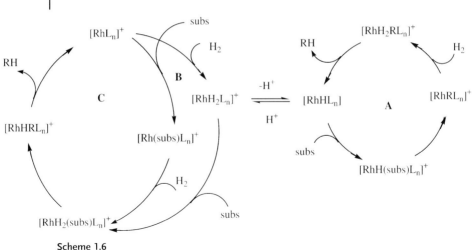

Scheme 1.6

catalysts than the dihydrides $[RhH_2S_xL_n]^+$. However, in order to hydrogenate an alkene without concomitant isomerization, path A must be suppressed; thus, hydrogenation under acidic conditions (path B) becomes operative due to the presence of the cationic $[RhH_2S_xL_n]^+$ dihydrides. The mechanism for hydrogenation via path B shows a close analogy to the proposed mechanism involving the Wilkinson catalyst (Scheme 1.5). The equilibrium illustrated in Eq. (7) offers considerable flexibility, and it is possible to identify appropriate conditions for the selective hydrogenation of alkynes to *cis* alkenes following pathways A and B. However, the rate of hydrogenation with the monohydride species is at least twice that with the dihydride species. It has been proposed that the origin of the selective reduction of 2-hexyne to *cis*-2-hexene seems to arise from the fact that the coordinated *cis*-2-hexene, formed by hydrogenation, is immediately displaced by 2-hexyne, before it can be isomerized or hydrogenated to hexane. Almost exclusive *cis*-addition of hydrogen to the alkyne was observed [28b].

The selective hydrogenation of dienes to monoenes can also be accomplished by using these rhodium cationic catalysts with proper choice of the phosphine ligand [28c]. Path C seems to be dominant, where the strongly coordinated dienes were hydrogenated selectively to monoenes, with hydrogen adding both 1,2 and 1,4 to conjugated dienes. The hydrogenation process follows the unsaturated route; in fact, the only detectable species are $[Rh(diene)L_2]^+$ which, after reaction with hydrogen, yield the alkyl–hydride complex. It should be noted that the latter species are common to both pathways B and C, the only difference being the sequence of substrate coordination and previous or posterior reaction with molecular hydrogen. The hydrogenation activity of catalysts derived from $[Rh(TFB)(P(C_6H_4\text{-}4\text{-}X)_3)_2]^+$ species depends on the electron-releasing ability of the X-substituent [29, 30].

Cationic $[Rh(diene)L_2]^+$ species are also catalyst precursors for the hydrogenation of ketones and aldehydes [19]. The mechanism presents some analogies with that proposed for alkene hydrogenation, with formation of alkoxy species by insertion.

The catalytic activity of cationic rhodium precursors of formula [Rh(diene)(di-phosphine)]⁺ was also explored by Schrock and Osborn [28]. Halpern and co-workers made very detailed mechanistic studies of olefin hydrogenation by [RhS₂(diphos)]⁺ species (diphos = 1,2-bis(diphenylphosphino)ethane; S = solvent) [31]. Significant differences have been observed in the reaction of the catalyst precursors [Rh(NBD)(PPh₃)₂]⁺ and [Rh(NBD)(diphos)]⁺ in methanol, as shown in Eqs. (8) and (9):

$$[Rh(NBD)(PPh_3)_2]^+ + 3\,H_2 \xrightarrow{\text{MeOH}} [RhH_2(MeOH)_2(PPh_3)_2]^+ + \text{norbornane}$$

$$(8)$$

$$[Rh(NBD)(diphos)]^+ + 3\,H_2 \xrightarrow{\text{MeOH}} [Rh(MeOH)_2(diphos)]^+ + \text{norbornane} \quad (9)$$
$$(NBD = 2,5\text{-norbornadiene}; \ diphos = 1,2\text{-bis(diphenylphosphino)ethane})$$

Although the latter product is a solvated mononuclear [Rh(MeOH)₂(diphos)]⁺ cation, in the solid state it is isolated as a binuclear complex of formula [Rh₂(diphos)₂](BF₄)₂, in which each rhodium center is bonded to two phosphorus atoms of a chelating bis(diphenylphosphino)ethane ligand, and to a phenyl ring of the bis(diphenylphosphino)ethane ligand of the other rhodium atom. This dimer reverts to a mononuclear species on redissolving. The mechanism of hydrogenation of the prochiral alkene methyl(Z)-α-acetamidocinnamate, studied in detail by Halpern [31], is depicted in Scheme 1.7.

The hydrogenation process follows the unsaturated route, the rate-determining step being the reaction of [Rh(alkene)(diphos)]⁺ with hydrogen. The proposed mechanism corresponds to pathway C of Scheme 1.6. Analogous cationic rhodium complexes with chiral diphosphines have allowed remarkable progress to be made on the subject of enantioselective hydrogenations, as will be discussed in the following section.

Several heteroaromatic compounds can be hydrogenated by [Rh(COD)(PPh₃)₂]⁺ species. Thus, this cationic complex has been reported to be a catalyst precursor for the homogeneous hydrogenation of heteroaromatic compounds such as quinoline [32] or benzothiophene [33]. Detailed mechanistic cycles have been proposed by Sánchez-Delgado and coworkers. The mechanism of hydrogenation of benzothiophene by the cationic rhodium(III) complex, [Rh(C₅Me₅)(MeCN)₃]²⁺, has been elucidated by Fish and coworkers [34].

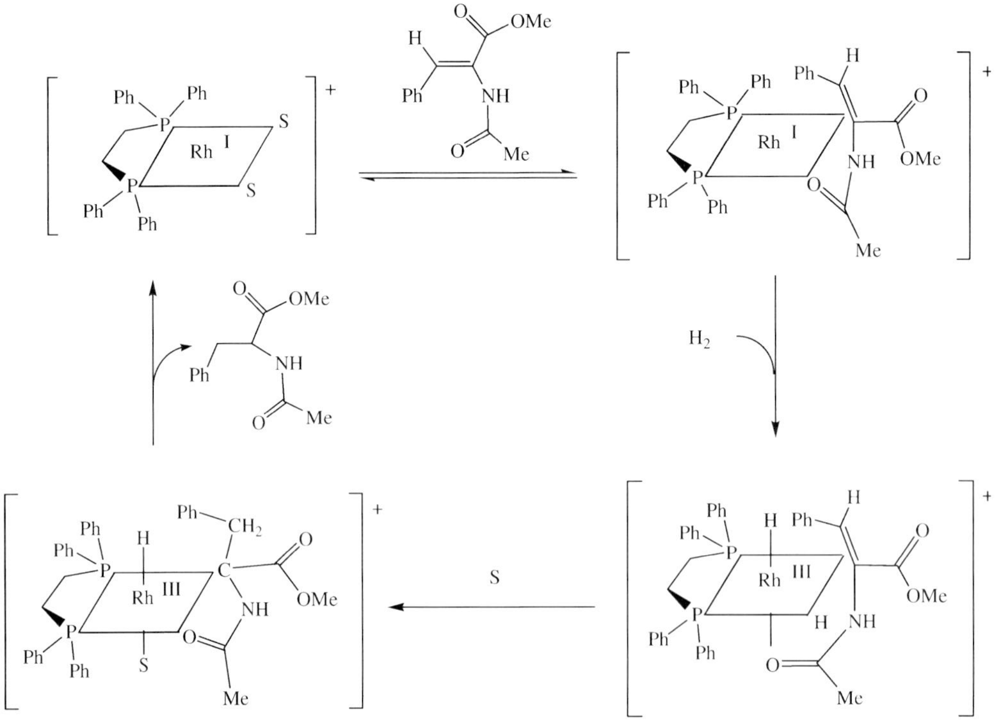

Scheme 1.7

1.6
Enantioselective Rhodium Catalysts

1.6.1
Hydrogenation of Alkenes

Following Wilkinson's discovery of [RhCl(PPh$_3$)$_3$] as an homogeneous hydrogenation catalyst for unhindered alkenes [14b, 35], and the development of methods to prepare chiral phosphines by Mislow [36] and Horner [37], Knowles [38] and Horner [15, 39] each showed that, with the use of optically active tertiary phosphines as ligands in complexes of rhodium, the enantioselective asymmetric hydrogenation of prochiral C=C double bonds is possible (Scheme 1.8).

Knowles reported the hydrogenation of α-phenylacrylic acid and itaconic acid with 15% and 3% optical purity, respectively, by using [RhCl$_3$(P*)$_3$] [P* = (R)-(–)-methyl-n-propylphenylphosphine] as homogeneous catalyst [38]. Horner found that α-ethylstyrene and α-methoxystyrene can be hydrogenated to (S)-(+)-2-phenylbutane (7–8% optical purity) and (R)-(+)-1-methoxy-1-phenylethane (3–4% optical purity), respectively, by using the complex formed *in situ* from [Rh(1,5-hexadiene)Cl]$_2$ and (S)-(–)-methyl-n-propylphenylphosphine as catalyst [39].

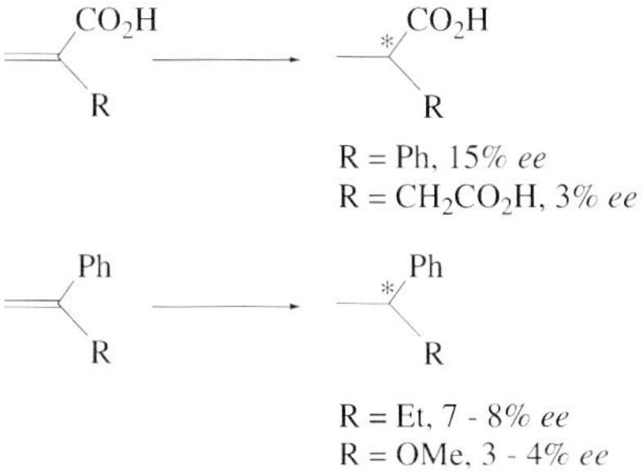

R = Ph, 15% ee
R = CH₂CO₂H, 3% ee

R = Et, 7 - 8% ee
R = OMe, 3 - 4% ee   **Scheme 1.8**

These were the first examples of metal-catalyzed enantioselective hydrogenation. The ee-values achieved were modest [40], but they established the validity of the hydrogenation method. Furthermore, the inherent generality of the method offered enormous opportunities for optimization of the results. It is interesting to note that, in these examples, the chirality of the ligands resides in the phosphorus atom.

The degree of enantioselective bias was improved shortly after this time. In 1971, Morrison et al. reported that the rhodium(I) complex [RhCl(NMDPP)₃] (NMDPP = neomethyldiphenylphosphine) reduces (E)-β-methylcinnamic and (E)-α-methylcinnamic acids with 61 and 52% ee, respectively (Scheme 1.9) [41]. NMDPP is a monodentate phosphine derived from (–)-menthol, and its asymmetry lies at carbon atoms.

An important breakthrough was the finding, by Kagan and Dang, that a rhodium complex containing the chiral diphosphine (–)-DIOP ((–)-2,3-O-isopropylidene-2,3-dihydroxy-1,4-bis(diphenylphosphino)butane, derived from (+)-ethyl tartrate) efficiently catalyzed the enantiomeric reduction of unsaturated prochiral acids. In particular, α-acetamidocinnamic acid and α-phenylacetamidoacrylic acid were reduced quantitatively with 72 and 68% ee, respectively (Scheme 1.10) [42]. Two assumptions were presented by Kagan as being necessary to achieve a high degree of stereoselectivity in enantioselective catalysis:
- the ligand conformations must have maximum rigidity;
- in order to avoid epimerization equilibria, ligands must stay firmly bonded to the metal.

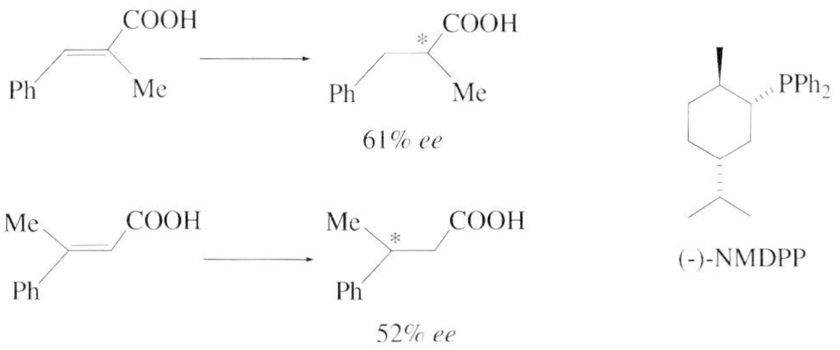

61% ee

52% ee

(-)-NMDPP

Scheme 1.9

Kagan concluded that diphosphines fulfill both of these conditions. In addition, it is desirable to avoid the possibility of geometric isomerism by choosing diphosphine having two equivalent phosphorus atoms [43]. These conditions strongly conditioned the subsequent development of the field, with the greater research effort being subsequently laid on metallic compounds with chiral diphosphines with C_2 symmetry as ligands. Additionally, Kagan introduced amino acid precursors as benchmark in enantioselective hydrogenation reactions.

At the same time, Knowles' team concentrated their efforts on the use of chiral at phosphorus ligands. Four years after the first report, these authors obtained 88% ee in the reduction of *α*-acylaminoacrylic acids by incorporating cyclohexyl and *ortho*-anisyl substituents [44] (Scheme 1.11).

A crucial achievement significantly stimulated the development of the investigation in the field of homogeneous enantioselective catalysis. The Knowles group established a method for the industrial synthesis of *L*-DOPA, a drug used for the treatment of Parkinson's disease. The key step of the process is the enantiomeric hydrogenation of a prochiral enamide, and this reaction is efficiently catalyzed by the air-stable rhodium complex $[Rh(COD)((R_P)\text{-}CAMP)_2]BF_4$ (Scheme 1.12).

This achievement was unique in two respects: 1) it was the first example of industrial application of a homogeneous enantioselective catalysis methodology; and 2) it represented a rare example of very quick convergence of basic knowledge into commercial application. The monophosphine ligand CAMP was shortly replaced by the related diphosphine ligand DIPAMP which improved the selectivity for the *L*-DOPA system up to 95% ee [45].

The following years witnessed the development of a plethora of new diphosphine ligands with chiral carbon backbones, and at a very impressive pace [46]. Among these, two examples were of particular interest.

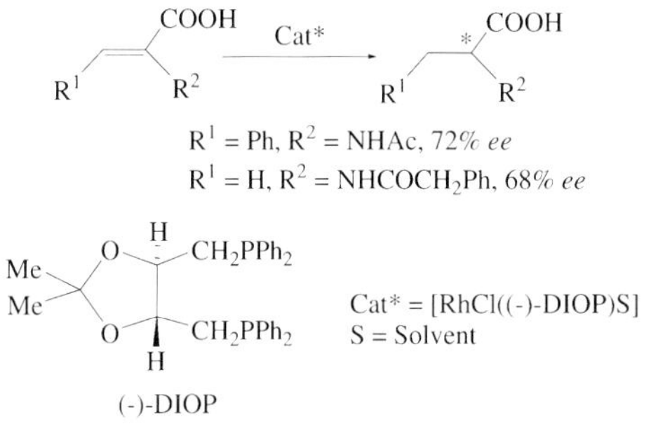
COOH
Cat*
COOH
R1
R2
R1
R2
R1 = Ph, R2 = NHAc, 72% ee
R1 = H, R2 = NHCOCH2Ph, 68% ee
Me
Me
O
O
H
H
CH2PPh2
CH2PPh2
Cat* = [RhCl((-)-DIOP)S]
S = Solvent
(-)-DIOP

Scheme 1.10

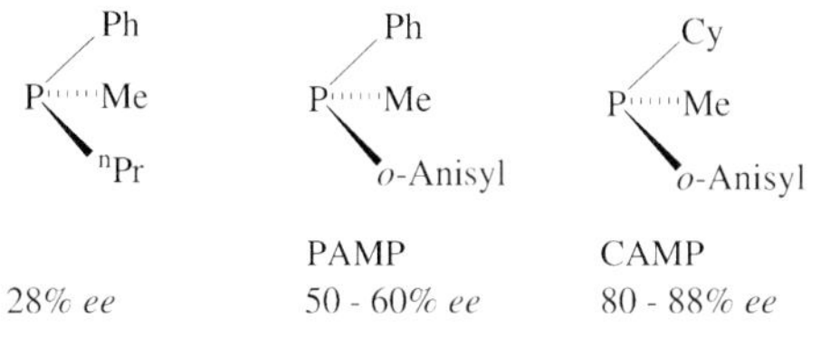
Ph
P
Me
nPr
Ph
P
Me
o-Anisyl
Cy
P
Me
o-Anisyl
PAMP
CAMP
28% ee
50 - 60% ee
80 - 88% ee

Scheme 1.11

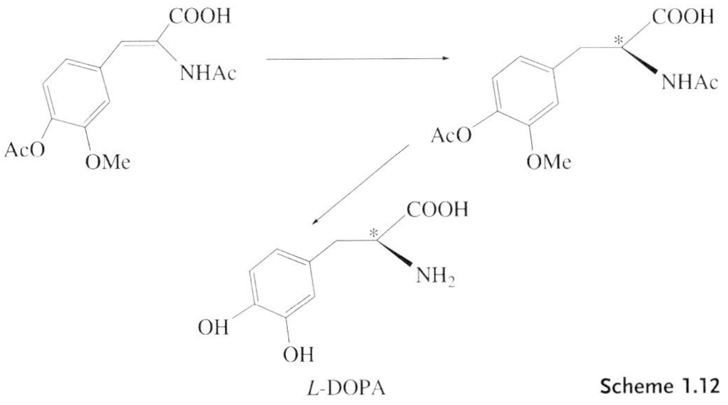

Scheme 1.12

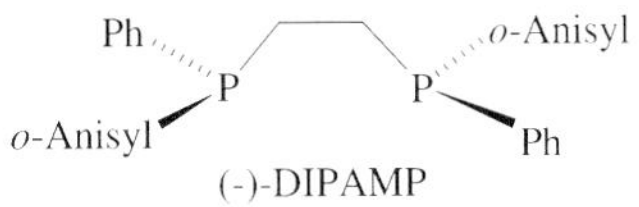

L-DOPA

(-)-DIPAMP

Scheme 1.13

The first example was the 2,2′-bis(diphenylphosphino)-1,1′-binaphthyl (BI-NAP) ligand synthesized by Noyori and Takaya in 1980 [47]. BINAP is an atro-poisomeric C_2 diphosphine that forms a seven-membered chelate ring through coordination to metals. The λ or δ conformations of this chelate determine the chiral disposition of the phenyl rings and, in this way, the chirality of BINAP is transmitted to the metal environment. BINAP complexes exhibit high enantioselectivity in various catalytic reactions, including hydrogenation. Thus, BINAP rhodium complexes [48] readily catalyze the hydrogenation of prochiral α-acylaminoacrylic acids with up to 100% ee (Scheme 1.14).

The second example was the bis(phospholanyl)ethanes (BPE) and bis(phospholanyl)benzenes (DuPHOS) ligands developed by Burk some years later [49]. These ligands are electron-rich diphosphines that contain two phospholanes *trans*-substituted in the 2,5 position, and in which the phosphorus atoms are connected by 1,2-ethano or 1,2-phenylene backbones. Chirality lies in the carbon atoms adjacent to the phosphorus atoms and therefore, in the derived catalysts, it lies in the immediacy of the rhodium atoms (Scheme 1.15). By choosing the appropriate R substituent, a large variety of substrates, including α-(acylamino) acrylic acids, enamides, enol acetates, β-keto esters, unsaturated carboxylic acids and itaconic acids, are efficiently hydrogenated by cationic rhodium complexes of the type [Rh(COD)(bisphospholane)]X (X = weakly coordinating anion) with exceedingly high enantioselectivities [50].

Subsequently, it was shown that rhodium complexes derived from diphosphinites, diphosphonites, and diphosphites – as well as hybrid ligands – are similarly active and selective to diphosphines [46 i].

Enantioselective hydrogenation catalysts with chiral nitrogen ligands have been rather neglected, however. Until now, it seems that ligands containing only

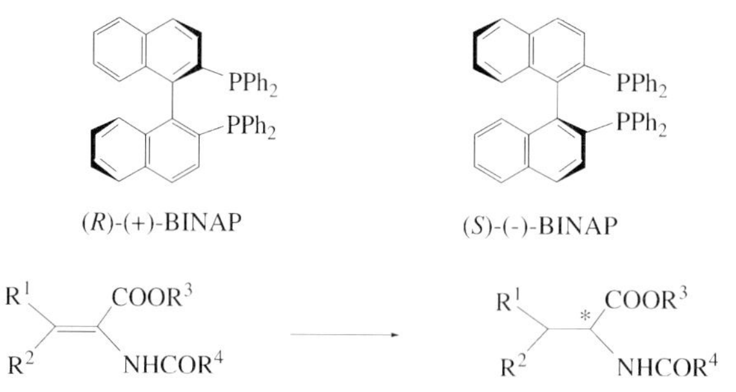

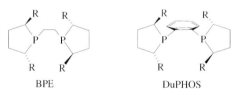

(R)-(+)-BINAP   (S)-(-)-BINAP

R¹	R²	R³	R⁴	ee (%)
H	Ph	H	Ph	100
Ph	H	H	Ph	87
H	Ph	Me	Ph	93
H	Ph	H	Me	84
H	m–OMe, p–OH– C₆H₂	H	Ph	79
H	H	H	Ph	98

Scheme 1.14

BPE   DuPHOS

Scheme 1.15

nitrogen donor atoms are not well suited for this purpose. Nevertheless, there are some examples of rhodium-based catalysts containing mixed P,N-chiral chelating ligands active for the reduction of C=C double bonds [51].

Since 1999 [52], the potential of monophosphorus ligands in the area has been reconsidered. The commonly accepted rule that bidentate ligands are necessary to achieve high enantioselectivity in hydrogenation reactions has been countered by the discovery that enantiopure monodentate P ligands may provide optical yields comparable to those obtained with chelating diphosphines [46i, 53]. This mostly occurs when two ligands on rhodium strongly restrict each other's conformational freedom [54]. In fact, monodentate P ligands containing biaryl, spirobiindane, oxaphosphinane, or carbohydrate building blocks have been used to efficiently hydrogenate α- and β-dehydroamino acids, itaconic acid derivatives and enamides with an ee systematically greater than 99% [46j, 53]. Recently, Reetz's group has demonstrated that mixtures of two different monodentate phosphorus ligands can be used in a combinatorial approach [53d–f]. Notably it has been shown that, in several cases, monodentate catalytic ligands induce higher enantioselectivity and lead to faster asymmetric hydrogenations than bidentate analogues [54, 55].

1.6.2
Hydrogenation of Ketones

Enantioselective hydrogenation of unfunctionalized ketones is promoted only by a limited number of transition metal catalysts. Among these, a few reports dealing with rhodium catalysts have appeared [46 a, b, e, g, i, j, m, 56]. In 1980, Markó et al. reported on the enantioselective catalytic reduction of unfunctionalized ketones, for the first time. In the presence of NEt$_3$, a neutral rhodium complex (prepared from [RhCl(NBD)]$_2$ and (+)-DIOP) hydrogenated methyl, α-naphthyl ketone to give the corresponding chiral alcohol in 84% ee [57]. Another important contribution to the field was the communication by Zhang et al. that alkyl methyl ketones can be efficiently hydrogenated by the *in-situ*-prepared catalyst from [RhCl(COD)]$_2$ and the conformational rigid chiral bisphosphines *P,P'*-1,2-phenylenebis(*endo*-2,5-dialkyl-7-phosphabicyclo[2.2.1] heptanes (PennPhos; Scheme 1.16). Enantiomeric excesses of up to 94% ee for the *tert*-butyl methyl ketone and 92% ee for cyclohexyl methyl ketone were observed, with the R=Me PennPhos ligand [58].

   The hydrogenation of ketones with O or N functions in the α- or β-position is accomplished by several rhodium compounds [46 a, b, e, g, i, j, m, 56]. Many of these examples have been applied in the synthesis of biologically active chiral products [59]. One of the first examples was the asymmetric synthesis of pantothenic acid, a member of the B complex vitamins and an important constituent of coenzyme A. Ojima et al. first described this synthesis in 1978, the most significant step being the enantioselective reduction of a cyclic α-keto ester, dihydro-4,4-dimethyl-2,3-furandione, to D-(–)-pantoyl lactone. A rhodium complex derived from [RhCl(COD)]$_2$ and the chiral pyrrolidino diphosphine, (2*S*,4*S*)-*N-tert*-butoxycarbonyl-4-diphenylphosphino-2-diphenylphosphinomethyl-pyrrolidine   ((*S,S*)-BPPM; Scheme 1.17) was used as catalyst [60]. The enantioselective hydrogenation of functionalized ketones was also efficiently achieved by a series of rhodium(I) aminophosphine- and amidophosphine-phosphinite complexes [61].

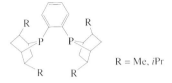

R = Me, *i*Pr

PennPhos   **Scheme 1.16**

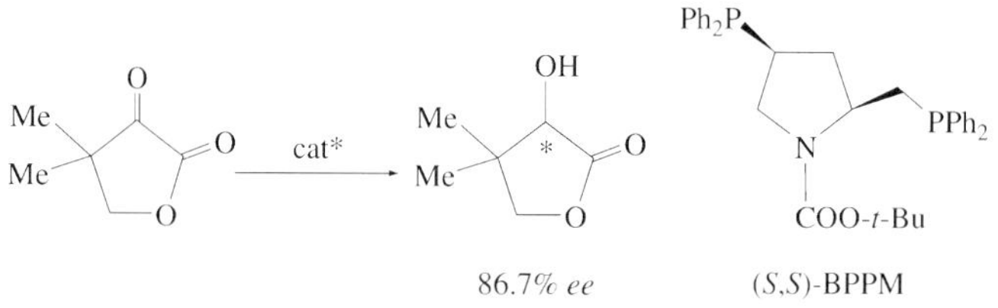
Me
Me
O
O
O
cat*
OH
Me
Me
*
O
O
86.7% ee
Ph2P
PPh2
N
COO-t-Bu
(S,S)-BPPM

Scheme 1.17

1.6.3 Hydrogenation of Imines

To date, catalyzed enantioselective reductions of C=N double bonds have only led to relatively limited success. Currently, only a few efficient chiral catalytic systems are available for the hydrogenation of imines. Hydrogenation of functionalized C=N double bonds such as *N*-acylhydrazones, sulfonimides, and *N*-diphenylphosphinylketimines has also been attempted, with some relevant results [46 b, g, i, j, 50, 62]. The first report on enantioselective homogeneous reduction of C=N double bonds mediated by rhodium complexes appeared in 1975. Kagan et al. reported that *N*-benzyl-*α*-phenyl ethylamine was prepared by hydrosilylation with 65% optical purity, using a chiral rhodium complex with (+)-DIOP [63]. Some years later, the Markó group reported that using catalysts prepared *in situ* from $[RhCl(NBD)]_2$ and chiral phosphines of the type $Ph_2PCHRCH_2PPh_2$ (R = Ph, *i*Pr, $PhCH_2$), optical yields up to 77% were achieved, although reproducibility of the results was poor [64]. Subsequently, some rhodium complexes have shown good ee values, among which two cases are remarkable. Acyclic *N*-alkylimines $ArC(Me){=}NCH_2Ph$ (Ar = Ph, 2-$MeOC_6H_4$, 3-$MeOC_6H_4$, 4-$MeOC_6H_4$) were hydrogenated in a two-phase system, with ee values up to 96%, using $[RhCl(COD)]_2$ together with sulfonated (*S*,*S*)-(–)-2,4-bis(diphenylphosphino)pentane ((*S*,*S*)-$BDPP_{sulf}$; Scheme 1.18) [65]. Similarly, the C=N group of *N*-acylhydrazones can be reduced by [Rh(COD)(DuPHOS)](CF_3SO_3) as catalyst precursor. Enantioselectivities up to 97% were achieved (Scheme 1.19). The N–N bond of the resulting *N*-benzoylhydrazines is cleaved by SmI_2 to afford the corresponding amines [66].

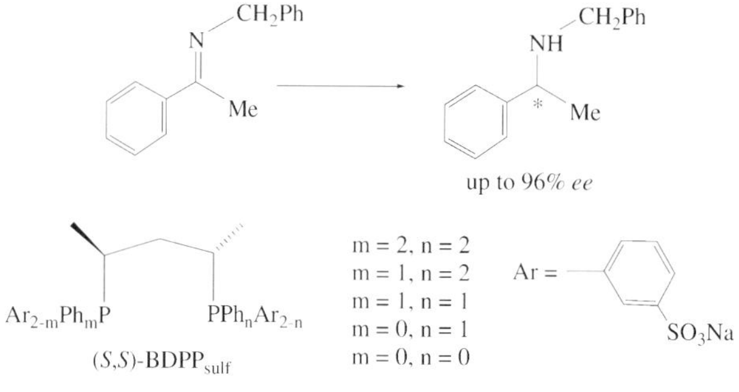
CH2Ph
N
Me
CH2Ph
NH
*
Me
up to 96% ee
Ar2-mPhmP
PPhnAr2-n
(S,S)-BDPPsulf
m = 2, n = 2
m = 1, n = 2
m = 1, n = 1
m = 0, n = 1
m = 0, n = 0
Ar =
SO3Na

Scheme 1.18

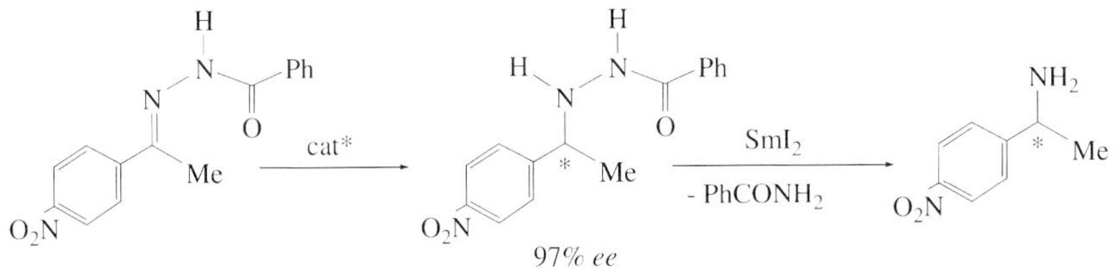

cat* = [Rh(COD)(Et-DuPHOS)](CF$_3$SO$_3$)

Scheme 1.19

1.6.4
Mechanism of Rhodium-Catalyzed Enantioselective Hydrogenation

The mechanism of rhodium-catalyzed enantioselective hydrogenation is one of the most thoroughly studied mechanisms of all metal-catalyzed processes [46b, c, h, i, k, 23, 50, 67]. Studies have been focused on cationic rhodium chiral diphosphine complexes as catalyst precursors, and on prochiral enamides as hydrogenation substrates. Early mechanistic studies were conditioned by the case of achiral hydrogenation of alkenes by Wilkinson's catalyst. In the catalytic cycle proposed for Wilkinson's catalyst, dihydride rhodium(III) complexes [14b, 35] are obtained by reversible oxidative addition of dihydrogen to the active species RhCl(PPh$_3$)$_2$ [23] (see Section 1.4). Thus, the formation of similar intermediates was expected upon hydrogenation of the catalyst precursors employed in asymmetric hydrogenation. However, Halpern et al. showed that [Rh(DIPHOS)(NBD)]$^+$ reacted with 2.0 mol of H$_2$/Rh yielding norbornane and the solvate complex [Rh(DIPHOS)S$_2$]$^+$ which was isolated as the dimer compound [Rh$_2$(DIPHOS)$_2$](BF$_4$)$_2$ (see Section 1.5) [31a, 68]. Further investigations showed that the formation of [Rh(PP*)S$_2$]$^+$ (PP*=chiral diphosphine; S=solvent molecule) complexes by hydrogenation of the catalyst precursors is a general behavior of *cis*-chelating diphosphine rhodium complexes [46c, 70]. It has been concluded that, under catalytic conditions, a molecule of the prochiral alkene coordinates to the rhodium atom, affording a bischelate catalyst–substrate complex (see Scheme 1.21). In fact, in 1978, for the first time, Brown and Chaloner provided ^{31}P-NMR-based evidence for the formation of such a type of compound [72], whilst shortly afterwards Halpern et al. reported the first X-ray molecular structure determination for a catalyst–substrate species, namely [Rh{(S,S)–CHIRAPHOS}{(Z)-EAC}]ClO$_4$ (EAC=ethyl-α-acetamidocinnamate) [73]. For C$_2$-symmetrical diphosphines – the most commonly used in enantioselective hydrogenation – two diastereomeric catalyst–substrate complexes can be formed, depending on the alkene enantioface that coordinates with the metal. However, only a single diastereomer of [Rh{(S,S)-CHIRAPHOS}{(Z)-EAC}]ClO$_4$ was detected in solution by NMR (Scheme 1.20); therefore, the second diastereomer must be present to the extent of less than 5%. Subsequently, numerous catalyst–substrate complexes have been described, and all have demonstrated stereodifferentiation in solution [67d].

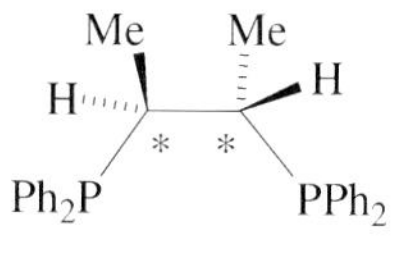

Me   Me

H........   .....H
         *   *
Ph₂P   PPh₂

(S,S)-CHIRAPHOS

Scheme 1.20

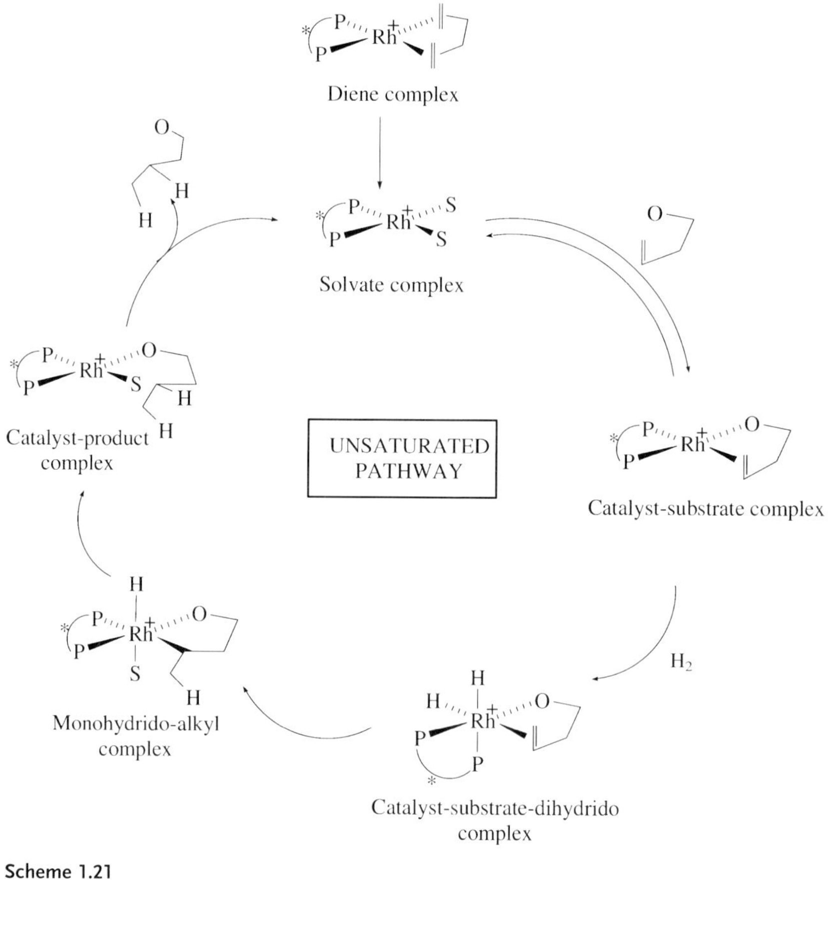

Diene complex

Solvate complex

Catalyst-product complex

UNSATURATED PATHWAY

Catalyst-substrate complex

Monohydrido-alkyl complex

Catalyst-substrate-dihydrido complex

H₂

Scheme 1.21

From all the above observations, it was concluded that, for diphosphine chelate complexes, the hydrogenation stage occurs after alkene association; thus, the unsaturated pathway depicted in Scheme 1.21 was proposed [31a,c, 74]. The monohydrido-alkyl complex is formed by addition of dihydrogen to the enamide complex, followed by transfer of a single hydride. Reductive elimination of the product regenerates the active catalysts and restarts the cycle. The monohydrido-alkyl intermediate was also observed and characterized spectroscopically [31c, 75], but the catalyst–substrate–dihydrido complex was not detected.

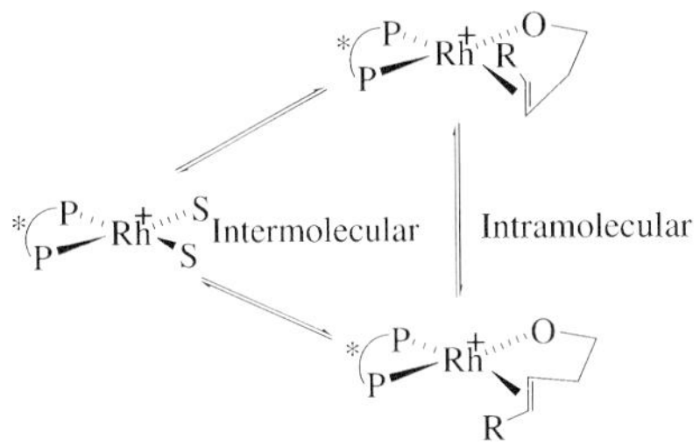

Scheme 1.22

The catalyst–substrate complexes deserve some additional comments. The two possible diastereomers for C_2-symmetrical diphosphines interconvert inter- and intramolecularly, the latter being the dominant mechanism [76] (Scheme 1.22). A second property – at least of some catalyst–substrate complexes – is that the reactivity of the minor diastereomer toward H_2 is notably higher than that of the major diastereomer.

Consequently, the stereochemistry of the hydrogenation is regulated by this relative reactivity rather than by the relative thermodynamic stability of the diastereomers [73, 75, 77]. Halpern and Landis accomplished detailed kinetic measurements on the hydrogenation of methyl-(Z)-α-acetamidocinnamate [(Z)-MAC] enamide catalyzed by Rh(DIPAMP) species. These authors studied the influence of temperature and pressure on the interconversion of catalyst–substrate complexes, on their reaction with H_2 and, as a consequence, on the ee achieved. The oxidative addition of H_2 is the enantio-determining and the turnover-limiting step, and it was concluded that the ee increased with decreasing temperature, because the concentration of the minor diastereomer increased, whereas it decreased with increasing pressure because hydrogenation of the major diastereomer became significant [78].

Extensive computational calculations have been performed by using molecular mechanics (MM) [79], quantum mechanics (QM) [80], or combined MM/QM methods [81]. As major contributions, these theoretical studies predict the greater stability of the major isomer, explain the higher reactivity of the minor diastereomer, introduce the formation of a dihydrogen adduct as intermediate in the oxidative addition of H_2 to the catalyst–substrate complexes, and propose the migratory insertion, instead of the oxidative addition, as a turnover-limiting step.

A new insight into the mechanism of enantioselective hydrogenation was achieved when the elusive dihydride intermediates were detected and characterized spectroscopically [82]. Bargon, Selke et al. detected, through parahydrogen-induced polarization (PHIP), NMR experiments [83, 84], the formation of a dihydride in the hydrogenation of dimethylitaconate with an enantiomerically pure bis(phosphinite)rhodium(I) catalyst (see Scheme 1.23). The structural assignment was not complete: it was not proven that the itaconate is in the coordination sphere of the metal (no ^{13}C-labeled experiments were performed) and there is a supposed *trans* $^2J_{PH}$ of only 4 Hz, far from the usual values for this coupling. However, the presence of

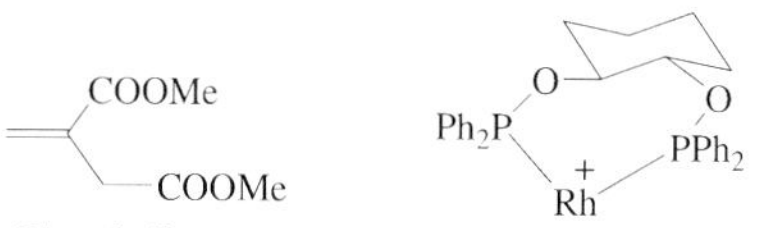

COOMe
COOMe
Dimethylitaconate

Ph₂P
O
O
PPh₂
+
Rh
Bis(phosphinite)rhodium complex

Scheme 1.23

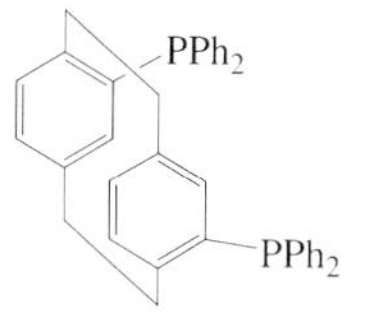

PPh₂
PPh₂

PHANEPHOS   **Scheme 1.24**

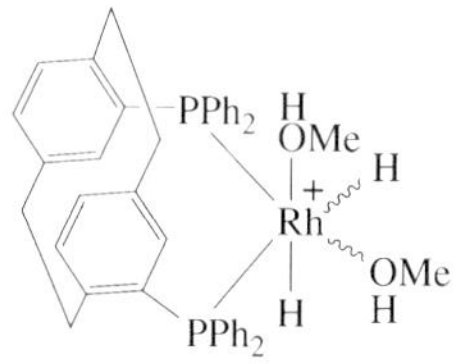

PPh₂ H
OMe
H
+
Rh
OMe
PPh₂ H   H

Scheme 1.25

some transient rhodium dihydride is definitive from the experimental evidences [84].

The use of the diphosphine PHANEPHOS (see Scheme 1.24) permitted Bargon, Brown and colleagues to detect and characterize a dihydrido intermediate in the hydrogenation of the enamide MAC by a rhodium-based catalyst. The PHIP NMR technique was employed, and showed one of the hydrogen atoms to be agostic between the rhodium center and the β-carbon of the substrate [85]. By using the same diphosphine and technique it was also possible to detect two diastereomers of the dihydride depicted in Scheme 1.25, which may also be detected using conventional NMR measurements [86].

Following a preparative method developed by Evans [87], Imamoto et al. prepared the C_2-symmetric electron-rich diphosphines BisP* with two stereogenic phosphorus centers [88] (Scheme 1.26). Further, Gridnev and Imamoto carried out detailed mechanistic investigations on the hydrogenation of α-dehydroamino acids and other unsaturated substrates such as enamides, (E)-β-dehydroamino acids or 1-benzoyloxyethenephosphonate using the rhodium complexes [Rh(diene)(t-Bu-BisP*)]⁺ (diene = COD, NBD) as catalyst precursors [46h, 67c, d, 89]. These authors showed that hydrogenation of the diene precursors at −20 °C gives the expected solvate complexes [Rh(t-Bu-BisP*)S₂]⁺, but at −90 °C affords two diastereomers of the catalysts–solvate dihydrido complex [RhH₂(t-Bu-BisP*)S₂]⁺. The formation of these dihydrides is reversible [89a]. Similar dihydrides were prepared for the other BisP* ligands [89b]. The catalysts–solvate–dihydrido reacts with α-dehydroamino acids, giving monohydride intermediates

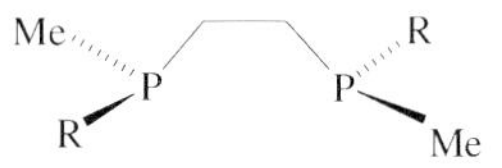

R = *t*-Bu, 1-adamantyl, 1-methylcyclohexyl,
   Et₃C, *c*-C₅H₉, *c*-C₆H₁₁, *i*-Pr

(*S,S*)-BisP*

Scheme 1.26

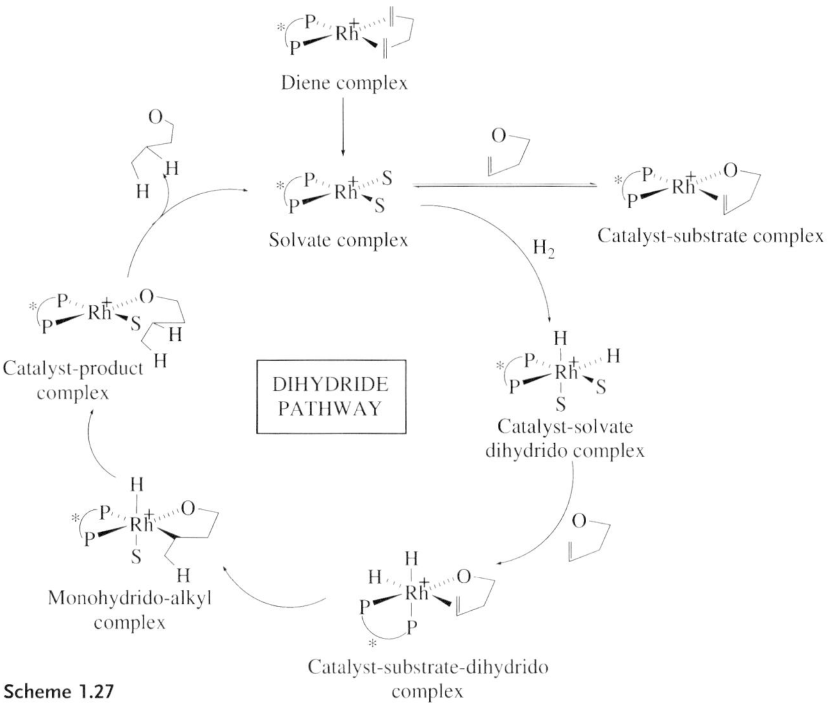

Diene complex

Solvate complex

Catalyst-substrate complex

Catalyst-product complex

DIHYDRIDE PATHWAY

Catalyst-solvate dihydrido complex

Monohydrido-alkyl complex

Catalyst-substrate-dihydrido complex

Scheme 1.27

and, eventually, the hydrogenation product with the same configuration and ee value (99%) as obtained in the catalytic reaction. On the other hand, the catalyst–substrate complexes were easily prepared by adding the substrates to the solvates. Hydrogenation of the catalyst-substrate complexes gives, in general, the hydrogenation product, but only at higher temperatures and with a lower ee. From the comparison of their results, the authors concluded that in this case the dihydride mechanism depicted in Scheme 1.27 was operating. The catalyst–substrate complexes dissociate, yielding the solvate complex that is hydrogenated to the solvate–dihydride complex. This dihydride reacts with the substrate, giving the monohydride–alkyl complex as the next observable species. The hydrogenation product is obtained after the reductive elimination step.

Notably, under catalytic conditions, reaction steps preceding catalyst–substrate–dihydrido formation are shown to be in rapid equilibrium. Therefore, stereoselection must occur at a later step in the catalytic cycle, and the authors suggested the migratory insertion as an enantio-determining and turnover-limiting irreversible step of enantioselective hydrogenation. As a major consequence, following the Gridnev and Imamoto contribution, the boundaries between unsaturated pathway and dihydride pathway are blurred. At least in some cases, both mechanisms are operating, joining in a single pathway before stereoselection occurs.

1.7
Some Dinuclear Catalyst Precursors

Several dinuclear rhodium complexes such as the above-mentioned [Rh$_2$(OAc)$_4$] have been used as hydrogenation catalysts [22, 23]. Maitlis and coworkers have studied the chemistry and catalytic activity of the [Rh(C$_5$Me$_5$)Cl$_2$]$_2$ complex and related complexes. Kinetic studies suggested that cleavage into monomer occurs in the most active catalysts [90].

Muetterties has suggested that the dimeric hydride [RhH(P{OiPr}$_3$)$_2$]$_2$ catalyzes alkene and alkyne hydrogenation *via* dinuclear intermediates [91]. However, no kinetic evidence has been reported to prove the integrity of the catalysts during the reactions. On the other hand, studies of the kinetics of the hydrogenation of cyclohexene catalyzed by the heterodinuclear complexes [H(CO)(PPh$_3$)$_2$Ru(μ-bim)M(diene)] (M = Rh, Ir; bim = 2,2'-biimidazolate) suggested that the full catalytic cycle involves dinuclear intermediates [92].

1.8
Concluding Remark

Throughout the history of homogeneous catalysis – and especially of homogeneous hydrogenation – rhodium complexes have played a crucial role due to the combination of excellent coordinative properties, remarkable activity and selectivity, the adequate redox characteristics of the couple Rh(I)/Rh(III) as well as NMR properties (100% abundance, spin 1/2). The selection of results presented in this chapter show that homogeneous hydrogenation by rhodium complexes represents not only a rich past, as indicated by the exponential growth of the past forty years, but also a bright future.

Abbreviations

BINAP   2,2'-bis(diphenylphosphino)-1,1'-binaphthyl
BPE   bis(phospholanyl)ethane
COD   1,5-cyclooctadiene
DIPHOS   1,2-bis(diphenyl)phosphinoethane
DuPHOS   bis(phospholanyl)benzene
MH   monohydride

MM	molecular mechanics
NBD	2,5-norbornadiene
NMDPP	neomethyldiphenylphosphine
PHIP	parahydrogen-induced polarization
QM	quantum mechanics
TFB	tetrafluorobenzobarrelene
(Z)-MAC	methyl-(Z)-α-acetamidocinnamate

References

1 B. R. James, *Homogeneous Hydrogenation*, Wiley, New York, **1973**.

2 P. A. Chaloner, M. A. Esteruelas, F. Joó, L. A. Oro, *Homogeneous Hydrogenation*, Kluwer Academic, Dordrecht, **1994**.

3 F. J. McQuilin, *Homogeneous Hydrogenation in Organic Chemistry*, D. Reidel, Dordrecht, **1976**.

4 (a) L. A. Oro, in: *Encyclopedia of Catalysis*, I. V. Horváth, E. Iglesia, M. T. Klein, J. A. Lercher, A. J. Russell, E. I. Stiefel (Eds.), John Wiley & Sons, **2003**, Vol. 4, p. 55; (b) R. Sánchez-Delgado, M. Rosales, *Coord. Chem. Rev.* **2000**, *196*, 249; (c) A. Spencer, in: *Comprehensive Coordination Chemistry*, G. Wilkinson, R. D. Guillard, J. A. McCleverty (Eds.), Pergamon, Oxford, **1987**, Vol. 6, Chapter 61.2, p. 229; (d) B. R. James, in: *Comprehensive Organometallic Chemistry*, G. Wilkinson, F. G. A. Stone, E. A. Abel (Eds.), Pergamon, Oxford, **1982**, vol. 8, Chapter 51, p. 285; (e) B. R. James, *Adv. Organomet. Chem.*, **1979**, *17*, 319.

5 (a) M. Calvin, *Trans. Faraday Society* **1938**, *34*, 1181; (b) M. Calvin, *J. Am. Chem. Soc.* **1939**, *61*, 2230.

6 M. Iguchi, *J. Chem. Soc. Japan* **1939**, *60*, 1287.

7 P. J. Brothers, *Progress Inorg. Chem.* **1981**, *28*, 1.

8 (a) M. Iguchi, *J. Chem. Soc. Japan* **1942**, *63*, 634; (b) M. Iguchi, *J. Chem. Soc. Japan* **1942**, *63*, 1752.

9 (a) S. S. Bath, L. Vaska, *J. Amer. Chem. Soc.* **1965**, *85*, 3500; (b) L. Vaska, *Inorg. Nucl. Chem. Lett.* **1965**, *1*, 89.

10 (a) D. Evans, G. Yagupsky, G. Wilkinson, *J. Chem. Soc. (A)* **1968**, 2660; (b) M. Yagupsky, C. K. Brown, G. Yagupsky, G. Wilkinson, *J. Chem. Soc. (A)* **1970**, 937.

11 (a) C. O'Connor, G. Yagupsky, D. Evans, G. Wilkinson, *Chem. Commun.* **1968**, 420; (b) C. O'Connor, G. Wilkinson, *J. Chem. Soc. (A)* **1968**, 2665.

12 G. Yagupsky, G. Wilkinson, *J. Chem. Soc. (A)* **1970**, 941.

13 (a) D. Evans, J. A. Osborn, G. Wilkinson, *J. Chem. Soc. (A)* **1968**, 3133; (b) G. Yagupsky, C. K. Brown, G. Wilkinson, *Chem. Commun.* **1969**, 1244; (c) G. Yagupsky, C. K. Brown, G. Wilkinson, *J. Chem. Soc. (A)* **1970**, 1392; (d) C. K. Brown, G. Wilkinson, *J. Chem. Soc. (A)* **1970**, 2753.

14 (a) J. A. Osborn, G. Wilkinson, J. F. Young, *Chem. Commun.* **1965**, 17; (b) J. A. Osborn, F. H. Jardine, J. F. Young, G. Wilkinson, *J. Chem. Soc. (A)* **1966**, 1711; (c) M. A. Bennett, P. A. Longstaff, *Chem. Ind. (London)* **1965**, 846; (d) R. S. Coffey, British Patent 1121642, **1965**; (e) L. Vaska, R. E. Rhodes, *J. Am. Chem. Soc.* **1965**, *87*, 4970.

15 L. Horner, H. Büthe, H. Siegel, *Tetrahedron Lett.* **1968**, 4023.

16 F. H. Jardine, *Progress Inorg. Chem.* **1981**, *28*, 63.

17 J. R. Shapley, R. R. Schrock, J. A. Osborn, *J. Am. Chem. Soc.* **1969**, *91*, 2816.

18 (a) B. F. G. Jonson, J. Lewis, D. A. White, *J. Am. Chem. Soc.* **1969**, *91*, 5186; (b) L. M. Haines, *Inorg. Nucl. Chem. Lett.* **1969**, *5*, 399; (c) M. Green, T. A. Kuc, S. Taylor, *Chem. Commun.* **1970**, 1553.

19 R. R. Schrock, J. A. Osborn, *Chem. Commun.* **1970**, 567.

20 B. C. Hui, G. L. Rempel, *Chem. Commun.* **1970**, 1195.

21 (a) P. Legzdins, G. Rempel, G. Wilkinson, *Chem. Commun.* **1969**, 825; (b) P. Legzdins, R. W. Mitchell, G. Rempel,

J. D. Ruddick, G. Wilkinson, *J. Chem. Soc. (A)* **1970**, 3322.

22 S. J. La Placa, J. Ibers, *Acta Cryst.* **1965**, *18*, 511.

23 J. Halpern, *Inorg. Chim. Acta* **1981**, *50*, 11, and references therein.

24 (a) S. B. Duckett, C. L. Newell, R. Eisenberg, *J. Am. Chem. Soc.* **1994**, *116*, 10548; (b) S. B. Duckett, C. L. Newell, R. Eisenberg, *J. Am. Chem. Soc.* **1997**, *119*, 2068.

25 (a) S. A. Colebrooke, S. B. Duckett, J. A. B. Lohman, *Chem. Commun.* **2000**, 695; (b) S. A. Colebrooke, S. B. Duckett, J. A. B. Lohman, R. Eisenberg, *Chem. Eur. J.* **2004**, *10*, 2459.

26 D. Blazina, S. B. Duckett, J. P. Dunne, C. Godard, *Dalton Trans.* **2004**, 2601.

27 F. Joó, P. Csiba, A. Bényei, *J. Chem. Soc., Chem. Commun.* **1993**, 1602.

28 (a) R. R. Schrock, J. A. Osborn, *J. Am. Chem. Soc.,* **1976**, *98*, 2134; (b) R. R. Schrock, J. A. Osborn, *J. Am. Chem. Soc.* **1976**, *98*, 2143; (c) R. R. Schrock, J. A. Osborn, *J. Am. Chem. Soc.* **1976**, *98*, 4450.

29 R. Usón, L. A. Oro, R. Sariego, M. Valderrama, C. Rebullida, *J. Organomet. Chem.* **1980**, *197*, 87.

30 M. A. Esteruelas, L. A. Oro, *Coord. Chem. Rev.* **1999**, *193*, 557.

31 (a) J. Halpern, D. P. Riley, A. S. C. Chan, J. J. Pluth, *J. Am. Chem. Soc.* **1977**, *98*, 8055; (b) J. Halpern, D. P. Riley, A. S. C. Chan, J. J. Pluth, *Adv. Chem. Ser.* **1977**, *173*, 16; (c) A. S. C. Chan, J. Halpern, *J. Am. Chem. Soc.* **1980**, *102*, 838.

32 R. A. Sánchez-Delgado, D. Rondón, A. Andriollo, V. Herrera, G. Martín, B. Chaudret, *Organometallics* **1993**, *12*, 4291.

33 (a) R. A. Sánchez-Delgado, V. Herrera, L. Rincón, A. Andriollo, G. Martín, *Organometallics* **1994**, *13*, 553; (b) V. Herrera, A. Fuentes, M. Rosales, R. A. Sánchez-Delgado, C. Bianchini, A. Meli, F. Vizza, *Organometallics* **1997**, *16*, 2465.

34 E. Baralt, S. J. Smith, J. Hurwitz, I. T. Horvath, R. H. Fish, *J. Am. Chem. Soc.* **1992**, *114*, 5187.

35 J. F. Young, J. A. Osborn, F. H. Jardine, G. Wilkinson, *Chem. Commun.* **1965**, 131.

36 O. Korpin, K. Mislow, *J. Am. Chem. Soc.* **1967**, *89*, 4784.

37 L. Horner, H. Winkler, A. Rapp, A. Mentrup, H. Hoffmann, P. Beck, *Tetrahedron Lett.* **1961**, 161.

38 W. S. Knowles, M. J. Sabacky, *Chem. Commun.* **1968**, 1445.

39 L. Horner, H. Siegel, H. Büthe, *Angew. Chem.* **1968**, *80*, 1034; *Angew. Chem., Int. Ed. Engl.* **1968**, *7*, 942.

40 The ee were improved up to 28% by increasing the optical purity of the phosphine (it was 69% ee at the beginning).

41 J. D. Morrison, R. E. Burnett, A. M. Aguiar, C. J. Morrow, C. Phillips, *J. Am. Chem. Soc.* **1971**, *93*, 1301.

42 T. P. Dang, H. B. Kagan, *Chem. Commun.* **1971**, 481.

43 H. B. Kagan, T.-P. Dang, *J. Am. Chem. Soc.* **1972**, *94*, 6429.

44 W. S. Kowles, M. J. Sabacky, B. D. Vineyard, *J. Chem. Soc., Chem. Commun.* **1972**, 10.

45 (a) W. S. Knowles, M. J. Sabacky, B. D. Vineyard, D. J. Weinkauff, *J. Am. Chem. Soc.* **1975**, *97*, 2567; (b) W. S. Knowles, *Acc. Chem. Res.* **1983**, *16*, 106; (c) W. S. Knowles, *Angew. Chem.* **2002**, *114*, 2096; *Angew. Chem. Int. Ed.* **2002**, *41*, 1998 (Nobel lecture).

46 (a) H. Takaya, T. Otha, R. Noyori, in: *Catalytic Asymmetric Synthesis*, I. Ojima (Ed.), VCH: New York, **1993**, p. 1; (b) R. Noyori, in: *Asymmetric Catalysis in Organic Synthesis*, Wiley: New York, **1994**, p. 16; (c) J. M. Brown, in: *Comprehensive Asymmetric Catalysis*, E. N. Jacobsen, A. Pfaltz, H. Yamamoto (Eds.), Springer: Berlin, **1999**, Vol. 1, p. 121; (d) R. L. Halterman, *Comprehensive Asymmetric Catalysis*, E. N. Jacobsen, A. Pfaltz, H. Yamamoto (Eds.), Springer: Berlin, **1999**, Vol. 1, p. 183; (e) T. Ohkuma, R. Noyori, *Comprehensive Asymmetric Catalysis*, E. N. Jacobsen, A. Pfaltz, H. Yamamoto (Eds.), Springer: Berlin, **1999**, Vol. 1, p. 199; (f) H.-U. Blaser, F. Spindler, *Comprehensive Asymmetric Catalysis*, E. N. Jacobsen, A. Pfaltz, H. Yamamoto (Eds.), Springer: Berlin, **1999**, Vol. 1, p. 247; (g) T. Okhuma, M. Kitamura, R. Noyori, in: *Catalytic Asymmetric Synthesis*, I. Ojima (Ed.), Wiley-VCH: New York, **2000**, p. 1; (h) K. V. L.

Crépy, T. Imamoto, *Adv. Synth. Catal.* **2003**, *345*, 79; (i) H.-U. Blaser, C. Malan, B. Pugin, F. Spindler, H. Steiner, M. Studer, *Adv. Synth. Catal.* **2003**, *345*, 103; (j) W. Tang, X. Zhang, *Chem. Rev.* **2003**, *103*, 3029; (k) L. Dahlenburg, *Eur. J. Inorg. Chem.* **2003**, 2733; (l) J.-P. Genet, *Acc. Chem. Res.* **2003**, *36*, 908; (m) P. Barbaro, C. Bianchini, G. Giambastiani, S. L. Parivel, *Coord. Chem. Rev.* **2004**, *248*, 2131.

47 A. Miyashita, A. Yasuda, H. Takaya, K. Toriumi, T. Ito, T. Souchi, R. Noyori, *J. Am. Chem. Soc.* **1980**, *102*, 7932.

48 The methanol complex [Rh(BINAP) (MeOH)$_2$]ClO$_4$ and the complex resulting from loss of MeOH from it are used as catalysts [47]. Both BINAP enantiomers were employed.

49 M. J. Burk, *J. Am. Chem. Soc.* **1991**, *113*, 8518.

50 M. J. Burk, *Acc. Chem. Res.* **2000**, *33*, 363.

51 (a) A. Togni, L. M. Venanzi, *Angew. Chem.* **1994**, *106*, 517; *Angew. Chem. Int. Ed. Engl.* **1994**, *33*, 497; (b) F. Fache, E. Schulz, M. L. Tommasino, M. Lemaire, *Chem. Rev.* **2000**, *100*, 2159.

52 F. Guillen, J. Fiaud, *Tetrahedron Lett.* **1999**, *40*, 2939.

53 (a) T. Hayashi, *Acc. Chem. Res.* **2000**, *33*, 354; (b) F. Lagasse, H. B. Kagan, *Chem. Pharm. Bull.* **2000**, *48*, 315; (c) I. V. Komarov, A. Börner, *Angew. Chem.* **2001**, *113*, 1237; *Angew. Chem. Int. Ed.* **2001**, *40*, 1197; (d) M. T. Reetz, T. Sell, A. Meiswinkel, G. Mehler, *Angew. Chem.* **2003**, *115*, 814; *Angew. Chem. Int. Ed.* **2003**, *42*, 790; (e) M. T. Reetz, G. Mehler, *Tetrahedron Lett.* **2003**, *44*, 4593; (f) M. T. Reetz, G. Mehler, A. Meiswinkel, *Tetrahedron: Asymmetry* **2004**, *15*, 2165; (g) M. T. Reetz, J.-A. Ma, R. Goddard, *Angew. Chem.* **2005**, *117*, 416; *Angew. Chem. Int. Ed.* **2005**, *44*, 412.

54 C. Claver, E. Fernández, A. Gillon, K. Heslop, D. J. Hyett, A. Martorell, A. G. Orpen, P. G. Pringle, *Chem. Commun.* **2000**, 961.

55 (a) M. van den Berg, A. J. Minnaard, E. P. Schudde, J. van Esch, A. H. M. de Vries, J. G. de Vries, B. Feringa, *J. Am. Chem. Soc.* **2000**, *122*, 11539; (b) D. Peña, A. J. Minnaard, A. H. M. de Vries, J. G. de Vries, B. L. Feringa, *Org. Lett.* **2003**, *5*, 475.

56 R. Noyori, T. Ohkuma, *Angew. Chem.* **2001**, *113*, 41; *Angew. Chem. Int. Ed. Engl.* **2001**, *40*, 40.

57 S. Törös, B. Heil, L. Kollár, L. Markó, *J. Organomet. Chem.* **1980**, *197*, 85.

58 Q. Jiang, Y. Jiang, D. Xiao, P. Cao, X. Zhang, *Angew. Chem.* **1998**, *110*, 1203; *Angew. Chem. Int. Ed. Engl.* **1998**, *37*, 1100.

59 (a) K. Inoguchi, S. Sakuraba, K. Achiwa, *Synlett* **1992**, 169; (b) H. P. Märki, Y. Crameri, R. Eigenmann, A. Krasso, H. Ramuz, K. Bernauer, M. Goodman, K. L. Melmon, *Helv. Chim. Acta* **1988**, *71*, 320; (c) S. Sakuraba, N. Nakajima, K. Achiwa, *Tetrahedron: Asymmetry* **1993**, *4*, 1457; (d) H.-U. Blaser, H.-P. Jalett, F. Spindler, *J. Mol. Catal. A: Chem.* **1996**, *107*, 85.

60 I. Ojima, T. Kogure, T. Terasaki, K. Achiwa, *J. Org. Chem.* **1978**, *43*, 3444.

61 C. Pasquier, S. Naili, A. Mortreux, F. Agbossou, L. Pélinski, J. Brocard, J. Eilers, I. Reiners, V. Peper, J. Martens, *Organometallics* **2000**, *19*, 5723 and references therein.

62 H.-U. Blaser, F. Spindler, *Comprehensive Asymmetric Catalysis*, E. N. Jacobsen, A. Pfaltz, H. Yamamoto (Eds.), Springer: Berlin, **1999**, Vol. 1, p. 247.

63 H. B. Kagan, N. Langlois, T. P. Dang, *J. Organomet. Chem.* **1975**, *90*, 353.

64 S. Vastag, J. Bakos, S. Törös, N. E. Takach, R. B. King, B. Heil, L. Markó, *J. Mol. Catal.* **1984**, *22*, 283.

65 J. Bakos, A. Orosz, B. Heil, M. Laghmari, P. Lhoste, D. Sinou, *J. Chem. Soc., Chem. Commun.* **1991**, 1684.

66 M. J. Burk, J. E. Feaster, *J. Am. Chem. Soc.* **1992**, *114*, 6266.

67 (a) J. Halpern, *Science* **1982**, *217*, 401; (b) J. M. Brown, *Chem. Soc. Rev.* **1993**, *22*, 25; (c) J. M. Brown, R. Giernoth, *Curr. Opinion Drug Discov. Devel.* **2000**, *3*, 825; (d) I. D. Gridnev, T. Imamoto, *Acc. Chem. Res.* **2004**, *37*, 633.

68 Some years later, *para*-enriched dihydrogen experiments indicated the existence of a non-detectable amount of a dihydride in equilibrium with [RhP$_2$S$_2$]$^+$ species [69].

69 J. M. Brown, L. R. Canning, A. J. Downs, A. M. Foster, *J. Organomet. Chem.* **1983**, *255*, 103.

70 At an early stage, only for diphosphine ligands capable of *trans*-coordination were detected hydride solvate intermediates [71], but a catalytic pathway based on diphosphine *trans*-chelation has not been developed to date (see [80b]).

71 J. M. Brown, P. A. Chaloner, A. G. Kent, B. A. Murrer, P. N. Nicholson, D. Parker, P. J. Sidebottom, *J. Organomet. Chem.* **1981**, *216*, 263.

72 J. M. Brown, P. A. Chaloner, *J. Chem. Soc., Chem. Commun.* **1978**, 321.

73 J. Halpern, A. S. C. Chan, J. J. Pluth, *J. Am. Chem. Soc.* **1980**, *102*, 5952.

74 A. S. C. Chan, J. J. Pluth, J. Halpern, *Inorg. Chim. Acta* **1979**, *37*, L477.

75 J. M. Brown, P. A. Chaloner, *J. Chem. Soc., Chem. Commun.* **1980**, 344.

76 J. A. Ramsden, T. D. W. Claridge, J. M. Brown, *J. Chem. Soc., Chem. Commun.* **1995**, 2469 and references therein.

77 P. S. Chua, N. K. Roberts, B. Bosnich, S. J. Okrasinski, J. Halpern, *J. Chem. Soc., Chem. Commun.* **1981**, 1278.

78 C. R. Landis, J. Halpern, *J. Am. Chem. Soc.* **1987**, *109*, 1746.

79 J. S. Giovannetti, C. M. Kelly, C. R. Landis, *J. Am. Chem. Soc.* **1993**, *115*, 4040.

80 (a) C. R. Landis, P. Hilfenhaus, S. Feldgus, *J. Am. Chem. Soc.* **1999**, *121*, 8741; (b) A. Kless, A. Börner, D. Heller, R. Selke, *Organometallics* **1997**, *16*, 2096.

81 (a) C. R. Landis, S. Feldgus, *Angew. Chem.* **2000**, *112*, 2985; *Angew. Chem. Int. Ed.* **2000**, *39*, 2863; (b) S. Feldgus, C. R. Landis, *J. Am. Chem. Soc.* **2000**,

122, 12714; (c) S. Feldgus, C. R. Landis, *Organometallics* **2001**, *20*, 2374.

82 K. Rosen, *Angew. Chem.* **2001**, *113*, 4747; *Angew. Chem. Int. Ed.* **2001**, *40*, 4611.

83 C. Russell Bowers, D. P. Weitekamp, *J. Am. Chem. Soc.* **1987**, *109*, 5541.

84 A. Harthun, R. Kadyrov, R. Selke, J. Bargon, *Angew. Chem.* **1997**, *109*, 1155; *Angew. Chem. Int. Ed. Engl.* **1997**, *36*, 1103.

85 R. Giernoth, H. Heinrich, N. J. Adams, R. J. Deeth, J. Bargon, J. M. Brown, *J. Am. Chem. Soc.* **2000**, *122*, 12381.

86 H. Heinrich, R. Giernoth, J. Bargon, J. M. Brown, *Chem. Commun.* **2001**, 1296.

87 A. R. Muci, K. R. Campos, D. A. Evans, *J. Am. Chem. Soc.* **1995**, *117*, 9075.

88 T. Imamoto, J. Watanabe, Y. Wada, H. Masuda, H. Yamada, H. Tsuruta, S. Matsukawa, K. Yamaguchi, *J. Am. Chem. Soc.* **1998**, *120*, 1635.

89 (a) I. D. Gridnev, N. Higashi, K. Asakura, T. Imamoto, *J. Am. Chem. Soc.* **2000**, *122*, 7183; (b) I. D. Gridnev, Y. Yamanoi, N. Higashi, H. Tsuruta, M. Yasutake, T. Imamoto, *Adv. Synth. Catal.* **2001**, *343*, 118.

90 P. M. Maitlis, *Acc. Chem. Res.* **1978**, *11*, 301.

91 (a) A. J. Sivak, E. L. Muetterties, *J. Am. Chem. Soc.* **1979**, *101*, 4878; (b) E. L. Muetterties, *Inorg. Chim. Acta* **1981**, *50*, 9.

92 (a) M. P. García, A. M. López, M. A. Esteruelas, F. J. Lahoz, L. A. Oro, *J. Chem. Soc., Chem. Commun.* **1988**, 793; (b) M. A. Esteruelas, M. P. García, A. M. López, L. A. Oro, *Organometallics* **1991**, *10*, 127.

2
Iridium

Robert H. Crabtree

2.1
Introduction

Today, iridium compounds find so many varied applications in contemporary homogeneous catalysis it is difficult to recall that, until the late 1970s, rhodium was one of only two metals considered likely to serve as useful catalysts, at that time typically for hydrogenation or hydroformylation. Indeed, catalyst/solvent combinations such as [IrCl(PPh$_3$)$_3$]/MeOH, which were modeled directly on what was previously successful for rhodium, failed for iridium. Although iridium was still considered potentially to be useful, this was only for the demonstration of stoichiometric reactions related to proposed catalytic cycles. Iridium tends to form stronger metal–ligand bonds (e.g., Cp(CO)Rh-CO, 46 kcal mol^{-1}; Cp(CO)Ir-CO, 57 kcal mol^{-1}), and consequently compounds which act as reactive intermediates for rhodium can sometimes be isolated in the case of iridium.

When low-coordinate iridium fragments in "non-coordinating" solvents (e.g., {Ir(PPh$_3$)$_2$}$^+$ in CH$_2$Cl$_2$) were found to be much more active than their rhodium analogues, it became clear that it is the *dissociation* of ligands or solvent – much slower for Ir versus Rh and for MeOH versus CH$_2$Cl$_2$ – that leads to low catalytic rates with [IrCl(PPh$_3$)$_3$]/MeOH. The other steps in the catalytic cycle are often very fast for Ir, so if the need for dissociation is avoided, then highly active Ir catalysts can be formed. However, a new consensus has now emerged: rhodium catalysts are often considered to be slower but more selective, whilst iridium catalysts are faster but less selective.

2.2
Historical Aspects

Iridium made its first major mark in 1965, in the arena of organometallic chemistry with the discovery of Vaska's complex, [IrCl(CO)(PPh$_3$)$_2$] (**1**) [1]. Only weakly catalytic itself, Vaska's complex is nevertheless highly relevant to cataly-

The Handbook of Homogeneous Hydrogenation.
Edited by J. G. de Vries and C. J. Elsevier
Copyright © 2007 WILEY-VCH Verlag GmbH & Co. KGaA, Weinheim
ISBN: 978-3-527-31161-3

sis in providing the classic examples of oxidative addition – normally a key step in almost any catalytic cycle. Equation (1) shows how a variety of molecules X-Y can oxidatively add in a concerted manner to this Ir(I) species to form a series of Ir(III) adducts. The H_2 adduct (X=Y=H) is only very weakly catalytically active for alkene hydrogenation because all the ligands in $[IrH_2Cl(CO)(PPh_3)_2]$ are firmly bound and do not dissociate to make way for substrate alkene. Without alkene binding, hydrogen transfer from the metal to the alkene cannot occur.

$$(1)$$

Following the discovery of Wilkinson's hydrogenation catalyst, $[RhCl(PPh_3)_3]$ (2) in 1964, the iridium analogue was naturally also investigated as a catalyst, but proved to be only very weakly active. Once again, the reason was that the adduct $[IrH_2Cl(PPh_3)_3]$ failed to lose PPh_3, unlike the Rh analogue, so that the alkenes were unable to bind and undergo reduction [2].

Schrock and Osborn [3] introduced the valuable idea that the reaction should be started with a PR_3 to Rh ratio of 2:1 in order to avoid the need for ligand dissociation. These authors used Chatt's diene-metal precursors, $[(nbd)RhCl]_2$ (nbd=norbornadiene), to form a series of very useful catalysts of the type $[(nbd)Rh(PR_3)_2]BF_4$. The nbd was shown to be lost during hydrogenation to form species based on the $\{Rh(PR_3)_2\}^+$ fragment, such as $[(MeOH)_2RhL_2]BF_4$. In the Rh series, MeOH was easily lost and catalytic alkene reduction was rapid. In the iridium analogues, however, the Ir(III) complexes $[IrH_2(solvent)_2(PPh_3)_2]^+$ (3, solvent=MeOH) were formed. These proved to be very much less labile and less active than the Rh series [4], and consequently attention was naturally focused on rhodium.

At this point, the initial intent of these investigations was to seek stable hydrides in iridium that were relevant to transient intermediates proposed in the rhodium series. With this aim in view, attention was focused on a series of complexes $[(cod)Ir(PR_3)_2]BF_4$, analogous to the Schrock-Osborn Rh catalysts; many of these had been synthesized previously, but had only been tested for catalysis in coordinating solvents and the results had been disappointing. The related mixed-ligand complexes, such as $[(cod)Ir(py)(PR_3)_2]BF_4$ (cod=1,5-cyclooctadiene; py=pyridine), were new [5, 6]. Since solvent dissociation from 3 was needed to generate a site for alkene binding, it seemed appropriate to examine the variation of the solvent, particularly the use of CH_2Cl_2; this was considered to be non-coordinating because, at the time, it was not known to be capable of binding to metals. Halocarbon solvents in general had been avoided for Rh catalysts, presumably because of the risk of C–Cl oxidative addition to Rh(I). The iridium complexes resisted such pathways, possibly because their resting state is Ir(III) (versus Rh(I)), and possibly also because of their cationic nature; many neutral Ir(I) species do add C–Cl bonds easily. Not only was the catalytic rate very greatly enhanced in CH_2Cl_2 but, more importantly, the substrate scope was also greatly expanded. At the time, no homogeneous hydrogenation catalysts were known which would reduce tri- and especially tetrasubstituted alkenes efficiently; even today, these are very rare. By using a low PR_3 to M ratio, a non-coordinating solvent, and Ir rather than Rh, very high activity was achieved for hindered alkenes [7].

If a PR_3 to M ratio of 2 was so good, then would a ratio of 1 be better? A catalyst of this type indeed proved to be the best of the whole series. $[Ir(cod)(PCy_3)$ $(py)]BF_4$ (**4**, Cy=cyclohexyl) is sometimes referred to as Crabtree's catalyst, although both Hugh Felkin and George Morris were also very closely associated with its initial development [5, 6]. The rates measured for reduction of various alkenes by **4** illustrate the high activity for hindered alkenes: t-BuCH=CH$_2$, 8300; 1-hexene, 6400; cyclohexene, 4500; 1-methylcyclohexene, 3800; $Me_2C=CMe_2$, 4000 h^{-1}. Even at 0.1% loading, the catalyst completely reduces all but the tetrasubstituted alkene, where 400 catalytic turnovers are seen ($Me_2C=CMe_2$, 0 °C, CH_2Cl_2) before catalyst deactivation. The deactivation product is a hydride-bridged polynuclear complex [7], presumably formed by intermolecular reaction of the catalyst when the depleted substrate is no longer able to compete effectively for binding to the metal. Hydrogenation tends to be favored over deactivation by operating at 0 °C rather than at room temperature.

4

The above-mentioned rates can usefully be compared with those for other catalysts under similar conditions [7]: $[RhCl(PPh_3)_3]$ at 0 °C (1-hexene, 60; cyclohexene, 70; $Me_2C=CMe_2$, 0 h^{-1}) is far slower and $[RuHCl(PPh_3)_3]$ at 25 °C in C_6H_6 (1-hexene, 9000; cyclohexene, 7; 1-methylcyclohexene, $Me_2C=CMe_2$, 0) is highly selective for terminal alkenes.

The initial studies on the catalyst did not attract the attention of the organic synthetic community, partly because the details were not published in an organic chemistry journal, and the substrates used were not "real" multifunctional organic compounds. On the basis of a suggestion made by Bill Suggs, the catalyst was used for more appropriate substrates, and the results obtained published [8]. More importantly, based on a further suggestion by Sarah Danishevsky, strong (99%) directing effects were also found in which the catalyst binds to a substrate OH or

C=O group and then delivers H_2 almost exclusively from the face of the substrate that contains the binding group [9]. This property of the catalyst, which was discovered independently by Stork [10], is illustrated in Eq. (2). Any of a variety of directing groups such as ether, ketone or ester is capable of binding to the catalyst before hydrogenation takes place. This sets the stereochemistry of as many as two new stereocenters in the reduction. Since Stork is a highly respected member of the organic chemistry community, his intervention was critical in first making the catalyst known, after which time it began to be used more generally.

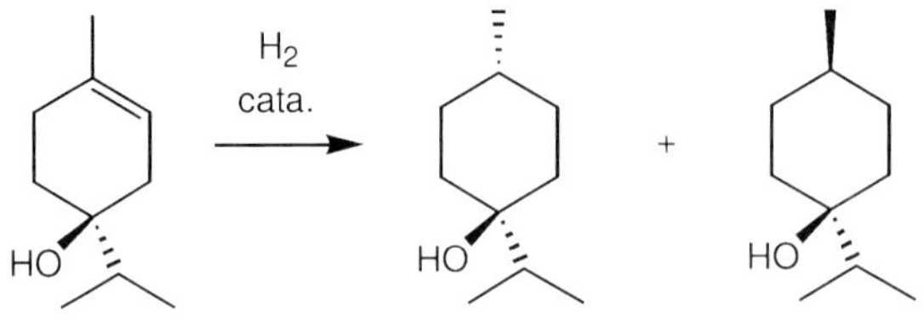

(2)

Pd/C	20%	80%
$[Ir(cod)(PCy_3)py]^+$	99.9%	0.1%

The reason that directing effects are so efficient is related to the low PR_3 to Ir ratio, which allows the directing group, the H_2 and the C=C bond all to bind to the metal at the same time. This was suggested by the detection of **5** at low temperature in the reaction of Eq. (3) [9].

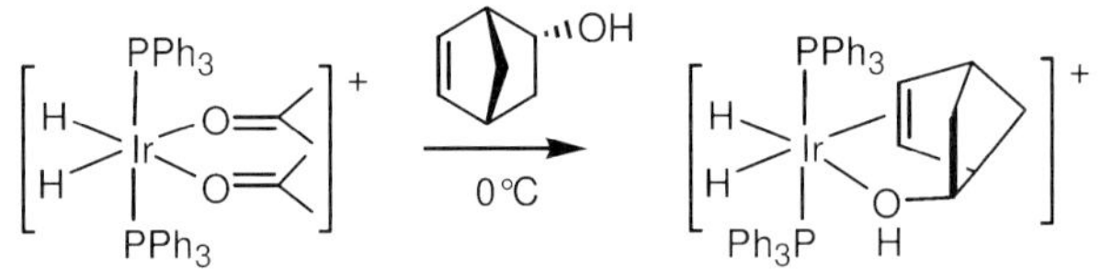

(3)

3   **5**

In the initial studies, the Ir system appeared to be less useful for enantioselective reduction because the *e.e.* values were never as high as seen for the Rh analogues. In commercial practice, however, rate can be more important than *e.e.* In this vein, Blaser [11] was able to equip the $\{(cod)Ir\}^+$ fragment with an asymmetric ligand of Togni's [12] to give a complex **6** that is used for the commercial production of the agrochemical metolachlor (Dual Magnum®). This is one of the few enantioselective hydrogenation systems that is in commercial use today.

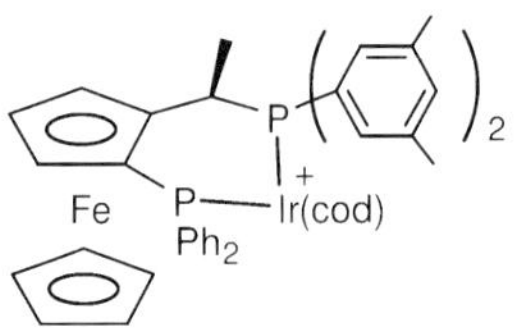

6

In a purely mechanistic experiment, the deuteration of 8-methylquinoline and related compounds by the Ir catalysts was examined, whereupon very rapid and selective isotope incorporation into the methyl CH bonds was found; once again, chelation control was operating [13]. Much later, the pharmaceutical industry developed this aspect of the catalyst for the tritiation of drug candidates, needed for metabolic studies. By introducing the radioactive tritium at the last step, a full organic synthesis involving radioactive intermediates was avoided; this also greatly minimized the production of radioactive organic waste. Catalysts **3**, **4** and $[Ir(cod)(dppb)]BF_4$ $(dppb = Ph_2P\{CH_2\}_4PPh_2)$ have all proved useful in this commercially important reaction, with each catalyst having a slightly different selectivity [14]. As before, pronounced directing effects caused exchange to occur at well-defined positions on the substrate, notably those immediately adjacent to the point on the compound where the catalyst binds. This is usually an O heteroatom, such as in an amide, ester, alcohol or ketone.

$$ \text{(4)} $$

A wide variety of iridium-based hydrogenation catalysts are currently under development, notably for organic syntheses including enantioselective synthesis. Hydrogenation by hydrogen transfer is well known [15], and the reduction of C=O and C=N double bonds is also possible [16, 17].

The hydroboration of terminal and internal alkenes with pinacolborane can be carried out at room temperature in the presence of an iridium(I) catalyst (3 mol.%) formed by the addition of dppm (2 equiv.) to $[Ir(cod)Cl]_2$ $(dppm = Ph_2PCH_2PPh_2)$, a mixture that presumably furnishes $[Ir(cod)(dppm)]Cl$ as the true catalyst precursor. Hydroboration results in the addition of the boron atom to the terminal carbon of 1-alkenes with more than 99% selectivity [18].

The reversal of hydrogenation is also possible, as evidenced by the many iridium catalysts for alkane dehydrogenation to alkenes or arenes, though to date this area is of mainly academic interest rather than practical importance [19].

One point of practical importance is the sensitivity of these catalysts to counterion and solvent; this is particularly the case in asymmetric hydrogenation, where significant changes in properties have been seen in several cases [20]. This implies that a range of solvents and counterions might usefully be examined in planning trials of the catalyst for a given reduction. In one case [20a], even the usually satisfactory triflate and tetrafluoroborate counterions almost completely inhibited a cationic iridium-PHOX catalyst. In that case, catalysts with $[Al\{OC(CF_3)_3\}_4]^-$, BArF$^-$, and $[B(C_6F_5)_4]^-$ counterions did not lose activity during the reaction, and even remained active after all of the substrate had been consumed. Tetraphenylborate is another undesirable anion as it tends to coordinate via an arene ring. In contrast to their sensitivity to anion and solvent, the Ir catalysts are air-stable, unlike typical Rh analogues.

2.3
Organometallic Aspects

The above-mentioned catalysts rely for their activity on losing the cod ligand via hydrogenation to give cyclooctane, thus liberating sites on the metal. The origin of cod as a ligand lies in some of Chatt's early studies [21] that were related to the development of the Dewar-Chatt model [21]. The intellectual roots of the concept go back to Langmuir and to Pauling in the 1920s and 1930s, who proposed that CO could form multiple bonds with metals such as Ni(0) [22].

Many useful iridium catalysts, such as those mentioned above, are synthetically accessible from $[Ir(cod)Cl]_2$, which is now commercially available. Treatments with PR_3 in a nonpolar solvent gives $[Ir(cod)PR_3Cl]$ for the less bulky members of the series, with PEt_3 marking the dividing line between the two types of pathway. Smaller ligands produce neutral bis-phosphine halo-complexes. In polar solvents (e.g., aqueous acetone), in contrast, the chloride ion can dissociate and ionic $[(cod)Ir(PR_3)_2]^+$ (**7**) or $[(cod)Ir(PR_3)_3]^+$ are obtained, again depending on the steric bulk, with smaller ligands yielding the tris-phosphine species. If $[IrCl(cod)PCy_3]$ is treated with pyridine in aqueous acetone, $[Ir(cod)(PCy_3)py]^+$ (**4**) is obtained. This species is not in equilibrium with $[Ir(cod)(PPh_3)_2]^+$ and $[Ir(cod)py_2]^+$ to any detectable extent (^1H- and ^{31}P-NMR spectroscopy). Variants of these routes can be made to provide chelate compounds of the type $[(cod)Ir(L-L)]^+$, where L-L are diphosphines, diamines, or mixed-donor ligands [5, 6, 23]. Typically, reactions are carried out at room temperature under N_2 or Ar.

A vast number of derivatives of these general types have been prepared by similar routes for catalytic applications, and at this point we can do no more than provide a series of recent references: some have P-donor ligands [24], some have N-heterocyclic carbenes [25], and others have mixed donors [26].

The hydrogenation product from $[Ir(cod)(PPh_3)_2]BF_4$ in various solvents is the readily isolable series $[IrH_2(solvent)_2(PPh_3)_2]BF_4$ [4], where the solvent can be Me_2CO, $MeOH$, and even H_2O. The acetone complex (**3**) has been characterized crystallographically [27]. These are precursors for the synthesis of a wide variety of unusual derivatives (Scheme 2.1). The first complexes of halocarbons were made by the route of Eq. (4), where $L=MeI$ [28]. For $L=H_2$, the products were the first bis-dihydrogen complexes [29]. Agostic species arise from reaction with 8-methylquinoline (Scheme 2.1). Instead, benzoquinoline undergoes cyclometalation.

Styrene yields a stable η^6-arene complex (Scheme 2.1), which explains why neither **3** nor **7** is an effective hydrogenation catalyst for styrene and related substrates. The formation of such stable adducts is highly disadvantageous for rapid catalysis, but not for the exploration of organometallic chemistry. No similar stable complexes have been obtained from the catalyst **4**; the faster catalytic rates seen for **4** may correlate with the presence of less stable intermediates in this case [30].

One of the limitations of both **4** and **7** in catalysis is their ready decomposition to inactive cluster hydride complexes in the absence of substrate. If the substrate is a weak ligand (e.g., $Me_2C=CMe_2$), this decomposition can be competitive with cluster formation. A high concentration of substrate favors catalysis

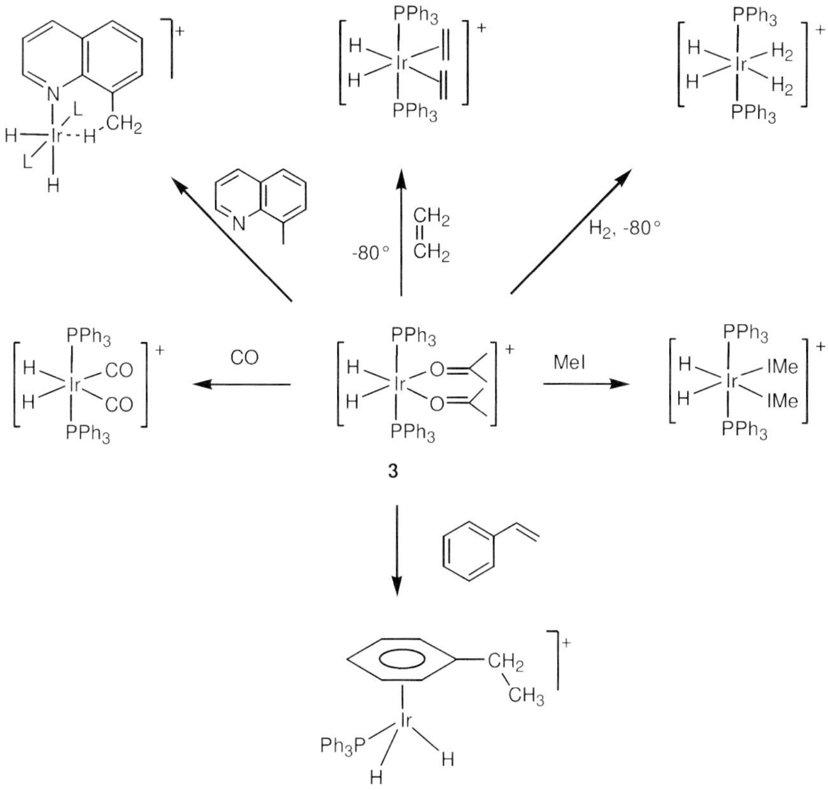

Scheme 2.1 Some reactions of [IrH$_2$(Me$_2$CO)$_2$(PPh$_3$)$_2$]$^+$.

by intercepting unsaturated metal-containing intermediates before they have a chance to cluster [31].

Moving to specific cases, [Ir(cod)(PPh$_3$)$_3$]BF$_4$ (**7**) yields the tris hydrogen-bridged cluster shown in Eq. (5).

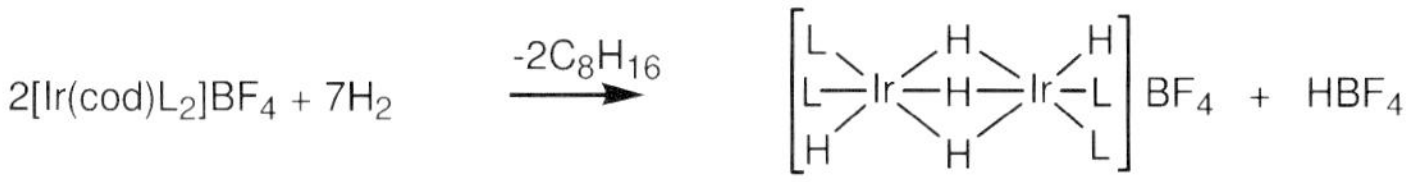

$$2[\text{Ir(cod)L}_2]\text{BF}_4 + 7\text{H}_2 \xrightarrow{-2\text{C}_8\text{H}_{16}} \left[\begin{array}{c} \text{L} \\ \text{L} \end{array}\!\!\!\text{Ir}\!\!\overset{\text{H}}{\underset{\text{H}}{\diagdown}}\!\!\overset{\text{H}}{\text{Ir}}\!\!\overset{\text{H}}{\underset{\text{L}}{\diagup}} \right] \text{BF}_4 + \text{HBF}_4 \qquad (5)$$

[Ir(cod)(PCy$_3$)(py)]BF$_4$ (**4**) forms the tri-nuclear cluster shown in Eq. (6):

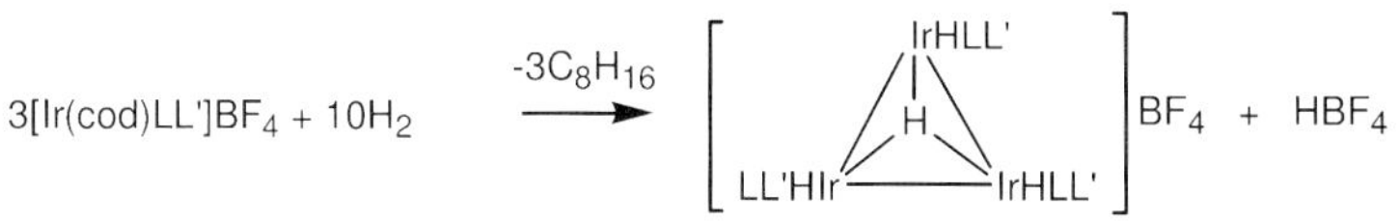

$$3[\text{Ir(cod)LL'}]\text{BF}_4 + 10\text{H}_2 \xrightarrow{-3\text{C}_8\text{H}_{16}} \left[\begin{array}{c} \text{IrHLL'} \\ \text{H} \\ \text{LL'HIr} \!-\! \text{IrHLL'} \end{array} \right] \text{BF}_4 + \text{HBF}_4 \qquad (6)$$

Rates of cluster formation are minimized by having the catalyst concentrations as low as possible. Successive additions of aliquots of catalyst can help in difficult cases. None of the cluster hydrides can be converted back to catalytically active or mononuclear complexes (H_2, 1 atm, $-80°$ to $+60°C$).

The addition of H_2 at $-80°C$ to $[Ir(cod)(PPh_3)_2]^+$ results in complete conversion to a detectable intermediate dihydride **8** (Scheme 2.2). On warming under H_2 to about $-20°C$, this produces cyclooctane and a trinuclear hydride cluster. If excess cod is present during the warming procedure, a new alkene complex (**9**) is formed. This is much more stable than species **8** and survives to room temperature. This explains why the $[Ir(cod)(PPh_3)_2]BF_4$ catalyst is ineffective for cod as substrate. The lack of reactivity of **8** can be explained by the C=C bond being coplanar with the cis hydride, allowing insertion. **9** also has C=C cis to an Ir–H, but the C=C bond is now orthogonal, forbidding insertion. **8** must be implicated in the activation of the catalyst by hydrogen. As before, catalyst **4** does not give rise to stable intermediates of similar structure, although they are assumed to be present [32].

At low temperatures ($-80°C$), $[IrH_2(solvent)_2(PPh_3)_2]^+$ (**3**) also reacts with small monoolefins such as ethylene in CH_2Cl_2 solution, to give $[IrH_2(olefin)_2(PPh_3)_2]^+$. These transfer coordinated H_2 to olefin on warming to $-20°C$, and so can be considered as probable intermediates in hydrogenation. Bulky alkenes such as tBuCH=CH produce $[IrH_2(olefin)(solvent)(PPh_3)_2]^+$.

Under similar conditions ($-80°C$, CD_2Cl_2) H_2 also reacts with **3** to give bis dihydrogen complex $[IrH_2(H_2)_2(PPh_3)_2]^+$; this is detected by ^1H-NMR spectroscopy, including T_1 relaxation measurements. This loses H_2 at $0°C$ when the H_2 is removed, to form the dinuclear hydride of Eq. (5).

These results suggest that the resting state of the catalyst is probably an $[IrH_2(L)_2(PPh_3)_2]^+$ species, where L can be solvent, substrate or H_2 depending on conditions, with L=substrate being predominant at the start of the reduction when the substrate concentration is highest.

Apparently similar Rh catalysts appear to have Rh(I) resting states of type $[Rh(PPh_3)_2L_2]^+$, which possibly accounts for their very different properties, for example their inability to reduce tri- and tetrasubstituted olefins.

Monoolefins containing coordinating groups often chelate, as in **5**. These also transfer coordinated H_2 to the C=C bond on warming to $-20°C$ and provide a rationalization for the directed hydrogenation mentioned earlier, in which hydrogenation occurs with almost exclusive H_2 addition from the face of the substrate that contains the coordinating group.

The presence of base such as NEt_3 in the system leads to conversion of the cationic $[IrH_2L_2(PPh_3)_2]^+$ forms to catalytically inactive neutral analogues. An ex-

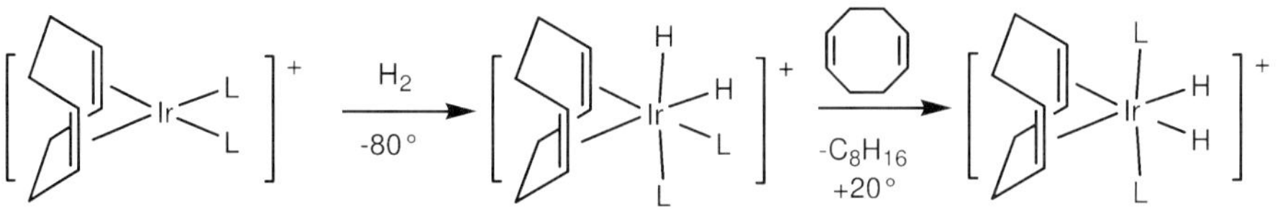

Scheme 2.2 Some intermediates in the hydrogenation of $[Ir(cod)(PPh_3)_2]BF_4$.

ample of a reaction of this type that gives an isolable neutral hydride is shown in Eq. (7):

$$[Ir(cod)(PPh_3)_2]BF_4 + NEt_3 + 6\,H_2 \rightarrow [IrH_5(PPh_3)_2] + [Et_3NH]BF_4 + C_8H_{16} \quad (7)$$

2.4
Catalysis

The above-mentioned reaction with base has relevance for catalytic chemistry in that substrates that are also bases may deactivate the catalyst by deprotonation; this can be avoided by addition of HOAc, HBF_4 or H_2SO_4 or use of the corresponding salt of the substrate. Coordinating anions react with the catalyst, again with deactivation of the catalyst, so any halide counterions should be replaced by BF_4 or PF_6. Carboxylate salts also react with the system to give inactive $[IrH_2(O_2CR)(PPh_3)_2]$, so carboxylates should be reduced in the protonated form (or as the ester). Amides bind via the carbonyl oxygen, albeit reversibly, so they can affect the rate of reaction and the stereochemistry of the product via directing effects, but are otherwise well tolerated. Esters and alcohols bind less strongly and have little effect on the rate, but still show directing effects. The Ir catalyst has been used for a wide variety of transformations in the organic synthesis of complex molecules. When attention is paid to the points mentioned above, the results have often proved very satisfactory.

2.4.1
Enantioselective Versions of the Iridium Catalyst

Despite extensive efforts, only a handful of enantioselective hydrogenations have as yet achieved the status of commercial processes. Among these is one that involves the enantiomeric reduction of imines by catalyst **6**: Syngenta's process for (S)-metolachlor [11]. The latter is now the largest scale industrial enantiomeric catalytic process, with annual sales of the product, Dual-Magnum®, now exceeding 10^4 tons. Imines tend to be difficult substrates because of the possibility of unproductive ligand binding via the imine lone pair. For reasons that are still not entirely clear, the Ir catalysts are less seriously affected by such binding as are the Rh analogues. It is possible that the high trans-effect of the hydrides in the Ir(III) resting state labilizes the substrate binding sites, located trans to the hydrides. Enhanced back-bonding by the third row metal may also enhance the relative stability of the η^2-bound form of the imine that leads to insertion and productive catalysis.

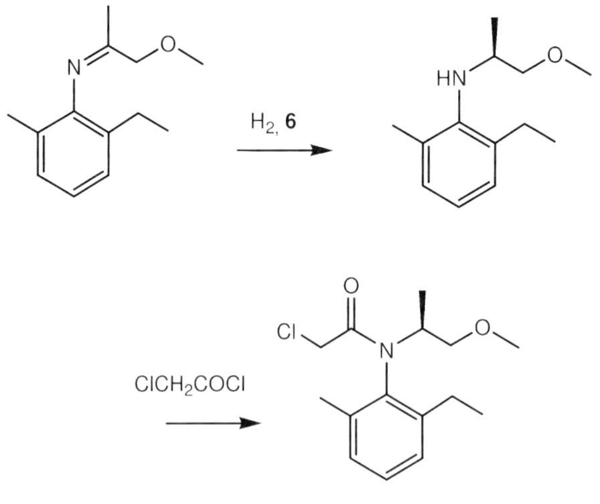

(8)

(S)-metolachlor

Bulky groups on the imine also help to disfavor η^1 binding. A ketimine is normally required for the reduction product to contain an asymmetric carbon α to nitrogen, as in the case of metolachlor (Eq. (8)). Finally, the presence of an acid of a non-coordinating anion helps to protonate the nitrogen lone pair and disfavor η^1 binding to the metal via this lone pair. The iodide additive leads to the formation of iodoiridium species that are beneficial for precatalyst **6**. Rates of up to $1.8 \times 10^6 \, h^{-1}$ are achieved (50 °C, 80 bar) allowing substrate/catalyst ratios of 10^6. This is said to be one of the fastest homogeneous catalysts of any type known. For economic success of the process, the rate is more significant than the ee (80%), whereas in reports made by academic contributors the ee-values often dominate the discussion. A more appropriate figure of merit (FOM) [11] might be obtained by multiplying the ee by the rate; hence, an FOM value for the metolachlor catalyst system is $1.45 \times 10^6 \, h^{-1}$.

2.4.2
Mechanism

The fastest $[Ir(cod)LL']BF_4$ systems have proved difficult to study from a mechanistic standpoint because they are so active that the rates are often limited by the mass transfer of hydrogen from the gas phase into solution. This implies that efficient stirring is desirable for the most effective use of the catalyst.

Perhaps the best data are available from Brandt's study of Pfaltz's asymmetric $[Ir(cod)(P-N)]^+$ catalyst [33], bearing a chelating phosphino-oxazolidene ligand. The rate is first order in catalyst and H_2, but zero order in substrate. Taken together with the density functional theory (DFT) calculations, this is consistent with the mechanism of Scheme 2.3, shown here in its essentials only (the interested reader is urged to consult the original paper for the complete story). Surprisingly, an Ir(III)/Ir(V) cycle is proposed, rather than the M(I)/M(III) cycle

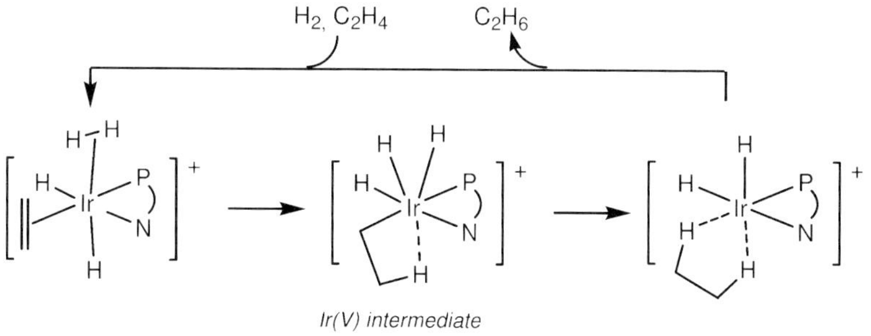

Scheme 2.3 The essential features of the Brandt mechanism for the Pfaltz catalyst [34].

that is usually considered for iridium and that is well established for rhodium. This explains the insensitivity of the iridium system to air and to oxidizing solvents, since Ir(III) and Ir(V) tend to be more stable than Ir(I) both to air and to oxidants in general. It also explains the markedly different catalytic selectivities of what are entirely analogous Rh(I) and Ir(I) catalyst precursors. It is very likely that a similar Ir(III)/Ir(V) cycle applies to typical [(cod)IrL$_2$]$^+$ catalysts. Related iridium species are effective alkane dehydrogenation catalysts, for which a similar reverse-hydrogenation mechanism could readily apply.

In other studies, imine reduction by [Ir(cod)(PPh$_3$)$_2$]BF$_4$ in THF has been shown to be first order in each of the catalyst, the H$_2$, and the substrate. Initial formation of [IrH$_2$(imine)$_2$(PPh$_3$)$_2$]$^+$ was proposed to lead to amine and [Ir(imine)$_2$(PPh$_3$)$_2$]. Oxidative addition regenerates the Ir(III) species [34].

Oro, Werner and coworkers found that alkyne reduction by the P,O chelated [Ir(cod)(PrPr$_2$CH$_2$CH$_2$OMe)]BF$_4$ in CH$_2$Cl$_2$ at 25 °C is also first order in each of catalyst, H$_2$ and substrate. Styrene is formed rapidly, whilst subsequent reduction to ethyl benzene is much slower. Stopping the reaction after the appropriate time led to essentially complete selectivity for styrene formation [35]. Surprisingly, the cod remains coordinated to Ir throughout the catalytic cycle, in contrast to every other case, where cod is proposed to be hydrogenated or the cyclooctane hydrogenation product is detected. In view of the case with which 6-alkynes rearrange to vinylidenes, such a pathway might easily be involved in 1-alkyne hydrogenation. The appropriate isotope labeling experiments seem to be carried out only rarely.

A detailed combined experimental computational mechanistic study, performed for isotope exchange in 2-dimethylamino pyridine, showed how the presence of hydrides in the Ir(III) intermediates helps to flatten the potential energy surface, accounting for the extremely high rates of exchange. In this case, carbene intermediates were also involved as a result of double C–H activation.

2.4.3
Practical Considerations

As the iridium catalysts are often somewhat thermally sensitive, synthetic procedures to prepare them should be carried out at room temperature, or below. These catalysts are normally stable to air as solids, but are somewhat air-sensitive in solution. An inert atmosphere (N_2 or Ar) is typically used for the storage of solids and to protect solutions, as the catalysts deactivate in the absence of substrate. The order of addition must be: substrate first, followed by H_2. Weakly coordinating solvents are required for optimum activity. Dichloromethane is typical, but tetrahydrofuran (THF) has also been used. $PhNO_2$, PhCl and $PhCF_3$ may also be satisfactory, but MeCN, pyridine and alcohols should be avoided. The presence of water is tolerated. Basic substrates should be neutralized by the addition of HOAc or HBF_4 in an amount equivalent to the number basic groups to be neutralized, though an excess does not seem to be detrimental. A catalyst loading of 0.1% is usually satisfactory, though very much lower loadings have been used in commercial processes. BF_4^- is the usual counterion, but PF_6^- can also be used. BPh_4^- is unsatisfactory because it tends to bind to the metal to produce catalytically inactive arene complexes. Coordinating anions such as halides are to be avoided in the substrate, but the presence of some iodide has proved beneficial in one case. In the relatively low-polarity solvents used, the complexes form tight ion pairs. In related systems, such as $[IrH_2(dipy)(PPh_3)_2]BF_4^-$, the ion pair has a definite structure, as shown by NMR spectroscopy [36]. Hydrogen is usually supplied at 1 atm pressure, although commercial applications use pressures up to 80 atm. Rates may also slow at low H_2 pressures, but the reaction still occurs. Reaction temperatures from 0 °C to 50 °C have been used successfully.

A variety of functional groups resist reduction: arene rings, NO_2, COOMe, $CONH_2$, sulfones, nitrile, and ArHal. Nitriles can bind to the metal, and the N lone pair is not effectively masked by acid addition so lower rates can be encountered if this group is present. Alkynes, alkenes, and imines are the best-studied substrates for which reduction is efficient.

The isolation of product is usually possible after evaporation of the solvent and extraction with hexane, ether, or toluene. Supported versions, for example on polystyrene grafted with PPh_2 groups, have proved unsatisfactory because the rate of deactivation is greatly enhanced under these conditions [37]. Asymmetric versions exist, but the ee-values tend to be lower than in the Rh series [38]. With acid to neutralize the basic N lone pair, imine reduction is fast. Should it be necessary to remove the catalyst from solutions in order to isolate a strictly metal-free product, a resin containing a thiol group should prove satisfactory. A thiol group in the substrate deactivates the catalyst, however.

Acknowledgments

The author thanks the U.S. Department of Energy and Johnson Matthey for the support of these studies, and also those coworkers mentioned in the references.

Abbreviations

cod	1,5-cyclooctadiene
Cy	cyclohexyl
DFT	density functional theory
ee	enantiomeric excess
FOM	figure of merit
nbd	norbornadiene
py	pyridine
THF	tetrahydrofuran

References

1 L. Vaska, D. Rhodes, *J. Am. Chem. Soc.* **1965**, *87*, 4970.

2 J. A. Osborn, F. H. Jardine, J. F. Young, G. Wilkinson, *J. Chem. Soc.* **1966**, 1711.

3 R. R. Schrock, J. A. Osborn, *J. Am. Chem. Soc.* **1976**, *98*, 2134, 2143, 4450.

4 J. R. Shapley, R. R. Schrock, J. A. Osborn, *J. Am. Chem. Soc.* **1969**, *91*, 2816.

5 R. H. Crabtree, H. Felkin, G. E. Morris, *J. Organomet. Chem.* **1977**, *141*, 205.

6 R. H. Crabtree, H. Felkin, T. Fillebeen-Khan, G. E. Morris, *J. Organomet. Chem.* **1979**, *168*, 183.

7 R. H. Crabtree, *Acc. Chem. Res.* **1979**, *12*, 331.

8 J. W. Suggs, S. D. Cox, R. H. Crabtree, J. M. Quirk, *Tetrahedron Lett.* **1981**, *22*, 303.

9 R. H. Crabtree, M. W. Davis, *Organometallics* **1983**, *2*, 681; *J. Org. Chem.* **1986**, *51*, 2655.

10 G. Stork, D. E. Kahne, *J. Am. Chem. Soc.* **1983**, *105*, 1072.

11 H. U. Blaser, H. P. Buser, K. Coers, R. Hanreich, H. P. Jalett, E. Jelsch, B. Pugin, H. D. Schneider, F. Spindler, A. Wegmann, *Chimia* **1999**, *53*, 275.

12 A. Togni, C. Breutel, A. Schnyder, F. Spindler, H. Landert, A. Tijani, *J. Am. Chem. Soc.* **1994**, *116*, 4062.

13 R. H. Crabtree, E. M. Holt, M. E. Lavin, S. M. Morehouse, *Inorg. Chem.* **1985**, *24*, 1986.

14 (a) J. S. Valsborg, L. Sorensen, C. Foged, *J. Label. Compds. Radiopharm.* **2001**, *44*, 209; (b) A. Y. L. Shu, D. Saunders, S. H. Levinson, S. W. Landvatter, A. Mahoney, S. G. Senderoff, J. F. Mack, J. R. Heys, *J. Label. Compds. Radiopharm.* **1999**, *42*, 797.

15 J. S. Chen, Y. Y. Li, Z. R. Dong, B. Z. Li, J. X. Gao, *Tetrahedron Lett.* **2004**, *45*, 8415.

16 R. Sablong, J. A. Osborn, *Tetrahedron Lett.* **1996**, *37*, 4937.

17 K. Tani, J. Onouchi, T. Yamagata, Y. Kataoka, *Chem. Lett.* **1995**, 955.

18 Y. Yamamoto, R. Fujikawa, T. Umemoto, N. Miyaura, *Tetrahedron* **2004**, *60*, 10695.

19 (a) R. H. Crabtree, C. P. Parnell, *Organometallics* **1985**, *4*, 519–523; (b) C. M. Jensen, *Chem. Commun.* **1999**, 2443; (c) P. Braunstein, Y. Chauvin, J. Nahring, A. DeCian, J. Fischer, A. Tiripicchio, F. Ugozzoli, *Organometallics* **1996**, *15*, 5551–5567; (d) F. C. Liu, A. S. Goldman, *Chem. Commun.* **1999**, 655; (e) D. M. Tellers, R. G. Bergman, *Organometallics* **2001**, *20*, 4819.

20 (a) S. P. Smidt, N. Zimmermann, M. Studer, A. Pfaltz, *Chem. Eur. J.* **2004**, *10*, 4685; (b) N. Kinoshita, K. H. Marx, K. Tanaka, K. Tsubaki, T. Kawabata, N. Yoshikai, E. Nakamura, K. Fuji, *J. Org. Chem.* **2004**, *69*, 7960 and references cited therein.

21 J. Chatt, R. G. Wilkins, *Nature* **1950**, *165*, 859; J. Chatt, L. M. Venanzi, *J. Chem. Soc.* **1957**, 4735.

22 R. H. Crabtree, *J. Organomet. Chem.* **2004**, *689*, 4083.

23 R. H. Crabtree, G. E. Morris, *J. Organomet. Chem.* **1977**, *135*, 395.

24 T. Focken, G. Raabe, C. Bolm, *Tetrahedron: Asymmetry* **2004**, *15*, 1693.

25 (a) C. Bolm, T. Focken, G. Raabe, *Tetrahedron: Asymmetry* **2003**, *14*, 1733; (b) A. C. Hillier, H. M. Lee, E. D. Stevens, S. P. Nolan, *Organometallics* **2001**, *20*, 4246.

26 (a) O. Pamies, M. Dieguez, G. Net, A. Ruiz, C. Claver, *Organometallics* **2000**, *19*, 1488; (b) X. Sava, N. Mezailles, L. Ricard, F. Mathey, P. Le Floch, *Organometallics* **1999**, *18*, 807; (c) M. Dieguez, A. Orejon, A. M. Masdeu-Bulto, R. Echarri, S. Castillon, C. Claver, A. Ruiz, *J. Chem. Soc. Dalton Trans.* **1997**, 4611; (d) C. Bianchini, L. Glendenning, M. Peruzzini, G. Purches, F. Zanobini, E. Farnetti, M. Graziani, G. Nardin, *Organometallics* **1997**, *16*, 4403.

27 R. H. Crabtree, G. G. Hlatky, C. A. Parnell, B. E. Segmuller, R. J. Uriarte, *Inorg. Chem.* **1984**, *23*, 354.

28 (a) M. J. Burk, R. H. Crabtree, E. M. Holt, *Organometallics* **1984**, *3*, 638; (b) M. J. Burk, B. Segmuller, R. H. Crabtree, *Organometallics* **1987**, *6*, 2241.

29 R. H. Crabtree, M. Lavin, *Chem. Commun.* **1985**, 1661.

30 (a) R. H. Crabtree, M. F. Mellea, J. M. Quirk, *Chem. Commun.* **1981**, 1217; (b) R. H. Crabtree, M. F. Mellea, J. M. Quirk, *J. Am. Chem. Soc.* **1984**, *106*, 2913.

31 (a) R. H. Crabtree, H. Felkin, G. E. Morris, T. J. Khan, J. A. Richards, *J. Organomet. Chem.* **1976**, *113*, C7; (b) D. F. Chodosh, R. H. Crabtree, H. Felkin, G. E. Morris, *J. Organomet. Chem.* **1978**, *161*, C67–C70; (c) D. F. Chodosh, R. H. Crabtree, H. Felkin, S. Morehouse, G. E. Morris, *Inorg. Chem.* **1982**, *21*, 1307.

32 R. H. Crabtree, H. Felkin, T. Fillebeen-Khan, G. E. Morris, *J. Organomet. Chem.* **1979**, *168*, 183.

33 (a) P. Brandt, E. Hedberg, P. G. Andersson, *Chem. Eur. J.* **2003**, *9*, 339; (b) A. Pfaltz, J. Blankenstein, R. Hilgraf, E. Hormann, S. McIntyre, F. Menges, M. Schonleber, S. P. Smidt, B. Wustenberg, N. Zimmermann, *Adv. Synth. Catal.* **2003**, *345*, 33.

34 V. Herrera, B. Munoz, V. Landaeta, N. Canudas, *J. Mol. Catal. A* **2001**, *174*, 141.

35 M. A. Esteruelas, A. M. Lopez, L. A. Oro, A. Perez, M. Schultz, H. Werner, *Organometallics* **1993**, *12*, 1823.

36 A. Macchioni; C. Zuccaccia, E. Clot, K. Gruet, R. H. Crabtree, *Organometallics* **2001**, *20*, 2367.

37 R. H. Crabtree, unpublished data.

38 P. Schnider, G. Koch, R. Pretôt, G. Z. Wang, F. M. Bohnen, C. Kruger, A. Pfaltz, *Chem. Eur. J.* **1997**, *3*, 887.

3
Ruthenium and Osmium

Robert H. Morris

3.1
Introduction

There is much current excitement and activity in the field of homogeneous hydrogenation using ruthenium catalysts. This is reflected in the recent, explosive
increase in the number of research publications in this area, now rivaling those
for rhodium catalysts (Fig. 3.1). Meanwhile, the price of rhodium metal has risen dramatically, becoming about ten times that of ruthenium, on a molar basis. The number of reports on the use of osmium catalysts has remained low,
partly because of the higher price of osmium compounds – about ten times that
of ruthenium – and partly because the activity of osmium catalysts is often lower.

   During the early years of catalyst development (1960–1980), rhodium chemistry dominated the scene, led by the investigations, for example, of Wilkinson,
Kagan, Osborn, and Knowles [1]. The more complex catalytic chemistry of
ruthenium was slower to develop, starting with studies by Halpern [2] and Wilkinson [3] during the 1960s. This continued with an exploration of the types of
ruthenium complexes that were active hydrogenation catalysts in the 1970s, as
reviewed by James [4, 5]. During the 1980s the search for new chemistry for
synthesis gas (CO, H_2) and coal utilization to combat petroleum shortages (the
"energy crisis") shifted attention to Ru and Os complexes, and promising activity was found for the hydrogenation of difficult substrates such as arenes, simple ketones, nitriles, and esters. For both economic and scientific reasons, attention then shifted to enantioselective hydrogenations using ruthenium complexes. Japanese scientists were on the crest of this new wave, with Noyori leading the way. Noyori was awarded the Nobel prize for this work in 2001 and his
lecture has subsequently been published [6, 7].

   The current research areas with ruthenium chemistry include the effective
asymmetric hydrogenation of other substrates such as imines and epoxides, the
synthesis of more chemoselective and enantioselective catalysts, CO_2 hydrogenation and utilization, new methods for recovering and recycling homogeneous
catalysts, new solvent systems, catalysis in two or three phases, and the replace-

The Handbook of Homogeneous Hydrogenation.
Edited by J. G. de Vries and C. J. Elsevier
Copyright © 2007 WILEY-VCH Verlag GmbH & Co. KGaA, Weinheim
ISBN: 978-3-527-31161-3

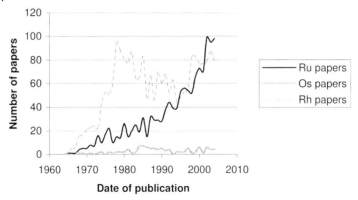

Fig. 3.1 Graphical illustration of numbers of reports per year versus date of publication. Data were obtained by searching the Chemical Abstracts Database using the term "hydrogenation catalyzed by ruthenium complexes" or osmium complexes or rhodium complexes. These are not comprehensive searches but are still representative of the activity in the field.

ment of phosphine ligands with other donors such as stable carbene and nitrogen donors.

3.2
Ruthenium

3.2.1
The First Catalysts for Alkene Hydrogenation: Mechanistic Considerations

In 1961, Halpern's group reported that the water-soluble, activated alkenes, fumaric, acrylic and maleic acid, could be catalytically hydrogenated in a solution containing chlororuthenium(II) species at 70 to 90 °C and 1 bar H_2 [2]. Interest in such chloro complexes grew out of reports about their electron-transfer behavior, a topic of interest at the time due to the extensive studies of Taube and others. Details of the hydrogenation of maleic acid are provided in Table 3.1. The kinetics of this system were thoroughly investigated by H_2 uptake measurements and spectroscopy, and the rate law was consistent with a mechanism where the alkene first binds to the metal in a pre-equilibrium followed by the turnover-limiting reaction of the alkene complex with dihydrogen where hydrogen is added *cis* on the double bond, as in Scheme 3.1.

Chatt and Hayter reported the first ruthenium and osmium hydride complexes of the type $MHCl(PR_2CH_2CH_2PR_2)_2$ in 1959, but these are not catalysts [9, 10]. Subsequently, in 1965, Wilkinson and coworkers found that the reaction of $RuCl_2(PPh_3)_3$ with hydrogen and a base gave the hydride complex $RuHCl(PPh_3)_3$, a very active hydrogenation catalyst [3]. A modern interpretation

Table 3.1 Representative conditions for the hydrogenation of alkenes.

Precatalyst (Ru)	Substrate (S)	S:C	Solvent	$p(H_2)$ [bar]	Product	Conversion [%]	Temp [°C]	TON	TOF [h^{-1}]	Reference
1 [RuCl$_6$]$^{4-}$ by reduction of [RuCl$_6$]$^{2-}$ with TiCl$_3$	Maleic acid (*cis*-HOOCCH=CHCOOH) (0.061 M)	100	3 M HCl	1	Succinic acid	100	80	100	3	8
2 RuHCl(PPh$_3$)$_2$	1-Octene (1.1 M)	1400	Toluene	0.66	Heptane		25	1400	<1×10^{4} [a]	12
3 RuCl$_2$(PPh$_3$)$_2$	Maleic acid	100	Dimethyl-acetamide	1	Succinic acid	100	35	100	10 [b]	4
4 Ru(1,5-COD)(1,3,5-COT)	1,3,5-Cycloheptatriene	46	THF	1	Cycloheptene	90	37	46	7	39
5 Os(η^2-H$_2$)HCl(CO)(P^{i}Pr$_3$)$_2$	Styrene	100	iPrOH	1	Ethylbenzene	100	60	100	1200	123

a) With other conditions constant as listed, the TOF varies with the alkene concentration as TOF $= 2.1 \times 10^4$ [octene]/(1.3+[octene]).
b) TOF varies depending on the concentration of several species; the rate law and kinetic parameters have been reported [4].

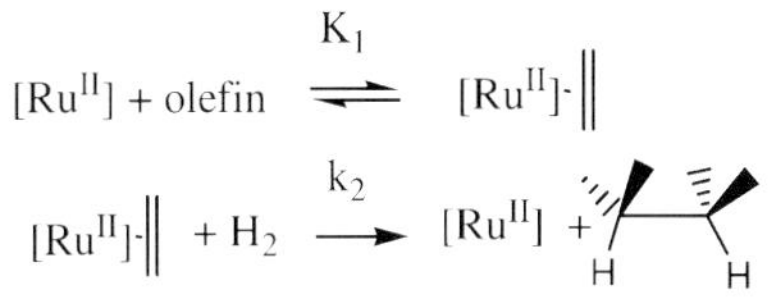

$$[Ru^{II}] + olefin \xrightleftharpoons{K_1} [Ru^{II}]\cdot\|$$

$$[Ru^{II}]\cdot\| + H_2 \xrightarrow{k_2} [Ru^{II}] +$$

Scheme 3.1

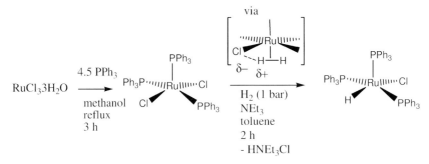

Scheme 3.2 Preparation of the alkene hydrogen catalyst $RuHCl(PPh_3)_3$.

of the formation of the hydride is that it proceeds via an acidic η^2-dihydrogen complex [11] (Scheme 3.2). This monohydride complex is an extremely active and selective catalyst for the hydrogenation of 1-alkenes in benzene at 25 °C [3, 12]. The turnover frequency (TOF) for 1-octene hydrogenation is about 10^4 h^{-1} for the mild conditions listed in Table 3.1, entry 1 (e.g., 0.66 bar H_2, 25 °C), and this changes with the alkene concentration, as listed. Disubstituted alkenes are hydrogenated about 1000-fold more slowly. The catalyst is only soluble to the extent of 10^{-4} M in toluene. It is about 20 times more active than the well-known alkene hydrogenation catalyst $RhCl(PPh_3)_3$ under similar conditions [12].

It has been a challenge to determine the mechanism of catalysis of this very oxygen-sensitive system (the current view is summarized in Scheme 3.3). $RuHCl(PPh_3)_3$ is an unusual case of a coordinatively unsaturated (5-coordinate) d^6 complex. The three bulky triphenylphosphine ligands prevent the coordination of other large ligands. In the catalytic reaction, this complex reacts with the alkene substrate to form an unstable alkyl intermediate by hydride addition to the double bond. In the turnover-limiting step, dihydrogen coordinates and becomes acidic. Proton transfer to the alkyl carbon releases the hydrogenated product with retention of configuration at carbon, and regenerates the starting hydride. The hydrogenolysis of a ruthenium–carbon bond via protonation by an acidic dihydrogen ligand *cis* to the alkyl has become a well-accepted mechanism [11, 13, 14], and would provide the observed *cis* stereochemistry of the addition of dihydrogen to the double bond. The formation of an alkyl intermediate is supported by the observation that the related complex $RuH(OC(O)CF_3)(PPh_3)_3$ reacts with ethylene in the absence of H_2 to give, reversibly, an ethyl complex $Ru(Et)(OC(O)CF_3)(PPh_3)_3$. Such a β-addition/elimination of hydride explains why such monohydride complexes are alkene isomerization catalysts. This po-

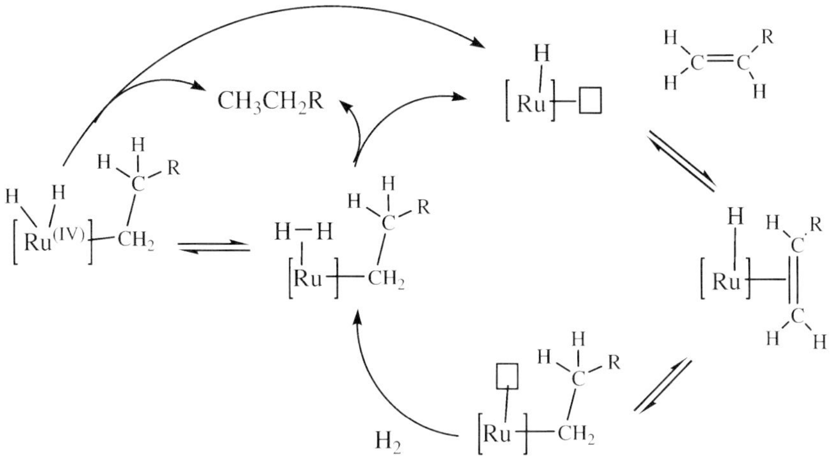

Scheme 3.3 Mechanism for the hydrogenation of 1-alkenes catalyzed by RuHCl(PPh$_3$)$_3$. [Ru] represents the RuCl(PPh$_3$)$_n$ fragment. The box represents an empty coordination site on ruthenium (II).

tentially undesirable side reaction may have been a reason why rhodium catalysts were favored over Ru(II) catalysts during the early days of these studies. Most rhodium catalysts proceed through a dihydride intermediate that hydrogenates, but does not isomerize, alkenes.

Quantitative rate measurements under a variety of conditions support such a mechanism [4, 15]. A complete kinetic analysis is available for the hydrogenation of acrylic acid derivatives using the precatalysts RuCl$_2$(PPh$_3$)$_3$ in the solvent dimethylacetamide, although the system is much less active in this more polar and coordinating solvent (e.g., entry 3, Table 3.1).

The triphenylphosphine complexes of the type RuCl$_2$(PPh$_3$)$_3$, RuHX(PPh$_3$)$_3$, X=Cl, O$_2$CR, etc., RuH$_2$(PPh$_3$)$_4$, RuH(CO)X(PPh$_3$)$_3$, RuCl$_2$(CO)$_2$(PPh$_3$)$_2$ all proved to be catalysts for a variety of reductions, although the carbonyl complexes tended to require higher temperatures [5]. For example, the last complex is a catalyst for the selective hydrogenation of 1,5,9-cyclododecatriene to cyclododecene in dimethylformamide (DMF) at 140 °C and 10 bar H$_2$ in the presence of PPh$_3$ [16]. The complex RuCl$_2$(PPh$_3$)$_3$ proved active in the hydrogenation of the C=C bond in α,β-unsaturated ketones by hydrogen transfer from formic acid or benzylalcohol [17]. Later, it was demonstrated that the addition of base greatly accelerates such transfer reactions by promoting the formation of hydride species, as reviewed elsewhere [18, 19]. Thus, RuCl$_2$(PPh$_3$)$_3$ in the presence of a base catalyzes the transfer of hydrogen to ketones or imines from *i*PrOH or formic acid [18]. Transfer hydrogenation reactions will be discussed further in Chapters 20 and 32.

3.2.2
Synthesis of Ruthenium Precatalysts and Catalysts

The modification of these precursor compounds with other ligands, including a vast array of chiral phosphorus-donors, has resulted in an ever-expanding list of useful ruthenium hydrogenation catalysts, as described in the following sections. Figure 3.2 illustrates how the PPh$_3$ ligands of RuCl$_2$(PPh$_3$)$_3$ are readily displaced by a wide range of ligands to produce new catalysts. The reaction with diphosphines with medium bite angles (dppb, diop, binap) (Fig. 3.3) produces complexes RuCl$_2$(diphosphine)(PPh$_3$) that are used as catalysts for the hydrogenation of 1,3-diketones [20], the hydrogenation of benzonitrile [21], and the hydrogenation of imines [22]. The dppb complex can be converted to the binuclear dihydrogen complex (η^2-H$_2$)(dppb)Ru(μ-Cl)$_3$Ru(dppb)Cl, which is a precatalyst for the hydrogenation of styrene and aldimines [23, 24]. The reactions with P-N ligands (chiral phosphinooxazolines [25] or phosphine-imines [26]) produce RuCl$_2$(PPh$_3$)(P-N) precatalysts for the enantioselective transfer hydrogenation of ketones. The reaction with diamines such as ethylene diamine produces RuCl$_2$(PPh$_3$)$_2$(diamine) complexes for the efficient H$_2$-hydrogenation of simple ketones [27] (see below). The reaction with 2 equiv. of chiral β-aminophosphine ligands produces RuCl$_2$(P-NH$_2$)$_2$, very active enantioselective hydrogenation cat-

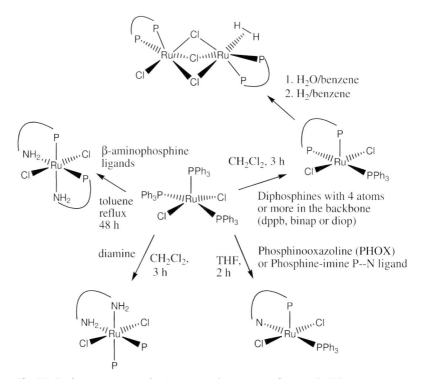

Fig. 3.2 Synthetic routes to ruthenium precatalysts starting from RuCl$_2$(PPh$_3$)$_3$.

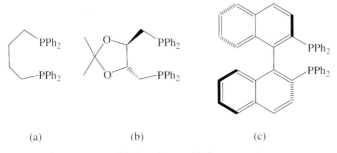

(a)   (b)   (c)

Fig. 3.3 The structures of diphosphines with four atoms in
the backbone: (a) dppb; (b) (–)-(R,R)-diop; (c) (R)-binap.

alysts for ketones and imines [28, 29]. Finally, the reaction with water-soluble
sulfonated tri-arylphosphines (not shown in Fig. 3.2) produces water-soluble
complexes such as $[RuCl_2(P(C_6H_4-m-SO_3Na)_3)_2]_2$ that catalyze the H_2-hydroge-
nation of aldehydes in water [30] and $[RuCl_2(PPh_2(C_6H_4-m-SO_3Na))_2]_2$ which, in
the presence of excess phosphine, selectively hydrogenates the C=C bond of a,β-
unsaturated aldehydes at pH 3 but switches to selectively hydrogenating the aldehyde C=O at pH 9 [31].

The PPh$_3$ ligands in RuHCl(PPh$_3$)$_3$ can be displaced in a similar fashion to
produce a range of analogous precatalysts such as RuHCl(diamine)(PPh$_3$)$_2$ and
trans-RuHCl(diamine)(diphosphine). When the former diamine compound is activated with alkoxide base under H_2, it is an active catalyst for ketone and imine
hydrogenation [32, 33], while the latter is a precatalyst for the asymmetric hydrogenation of imines and ketones under mild conditions [34, 35].

The compounds $[RuCl_2(C_6H_6)]_2$ [36] (Fig. 3.4a), Ru(η^3-methylallyl)$_2$(COD) [37]
(Fig. 3.4b), COD=η^4-1,5-cyclooctadiene and $[RuCl_2(COD)]_n$ [38, 39] are also
very useful starting materials that are commercially available. The complex
RuCl$_2$(dmso)$_4$ [40] in Figure 3.4c has relatively labile ligands. The starting material Ru(COD)(COT) [38] (Fig. 3.4d) is a source of Ru^0 complexes and the dihydrogen complex RuH$_2$(H$_2$)$_2$(PCy$_3$)$_2$ (see Fig. 3.6). The complex Ru(COD)(COT)
is also a useful catalyst for the hydrogenation of trienes to monoenes (see Table
3.1, entry 4) [39].

The structure of $[RuCl_2(COD)]_n$ is not well defined, but it is a very useful
starting material to catalysts (Fig. 3.5). Its reaction with binap (see Fig. 3.3) and
NEt$_3$ can lead to the chloride-bridged dimer $[NEt_2H_2][Ru_2Cl_5(binap)_2]$, or with
sodium acetate to the excellent catalyst precursor Ru(binap)(OAc)$_2$ (see below).
The former complex [41] was originally thought to be Ru$_2$Cl$_4$(binap)$_2$(NEt$_3$) [42];
however, the ethyl group in NEt$_3$ appears to undergo an interesting fragmentation reaction. It is an excellent precatalyst for the enantioselective hydrogenation
of dehydroamino acids [24, 41–43]. The reaction of the Ru(η^3-methylallyl)$_2$(COD)
complex with enantiopure diphosphines, and then with HBr, yields catalyst solutions thought to contain a solvated form of RuBr$_2$(diphosphine) that are useful for the asymmetric hydrogenation of functionalized alkenes and ketones in-

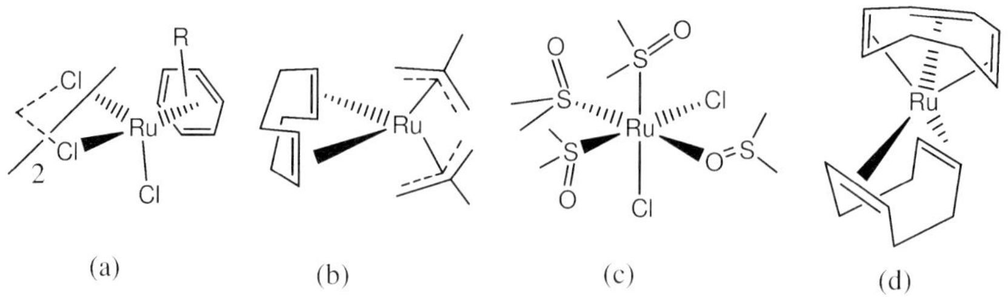

Fig. 3.4 Useful starting ruthenium complexes.

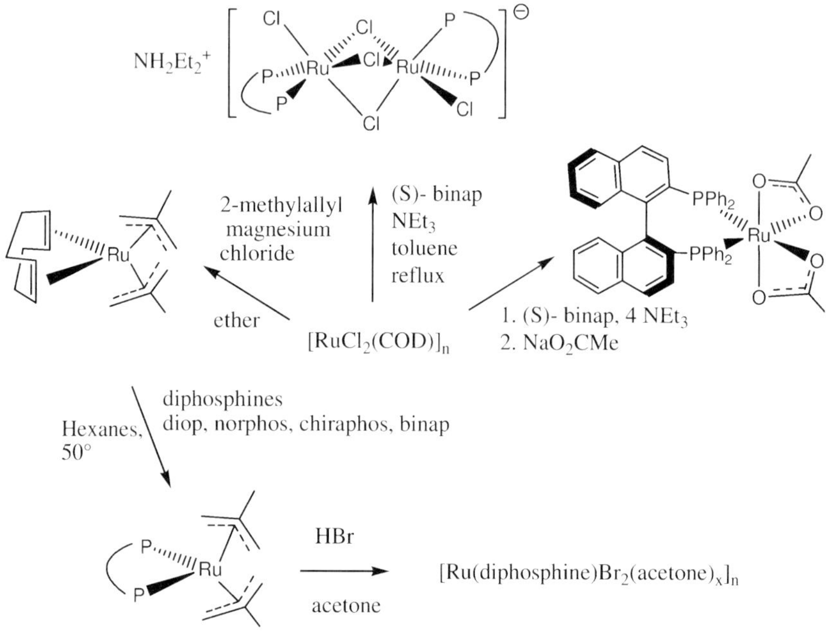

Fig. 3.5 Reactions starting with [RuCl$_2$(1,5-cyclooctadiene)]$_n$.

cluding unsaturated acids, β-ketoesters, and allylic alcohols [44, 45]. The π-allyl complex can also be reacted with chiral diphosphines and HBF$_4$/BF$_3$ to generate a very active hydrogenation catalyst for tetrasubstituted alkenes that are precursors to fragrances [46].

3.2.3
Dihydrogen Complexes and Non-Classical Hydrogen Bonding in Catalysis

Schemes 3.2 and 3.3 show intermediates containing dihydrogen ligands with the H–H bond intact. It has only been appreciated since the discovery of the first dihydrogen complexes by Kubas and coworkers in 1984 [14] that such complexes are key intermediates in catalytic cycles [11, 13, 14].

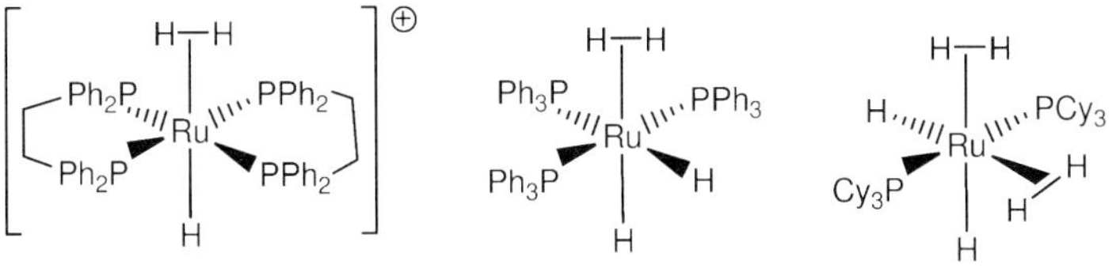

Fig. 3.6 The dihydrogen complexes $[Ru^{II}(\eta^2\text{-}H_2)H(dppe)_2]^+$, $Ru^{II}(\eta^2\text{-}H_2)(H)_2(PPh_3)_3$ and $Ru^{II}(\eta^2\text{-}H_2)_2(H)_2(PCy_3)_2$.

Before 1984, the oxidative addition of H_2 to square-planar Ru^{II} to produce octahedral Ru^{IV} ($H_2 + [Ru^{II}] \rightarrow [Ru^{IV}](H)_2$) was thought to be the turnover-limiting step in this cycle (c.f., the left equilibrium of Scheme 3.3) by analogy to rhodium systems. The discovery that the complexes $[RuH_3(diphosphine)_2]^+$ [47] and $RuH_4(PPh_3)_3$ [48] are not seven-coordinate Ru^{IV} structures but instead are octahedral, Ru^{II} complexes $[Ru(\eta^2\text{-}H_2)H(diphosphine)_2]^+$ and $Ru(\eta^2\text{-}H_2)(H)_2(PPh_3)_3$ (Fig. 3.6) supports the inner pathway of Scheme 3.3. The dihydrogen ligands in these complexes have H–H distances of 0.94 Å [49] and about 1.1 Å, respectively, longer than that of free H_2 at 0.74 Å. Even $RuH_6(PCy_3)_2$ [50, 51] retains an octahedral, Ru^{II} configuration.

Dihydrogen complexes display a wide range of acidity or, in other words, a propensity to undergo heterolytic splitting. The neutral dihydrogen complexes of Figure 3.6 have approximate pK_a^{THF} values of about 36–40 [52] (similar to cyclohexanol in THF), while the cationic complex has a value of about 14 [53]. Dicationic complexes in CH_2Cl_2 containing a π-acid ligand become very acidic; for example, $trans$-$[Ru(\eta^2\text{-}H_2)(CO)(PPh_2CH_2CH_2CH_2PPh_2)_2]^{2+}$ has a $pK_a^{CH_2Cl_2}$ value of –7 relative to $HPCy_3^+/PCy_3$ defined as 9 [54]. Such values are determined by measuring an equilibrium constant, usually by use of nuclear magnetic resonance (NMR), for a reaction of the dihydrogen complex with a base, the conjugate acid of which has a known pK_a value [52]. For example, the dihydrogen complex $[Ru(\eta^2\text{-}H_2)(\eta^2\text{-}C_5H_5)(dppm)]^+$, $dppm = PPh_2CH_2PPh_2$, has an approximate pK_a^{THF} of about 7.3 as determined from the equilibrium constant of Eq. (1) [52].

$$[Ru(\eta^2H_2)(C_5H_5)(PPh_2CH_2PPh_2)]^+ + PBu_3 \rightleftharpoons$$
$$RuH(C_5H_5)(PPh_2CH_2PPh_2) + HPBu_3^+ \qquad (1)$$

The easy heterolytic splitting of dihydrogen in such cationic cyclopentadienyl complexes can be exploited in the hydrogenation of CO_2. Lau and coworkers found that heating solutions of $[(\eta^5,\eta^1\text{-}C_5H_4(CH_2)_3NMe_2)Ru(dppm)]BF_4$, under H_2/CO_2 (40 bar/40 bar) at 80 °C for 16 h gave formic acid in low yields (TON = 8) [55]. These authors proposed that dihydrogen undergoes heterolytic splitting into a hydride and a proton on the amine as shown in Scheme 3.4, and that the hydride and proton then react with the CO_2 to produce formic acid. This ligand-assisted splitting of dihydrogen is also observed in the enantioselective hydrogenation of tiglic acid and in the Noyori ketone hydrogenation catalysts (see below). A feature of such a reaction is that when the dihydrogen is deprotonated by the base in a

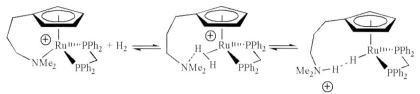

Scheme 3.4 The heterolytic splitting of dihydrogen at Ru(II)
to give a hydridic-protonic bond, as proposed by Chu et al.
[55] in the mechanism of the homogeneous hydrogenation of
carbon dioxide.

low-dielectric solvent such as toluene or THF, the protonated base can donate a
non-classical hydrogen bond (also referred to as a dihydrogen bond [56]) to the hydride, as shown in Scheme 3.4. This type of $MH \cdots HN$ or $MH \cdots HO$ hydrogen
bond was discovered by Crabtree's group [58] and Morris' group [59] in 1994, and
can have an energy of several kcal mol^{-1} and have an $H \cdots H$ distance of 1.8–2.3 Å.
These are now known to be important features of mechanisms of reactions involving transition metal hydrides.

A related chiral complex $[Ru(\eta^2\text{-}H_2)(\eta^5\text{-}C_5H_5)(chiraphos)]^+$ has been used for
the enantioselective outer-sphere hydrogenation of iminum salts [60].

The cationic complexes $[RuH(\eta^2\text{-}H_2)(PP_3)]BPh_4$, $PP_3 = P(CH_2CH_2PPh_2)_3$ [61]
and $[RuH(L)(PMe_2Ph)_4]PF_6$, $L = PMe_2Ph$ [62] or $\eta^2\text{-}H_2$ [63], are catalysts for the
selective hydrogenation of alkynes to alkenes, even in the presence of added alkenes. The PMe_2Ph compounds are sources of $[RuH(PMe_2Ph)_4]^+$ that hydrogenates terminal and internal alkynes in the presence of excess PMe_2Ph, probably
as shown in Scheme 3.5. The alkyne coordinates to ruthenium and is attacked
by the hydride to give an intermediate vinyl species. This is hydrogenolyzed,
probably via proton transfer from an acidic η^2-dihydrogen ligand situated cis to
the vinyl. However, the alternative oxidative addition of dihydrogen and reductive elimination of the hydrogenolyzed product has not been ruled out. 1-Hex-
yne is hydrogenated to 1-hexene with an initial TOF of 4 h^{-1} at 1 bar H_2, 30 °C.
Steric effects of the phosphine ligands in $[RuHL_5]^+$ are very important. The rate
is smaller, the smaller the cone angle of the phosphine used ($PMe_2Ph > PMe_3 >$
$P(OMe)_3$) [64].

The $[RuH(\eta^2\text{-}H_2)(PP_3)]BPh_4$ complex is also thought to operate by the mechanism of Scheme 3.5, and the hydrogenolysis step is shown to be turnover-limiting
[61]. A representative TOF for 94% conversion of phenylacetylene to styrene is
376 h^{-1} at 40 °C, 5 bar H_2 with a turnover number (TON) of 940 [61]. At higher
pressures the TOF is reduced, probably because the dissociation of H_2 from the
starting dihydrogen complex is quickly reversed. Terminal alkynes can undergo
a side reaction where they couple to form other complexes that are inactive or
less active as hydrogenation catalysts. This coupling is prevented in the case of
the PMe_2Ph systems by adding excess PMe_2Ph. The complex $[Ru(COD)(H)$
$(PMe_2Ph)_3]PF_6$ is, under H_2 gas, a source of $[RuH(PMe_2Ph)_3(solvent)_2]^+$; this
species is a very active hydrogenation catalyst for alkynes and alkenes, although

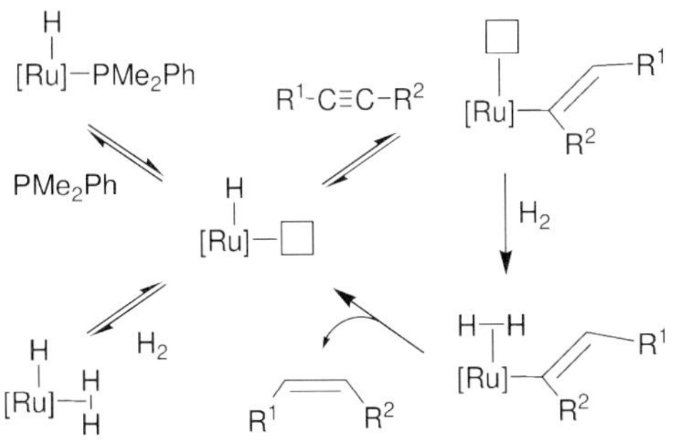

Scheme 3.5 Hydrogenation of alkynes to alkenes catalyzed by [RuH(η^2-H$_2$)(P(CH$_2$CH$_2$PPh$_2$)$_3$)]BPh$_4$ ([Ru]= [Ru(P(CH$_2$CH$_2$PPh$_2$)$_3$)]$^+$) or [RuH(PMe$_2$Ph)$_5$]PF$_6$ or [RuH(η^2-H$_2$)(PMe$_2$Ph)$_4$]PF$_6$ ([Ru]=[Ru(PMe$_2$Ph)$_4$]$^+$). The square represents a vacant site on ruthenium.

the system deactivates rapidly for terminal alkynes [64]. The rate of hydrogenation to cis alkenes increased as 1-hexyne < 2-hexyne < 3-hexyne.

3.2.4
Toward the Reduction of Simple Ketones, Nitriles, Esters and Aromatics with Monodentate Phosphine Systems

At the end of the 1970s, chemists were focusing on applying ruthenium catalysts to enantioselective hydrogenation reactions (see below), and to the hydrogenation of more difficult substrates such as simple ketones, nitriles and esters and reactions related to coal and synthesis gas (H$_2$/CO) chemistry. Important to the utilization of coal (and lignin [65]) is the hydrogenation of arenes and polycyclic aromatics. The very oxygen- and water-sensitive anionic hydride complexes K[RuH$_2$((C$_6$H$_4$)PPh$_2$)(PPh$_3$)$_2$] and K$_2$[Ru$_2$H$_4$(PPh$_2$)(PPh$_3$)$_3$] were reported by Pez and coworkers to catalyze a variety of difficult hydrogenations, including simple ketones to alcohols (e.g., acetone to iPrOH in toluene, 80 °C, 6 bar, TON 380, TOF 24 h^{-1}), esters activated with CF$_3$ groups to the alcohols (90 °C, 6 bar, toluene), nitriles to amines with selectivities up to 90% for the primary amine (acetonitrile to ethylamine in toluene, 90 °C, 6 bar, TON 150, TOF 8 h^{-1}) [66], and anthracenes to 1,2,3,4-tetrahydroanthracenes. The rate of ketone hydrogenation tripled when 18-crown-6 was added to complex the potassium.

Linn and Halpern later found that the active catalyst in the ketone and anthracene hydrogenation reactions of Pez was likely to be Ru(η^2-H$_2$)(H)$_2$(PPh$_3$)$_3$ (Fig. 3.6) [67]. For example, cyclohexanone is converted to cyclohexanol under mild conditions in toluene (see Table 3.3). The TOF depends on the substrate concentration, and the rate law for the catalytic reaction was determined to be given by Eq. (2), with $k=1.3\times10^{-3}$ M^{-1} s^{-1} at 20 °C.

$$\text{Rate} = k[\text{RuH}_4(\text{PPh}_3)_3][\text{ketone}] \qquad (2)$$

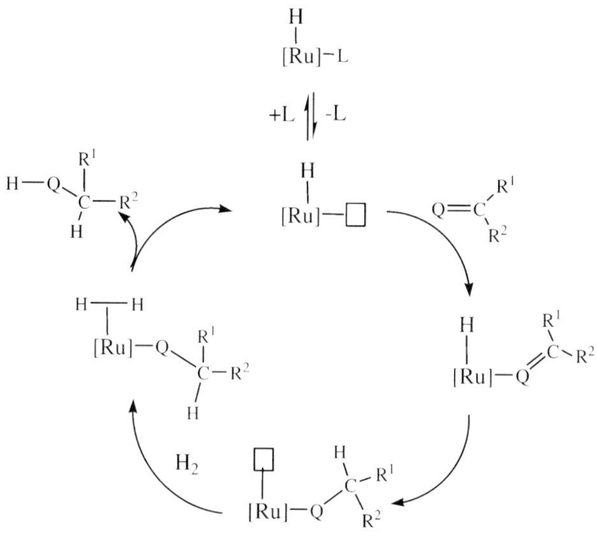

Scheme 3.6 Conventional mechanism for the H_2-hydrogenation of aldehydes, ketones (Q=O) and imines (Q=NR). Ruthenium remains as Ru^{II} throughout the cycle. The square represents a vacant site on ruthenium.

Linn and Halpern proposed a mechanism where the lack of a dihydrogen concentration dependence in the rate law of Eq. (2) was rationalized by the canceling effects of a pre-equilibrium H_2 dissociation and then rate-determining re-addition step. In this mechanism, H_2 dissociates from $RuH_4(PPh_3)_3$ when the ketone coordinates, an alkoxide intermediate $RuH(OR)(PPh_3)_3$ forms, and then H_2 re-coordinates to this intermediate in the rate-determining step. This is followed by the rapid elimination of alcohol and reaction with H_2 to reform $RuH_4(PPh_3)_3$. These steps are commonly proposed for inner-sphere hydrogenation mechanisms (HI) of carbonyl compounds (Scheme 3.6, $[Ru]=RuH(PPh_3)_3$, Q=O, L=H_2) [19]. Note the striking similarities between Schemes 3.3 and 3.6.

Directly related to the cycle shown in Scheme 3.6 is the mechanism of transfer-hydrogenation of ketones and imines catalyzed by, for example, $RuCl_2(PPh_3)_3/$ base or $RuH_2(PPh_3)_4$ solutions in iPrOH. Here, instead of the H_2 in Scheme 3.6, the iPrOH solvent, formic acid or formate is the source of H^+/H^- for regeneration of the starting hydride catalyst, as shown in Scheme 3.7. In the case of dihydride catalysts, Scheme 3.8 has been proposed [18]. Note that the former mechanism involves β-hydride elimination from formate or alkoxide that maintains a Ru^{II} oxidation state, while the later mechanism involves reductive elimination of an alkoxide and hydride with a resulting reduction of the metal to Ru^0.

More recently, dihydrogen complexes have been patented for nitrile hydrogenation. For example, the complex $Ru(\eta^2\text{-}H_2)_2(H)_2(PCy_3)_2$ (Fig. 3.6) catalyzes the hydrogenation of adiponitrile to hexamethylenediamine (HMD) in toluene at $90\,°C$, 70 bar H_2 with TON 52, TOF 5 h^{-1} [68]. At intermediate conversions, the

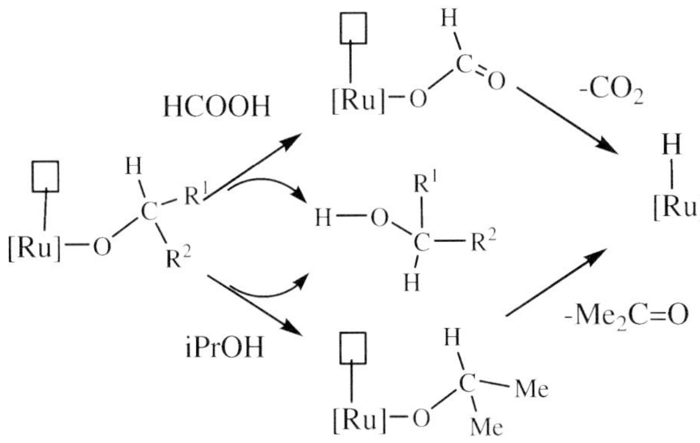

HCOOH

[Ru]—O

H
|
C=O

−CO₂

$[Ru]-\square$

H
|
[Ru]—□

H—O
R¹
|
C—R²
|
H

−Me₂C=O

[Ru]—O
H
|
C—R¹
R²

iPrOH

[Ru]—O
H
|
C—Me
Me

Scheme 3.7 Generation of the active hydride catalyst by hydrogen transfer from formic acid or iso-propanol via β-hydride elimination from formate or alkoxide intermediates. The square represents a vacant site on ruthenium.

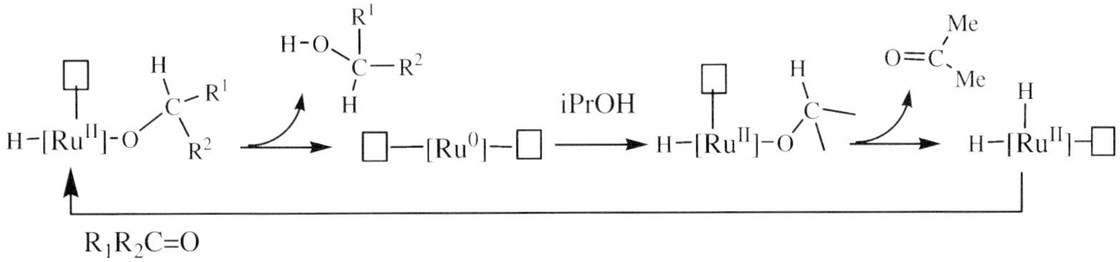

H—[Ru^{II}]—O
H
|
C—R¹
R²

H—O
R¹
|
C—R²
|
H

□—[Ru⁰]—□

iPrOH

H—[Ru^{II}]—O
H
|
C—

O=C
Me
Me

H—[Ru^{II}]—□

R₁R₂C=O

Scheme 3.8 Generation of the active dihydride catalyst by transfer hydrogenation by reductive elimination of the product to give a ruthenium(0) intermediate ([Ru] = Ru(PPh₃)₃).

system displays an interesting, non-statistical reduction of the two CN groups, giving a higher ratio of aminocapronitrile to HMD than expected.

Several ruthenium systems catalyze the hydrogenation of aromatic rings, and this topic is detailed in Chapter 16. An early example reported by Bennett and coworkers was that of RuHCl(η^6-C₆Me₆)(PPh₃), which catalyzed the hydrogenation of benzene to cyclohexane at 25 °C, 1 bar H₂ [69]. Since ruthenium colloids are very active for this reaction under certain conditions, there is evidence that at least some of the reported catalysts are heterogeneous [70].

The hydrogenation of esters remains a challenge. Some recent progress has been reported by Teunissen and Elsevier [71, 72] where a mixture of Ru(acac)₃ and MeC(CH₂PPh₂)₃ was used to hydrogenate aromatic and aliphatic esters to the alcohols in MeOH at 100–120 °C with 85 bar H₂.

The use of Ru(acac)₃ under very high temperature (268 °C) and pressure (1300 bar of H₂/CO) in THF provides a catalyst for the hydrogenation of carbon monoxide to methanol and methyl formate [73]. The active species is derived from Ru(CO)₅.

3.2.5
Enantiomeric Hydrogenation of Alkenes with Bidentate Ligand Systems

More than one-half of the reports in Figure 3.1 are associated with asymmetric hydrogenation and its application in organic synthesis. The first studies from the groups of James and Bianchi in the 1970s involved Kagan's readily prepared chiral, chelating ligand (–)-diop (see Fig. 3.3), in ruthenium complexes such as $Ru_2Cl_4(diop)_3$ [74], trans-$RuHCl(diop)_2$ [5], and $Ru_4H_4(CO)_8(diop)_2$ [75]. The chloro complexes were moderately active and selective for the hydrogenation of acrylic acid derivatives (Table 3.2). A kinetic study revealed that the active catalyst contained only one diop ligand per ruthenium [76].

Complexes containing one binap ligand per ruthenium (Fig. 3.5) turned out to be remarkably effective for a wide range of chemical processes of industrial importance. During the 1980s, such complexes were shown to be very effective, not only for the asymmetric hydrogenation of dehydroamino acids [42] – which previously was rhodium's domain – but also of allylic alcohols [77], unsaturated acids [78], cyclic enamides [79], and functionalized ketones [80, 81] – domains where rhodium complexes were not as effective. Table 3.2 (entries 3–5) lists impressive TOF values and excellent ee-values for the products of such reactions. The catalysts were rapidly put to use in industry to prepare, for example, the perfume additive citronellol from geraniol (Table 3.2, entry 5) and alkaloids from cyclic enamides. These developments have been reviewed by Noyori and Takaya [82, 83].

Ashby and Halpern deduced the mechanism of the hydrogenation of tiglic acid catalyzed by $Ru(binap)(OAc)_2$ in MeOD [84]. This is shown in Scheme 3.9, with some modification to accommodate more recent knowledge of the heterolytic splitting of dihydrogen assisted by a ligand [57]. In the turnover-limiting addition of dihydrogen, this molecule splits into a hydride on the metal and a proton on the carboxylate ligand. The enantioselectivity of the process is directed by the binap ligand ((S)-binap in this case) that sets the chirality at the metal (Δ in this case) and the carbon on the C=C double bond to which the hydride adds. The difference from the classical alkene hydrogenation mechanism of Scheme 3.3 is that the alkyl intermediate is protonated by the carboxylic acid and not by a dihydrogen ligand. The evidence for this is the selective formation of (S)-3-deutero-2-methylbutanoic acid when MeOD is used as the solvent.

By contrast, a recent, detailed mechanism of the enantiomeric hydrogenation of α-(acylamino)acrylic esters catalyzed by $Ru((S)-binap)(OAc)_2$ follows that of Scheme 3.3, where both H atoms from the dihydrogen add to the C=C double bond [85]. The high enantioselectivity of the process is produced, in part, by the chelation of the alkene substrate via the C=C double bond and by a carbonyl oxygen of the substrate [86].

Table 3.2 Representative conditions for the enantiomeric hydrogenation of alkenes.

	Precatalyst (Ru)	Substrate (S)	S:C	Solvent	p(H$_2$) [bar]	Product	Conversion [%]	ee [%]	Time [h]	Temp. [C]	TON	TOF [h^{-1}]	Reference
1	Ru$_2$Cl$_4$((−)-diop)$_3$	Atropic acid H$_2$C=CPh(COOH) (0.2 M)	50	Dimethyl-acetamide	1	(R)-2-phenyl-propionic acid	100	40		60	50	8	127
2	Ru$_2$Cl$_4$((−)-diop)$_3$	2-Acetamidoacrylic acid	50	Dimethyl-acetamide	1	(S)-acetylalanine	100	59		60	50	1	127
3	"Ru$_2$Cl$_4$((S)-binap)$_2$-NEt$_3$" (now thought to be NEt$_2$H$_2$-[Ru$_2$Cl$_5$((S)-binap)$_2$])	PhHC=C(COOH) (NHCOPh)	80	EtOH/THF (NEt$_3$ added)	2	(R)-phenylalanine derivative	100	>90	<24	35	80	>3	42
4	Λ-Ru((R)-binap)-(O$_2$CMe)$_2$	Tiglic acid MeHC=C-MeCOOH (0.05 M)	500	MeOH	1	(R)-EtCMeH-(COOH)	100	93 (R)	0.3	21	500	<4000 [a]	128
5	Δ-Ru((S)-bina-p)(O$_2$CCF$_3$)$_2$	Geraniol (5.8 M)	20000	MeOH	30	(R)-citronellol	100	92 (R)	13	20	20000	1500	77
6	[RuH((R)-binap)-(NCMe)$_{3-n}$(sol.)$_n$]$^+$	(Z)-methyl-R-acetamidocinnamate (0.13 M)	50	Acetone	4	(R)-PhCH$_2$CH-(COOMe) (NHCOMe)	100	92 (R)	96	30	980	54	129

a) TOF varies with alkene concentration as TOF = 8 × 10^4 [alkene] h^{-1}.

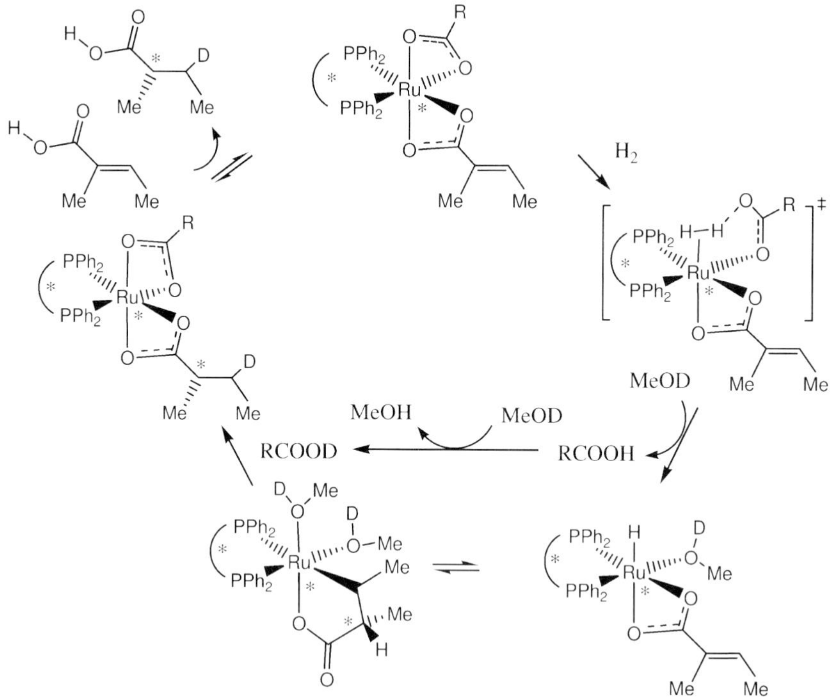

Scheme 3.9 A possible mechanism of the hydrogenation of tiglic acid catalyzed by Ru((S)-binap)(OAc)₂ (as adapted from [84]). The stereochemistry of the metal center and coordination geometries are speculative at this stage.

3.2.6
Enantiomeric Hydrogenation of Carbonyl Compounds

Complexes of the type RuX₂(diphosphine), where X is a halogen or carboxylic acid (see Fig. 3.5), are precatalysts for the hydrogenation of ketones that have a functional group such as an ester carbonyl or amino group in the vicinity of the C=O bond so that the two groups can chelate to the metal [45, 80, 81]. The mechanism is thought to involve a monohydride route (as shown in Scheme 3.6), with a step that involves an inner-sphere transfer of hydride to the carbonyl of the ketone (Scheme 3.10). Similarly, the cationic catalyst [RuH((R)-binap) (NCMe)₃₋ₙ(sol.)ₙ]⁺, sol.=solvent, is very active for the hydrogenation of ketoesters (Table 3.3) and in this case, the intermediate alkoxide complex, where the hydride has added to the carbonyl group, has been completely characterized [87].

In a series of breakthroughs during the 1990s, Noyori's group discovered that simple prochiral ketones that do not contain such functional groups are hydrogenated to pure, optically active alcohols by use of extremely active ruthenium complexes containing primary or secondary amine groups [88, 89]. These cata-

inner-sphere
hydride transfer

outer-sphere
hydride transfer

Scheme 3.10 Inner-sphere versus outer-sphere hydride transfer to the ketone.

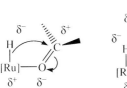

lysts follow a fundamentally different, newly discovered mechanism, involving the outer-sphere transfer of the hydride to the carbonyl assisted by an N–H group (Scheme 3.10). Noyori has called this "metal–ligand bifunctional cataly-sis", where both the ruthenium and the amine are involved in the hydrogena-tion of the ketone and also in the dihydrogen activation (see below). First, they reported that the presence of a diamine with at least one N–H group in Ru^{II} precatalysts of the type RuCl$_2$(diamine)(PR$_3$)$_2$ and RuCl$_2$(diamine)(diphosphine) spectacularly increased the activity of ruthenium complexes toward the hydroge-nation of simple ketones [90]. The chirality of the diamine, such as (R,R)-NH$_2$CHPhCHPhNH$_2$ ((R,R)-dpen), and the diphosphine, such as (R)-binap, must be properly matched to obtained high ee-values in the hydrogenation of a wide range of ketones [89]. The precatalysts are activated by reaction with dihy-drogen and base to give the active catalyst solution. The example in Table 3.3 for the hydrogenation of acetophenone catalyzed by the Ru(Cl)$_2$((S)-tolbinap) ((S,S)-dpen)/KO^{t}Bu system shows an astounding TOF of 2×10^5 h^{-1} at 30 °C, 45 bar H$_2$ (TOF increases as the hydrogen pressure increases). This illustrates the orders of magnitude effect of the N–H group compared to the first two en-tries of Table 3.3 that probably involve inner-sphere hydride transfer. Clapham et al. [19] have reviewed the mechanisms of ruthenium hydrides in catalytic hy-drogenation proposed in the literature up to 2004, and have systematized them according to the inner-sphere and outer-sphere classification.

Recent mechanistic studies conducted by the present author and colleagues [32, 33, 91, 92] and Noyori and colleagues [93] suggest that a *trans*-dihydride complex and an amineamido complex are the active catalysts in the main cycle (Scheme 3.11). The dihydride forms a six-member RuH$\cdots$C-O$\cdots$HN ring with the aryl ketone in the transition state, while simultaneous outer-sphere hydride and proton transfer gives the alcohol and an amineamido complex with a dis-torted trigonal bipyramidal geometry about ruthenium. Addition of dihydrogen to the ruthenium-amido bond via an unstable dihydrogen complex regenerates the *trans*-dihydride. The amido ligand assists in the heterolytic splitting of the dihydrogen. There is evidence that the alcohol solvent also assists in this split-ting process. The lack of coordination sites *cis* to the hydride means that C=C bonds cannot be hydrogenated by an inner-sphere mechanism, and so these cat-alysts are selective for the hydrogenation of polar bonds (C=O) or (C=N) [34] over C=C bonds.

Table 3.3 Representative conditions for the hydrogenation of carbonyl compounds including enantiomeric reactions.

Precatalyst (Ru)	Substrate (S)	S:C	Solvent	p(H$_2$) [bar]	Conversion [%]	ee [%]	Time [h]	Temp. [°C]	TOF [h^{-1}]	Reference
1 Ru(H)$_2$(H$_2$)(PPh$_3$)$_3$	Cyclohexanone	36	Toluene	0.6	3		1	20	1	67
2 [RuH((R)-binap)(NCMe)$_{3-n}$(sol.)$_n$]$^+$	MeOOCCMe$_2$C(=O)COOMe	200	MeOH	50	100	59 (R)	50	50	4	87
3 Ru(Cl)$_2$((S)-tolbinap)((S,S)-dpen)/KO^{t}Bu	PhMeC=O	2,400,000[a]	iPrOH	45	100	80 (R)	48	30	2×10^5	89
4 RuH$_2$((R)-binap)(NH$_2$CMe$_2$CMe$_2$NH$_2$)	PhMeC=O	400	Benzene	8	100	62–68 (R)	2	20	200	91
5 OsH(NHCMe$_2$CMe$_2$NH$_2$)(PPh$_3$)$_2$	PhMeC=O	346	Benzene	5	100	(R)	0.3	20	1400	125

a) Substrate : base = 100 : 1.

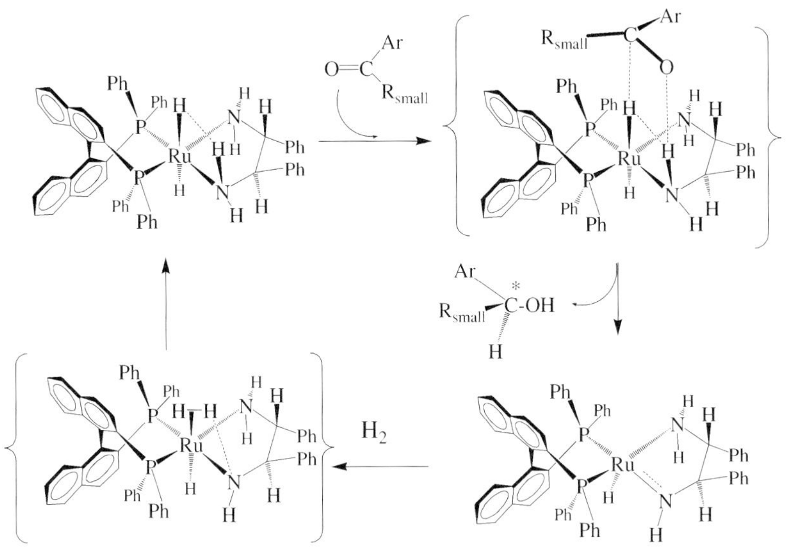

Scheme 3.11 Partial mechanistic scheme for the hydrogenation of aryl ketones to give the (S)-alcohol catalyzed by $RuCl_2((R)\text{-binap})((R,R)\text{-dpen})/KO^tBu/H_2$ as based on the observed mechanism for $RuH_2((R)\text{-binap})(NH_2CMe_2CMe_2NH_2)$.

Noyori and coworkers reported well-defined ruthenium(II) catalyst systems of the type $RuH(\eta^6\text{-arene})(NH_2CHPhCHPhNTs)$ for the asymmetric transfer hydrogenation of ketones and imines [94]. These also act via an outer-sphere hydride transfer mechanism shown in Scheme 3.12. The hydride transfer from ruthenium and proton transfer from the amino group to the C=O bond of a ketone or C=N bond of an imine produces the alcohol or amine product, respectively. The amido complex that is produced is unreactive to H_2 (except at high pressures), but readily reacts with iPrOH or formate to regenerate the hydride catalyst.

An interesting catalytic ruthenium system, $Ru(\eta^5\text{-}C_5Ar_4OH)(CO)_2H$ based on substituted cyclopentadienyl ligands was discovered by Shvo and coworkers [95–98]. This operates in a similar fashion to the Noyori system of Scheme 3.12, but transfers hydride from the ruthenium and proton from the hydroxyl group on the ring in an outer-sphere hydrogenation mechanism. The source of hydrogen can be H_2 or formic acid. Casey and coworkers have recently shown, on the basis of kinetic isotope effects, that the transfer of H^+ and H^- equivalents to the ketone for the Shvo system and the Noyori system (Scheme 3.12) is a concerted process [99, 100].

Palmer and Wills in 1999 reviewed other ruthenium catalysts for the asymmetric transfer hydrogenation of ketones and imines [101]. Gladiali and Mestroni reviewed the use of such catalysts in organic synthesis up to 1998 [102]. Review articles that include the use of ruthenium asymmetric hydrogenation catalysts cover the literature from 1981 to 1994 [103, 104], the major contributions

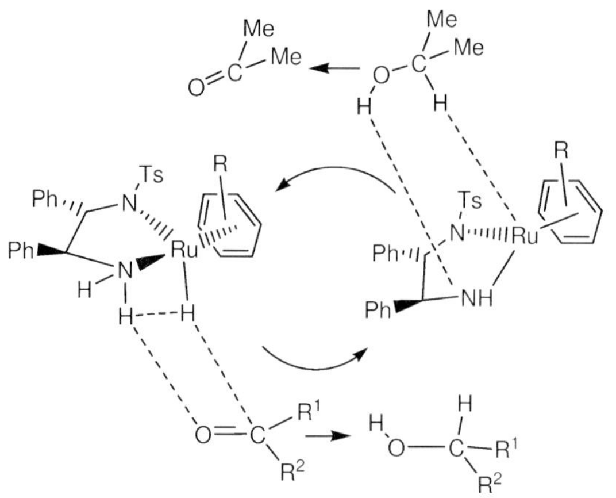

Scheme 3.12 Enantioselective hydrogenation of a ketone by transfer from *iso*-propanol catalyzed by the hydride complex RuH(η^6-arene)(NH$_2$CHPhCHPhNTs) and the amido complex Ru(η^6-arene)(NHCHPhCHPhNTs) [94].

by the group of Genêt until 2003 [45], and the field from an industrial perspective to 2003 [105] (see also Chapter 25). The field of asymmetric imine hydrogenation, that includes ruthenium catalysts, has been reviewed both in 1997 [106] and 2001 [107]. The specific use of the following ligand systems in ruthenium H$_2$-hydrogenation catalysts has been summarized: aminophosphine-phosphinite ligands in 1998 [108], P-chirogenic diphosphine ligands in 2003 [109], chiral ferrocenyl phosphines [110], and a range of new chiral ligand systems in 2003 [111]. Much current research effort is directed at immobilizing these valuable chiral catalysts [112] or keeping them in the aqueous phase [113] so that they can be recovered and recycled. Aqueous-phase and biphasic catalysis involving ruthenium complexes is an active area that was reviewed in 2002 [31, 114].

3.3
Osmium

Complexes of Os^{II} have similar properties to those of Ru^{II}, and can often be prepared in analogous fashions. However, fewer exploratory investigations have been conducted into the starting materials for osmium chemistry than for ruthenium chemistry. In a review of the few osmium hydrogenation catalysts known up to 1995, Sanchez-Delgado et al. [115] point out that the stronger bonding of this 5d metal results in catalysts with higher thermal and oxidative stability than its 4d counterpart, ruthenium, and this – along with other interesting properties – may counter the high cost of using osmium. These authors have since discussed the mechanism of related ruthenium and osmium systems to 2000 [116]. Esteruelas and Oro have described the catalysts based on dihydrogen complexes of osmium [13], and specifically on the derivatives of the five-co-ordinate compound $OsHCl(CO)(P^iPr_3)_2$ [117].

The investigation of osmium hydrogenation catalysts began with a brief report by Vaska in 1965 that the six-coordinate trisphosphine complex $OsHCl(CO)(PPh_3)_3$ could catalyze the hydrogenation of acetylene to ethylene and ethane [118]. Activity during the 1970s and early 1980s focused mainly on the potential of osmium carbonyl clusters as catalysts for the hydrogenation of CO [119]. An interest here is whether a molecule that is made up of a well-defined multimetallic cluster could act like a metal surface found in a Fisher-Tropsch catalyst. The activity of such clusters is relatively low, even for the catalytic hydrogenation of alkenes, as reported, for example, for $Os_3(H)_2(CO)_{10}$ by Keister and Shapely in 1976 [120]. At 50 °C and 3 bar H_2, the hydrogenation of 1-hexene to n-hexane proceeded at a TOF of 1 h^{-1} for a TON of 31, but at the same time the isomerization of some of the hexene to internal alkenes proceeded at a TOF of 2 h^{-1} with a TON of 69. The observation of triosmium intermediates in the reaction indicated that the triangular cluster remains intact throughout the cycle. The catalyst $OsHBr(CO)(PPh_3)_3$ is somewhat less active, isomerizing hexene in the same way, but eventually hydrogenating the intermediates to hexane with a TON of about 60 and a TOF of 5 h^{-1} at 100 °C, 1 bar H_2. Under similar conditions, cyclohexene was hydrogenated to cyclohexane at a TOF of 0.5 h^{-1}, while the C=C bond of cyclohex-2-en-1-one was reduced with a TOF of 24 h^{-1} with a TON of 80. Osmium and ruthenium complexes of the type $MHX(CO)(PR_3)_3$, X=halogen, carboxylate, showed similar, low activity of about TOF 0.5 to 3 h^{-1} for the hydrogenation of propionaldehyde in toluene at 150 °C, 30 bar H_2. Acetone was hydrogenated at a slow rate at 150 °C, 65 bar H_2 [121] until the catalyst decomposed to metal, at which point the rate increased and also the solvent, toluene, was hydrogenated [122]. Several other substrates were investigated as described elsewhere [115].

The five-coordinate bisphosphine complexes $MHCl(CO)(PR_3)_2$, M = Ru, Os, $PR_3 = PMe^tBu_2$, P^iPr_3, PCy_3 and their air-stable precatalysts forms such as $OsHCl(CO)(\eta^2\text{-}O_2)(PR_3)_2$ or $RuHCl(CO)(styrene)(PR_3)_2$ are active alkene hydrogenation catalysts and ketone transfer hydrogenation catalysts in the presence of $NaBH_4$. The dihydrogen complex $OsHCl(CO)(\eta^2\text{-}H_2)(P^iPr_3)_2$, presumably a source of $OsHCl(CO)(P^iPr_3)_2$ by loss of H_2, catalyzes the H_2-hydrogenation of styrene in iPrOH at 60 °C, 1 bar H_2 with a TON of 100 and a TOF of 1200 h^{-1} [123]. Phenyl acetylene is hydrogenated slowly by $OsHCl(CO)(\eta^2\text{-}H_2)(P^iPr_3)_2$, first completely to styrene, because a stable styryl intermediate $OsCl(CO)(CH=CHPh)(P^iPr_3)_2$ ties up all of the osmium and prevents reactions with styrene (Scheme 3.13). This styryl complex is hydrogenolyzed in the turnover-limiting step. The styrene that is produced cannot be hydrogenated until this compound is consumed, after which the hydrogenation to ethylbenzene is rapid [117]. The catalyst precursor $OsHCl(CO)(\eta^2\text{-}O_2)(PCy_3)_2$ is effective, and more active than $RhCl(PPh_3)_3$, for the selective hydrogenation of the disubstituted C=C bonds instead of the C≡N triple bonds of nitrile-butadiene rubbers at 5–40 bar H_2, 130 °C in monochlorobenzene [124].

The mildest conditions for the osmium-catalyzed hydrogenation of a simple ketone (in this case acetophenone) were reported recently by Clapham and Mor-

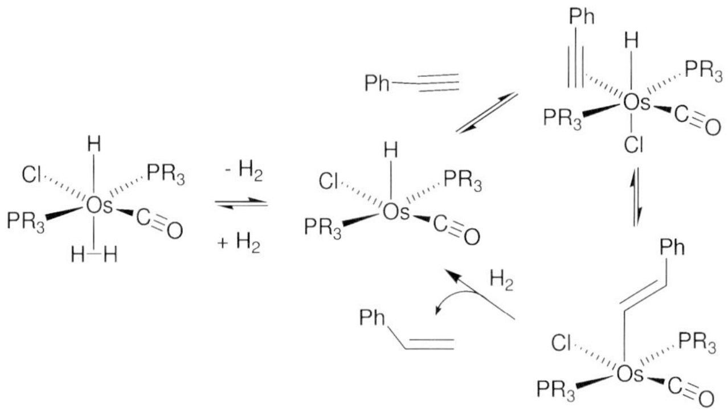

Scheme 3.13 Proposed mechanism for the hydrogenation of phenyl acetylene catalyzed by OsHCl(CO)(P^{i}Pr$_3$)$_2$ [115].

ris [125] by use of the catalyst OsH(NHCMe$_2$CMe$_2$NH$_2$)(PPh$_3$)$_2$ in benzene with a maximum TOF of 1400 h^{-1} and TON of 346 at 20 °C, 5 bar H$_2$. This reaction is thought to proceed through a mechanism analogous to the one shown in Scheme 3.11. Here, the osmium complex appears to be as active as the ruthenium analogue.

Bianchini and coworkers [126] found a difference in the chemoselectivity between the metals Fe, Ru, and Os in the complexes [M(H$_2$)H(P(CH$_2$CH$_2$PPh$_2$)$_3$)]-BPh$_4$ in the hydrogenation of benzylideneacetone by transfer from *iso*-propanol. The Fe and Ru catalysts reduced the C=O bond to give the allyl alcohol, with Ru more active than iron (TOF 79 h^{-1} at 60 °C for Ru versus 13 h^{-1} at 80 °C for Fe), while the Os catalyst first reduced the C=O bond but then catalyzed isomerization of the allyl alcohol to give the saturated ketone (TOF 55 h^{-1} at 80 °C). The difference in reactivity was attributed to the weak binding of the alkene of the allyl alcohol to Fe and Ru relative to Os in these complexes. A variety of selectivities was noted for other unsaturated ketones, whereas unsaturated aldehydes were not hydrogenated.

In future, it will be interesting to identify a catalytic hydrogenation process that justifies the use of osmium over ruthenium, though one possibility might be a high temperature application such as that required in the hydrogenation of unsaturated rubbers.

Acknowledgment

The author wishes to thank NSERC Canada for the provision of a discovery grant.

Abbreviations

DMF   dimethylformamide
ee   enantiomeric excess
HMD   hexamethylenediamine
SCR   substrate catalyst ratio
TOF   turnover frequency
TON   turnover number

References

1 Brown, J. M., Chaloner, P. A. In: *Homogeneous Catalysis with Metal Phosphine Complexes*, Pignolet, L. H. (Ed.), Plenum Press, New York, **1983**, Chapter 4.

2 Halpern, J., Harrod, J. F., James, B. R., *J. Am. Chem. Soc.* **1961**, *83*, 753.

3 Evans, D., Osborn, J. A., Jardine, F. H., Wilkinson, G., *Nature* **1965**, *208*, 1203.

4 James, B. R., *Homogeneous Hydrogenation*, John Wiley, New York, **1973**.

5 James, B. R., *Adv. Organomet. Chem.* **1979**, *17*, 319.

6 Noyori, R., *Angew. Chem. Int. Ed. Engl.* **2002**, *41*, 2008.

7 Noyori, R., *Adv. Synth. Cat.* **2003**, *345*, 15.

8 Halpern, J., Harrod, J. F., James, B. R., *J. Am. Chem. Soc.* **1966**, *88*, 5150.

9 Chatt, J., Hayter, R. G., *Proc. Chem. Soc.* **1959**, 153.

10 Chatt, J., Hayter, R. G., *J. Chem. Soc.* **1961**, 2605.

11 Jessop, P. G., Morris, R. H., *Coord. Chem. Rev.* **1992**, *121*, 155.

12 Hallman, P. S., McGarvey, B. R., Wilkinson, G., *J. Chem. Soc. A* **1968**, 31430.

13 Esteruelas, M. A., Oro, L. A., *Chem. Rev.* **1998**, *98*, 577.

14 Kubas, G. J., *Metal Dihydrogen and Sigma-Bond Complexes.* Kluwer Academic/Plenum, New York, **2001**.

15 Rose, D., Gilbert, J. D., Richardson, R. P., Wilkinson, G., *J. Chem. Soc. A* **1969**, 2610.

16 Fahey, D. R., *J. Org. Chem.* **1973**, *38*, 80–87.

17 Blum, J., Sasson, Y., Iflah, S., *Tetrahedron Lett.* **1972**, 1015.

18 Bäckvall, J.-E., *J. Organomet. Chem.* **2002**, *652*, 105.

19 Clapham, S. E., Hadzovic, A., Morris, R. H., *Coord. Chem. Rev.* **2004**, *248*, 2201.

20 Maienza, F., Santoro, F., Spindler, F., Malan, C., Mezzetti, A., *Tetrahedron Asymm.* **2002**, *13*, 1817.

21 Joshi, A. M., MacFarlane, K. S., James, B. R., Frediani, P., *Stud. Surf. Sci. Catal.* **1992**, *73*, 143.

22 Fogg, D. E., James, B. R. In: *Catalysis of Organic Reactions of the Chemical Industry.* Dekker, **1995**, Vol. 62, pp. 435.

23 Joshi, A. M., Macfarlane, K. S., James, B. R., *J. Organomet. Chem.* **1995**, *488*, 161.

24 Macfarlane, K. S., Thorburn, I. S., Cyr, P. W., Chau, D., Rettig, S. J., James, B. R., *Inorg. Chim. Acta* **1998**, *270*, 130.

25 Langer, T., Helmchen, G., *Tetrahedron Lett.* **1996**, *37*, 1381.

26 Crochet, P., Gimeno, J., Garcia-Granda, S., Borge, J., *Organometallics* **2001**, *20*, 4369.

27 Doucet, H., Ohkuma, T., Murata, K., Yokozawa, T., Kozawa, M., Katayama, E., England, A. F., Ikariya, T., Noyori, R., *Angew. Chem. Int. Ed. Eng.* **1998**, *37*, 1703.

28 Abdur-Rashid, K., Guo, R., Lough, A. J., Morris, R. H., Song, D., *Adv. Synth. Catal.* **2005**, *347*, 571.

29 Guo, R., Lough, A. J., Morris, R. H., Song, D., *Organometallics* **2004**, *23*, 5524.

30 Fache, E., Santini, C., Senocq, F., Basset, J. M., *J. Mol. Catal.* **1992**, *72*, 337.

31 Joó, F., *Acc. Chem. Res.* **2002**, *35*, 738.

32 Abdur-Rashid, K., Lough, A. J., Morris, R. H., *Organometallics* **2000**, *19*, 2655.

33 Abbel, R., Abdur-Rashid, K., Faatz, M., Hadzovic, A., Lough, A. J., Morris, R. H., *J. Am. Chem. Soc.* **2005**, *127*, 1870.

34 Abdur-Rashid, K., Lough, A. J., Morris, R. H., *Organometallics* **2001**, *20*, 1047.

35 Guo, R., Elpelt, C., Chen, X., Song, D., Morris, R. H., *Chem. Commun.* **2005**, 3050.

36 Bennett, M. A., Smith, A. K., *J. Chem. Soc., Dalton Trans.* **1974**, 233.

37 Schrock, R. R., Johnson, B. F. G., Lewis, J., *J. Chem. Soc., Dalton Trans.* **1974**, 951.

38 Frosin, K.-M., Dahlenburg, L., *Inorg. Chim. Acta* **1990**, *167*, 83.

39 Airoldi, M., Deganello, G., Dia, G., Gennaro, G., *J. Organomet. Chem.* **1980**, *187*, 391.

40 Evans, I. P., Spencer, A., Wilkinson, G., *J. Chem. Soc., Dalton Trans.* **1973**, 204.

41 DiMichele, L., King, S. A., Douglas, A. W., *Tetrahedron Asymm.* **2003**, *14*, 3427.

42 Ikariya, T., Ishii, Y., Kawano, H., Arai, T., Saburi, M., Yoshikawa, S., Akutagawa, S., *J. Chem. Soc., Chem. Commun.* **1985**, 922.

43 Ohta, T., Tonomura, Y., Nozaki, K., Takaya, H., Mashima, K. *Organometallics* **1996**, *15*, 1521.

44 Genêt, J.-P., Pinel, C., Ratovelomanana-Vidal, V., Mallart, S., Pfister, X., Bischoff, L., Deandrade, M. C. C., Darses, S., Galopin, C., Laffitte, J.A., *Tetrahedron Asymm.* **1994**, *5*, 675.

45 Genêt, J. P., *Acc. Chem. Res.* **2003**, *36*, 908.

46 Dobbs, D. A., Vanhessche, K. P. M., Brazi, E., Rautenstrauch, V., Lenoir, J.-Y., Genêt, J.-P., Wiles, J., Bergens, S. H., *Angew. Chem. Int. Ed. Engl.* **2000**, *39*, 1992.

47 Morris, R. H., Sawyer, J. F., Shiralian, M., Zubkowski, J., *J. Am. Chem. Soc.* **1985**, *107*, 5581.

48 Crabtree, R. H., Hamilton, D. G., *J. Am. Chem. Soc.* **1986**, *108*, 3124.

49 Albinati, A., Klooster, W., Koetzle, T. F., Fortin, J. B., Ricci, J. S., Eckert, J., Fong, T. P., Lough, A. J., Morris, R. H., Golombek, A., *Inorg. Chim. Acta* **1997**, *259*, 351.

50 Arliguie, T., Chaudret, B., Morris, R. H., Sella, A., *Inorg. Chem.* **1988**, *27*, 598.

51 Sabo-Etienne, S., Chaudret, B., *Coord. Chem. Rev.* **1998**, *180*, 381.

52 Abdur-Rashid, K., Fong, T. P., Greaves, B., Gusev, D. G., Hinman, J. G., Landau, S. E., Morris, R. H., *J. Am. Chem. Soc.* **2000**, *122*, 9155.

53 Cappellani, E. P., Drouin, S. D., Jia, G., Maltby, P. A., Morris, R. H., Schweitzer, C. T., *J. Am. Chem. Soc.* **1994**, *116*, 3375.

54 Rocchini, E., Mezzetti, A., Ruegger, H., Burckhardt, U., Gramlich, V., Del Zotto, A., Martinuzzi, P., Rigo, P., *Inorg. Chem.* **1997**, *36*, 711.

55 Chu, H. S., Lau, C. P., Wong, K. Y., Wong, W. T., *Organometallics* **1998**, *17*, 2768.

56 Crabtree, R. H., Eisenstein, O., Sini, G., Peris, E., *J. Organomet. Chem.* **1998**, *567*, 7.

57 Morris, R. H. In: *Recent Advances in Hydride Chemistry*. Peruzzini, M., Poli, R. (Eds.). Elsevier, Amsterdam, **2001**, pp. 1.

58 Lee, Jr, J. C., Rheingold, A. L., Muller, B., Pregosin, P. S., Crabtree, R. H., *J. Chem. Soc. Chem. Commun.* **1994**, 1021.

59 Lough, A. J., Park, S., Ramachandran, R., Morris, R. H., *J. Am. Chem. Soc.* **1994**, *116*, 8356.

60 Guan, H. R., Iimura, M., Magee, M. P., Norton, J. R., Zhu, G., *J. Am. Chem. Soc.* **2005**, *127*, 7805.

61 Bianchini, C., Bohanna, C., Esteruelas, M. A., Frediani, P., Meli, A., Oro, L. A., Peruzzini, M., *Organometallics* **1992**, *11*, 3837.

62 Albers, M. O., Singleton, E., Viney, M. M., *J. Mol. Catal.* **1985**, *30*, 213–217.

63 Lough, A. J., Morris, R. H., Ricciuto, L., Schleis, T., *Inorg. Chim. Acta* **1998**, *270*, 238.

64 Nkosi, B. S., Coville, N. J., Albers, M. O., Gordon, C., Viney, M. M., Singleton, E., *J. Organomet. Chem.* **1990**, *386*, 111.

65 Wong, T. Y. H., Pratt, R., Leong, C. G., James, B. R., Hu, T.Q. In: *Catalysis of Organic Reactions, Chemical Industries*. Dekker, **2001**, Vol. 82, pp. 255.

66 Grey, R. A., Pez, G., Wallo, A., *J. Am. Chem. Soc.* **1981**, *103*, 7536.

67 Linn, D. E., Halpern, J., *J. Am. Chem. Soc.* **1987**, *109*, 2969.

68 Beatty, R. P., Paciello, R. A. **1998**, US Patent 5726334.

69 Bennett, M. A., Huang, T.-N., Smith, A. K., Turney, T. W., *J. Chem. Soc., Chem. Commun.* **1978**, 582.

70 Widegren, J. A., Bennett, M. A., Finke, R. G., *J. Am. Chem. Soc.* **2003**, *125*, 10301.

71 Teunissen, H. T., Elsevier, C., *J. Chem. Commun.* **1997**, 667.

72 Teunissen, H. T., Elsevier, C., *J. Chem. Commun.* **1998**, 1367.

73 Bradley, J. S., *J. Am. Chem. Soc.* **1979**, *101*, 7419.

74 James, B. R., Wang, D. K. W., Voigt, R. F., *J. Chem. Soc., Chem. Commun.* **1975**, 574.

75 Botteghi, C., Gladiali, S., Bianchi, M., Matteoli, U., Frediani, P., Vergamini, P. G., Benedetti, E., *J. Organomet. Chem.* **1977**, *140*, 221.

76 James, B. R., Wang, D. K. W., *Can. J. Chem.* **1980**, *58*, 245.

77 Takaya, H., Ohta, T., Sayo, N., Kumobayashi, H., Akutagawa, S., Inoue, S., Kasahara, I., Noyori, R., *J. Am. Chem. Soc.* **1987**, *109*, 1596.

78 Ohta, T., Takaya, H., Kitamura, M., Nagai, K., Noyori, R., *J. Org. Chem.* **1987**, *52*, 3174.

79 Noyori, R., Ohta, M., Hsiao, Y., Kitamura, M., Ohta, T., Takaya, H., *J. Am. Chem. Soc.* **1986**, *108*, 7117.

80 Kitamura, M., Ohkuma, T., Inoue, S.-I., Sayo, N., Kumobayashi, H., Akutagawa, S., Ohta, T., Takaya, H., Noyori, R., *J. Am. Chem. Soc.* **1988**, *110*, 629.

81 Noyori, R., Ohkuma, T., Kitamura, M., Takaya, H., Sayo, N., Kumobayashi, H., Akutagawa, S., *J. Am. Chem. Soc.* **1987**, *109*, 5856.

82 Noyori, R., Takaya, H., *Acc. Chem. Res.* **1990**, *23*, 345.

83 Takaya, H., Ohta, T., Mashima, K. In: *Homogeneous Transition Metal Catalyzed Reactions.* Moser, W. R., Slocum, D. W. (Eds.), American Chemical Society, Washington, **1992**, Chapter 8.

84 Ashby, M. T., Halpern, J., *J. Am. Chem. Soc.* **1991**, *113*, 589.

85 Kitamura, M., Tsukamoto, M., Bessho, Y., Yoshimura, M., Kobs, U., Widhalm, M., Noyori, R., *J. Am. Chem. Soc.* **2002**, *124*, 6649.

86 Wiles, J. A., Bergens, S. H., *Organometallics* **1999**, *18*, 3709.

87 Daley, C. J. A., Bergens, S. H., *J. Am. Chem. Soc.* **2002**, *124*, 3680.

88 Noyori, R., Hashiguchi, S., *Acc. Chem. Res.* **1997**, *30*, 97.

89 Noyori, R., Ohkuma, T., *Angew. Chem. Int. Ed. Engl.* **2001**, *40*, 40.

90 Ohkuma, T., Ooka, H., Hashiguchi, S., Ikariya, T., Noyori, R., *J. Am. Chem. Soc.* **1995**, *117*, 2675.

91 Abdur-Rashid, K., Clapham, S. E., Hadzovic, A., Harvey, J. N., Lough, A. J., Morris, R. H., *J. Am. Chem. Soc.* **2002**, *124*, 15104.

92 Abdur-Rashid, K., Faatz, M., Lough, A. J., Morris, R. H., *J. Am. Chem. Soc.* **2001**, *123*, 7473.

93 Sandoval, C. A., Ohkuma, T., Muñiz, K., Noyori, R., *J. Am. Chem. Soc.* **2003**, *125*, 13490.

94 Haack, K. J., Hashiguchi, S., Fujii, A., Ikariya, T., Noyori, R., *Angew. Chem. Int. Ed. Engl.* **1997**, *36*, 285.

95 Shvo, Y., Czarkie, D., Rahamim, Y., Chodosh, D. F., *J. Am. Chem. Soc.* **1986**, *108*, 7400.

96 Abed, M., Goldberg, I., Stein, Z., Shvo, Y., *Organometallics* **1988**, *7*, 2054.

97 Menashe, N., Shvo, Y., *Organometallics* **1991**, *10*, 3885.

98 Menashe, N., Salant, E., Shvo, Y., *J. Organomet. Chem.* **1996**, *514*, 97.

99 Casey, C. P., Johnson, J. B., *J. Org. Chem.* **2003**, *68*, 1998.

100 Casey, C. P., Singer, S. W., Powell, D. R., Hayashi, R. K., Kavana, M., *J. Am. Chem. Soc.* **2001**, *123*, 1090.

101 Palmer, M. J., Wills, M., *Tetrahedron Asymm.* **1999**, *10*, 2045.

102 Gladiali, S., Mestroni, G., *Transition Metals in Organic Synthesis.* Beller, M., Bolm, C. (Eds.), Wiley-VCH, Weinheim, **1998**, Volume 2, pp. 97.

103 Trost, B. M., Fleming, I., *Comprehensive Organic Synthesis: Selectivity, Strategy, and Efficiency in Modern Organic Chemistry.* Pergamon Press, New York, **1991**, Vol. 8.

104 Ojima, I., Eguchi, M., Tzamarioudaki, M. In: *Comprehensive Organometallic Chemistry.* Abel, E. W., Stone, F. G. A.,

Wilkinson, G. (Eds.), Pergamon, New York, **1995**, Vol. 12, pp. 9.

105 Blaser, H. U., Malan, C., Pugin, B., Spindler, F., Steiner, H., Studer, M., *Adv. Synth. Catal.* **2003**, *345*, 103.

106 James, B. R., *Catalysis Today* **1997**, *37*, 209.

107 Dai, X., Qin, Z., *Prog. Chem.* **2001**, *13*, 183.

108 Agbossou, F., Carpentier, J. F., Hapiot, F., Suisse, I., Mortreux, A., *Coord. Chem. Rev.* **1998**, *180*, 1615.

109 Crepy, K. V. L., Imamoto, T., *Adv. Synth. Catal.* **2003**, *345*, 79.

110 Barbaro, P., Bianchini, C., Giambastiani, G., Parisel, S. L., *Coord. Chem. Rev.* **2004**, *248*, 2131.

111 Tang, W., Zhang, X., *Chem. Rev.* **2003**, *103*, 3029.

112 Bianchini, C., Barbaro, P., *Top. Catal.* **2002**, *19*, 17.

113 Sinou, D., *Adv. Synth. Catal.* **2002**, *344*, 221.

114 Dwars, T., Oehme, G., *Adv. Synth. Catal.* **2002**, *344*, 239.

115 Sanchez-Delgado, R. A., Rosales, M., Esteruelas, M. A., Oro, L. A., *J. Mol. Cat. A* **1995**, *96*, 231.

116 Sanchez-Delgado, R. A., Rosales, M., *Coord. Chem. Rev.* **2000**, *196*, 249.

117 Esteruelas, M. A., Oro, L. A., *Adv. Organomet. Chem.* **2001**, *47*, 1.

118 Vaska, L., *Inorg. Nucl. Chem. Lett.* **1965**, *1*, 89.

119 Choi, H. W., Muetterties, E. L., *Inorg. Chem.* **1981**, *20*, 2664.

120 Keister, J. B., Shapely, J. R., *J. Am. Chem. Soc.* **1976**, *98*, 1056.

121 Sanchez-Delgado, R. A., Andriollo, A., Gonzalez, E., Valencia, N., Leon, V., Espidel, J., *J. Chem. Soc., Dalton Trans.* **1985**, 1859.

122 Sanchez-Delgado, R. A., Valencia, N., Marquez-Silva, R., Andriollo, A., Medina, M., *Inorg. Chem.* **1986**, *25*, 1106.

123 Esteruelas, M. A., Sola, E., Oro, L. A., Meyer, U., Werner, H., *Angew. Chem.* **1988**, *100*, 1621.

124 Parent, J. S., McManus, N. T., Rempel, G. L., *J. Appl. Polymer Sci.* **2001**, *79*, 1618.

125 Clapham, S. E., Morris, R. H., *Organometallics* **2005**, *24*, 479.

126 Bianchini, C., Farnetti, E., Graziani, M., Peruzzini, M., Polo, A., *Organometallics* **1993**, *12*, 3753.

127 James, B. R., McMillan, R. S., Morris, R. H., Wang, D. K. W. In: *Advances in Chemistry Series.* Bau, R. (Ed.), American Chemical Society, Washington, DC, **1978**, Vol. 167, pp. 122.

128 Ashby, M. T., Halpern, J., *J. Am. Chem. Soc.* **1991**, *113*, 589.

129 Wiles, J. A., Bergens, S. H., *Organometallics* **1998**, *17*, 2228.

4
Palladium and Platinum

Paolo Pelagatti

4.1
Introduction

Studies of homogeneous hydrogenation catalyzed by soluble palladium (Pd) or platinum (Pt) catalysts first began when the process of heterogeneous hydrogenation was already known. The reduction of $[Pt(C_2H_4)Cl_2]_2$ with H_2 to produce Pt(0), C_2H_6 and HCl was first reported in 1954, and can be considered as one of the events to have opened up the field of homogeneous hydrogenation [1]. The homogeneity of the reaction appeared to be dependent upon the temperature and on the amount of hydrogen employed [2]. The discovery that H_2 could be homogeneously activated by Pd(II) was first observed by Halpern's group in the 1950s and 1960s, as part of a series of hydrometallurgical investigations involving the precipitation of metals from solution with H_2 [3]. The first reports relating to homogeneous catalysts appeared in the literature during the 1960s, their aim being mainly to identify the reaction mechanisms of heterogeneous hydrogenations [2]. The use of Pd(II) as a hydrogenation catalyst was retarded by its instability under H_2 atmosphere, as demonstrated during alkene isomerization reactions [4, 5], in favor of the more stable catalyst, Pt(II). However, interest towards Pd- and Pt-catalyzed homogeneous reductions of unsaturated functions was renewed when it was found that the selectivities were usually higher than those obtained under heterogeneous conditions [6]. Today, the use of soluble Pd and Pt hydrogenating catalysts remains the subject of intense academic and industrial research, as evidenced by the numerous publications on the subject, and the many deposited patents.

An analysis of the most significant homogeneous catalytic systems reported in the literature reveals a structural variety for Pd which is not found for Pt. In fact, although in most cases Pd is incorporated into the (pre)catalyst as divalent ion, active Pd(0)-catalysts have also been reported. By contrast, Pt(0)-catalysts are a rarity. Moreover, Pd complexes containing mono-, di-, tri-, and even tetradentate ligands have found application as hydrogenation catalysts, and often their activity and selectivity is governed by the steric and electronic features of

The Handbook of Homogeneous Hydrogenation.
Edited by J. G. de Vries and C. J. Elsevier
Copyright © 2007 WILEY-VCH Verlag GmbH & Co. KGaA, Weinheim
ISBN: 978-3-527-31161-3

the chelating systems. The most thoroughly studied Pt-catalysts are instead simple phosphine-containing Pt(II) complexes, usually activated with stannous chloride, $SnCl_2$. Tin(II) salts have instead found scarce application with Pd, and in some cases have turned out to be poisons of the catalytic processes. Another important aspect which differentiates the two metals has practical consequences: the well-known higher reactivity of Pd with respect to Pt [7] allows the Pd-promoted hydrogenations to be carried out under much milder conditions (room temperature and atmospheric pressure of H_2) than those usually required for activating Pt-catalysts. However, Pd-based catalysts are usually more subject to decomposition than are Pt-based catalysts.

As unsaturated C–C bonds have certainly been the most thoroughly investigated substrates, this chapter focuses on the hydrogenation of alkenes and alkynes, and the hydrogenation of other functional groups such as nitro, nitrile, and carbonyl will be excluded. Particular emphasis will be given to those catalytic systems for which a mechanistic study has been carried out. Where possible, a brief discussion of the homogeneous character of the catalytic processes will be given.

4.2
Palladium

4.2.1
Phosphorus-Containing Catalysts

In 1963, in a report which focused mainly on the use of first-row transition-metal catalysts combined with organoaluminum compounds, the hydrogenating activity of the catalytic system $[PdCl_2(Pn\text{-}Bu_3)_2/Al(i\text{-}Bu)_3]$ towards 1-hexene under mild conditions (heptane, 25 °C, 3.5–3.7 atm of H_2 pressure, Pd:alkene ratio $\sim 1:80$) was briefly addressed [8]. After 19 h of reaction, a 25.5% conversion to hexane was obtained (TOF $\sim 1.1\ h^{-1}$). In 1967, several Pd(II) complexes of the general formula $[PdX_2(Ph_3Q)_2]$ (Q = P or As, X = Cl or CN) were used to homogeneously hydrogenate soybean oil Me-ester in the presence or absence of co-catalysts, such as $SnCl_2$ or $GeCl_2$ [9]. *Cis-trans* double-bond isomerization, migration of isolated double bonds to conjugated dienes, and selective hydrogenation of polyenes to monoenes without further reduction were observed. The most active system was found to be $PdCl_2(PPh_3)_2/SnCl_2 \cdot 2\,H_2O$ which, after 3 h, converted both linoleate and linolenate to monoenes almost completely, but not at all to stearate (benzene/methanol, 90 °C, 39.1 atm H_2 pressure); Me-oleate was selectively converted to the corresponding monoene under the same experimental conditions and reaction time. In both cases high catalyst loadings were applied. Traces of Pd black were observed at the end of the reaction. Lowering of the temperature precluded the formation of Pd black, but slowed down the hydrogenations.

By the end of the 1970s, $PdCl_2(PPh_3)_2$ was being used to hydrogenate 1,5-cyclooctadiene [10]. The substrate isomerization to 1,3-cyclooctadiene preceded its reduction to cyclooctene. For example, after 5 h of reaction (CH_2Cl_2/MeOH, 90 °C, 51 atm H_2 pressure, alkene:Pd ratio=1:15) the products distribution was 4% of 1,3-cyclooctadiene, 93% of cyclooctene, and 3% of cyclooctane. The reactivity and selectivity remained quite high up to an alkene:Pd ratio of 250. However, under the same experimental conditions, $SnCl_2$ proved to be a poison of the reaction. The π-allylic reaction intermediate [PdCl(π-cyclooctenyl)(PPh₃)] was isolated from the reactant solutions and resulted in a much more active catalyst than its precursor $PdCl_2(PPh_3)_2$, although it was less selective. After 3 h of reaction, under 34 atm H_2 pressure, the product distribution was 10% of 1,3-cyclooctadiene, 83% of cyclooctene, and 6% of cyclooctane. During the same period, other π-allyl-Pd(II) complexes that were effective in the selective hydrogenation of allene to propene (THF, 15 °C, 1 atm total pressure) were reported [11]. Turnover numbers (TONs) ranging from 18 to 75 were obtained with [(η^3-allyl)-PdCl(PR₃)] (allyl=C_3H_5, 1-Me-C_3H_4, 2-Me-C_3H_4; R=PPh₃, PPh₂Me, Pt-Bu₃) or [Pd(η^3-C_3H_5)₂] ([Pd(η^3-C_3H_5)₂] was used at 0 °C). The TOFs ranged from 0.19 to 75 h^{-1}, the highest being obtained with the bis-allene complex [Pd(C_3H_5)₂], though this was significantly decomposed. Since the stability of the complexes under H_2 atmosphere were shown to differ, a comparison of the catalytic results was possible only for the strictly similar allyl-complexes [(η^3-C_3H_5)PdCl(PPh₃)], [(η^3-1-Me-C_3H_4)PdCl(PPh₃)] and [(η^3-2-Me-C_3H_4)PdCl(PPh₃)]. These showed similar turnover frequency (TOF) values of 0.21, 0.23, and 0.19 h^{-1}, respectively, and comparable stability. The same complexes slowly catalyzed the selective hydrogenation of 1,5-cyclooctadiene and 1,3-cyclooctadiene to the corresponding monoenes (Pd:diene ratio=1:31) [12]; 1,5-cyclooctadiene was first isomerized to 1,3-cyclooctadiene and then hydrogenated to cyclooctene. In several cases, however, Pd-black was detected at the end of the catalytic reactions.

The catalytic activity of other Pd-complexes containing mono- or chelating phosphines was studied by Stern and Maples [13]. In the hydrogenation of butadiene (toluene, r.t., 6.8 atm H_2 pressure), the dinuclear Pd(0) complex $Pd_2(dppm)_3$ showed the highest TOF (0.25 h^{-1}), leading to an excess of 1-butene with respect to *cis*- or *trans*-2-butene. (Since the complex was handled in plain air, caution must be taken about its nature.) The Pd(II) complexes [$PdCl_2(dppm)$], [$PdCl_2(1,1\text{-}dppe)$] and [$PdCl_2(dppe)$], although not completely soluble in toluene, also promoted the hydrogenation, albeit with lower TOF values and selectivities. $PdCl_2(PhCN)_2$ was practically inactive. The hydrogenating activity of $Pd_2(dppm)_3$ towards a variety of other unsaturated substrates (alkenes, dienes, trienes, and acetylenes) was reported in the same work. The nature of the substrate appeared to regulate the catalyst activity, in the sense that complexing substrates led to faster reactions. For example, the TOFs varied in the following order: 0.012 h^{-1} (1,9-decadiene), 0.014 h^{-1} (1,5-hexadiene), 0.16 h^{-1} (1,4-pentadiene), 0.25 h^{-1} (butadiene), 0.44 h^{-1} (*cis*-1,3-pentadiene), 0.58 h^{-1} (1,3-cyclohexadiene), 0.80 h^{-1} (*trans*-1,3-pentadiene), and 1.3 h^{-1} (norbornadiene). Studies on the catalyst pretreatment showed that a dissociation equilibrium in solution is necessary in order to enable a coordinatively unsaturated species to bind the substrate, according to Eqs. (1) and (2) (L=ligand, S=solvent, ol=alkene).

$$PdL_4 + S \leftrightarrow PdL_3S + L \tag{1}$$

$$PdL_3S + ol \leftrightarrow PdL_3ol + S \tag{2}$$

Treatment of the precatalytic solutions with oxygen brought about a remarkable enhancement of the catalysis, and an improved selectivity. The effect of oxygen was tentatively rationalized in terms of oxidation of a phosphine ligand with subsequent dissociation of the so-formed phosphine oxide or, alternatively, with formation of oxygen complexes able to promote the substrate/hydrogen activation via hydroperoxide intermediates or the formation of either hydrogen or substrate complexes. The effect of the activating oxygen pretreatment was, at least partially, clarified by Alper et al. for the dinuclear Pd(II) complex [{(t-Bu)$_2$HP}PdP(t-Bu)$_2$]$_2$. This alone was a completely inactive hydrogenating catalyst towards a,β-unsaturated compounds. Once exposed to oxygen for some minutes, it transformed into the mononuclear species [Pd{O$_2$P(t-Bu)$_2$}{OP(t-Bu)$_2$}{OHP(t-Bu)$_2$}] [14] containing a η^2-phosphinate ligand. This species was an active pre-catalyst for the selective reduction of the double C–C bond of several a,β-unsaturated ketones and aldehydes under mild conditions (THF, r.t., 1 atm H$_2$ pressure, 1–2% catalyst loading) [15]. A TOF ≈ 5 h^{-1} was achieved in the hydrogenation of 3-nonen-2-one. The same system was applied in the chemoselective hydrogenation of several substrates, such as a,β-unsaturated sulfones and phosphonates [16], simple and functionalized conjugated dienes [17], and vinyl epoxides [18]. In all cases good catalytic activities and selectivities were obtained (TOFs up to 100 h^{-1}).

In recent years, a number of other polynuclear complexes have been investigated in addition to [Pd$_2$(dppm)$_3$]. In 1989, Eisenberg reported that the reaction between Pd$_2$Cl$_2$(dppm)$_2$ with an excess of NaBH$_4$ led to the formation of a palladium hydride species of approximate stoichiometry [Pd$_2$H$_x$(dppm)$_2$] [19]. This hydride was effective in small-scale hydrogenations of alkynes and alkenes. Parahydrogen-induced polarization was observed in styrene formed during the hydrogenation of phenylacetylene, indicating that the transfer of hydrogen to the substrate occurred pairwise and rapidly relative to proton relaxation. In 1998, the structure of [Pd$_2$H$_x$(dppm)$_2$] was inferred by using a variety of spectroscopic tools [20], and revealed to be a cluster of formula [Pd$_4$(dppm)$_4$(H$_2$)]$^{2+}$. The hydrogenating capability of this material was further investigated quite recently [21]. The [Pd$_4$(dppm)$_4$(H$_2$)](BPh$_4$)$_2$ cluster catalyzed the homogeneous hydrogenation of phenylacetylene, diphenylethyne and 1-phenyl-1-propyne (THF, 20 °C, 1 atm H$_2$ pressure), with TOFs of 500, 200, and 500 h^{-1}, respectively. The products distribution was seen to be time-dependent; after 3 h the cis-alkenes were in the range 75–90%, whereas after 24 h the over-reduced products were predominant, at least with phenylacetylene and diphenylethyne. Strongly coordinating solvents such as dimethylformamide (DMF) led to high TOFs (up to 1800 h^{-1} under 41 atm H$_2$ pressure), while less-coordinating solvents such as tetrahydrofuran (THF), acetone (Me$_2$CO), and acetonitrile (MeCN) led to lower TOF values (1240, 1130, and 1060 h^{-1}, respectively, under the same pressure); pyridine inhibited the reaction (TOF = 660 h^{-1}), most likely due to the presence of reactivity with $[Pd_4(dppm)_4(H_2)]^{2+}$. The polymeric low-valent complex $[Pd_5(PPh_2)]_n$ ($n \approx 4$) was also reported to be a highly active catalyst in the semi-hydrogenation of phenylacetylene and 1,3-pentadiene, as well as in the hydrogenation of 1- and 2-pentene (DMF, 20 °C, 1 atm H_2 pressure) [22]. The TOFs reached with the different substrates were 7200, 60 000, and 6000 (for both simple alkenes) h^{-1}, respectively.

Palladium(II) complexes containing tridentate PNP ligands first appeared during the late 1980s. A complex of general formula [Pd(PNP)Cl]Cl containing the pincer ligand N,N'-bis(diphenylphosphino)-2,6-diaminopyridine was briefly mentioned in 1987 as a catalyst for styrene hydrogenation (ethanol, 60 °C, 10 atm H_2 pressure) [23]; the platinum version was also active. In 1988, a kinetic analysis of the hydrogenation of cyclohexene catalyzed by different [Pd(PNP)Cl]Cl complexes (PNP = bis-2-(diphenylphosphino)ethyl benzylamine [24], bis-2-(diphenylphosphino)ethyl amine [25] or tris-2-(diphenylphosphino)ethyl amine [25]) under mild conditions (ethanol, 10–40 °C, 0.1–1 atm H_2 pressure) was reported (see Section 4.2.4).

During the 1990s, [Pd(PNO)X] (X = OAc, Cl, I) complexes derived from protic HPNO acyl-hydrazones (**1** in Scheme 4.1) were reported to be effective in the homogeneous hydrogenation of styrene and phenylacetylene under mild conditions (MeOH, r.t. or 40 °C, 1 atm H_2 pressure, 1% catalyst loading) [26–28]. With styrene, the acetate-complexes showed appreciable activity (TOFs up to 67 h^{-1}), the chloride-complexes reacted sluggishly (complete conversion after 48 h), and the iodide-complexes were completely inactive. With phenylacetylene as the substrate, the hydrogenations catalyzed by the chloride-complexes proceeded slowly (conversions not complete after 24 h), although with good selectivity to styrene, whilst the iodide complexes were, again, not active. With the acetate complexes, hydrogenation of the alkyne was poisoned by the precipitation of phenylethynyl palladium(II) complexes of the type [Pd(PNO)(C≡C–Ph)], which formed by the elimination of acetic acid. The formation of these organometallic species is independent of the catalytic conditions, as shown by the possibility of synthesizing them by reaction between the acetate complexes [Pd(PNO)(OAc)] and an excess of alkyne in methanol. Catalytic hydrogenations promoted by the phenylethynyl palladium(II) complexes in solvents in which these are completely soluble (CH_2Cl_2 or THF), led to much lower conversions than those reached with the acetate precursors, which exemplifies the pollutant nature of this *in-situ*-formed species.

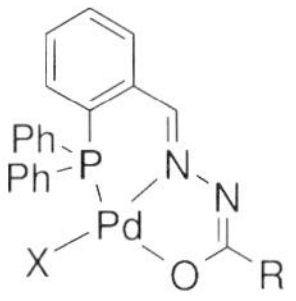

X = OAc, Cl, I

R = Me, C_6H_5, p-Me-C_6H_4
p-Br-C_6H_4, p-NO_2-C_6H_4, py

1

Scheme 4.1 The Pd(II) complexes containing acyl-hydrazone ligands, Pd(PNO)X.

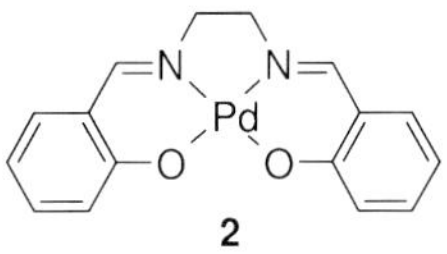

2   **Scheme 4.2** The Pd-Salen complex.

4.2.2
Nitrogen-Containing Catalysts

Nitrogen ligand-containing complexes were first reported in the literature during the 1970s. Pd-Salen (**2** in Scheme 4.2) was reported as being a suitable catalyst for the hydrogenation of 1-hexene, and a good enzyme hydrogenase model [29, 30]. The complex was active in heterogeneous as well as homogeneous conditions, depending on the solvent employed (ethanol or DMF, respectively). Modified versions of the present catalyst appeared later in the literature [31, 32]; low reactivities and selectivities were observed in the hydrogenations of 1-hexene and phenylacetylene.

Ferrocenyl sulfide palladium(II) complexes of formula $[(\eta^5\text{-}C_5H_5)Fe(\eta^5\text{-}C_5H_3\text{-}1\text{-}CHRNR'_2\text{-}2\text{-}SR'')PdCl_2]$ or $[(R''S)(\eta^5\text{-}C_5H_4)Fe(\eta^5\text{-}C_5H_3\text{-}1\text{-}CHRNR'_2\text{-}2\text{-}SR'')PdCl_2]$ were employed in the selective hydrogenation of conjugated dienes to monoenes and in the hydrogenation of styrene derivatives [33–38]. The speed and selectivity were governed mainly by the nature of the R'' group, with 4-Cl-C$_6$H$_4$ leading to the best results in both series of complexes (TOFs up to 722 h^{-1} and almost complete selectivity to cyclooctene were achieved in the hydrogenation of 1,3-cyclooctadiene conducted at r.t. under 7 atm H$_2$ pressure). The solvent of choice was acetone, while dichloromethane, THF, and DMF led to much poorer results. Replacement of the Pd–S bond with a Pd–Se bond, or substitution of Pd with Pt, led to inactive systems. On the basis of these observations the authors argued that rupture of the Pd–S bond was a necessary prerequisite to form the active species. Nevertheless, induction times observed in some catalytic reactions (up to 49.7 h) cannot completely rule out – at least in those cases – a heterogeneous contribution to the reaction. Systems containing two sulfide moieties were almost inactive under the same experimental conditions [38].

Orthometallated dimer palladium(II) complexes [Pd(NC)X]$_2$ (**3** in Scheme 4.3) were reported by Saha as effective catalysts for the hydrogenation of alkenes and alkynes under mild conditions (25 °C, 1 atm H$_2$ pressure) [39, 40]. The reactions were efficient in either DMF or DMSO, reaching initial TOFs higher than 10 000 h^{-1} in the case of styrene, while in less-coordinating or non-coordinating solvents no activity was observed. Selectivity was never obtained, and the overreduced products (or a mixture of different isomers) were always obtained. This, and the fact that the authors reported for a set of experiments that "... the yellow DMF solution of the catalyst turned deep greenish brown within 10 min on stirring under hydrogen at 20 °C" before the addition of the substrate [40], cast serious doubt on the homogeneous character of the reaction.

In 1991, Elsevier reported other nitrogen ligand-containing complexes as active hydrogenating catalysts. Palladium(0) complexes containing the Ar-bian bis-imine

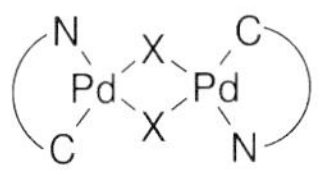

3

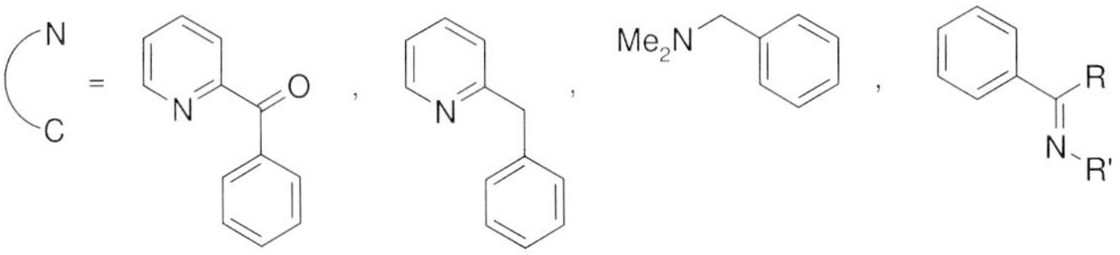

X = OAc or Cl; R = H or Me; R' = Me or aryl

Scheme 4.3 The orthometallated Pd(II) complexes, [Pd(NC)₂X]₂.

Ar = C_6H_5, p-Me-C_6H_4, p-MeO-C_6H_4
E = CO_2Me, CN, C(O)–O–C(O)

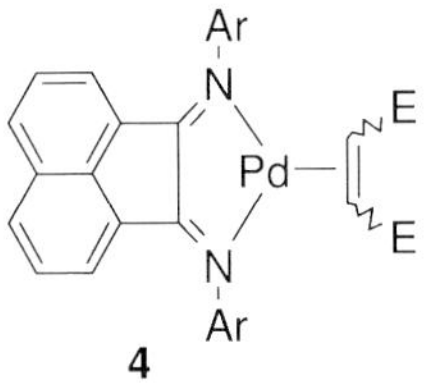

4

Scheme 4.4 The [Pd(Ar-bian)(dmf)] complexes.

ligand bis(arylimino)acenaphthene (**4** in Scheme 4.4), were successful in the reduction of electron-deficient alkenes [41]. The hydrogenations proceed under mild conditions (THF, 20 °C, 1–1.5 atm H_2 pressure, 1% catalyst loading, complete conversion within 8–16 h), with high chemoselectivity for C=C double bonds. Thus, nitro and cyano groups were not reduced to the corresponding amines, and the alkene functions of unsaturated esters, ketones and anhydrides were selectively hydrogenated. High selectivity (up to 90%) was also observed in the hydrogenation of α,β-unsaturated aldehydes. In the case of alkenes not containing electron-withdrawing substituents, the hydrogenations began only after an induction period, and for these instances a heterogeneous contribution was evidenced. [Pd(Ar-bian)(dmf)] (dmf=dimethylfumarate) complexes were also efficient in the highly selective hydrogenation of alkynes to the corresponding Z-alkenes [42]. The complex with Ar=p-Me-C_6H_4 was able to hydrogenate several linear and cyclic alkynes with selectivity up to 99% to the corresponding alkenes, and with high tolerance towards different functional groups. Substitution of the Ar-bian ligand with other nitrogen (bipy or dab=p-anisyldiazabutadiene) or phosphine (dppe) ligands led to poorer catalysts. More recently, other bis-imine ligands containing Pd(0) complexes of the general formula 6-R″-C_5H_3N-(C(R′)=NR)-2 (R″=H, Me; R′=H, Me; R=alkyl, aryl) were used in the stereoselective hydrogenation of 1-phenyl-1-propyne to Z-1-phenyl-1-propene [43]. The experimental conditions were identical to those applied with the [Pd(Ar-bian)(dmf)] complexes. The stability and behavior of the complexes under H_2 atmosphere were strongly dependent on the substitu-

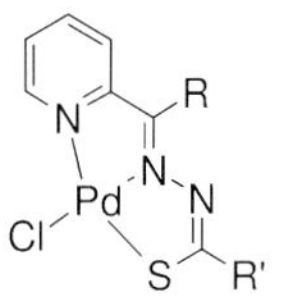

Scheme 4.5 The Pd(II) complexes with thiosemicarbazone or thiobenzoylhydrazone ligands, Pd(NN'S)Cl.

5

R = Me or Ph
R' = NH$_2$ or Ph

ents R, R', and R''. Moderately bulky, σ-donating substituents were necessary in order to have hydrogenating activity without loss of palladium. The best compromise was reached with R'' = H, R' = H and R = i-Pr, which led to complete consumption of the alkyne within 2 h of reaction, and with the following products distribution: 87% Z-alkene, 3% E-alkene, and 10% alkane. Water-soluble bis-imine Pd(0) complexes of the general formula 6-R'-C$_5$H$_3$N-(C(H)=NR)-2 (R' = H, Me; R = 2,3,4,6-tetra-O-acetyl-β-D-glucopyranose residue) were used to hydrogenate unsaturated nitriles in water (r.t., 1 atm H$_2$ pressure, Pd:substrate ratio = 1:50) [44]. The complexes displayed high activity in distilled water, but the reactions were substantially heterogeneous. In 0.1 M KOH solution, the hydrogenations of acrylonitrile, methacrylonitrile, and crotonitrile were homogeneous and proceeded fairly rapidly (after 2 h the yields of the saturated products were 70, 50, and 50%, respectively); however, undesired side reactions also occurred. In 0.25 M KOH solution only the hydrogenation of acrylonitrile was homogeneous and complete within 2 h. No enantioselectivity was observed in the hydrogenation of α-ethylacrylonitrile.

High chemoselectivities (up to 96%) were reached in the hydrogenation of phenylacetylene to styrene (DMF, 30 °C, 1 atm H$_2$ pressure, 1% catalyst loading, TOFs up to 4.2 h^{-1}) catalyzed by chloride Pd(II) complexes [Pd(NN'S)Cl] containing thiosemicarbazone or thiobenzoylhydrazone ligands (**5** in Scheme 4.5) [45]. Instead, minor reactivities and selectivities were obtained with NN'N'' pyridyl-hydrazone-containing Pd(II) complexes in the hydrogenation of phenylacetylene [46].

4.2.3
Other Catalysts

Pd(acac)$_2$ has been reported to be an active catalyst in soybean oil hydrogenation [47]. The reactions were conducted in bulk with low catalyst loadings (1–60 ppm) and without any co-catalyst. Under 10 atm H$_2$ pressure and at 80–120 °C, optimum linolenate selectivity and high $trans$-isomers content were obtained. Decomposition of the catalyst occurred at temperatures above 120 °C.

Simple inorganic salts, such as PdCl$_2$, were also studied. PdCl$_2$/SnCl$_2$ · 2 H$_2$O and K$_2$PdCl$_4$/SnCl$_2$ · 2 H$_2$O turned out to be completely inactive in the hydrogenation of soybean oil methylester [9], while PdCl$_2$ in DMF was active towards conjugated dienes and alkynes (DMF, 25 °C, 25 atm H$_2$ pressure, 0.31–0.5% catalyst loading) [48]. 1-Heptyne and 2-pentyne were selectively reduced to the corresponding alkenes within 25 minutes, while the hydrogenation of dienes was slower. PdCl$_2$ dissolved in a CH$_2$Cl$_2$-polyethylene glycol (PEG) mixture was employed in the hydrogenation of diphenylacetylene (10 °C, 1 atm H$_2$ pressure, Pd : substrate ratio $\approx$ 1 : 10) [49]. The rate and selectivity of the reaction were dependent on the PEG molecular weight; with PEG-400, 90% conversion was achieved after 80 minutes, with 80% of cis-stilbene.

4.2.4
Mechanistic Aspects

Reaction mechanisms based on kinetic studies have been proposed for most of the aforementioned catalytic hydrogenations. For the Pd(II)-catalysts, the heterolytic activation of molecular hydrogen results generally favored compared to oxidative addition which, in contrast, is common for low-valent Pd-catalysts. For example, a molecular orbital analysis of the hydrogenation of styrene catalyzed by the [PdCl$_4$]$^{2-}$ anion [50] showed that H$_2$ was activated by heterolytic splitting (in these experiments the catalyst was supported on a solid, but no profound differences were believed to exist between the homogeneous and heterogeneous cases). The four successive steps depicted in Eqs. (3) to (6) were defined as:

- Splitting of H$_2$ with formation of the palladium hydride [HPdCl$_3$]$^{2-}$ and HCl (Eq. (3)).
- π-coordination of styrene through a pentacoordinate intermediate (Eq. (4)).
- Alkene insertion into the Pd–H bond with formation of an alkyl intermediate (Eq. (5)).
- Reaction with HCl to eliminate ethylbenzene and re-form the anion [PdCl$_4$]$^{2-}$ (Eq. (6)).

$$[\text{PdCl}_4]^{2-} + \text{H}_2 \ \leftrightarrow \ [\text{HPdCl}_3]^{2-} + \text{HCl} \tag{3}$$

$$[\text{HPdCl}_3]^{2-} + \text{C}_6\text{H}_5\text{-CH} = \text{CH}_2 \ \leftrightarrow \ [\text{HPdCl}_3(\text{C}_6\text{H}_5\text{-CH} = \text{CH}_2)]^{2-} \tag{4}$$

$$[\text{HPdCl}_3(\text{C}_6\text{H}_5\text{-CH} = \text{CH}_2)]^{2-} \ \leftrightarrow \ [(\text{C}_6\text{H}_5\text{-CH}_2\text{-CH}_2)\text{PdCl}_3]^{2-} \tag{5}$$

$$[(\text{C}_6\text{H}_5\text{-CH}_2\text{-CH}_2)\text{PdCl}_3]^{2-} + \text{HCl} \ \rightarrow \ [\text{PdCl}_4]^{2-} + \text{C}_6\text{H}_5\text{CH}_2\text{CH}_3 \tag{6}$$

The elimination of HCl was proposed to occur also during the H$_2$ activation with the [Pd(PNP)Cl]Cl complexes (PNP = bis-2-(diphenylphosphino)ethyl benzylamine, bis-2-(diphenylphosphino)ethyl amine or tris-2-(diphenylphosphino)ethyl amine) [24, 25]. Based on the findings of ^{31}P$\{^1$H$\}$- and ^{1}H-NMR investigations, the hydride [HPd(PNP)]Cl was detected under H$_2$ atmosphere. The alternative mechanism which involves the oxidative addition of H$_2$ with formation of a Pd(IV)-dihydride intermediate, appeared less likely on the basis of thermodynamic considerations.

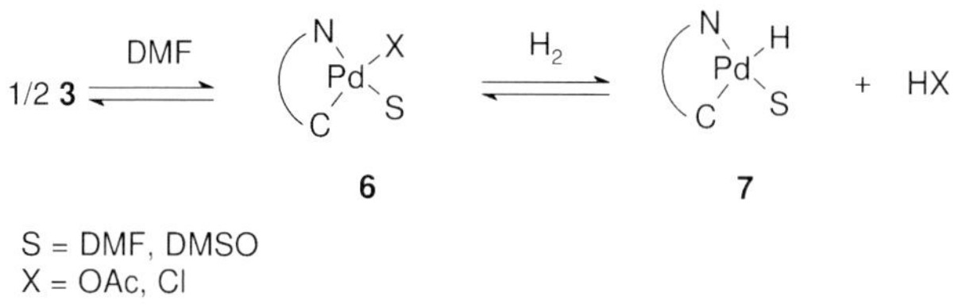

Scheme 4.6 Heterolytic splitting of H$_2$ with [Pd(NC)$_2$X]$_2$ complexes.

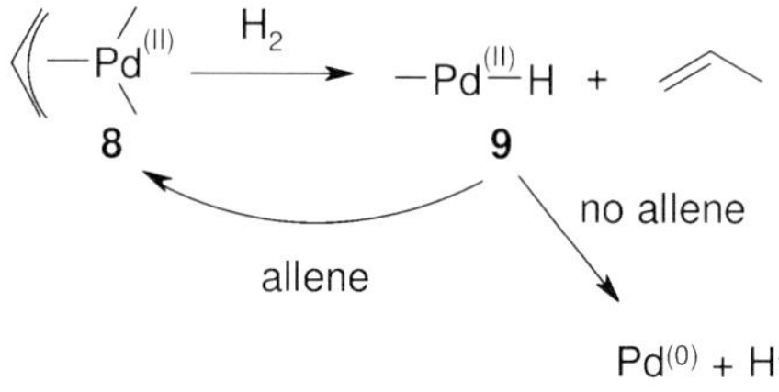

Scheme 4.7 Hydrogenation of allene catalyzed by [(η^3-allyl)Pd(PR$_3$)Cl] complexes.

The efficiency of the orthometallated Pd(II) complexes [Pd(NC)X]$_2$ [39, 40] (**3** in Scheme 4.3) in coordinating solvents such as DMF or DMSO was considered to be in favor of the initial dissociation of the dimers **3** into the mononuclear species **6** (Scheme 4.6). The heterolytic splitting of H$_2$ leads to the formation of the hydride **7** plus HCl or AcOH. The solvato-hydride intermediate **7** was characterized spectroscopically and its elemental analysis furnished [39]; regrettably, no data regarding its catalytic activity were reported.

The dissociation of a coordinated allene by hydrogen was evidenced for [(η^3-allyl)Pd(PR$_3$)Cl] complexes [11]. The hydrogenation of allene to propene was then invoked to follow the pathway depicted in Scheme 4.7:

- H$_2$ activation by protonation of the coordinated allene in **8** with formation of the palladium hydride intermediate **9** and propene.
- Insertion of an allene molecule into the Pd–H bond, with regeneration of the starting allene complex **8** which then re-enters the catalytic cycle. In the absence of allene, extensive Pd black formation was observed.

In the hydrogenation of styrene catalyzed by the [Pd(PNO)(OAc)] complexes (**1** in Scheme 4.1) [26–28], a clear correlation between the activity of the complexes and the basicity of the hydrazonic ligand, in turn governed by the nature of the R group, was established [28]. Indeed, the higher the ligand basicity, the faster the hydrogenation reactions. This result, combined with a kinetic analysis, led to the mechanism depicted in Scheme 4.8. The first step is the heterolytic activation of H$_2$ with protonation of the hydrazonic arm in **1** and formation of the hydride intermediate **10**; **10** can be considered at equilibrium with **11**, where the acetate anion re-enters in the coordination sphere by displacement of the C=O amide group of the ligand. By rupture of the Pd–O(amide) (**10**) or Pd-O-

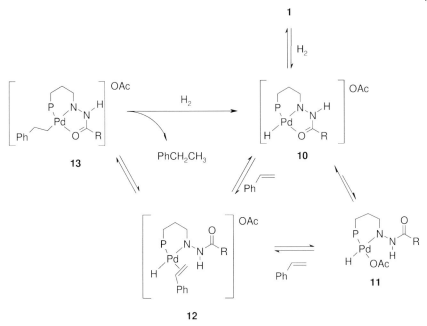

Scheme 4.8 Proposed pathway for the hydrogenation
of styrene catalyzed by Pd(PNO)(OAc) complexes.

(acetate) (**11**) bond, a styrene molecule coordinates to palladium with formation
of **12**, and after the alkene insertion into the Pd–H bond the alkyl-intermediate
13 forms. The involvement of **10** in the catalytic cycle appeared more likely than
that of **11**, on the basis of the inertness shown by the acetate complex
[Pd(PNS)(OAc)] (PNS = 2-(diphenylphosphino)benzaldehyde thiosemicarbazone),
where the two soft donors P and S make the complex too stable [27]. The cycle
closes with the intervention of a H_2 molecule, which leads to the restoration of
10 (or alternatively **11**) and elimination of ethylbenzene.

A similar H_2 activation mechanism was proposed for the [Pd(NN′S)Cl] complexes (**5** in Scheme 4.5) in the semi-hydrogenation of phenylacetylene [45];
after formation of the hydride **14** (Scheme 4.9), coordination of the alkyne occurs by displacement of the chloride ligand from Pd (**15**). The observed chemoselectivity (up to 96% to styrene) was indeed ascribed to the chloride anion,
which can be removed from the coordination sphere by phenylacetylene, but
not by the poorer coordinating styrene. This was substantiated by the lower chemoselectivities observed in the presence of halogen scavengers, or in the hydrogenations catalyzed by acetate complexes of formula [Pd(NN′S)(OAc)]. Here, the
acetate anion can be easily removed by either phenylacetylene or styrene.

A net heterolytic H_2 cleavage was evidenced for [Pd(Salen)] (**2** in Scheme 4.2)
[29, 30]. Its catalytic activity was pH-dependent; namely, it was inhibited by
acids and enhanced by bases. A kinetic study of the hydrogenation of 1-hexene
in DMF led to the definition of the following mechanism (Scheme 4.10):

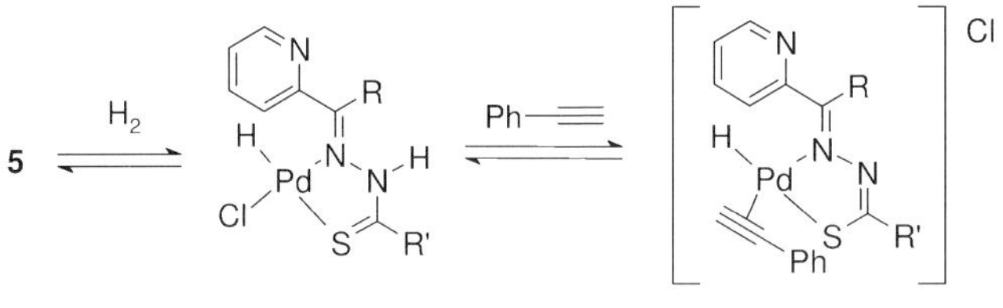

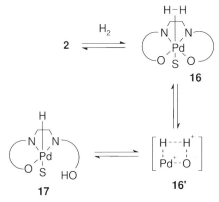

Scheme 4.9 Heterolytic splitting of H_2 and coordination of phenylacetylene in Pd(NN′S)Cl complexes.

Scheme 4.10 Heterolytic H_2 splitting promoted by Pd–Salen complex; $S = DMF$.

- solvent-assisted hydrogen coordination to the metal (**16**);
- heterolytic activation of the molecular hydrogen through a highly polarized, four-center transition state (**16′**) which involves a coordinate phenolate group. The so-formed OH group leaves a vacant coordination site on the metal (**17**), which can now accommodate the entering alkene, the subsequent alkene insertion into the Pd–H bond forming an alkyl intermediate. Finally, hydrogen transfer from the OH group of the ligand to the coordinated alkyl group gives the final alkane, with restoration of **2**.

Oxidative addition of molecular hydrogen was considered to be involved in the alkyne hydrogenations catalyzed by [Pd(Ar-bian)(dmf)] complexes (**4** in Scheme 4.4) [41, 42]. Although the mechanism was not completely addressed, **4** was considered to be the pre-catalyst, the real catalyst most likely being the [Pd(Ar-bian)(alkyne)] complex **18** in Scheme 4.11. Alkyne complex **18** was then invoked to undergo oxidative addition of H_2 followed by insertion/elimination or pairwise transfer of hydrogen atoms, giving rise to the alkene-complex **19**.

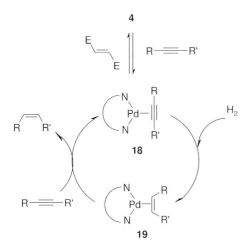

Scheme 4.11 Proposed catalytic cycle for the hydrogenation of alkynes promoted by Pd(Ar-bian)(dmf) complexes.

Displacement of the alkene by an incoming alkyne molecule leads to the elimination of the product and restoration of **18** (see also chapter 14).

The higher catalytic activity of the cluster compound $[Pd_4(dppm)_4(H_2)](BPh_4)_2$ [21] (**20** in Scheme 4.12) in DMF with respect to less coordinating solvents (e.g., THF, acetone, acetonitrile), combined with a kinetic analysis, led to the mechanism depicted in Scheme 4.12. Initially, **20** dissociates into the less sterically demanding d^9-d^9 solvento-dimer **21**, which is the active catalyst. An alkyne molecule then inserts into the Pd–Pd bond to yield **22** and, after migratory insertion into the Pd–H bond, the d^9-d^9 intermediate **23** forms. Now, H_2 can oxidatively add to **23** giving rise to **24** which, upon reductive elimination, results in the formation of the alkene and regenerates **21**.

The high activity of the polymeric compound $[Pd_5(PPh_2)]_n$ ($n \approx 4$) [22] in the hydrogenation of alkenes and alkynes was ascribed to the initial DMF-assisted dissociation of the polymer (Eq. (7)); the resulting $[Pd_5(PPh)_2]$ species forms an adduct with the substrate molecule (Sub) (Eq. (8)) which, after reaction with H_2, leads to the hydrogenation product and gives back $[Pd_5(PPh)_2]$ (Eq. (9)), which re-enters the catalytic cycle.

$$[Pd_5(PPh)_2]_n \ \leftrightarrow \ n[Pd_5(PPh)_2] \tag{7}$$

$$[Pd_5(PPh)_2] + Sub \ \leftrightarrow \ [Pd_5(PPh)_2](Sub) \tag{8}$$

$$[Pd_5(PPh)_2](Sub) + H_2 \ \leftrightarrow \ [Pd_5(PPh)_2] + SubH_2 \tag{9}$$

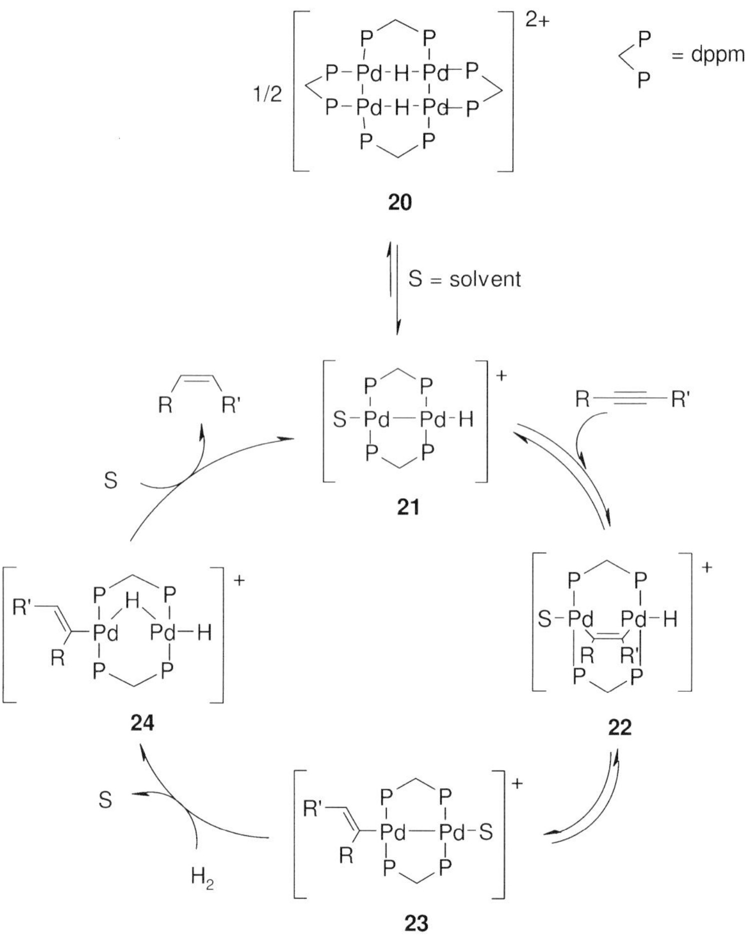

Scheme 4.12 Proposed mechanism for the hydrogenation of alkynes catalyzed by $[Pd_4(dppm)_4(H_2)]^{2+}$.

4.3
Platinum

4.3.1
Platinum Complexes Activated with Sn(II) Salts

4.3.1.1 Phosphorus-Containing Catalysts

The catalytic activity of $[PtX_2(QPh_3)_2]/SnCl_2$ complexes (Q = P or As, X = halogen or pseudo-halogen) was intensively studied, starting during the late 1960s. Such catalytic systems promoted the isomerization and hydrogenation of methyl oleate [51], methyl linolenate [52], and polyenes [53], as well as short-chain alkenes

[54]. The Pt:substrate ratios ranged from 1:10 to 1:180. The formation of conjugated isomers which were slowly hydrogenated to the corresponding monoene was observed (see Section 4.3.3). Cis-[PtCl$_2$L(PR$_3$)]/SnCl$_2$ complexes (L=SR$_2'$, p-X-C$_6$H$_4$NH$_2$; R=aryl) also showed activity in the hydrogenation of alkenes, such as styrene (acetone, 60 °C, 41 atm H$_2$ pressure) [55]. TOF values up to 4400 h^{-1} were reached, depending on the nature of ligands L and PR$_3$. These complexes resulted in more active catalysts than the parent complexes cis-[PtCl$_2$L$_2$]/SnCl$_2$ or cis-[PtCl$_2$(PR$_3$)$_2$]/SnCl$_2$, which brought about a maximum TOF of 910 h^{-1}. Rationalization of the results, in terms of a plausible reaction mechanism, was far from straightforward because of the contemporary presence, under catalytic conditions, of several Pt and Pt-Sn species, both neutral and ionic, all of which played a catalytic role. Indeed, the authors reported that the multicomponent system was necessary to establish the catalytic cycle, as evidenced by catalytic runs involving isolated species. Dimeric versions of the aforementioned complexes of the type [Pt$_2$(μ-Cl)$_2$(PR$_3$)$_2$]/SnCl$_2$ [55] or cis-[Pt$_2$(μ-SR$'$)(μ-Cl)Cl$_2$(PR$_3$)$_2$] [56] led to reactivities similar to those observed with the mononuclear parents, thus indicating the necessity of dimer dissociation to bring about the formation of active mononuclear species.

4.3.1.2 Other Catalysts

In 1963, the hydrogenation of ethylene and acetylene under mild conditions (methanol, r.t., 1 atm total pressure) was readily carried out with the catalytic system H$_2$PtCl$_6$/SnCl$_2$ (Pt/Sn ratio=1:10) [57]. With higher Pt/Sn ratios and higher alcohols, the hydrogenation of higher alkenes became feasible. For example, the hydrogenation of cyclohexene in i-PrOH proceeded with a TOF of 5 h^{-1} [58]. The addition of chloride or bromide ions and water strongly increased the hydrogenation rate (TOFs up to 94 h^{-1}). Carboxylic acids or esters also turned out to be suitable solvents: with a Pt/Sn ratio of 1:5, the hydrogenation of 1-hexene in glacial acetic acid led, after 2 h of reaction, to a mixture of hexane ($\approx$ 50%), 2-$trans$-hexene ($\approx$ 25%), 3-$trans$-hexene ($\approx$ 5%), traces of 2-cis- and 3-cis-hexene, and unreacted 1-hexene [59]. As expected, the hydrogenation of 2-hexene proceeded somewhat sluggishly, whilst in the hydrogenation of soybean oil a remarkable preference for the reduction of linoleic acid was observed. Again, H$_2$PtCl$_6$/SnCl$_2$ was employed in the hydrogenation of Dewar-benzene derivatives, such as dimethyl tetramethylbicyclo[2.2.0]hexa-2,5-diene-2,3-dicarboxylate (**25** in Scheme 4.13) and its di-$tert$-Bu-ester under mild conditions (i-PrOH, 25 °C, 1 atm H$_2$ pressure, Pt:substrate ratio 1:10) [60]. Contrary to heterogeneous systems, the hydrogenation of **25** was restricted to the ester-substituted double bond, giving rise to dimethyl tetramethylbicyclo[2.2.0]hex-5-ene-2-endo, 3-endo-dicarboxylate (**26**); the exo,endo hydrogenation (**27**) was involved for 16%. In 1972, Parshall showed that PtCl$_2$ could be used as hydrogenation catalyst for alkenes in molten [R$_4$N][SnCl$_3$] [61]. With PtCl$_2$/[Et$_4$N][SnCl$_3$], a 50% conversion was obtained in the hydrogenation of ethylene after 5 h of reaction (100 °C, 1.7 atm total pressure, Pt:ethylene ratio 1:13). At higher temperatures

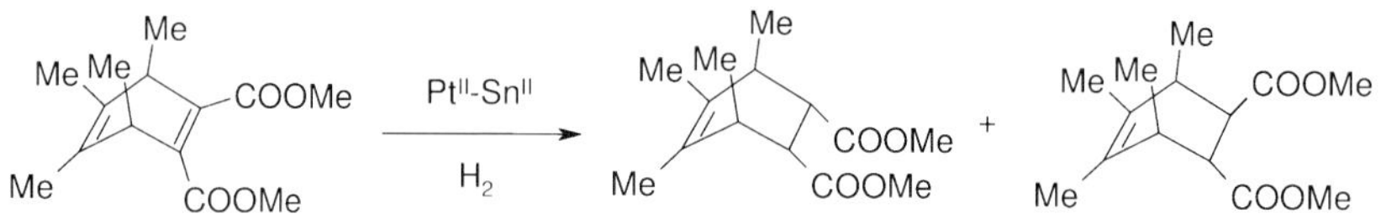

Scheme 4.13 Hydrogenation of Dewar-benzene derivatives.

and pressures, 1,5,9-cyclododecatriene was hydrogenated to cyclododecene with considerable selectivity: 87% monoene, 10% diene, 2% of unreacted triene, and traces of over-reduced product were obtained at 160 °C and 100 atm H_2 pressure after 8 h of reaction (Pt:substrate ratio = 1:15).

4.3.2
Platinum Complexes not Activated with Sn(II) Salts

Platinum complexes that display hydrogenating activity without the addition of Sn(II) salts are scarce in the literature. The dihydride Pt(II) complex [PtH$_2${(t-Bu)$_2$PCH$_2$CH$_2$P(t-Bu)$_2$}] hydrogenated cyclohexene to cyclohexane only under drastic conditions (benzene, 100 °C, 77.4 atm H_2 pressure, 1% catalyst loading), with a TOF of 0.26 h^{-1} [62]. The chloro-bridged complex [PtCl$_2$(2,4,6-Me$_3$-pyridine)]$_2$ was active at r.t. and 1 atm H_2 pressure toward mono- and dienes (a TOF of 270 h^{-1} was obtained with styrene) [63]; however, when racemization was possible all the isomers were detected in the final solutions. The hydride complex *trans*-[PtH(NO$_3$)(PEt$_3$)$_2$] [64] was active towards both internal and terminal alkenes, but was inactive toward alkenes bearing electron-withdrawing substituents. The hydrogenation rate was dependent on the solvent employed, with methanol leading to the highest TOF values. With styrene as substrate, a TOF of 115 h^{-1} was reached at 60 °C and 41 atm H_2 pressure. The methoxide intermediate PtH(OMe)(PEt$_3$)$_2$ was considered to be involved in the reaction mechanism. The activity towards internal alkenes was slower.

Pt(0) complexes of the type [Pt(P-P)(C$_2$H$_4$)] (P-P = dppb, 1,2-bis[(diphenylphosphino)methyl]benzene (dpmb) and (+)-2,3-O-isopropylidene-2,3-dihydroxy-1,4-bis(diphenylphosphino)butane ((+)-diop)) were used in the hydrogenation of several alkenes in combination with CH$_3$SO$_3$H (toluene, 80 °C, 50 atm H_2 pressure, Pt:substrate ratio 1:320) [65]. With the dppb-containing complex, the reduction of terminal alkenes was accompanied by extensive isomerization, and the resulting internal alkenes underwent very little hydrogenation (1-hexene led to 29.3% yield of hexane and 68.8% yield of hexenes after 22 h). Alkenes bearing electron-withdrawing substituents were hydrogenated much more easily (3-buten-2-one and 2-cyclohexen-1-one were reduced to the corresponding saturated ketones with TOFs of 68 and 282 h^{-1}, respectively). The activity of [Pt(dppb)(C$_2$H$_4$)] was usually higher than that of [Pt(dpmb)(C$_2$H$_4$)], both with simple olefins (styrene, TOFs of 6.2 and 3.7 h^{-1}, respectively) and unsaturated ketones (2-cyclohexen-1-one, TOFs of 80 and 48 h^{-1}, respectively). Poor chiral induction (8% *ee*, *R*-isomer) was observed in the hydrogenation of *α*-ethylstyrene catalyzed by [Pt((+)-diop)(C$_2$H$_4$)].

Finally, the only example of a polynuclear homogeneous catalyst is the dinuclear complex [Pt$_2$(P$_2$O$_5$H$_2$)$_4$]$^{4-}$ [66], which catalyzed the hydrogenation of styrene, phenylacetylene, 1-octyne, and 1-hexyne (*i*-PrOH, 60 °C, 20.7 atm H$_2$ pressure, Pd:substrate ratio 1:1800) to the corresponding alkanes within 10 h of reaction.

4.3.3
Mechanistic Aspects

Because of its industrial importance, many studies have conducted in order to establish the mechanism which regulates the hydrogenation of polyunsaturated fatty acid derivatives catalyzed by the [PtX$_2$(QPh$_3$)$_2$]/SnCl$_2$ complexes (Q=P or As, X=halogen or pseudo-halogen) [51, 52]. The following observations were made:

- Initial, rapid step-wise migration of double bonds takes place to give a conjugated isomer [67], which is then reduced more slowly to the corresponding monoene. The necessary formation of conjugated isomers was highlighted by the inertness of polyenes where the double bonds are separated by several methylene groups [52]. Isomerization occurred before, as well as after, the hydrogenation. *Cis-trans* isomerization also occurred. High H$_2$ pressures (usually higher than 30 atm) were required, otherwise only isomerization was observed.
- Hydride formation was a fundamental step in the mechanism, and indeed PtHCl(PPh$_3$)$_2$ species were isolated from the reactant solutions. However, their formation was not the rate-determining step, since the same rate of hydrogenation was observed with either PtCl$_2$(PPh$_3$)$_2$ or PtHCl(PPh$_3$)$_2$ complexes.
- Metal–alkene complex formation was also necessary, but again was not rate-determining.
- The formation of hydride-Pt(II)-alkene complexes was thought to be the rate-determining step, as evidenced by the isolation of a number of such species.
- The addition of an excess of SnCl$_2$ (a Pt:Sn ratio of 1:5 was found to be the best) was necessary as co-catalyst; SnCl$_2$ reacts with chloride ions to give the poor *σ*-donor and strong *π*-acceptor SnCl$_3^-$ ligand [68] that, decreasing the electron density on Pt, favors the formation of Pt–H or Pt–alkene bonds.
- Triphenylphosphine and related ligands stabilized the hydride intermediate once formed, besides rendering Pt soluble in non-polar organic solvents. These observations have been condensed in the mechanism depicted in Scheme 4.14.

Penta-coordinate Pt–Sn anions of the type [Pt(SnCl$_3$)$_5$]$^{3-}$ [69] appeared also to be involved in the hydrogenations catalyzed by the phosphine-free H$_2$PtCl$_6$/SnCl$_2$ [57] and PtCl$_2$/[R$_4$N][SnCl$_3$] [61] systems. The anion [Pt(SnCl$_3$)$_5$]$^{3-}$ transforms into the active species [PtH(SnCl$_3$)$_3$]$^{2-}$ upon heterolytic hydrogen activation and loss of two SnCl$_3^-$ ligands (Eqs. (10) to (12)).

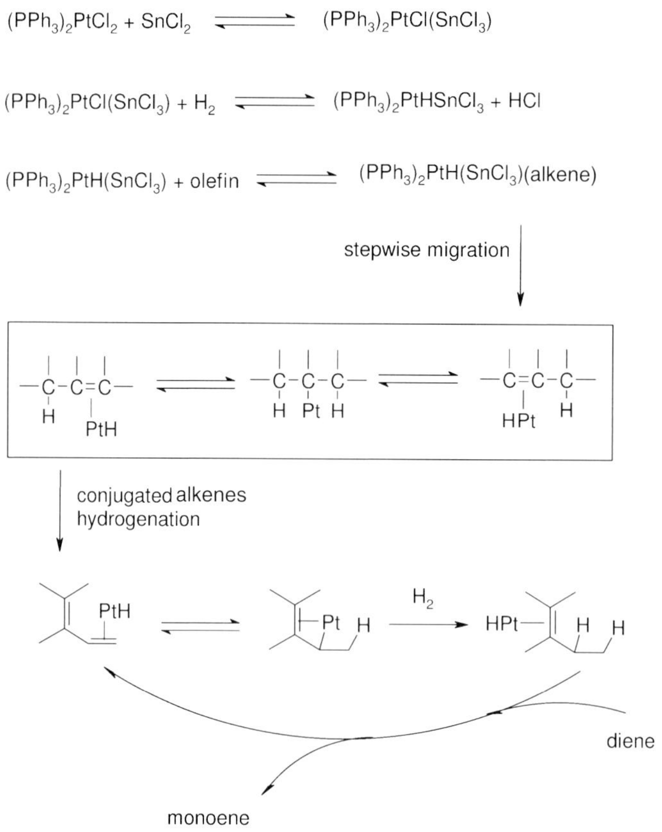

$$(PPh_3)_2PtCl_2 + SnCl_2 \quad \rightleftharpoons \quad (PPh_3)_2PtCl(SnCl_3)$$

$$(PPh_3)_2PtCl(SnCl_3) + H_2 \quad \rightleftharpoons \quad (PPh_3)_2PtHSnCl_3 + HCl$$

$$(PPh_3)_2PtH(SnCl_3) + \text{olefin} \quad \rightleftharpoons \quad (PPh_3)_2PtH(SnCl_3)(\text{alkene})$$

stepwise migration

conjugated alkenes
hydrogenation

diene

monoene

Scheme 4.14 Pathway for the selective hydrogenation of polyenes to monoenes catalyzed by $PtX_2(QPh_3)_2/SnCl_2$; Q=P or As, X=halogen or pseudo-halogen.

$$[Pt(SnCl_3)_5]^{3-} \leftrightarrow [Pt(SnCl_3)_4]^{2-} + SnCl_3^- \tag{10}$$

$$[Pt(SnCl_3)_4]^{2-} + H_2 \leftrightarrow [PtH(SnCl_3)_4]^{3-} + H^+ \tag{11}$$

$$[PtH(SnCl_3)_4]^{3-} \leftrightarrow [PtH(SnCl_3)_3]^{2-} + SnCl_3^- \tag{12}$$

In the case of ethylene hydrogenation, the mechanism proposed by Parshall [61] involves the coordination of an alkene molecule through a five-coordinate intermediate (Eq. (13)); the subsequent alkene insertion into the Pt–H bond (Eq. (14)) and intervention of a second molecule of H_2 (Eq. (15)) leads to the elimination of ethane and restoration of the catalytic active species $[PtH(SnCl_3)]^{2-}$. However, in 1976 Yasumori and coworkers reported a kinetic analysis conducted on the hydrogenation of ethylene catalyzed by the Pt–Sn complex $[(Me)_4N]_3[Pt(SnCl_3)_5]$ [70], under much milder conditions than those applied by Parshall (acetone, 0–15 °C, partial H_2 pressure up to 0.13 atm). The collected data were in agreement with the formation of Pt_n clusters of different sizes (n = 1 to 6, depending on Pt concentration), which were invoked as being responsible for the observed catalytic activity. The possibility that the catalyses described by Parshall might have a heterogeneous character is further supported by the well-known ability of tetraalkylammonium salts to stabilize metal-clusters, and by the harsh experimental conditions applied [71].

$$[PtH(SnCl_3)_3]^{2-} + C_2H_4 \;\leftrightarrow\; [PtH(C_2H_4)(SnCl_3)_3]^{2-} \tag{13}$$

$$[PtH(C_2H_4)(SnCl_3)_3]^{2-} \;\leftrightarrow\; [Pt(C_2H_5)(SnCl_3)_3]^{2-} \tag{14}$$

$$[Pt(C_2H_5)(SnCl_3)_3]^{2-} + H_2 \;\rightarrow\; [PtH(SnCl_3)_3]^{2-} + C_2H_6 \tag{15}$$

A kinetic analysis of the styrene hydrogenation catalyzed by $[Pt_2(P_2O_5H_2)_4]^{4-}$ [66] was indicative of the fact that the dinuclear core of the catalyst was maintained during hydrogenation. However, three speculative mechanisms were in agreement with the kinetic data, which mainly differ in the H_2 activation step. This in fact can occur through the formation of two Pt–monohydrides, still connected by a Pt–Pt bond, or through the formation of two independent Pt–monohydrides. The third mechanism involves the dissociation of a phosphine from one Pt center, with subsequent oxidative addition of H_2 to produce a Pt–dihydride intermediate.

Abbreviations

dmf	dimethylfumarate
DMF	dimethylformamide
DMSO	dimethylsulfoxide
PEG	polyethylene glycol
THF	tetrahydrofuran
TOF	turnover frequency
TON	turnover number

References

1 J. H. Flynn, H. M. Hulburt, *J. Am. Chem. Soc.* **1954**, *76*, 3393.
2 A. S. Gow, H. Heinemann, *J. Phys. Chem.* **1960**, *64*, 1574.
3 J. Halpern, *J. Organomet. Chem.* **1980**, *200*, 133.
4 J. F. Harrod, A. J. Chalk, *J. Am. Chem. Soc.* **1964**, *86*, 1776.
5 R. Cramer, R. V. Lindsey, *J. Am. Chem. Soc.* **1966**, *88*, 3534.
6 B. R. James, *Homogeneous Hydrogenation*, Wiley, New York, **1975**.
7 F. Basolo, H. B. Gary, R. G. Pearson, *J. Am. Chem. Soc.* **1960**, *82*, 4200.
8 M. F. Sloan, A. S. Matlack, D. S. Breslow, *J. Am. Chem. Soc.* **1963**, *85*, 4014.

9 H. Itatani, J.C. Bailar Jr., *J. Am. Oil Chemists' Soc.* **1967**, *44*, 147.

10 Y. Fujii, J.C. Bailar Jr., *J. Catal.* **1978**, *55*, 146.

11 G. Carturan, G. Strukul, *J. Organomet. Chem.* **1978**, *157*, 475.

12 G. Strukul, G. Carturan, *Inorg. Chim. Acta* **1979**, *35*, 99.

13 E.W. Stern, P.K. Maples, *J. Catal.* **1972**, *27*, 120.

14 P. Leoni, F. Marchetti, M. Pasquali, *J. Organomet. Chem.* **1993**, *451*, C25.

15 M. Sommovigo, H. Alper, *Tetrahedron Lett.* **1993**, *34*, 59.

16 I.S. Cho, H. Alper, *J. Org. Chem.* **1994**, *59*, 4027.

17 I.S. Cho, H. Alper, *Tetrahedron Lett.* **1995**, *36*, 5673.

18 I.S. Cho, B. Lee, H. Alper, *Tetrahedron Lett.* **1995**, *36*, 6009.

19 R.U. Kirss, R. Eisenberg, *Inorg. Chem.* **1989**, *28*, 3372.

20 I. Gauthron, J. Gagnon, T. Zhang, D. Rivard, D. Lucas, Y. Mugnier, P.D. Harvey, *Inorg. Chem.* **1998**, *37*, 1112.

21 D. Evrard, K. Groison, Y. Mugnier, P.D. Harvey, *Inorg. Chem.* **2004**, *43*, 790.

22 I.I. Moiseev, M.N. Vargaftic, *New J. Chem.* **1998**, 1217.

23 W. Schirmer, U. Flörke, H.-J. Haupt, *Z. Anorg. Allg. Chem.* **1987**, *545*, 83.

24 V.V.S. Reddy, *J. Mol. Catal.* **1988**, *45*, 73.

25 M.M. Taqui Khan, B. Taqui Khan, S. Begum, *J. Mol. Catal.* **1988**, *45*, 305.

26 A. Bacchi, M. Carcelli, M. Costa, P. Pelagatti, C. Pelizzi, G. Pelizzi, *Gazz. Chim. Ital.* **1994**, *124*, 429.

27 A. Bacchi, M. Carcelli, M. Costa, A. Leporati, E. Leporati, P. Pelagatti, C. Pelizzi, G. Pelizzi, *J. Organomet. Chem.* **1997**, *535*, 107.

28 P. Pelagatti, A. Bacchi, M. Carcelli, M. Costa, A. Fochi, P. Ghidini, E. Leporati, M. Masi, C. Pelizzi, G. Pelizzi, *J. Organomet. Chem.* **1999**, *583*, 94.

29 G.H. Olivé, S. Olivé, *Angew. Chem. Int. Ed. Engl.* **1974**, *13*, 549.

30 G.H. Olivé, S. Olivé, *J. Mol. Catal.* **1975/76**, *1*, 121.

31 A.El-M.M. Ramadan, *Transition Met. Chem.* **1996**, *21*, 536.

32 A. Bacchi, M. Carcelli, L. Gabba, S. Ianelli, P. Pelagatti, G. Pelizzi, D. Rogolino, *Inorg. Chim. Acta* **2003**, *342*, 229.

33 R.V. Honeychuck, M.O. Okoroafor, L.-H. Shen, C.H. Brubaker, Jr., *Organometallics* **1986**, *5*, 482.

34 M.O. Okoroafor, L.-H. Shen, R.V. Honeychuck, C.H. Brubaker, Jr., *Organometallics* **1988**, *7*, 1297.

35 C.-K. Lai, A.A. Naiini, C.H. Brubaker, Jr., *Inorg. Chim. Acta* **1989**, *164*, 205.

36 A.A. Naiini, C.-K. Lai, D.L. Ward, C.H. Brubaker, Jr., *J. Organomet. Chem.* **1990**, *390*, 73.

37 A.A. Naiini, H.M. Ali, C.H. Brubaker, Jr., *J. Mol. Catal.* **1991**, *67*, 47.

38 C.-H. Wang, C.H. Brubaker, Jr., *J. Mol. Catal.* **1992**, *75*, 221.

39 A. Bose, C.R. Saha, *Indian J. Chem.* **1990**, *29A*, 461.

40 D.K. Mukherjee, B.K. Palit, C.R. Saha, *Indian J. Chem.* **1992**, *31A*, 243.

41 R. van Asselt, C.J. Elsevier, *J. Mol. Catal.* **1991**, *65*, L13.

42 M.W. van Laren, C.J. Elsevier, *Angew. Chem. Int. Ed. Engl.* **1999**, *38*, 3715.

43 M.W. van Laren, M.A. Duin, C. Klerk, M. Naglia, D. Rogolino, P. Pelagatti, A. Bacchi, C. Pelizzi, C.J. Elsevier, *Organometallics* **2002**, *21*, 1546.

44 C. Borriello, M.L. Ferrara, I. Orabona, A. Panunzi, F. Ruffo, *J. Chem. Soc., Dalton Trans.* **2000**, 2545.

45 P. Pelagatti, A. Venturini, A. Leporati, M. Carcelli, M. Costa, A. Bacchi, G. Pelizzi, C. Pelizzi, *J. Chem. Soc., Dalton Trans.* **1998**, 2715.

46 M. Costa, P. Pelagatti, C. Pelizzi, D. Rogolino, *J. Mol. Catal. A: Chem.* **2002**, *178*, 21.

47 S. Koritala, *J. Am. Oil Chemist's Soc.* **1985**, *62*, 517.

48 A. Sisak, F. Ungváry, *Chem. Ber.* **1976**, *109*, 531.

49 N. Suzuki, Y. Ayaguchi, Y. Izawa, *Chem. Ind.* **1983**, *4*, 166.

50 D.R. Armstrong, O. Novaro, M.E. Ruiz-Vizcaya, R. Linarte, *J. Catal.* **1977**, *48*, 8.

51 J.C. Bailar, Jr., H. Itatani, *J. Am. Chem. Soc.* **1967**, *89*, 1592.

52 E.N. Frankel, E.A. Emken, H. Itatani, J.C. Bailar, Jr., *J. Org. Chem.* **1967**, *32*, 1447.

53 H.A. Tayim, J.C. Bailar, Jr., *J. Am. Chem. Soc.* **1967**, *89*, 4330.

54 R.W. Adams, G.E. Batley, J.C. Bailar, Jr., *J. Am. Chem. Soc.* **1968**, *90*, 6051.

55 G. K. Anderson, C. Billard, H. C. Clark, J. A. Davies, C. S. Wong, *Inorg. Chem.* **1983**, *22*, 439.

56 V. K. Jain, G. S. Rao, *Inorg. Chim. Acta* **1987**, *127*, 161.

57 R. D. Cramer, E. L. Jenner, R. V. Lindsey, Jr., U. G. Stolberg, *J. Am. Chem. Soc.* **1963**, *85*, 1691.

58 H. van Bekkum, J. van Gogh, G. van Minnen-Pathuis, *J. Catal.* **1967**, *7*, 292.

59 L. P. van 't Hoff, B. G. Linsen, *J. Catal.* **1967**, *7*, 295.

60 F. van Rantwijk, G. J. Timmermans, H. van Bekkum, *Recl. Trav. Chim. Pays-Bas* **1976**, *95*, 39.

61 G. W. Parshall, *J. Am. Chem. Soc.* **1972**, *94*, 8716.

62 T. Yoshida, T. Yamagata, T. H. Tulip, J. A. Ibers, S. Otsuka, *J. Am. Chem. Soc.* **1978**, *100*, 2063.

63 R. Rumin, *J. Organomet. Chem.* **1983**, *247*, 351.

64 H. C. Clark, C. Billard, C. S. Wong, *J. Organomet. Chem.* **1979**, *173*, 341.

65 S. Paganelli, U. Matteoli, A. Scrivanti, C. Botteghi, *J. Organomet. Chem.* **1990**, *397*, 375.

66 J. Lin, C. U. Pittman, Jr., *J. Organomet. Chem.* **1996**, *512*, 69.

67 H. A. Tayim, J. C. Bailar, Jr., *J. Am. Chem. Soc.* **1967**, *89*, 3420.

68 G. W. Parshall, *J. Am. Chem. Soc.* **1966**, *88*, 704.

69 R. D. Cramer, R. V. Lindsey, Jr., C. T. Prewitt, U. G. Stolberg, *J. Am. Chem. Soc.* **1965**, *87*, 658.

70 H. Nowatary, K. Hirabayashi, I. Yasumori, *J. Chem. Soc., Faraday Trans.* **1976**, *72*, 2785.

71 J. A. Widegren, R. G. Finke, *J. Mol. Catal. A: Chem.* **2003**, *198*, 317.

5
Nickel

Elisabeth Bouwman

5.1
Introduction

Nickel is frequently used in industrial homogeneous catalysis. Many carbon–carbon bond-formation reactions can be carried out with high selectivity when catalyzed by organonickel complexes. Such reactions include linear and cyclic oligomerization and polymerization reactions of monoenes and dienes, and hydrocyanation reactions [1]. Many of the complexes that are active catalysts for oligomerization and isomerization reactions are supposed also to be active as hydrogenation catalysts.

The choice of nickel as a homogeneous hydrogenation catalyst is self-evident if one considers Nature. The hydrogenase enzymes make use of the abundantly available transition metals nickel and iron for the activation of dihydrogen or the reduction of protons. The nickel-containing hydrogenase enzyme is not able to hydrogenate organic substrates, but it can activate dihydrogen at atmospheric pressure and ambient temperatures [2]. Dihydrogen oxidation in living cells is coupled to the reduction of electron acceptors such as oxygen, nitrate, sulfate, carbon dioxide and fumarate, whereas proton reduction (dihydrogen evolution) is an essential element of pyruvate fermentation or the disposal of excess electrons. Studies of either *para*- to *ortho*-hydrogen conversion or H_2/D^+ (or D_2/H^+) exchange reactions indicate that in the hydrogenase active site dihydrogen is activated *via* a heterolytic pathway, forming a hydride and a proton, as shown schematically in Scheme 5.1 [2]. The rational design and synthesis of structural and functional models for hydrogenases [3], as well as their characterization and reactivity studies, might contribute to the development of new catalysts for practical use in, for example, fuel cells, as has been expertly described by Cammack et al. [4]. The advantage of the use of inexpensive nickel over costly metals such as rhodium or ruthenium is evident.

Despite the fact that Nature uses nickel for the activation of dihydrogen, and that Raney-Ni is one of the oldest and the most important heterogeneous hydrogenation catalysts, very few nickel complexes are known to catalyze the homoge-

The Handbook of Homogeneous Hydrogenation.
Edited by J. G. de Vries and C. J. Elsevier
Copyright © 2007 WILEY-VCH Verlag GmbH & Co. KGaA, Weinheim
ISBN: 978-3-527-31161-3

(a) $M^{n+} + H_2$ $\rightleftharpoons$ $M^{(n+2)+}(H)_2$

(b) $2\,M^{n+} + H_2$ $\rightleftharpoons$ $2\,M^{(n+1)+}H$

(c) $M^{n+}\!-\!X + H_2$ $\rightleftharpoons$ $M^{n+}\!-\!H + HX$

Scheme 5.1 Different modes of dihydrogen activation.
(a) Oxidative addition, forming a metal dihydride species;
(b) homolytic activation, forming two metal monohydride
species; (c) heterolytic splitting, forming a metal hydride
species and a proton.

neous hydrogenation reaction. In this section, a brief overview will be provided
of mononuclear nickel complexes, which have been reported to catalyze homogeneously the hydrogenation of various olefins. Many of the reported nickel-containing hydrogenation catalysts are of the Ziegler type (i.e., a nickel salt in combination with a reducing agent such as trialkylaluminum), and in many cases it
is questionable whether the reported complexes yield really homogeneous hydrogenation catalysts or that the observed catalytic hydrogenation activity is due
to formation of colloidal (heterogeneous) nickel particles [5].

5.2
Coordination Chemistry and Organometallic Aspects of Nickel

5.2.1
Nickel–Hydride Complexes

The existence of hydride complexes of nickel has long been recognized, with
the first reports of these compounds relying mainly on the evidence of ^{1}H-NMR
investigations. The number of X-ray structures containing at least one nickel–
hydride bond is rather limited, and the reported structures include a relatively
large number of organometallic clusters. Only a limited number of mononuclear or dinuclear structures possibly relevant for hydrogenation catalysis have
been reported. These complexes typically contain electron-donating phosphane
ligands with cyclohexyl, isopropyl or tertiary-butyl groups at phosphorus. The
synthesis of the nickel hydride complexes is usually achieved by reaction of
nickel(II) halide complexes with the mild reducing agents such as $HBMe_3^-$ and
BH_4^-, or by addition of H^+ to a nickel(0) complex.

An interesting nickel–hydride complex with a biologically relevant ligand arising from enzyme modeling has been reported by the group of Holm [6]. Reaction
of the complex $[Ni(NS_3)Cl]^+$ (NS_3 = tris(t-butylsulfanylethyl)amine) with $NaBH_4$ results in the trigonal bipyramidal complex $[Ni(NS_3)H]^+$; the presence of the hydride
was confirmed by a high-field ^{1}H-NMR resonance at −37.75 ppm. The latter complex reacts reversibly with ethene to form the ethyl complex, and it catalyzes the
isomerization of 1-hexene [6]. A similar complex has been reported with the tripo-

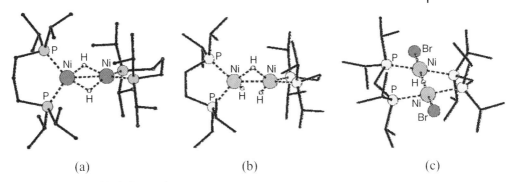

Fig. 5.1 Projection of hydride complexes:
(a) [Ni(dippp)(μ-H)]$_2$ [15]; (b) [{HNi(dippe)}$_2$(μ-H)]$^+$ [16];
(c) [Ni$_2$Br$_2$(dippm)$_2$(μ-H)] [17].

dal phosphorus ligand tris(2-diphenylphosphanylethyl)amine, but in this case substoichiometric amounts of hydride were found to be present in the complex [7].

Using monodentate phosphane ligands, only a few square-planar nickel(II) complexes have been reported, containing a hydride coordinated *trans* to a halide ion (based only on NMR evidence) [8], a phenolate group [9], an imine nitrogen [10], or a tetrahydridoborate anion [11]. The structures show significant distortion from the ideal square-planar geometry due to the bulky phosphane ligands. The hydride is visible in the ^{1}H-NMR spectra around –25 ppm as a triplet resulting from the coupling with two *cis* phosphane donors.

The use of chelating didentate phosphane ligands with an ethyl or propyl bridge often results in the formation of dinuclear nickel complexes with two bridging hydrides [12–15]. The complexes are readily formed by the reaction of the corresponding mononuclear nickel dichloride complex with mild reducing agents such as HBMe$_3^-$ or HBEt$_3^-$. In the resulting complexes of general formula [P$_2$Ni(μ-H)]$_2$, the metal ion is formally nickel(I) with a nickel–nickel bond length of 2.4 Å. A projection of the structure of [Ni(dippp)(μ-H)]$_2$ is shown in Figure 5.1 (a) (dippp = 1,3-bis(diisopropylphosphanyl)propane). Despite earlier predictions and theoretical calculations, the dinuclear core is nonplanar; the complexes are diamagnetic, showing the hydride resonance as a quintet around –10 ppm [12–15]. Bach et al. have studied the reactivity of the complex [(dbpe)-Ni(μ-H)]$_2$ (dbpe = *t*Bu$_2$PCH$_2$CH$_2$P*t*Bu$_2$) with deuterium gas; scrambling was observed resulting in the formation of the monodeuteride and dideuteride complexes [14]. The ^{31}P-NMR spectrum of the complex [(dbpe)Ni(μ-H)]$_2$ shows a singlet at +94.6 ppm; when a spectrum is recorded in an H$_2$ atmosphere it shows an additional singlet at +108 ppm which has been tentatively ascribed to a mononuclear nickel(II) dihydride complex, or to a nickel(IV) tetrahydride complex as a result of oxidative addition of dihydrogen, and which would be intermediates in the scrambling process [14].

The reaction of [Ni(dippe)Br$_2$] (dippe = 1,2-bis(diisopropylphosphanyl)ethane) with NaBH$_4$ results in the formation of a dinuclear trihydride complex [{HNi-

(dippe)}$_2$(μ-H)]$^+$ (see Fig. 5.1b) [16]. The structure shows a bridging hydride at a bent angle as well as a Ni–Ni bond with a distance of 2.3 Å. The complex is diamagnetic and displays one quintet at –13.4 ppm, indicating rapid interconversion of the bridging hydride with the terminal hydrides. The structure can be regarded as a protonated form of the dihydrides described above, but attempts to deprotonate the complex have failed, even using a strong base [16].

The small bite-angle ligands dcpm (bis(dicyclohexylphosphanyl)methane) or dippm (bis(diisopropylphosphanyl)methane) yield dinuclear nickel complexes of formula [Ni$_2$X$_2$(μ-P$_2$)$_2$(μ-H)] (X=halide) with one hydride bridging the two nickel centers [17, 18]. The ligands in these complexes are not chelating, but are binding two nickel ions. The compounds are mixed-valent, containing a nickel-(I) and a nickel(II) ion. The complexes are therefore paramagnetic and for [Ni$_2$Cl$_2$(μ-dcpm)$_2$(μ-H)] a g-value of 2.139 at room temperature was reported [18]. The complex [Ni$_2$Cl$_2$(μ-dcpm)$_2$(μ-H)] is bent, with a Ni–H–Ni angle of 128° and a Ni–Ni distance of 2.9 Å [18], whereas the complex [Ni$_2$Br$_2$(μ-dippm)$_2$-(μ-H)] (see Fig. 5.1c) is nearly planar with a Ni–H–Ni angle of 178° and a Ni–Ni distance of 3.2 Å [17].

A mononuclear nickel hydride complex with three N-heterocyclic carbene ligands has been reported; the compound was formed by oxidative addition of an imidazolium salt to the Ni(0) bis(carbene) complex [19]. The hydride signal of this nickel(II) complex appears at –15 ppm.

Recently, the crystal structure of a nickel(II) complex with a tridentate silyl ligand has been reported [20]. The structure in the solid state shows an η^2-(Si–H) binding to nickel, with a Ni–H distance of 1.47 Å; NMR spectra of the complex in solution at –80 °C suggest the formation of a nickel(IV) hydride species through oxidative addition of the silyl-hydrogen to nickel [20].

5.2.2
Nickel–Alkene and Nickel–Alkyl Complexes

A large number of (mostly zero-valent) nickel–alkene complexes has been reported. Although these complexes have not been recently reviewed, their general properties and structures were expertly described in 1982 [21]. A complete overview of the reported nickel–alkene and nickel–alkyl complexes is beyond the scope of this section, in which a selection of nickel–alkene and nickel–alkyl complexes is described, mostly related to possible intermediates in hydrogenation catalysis.

The reaction of [Ni(ethene)$_3$] with a hydride donor such as trialkyl(hydrido)-aluminate results in the formation of the dinuclear anionic complex [{Ni(ethene)$_2$}$_2$H]$^-$ [22]. The nickel(0) centers in this complex are in a trigonal planar environment of two ethene molecules and a bridging hydride ion, with the ethene carbons in the plane of coordination. The two planes of coordination within the dinuclear complex are almost perpendicular to each other, and the Ni–H–Ni unit is significantly bent, with an angle of 125° and a Ni–Ni distance of 2.6 Å [22].

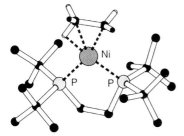

Fig. 5.2 Projection of [Ni(dbpe)(ethyl)] [23].

Protonation of (diphosphane)nickel(0)–ethene complexes with $HBF_4 \cdot OEt_2$ results in the formation of the corresponding cationic three-coordinate nickel(II)–alkyl complexes, in which a β-agostic hydrogen is present [23]. Figure 5.2 illustrates the structure of [Ni(dbpe)(ethyl)] with an agostic Ni–H distance of 1.64 Å [23]. For these nickel complexes, β-elimination and alkene rotation appeared to be slow on the NMR timescale at temperatures below 27 °C. NMR spectroscopy at 20 °C revealed a multiplet at –1.24 ppm for the three protons at the β-carbon, whereas at –100 °C a broad signal at –5.8 ppm is assigned to the agostic hydrogen [23].

In contrast, the fluxionality of neutral three-coordinate nickel(II)–alkyl complexes with β-diketiminato ligands is described in a recent publication [24]. The β-agostic complexes undergo facile β-elimination and reinsertion resulting in a mixture of primary and secondary nickel(II)–propyl complexes [24]. The corresponding four-coordinate, square-planar complexes with an additional nitrogen-donor ligand do not show β-agostic interactions in the solid state. In solution, however, dissociation of the nitrogen-donor ligand readily occurs and in the NMR spectra resonances are observed that are due to a β-agostic hydrogen interaction of the coordinated alkyl group [25]. Similarly, the neutral (acetylacetonato)(triphenylphosphane)nickel(II)–ethyl complex does not show agostic interactions in the square-planar solid-state structure, whereas in solution scrambling of the alkyl protons is observed [26]. In another recent report a remarkably stable dinuclear nickel(II)–hexyl complex has been reported; the compound does not show any agostic interactions, neither in the solid state nor in solution [27].

5.2.3
Mechanistic Aspects of Hydrogen Activation

For catalytic hydrogenation the activation of a dihydrogen molecule, the binding of an alkene, and the formation of a metal–alkyl intermediate are the most crucial aspects of a plausible catalytic mechanism. Support for a nickel–hydride mechanism in the oligomerization of ethene by nickel complexes with a chelating phosphorus–oxygen ligand was obtained through the crystallization of a complex trapped with triphenylphosphane [28]. The proton NMR spectrum of this complex shows the hydride resonance at –24.85 ppm as a double doublet due to the two *cis* phosphorus atoms [28].

Protonation mechanisms of nickel complexes and the relevance to hydrogenases, as well as to industrial catalysts, have recently been reviewed [29]. The huge interest in the possibilities of a hydrogen economy, based on the cheap and sustainable generation of H_2, and the clean and efficient oxidation of H_2 in fuel cells has given impetus to the research for functional models of hydrogenases. In the search for active functional models for [NiFe] hydrogenases, nickel complexes are tested on hydrogenase activity and three types of reactivity can be distinguished: 1) H_2/D^+ exchange (scrambling); 2) activation of H_2 (oxidation to H^+); and 3) reduction of H^+ to H_2. For functional models of the [NiFe] hydrogenases the first two activities are of importance.

Bis(diphosphane)nickel complexes have been reported to show heterolytic splitting of dihydrogen [30]. The complex $[Ni(PNP)_2](BF_4)_2$, in which PNP is the ligand $Et_2PCH_2N(Me)CH_2PEt_2$, possesses both hydride and proton acceptor sites. When dihydrogen gas is passed through a solution of $[Ni(PNP)_2]^{2+}$ in acetonitrile or dichloromethane, the solution bleaches during formation of the hydride species $[HNi(PNHP)(PNP)]^{2+}$, in which one of the central nitrogen atoms of the ligands is protonated. In the ^{1}H-NMR of the complex $[HNi(PNHP)(PNP)]^{2+}$ at low temperatures, two resonances are observed at -15.2 and 7.4 ppm, that have been assigned to the hydride and the NH-proton, respectively. The hydride and the NH-proton undergo rapid intramolecular exchange with each other and with protons in solution [30].

The structures and reactivity of the nickel complexes described above seem to indicate that, for nickel-containing hydrogenation catalysts, heterolytic activation of dihydrogen may be the most likely mechanistic pathway. The nickel–hydride complexes are stabilized by the presence of electron-donating phosphane ligands, making the nickel complexes with these ligands the most likely candidates for active homogeneous hydrogenation catalysts.

5.3
Hydrogenation Catalysis

5.3.1
Ziegler Systems

The so-called Ziegler-type catalysts have been studied rather extensively. This type of catalyst is formed by treating a transition-metal salt with a reducing agent. The most frequently used reducing agents include trialkylaluminum, alkylaluminum chlorides, alkali metals, $LiAlH_4$, and $NaBH_4$. The mixtures yield dark, air-sensitive solutions that appear homogeneous to the eye, but most likely contain colloidal metal [5]. Ziegler-type catalysts made from cobalt or nickel salts with triethylaluminum can be used for the hydrogenation of benzene to cyclohexane under rather mild conditions (155 °C, 10 bar). In a typical patent example, benzene is hydrogenated to cyclohexane (99% conversion) with more than 2500 turnovers after 1 h, using a catalyst prepared from a mixture of nickel and zinc octoates and excess triethylaluminum [31]. A process has been developed by Institut Français du Petrole [32].

Ziegler-type hydrogenation catalysts have been used for the hydrogenation of vegetable oils [33, 34], cyclohexadienes [35], and butadiene rubbers [36–38]. Under rather mild (1 bar H_2, 40 °C) conditions (substituted) 1,4-cyclohexadienes can be hydrogenated selectively to the cyclohexenes in yields of up to 80%; trace amounts of benzene were detected as a result of disproportionation reactions [35]. The active catalyst was prepared from [Ni(acac)$_2$] (Hacac = acetylacetone), $Et_3Al_2Cl_3$ (10 equiv. to Ni) and PPh$_3$. The activity is rather low, and with a substrate:catalyst ratio (SCR) of 100 complete conversion is reached after 6 to 28 h. Isomerization of the substituted cyclohexadienes occurred, and a mechanistic pathway involving η^3-allyl coordination of the substrate was proposed. Dihydrogen activation was proposed to occur via oxidative addition [35].

Recent investigations into the hydrogenation of rubbers include catalytic systems prepared from nickel(II) 2-ethylhexanoate or nickel(II) acetylacetonate activated with triisobutylaluminum [36], triethylaluminum [38], or butyl lithium [37, 38]. Large amounts of co-catalysts (initiator) are necessary. Hydroxyl-containing polybutadienes were hydrogenated at low H_2 pressures, affording the saturated polymers with full retention of the OH functionality [38]. Again, the activity of the catalysts is rather low, with SCRs of approximately 20 having been used.

5.3.2
Nickel Complexes of Oxygen- or Nitrogen-Containing Ligands

Catalytic homogeneous hydrogenation of cyclohexene has been claimed for simple systems such as nickel(II) acetylacetonate [39] or a nickel–chloride complex with two monodentate amines [40]. The latter complex was used as comparison for a heterogeneous catalyst obtained by impregnation of the complex on γ-alumina [40]. SCRs of 100 were used at 30 atm. H_2 and temperatures up to 100 °C, resulting in conversions of only 20–35% after 1 h.

Nickel complexes of simple tetradentate ligands containing nitrogen- and oxygen-donor atoms have also been investigated for their catalytic activity in homogeneous hydrogenation [41–43]. The complex [Ni(saloph)] (saloph=bis(salicylidene)phenylenediamine) has been reported to catalyze the hydrogenation of cyclohexene to cyclohexane and cyclooctene to cyclooctane at a moderately high dihydrogen pressure of 60 bar and a temperature of 50 °C [42]. The catalyst is more reactive in the hydrogenation of cyclohexene than in the hydrogenation of cyclooctene, the turnover frequency (TOF) being 13 h^{-1} and 7 h^{-1}, respectively (SCR 100, t=5 h, in ethanol). The reaction is suggested to proceed *via* the high-valent [Ni^{IV}(saloph)(H)$_2$] species, which supposedly is formed through the oxidative addition of dihydrogen to the divalent [Ni(saloph)] complex [42]. It has been claimed that encapsulation of the analogous nickel complex [Ni(salen)] (salen=bis(salicylidene)ethylenediamine) in zeolite Y also results in a homogeneous catalyst for hydrogenation reactions [44]. The hydrogenation of 1,5,9-cyclododecatriene has been investigated using similar salen-type nickel complexes under rather harsh conditions (150 °C, 50 bar), but giving rather low activities (TOF < 14 h^{-1}) [41]. In a recent study, nickel complexes of similar tetradentate ligands derived from pyrrole 2-carboxaldehyde and various diamines were used as catalysts in the hydrogenation of phenylacetylene [43]. The activity of these systems is also rather low, and complete conversion was not even reached after 24 h (SCR 150, at 40 °C and 1 bar). Also in this case activation of dihydrogen is believed to proceed via oxidative addition to give a Ni(IV) dihydride intermediate [43].

[Ni(acac)$_2$] has been reported to be an active catalyst for the hydrogenation of cyclohexene to cyclohexane in methanol at moderate pressures and high temperature. At 100 °C and a H$_2$ pressure of 30 bar, a TOF of 450 h^{-1} is reported (SCR 1000, t = 1 h, in methanol). From a study including pretreatment of the catalyst under a nitrogen atmosphere, it was concluded that alkene coordination precedes hydrogen activation [39]. [Ni-(acac)$_3$] has been used for the catalytic hydrogenation of unsaturated fatty acids and esters [45]. No hydrogenation activity is observed at temperatures below 100 °C. After an induction period of 3 h, methyl linolenate can be hydrogenated with a TOF of 10 h^{-1} (SCR 10, t = 1 h, 100 °C, 7 bar, in methanol). It is quite likely that the actual catalyst is heterogeneous, as an induction period is needed and black precipitates are formed during the hydrogenation reaction.

5.3.3
Nickel Complexes of Triphenylphosphane

The tetrahedral nickel complexes [NiX$_2$(PPh$_3$)$_2$] (X = Cl, Br, or I) have been reported to show catalytic activity in the hydrogenation of linear, cyclic, terminal, and sterically hindered olefins; a summary of the reported activities in the hydrogenation of various substrates is provided in Table 5.1 [46–50]. The catalytic activity of the different halide complexes in the hydrogenation of methyl linoleate decreases in the order I > Br > Cl, but also when the iodide complex is used the activities remain extremely low, even at elevated temperatures and pressures [46]. Selective hydrogenation of polyunsaturated compounds to monounsaturated compounds is observed, as well as migration of the double bond and the isomerization of *cis*- to *trans*-alkenes. Nonconjugated dienes are assumed to be converted into conjugated dienes through migration of the double bonds. Subsequently, these conjugated dienes are hydrogenated to monoenes. Monoolefins are hardly further hydrogenated. The isomerization reaction was found not to occur in the absence of dihydrogen. Therefore, it was suggested that both the hydrogenation and isomerization reaction proceed *via* a monohydride species and that dihydrogen is activated via a heterolytic pathway [46].

Similar low activities were found in the hydrogenation of 1-octene [47]. The use of [Ni(PPh$_3$)$_2$I$_2$] in the hydrogenation of norbornadiene resulted in considerable amounts of nortricyclene, *via* transannular ring closure, whereas 1,5-cyclooctadiene yielded bis-cyclo-[3.3.0]oct-2-ene. According to these authors, the re-

Table 5.1 Hydrogenation of several substrates in the presence of the nickel complex [Ni(PPh$_3$)$_2$I$_2$]. [a]

Entry	Substrate	pH$_2$ [bar]	T [°C]	TOF[b]	Main product	Reference
1	1,5,9-cyclododecatriene	78.5	175	165	cyclododecene	50
2	1,5,9-cyclododecatriene	50	100	1	cyclododecene	49
3	methyl linoleate	40	90	2	methyl oleate	46
4	1,3-cyclooctadiene	50	100	8	cyclooctene	49
5	1,5-cyclooctadiene	50	100	9	cyclooctene[c]	49
6	2,5-dimethyl-1,5-hexadiene	50	100	7	2,5-dimethyl-2-hexene	49
7	2,5-dimethyl-2,4-hexadiene	50	100	8	2,5-dimethyl-2-hexene	49
8	isoprene	50	100	0		49
9	norbornadiene	50	100	29	nortricyclene[d]	49
10	methyl oleate	40	90	0		46
11	methyl cis-9,cis-15-octadecadienoate	40	90	trace	monoene	48
12	1-octene	1	20	trace	n-octane	47

a) Reaction conditions for entry 1: SCR = 100, t = 30 min, no solvent added; entries 2, 4–9: SCR = 15 (59 for entry 8), t = 1 h (30 min for entry 9), in acetic acid; entries 3, 10, 11: SCR = 5 (9 for entry 11), t = 3 h, in benzene.
b) Turnover frequency in mol converted substrate per mol Ni per hour, calculated from data in [46–50].
c) Transannular ring closure was also observed.
d) Via transannular ring closure, nortricyclene is further hydrogenated to norbornane.

sults suggested the involvement of a η^3-allyl-nickel intermediate in the catalytic cycle [49]. Kinetic studies have been undertaken for the hydrogenation of 1,5,9-cyclododecatriene to cyclododecene, using the catalyst [Ni(PPh$_3$)$_2$I$_2$] under rather harsh conditions (155–175 °C, 80 bar) [50]. It was proposed that a monohydride–nickel complex is the active species, and dihydrogen activation was proposed to proceed via oxidative addition [50].

5.3.4
Nickel Complexes of Didentate Phosphane Ligands

The catalytic activity of nickel(II) acetate in combination with a number of chelating phosphane ligands in the hydrogenation of 1-octene has been studied extensively during the past decade [51–56]. It was found that in-situ-prepared mixtures of nickel(II) acetate with the electron-donating ligands o-MeO-dppe, o-MeO-dppp, dcpe, or dcpp (see Scheme 5.2) achieved quite high activities in the hydrogenation of 1-octene, the nonoptimized catalytic systems showing turnover numbers (TONs) of up to 460 per hour at 25 °C and 50 bar [51]. In

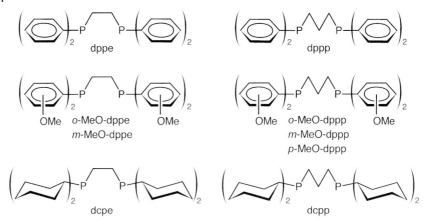

Scheme 5.2 Structure of the ligands used. dppe: 1,2-*bis*(diphenylphosphanyl)ethane; *o*-MeO-dppe: 1,2-*bis*(di(*o*-methoxyphenyl) phosphanyl)ethane; dcpe: 1,2-*bis*(dicyclohexylphosphanyl)ethane; *m*-MeO-dppe: 1,2-*bis* (di(*m*-methoxyphenyl)phosphanyl)ethane; dppp: 1,3-*bis*(diphenylphosphanyl)propane;

o-MeO-dppp: 1,3-*bis*(di-*o*-methoxyphenyl-phosphanyl)propane; dcpp: 1,3-*bis*(dicyclohexylphosphanyl)propane; *m*-MeO-dppp: 1,3-*bis*(di(*m*-methoxyphenyl)phosphanyl)propane; *p*-MeO-dppp: 1,3-*bis*(di(*p*-methoxyphenyl)phosphanyl)propane.

contrast, use of the simple ligands dppe or dppp resulted in inactive systems; the results are summarized in Table 5.2 [51].

In a first attempt to separate electronic from steric effects of the *ortho*-methoxy group, the *meta* and *para* analogues of the methoxy-substituted ligands were also tested for hydrogenation activity. None of these showed any hydrogenation activity, however, seemingly indicating that the positive effect of the *ortho*-methoxy group should be attributed mainly to steric effects [51].

In order to explain the differences in catalytic activity of nickel compounds containing the various ligands, the behavior of the nickel complexes in solution was investigated in more detail, for which NMR spectroscopy appeared to be a valuable tool [54]. Using ^{1}H- and ^{31}P$\{^1$H$\}$-NMR spectroscopy, the behavior of the complexes has been studied in several solvents. It appeared that, in solutions, the nickel complexes of the ligand *o*-MeO-dppe are involved in ligand-exchange reactions; the mono(ligand) complexes [Ni(*o*-MeO-dppe)X$_2$] are in equilibrium with bis(ligand) complexes [Ni(*o*-MeO-dppe)$_2$]$^{2+}$ and "naked" (solvated) nickel ions ["NiX$_4$"]$^{2-}$ (see Scheme 5.3) [54]. The exact nature of the species "NiX$_4^{2-}$" is dependent on the solvent and the anions; in coordinating solvents it may exist as [Ni(solvent)$_6$]$^{2+}$. The position of the equilibrium appeared to be highly dependent on the anion X and the solvent used. In a polar solvent in combination with weakly coordinating anions, only the bis(ligand) complex was observed, whereas in an apolar solvent in combination with coordinating anions only the mono(ligand) complex was present. From these studies it appeared that the differences in catalytic hydrogenation activity of nickel(II) acetate in combi-

Table 5.2 Nickel-catalyzed hydrogenation of 1-octene.[a]

Entry	Ligand	TOF$_{1\text{-octene}}$[b]	S$_{n\text{-octane}}$[c]	[Ni] [mM]	SCR
At 25 °C					
1	dppe	0		1.8	1000
2	dppp	2	100	1.8	1000
3	o-MeO-dppe	8	100	1.4	2000
4	o-MeO-dppp	500	100	1.4	2000
5	dcpe	2000[d]	98	3.6	500
6	dcpe	1740	80	1.4	2000
7	dcpp	360	97	3.6	500
At 50 °C					
8	dppe	0		1.8	1000
9	dppp	0		1.8	1000
10	o-MeO-dppe	400	94	1.4	2000
11	o-MeO-dppp	1160	99	1.4	2000
12	dcpe	1990	83	1.4	2000
13	m-MeO-dppe[e]	0		3.6	500
14	m-MeO-dppp[e, f]	0		3.6	500
15	p-MeO-dppp[f]	10	0	3.6	500

a) Reaction conditions: Ni:ligand ratio=1, t=1 h, pH$_2$=50 bar, in 20 mL methanol.
b) Turnover number in mol converted 1-octene per mol Ni(OAc)$_2$ after 1 h.
c) Selectivity towards n-octane (mol n-octane per mol converted 1-octene).
d) Complete conversion after 15 min.
e) No clear solution could be obtained after mixing nickel(II) acetate and the ligand.
f) Reaction performed in ethanol.
All data taken from [57].

nation with dppe or o-MeO-dppe could not be rationalized by the assumption that the bulkier ligand would prevent the formation of inactive bis(ligand) complexes. A comparison of the behavior of the nickel complexes of the ligand o-MeO-dppe with the unsubstituted analogue dppe showed that the latter ligand was more readily oxidized [54]. The source of the oxygen was unknown; as the experiments were carried out in an inert atmosphere the oxidation seemed to be related to the nickel(II) center, as observed previously [58]. It has been reported earlier that Pd(OAc)$_2$ is able to oxidize tertiary phosphanes [59]. This possibility of oxidation was found to play a major role in the case of the nickel complexes of the propane-bridged ligands o-MeO-dppp and dppp, as for both of these ligands the formation of bis(ligand) complexes is not possible, but only the substituted ligand o-MeO-dppp yields an active catalyst [53]. Prevention of oxidation of the phosphorus atom by the *ortho*-methoxy group is most likely due to steric effects; nickel(II) acetate in combination with p-MeO-dppp did not

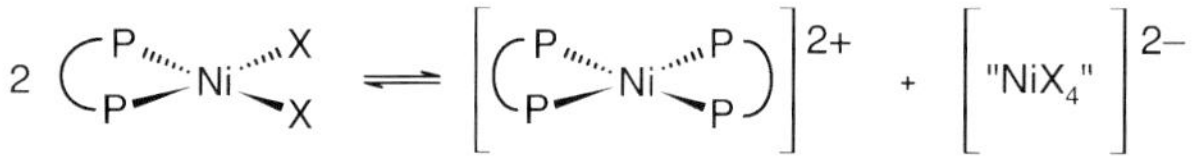

2 $\left(\begin{smallmatrix} P_{\prime\prime\prime} & \ & \ _{\prime\prime\prime}X \\ & Ni & \\ P^{\prime} & \ & \ ^{\prime}X \end{smallmatrix} \right)$ $\rightleftharpoons$ $\left[\left(\begin{smallmatrix} P_{\prime\prime\prime} & \ & \ _{\prime\prime\prime}P \\ & Ni & \\ P^{\prime} & \ & \ ^{\prime}P \end{smallmatrix} \right) \right]^{2+}$ + $\left[\text{"NiX}_4\text{"} \right]^{2-}$

Scheme 5.3 Schematic representation of the ligand-exchange equilibrium.

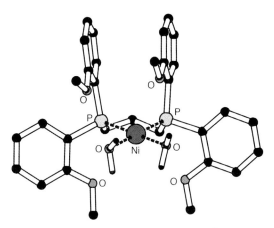

Fig. 5.3 Projection of $[\text{Ni}(o\text{-MeO-dppp})(\text{H}_2\text{O})_2]^{2+}$.

yield an active hydrogenation catalyst [51], and $^{31}\text{P}\{^1\text{H}\}$-NMR spectroscopy showed that also in this case the ligand is rapidly oxidized [54].

The inability to form a *bis*(ligand) complex with the ligand *o*-MeO-dppp is most likely due to a combination of the large bite angle caused by the propylene-bridge in the ligand, and the rather small ionic radius of the nickel(II) ion. Attempts to synthesize a *bis*(ligand) complex using the ligand *o*-MeO-dppp were unsuccessful. Instead, an unprecedented low-spin nickel complex containing two coordinated water molecules was formed – that is the complex [Ni(*o*-MeO-dppp)(H$_2$O)$_2$](PF$_6$)$_2$ (Fig. 5.3) [53]. This complex yields a highly active and selective hydrogenation catalyst, with a TOF of 2800 h^{-1} in the hydrogenation of 1-octene [55]. The impossibility to form a *bis*(ligand) complex might be the most important reason for the considerably higher hydrogenation activity of catalysts based on this ligand, compared to those based on *o*-MeO-dppe [53].

The catalytic activity of nickel(II) complexes of didentate phosphane ligands containing *ortho*-methoxy phenyl, *ortho*-ethoxy phenyl or cyclohexyl groups in the hydrogenation of 1-octene has been studied in more detail [55, 56]. The only observed side reaction in the hydrogenation of 1-octene is the isomerization to internal alkenes. The factors which influence the availability and accessibility of the coordination sites in the catalytically active species are solvent, anion, and amount and nature of ligand. The influence of these factors has been investigated using catalysts containing *o*-MeO-dppp, as these are not prone to ligand-exchange equilibria [55]. The presence of a solvent that is both polar and protic has been found to be essential in order to observe catalytic activity. This catalytic activity is furthermore inversely related to the coordination ability of the anion. In the case of *o*-MeO-dppe, the catalyst is not active at all at a ligand-to-metal ratio higher than 2. In the case of *o*-MeO-dppp, the catalytic activity gradually decreases with increasing ligand-to-metal ratio. At a ratio of 5:1 an activity of 77 turnovers after 1 h is still observed, though the selectivity for hydrogenation has fallen considerably (70%). Although the ligand-exchange equilibrium occurs, the nickel complexes containing the ligands *o*-MeO-dppe and *o*-EtO-dppe still yield active hydrogenation catalysts. The presence of the more bulky ethoxy groups instead of the methoxy groups increases the hydrogenation activity by a factor of three, with the TOF rising from 400 h^{-1} to 1200 h^{-1}; the selectivity for *n*-octane, however, is decreased from 99 to 65% [55]. The type of ligand merely influences the accessibility of the axial or equatorial coordination sites for the reactants, and hence the selectivity towards the hydrogenated product *n*-octane [55]. With the ligand dcpe, very active hydrogenation catalysts are obtained; at 50 °C and 50 bar H_2, TONs of up to 3000 h^{-1} have been reached, although also in this case the selectivity was decreased to 88% [56].

The kinetics of the nickel-catalyzed hydrogenation of 1-octene have been studied using nickel(II) acetate in combination with the ligand *o*-MeO-dppp [52]. The effects of temperature and dihydrogen pressure, as well as the influence of nickel and substrate concentrations have been studied (T = 30–60 °C, pH_2 = 30–60 bar, [Ni] = 0.35–2.81 mM, [1-octene] = 0.98–3.91 M, ligand:Ni ratio = 1.1 in a mixture of dichloromethane and methanol). The kinetic study resulted in the surprisingly simple rate law: $r = k_{cat}[Ni][1\text{-octene}]pH_2$. These kinetic data are consistent with a mechanism in which the terminating hydrogenolysis step of the nickel–alkyl complex is the rate-determining step of the catalytic cycle with all preceding steps at equilibrium. The first-order dependence in the nickel concentration indicates that dinuclear nickel complexes with bridging hydrides are not present during catalysis, and that thus only mononuclear nickel complexes are present [52]. This absence of dinuclear species is remarkable, as these dinuclear complexes are known for a number of didentate phosphane ligands (see Section 5.2.1). The first-order dependence of the substrate was confirmed by the results with other substrates, as the hydrogenation rate was seen to depend strongly on the nature of the substrate. The results for some different substrates are listed in Table 5.3 [52]. The rate of hydrogenation of the internal alkene cyclooctene was much lower than that of the terminal alkene 1-octene. The hydrogenation of ethene was found to be very rapid, with an estimated TOF of 9000 h^{-1}; this is quite unusual, as ethene is not hydrogenated by the generally very active Wilkinson's catalyst [60].

The homogeneous nature of the catalysts is confirmed by the linear dependence of the catalytic activity on the concentration of nickel(II). Only in the case of a truly homogeneous catalyst is the activity expected to be directly proportional to the catalyst concentration. In the case of the formation of nanoclusters, larger agglomerates – and, therefore, a comparatively lower number of active sites – will be formed at higher concentrations of the nickel salt. Furthermore, a sigmoidal curve for the rate of consumption of substrate has been proposed

Table 5.3 Different substrates in the homogeneous hydrogenation catalyzed by *in-situ* mixtures of Ni(OAc)$_2$ and *o*-MeO-dppp. [a)]

Entry	Substrate	TOF [b)]	SCR	Main product
1	1-octene	1480	2000	*n*-octane
2	ethene [c)]	9000	2000	ethane
3	cyclooctene	30	400	cyclooctane
4	1,7-octadiene	30	1000	1-octene
5	styrene	0	200	
6	1-octyne	0	250	

a) Reaction conditions: [Ni]=2.5 mM (before substrate is added), t=1 h, pH$_2$=50 bar, T=50 °C, in 20 mL dichloromethane/methanol (1:1, v/v).
b) Turnover is mol converted substrate per mol nickel in 1 h.
c) T=40 °C and t=10 min.
All data taken from [52].

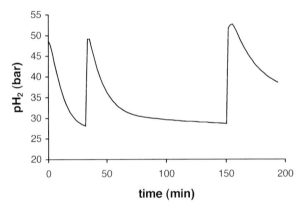

Fig. 5.4 Dihydrogen pressure drop of the hydrogenation of 1-octene using an *in-situ*-formed catalyst containing Ni(OAc)$_2$ and the ligand *o*-MeO-dppe. After 30 and 150 min, fresh 1-octene is added to the reaction mixture and the H$_2$ pressure is reset to 50 bar. (Reproduced from [57])

as being characteristic for the nanocluster formation reaction [5]. For the nickel-(II) catalysts with didentate phosphane ligands the hydrogenation reaction starts immediately without any induction period (Fig. 5.4) [57]. In addition, it is possible to regenerate the catalysts from the reaction mixtures after the hydrogenation experiments. These results – in combination with the observation that pure nickel(II) acetate does not lead to an active catalyst under the applied reaction conditions (50 °C) and the excellent reproducibility of the results – lead to conclusion that the reported nickel(II) complexes of didentate phosphane ligands yield truly homogeneous hydrogenation catalysts.

5.4
Concluding Remarks

Many Ziegler-type catalytic systems, in addition to a few mononuclear nickel complexes of nitrogen- or oxygen-containing ligands, have been reported as being homogeneous hydrogenation catalysts for a number of substrates. In most of these cases, it is doubtful whether the reported nickel complexes really yield homogeneous hydrogenation catalysts; it is more likely that catalytic hydrogenation activity is due to formation of colloidal (heterogeneous) nickel particles. Despite the recent report of a nickel(IV) hydride species [20], the assumption that dihydrogen activation in these catalytic systems would proceed via oxidative addition remains dubious.

Nickel(II) complexes with the correct choice of a didentate phosphane ligand do yield very active homogeneous hydrogenation catalysts. It has been found that nickel complexes with chelating diphosphane ligands may be involved in three different processes, namely ligand-exchange reactions, ligand oxidation, and the actual hydrogenation reaction (see Scheme 5.4).

The larger bite angles of the C_3-bridged ligands preclude the ligand-exchange equilibrium, the ionic radius of the nickel(II) ion being too small to accommo-

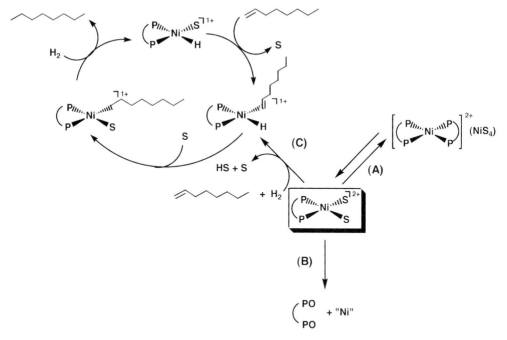

Scheme 5.4 Three processes in which nickel diphosphane complexes can be involved under catalytic hydrogenation conditions. (A) ligand exchange; (B) oxidation; and (C) hydrogenation of 1-octene. For simplicity, the reaction steps are not depicted as equilibria and the isomerization reaction has been omitted. In this scheme, S depicts a solvent molecule, but it might also be an anion.

date two of these didentate ligands in a square-planar geometry. Considering the C$_2$-bridged ligands, only dcpe (see Scheme 5.2) has a ligand cone that is large enough to prevent coordination of a second didentate ligand. It is assumed that the nickel(II) ion catalyzes the oxidation of tertiary phosphane ligands. This oxidation appeared to be very rapid in the case of the unsubstituted ligands dppe and dppp, whereas it is relatively slow for *o*-MeO-dppe and *o*-MeO-dppp. Most probably, the electronic properties of the ligand influence the oxidation sensitivity of the ligand in the presence of nickel.

If the ligand-exchange reaction, as well as oxidation of the ligand, could be prevented, then the formation of highly active hydrogenation catalysts with simple diphosphane ligands might be possible. Consequently, the bulk application of inexpensive, nickel-containing homogeneous hydrogenation catalysts might come within reach.

Abbreviations

SCR substrate:catalyst ratio
TOF turnover frequency
TON turnover number

References

1 W. Keim, *Angew. Chem., Int. Ed. Engl.* **1990**, *29*, 235.
2 R. Cammack, P. van Vliet. In: *Bioinorganic Catalysis*, 2nd edn. J. Reedijk, E. Bouwman (Eds.), Marcel Dekker, Inc., New York, **1999**.
3 E. Bouwman, J. Reedijk, *Coord. Chem. Rev.* **2005**, *249*, 1555.
4 R. Cammack, M. Frey, R. Robson, *Hydrogen as a Fuel: Learning from Nature.* Taylor & Francis, London and New York, **2001**.
5 J.A. Widegren, R.G. Finke, *J. Mol. Catal.* **2003**, *198*, 317.
6 P. Stavropoulos, M.C. Muetterties, M. Carrié, R.H. Holm, *J. Am. Chem. Soc.* **1991**, *113*, 8485.
7 L. Sacconi, A. Orlandini, S. Midollini, *Inorg. Chem.* **1974**, *13*, 2850.
8 K. Jonas, G. Wilke, *Angew. Chem., Int. Ed. Engl.* **1969**, *8*, 519.
9 A.L. Seligson, R.L. Cowan, W.C. Trogler, *Inorg. Chem.* **1991**, *30*, 3371.
10 M. Aresta, E. Quaranta, A. Dibenedetto, P. Giannoccaro, I. Tommasi, M. Lanfran-

chi, A. Tiripicchio, *Organometallics* **1997**, *16*, 834.
11 T. Saito, M. Nakajima, A. Kobayashi, Y. Sasaki, *J. Chem. Soc., Dalton Trans.* **1978**, 482.
12 K. Jonas, G. Wilke, *Angew. Chem.* **1970**, *82*, 295.
13 B.L. Barnett, C. Kruger, Y.H. Tsay, R.H. Summerville, R. Hoffmann, *Chem. Ber.* **1977**, *110*, 3900.
14 I. Bach, R. Goddard, C. Kopiske, K. Seevogel, K.R. Pörschke, *Organometallics* **1999**, *18*, 10.
15 M.D. Fryzuk, G.K.B. Clentsmith, D.B. Leznoff, S.J. Rettig, S.J. Geib, *Inorg. Chim. Acta* **1997**, *265*, 169.
16 M. Jiménez Tenorio, M.C. Puerta, P. Valerga, *J. Chem. Soc., Dalton Trans.* **1996**, 1305.
17 D.A. Vicic, T.J. Anderson, J.A. Cowan, A.J. Schultz, *J. Am. Chem. Soc.* **2004**, *126*, 8132.
18 C.E. Kriley, C.J. Woolley, M.K. Krepps, E.M. Popa, P.E. Fanwick, I.P. Rothwell, *Inorg. Chim. Acta* **2000**, *300*, 200.

19 N. D. Clement, K. J. Cavell, C. Jones, C. J. Elsevier, *Angew. Chem. Int. Ed.* **2004**, *43*, 1277.

20 W. Z. Chen, S. Shimada, M. Tanaka, Y. Kobayashi, K. Saigo, *J. Am. Chem. Soc.* **2004**, *126*, 8072.

21 P. W. Jolly. In: *Comprehensive Organometallic Chemistry*, Vol. 8. G. Wilkinson, F. G. A. Stone, E. W. Abel (Eds.), Pergamon Press, Oxford, **1982**, Section 56.2.

22 K. R. Pörschke, W. Kleimann, G. Wilke, K. H. Claus, C. Krüger, *Angew. Chem., Int. Ed. Engl.* **1983**, *22*, 991.

23 F. M. Conroy-Lewis, L. Mole, A. D. Redhouse, S. A. Litster, J. L. Spencer, *J. Chem. Soc., Chem. Commun.* **1991**, 1601.

24 E. Kogut, A. Zeller, T. H. Warren, T. Strassner, *J. Am. Chem. Soc.* **2004**, *126*, 11984.

25 H. L. Wiencko, E. Kogut, T. H. Warren, *Inorg. Chim. Acta* **2003**, *345*, 199.

26 F. A. Cotton, B. A. Frenz, D. L. Hunter, *J. Am. Chem. Soc.* **1974**, *96*, 4820.

27 M. Stollenz, M. Rudolph, H. Görls, D. Walther, *J. Organometal. Chem.* **2003**, *687*, 153.

28 U. Müller, W. Keim, C. Krüger, P. Betz, *Angew. Chem., Int. Ed. Engl.* **1989**, *28*, 1011.

29 R. A. Henderson, *J. Chem. Research (S)* **2002**, 407.

30 C. J. Curtis, A. Miedaner, R. Ciancanelli, W. W. Ellis, B. C. Noll, M. R. DuBois, D. L. DuBois, *Inorg. Chem.* **2003**, *42*, 216.

31 D. Durand, G. Hillion, C. Lassau, L. Sajus, US Patent 4271323, **1981**.

32 B. Cornils, W. A. Herrmann, *Applied Homogeneous Catalysis with Organometallic Compounds*. VCH, Weinheim, **1996**.

33 A. G. Hinze, D. J. Frost, *J. Catal.* **1972**, *24*, 541.

34 P. Abley, F. J. McQuillin, *J. Catal.* **1972**, *24*, 536.

35 M. Sakai, N. Hirano, F. Harada, Y. Sakakibara, N. Uchino, *Bull. Chem. Soc. Jpn.* **1987**, *60*, 2923.

36 S. N. Gan, N. Subramaniam, R. Yahya, *J. Appl. Polym. Sci.* **1996**, *59*, 63.

37 V. A. Escobar Barrios, R. Herrera Nájera, A. Petit, F. Pla, *Eur. Polym. J.* **2000**, *36*, 1817.

38 S. Sabata, J. Hetflejs, *J. Appl. Polym. Sci.* **2002**, *85*, 1185.

39 T. Thangaraj, S. Vancheesan, J. Rajaram, J. C. Kuriacose, *Indian J. Chem. Sect A-Inorg. Phys. Theor. Anal. Chem.* **1980**, *19*, 404.

40 P. C. L'Argentière, E. A. Cagnola, M. G. Canon, D. A. Liprandi, D. V. Marconetti, *J. Chem. Technol. Biotechnol.* **1998**, *71*, 285.

41 T. E. Zhesko, N. S. Barinov, A. G. Nikitina, I. Y. Kvitko, L. V. Alam, N. P. Smirnova, *Zhurnal Obshchei Khimii* **1980**, *50*, 2301.

42 D. Chatterjee, H. C. Bajaj, S. B. Halligudi, K. N. Bhatt, *J. Mol. Catal.* **1993**, *84*, L1.

43 A. Bacchi, M. Carcelli, L. Gabba, S. Ianelli, P. Pelagatti, G. Pelizzi, D. Rogolino, *Inorg. Chim. Acta* **2003**, *342*, 229.

44 D. Chatterjee, H. C. Bajaj, A. Das, K. Bhatt, *J. Mol. Catal.* **1994**, *92*, L235.

45 E. A. Emken, E. N. Frankel, R. O. Butterfield, *J. Am. Oil Chemists Soc.* **1966**, *43*, 14.

46 H. Itatani, J. C. Bailar, *J. Am. Chem. Soc.* **1967**, *89*, 1600.

47 P. Abley, F. J. McQuillin, *Discuss. Faraday Soc.* **1968**, 31.

48 E. N. Frankel, H. Itatani, J. C. Bailar, *J. Am. Oil Chemists Soc.* **1972**, *49*, 132.

49 H. Itatani, J. C. Bailar, *Ind. Eng. Chem. Prod. Res. Develop.* **1972**, *11*, 146.

50 T. E. Zhesko, Y. N. Kukushkin, A. G. Nikitina, V. P. Kotelnikov, N. S. Barinov, *Zhurnal Obshchei Khimii* **1979**, *49*, 2254.

51 I. M. Angulo, A. M. Kluwer, E. Bouwman, *Chem. Commun.* **1998**, 2689.

52 I. M. Angulo, E. Bouwman, *J. Mol. Catal.* **2001**, *175*, 65.

53 I. M. Angulo, E. Bouwman, S. M. Lok, M. Lutz, W. P. Mul, A. L. Spek, *Eur. J. Inorg. Chem.* **2001**, 1465.

54 I. M. Angulo, E. Bouwman, M. Lutz, W. P. Mul, A. L. Spek, *Inorg. Chem.* **2001**, *40*, 2073.

55 Angulo, S. M. Lok, V. F. Q. Norambuena, M. Lutz, A. L. Spek, E. Bouwman, *J. Mol. Catal.* **2002**, *187*, 55.

56 I. M. Angulo, E. Bouwman, R. van Gorkum, S. M. Lok, M. Lutz, A. L. Spek, *J. Mol. Catal.* **2003**, *202*, 97.

57 I. M. Angulo, PhD Thesis, Leiden University, Leiden, The Netherlands, **2001**.

58 P. S. Jarrett, P. J. Sadler, *Inorg. Chem.* **1991**, *30*, 2098.

59 C. Amatore, E. Carré, A. Jutland, M. A. M'Barki, *Organometallics* **1995**, *14*, 1818.

60 J. A. Osborn, F. H. Jardine, J. F. Young, G. Wilkinson, *J. Chem. Soc. A* **1966**, 1711.

6

Hydrogenation with Early Transition Metal, Lanthanide and Actinide Complexes

Christophe Copéret

6.1
Introduction

Homogeneous hydrogenation catalysts have been mainly based on late transition-metal complexes since the discovery of Wilkinson's catalyst, [(Ph$_3$P)$_3$RhCl] [1]. Nonetheless, some of the first homogeneous catalysts to emerge were based on early transition metals. In fact, soon after the discovery of metallocene complexes [2, 3], which were rapidly exploited as the soluble equivalent of Ziegler-Natta olefin polymerization catalysts [4], these systems were introduced as homogeneous hydrogenation catalysts [5–11]. Other Ziegler-Natta-type olefin polymerization catalysts, [L$_n$MX$_n$/M′R] {M = Ti, Zr, V, Cr, Mo, Mn, Fe, Co, Ni, Pd, Ru; X = Cl, OR; M′ = Al or Li, R = H, Et, Bu or *i*Bu} were also investigated, and the results obtained showed that all metals could indeed activate H$_2$ and hydrogenate olefins. This pioneering chemistry most likely formed the basis for the discovery of most homogeneous catalyzed processes known to date, and has led to the development of well-defined systems and to the success of molecular organometallic chemistry and homogeneous catalysis [12, 13].

In hydrogenation, early transition-metal catalysts are mainly based on metallocene complexes, and particularly the Group IV metallocenes. Nonetheless, Group III, lanthanide and even actinide complexes as well as later metals (Groups V–VII) have also been used. The active species can be stabilized by other bulky ligands such as those derived from 2,6-disubstituted phenols (aryloxy) or silica (siloxy) (*vide infra*). Moreover, the catalytic activity of these systems is not limited to the hydrogenation of alkenes, but can be used for the hydrogenation of aromatics, alkynes and imines. These systems have also been developed very successfully into their enantioselective versions.

This chapter will provide an overview of the development and use of early transition-metal complexes in hydrogenation, and in consequence has been divided into several sections. Section 6.2 will focus on the mechanistic differences in the hydrogenation reaction between early and late transition metals. The following three sections will describe the various systems based on Group IV (Section 6.3), earlier (Section 6.4: Group III, lanthanides and actinides) and later metals (Section 6.5: Groups V–VII). In each section, the hydrogenation of alkenes, dienes, alkynes, aromatics and imines will be described, and the development of enantioselective hydrogenation catalysts will be discussed. Section 6.6 will be devoted to their heterogeneous equivalents. Finally, Section 6.7 will provide a brief conclusion of the current state of the art in this area, its scope, and its limitations.

6.2
Mechanistic Considerations

For late transition metals, one of the key elementary steps is the oxidative addition of H_2 (see Chapter 1) [12, 13], and therefore the process requires transition-metal complexes which can readily undergo oxido-reduction processes (low oxidation state, d^n configuration). In contrast, most early transition-metal hydrogenation catalysts are based on d^0 metal complexes, having ligands, which often excludes the possibility to change the oxidation of the metal center. Therefore, the activation of H_2 and formation of the alkane must proceed through elementary steps other than oxidative addition and reductive elimination. Additionally, because these metal complexes have a d^0 configuration, they cannot generate stable π-alkene complexes. Three types of catalyst precursors can be used: L_nM-R, L_nM-H, or L_nM-X, the latter being used in combination with an alkylated agent or a metal hydride to generate the L_nM-R or the L_nM-H catalytically active species (Scheme 6.1).

Starting from L_nM-H, the first step is insertion of the alkene in the M–H bond to generate the corresponding alkyl complex [12]. Note that successive insertion and β-H transfer steps allow the starting alkene to be isomerized, and it can be

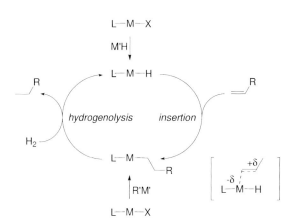

Scheme 6.1 Elementary steps for the hydrogenation of olefins with d^0 transition-metal complexes.

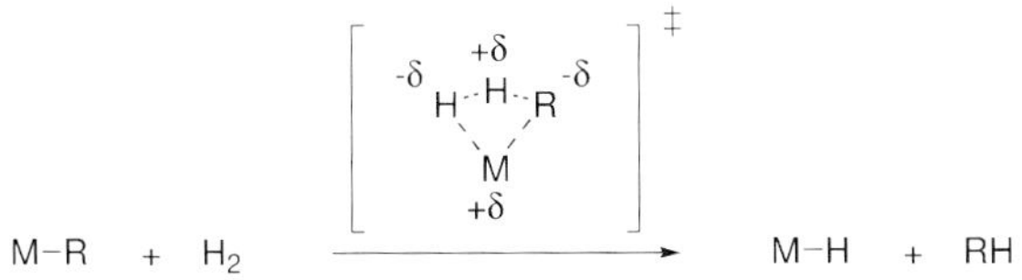

$$\left[\begin{array}{c} {}^{-\delta}\text{H}\cdots\overset{+\delta}{\text{H}}\cdots\text{R}^{-\delta} \\ \diagdown \; \diagup \\ \text{M} \\ {}^{+\delta} \end{array} \right]^{\ddagger}$$

$$\text{M–R} \quad + \quad \text{H}_2 \qquad \xrightarrow{\hspace{3cm}} \qquad \text{M–H} \quad + \quad \text{RH}$$

Scheme 6.2 σ-Bond metathesis transition state.

observed when hydrogenation is carried out under H_2-limiting conditions (low pressure, diffusion control experiments, poor stirring, etc.). The insertion step is probably preceded by a polarization of the alkene by the metal complex. While such types of intermediates have not been observed on d^0 L_nM-H complexes, they have been found on the corresponding d^0 L_nM-X complexes (X=Cl, OR, CH_2R) [14–28]. The following step is the direct hydrogenolysis of the metal–carbon bond by H_2 through a four-centered σ-bond metathesis transition state [29–34] which does not require a change of oxidation state, thereby regenerating the M–H species. While this step corresponds to an exchange of four atoms between two molecules, the geometry of this transition state does not have a kite-shape geometry, but rather a triangular shape, where three atoms are almost co-linear. Overall, it corresponds to a transfer of a proton between the H_2 molecule coordinated to the metal center and an alkyl ligand (Scheme 6.2). The reverse step, which is usually slightly endothermic for early transition metals, corresponds to a C–H activation on a metal–hydride. Successive C–H bond activation and hydrogenolysis will be responsible for H-scrambling, which can be possible under H_2-deprived conditions and can be observed by using D_2 in place of H_2.

Finally, the hydrogenation of other substrates such as imines involve similar elementary steps.

6.3
Group IV Metal Hydrogenation Catalysts

6.3.1
Hydrogenation of Alkenes

In Group IV metal complexes, metallocene complexes are the main catalyst precursors for hydrogenation. Two major catalytic systems have been used: 1) Cp_2MR_2 (R=H, Alkyl, Aryl); and 2) Cp_2MX_2 in combination with alkylating agent or an hydride (Table 6.1). The catalytic tests are typically run with 50 equiv. of substrate per metal, but in some cases turnover numbers (TONs) exceeding 1000 can be achieved [35].

The hydrogenation is usually limited to nonpolar alkenes (terminal and internal cyclic and acyclic alkenes), even though Ti systems have been used to hydrogenate alkenes containing ether and ester functionalities such as vinyl ethers or methyl oleate [42, 45, 59, 62].

Table 6.1 Catalytic systems based on Group IV dicyclopentadienyl complexes.

Ti Catalytic systems	Reference(s)	Zr and Hf catalytic systems	Reference(s)
Ti(OiPr)$_4$/R$_3$Al	5	Cp$_2$ZrCl$_2$/BuLi (Cp$_2$Zr-olefin)	36, 37
Ti(OiPr)$_4$/RLi	5	Cp$_2$ZrCl$_2$/R$_3$Al	5
Cp$_2$TiCl$_2$/BuLi	6	(RC$_5$H$_4$)$_2$ZrH$_2$	7, 38, 39
		R = H, Me, iPr, Bn	
AnsaCp$_2$TiCl$_2$/EtLi	40	(RC$_5$H$_4$)$_2$HfH$_2$	38, 39
		R = Me, iPr, Bn	
Cp$_2$TiCl$_2$/PhMgBr	6	[Cp*(carboranyl)HfH]$_2$	25
Cp$_2$TiCl$_2$/RMgBr	6, 41, 42		
Cp$_2$TiCl$_2$/R$_2$Mg	43		
Cp$_2$Ti(CO)$_2$	44		
Cp$_2$TiCl$_2$/Mg	45, 46		
Cp$_2$Ti(CO)(Ph$_2$C$_2$)	10, 11		
CpCp'Ti(PMe$_3$)(Ph$_2$C$_2$)	47		
Cp' = Cp*, MeC$_5$H$_4$			
Cp$_2$Ti(Ph$_2$C$_2$)	48, 49		
Cp$_2$Ti(AlH$_4$)$_2$	50–52		
Cp$_2$TiCl$_2$/NaH	35, 53–57		
Cp$_2$TiPh$_2$	9, 58		
Cp$_2$TiR$_2$ + hv	59		
R = Me, CH$_2$Ph, Ph			
Cp$_2$TiMe$_2$/RSiH$_3$	60, 61		

In the Cp$_2$Ti system, Brintzinger et al. have shown that the angle between
the two cyclopentadienyl ligands is important. For example, the activity in the
hydrogenation of cyclohexene varies as follows: [(CH$_2$)$_2$(C$_5$H$_4$)$_2$TiCl$_2$] (turnover
frequency (TOF) = 266 000 mol H$_2$ cons. mol^{-1} h^{-1}) > [(CH$_2$)$_3$(C$_5$H$_4$)$_2$TiCl$_2$] (TOF =
223 000 mol H$_2$ cons. mol^{-1} h^{-1}) > [(CH$_2$)(C$_5$H$_4$)$_2$TiCl$_2$] (TOF = 118 000 mol H$_2$
cons. mol^{-1} h^{-1}) > [(C$_5$H$_5$)$_2$TiCl$_2$] (TOF = 90 000 mol H$_2$ cons. mol^{-1} h^{-1}) [40].

This field is still active today: numerous recent patents [46, 63–89] and several
reports [90–92] have detailed the efficiency of such types of systems for the hydrogenation of unsaturated polymers resulting from the polymerization of
dienes (butadiene, isoprene, 1,3-cyclohexadiene) or the co-polymerization of
dienes and styrenes (block co-polymers of butadiene and styrene: SB, SBS,
SBSB polymers).

6.3.2
Hydrogenation of Alkynes and Dienes

Alkynes are hydrogenated to *cis* olefins with the same catalytic systems, and
subsequently undergo hydrogenation to yield the corresponding alkanes [7, 42,
45, 47, 49, 59, 93]. For example, Jordan et al. reported the selective hydrogenation of 3-hexyne into *cis* 3-hexene with a TOF of 25 h^{-1} [25], and *cis* 3-hexene is

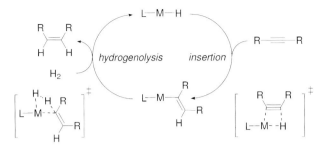

Scheme 6.3

Scheme 6.4

subsequently hydrogenated at 12 h^{-1} (Scheme 6.3). Typically, the hydrogenation of alkynes is slower than that of alkenes (hydrogenolysis is probably rate determining in this case), but because alkynes react more rapidly with the metal hydride intermediate than alkenes do, they can be hydrogenated selectively, especially at low conversions (*vide infra* for a thorough study of reaction rates with lanthanide and actinide complexes).

Similarly, in some cases, dienes can be selectively hydrogenated into the corresponding alkenes, but they usually provide the corresponding alkane or a mixture of alkanes and alkenes [6, 45, 49, 59]. For example, the hydrogenation of 50 equiv. of 1,3- or 1,4-cyclohexadiene in the presence of [Cp$_2$TiCl$_2$]-*i*PrMgBr

(1:6 ratio) gives selectively cyclohexene (>98%; Scheme 6.4) [42]. On the other hand, using the same experimental procedure, cyclopentadiene is converted (95% conv.) into a mixture of cyclopentene (85%) and cyclopentane (15%), and 1,3- or 1,5-cyclooctadienes give only cyclooctane. Additionally, linear nonconjugated dienes such as 1,4-pentadiene and 1,5-hexadiene are directly converted into the corresponding alkanes.

6.3.3
Enantioselective Hydrogenation of Olefins

Kagan et al. were the first to report the corresponding enantioselective catalytic hydrogenation using chiral metallocene derivatives [94, 95]. By using menthyl- and neomenthyl-substituted cyclopentadienyl titanium derivatives in the presence of activators (Scheme 6.5) [96], these authors observed low ee-values (7–14.9%) for the catalytic hydrogenation of 2-phenyl-1-butene into 2-phenylbutane. In contrast, no enantiomeric excess was obtained with the corresponding zirconocene derivatives.

The use of other simple chiral cyclopentadienyl systems proved to be unsuccessful [39]. The design of better catalysts was directed first at preparing bulkier chiral cyclopentadienyl ligands. The chirality was introduced by preparing substituted cyclopentadienes derived from pinene and camphor (Scheme 6.6), but the ee-values were still low (<35%) [97, 98]. The ee-values have been improved to 61–69% by introducing more steric bulk in the system (Scheme 6.6; R=H versus Me) [99]. The first high ee-values were obtained using a C_2 symmetric cyclopentadienyl ligand, for which both enantiomers are accessible in an enantiomerically pure form [100]. This system hydrogenates 2-phenyl-1-butene at −78 °C with enantiomeric excess up to 96%, albeit with a low TON (e.g., 10). It should be noted that *ansa* titanocene derivatives prepared by White et al. give low ee-values compared to the Group III and lanthanide derivatives developed by Marks et al. (*vide infra*, Section 6.4.1) [101].

The best results were obtained with the Brintzinger indenyl zirconene and titanocene derivatives [(EBTHI)MX$_2$], developed earlier for the stereocontrolled po-

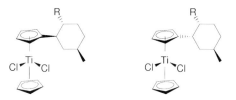

R = CMe$_2$Ph, CHMe$_2$ (menthyl)   R = CMe$_2$Ph, CHMe$_2$ (neomenthyl)

Scheme 6.5 First enantioselective hydrogenation by a Group IV metallocene catalysts.

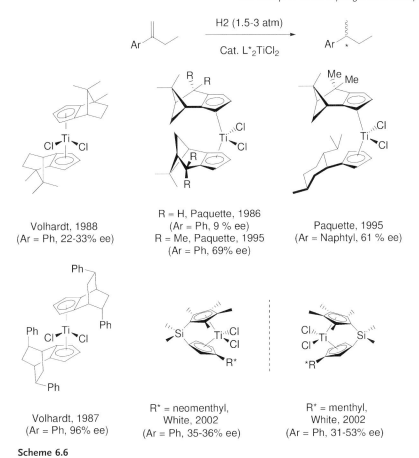

R = H, Paquette, 1986
(Ar = Ph, 9 % ee)
R = Me, Paquette, 1995
(Ar = Ph, 69% ee)

Volhardt, 1988
(Ar = Ph, 22-33% ee)

Paquette, 1995
(Ar = Naphtyl, 61 % ee)

Volhardt, 1987
(Ar = Ph, 96% ee)

R* = neomenthyl,
White, 2002
(Ar = Ph, 35-36% ee)

R* = menthyl,
White, 2002
(Ar = Ph, 31-53% ee)

Scheme 6.6

lymerization of propene (Scheme 6.7) [102, 103]. When 2-phenyl-1-butene is used as the standard test substrates, low ee% are still obtained, despite high activity [104, 105]. However, tri-substituted [106] and tetra-substituted [107] alkenes are hydrogenated with very high enantioselectivities. In the case of tri-substituted alkenes [106], the catalytic system is generated by the treatment of [(EBTHI)TiX$_2$] (5 mol.%) (X$_2$ = binolate) with BuLi in the presence of 2.5 equiv. PhSiH$_3$, which is used to stabilize the catalytic system. The active species is probably [(EBTHI)Ti^{III}H]. Acyclic and cyclic tri-substituted alkenes are hydrogenated in typically 70–90% yields, with enantiomeric excesses ranging from 83% to >99% (Table 6.2). The favored enantiomer is formed through the pathway in which the substituent R points as far away as possible from the indenyl ligand (Scheme 6.7).

When the hydrogenation of (E)-1,2-diphenylpropene is performed under D$_2$, 1,2-diphenylpropane is selectively deuterated in positions 1 and 2, which shows that no isomerization of the alken takes place under these conditions. Noteworthy, the reaction rate is highly dependent on the substrate, and typically

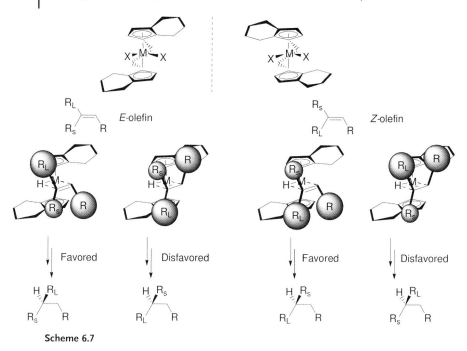

Scheme 6.7

Z-alkenes are reduced much more slowly than are *E*-isomers. For instance, the (*E*)-(1,2)-diphenylpropene is reduced in 9 h at 65 °C under 5.3 bar, while the *Z*-isomer reaches only 3% conversion after 48 h at 70 °C under 133 bar of H_2. These data are fully consistent with the model proposed to predict the stereochemical outcome of the reaction (see Scheme 6.7), which shows that, for the *Z*-isomers, the R_L substituent points towards the indenyl ligands in the favored pathway, thus decreasing the reaction rate.

For tetra-substituted alkenes [107], it was necessary to rely on the more reactive cationic Zr equivalent generated from [(EBTHI)ZrMe$_2$] and either methylaluminoxane or [PhMe$_2$NH]$^+$[Co(C$_2$B$_9$H$_{11}$)$_2$]$^-$ developed earlier by Waymouth et al. [104]. Using H_2 pressure ranging between 5 and 133 bar, it was possible to obtain the hydrogenated products with 80–98% ee in most cases (Table 6.3).

6.3.4
Enantioselective Hydrogenation of Imines and Enanimes

One of the best enantioselective olefin hydrogenation catalysts, (EBTHI)TiH, was originally developed for the enantioselective hydrogenation of imines [108, 109]. The catalytic system is generated by the treatment of [(EBTHI)TiX$_2$] (5 mol%, X=binolate) with BuLi in the presence of 2.5 equiv. PhSiH$_3$, and the hydrogenation is typically performed with 5 mol% of catalyst at 65 °C under 133 bar H_2. High pressure is usually required to obtain high ee-values with this

Table 6.2 Enantioselective hydrogenation of tri-substituted alkenes catalyzed by [(S,S,S)-(EBTHI)TiX$_2$]. [a]

Substrate	Product [b]	Time [h]	Yield [%]	ee [%]
(Me, diphenyl alkene structure)	(Me, diphenyl alkane structure)	48	91	99
(Me, Me; MeO-aryl structure)	(Me, Me; MeO-aryl structure)	48	79	95
(Me, Me; MeO-aryl structure)	(Me, Me; MeO-aryl structure)	146	80	31
(Me; MeO-naphthalenyl structure)	(Me; MeO-naphthalenyl structure)	44	77	92
(Me; MeO-naphthalenyl structure)	(Me; MeO-naphthalenyl structure)	132	70	93
(Me; naphthalenyl structure)	(Me; naphthalenyl structure)	184	70	83
(Ph; MeO-naphthalenyl structure)	(Ph; MeO-naphthalenyl structure)	169	87	83
(Me, NBn$_2$; phenyl structure)	(Me, NBn$_2$; phenyl structure)	43	75	95
(Me, OMe; MeO-aryl structure)	(Me, OMe; MeO-aryl structure)	48	80–86	93–94

a) Reaction conditions: 65 °C, 133 bar H$_2$ in THF with 5 mol% [(S,S,S)-(EBTHI)TiX$_2$].
b) No absolute configuration given when unknown.

system. Acyclic benzyl protected imines, used as mixture of diastereomers, are converted to their corresponding amines in good yields (66–93%) and moderate enantioselectivities (58–85% ee). Better results are obtained for cyclic imines, which are hydrogenated in 97–99% ee (Tables 6.4 and 6.5). The absolute configuration of the amine can be predicted using the following model and rules

Table 6.3 Enantioselective hydrogenation of tetra-substituted alkenes catalyzed by [(S,S)-(EBTHI)ZrMe$_2$]/[PhMe$_2$NH$^+$B(C$_6$F$_5$)$_4^-$].[a]

Substrate	Product	P$_{H2}$ [bar] (% cat)	Yield [%]	ee [%] (cis:trans)
F-phenyl C(Me)=C(Me)Me	*F-phenyl CH(Me)-CH(Me)Me*	5.3 (8) / 110 (8)	79 / 77	84 / 96
1-Me-indene, 2-Me	*1-Me-indane, 2-Me*	5.3 (5) / 110 (8)	76 / 87	86 (95:5) / 93 (>99:1)
1-Me-indene, 2-Bu	*1-Me-indane, 2-Bu*	5.3 (8)	96	92 (99:1)
1-Me-indene, 2-Ph	*1-Me-indane, 2-Ph*	5.3 (5) / 67 (8)	34 / 89	97 / 98 (>99:1)
1-Et-indene, 2-Me	*1-Et-indane, 2-Me*	5.3 (5) / 110 (8)	57 / 95	5 (9:1) / 52 (95:5)
1-Ph-indene, 2-Me	*1-Ph-indane, 2-Me*	5.3 (5) / 133 (8)	44 / 94	29 (>99:1) / 78 (>99:1)
1-Me-dihydronaphthalene, 2-Me	*1-Me-tetralin, 2-Me*	133 (5)	91	92 (>99:1)

a) Reaction conditions: 25 °C, 5.3–133 bar H$_2$ with 5–8 mol% [(S,S,S)-(EBTHI)ZrMe$_2$].

(Scheme 6.8): the substituent (R) on the nitrogen should point as far away as possible from the indenyl ligand. This model predicts that hydrogenation of *syn*- and *anti*-imines give rise to enantiomers of opposite stereochemistry as observed experimentally. The model also predicts that if one takes into consideration the influence of R$_S$ and R$_L$ substituents, the energy difference of the two possible pathways should be lower for the *syn* imines, giving rise to lower enantioselectivities.

As illustrated in the hydrogenation of cyclic imines, the system is compatible with a wide range of functional groups, such as olefins, protected or unpro-

Table 6.4 Enantioselective hydrogenation of imines catalyzed
by [(*R,R,R*)-(EBTHI)TiX$_2$] [108, 109]. [a]

Substrate	Product [k]	Yield [%] [a]	ee [%]
(Me, Bu, N, Ph) [b]	(Me, Bu, N, H, Ph)	68	56
(Me, N, Ph) [c]	(Me, N, H, Ph)	64	62
(Me, N, Ph) [d]	(Me, N, H, Ph)	66	75
(Me, cyclohexyl, N, Ph) [e]	(Me, cyclohexyl, N, H, Ph)	93	76
(Me, cyclohexyl, N) [f]	(Me, cyclohexyl, N, H)	70	79
(phenyl, N, Ph) [g]	(phenyl, N, H, Ph)	93	85
(MeO, N, Ph) [h]	(MeO, N, H, Ph)	86	86
(furan, N, Ph) [i]	(furan, N, H, Ph)	70	83
(naphthyl, N, Ph) [j]	(naphthyl, N, H, Ph)	82	70

a) Reaction conditions: 65 C, 133 bar H$_2$ in THF with 5 mol.% [(*R,R,R*)-(EBTHI)TiX$_2$].
b) *anti/syn* = 3.3.
c) *anti/syn* = 3.
d) *anti/syn* = 13.
e) *anti/syn* = 9.
f) *anti/syn* = 17.
g) *anti/syn* = 17.
h) *anti/syn* = 17.
i) *anti/syn* = 10.
j) *anti/syn* = 3.3.
k) No absolute configuration given when unknown.

Table 6.5 Enantioselective hydrogenation of cyclic imines catalyzed by [(R,R,R)-(EBTHI)TiX$_2$].

Substrate	Product[b]	Conditions [P/T/t][a]	Yield [%]	ee [%]
		5.3/65/42	83	99
		33/65/24	70	97
		5.3/65/30	74	97
		5.3/65/50	79–82	98
		5.3/65/6	72	99
		5.3/50/23	79	99
		5.3/50/27	73	99
		5.3/45/23	72	99
		5.3/65/16	82	99
		5.3/65/8	84	99
		5.3/65/10	72	99

a) P=pressure (bar); T=temperature (C); t=time (h).
b) No absolute configuration given when unknown.

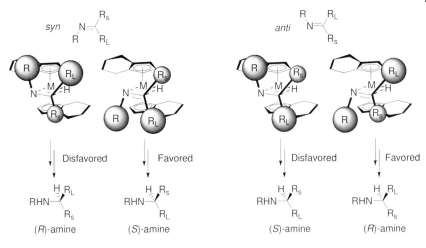

Scheme 6.8

Table 6.6 Kinetic resolution of di-substituted 1-pyrrolines catalyzed by [(R,R,R)-(EBTHI)TiX₂] [111].[a]

Substrate	Yield [% recovered S]	ee [%]	Yield [% product]	ee [%]
Me—⟨pyrroline⟩—Ph	37	99	34	99
Me—⟨pyrroline⟩—(N-Ph—CH₂ pyrrole)	42	96	44	98
iPr₃SiO—⟨pyrroline⟩—Ph	43	98	41	98
Me—⟨pyrroline⟩—C₁₁H₂₃	41	>95	41	>95
⟨pyrroline⟩—Ph, Ph	–	75	42	>95
Ph—⟨pyrroline⟩—Ph	33	49	44	99

a) 5 mol% cat., THF, 65–75 °C, 5.3 bar H₂, reaction run to 50% completion.

Table 6.7 Enantioselective hydrogenation of enamines catalyzed by [(R,R,R)-(EBTHI)TiX$_2$] [112]. [a]

Substrate	Product [b]	Pressure [bar]	Yield [%] [a]	ee [%]
		1	75	92
		1	72	92
		1	89	89
		1	77	96
		5.3	87	98
		5.3	83	96
		5.3	88	91

Table 6.7 (contributed)

Substrate	Product	Pressure [bar]	Yield [%] [a]	ee [%]
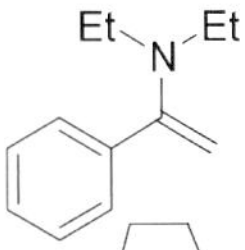	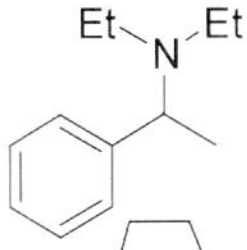	5.3	72	99
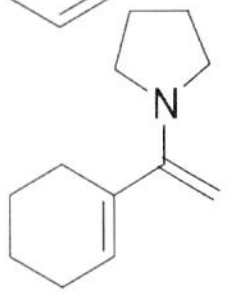	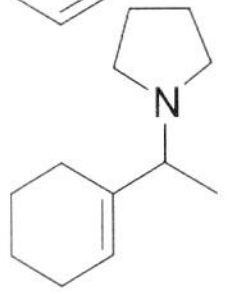	5.3	72	95

a)   Using 5 mol% of catalyst.
b)   No absolute configuration given when unknown.

tected alcohols, acetals and aromatic groups [109, 110]. The same model can be used to predict the absolute stereochemistry of the amine product – that is, formation of the *S*-amine from the (*R,R,R*)-catalyst (Scheme 6.8).

This catalytic system can be used for the kinetic resolution of di-substituted 1-pyrrolines, for which high ee-values are achieved for both the amine and the recovered materials, especially when they are substituted in positions 2 and 5 (Table 6.6) [111]. Moreover, it should be noted that acyclic enamines are converted with high ee-values into their corresponding amines (89–98% ee; Table 6.7), which is in sharp contrast to what is obtained for acyclic imines (*vide supra*) [112].

Hydrogenation is not limited to the use of (EBTHI)MX$_2$-type catalysts. In polymerization, linked amido-cyclopentadienyl ligands have emerged as very important systems, and the corresponding chiral derivatives have been prepared (Scheme 6.9) [113–116]. Nonetheless, whilst high TON can be achieved (500–1000), the ee-values are quite low (<25%).

Finally, Brintzinger et al. have reported a different type of *ansa*-metallocene, which can be readily prepared enantiomerically pure, and which catalyzes the

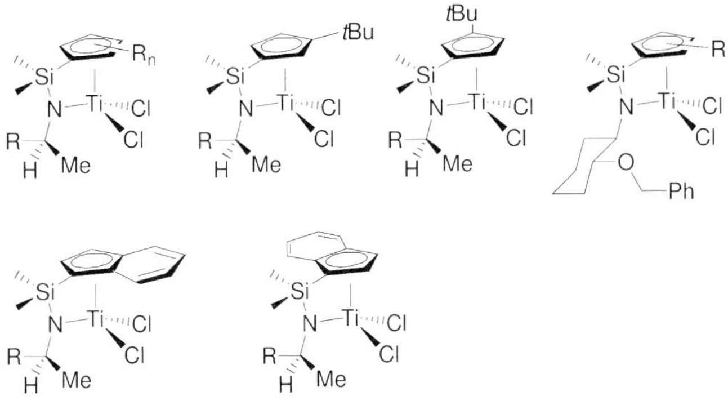

Scheme 6.9

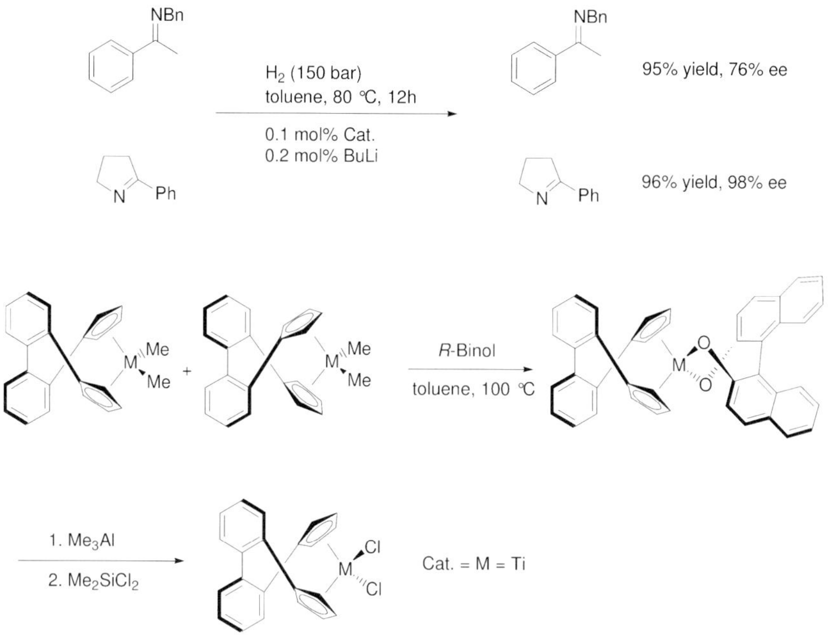

Scheme 6.10

hydrogenation of imines very efficiently (TON of 1000) with enantiomeric excesses comparable to those reported with (EBTHI)MX$_2$ [117, 118]. The synthesis of the enantiomerically pure catalyst is noteworthy because it relies on the preparation of a racemate. After reaction of this racemate with pure (*R*)-binol, two diastereomers are formed, but after heating at 100 °C a single diastereomer is obtained. This single diastereomer is converted into the enantiomerically pure dichloro complex by treatment of the binolate complex with Me$_3$Al and then Me$_3$SiCl (Scheme 6.10).

6.4
Hydrogenation Catalysts Based on Group III, Lanthanide, and Actinide Complexes

6.4.1
Hydrogenation of Alkenes with Group III Metal and Lanthanide Complexes

The hydrogenation of unfunctionalized alkenes is readily performed by Group III and lanthanide cyclopentadienyl hydride derivatives, one key feature being the high TOFs of these systems (up to 120 000 h^{-1} for hydrogenations catalyzed by Lu, Tables 6.8 and 6.9) [119, 120]. The reaction rate depends heavily on the metal and the ligands. It is inversely proportional to the metal radius (Lu > Sm > Nd > La), and it is faster for the Cp$_2^*$M derivatives than for the *ansa* di-

Table 6.8 Rate constants for the catalytic hydrogenation of
alkenes catalyzed by Group III and lanthanide complexes.[a]

	$[Cp_2^*LaH]_2$	$[Cp_2^*NdH]_2$	$[Cp_2^*SmH]_2$	$[Cp_2^*LuH]_2$
1-Hexene	6.2	21.6	23.8	34.5
(E) 2-Hexene	1.7	7.1	7.9	9.0
(E) 3-Hexene	7.3	33.4	51	240
(Z) 2-Hexene	–	6.6 [b]	–	13.4 [c]
(Z) 3-Hexene	–	2.2 [b]	3.8 [b]	8.1 [c]
Cyclohexene	0.015 [b]	0.014 [b]	0.005 [b]	0.023 [c]

a) Based on $r=k[M][H_2]$ unless otherwise specified, in units of
   $atm^{-1} s^{-1}$ (multiply by 224 to convert to $M^{-1} s^{-1}$).
b) $r=k[M]^{1/2}[alkene]$, in units of $M^{-1/2} s^{-1}$.
c) $r=k[M][alkene]$, in units of $M^{-1} s^{-1}$.

Table 6.9 Hydrogenation of alkenes catalyzed by Group III
and lanthanide complexes.

	$[Me_2SiCp_2''NdH]_n$	$[Me_2SiCp_2''SmH]_n$	$[Me_2SiCp_2''LuH]_n$
1-Hexene	21.6 [a]	23.8 [a]	34.5 [a]
Cyclohexene	0.014 [b]	0.005 [b]	0.023 [b]

a) Based on $r=k[M][H_2]$ unless otherwise specified, in units of
   $atm^{-1} s^{-1}$ (multiply by 224 to convert to $M^{-1} s^{-1}$).
b) $r=k[M]^{1/2}[alkene]$, in units of $M^{-1/2} s^{-1}$.

$Me_2SiCp_2''M-X = X-M$

cyclopentadienyl derivatives, $Me_2SiCp_2''MR$ (Tables 6.8 and 6.9). Kinetic studies
show that the rate-determining step depends on the alkene. In terms of elementary steps, these hydrides are dimers, and need to dissociate (Eq. (1)) prior to insertion of the alkenes (Eq. (2)). Hydrogenolysis subsequently liberates the alkane and re-forms the hydride (Eq. (3)).

$$(LMH)_2 \underset{k_{-1}}{\overset{k_1}{\rightleftarrows}} 2\ LMH \tag{1}$$

$$LM\text{-}H + alkene \underset{k_{-2}}{\overset{k_2}{\rightleftarrows}} LM\text{-}R \tag{2}$$

$$LM\text{-}R + H_2 \overset{k_3}{\longrightarrow} LM\text{-}H + RH \tag{3}$$

For 1-hexene, the rate can be expressed as: $r=k[L_nM][H_2]$, which is consistent
with hydrogenolysis being rate-determining (Eq. (3)). In contrast, for bulkier alkenes the rate law is as follows: $k[L_nM]^{1/2}[olefin]$, which is consistent with inser-

tion being rate-determining (Eq. (2); cyclohexene and internal *cis* acyclic alkenes). In the case of Lu, the dimer readily dissociates, and therefore the rate law is independent of the mechanism and the substrate, as observed experimentally $\{k[L_nM][alkene]\}$. Moreover, independently of the catalyst, hydrogenation of 1-hexene under D_2 generates 1,2-d_2-hexane, along with small amount of d_1-2-hexene, suggesting that β-H elimination can be competitive (the reverse step of insertion, Eq. (2)). In the case of cyclohexene however, Lu-based catalysts generate large quantities of various polydeuterated cyclohexane, suggesting that successive β-H elimination–insertion processes occur. Finally, when operating under deprived D_2 conditions, the amount of polydeuterated compounds increases, suggesting that the β-H elimination process becomes competitive with hydrogenolysis.

It should be noted that, in the case of $[(R)(Me)SiCp_2''YR]$, the TOF of the hydrogenation of 1-hexene decreases dramatically, from $11\,100\ h^{-1}$ ($R=Me$) to $200\ h^{-1}$ when an ether functional group is bound on the *ansa* bridge ($R=(CH_2)_5OMe$), which shows the sensitivity of Group III metal complexes towards polar functionalities [121].

These types of catalysts, $[Cp_2^*LnCH(SiMe_3)_2]$, are also used to hydrogenate substituted methylenecyclopentenes and cyclohexenes in good to very good diastereoselectivities, especially when the substituent is in the α-position to the alkene (Tables 6.10 and 6.11). However, the presence of functional groups such as amine or ether is detrimental to catalysis.

Table 6.10 Hydrogenation of substituted methylenecyclopentene catalyzed by $Cp_2^*LnCH(SiMe_3)_2$ [122].

Substrate	Catalyst	SCR (temp., °C)	Product	Yield [%] (dr, %)
iBu	Ln = Sm	10 (25)	iBu	84 (>99/1)
CH₂Ar Ar=2-Cl-C₆H₄	Ln = Sm	10 (25)	CH₂Ar Ar=2-Cl-C₆H₄	32 (>99/1)
nBu	Ln = Sm	33 (0)	nBu	32 (60/40)

dr = diastereomeric ratio; SCR = substrate : catalyst ratio.

Table 6.11 Hydrogenation of mono-substituted methylenecyclohexene catalyzed by $Cp_2^*LnCH(SiMe_3)_2$ [122].

Substrate	Catalyst	SCR (temp., °C)	Product	Yield [%] (dr, %)
R=Me	Ln=Sm	33 (−20)	R=Me	77 (93/7)
R=iBu	Ln=Sm	33 (−20)	R=iBu	90 (95/5)
R=tBu	Ln=Sm	20 (25)	R=tBu	95 (>99/1)
R=CH₂Ph	Ln=Sm	33 (−20)	R=CH₂Ph	95 (93/7)
R=(CH₂)₃NMe₂	Ln=Sm	33 (50)	(CH₂)₃NMe₂	76 (91/9)
R=OMe	Ln=Sm	20 (70)	R=OMe	0 (−)
	Ln=Sm	20 (70)		0 (−)
	Ln=Yb	33 (−20)		73 (61/39)
	Ln=Yb	33 (−20)		79 (73/27)
	Ln=Yb	33 (−20)		73 (77/23)

dr=diastereomeric ratio; SCR=substrate:catalyst ratio.

6.4.2
Hydrogenation of Dienes and Alkynes with Group III and Lanthanide Complexes

It is possible to selectively hydrogenate dienes into monoolefins when there is a large difference in reactivity between the two olefins. For example, vinylcyclohexene and 4-vinylnorbornene undergo a selective hydrogenation of the acyclic olefins in the presence of catalytic amount of [Cp₂*YMe] (Table 6.12). Note that α-allylmethylenecyclohexene gives a mixture of α-propylmethylenecyclohexene and the reductive cyclization product, which shows that the intramolecular insertion of a second alkene can be competitive with hydrogenolysis (Table 6.12).

Table 6.12 Hydrogenation of dienes catalyzed by Cp$_2^*$YMe [123].

Substrate	SCR (time, h)	Product	Yield [%]	Byproduct [%]
	100 (2)		70	–
	25 (1)		67	20
	100 (1)		72	–

SCR = substrate : catalyst ratio.

Similarly, when a diene is constituted of both terminal and tri-substituted olefins, the terminal olefin is selectively hydrogenated. Note that ether substituted systems could be hydrogenated in these cases. Moreover, for 3-substituted 1,5-hexadienes, the less-hindered olefins can be selectively hydrogenated into the corresponding 2-substituted 1-hexene, especially when the substituent in position 2 is large (Table 6.13). However, when no substituent is present on the alkyl chain, 1,5-dienes undergo reductive cyclization in the presence of *ansa* Cp derivatives of Y and Lu to give the corresponding cyclic alkanes in quantitative yields (Table 6.14) [124]. Note that no conversion is observed when an ether group is present in the starting material.

An original report by Evans reports the catalytic hydrogenation of 3-hexyne and diphenylacetylene into the corresponding alkanes by generating *in situ* [Cp$_2^*$SmH] from [Cp$_2^*$Sm] [120, 125]. Later, Evans showed that *cis*-3-hexene was formed selectively at the beginning of the reaction [126]. In a comprehensive study, Marks et al. have shown that the reaction rate of hydrogenation of alkynes, namely 3-hexyne, is slower (Table 6.15) than that of olefins (*vide supra*) [120, 125]. The rate law is as follows: $r = k[L_nMH][H_2]$, and it is fully consistent with hydrogenolysis being rate-determining. Since the rate of hydrogenation is faster for *cis* alkenes in the absence of alkynes, it shows that alkynes react faster with the hydride than alkenes.

Table 6.13 Hydrogenation of acyclic dienes catalyzed by Cp$_2^*$YMe [123].

Substrate	SCR (time, h)	Product	Yield [%]	Byproduct [%]
R=OSiMe$_2$tBu	50 (1)	R=OSiMe$_2$tBu	98	–
R=OMe	50 (2)	R=OMe	74	–
R=OCH$_2$Ph	50 (2)	R=OCH$_2$Ph	99	–
R=OSiMe$_2$tBu	33 (1)		85	10
R=OSiMe$_2$tBu	25 (1.5)		99	–
R=Ph	33 (1)		64	23
R=Ph	25 (1.5)		92	–
iPr	33 (1.1)		70	16
R=OCPh$_3$	33 (1)		96	1
	33 (1)		42	20

SCR = substrate : catalyst ratio.

6.4.3
Hydrogenation of Imines with Group III and Lanthanide Complexes

The hydrogenation of imines is typically carried out with 1 mol% of the lanthanocene catalyst under an H$_2$ pressure of 13 bar at 90 °C [127]. The best catalysts are based on Sm having Cp* ligands, the *ansa* systems being unreactive. The rate and the total conversion are improved by the addition of PhSiH$_3$, probably because it stabilizes the system (Table 6.16), and both are very sensitive to the

Table 6.14 Reductive cyclization of dienes catalyzed by $Me_2SiCp^*CpLnCH(SiMe_3)_2$ [124].

Substrate	Ln	SCR (P_{H_2}, atm)	Product	Yield [%]
	Y	200 (1)		100
	Lu	200 (1)		100
	Y	200 (1)		100
	Lu	200 (1)		100
OSiMe₂*t*Bu	Y	200 (1)	–	0
	Lu	200 (1)	–	0

SCR = substrate : catalyst ratio.

Table 6.15 Rate constants for the catalytic hydrogenation of olefins catalyzed by Group III and lanthanide complexes. [a]

	$[Cp_2^*LaH]_2$	$[Cp_2^*NdH]_2$	$[Cp_2^*SmH]_2$	$[Cp_2^*LuH]_2$
3-Hexyne	0.2	0.46	0.82	2.3

a) Based on $r = k[M][H_2]$, in units of $bar^{-1} s^{-1}$ (multiply by 224 to convert to $M^{-1} s^{-1}$).

substituents on the imine. Acyclic imines are fully converted into the corresponding amines under these conditions, while the cyclic ones do not react. Additionally, di-substituted imines at the carbon atom are less reactive than their mono-substituted counterparts. Finally, substitution at the nitrogen atom is also important, with their reactivity varying as follows: = N-Alkyl > = N-Aryl > = N-SiR₃.

6.4.4
Hydrogenation of Alkenes with Actinide Complexes

Biscyclopentadienyl actinide complexes catalyze the hydrogenation of alkenes. In the hydrogenation of 1-hexene, $[Cp_2^*UH_2]$ is more efficient (higher TOF) than the corresponding Th complex, and up to 812 TON can be achieved with the U complex (Table 6.17) [128]. Using cationic Th complexes improves the rate by an order of magnitude [129], but the strongest positive effect is obtained when Cp* ligands are replaced by a Si-tethered *ansa* dicyclopentadienyl ligand,

Table 6.16 Hydrogenation of imines catalyzed by Group III and lanthanide complexes.

Substrate	Product	Catalyst	Conditions [R/P/T/t] [a]	Conv. [%] [a]	Rate [h⁻¹]
PhCH=N–Me	PhCH$_2$–N(H)–Me	Cp$_2^*$SmR	100/13/90/92	83	1.0
		Cp$_2^*$SmR	100/13/50/122	57	0.5
		Cp$_2^*$SmR	100/14/25/51	4	0.04
		Cp$_2^*$SmR + PhSiH$_3$	100/13/90/44	98	2.2
		Cp$_2^*$LaR	100/10/25/50	11	0.05
		Cp$_2^*$LuR	100/13/25/90	51	0.60
PhCH=N–Ph	PhCH$_2$–N(H)–Ph	Cp$_2^*$SmR	100/13/90/120	16	0.10
		Cp$_2^*$SmR + PhSiH$_3$	100/13/90/120	10	–
PhCH=N–SiMe$_3$	PhCH$_2$–N(H)–SiMe$_3$	Cp$_2^*$SmR	100/13/90/58	21	0.40
PhC(Me)=N–Me	PhCH(Me)–N(H)–Me	Cp$_2^*$SmR	100/13/90/144	26	0.20
PhC(Me)=N–Bn	PhCH(Me)–N(H)–Bn	Cp$_2^*$SmR + PhSiH$_3$	100/13/90/134	98	0.70

a) SCR = substrate : catalyst ratio; P = pressure (bar); T = temperature (°C); t = time (h).

Table 6.17 Hydrogenation of alkenes catalyzed by actinide complexes.

Catalyst	Substrate	SCR (P_{H_2}) [a]	TOF [h⁻¹]
[Cp$_2^*$ThH$_2$]	1-hexene	14 (1)	0.5
[Cp$_2^*$UH$_2$]	1-hexene	14 (1)	70
[Cp$_2^*$ThMe$^+$B(C$_6$F$_5$)$_4^-$]	1-hexene	330 (1)	5.2
[Cp$_2^*$ThMe$_2$]/ {tBuCH[B(C$_6$F$_5$)$_2$]$_2$HNBu$_3$}	1-hexene	330 (1)	6
[Me$_2$SiCp$_2''$ThH$_2$]	1-hexene	– (1)	610
[Cp$_2^*$ThH$_2$]	E 2-hexene	14 (1)	0.086
[Me$_2$SiCp$_2''$ThH$_2$]	E 2-hexene	– (1)	2.45

a) SCR = substrate : catalyst ratio; P_{H_2} = H$_2$ pressure (bar).

Me$_2$SiCp$_2''$MX$_2$ =

for which TOFs up to $610\ h^{-1}$ are obtained [130]. For all these systems, the rate of hydrogenation decreases sharply (10- to 100-fold) in going from 1-hexene to *trans*-2-hexene, showing that the insertion step is highly sensitive to the substitution on the olefin [128, 130].

6.4.5
Enantiomeric Hydrogenation of Alkenes

Enantiomeric hydrogenation of alkenes can be performed with chiral lanthanide complexes. The design of an enantiomerically pure catalyst has been based on the introduction of the chirality (R^*) through the preparation of menthyl- or neomenthyl-substituted cyclopentadienyl derivatives $[R^*CpSiMe_2(C_5Me_4)]$ (Scheme 6.11) [131]. Formation of the *ansa* dichloro lanthanide complexes generates two diastereomers, usually with a high level of diastereoselection, and these can be separated by simple crystallization. Access to two enantiomorphous families as pure diastereomers is possible by using the menthyl and neomenthyl derivatives, respectively [121, 124, 132]. The catalyst precursor is then formed by a simple alkylation step, which takes place with retention of configuration.

In the Sm series, the two diastereomers of one enantiomorphous family give rise to very different ee-values (Table 6.18). For example, the hydrogenation of

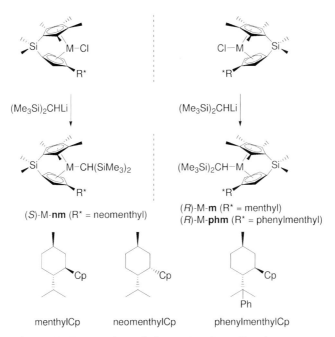

Scheme 6.11 Enantioselective hydrogenation Group III and lanthanide metallocene catalysts.

Table 6.18 Enantioselective hydrogenation of alkenes
catalyzed by Group III and lanthanide complexes.

Catalyst [a]	Substrate	SCR (P_{H2}) [b]	Temp. [C]	ee [%]
(R)-[Me$_2$SiCp'Cp^{nm}SmH]$_2$	2-phenyl-1-butene	100–500 (1)	25	71 (–)
(S)-[Me$_2$SiCp'Cp^{nm}SmH]$_2$	2-phenyl-1-butene	100–500 (1)	25	19 (+)
(R)-[Me$_2$SiCp'Cp^{nm}SmH]$_2$	Styrene [c]	100 (1)	25	43 (–)
(R)-[Me$_2$SiCp'Cp^{nm}SmR]	2-phenyl-1-butene	100–1000 (1)	25	71 (–)
(S)-[Me$_2$SiCp'Cp^{nm}SmR]	2-phenyl-1-butene	100–1000 (1)	25	19 (+)
(R)-[Me$_2$SiCp'Cp^{m}SmR]	2-phenyl-1-butene	100–1000 (1)	25	8 (–)
(S)-[Me$_2$SiCp'Cp^{m}SmR]	2-phenyl-1-butene	100–1000 (1)	25	19 (–)
(R)-[Me$_2$SiCp'Cp^{m}SmR]	2-phenyl-1-butene	100–1000 (1)	25	43 (–)
(S)-[Me$_2$SiCp'Cp^{m}YR]	2-phenyl-1-butene	100–1000 (1)	25	3 (–)
(S)-[Me$_2$SiCp'Cp^{m}YR]	2-phenyl-1-butene	100–1000 (1)	25	3 (–)

a) nm = neomenthyl, M-**nm**; m = menthyl, M-**m**.

Me$_2$SiCp'Cp^{x}M-R =

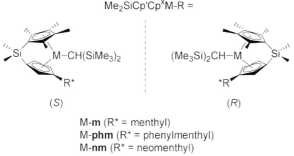

(S)   (R)

M-**m** (R* = menthyl)
M-**phm** (R* = phenylmenthyl)
M-**nm** (R* = neomenthyl)

b) SCR = substrate : catalyst ratio; P_{H2} = H$_2$ pressure (bar).
c) Using D$_2$ instead of H$_2$.

2-phenyl-1-butene at 25 °C gives (R)-(–)-2-phenylbutane in 70% ee with the (R)-Sm-**nm**, while it is (S)-(+)-2-phenylbutane in 19% ee with the (S)-Sm-**nm**. When using the other enantiomorphous family, opposite enantioselection is usually observed: the (S)-Sm-**m** gives the (R)-(–)-2-phenylbutane in 8% ee under the same conditions. Moreover, when a 70/30 mixture of the (S) and (R)-Sm-**m** is used, 64% ee in (S)-(+)-2-phenylbutane is observed. Reducing the temperature of the reaction to –78 °C increases the level of enantioselection to 96% e.e. Finally, using D$_2$ instead of H$_2$ provides exclusively the corresponding deuterated product, arising from a *cis*-addition. The best model to explain the observed stereochemistry corresponds to a frontal approach of the alkene towards the metal center (Scheme 6.12).

Kinetic studies show that insertion (the enantioselection step) is very rapid, and that the rate-determining step is the hydrogenolysis of the M–C bond. Nonetheless, under H$_2$-starving conditions, there is evidence that β-H elimination can be competitive with hydrogenolysis. β-H elimination of the alkyl intermediate gives back the starting alkene and, through an equilibration process, it

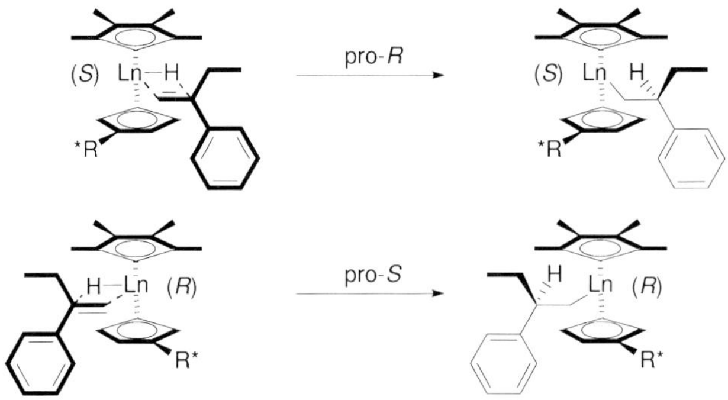

Scheme 6.12

induces an erosion of the enantioselectivity [132]. Noteworthy is the finding that, when Sm is replaced by Group III or other lanthanide metals, the following trend for ee-values is observed: La > Nd > Sm > Y > Lu, which parallels the decrease in ionic radius on going from La to Lu [132]. Other *ansa* dicyclopentadienyl systems have been designed, but this was only applicable to Y and Lu. Whilst the catalysts could be prepared as stable enantiomers, they show only low ee-values in the hydrogenation of styrene derivatives [124].

6.5
Hydrogenation Catalysts Based on Groups V–VII Transition-Metal Complexes

6.5.1
Hydrogenation of Alkenes and Dienes with Groups V–VII Transition-Metal Complexes

As shown by Breslow et al. during the mid-1960s, most transition-metal alkoxide or acetylacetonate complexes catalyze the hydrogenation of alkenes in the presence of an activator (Table 6.19) [5]. Other precursors have been used such as $[CpCr(CO)_3]_2$, but it is more difficult to understand how the active species are formed [133].

Moreover, systems based on Group V and VI transition metals can be used to selectively hydrogenate dienes into mono-enes. For example, Cp_2VCl_2 in combination with 2 equiv. BuLi or PhMgBr selectively hydrogenate butadiene into butenes, while the system based on $[Cp_2TiCl_2]/R'M$ gives butane [6]. Similarly, the V system hydrogenates isopropene selectively into 2-methyl-2-butene (92–93% selectivity), along with the other branched pentene isomers as well as traces of 2-methylbutane. Similarly, a catalytic amount of $[Cp_2MoH_2]$ (0.2 mol%) converts dienes and trienes selectively into the corresponding mono-enes in 50–90% yields at 180 °C (Scheme 6.13) [134, 135]. It has been reported that methyl acrylate, methyl crotonate, crotonaldehyde, and mesityl oxide can be hydrogenated under the same

Table 6.19 Hydrogenation catalysts based on Group V–VII transition-metal complexes activated with AliBu$_3$.

M complex	Activator [a]	Temp. [°C]	Olefin (SCR) [b]	Conv. [%] (time, h)
[VO(OiPr)$_3$]	iBu$_3$Al (3.9)	40	Cyclohexene (52)	100 (<20)
[VO(OiPr)$_3$]	iBu$_3$Al (3.9)	40	1-Octene (52)	100 (<20)
[Cr(acac)$_3$]	iBu$_3$Al (6.0)	30	Cyclohexene (105)	100 (2)
[Cr(acac)$_3$]	iBu$_3$Al (6.0)	30	1-Octene (105)	55 (1.2)
[MoO$_2$(acac)$_2$]	iBu$_3$Al (7.1)	30	Cyclohexene (63)	100 (<16)
[MoO$_2$(acac)$_2$]	iBu$_3$Al (7.1)	30	1-Octene (67)	100 (<21)
[Mn(acac)$_2$]	iBu$_3$Al (6.0)	30	Cyclohexene (63)	100 (<16)

a) Values in parentheses correspond to the number of equivalents of activator per transition-metal complex.
b) SCR = substrate : catalyst ratio.

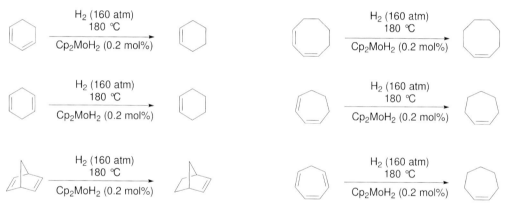

Scheme 6.13

reaction conditions, but no yields were reported. In sharp contrast, [Cp$_2$Cr, Cp$_2$WH$_2$] and [Cp$_2$ReH] are completely inactive under the same reaction conditions.

Finally, whilst rhenium hydride complexes have not been reported to hydrogenate alkenes, there are several reports of the dehydrogenation of alkanes in the presence of tBuCH=CH$_2$ as an hydrogen acceptor (Scheme 6.14) [136–142]. For example, cycloalkanes are transformed catalytically into the corresponding cyclic alkene, which shows that, in principle, a Re-based catalyst could be designed.

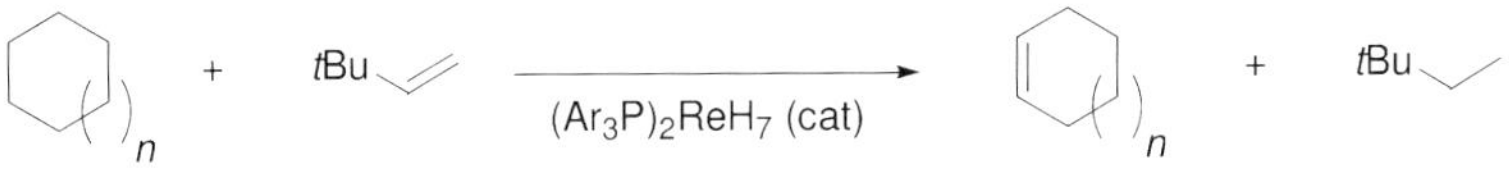

Scheme 6.14

6.5.2
Hydrogenation of Aromatics with Well-Defined Nb and Ta Aryloxide Complexes

Treatment under H_2 of $[(2,6\text{-diPhC}_6H_3O)_2Ta(CH_2R)_3]$ $(R=CH_2Ar)$ in the presence of different arylphosphine ligands generated well-defined monomeric Ta hydride complexes, $[(2,6\text{-diCyC}_6H_3O)_2Ta(H_3)(PCy_3)_2$ (Scheme 6.15) [143, 144]. During this treatment, the four phenyl substituents of the aryloxide ligands are hydrogenated into cyclohexyl units, which demonstrates the catalytic potential of these systems for the hydrogenation of aromatics.

In fact, this Ta complex catalyzes the hydrogenation of naphthalene into tetraline, and that of anthracene into 1,2,3,4-decahydroanthracene [145]. The corresponding dihydride, $[(2,6\text{-di-}i\text{PrC}_6H_3O)_3Ta(H_2)(PMe_2Ph)_2]$, prepared by treatment under H_2 of $[(2,6\text{-di-}i\text{PrC}_6H_3O)_3Ta(CH_2Ar)_2]$ in the presence of PMe_2Ph, is an efficient hydrogenation catalyst of aromatics such as naphthalene [146].

Similarly, $[(2,6\text{-diPhC}_6H_3O)_2NbR_3]$ $(R=CH_2Ph\text{-}4\text{-Me})$ is an hydrogenation catalyst precursor for aromatics. It catalyzes the hydrogenation of naphthalenes into the corresponding tetraline, giving selectively the product of hydrogenation of the nonsubstituted ring (Table 6.20). It is noteworthy that anthracene is converted exclusively into 1,2,3,4,5,6,7,8-octahydroanthracene, with no trace of 9,10-dihydroanthracene, whilst phenanthrene yields 9,10-dihydroanthracene (78%) along

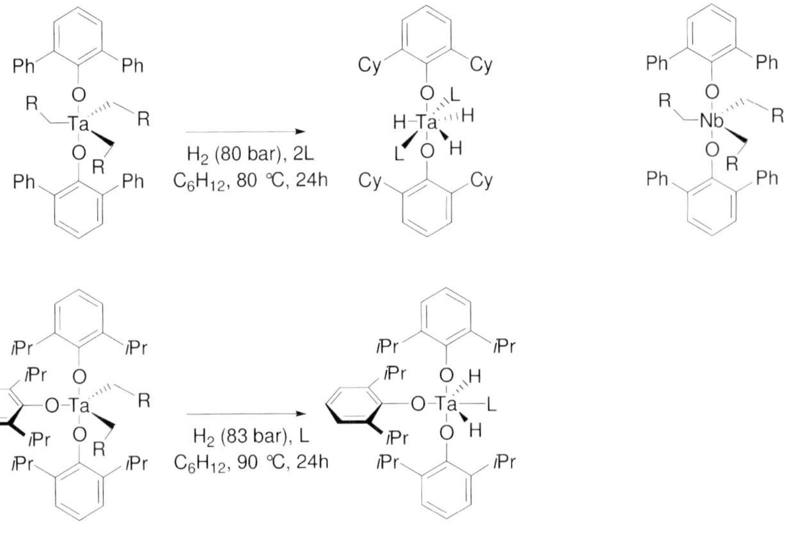

Scheme 6.15

Table 6.20 Hydrogenation of aromatics catalyzed by $[(2,6\text{-diPhC}_6\text{H}_3\text{O})_2\text{NbR}_3]$ and $[(2,6\text{-diCyC}_6\text{H}_3\text{O})_2\text{Ta}(\text{H})_3(\text{PCy}_3)_2]$ [145].

Substrate	Catalyst	SCR (P)[a]	Product	Yield [%]	Byproduct [%]
	Nb	20 (80)		>95	–
	Ta	20 (80)		>95	–
	Nb	20 (80)		>95	–
	Nb	20 (80)		>95	–
	Nb	20 (80)		>95	–
	Ta	20 (80)		>95	–
	Nb	20 (80)		78	22
	Nb	20 (80)		>95	–
	Nb	20 (80)		>95	–

a) SCR = substrate : catalyst ratio; P = pressure (bar).

with 1,2,3,4,5,6,7,8-octahydroanthracene (22%) [143–156]. Both, acenaphthylene and anthracene are transformed into 1,2,2a,3,4,5-hexahydronaphthalene. For both catalyst precursors, the hydrogenation of perdeuterated naphthalene gives exclusively the tetraline product with six hydrogens *cis*, all on one face.

Moreover, the Nb complex hydrogenates catalytically aryl- and benzyl-substituted phosphine under similar conditions (Scheme 6.16) [149]. Kinetic studies show that the hydrogenation of triphenylphosphine into the monocyclohexyl, dicyclohexyl, and tricyclohexylphosphine are successive reactions, and the rate of hydrogenation of the arylphosphine decreases as the number of cyclohexyl substituents increases [153].

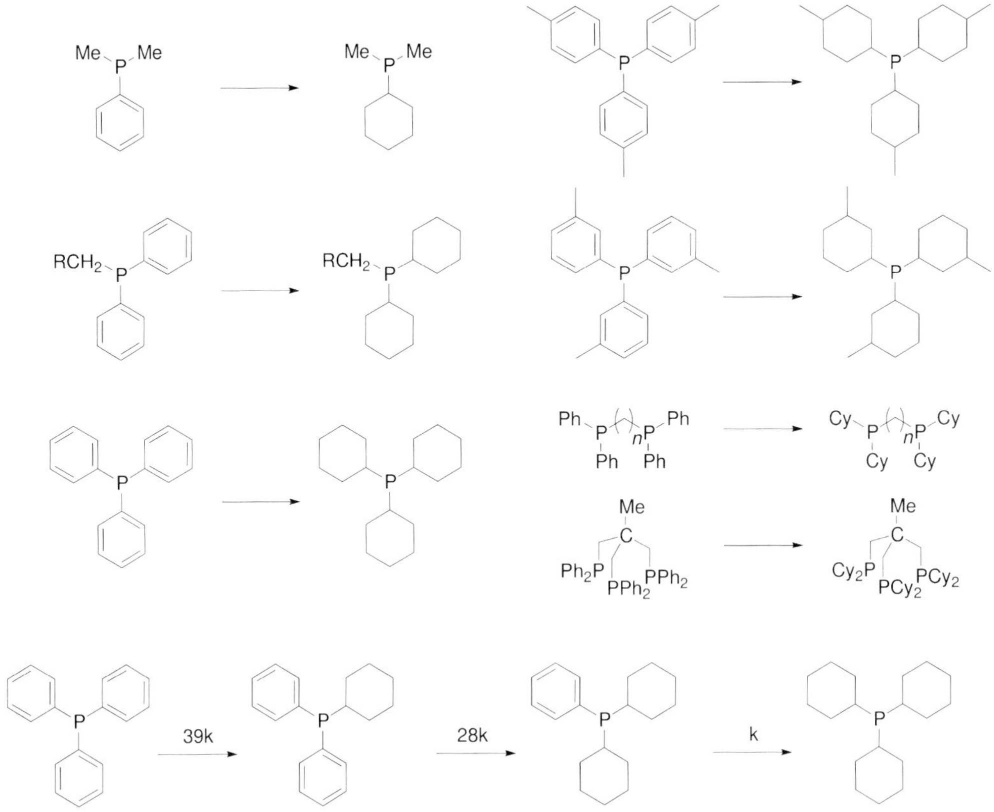

Scheme 6.16 Relative rates indicated above arrows.

6.6
Supported Early Transition-Metal Complexes as Heterogeneous Hydrogenation Catalysts

6.6.1
Supported Homogeneous Catalysts

Several strategies have been adopted to prepare the heterogeneous equivalents of the metallocene hydrogenation catalysts. One approach centers on the preparation of cyclopentadienyl-containing polymers based on either co-polymers of styrene and divinylbenzene (PSDVB) [157–164] or poly(phenylene oxide) (PPO) (Scheme 6.17) [165]. For the PSDVB system, the rate of hydrogenation of the supported systems is much greater than the molecular complex. For example, the TOF for the hydrogenation of cyclohexene for the polymer-supported Ti complex and [Cp$_2$TiCl$_2$] activated by BuLi are 1900 and 18 h^{-1}, respectively [160]. Note that critical factors are low metal loadings and small particle sizes (grind-

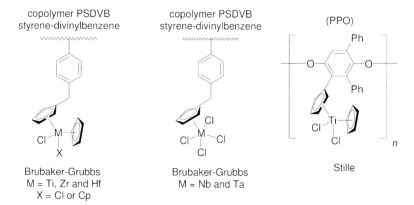

copolymer PSDVB
styrene-divinylbenzene

copolymer PSDVB
styrene-divinylbenzene

(PPO)

Brubaker-Grubbs
M = Ti, Zr and Hf
X = Cl or Cp

Brubaker-Grubbs
M = Nb and Ta

Stille

Scheme 6.17

ing of the polymer-supported catalysts prior to activation with BuLi is necessary for good activities). This system hydrogenates 1-hexene, styrene, 1,3- and 1,5-cyclooctadienes with comparable TOF, while tri-, tetra-substituted and functionalized alkenes do not undergo hydrogenation [158]. Several other metals have been used, but Ti is by far the most active [159, 163, 164]. Similarly, PPO-supported Ti complexes are more active (e.g., 10- to 70-fold) than their molecular equivalents [165].

Another approach has consisted of preparing a cyclopentadiene having an alkoxy silane substituent, which is grafted onto a silica support and then transformed into the corresponding metal cyclopentadiene complex (Scheme 6.18) [166]. After activation with BuLi or PrMgBr, their performances in the hydrogenation of 1-octene are about three- to five-fold better than the corresponding homogeneous catalysts (0.24–0.4 versus 0.08 h^{-1}, [CpTiCl$_3$]). Finally, a cyclopentadiene-containing silica has been prepared by sol-gel condensation of trisalkoxysilyl-substituted cyclopentadiene and then used to anchor the Ti center [167]. This system catalyzes the hydrogenation of 1-octene in the presence of BuLi,

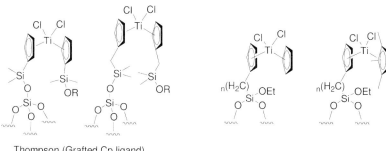

Thompson (Grafted Cp ligand)
R = Bu, Et or Si from silica

Cermak (Sol-Gel)
n = 1-3

Scheme 6.18

with activities comparable to those of the homogeneous systems (2 mol% Ti): [CpTiCl$_3$]   (5640 h^{-1}) > [($\equiv$ SiO-(CH$_2$)$_n$C$_5$H$_4$)Cp*TiCl$_2$]   (1730 h^{-1}) > [Cp*TiCl$_3$] (1300 h^{-1}) > [($\equiv$ SiO-(CH$_2$)$_n$C$_5$H$_4$)CpTiCl$_2$] (865 h^{-1}).

6.6.2
Heterogeneous Catalysts Prepared via Surface Organometallic Chemistry

In contrast to supported homogeneous catalysis, surface organometallic chemistry (SOMC) uses an inorganic oxide (E$_x$O$_y$) as a solid ligand, on which the metal is directly attached by at least a bond with a surface atom, usually an oxygen, through a M–OE bond.

During the late 1960s, polymer chemists were interested in developing well-defined equivalents of the Ziegler-Natta catalysts [168, 169], and prepared supported Group IV complexes by reacting organometallic complexes with oxide supports. Yermakov et al. showed that, under treatment under H$_2$, these systems evolved into the corresponding hydrides. All of these systems are efficient polymerization catalysts [169–172]. Moreover, the Ti- and Zr- hydrides have turned out to be efficient hydrogenation catalysts of alkenes and aromatics (TOF = 3600 to 360 000 h^{-1} for the hydrogenation of cyclohexene or benzene) [173–176]. The structure of these systems has required a detailed understanding of the surface chemistry. In short, as a general rule, the structure of the surface organometallic complex depends on the temperature of pretreatment of silica [177]. When silica is treated at a low temperature under vacuum (e.g., 200 °C), the reaction with an organometallic complex yields a bis(siloxy) surface complex. On the other hand, when silica is treated at a high temperature under vacuum (e.g., 700 °C), a mono(siloxy) surface complex is formed (Scheme 6.19). Under H$_2$, Group IV and V surface complexes generate well-defined surface metal hydrides (Ti [178], Zr [179, 180], Hf [181], and Ta [182]).

The hydrogenation catalysts can be prepared in situ, starting from the surface alkyl complex. In terms of catalytic performances, these catalysts are highly effective (Table 6.21) [150]. The best hydrogenation systems are based on silica supported dinuclear complexes, for which the structures of the active sites have not been investigated. Hydrogenation of toluene and xylenes can be achieved under similar conditions.

Similarly, the cyclopentadienyl Zr derivatives supported on alumina are highly active olefin hydrogenation catalysts [183, 184]. The initial rates of propene hydrogenation depending on the cyclopentadienyl derivatives and the level of dehydroxylation of alumina are as follows: [Cp*ZrMe$_3$]/Al$_2$O$_{3-(1000)}$ (3960 h^{-1}) > [Cp*ZrMe$_3$]/Al$_2$O$_{3-(500)}$ (1080 h^{-1}) ≈ [Cp$_2$ZrMe$_2$]/Al$_2$O$_{3-(1000)}$ (1080 h^{-1}) > [Cp*CpZrMe$_2$]/Al$_2$O$_{3-(1000)}$ (720 h^{-1}) > [Cp$_2$ZrMe$_2$]/Al$_2$O$_{3-(500)}$ (360 h^{-1}) > [Cp*CpZrMe$_2$]/Al$_2$O$_{3-(500)}$ (216 h^{-1}) [185]. Typically, highly dehydroxylated alumina (Al$_2$O$_{3-(1000)}$) is a better support, and it has been associated with the formation of cationic surface species.

Several other supports have been used in order to generate more-electrophilic Zr systems, including sulfated zirconia [186], sulfated alumina [187], or other sulfated oxide supports [188], though the surface species are quite complex for these systems. The catalytic activities in the hydrogenation of benzene are as follows: $[Cp^*Zr(CH_3)_3]/ZrO_{2\text{-sulfated}}$ $(970\,h^{-1})$ > $[Cp^*Zr(CH_3)_3]/Al_2O_{3\text{-sulfated}}$ $(360\,h^{-1})$> $[Cp^*Zr(CH_3)_3]/SnO_{2\text{-sulfated}}$ $(10\,h^{-1})$>$[Cp^*Zr(CH_3)_3]/SnO_{2\text{-sulfated}}$ $(5\,h^{-1})$ > $[Cp^*Zr(CH_3)_3]/TiO_{2\text{-sulfated}}$ $(<3\,h^{-1})$, and they are correlated with the surface Brönsted acidity [188].

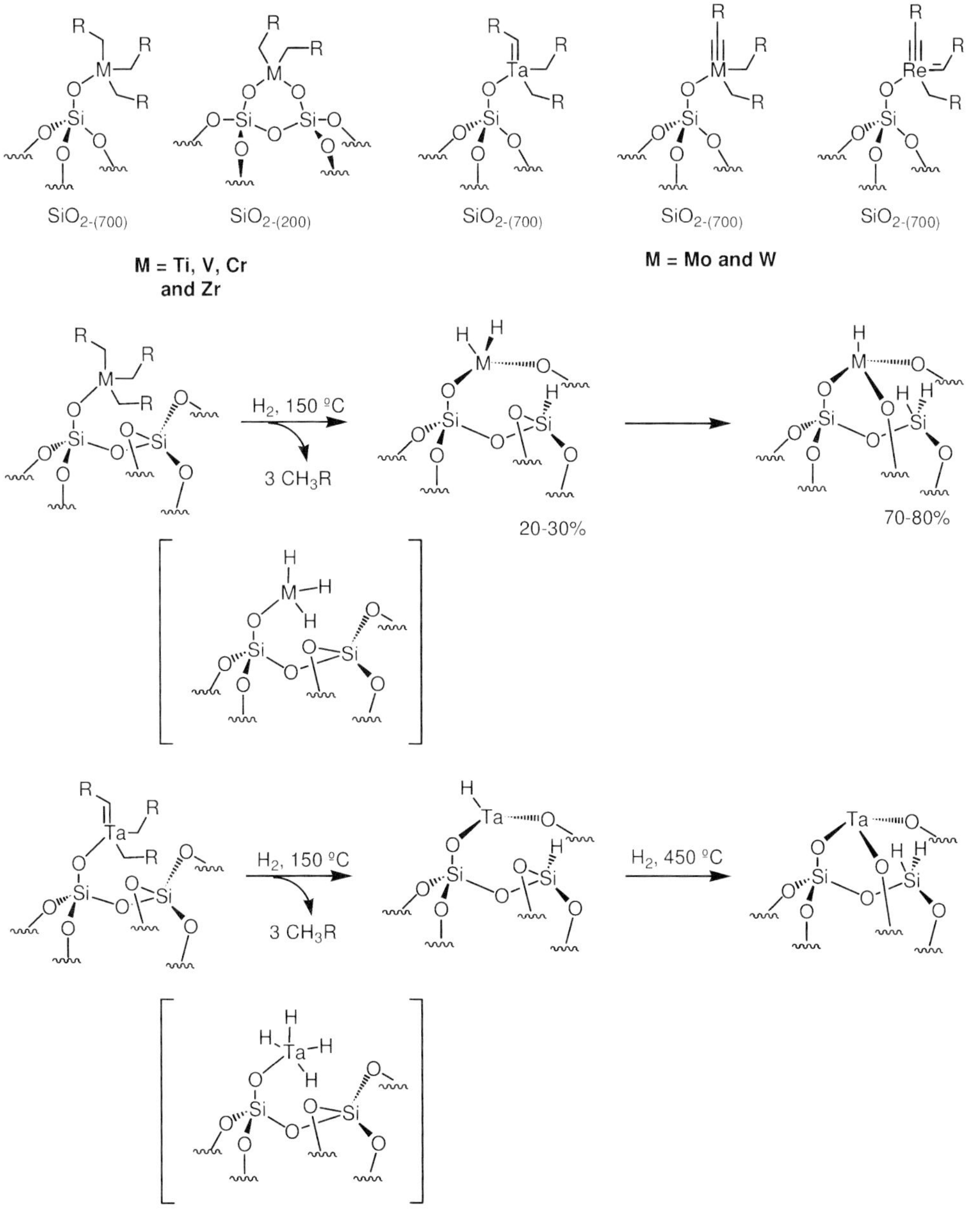

Scheme 6.19

Table 6.21 Hydrogenation of benzene and naphthalene catalyzed by silica-supported Group IV–VI transition-metal complexes.

Catalytic system	Substrate	Time [min]	Conversion [%]	Selectivity [%]
$Ti(CH_2SiMe_3)_4/SiO_2$	Benzene[a]	360	39	100
	Naphthalene[b]	150	100	$55:45$[d]
$Zr(CH_2Ph)_4/SiO_2$	Benzene[a]	240	100	100
	Naphthalene[b]	30	100	$25:75$[d]
$Hf(CH_2Ph)_4/SiO_2$	Benzene[a]	600	100	100
	Naphthalene[b]	60	100	$19:81$[d]
$Ta(CH_2Ar)_5/SiO_2$	Benzene[a]	24 h	100	100
$Ar = 4\text{-Me-C}_6H_4$	Naphthalene[b]	160	100	$20:80$[d]
$[Nb(CSiMe_3)(CH_2SiMe_3)_2]_2/SiO_2$	Benzene[a]	300	100	100
	Naphthalene[b]	20	100	$15:85$[d]
	Benzene[c]	25	100	100
	Toluene[c]	33	100	100
	o-Xylene[c]	30	100	$57:43$[d]
	m-Xylene[c]	20	100	$78:22$[d]
	p-Xylene[c]	25	100	$57:43$[d]
$[Ta(CSiMe_3)(CH_2SiMe_3)_2]_2/SiO_2$	Benzene[a]	720	100	100
	Naphthalene[b]	70	100	$20:80$[d]
	Benzene[c]	55	100	100
	Toluene[c]	64	100	100
	o-Xylene[c]	75	100	$91:9$[d]
	m-Xylene[c]	70	100	$92:8$[d]
	p-Xylene[c]	65	100	$82:18$[d]
$[Mo_2(CH_2SiMe_3)_6]/SiO_2$	Benzene[a]	720	100	100
	Naphthalene[b]	70	100	$20:80$[d]

a) 0.056 mmol M loaded on 5 g silica treated at 200 °C under vacuum, 17.5 g C_6H_6 (4000 equiv.), 80–100 bar H_2 at 120 °C.
b) 0.056 mmol M loaded on 5 g silica treated at 200 °C under vacuum, 2.0 g naphthalene (280 equiv.), 25 mL hexane, 100 bar H_2 at 120 °C.
c) 0.056 mmol M loaded on 5 g silica treated at 200 °C under vacuum, 26 mmol substrate (280 equiv.), 80–100 bar H_2 at 120 °C.
d) *cis/trans* ratio.

Soon after the discovery that dimeric actinide hydride complexes were active hydrogenation catalysts, Marks et al. investigated the formation of monomeric actinide active species by supporting them on an oxide support [184, 189]. The supported complexes on alumina are much more active than the corresponding molecular complexes. For comparison, the rate of hydrogenation of propene at 25 °C can be ranked as follows: $[Cp_2^*U(CH_3)_2]/Al_2O_{3\text{-}(1000)}$ (1080 h^{-1}) > $[Cp_2^*Th(CH_3)_2]/Al_2O_{3\text{-}(1000)}$ (580 h^{-1}) > $[Cp_2^*U(CH_3)_2]$ (68 h^{-1}) > $[Cp_2^*Th(CH_3)_2]$ (0.54 h^{-1}). Using partially dehydroxylated silica or an alumina dehydroxylated at a lower temperature as supports produces poorer catalysts [190]. Other supports have been used, but highly dehydroxylated alumina remains the best available to date. For example, the TOFs for the hydrogenation of propene at 25 °C are: $[Cp^*_2Th(CH_3)_2]/$ $Al_2O_{3\text{-}(1000)}$ (580 h^{-1}) > $[Cp^*_2Th(CH_3)_2]/SiO_2\text{-}Al_2O_3$ (160–230 h^{-1}) > $[Cp^*_2Th(CH_3)_2]/$ $MgCl_2$ (25–43 h^{-1}) > $[Cp^*_2Th(CH_3)_2]/SiO_2\text{-}MgO$ (0 h^{-1}) [191]. Note also that the TOF depends on the alkenes as measured by the hydrogenation at –45 °C: *cis*-butene (3960 h^{-1}) > *trans*-2-butene (2900 h^{-1}) ≫ propene (470 h^{-1}) ≫ isobutene (13 h^{-1}). Finally, some of these systems are highly active aromatic hydrogenation catalysts. At 90 °C and 13 bar, the TOF of hydrogenation is as follows: $[Th(allyl)_4]/Al_2O_{3\text{-}(1000)}$ (1970 h^{-1}) ≫ $[Th(CH_2Ar)_4]/Al_2O_{3\text{-}(1000)}$ (825 h^{-1}) > $[Cp^*Th(CH_2Ar)_3]$ (765 h^{-1}) [192, 193]. Noteworthy, the TOF for the hydrogenation of benzene with $[Th(allyl)_4]/$ $Al_2O_{3\text{-}(1000)}$ is comparable to Rh or Pt heterogeneous catalysts. The hydrogenation TOF is also function of the aromatic compounds: benzene (6850 k) > toluene (4100 k) > xylene (1450 k) ≫ naphthalene (1 k). Using D$_2$ in place of H$_2$ shows that hydrogenation corresponds to a random 1,2-*cis* addition of H$_2$.

6.7
Conclusions

Early transition-metal complexes have been some of the first well-defined catalyst precursors used in the homogeneous hydrogenation of alkenes. Of the various systems developed, the biscyclopentadienyl Group IV metal complexes are probably the most effective, especially those based on Ti. The most recent development in this field has shown that enantiomerically pure *ansa* zirconene and titanocene derivatives are highly effective enantioselective hydrogenation catalysts for alkenes, imines, and enamines (up to 99% ee in all cases), whilst in some cases TON of up to 1000 have been achieved.

Moreover, Group III, lanthanide and actinide biscyclopentadienyl complexes are efficient hydrogenation catalysts exhibiting a very rapid rate of hydrogenation. These catalysts are, nonetheless, somewhat more sensitive to the presence of functional groups, even though they can be used to hydrogenate imines. Enantiomerically pure chiral catalysts have also been developed for the enantiomeric hydrogenation of olefins in good enantiomeric excess (up to 71% ee).

Hydrogenation with Group V–VII transition metals has not yet been investigated in detail, despite the early discoveries of Breslow et al. Nonetheless, the findings of Rothwell et al. are noteworthy as Group V aryloxy systems can hydrogenate aromatics, including triphenylphosphine.

Finally, some of these systems have been supported on polymers or oxides, and these are typically more active than their homogeneous equivalents.

In general, the rate of alkene hydrogenation is typically ordered as follows: terminal > di-substituted > tri-substituted ≫ tetra-substituted. In fact, this allows terminal or di-substituted olefins to be hydrogenated selectively in the presence of tri- or tetra-substituted ones. Additionally, the rate of hydrogenation of alkynes is much slower than that of alkenes, although the *cis*-alkene intermediate can be observed and in some cases formed selectively, before its hydrogenation into the corresponding alkanes.

Whilst hydrogenation catalysts based on early transition metals are as active and selective as those based on late transition metals, they are usually not as compatible with functional groups, and this represents the major difficulty for their use in organic synthesis. Nonetheless, titanocene derivatives have been used in industry to hydrogenate unsaturated polymers.

Undoubtedly, further studies in this area are required, and it is most likely that, based on such investigations, new systems based on Group III–VII metals will emerge during the coming years.

Acknowledgments

These studies are dedicated to I.P. Rothwell for his pioneering contributions towards the development of early transition-metal hydrogenation catalysts. The author thanks the CNRS for financial support.

Abbreviations

ee	enantiomeric excess
PPO	poly(phenylene oxide)
PSDVB	co-polymers of styrene and divinylbenzene
SOMC	surface organometallic chemistry
TOF	turnover frequency
TON	turnover number

References

1 J. F. Young, J. A. Osborn, F. H. Jardine, G. Wilkinson, *J. Chem. Soc., Chem. Commun.* **1965**, 131.

2 G. Wilkinson, P. L. Pauson, J. M. Birmingham, F. A. Cotton, *J. Am. Chem. Soc.* **1953**, *75*, 1011.

3 G. Wilkinson, J. M. Birmingham, *J. Am. Chem. Soc.* **1954**, *76*, 4281.

4 D. S. Breslow, N. R. Newburg, *J. Am. Chem. Soc.* **1959**, *81*, 81.

5 M. F. Sloan, A. S. Matlack, D. S. Breslow, *J. Am. Chem. Soc.* **1963**, *85*, 4014.

6 Y. Tajima, E. Kunioka, *J. Org. Chem.* **1968**, *33*, 1689.

7 P. C. Wailes, H. Weigold, A. P. Bell, *J. Organometal. Chem.* **1972**, *43*, C32.

8 E. E. Van Tamelen, W. Cretney, N. Klaentschi, J. S. Miller, *J. Chem. Soc., Chem. Soc.* **1972**, 481.

9 J. E. Bercaw, R. H. Marvich, L. G. Bell, H. H. Brintzinger, *J. Am. Chem. Soc.* **1972**, *94*, 1219.

10 G. Fachinetti, C. Floriani, *J. Chem. Soc., Chem. Soc.* **1974**, 66.

11 C. Floriani, G. F. Fachinetti (Snamprogetti SpA, Italy), De2449257, **1975**.

12 G. W. Parshall, S. D. Ittel. The Applications and Chemistry of Catalysis by Soluble Transition Metal Complexes. In: *Homogeneous Catalysis*, 2nd edn., Wiley-VCH, Weinheim, **2000**.

13 D. Astruc. In: *Chimie Organométallique*, PUF, Grenoble, **2001**.

14 C.P. Casey, S.L. Hallenbeck, D.W. Pollock, C.R. Landis, *J. Am. Chem. Soc.* **1995**, *117*, 9770.

15 C.P. Casey, S.L. Hallenbeck, J.M. Wright, C.R. Landis, *J. Am. Chem. Soc.* **1997**, *119*, 9680.

16 C.P. Casey, J.J. Fisher, *Inorg. Chim. Act.* **1998**, *270*, 5.

17 C.P. Casey, M.A. Fagan, S.L. Hallenbeck, *Organometallics* **1998**, *17*, 287.

18 C.P. Casey, D.W. Carpenetti, II, H. Sakurai, *J. Am. Chem. Soc.* **1999**, *121*, 9483.

19 C.P. Casey, D.W. Carpenetti, II, *Organometallics* **2000**, *19*, 3970.

20 C.P. Casey, J.F. Klein, M.A. Fagan, *J. Am. Chem. Soc.* **2000**, *122*, 4320.

21 C.P. Casey, T.-Y. Lee, J.A. Tunge, D.W. Carpenetti, II, *J. Am. Chem. Soc.* **2001**, *123*, 10762.

22 C.P. Casey, D.W. Carpenetti, II, H. Sakurai, *Organometallics* **2001**, *20*, 4262.

23 C.P. Casey, J.A. Tunge, T.-Y. Lee, M.A. Fagan, *J. Am. Chem. Soc.* **2003**, *125*, 2641.

24 Z. Wu, R.F. Jordan, J.L. Petersen, *J. Am. Chem. Soc.* **1995**, *117*, 5867.

25 M. Yoshida, D.J. Crowther, R.F. Jordan, *Organometallics* **1997**, *16*, 1349.

26 J.-F. Carpentier, Z. Wu, C.W. Lee, S. Stroemberg, J.N. Christopher, R.F. Jordan, *J. Am. Chem. Soc.* **2000**, *122*, 7750.

27 E.J. Stoebenau, III, R.F. Jordan, *J. Am. Chem. Soc.* **2003**, *125*, 3222.

28 E.J. Stoebenau, III, R.F. Jordan, *J. Am. Chem. Soc.* **2004**, *126*, 11170.

29 P.L. Watson, *J. Am. Chem. Soc.* **1983**, *105*, 6491.

30 P.L. Watson, *J. Chem. Soc., Chem. Soc.* **1983**, 276.

31 M.E. Thompson, S.M. Baxter, A.R. Bulls, B.J. Burger, M.C. Nolan, B.D. Santarsiero, W.P. Schaefer, J.E. Bercaw, *J. Am. Chem. Soc.* **1987**, *109*, 203.

32 B.-J. Deelman, J.H. Teuben, S.A. Mac-Gregor, O. Eisenstein, *New J. Chem.* **1995**, *19*, 691.

33 L. Maron, L. Perrin, O. Eisenstein, *J. Chem. Soc., Dalton Trans.* **2002**, 534.

34 C. Copéret, A. Grouiller, M. Basset, H. Chermette, *Chem. Phys. Chem.* **2003**, *4*, 608.

35 Y.-H. Fan, S.-J. Liao, J. Xu, F.-D. Wang, Y.-L. Qian, J.-L. Huang, *J. Catal.* **2002**, *205*, 294.

36 T. Takahashi, N. Suzuki, M. Kageyama, Y. Nitto, M. Saburi, E. Negishi, *Chem. Lett.* **1991**, 1579.

37 S.J. Liao, Y. Xu, Y.P. Zhang, Q. Sun, R.A. Sun, R.W. Yang, *Chin. Chem. Lett.* **1994**, *5*, 689.

38 S. Couturier, B. Gautheron, *J. Organometal. Chem.* **1978**, *157*, C61.

39 S. Couturier, G. Tainturier, B. Gautheron, *J. Organometal. Chem.* **1980**, *195*, 291.

40 J.A. Smith, H.H. Brintzinger, *J. Organometal. Chem.* **1981**, *218*, 159.

41 K. Shikata, K. Nishino, K. Azuma, Y. Takegami, *Kogyo Kagaku Zasshi* **1965**, *68*, 358.

42 Y. Qian, G. Li, Y. Huang, *J. Mol. Cat.* **1989**, *54*, L19.

43 F. Masi, R. Santi, G. Longhini, A. Vallieri (Enichem S.p.A., Italy), Ep908234, **1999**.

44 K. Sonogashira, N. Hagihara, *Bull. Chem. Soc. Jap.* **1966**, *39*, 1178.

45 F. Scott, H.G. Raubenheimer, G. Pretorius, A.M. Hamese, *J. Organometal. Chem.* **1990**, *384*, C17.

46 K. Iwase, K. Kato, G. Yamamoto (Asahi Chemical Ind, Japan), Jp08041081, **1996**.

47 B. Demerseman, P. Le Coupanec, P.H. Dixneuf, *J. Organometal. Chem.* **1985**, *287*, C35.

48 V.B. Shur, V.V. Burlakov, M.E. Vol'pin, *Izv. Akad. Nauk SSSR, Ser. Khim.* **1986**, 728.

49 V.B. Shur, V.V. Burlakov, M.E. Vol'pin, *J. Organometal. Chem.* **1992**, *439*, 303.

50 E.B. Lobkovskii, L. Soloveichik, A.I. Sizov, B.M. Bulychev, *J. Organometal. Chem.* **1985**, *280*, 53.

51 V.K. Bel'skii, A.I. Sizov, B.M. Bulychev, G.L. Soloveichik, *J. Organometal. Chem.* **1985**, *280*, 67.

52 A.I. Sizov, T.M. Zvukova, V.K. Bel'sky, B.M. Bulychev, *Russ. Chem. Bull.* (Translation of Izvestiya Akademii Nauk, Seriya Khimicheskaya) **1998**, *47*, 1186.

53 Y. Zhang, S. Liao, Y. Xu, S. Chen, *J. Organometal. Chem.* **1990**, *382*, 69.

54 Q. Sun, S. Liao, Y. Xu, Y. Qian, J. Huang, *Cuihua Xuebao* **1996**, *17*, 495.

55 Y.-H. Fan, S.-J. Liao, J. Xu, Y.-L. Qian, J.-L. Huang, *Gaodeng Xuexiao Huaxue Xuebao* **1997**, *18*, 1683.

56 M. Zhang, Y. Fan, S. Liao, Y. Qian, *Huadong Ligong Daxue Xuebao* **1998**, *24*, 580.

57 Y. Fan, S. Liao, Y. Xu, Y. Qian, J. Huang, *Cuihua Xuebao* **1998**, *19*, 389.

58 J.-G. Lee, H. Y. Jeong, Y. H. Ko, J. H. Jang, H. Lee, *J. Am. Chem. Soc.* **2000**, *122*, 6476.

59 E. Samuel, *J. Organometal. Chem.* **1980**, *198*, C65.

60 J. F. Harrod, S. S. Yun, *Organometallics* **1987**, *6*, 1381.

61 J. F. Harrod, R. Shu, H.-G. Woo, E. Samuel, *Can. J. Chem.* **2001**, *79*, 1075.

62 Y. Tajima, E. Kunioka, *J. Am. Oil Chem. Soc.* **1968**, 45, 478.

63 F. Hayano, Y. Nakafutami, Y. Kishimoto (Asahi Chemical Industry Co., Ltd., Japan), Jp61034050, **1986**.

64 T. Masubuchi, Y. Kishimoto (Asahi Chemical Industry Co., Ltd., Japan), Jp61033132, **1986**.

65 T. Masubuchi, Y. Kishimoto (Asahi Chemical Industry Co., Ltd., Japan), Jp61057524, **1986**.

66 T. Teramoto, K. Goshima, M. Takeuchi (Japan Synthetic Rubber Co., Ltd., Japan), Ep339986, **1989**.

67 I. Hattori, A. Takashima, T. Imamura (Japan Synthetic Rubber Co., Ltd., Japan), Jp02272004, **1990**.

68 Y. Hashiguchi, H. Katsumata, K. Goshima, T. Teramoto, Y. Takemura (Japan Synthetic Rubber Co., Ltd., Japan), Ep434469, **1991**.

69 L. R. Chamberlain, C. J. Gibler (Shell Oil Co., USA), US5039755, **1991**.

70 L. R. Chamberlain, C. J. Gibler, R. A. Kemp, S. E. Wilson (Shell International Research Maatschappij B. V., Neth.), Ep471415, **1992**.

71 L. R. Chamberlain, C. J. Gibler, R. A. Kemp, S. E. Wilson, T. F. Brownscombe (Shell Oil Co., USA), US5177155, **1993**.

72 K. Kato, G. Yamamoto (Asahi Chemical Ind., Japan; Asahi Kasei Chemical Corporation), Jp08033846, **1996**.

73 H. Nakafutami, K. Kato (Asahi Chemical Ind, Japan), Jp08027216, **1996**.

74 H. Nakafutami, K. Kato (Asahi Chemical Ind, Japan), Jp08027231, **1996**.

75 K. Iwase, K. Kato, K. Miyamoto (Asahi Chemical Industry Co., Ltd., Japan), Jp09278677, **1997**.

76 E. J. M. de Boer, B. Hessen, A. A. van der Huizen, W. de Jong, A. J. van der Linden, B. J. Ruisch, L. Schoon, H. J. A. de Smet, F. H. van der Steen, H. C. T. L. van Strien, A. Villena, J. J. B. Walhof (Shell Internationale Research Maatschappij BV, Neth.), Ep801079, **1997**.

77 E. J. M. de Boer, B. Hessen, A. A. van der Huizen, W. de Jong, A. J. van der Linden, B. J. Ruisch, L. Schoon, H. J. A. de Smet, F. H. van der Steen, H. C. T. van Strien, A. Villena, J. J. B. Walhof (Shell Internationale Research Maatschappij BV, Neth.), Ep795564, **1997**.

78 K. Kato, Y. Kusanose (Asahi Chemical Industry Co., Ltd., Japan), Jp10259221, **1998**.

79 A. Vallieri, C. Cavallo, G. T. Viola (Enichem S. P. A., Italy), Ep816382, **1998**.

80 J. Yonezawa, K. Kato, E. Sasaya, T. Sato (Asahi Kasei Kogyo K. K., Japan), De19815895, **1998**.

81 J. A. Barrio Calle, M. D. Parellada Ferrer, M. J. Espinosa Soriano (Repsol Quimica S. A., Spain), Ep885905, **1998**.

82 K. Miyamoto, Y. Kitagawa, S. Sasaki (Asahi Kasei Kogyo Kabushiki Kaisha, Japan), Ep889057, **1999**.

83 S. Sasaki, M. Kamatani, H. Shinjo (Asahi Chemical Industry Co., Ltd., Japan), Jp2000095804, **2000**.

84 S. Sasaki, M. Kamatani, H. Shinjo (Asahi Chemical Industry Co., Ltd., Japan), Jp2000095814, **2000**.

85 J. Cai, W. Zhang, Y. Zhao, C. Huang, H. Xiao, Z. Zhu (Inst. of Industrial Technology, Financial Group, Peop. Rep. China; Qimei Industrial Co., Ltd.), Cn1324867, **2001**.

86 J.-C. Tsai, W.-S. Chang, Y.-S. Chao, C.-N. Chu, C.-P. Huang, H.-Y. Hsiao (Industrial Technology Research Institute, Taiwan; Chi Mei Corporation), Us6313230, **2001**.

87 H. Zhang, W. Li, H. Liang, J. Zhang, X. Peng, H. Fang, H. Luo (Yueyang Petrochemical General Plant, Baling Petrochemical Corp., Peop. Rep. China), Cn1332183, **2002**.

88 S. Sasaki, T. Kasai, Y. Kusanose (Asahi Kasei Chemical Corporation, Japan), Jp2004269665, **2004**.

89 A. J. Calle Barrio, M. D. Ferrer Parellada, M. J. E. Soriano (Epsol Quimica
S. A., Spain), Ro115527, **2000**.

90 R. C.-C. Tsiang, W.-S. Yang, M.-D. Tsai,
Polymer **1999**, *40*, 6351.

91 R. C.-C. Tsiang, W.-S. Yang, M.-D. Tsai,
ACS Symp. Ser. **2000**, *760*, 108.

92 S.-s. Xu, L. Yang, K. Yuan, B.-q. Wang,
X.-z. Zhou, W. Li, X.-j. He, *Gaodeng
Xuexiao Huaxue Xuebao* **2001**, *22*, 2022.

93 P. Meunier, B. Gautheron, S. Couturier,
J. Organometal. Chem. **1982**, *231*, C1.

94 V. E. Cesarotti, R. Ugo, H. B. Kagan,
Angew. Chem. Int. Ed. **1979**, *18*, 779.

95 E. Cesarotti, R. Ugo, R. Vitiello, *J. Mol.
Cat.* **1981**, *12*, 63.

96 E. Cesarotti, H. B. Kagan, R. Goddard,
C. Krueger, *J. Organometal. Chem.*
1978, *162*, 297.

97 L. A. Paquette, J. A. McKinney, M. L.
McLaughlin, A. L. Rheingold, *Tetrahedron Lett.* **1986**, *27*, 5599.

98 R. L. Halterman, K. P. C. Vollhardt,
Organometallics **1988**, *7*, 883.

99 L. A. Paquette, M. R. Sivik, E. I. Bzowej,
K. J. Stanton, *Organometallics* **1995**, *14*,
4865.

100 R. L. Halterman, K. P. C. Vollhardt,
M. E. Welker, D. Blaeser, R. Boese,
J. Am. Chem. Soc. **1987**, *109*, 8105.

101 P. Beagley, P. J. Davies, A. J. Blacker, C.
White, *Organometallics* **2002**, *21*, 5852.

102 F. R. W. P. Wild, L. Zsolnai, G. Huttner,
H. H. Brintzinger, *J. Organometal.
Chem.* **1982**, *232*, 233.

103 F. R. W. P. Wild, M. Wasiucionek, G.
Huttner, H. H. Brintzinger, *J. Organometal. Chem.* **1985**, *288*, 63.

104 R. Waymouth, P. Pino, *J. Am. Chem.
Soc.* **1990**, *112*, 4911.

105 R. B. Grossman, R. A. Doyle, S. L. Buchwald, *Organometallics* **1991**, *10*, 1501.

106 R. D. Broene, S. L. Buchwald, *J. Am.
Chem. Soc.* **1993**, *115*, 12569.

107 M. V. Troutman, D. H. Appella, S. L.
Buchwald, *J. Am. Chem. Soc.* **1999**,
121, 4916.

108 C. A. Willoughby, S. L. Buchwald,
J. Am. Chem. Soc. **1992**, *114*, 7562.

109 C. A. Willoughby, S. L. Buchwald,
J. Am. Chem. Soc. **1994**, *116*, 8952.

110 C. A. Willoughby, S. L. Buchwald,
J. Org. Chem. **1993**, *58*, 7627.

111 A. Viso, N. E. Lee, S. L. Buchwald,
J. Am. Chem. Soc. **1994**, *116*, 9373.

112 N. E. Lee, S. L. Buchwald, *J. Am. Chem.
Soc.* **1994**, *116*, 5985.

113 J. Okuda, S. Verch, T. P. Spaniol,
R. Stuermer, *Chem. Ber.* **1996**, *129*, 1429.

114 R. Stürmer, J. Okuda, K. Ritter (BASF
AG, Germany), De19622271, **1997**.

115 J. Okuda, S. Verch, R. Stürmer, T. P.
Spaniol, *J. Organometal. Chem.* **2000**,
605, 55.

116 J. Okuda, S. Verch, R. Stürmer, T. P.
Spaniol, *Chirality* **2000**, *12*, 472.

117 M. Ringwald, R. Stürmer, H. H. Brintzinger, *J. Am. Chem. Soc.* **1999**, *121*,
1524.

118 M. Ringwald, R. Stuermer, H. H. Brintzinger, *J. Am. Chem. Soc.* **1999**, *121*,
7278.

119 H. Mauermann, P. N. Swepston, T. J.
Marks, *Organometallics* **1985**, *4*, 200.

120 G. Jeske, H. Lauke, H. Mauermann,
H. Schumann, T. J. Marks, *J. Am.
Chem. Soc.* **1985**, *107*, 8111.

121 P. W. Roesky, C. L. Stern, T. J. Marks,
Organometallics **1997**, *16*, 4705.

122 G. A. Molander, J. Winterfeld, *J. Organometal. Chem.* **1996**, *524*, 275.

123 G. A. Molander, J. O. Hoberg, *J. Org.
Chem.* **1992**, *57*, 3266.

124 C. M. Haar, C. L. Stern, T. J. Marks,
Organometallics **1996**, *15*, 1765.

125 W. J. Evans, I. Bloom, W. E. Hunter,
J. L. Atwood, *J. Am. Chem. Soc.* **1983**,
105, 1401.

126 W. J. Evans, J. H. Meadows, W. E. Hunter, J. L. Atwood, *J. Am. Chem. Soc.*
1984, *106*, 1291.

127 Y. Obora, T. Ohta, C. L. Stern, T. J. Marks,
J. Am. Chem. Soc. **1997**, *119*, 3745.

128 P. J. Fagan, J. M. Manriquez, E. A. Maat-
ta, A. M. Seyam, T. J. Marks, *J. Am.
Chem. Soc.* **1981**, *103*, 6650.

129 L. Jia, X. Yang, C. Stern, T. J. Marks,
Organometallics **1994**, *13*, 3755.

130 C. M. Fendrick, L. D. Schertz, V. W. Day,
T. J. Marks, *Organometallics* **1988**, *7*, 1828.

131 V. P. Conticello, L. Brard, M. A. Giardello,
Y. Tsuji, M. Sabat, C. L. Stern, T. J. Marks,
J. Am. Chem. Soc. **1992**, *114*, 2761.

132 M. A. Giardello, V. P. Conticello,
L. Brard, M. R. Gagne, T. J. Marks,
J. Am. Chem. Soc. **1994**, *116*, 10241.

133 A. Miyake, H. Kondo, *Angew. Chem. Int. Ed.* **1968**, *7*, 631.

134 A. Nakamura, S. Otsuka, *J. Am. Chem. Soc.* **1973**, *95*, 7262.

135 A. Nakamura, S. Otsuka, *Tetrahedron Lett.* **1973**, 4529.

136 D. Baudry, M. Ephritikhine, H. Felkin, *J. Chem. Soc., Chem. Soc.* **1980**, 1243.

137 D. Baudry, M. Ephritikhine, H. Felkin, Y. Jeannin, F. Robert, *J. Organometal. Chem.* **1981**, *220*, C7.

138 D. Baudry, M. Ephritikhine, H. Felkin, J. Zakrzewski, *J. Chem. Soc., Chem. Soc.* **1982**, 1235.

139 D. Baudry, M. Ephritikhine, H. Felkin, *J. Chem. Soc., Chem. Soc.* **1982**, 606.

140 D. Baudry, M. Ephritikhine, H. Felkin, *J. Organometal. Chem.* **1982**, *224*, 363.

141 D. Baudry, M. Ephritikhine, H. Felkin, R. Holmes-Smith, *J. Chem. Soc., Chem. Soc.* **1983**, 788.

142 T. Burchard, H. Felkin, *New J. Chem.* **1986**, *10*, 673.

143 B.C. Ankianiec, P.E. Fanwick, I.P. Rothwell, *J. Am. Chem. Soc.* **1991**, *113*, 4710.

144 I.P. Rothwell, *Chem. Commun.* **1997**, 1331.

145 J.S. Yu, B.C. Ankianiec, I.P. Rothwell, M.T. Nguyen, *J. Am. Chem. Soc.* **1992**, *114*, 1927.

146 V.M. Visciglio, P.E. Fanwick, I.P. Rothwell, *J. Chem. Soc., Chem. Soc.* **1992**, 1505.

147 D.R. Mulford, J.R. Clark, S.W. Schweiger, P.E. Fanwick, I.P. Rothwell, *Organometallics* **1999**, *18*, 4448.

148 R.W. Chesnut, B.D. Steffey, I.P. Rothwell, *Polyhedron* **1989**, *8*, 1607.

149 J.S. Yu, I.P. Rothwell, *J. Chem. Soc., Chem. Soc.* **1992**, 632.

150 R.D. Profilet, A.P. Rothwell, I.P. Rothwell, *J. Chem. Soc., Chem. Soc.* **1993**, 42.

151 I.P. Rothwell, S.J. Yu (Research Corp. Technologies, Inc., USA), Wo9321192, **1993**.

152 M.A. Lockwood, M.C. Potyen, B.D. Steffey, P.E. Fanwick, I.P. Rothwell, *Polyhedron* **1995**, *14*, 3293.

153 M.C. Potyen, I.P. Rothwell, *J. Chem. Soc., Chem. Soc.* **1995**, 849.

154 B.C. Parkin, J.R. Clark, V.M. Visciglio, P.E. Fanwick, I.P. Rothwell, *Organometallics* **1995**, *14*, 3002.

155 V.M. Visciglio, M.T. Nguyen, J.R. Clark, P.E. Fanwick, I.P. Rothwell, *Polyhedron* **1996**, *15*, 551.

156 V.M. Visciglio, J.R. Clark, M.T. Nguyen, D.R. Mulford, P.E. Fanwick, I.P. Rothwell, *J. Am. Chem. Soc.* **1997**, *119*, 3490.

157 R.H. Grubbs, C. Gibbons, L.C. Kroll, W.D. Bonds, Jr., C.H. Brubaker, Jr., *J. Am. Chem. Soc.* **1973**, *95*, 2373.

158 W.D. Bonds, Jr., C.H. Brubaker, Jr., E.S. Chandrasekaran, C. Gibbons, R.H. Grubbs, L.C. Kroll, *J. Am. Chem. Soc.* **1975**, *97*, 2128.

159 E.S. Chandrasekaran, R.H. Grubbs, C.H. Brubaker, Jr., *J. Organometal. Chem.* **1976**, *120*, 49.

160 R. Grubbs, C.P. Lau, R. Cukier, C. Brubaker, Jr., *J. Am. Chem. Soc.* **1977**, *99*, 4517.

161 J.G.-S. Lee, C.H. Brubaker, Jr., *J. Organometal. Chem.* **1977**, *135*, 115.

162 C.-P. Lau, B.-H. Chang, R.H. Grubbs, C.H. Brubaker, Jr., *J. Organometal. Chem.* **1981**, *214*, 325.

163 B.H. Chang, R.H. Grubbs, C.H. Brubaker, Jr., *J. Organometal. Chem.* **1985**, *280*, 365.

164 B.H. Chang, C.P. Lau, R.H. Grubbs, C.H. Brubaker, Jr., *J. Organometal. Chem.* **1985**, *281*, 213.

165 L. Verdet, J.K. Stille, *Organometallics* **1982**, *1*, 380.

166 R. Jackson, J. Ruddlesden, D.J. Thompson, R. Whelan, *J. Organometal. Chem.* **1977**, *125*, 57.

167 J. Cermak, M. Kvicalova, V. Blechta, M. Capka, Z. Bastl, *J. Organometal. Chem.* **1996**, *509*, 77.

168 D.G.H. Ballard, *Adv. Catal.* **1973**, *23*, 263.

169 Y.I. Yermakov, B.N. Kuznetsov, V.A. Zakharov, *Stud. Surf. Sci. Catal.* **1981**, *8*, 1.

170 C.W. Tullock, F.N. Tebbe, R. Mulhaupt, D.W. Ovenall, R.A. Setterquist, S.D. Ittel, *J. Polym. Sci., A: Polym. Chem.* **1989**, *27*, 3063.

171 C. W. Tullock, R. Mulhaupt, S. D. Ittel, *Makromol. Chem., Rapid Commun.* **1989**, *10*, 19.

172 S. D. Ittel, *J. Macromol. Sci., Chem.* **1990**, *A27*, 1133.

173 J. Schwartz, M. D. Ward, *J. Mol. Cat.* **1980**, *8*, 465.

174 S. A. King, J. Schwartz, *Inorg. Chem.* **1991**, *30*, 3771.

175 Y. I. Yermakov, Y. A. Ryndin, O. S. Alekseev, D. I. Kochubei, V. A. Shmachkov, N. I. Gergert, *J. Mol. Cat.* **1989**, *49*, 121.

176 V. A. Zakharov, Y. A. Ryndin, *J. Mol. Cat.* **1989**, *56*, 183.

177 C. Copéret, M. Chabanas, R. P. Saint-Arroman, J.-M. Basset, *Angew. Chem. Int. Ed.* **2003**, *42*, 156.

178 C. Rosier, G. P. Niccolai, J.-M. Basset, *J. Am. Chem. Soc.* **1997**, *119*, 12408.

179 J. Corker, F. Lefebvre, C. Lecuyer, V. Dufaud, F. Quignard, A. Choplin, J. Evans, J.-M. Basset, *Science* **1996**, *271*, 966.

180 F. Rataboul, A. Baudouin, C. Thieuleux, L. Veyre, C. Copéret, J. Thivolle-Cazat, J.-M. Basset, A. Lesage, L. Emsley, *J. Am. Chem. Soc.* **2004**, *126*, 12541.

181 L. d'Ornelas, S. Reyes, F. Quignard, A. Choplin, J. M. Basset, *Chem. Lett.* **1993**, 1931.

182 V. Vidal, A. Theolier, J. Thivolle-Cazat, J. M. Basset, J. Corker, *J. Am. Chem. Soc.* **1996**, *118*, 4595.

183 R. L. Burwell, Jr., T. J. Marks, *Chem. Ind.* **1985**, *22*, 207.

184 T. J. Marks, *Acc. Chem. Res.* **1992**, *25*, 57.

185 K. H. Dahmen, D. Hedden, R. L. Burwell, Jr., T. J. Marks, *Langmuir* **1988**, *4*, 1212.

186 H. Ahn, C. P. Nicholas, T. J. Marks, *Organometallics* **2002**, *21*, 1788.

187 C. P. Nicholas, H. Ahn, T. J. Marks, *J. Am. Chem. Soc.* **2003**, *125*, 4325.

188 C. P. Nicholas, T. J. Marks, *Langmuir* **2004**, *20*, 9456.

189 R. G. Bowman, R. Nakamura, P. J. Fagan, R. L. Burwell, Jr., T. J. Marks, *J. Chem. Soc., Chem. Soc.* **1981**, 257.

190 M. Y. He, G. Xiong, P. J. Toscano, R. L. Burwell, Jr., T. J. Marks, *J. Am. Chem. Soc.* **1985**, *107*, 641.

191 R. D. Gillespie, R. L. Burwell, Jr., T. J. Marks, *Langmuir* **1990**, *6*, 1465.

192 M. S. Eisen, T. J. Marks, *J. Am. Chem. Soc.* **1992**, *114*, 10358.

193 M. S. Eisen, T. J. Marks, *J. Mol. Cat.* **1994**, *86*, 23.

7
Ionic Hydrogenations

R. Morris Bullock

7.1
Introduction

One pervasive mechanistic feature of many of the hydrogenations described in other chapters of this handbook concerns the bonding of the unsaturated substrate to a metal center. As illustrated in generalized form in Eq. (1) for the hydrogenation of a ketone, a key step in the traditional mechanism of hydrogenation is migratory insertion of the bound substrate into a metal hydride bond (M–H).

$$\tag{1}$$

Since this can formally be viewed as an addition of M–H to C=O, the bond order is reduced in this case to a C–O single bond. Reductive elimination generates the hydrogenated product and an unsaturated metal complex that subsequently re-enters the catalytic cycle. Many subtleties of this mechanism have been delineated in studies of hydrogenations of C=C and C=O bonds, and catalysts that follow this mechanism have been very successful.

Most homogeneous catalysts that proceed by traditional insertion mechanisms use precious metals. If the requirement for substrate binding and insertion is removed, then alternative mechanisms would be possible. Such alternative mechanisms could exploit other reactivity patterns accessible to metal hydrides, thus removing the requirement for precious metals. The use of inexpensive metals potentially offers several advantages, if catalysts containing them could be developed with sufficient turnover frequencies (TOFs) and lifetimes. A substantially lower cost of the metal is an obvious advantage, though it is recognized that many factors influence the overall costs of a process, and phosphines or

The Handbook of Homogeneous Hydrogenation.
Edited by J.G. de Vries and C.J. Elsevier
Copyright © 2007 WILEY-VCH Verlag GmbH & Co. KGaA, Weinheim
ISBN: 978-3-527-31161-3

Scheme 7.1

other ligands can substantially add to the cost of synthesis of an organometallic catalyst precursor. In addition to the initial cost of the metal, other considerations include less stringent requirements for recovery of the catalyst in an industrial process. For processes that might be used in the manufacture of pharmaceutical products, catalysts using abundant metals might be permitted at a higher residual level than other metals, owing to toxicity considerations. The premise of using "Cheap Metals for Noble Tasks" thus has appeal despite these caveats; fundamentally new mechanistic information and novel reactivity patterns have resulted from research directed towards this long-term goal.

Ionic hydrogenations involve addition of H_2 in the form of H^+ followed by H^-, as shown in Scheme 7.1; the proton and hydride transfers may be either sequential or concerted. A potential advantage of ionic hydrogenations is that the nature of the mechanism would tend to favor hydrogenation of polar bonds such as C=O over less-polar C=C bonds. Kinetic and mechanistic studies have played a key role in the development of ionic hydrogenations. In many cases to be discussed in this chapter, the individual proton-transfer and hydride-transfer steps that comprise the key steps in catalytic cycles can be separately studied in stoichiometric reactions. Mechanistic information can then be used to guide the rational design of new catalysts or the improvement and optimization of initially discovered ionic hydrogenation catalysts.

7.2
Stoichiometric Ionic Hydrogenations

7.2.1
Stoichiometric Ionic Hydrogenations using CF_3CO_2H and $HSiEt_3$

Ionic hydrogenations of C=C and C=O bonds were reported prior to the development of ionic hydrogenations mediated or catalyzed by transition metals. Trifluoroacetic acid (CF_3CO_2H) as the proton donor and triethylsilane ($HSiEt_3$) as the hydride donor are most commonly used, though a variety of other acids and several other hydride donors have also been shown to be effective. A review [1] by Kursanov et al. of the applications of ionic hydrogenations in organic synthesis documents the early progress in this field; a book gives further details [2]. As shown in Eq. (2), proton transfer to the alkene generates a carbenium ion, and hydride transfer from the hydrosilane generates the product.

$$(2)$$

Ionic hydrogenations of C=C bonds generally work well only in cases where a tertiary or aryl-substituted carbenium ion can be formed through protonation of the C=C bond. Alkenes that give a tertiary carbenium ion upon protonation include 1,1-disubstituted, tri-substituted and tetra-substituted alkenes, and each of these are usually hydrogenated by ionic hydrogenation methods in high yields.

The success of stoichiometric ionic hydrogenations is due to achieving a fine balance that favors the intended reactivity rather than any of several possible alternative reactions. The acid must be strong enough to protonate the unsaturated substrate, yet the reaction of the acid and the hydride should avoid producing H_2 too quickly under the reaction conditions. The commonly used pair of CF_3CO_2H and $HSiEt_3$ meets all these criteria.

The very strong acid, CF_3SO_3H (triflic acid, abbreviated as HOTf) can be used in conjunction with $HSiEt_3$ for the hydrogenation of certain alkenes [3]. These reactions proceed cleanly in 5 minutes at $-50\,°C$. This discovery was surprising, since a review of the use of CF_3CO_2H and $HSiEt_3$ had stated that "... stronger acids cannot be used in conjunction with silanes because they react" [1]. Indeed, rapid evolution of H_2 does occur when HOTf is added to $HSiEt_3$ in the absence of an alkene. The order of addition is important in the use of $HOTf/HSiEt_3$ for hydrogenation of C=C bonds, to ensure that acid-induced formation of H_2 is minimized. The addition of HOTf to a solution containing the alkene and the hydrosilane results in rapid and clean hydrogenation, but the reaction is still subject to the limitation of forming a tertiary carbenium ion.

Another potential mechanistic complication is capture of the intermediate carbenium ion by the conjugate base of the acid. When CF_3CO_2H is used as the acid, this would lead to trifluoroacetate esters. Kursanov et al. showed that, under the reaction conditions for ionic hydrogenations, trifluoroacetate esters can be converted to the hydrocarbon product (Eq. (3)).

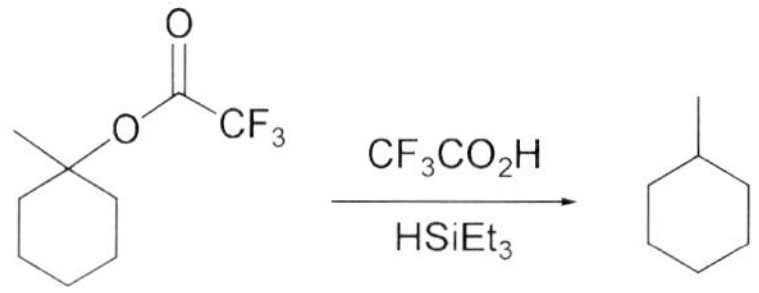

(3)

The intermediacy of the trifluoroacetate ester does not undermine the efficacy of the overall hydrogenation reaction, since the ionizing solvent CF_3CO_2H converts the ester back to the carbenium ion under the reaction conditions, resulting in its ultimate conversion to the hydrogenation product.

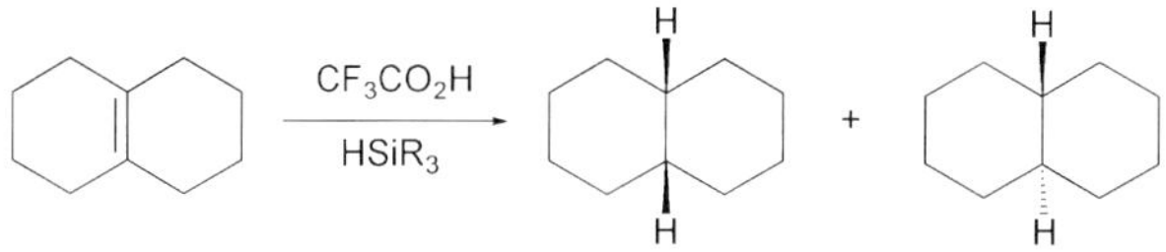

(4)

Studies of the ionic hydrogenation of $\Delta^{9(10)}$-octalin using CF_3CO_2H and a variety of hydrosilanes demonstrated that a considerable degree of stereoselectivity can be obtained (Eq. (4)) [4]. Use of $^nBuSiH_3$ as the hydride donor led to a 22:78 ratio of *cis*:*trans* decalin. Hydrogenation using the sterically demanding hydrosilane tBu_3SiH, in contrast, led to predominant formation of the opposite isomer, with 93% of the decahydronaphthalene product being the *cis* isomer. Steric factors of the silane hydride donor appear to dominate the stereoselectivity in this example, though in other examples the effect is much less, suggesting that additional factors can influence the product distribution. Ionic hydrogenation of 4-*tert*-butylmethylenecyclohexane with the same series of hydrosilanes invariably produced a predominance of the *trans* isomer [4].

Ionic hydrogenation of benzophenone using CF_3CO_2H and $HSiEt_3$ proceeds at room temperature and gives high yields of diphenylmethane (Eq. (5)) [1, 5]. This reaction presumably proceeds through the alcohol, as indicated in Eq. (5), but deoxygenation of the alcohol proceeds faster than hydrogenation of the C=O function, so the alcohol is not detected. Similar hydrogenations of the C=O group to CH_2 were found for aryl alkyl ketones. Dialkyl ketones react with CF_3CO_2H and $HSiEt_3$ to produce trifluoroacetate esters of the secondary alcohols, so conversion to the alcohol requires a subsequent hydrolysis reaction [1].

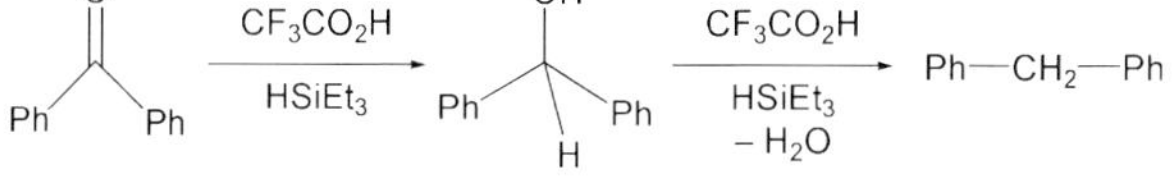

(5)

The hydrogenation of aldehydes by $CF_3CO_2H/HSiEt_3$ is often complicated by the formation of ethers. Doyle et al. found that the use of aqueous acids such as H_2SO_4, together with nonreactive solvents (CH_3CN), allowed some aldehydes and dialkyl ketones to be reduced to alcohols using $HSiEt_3$ [6].

7.2.2
Stoichiometric Ionic Hydrogenations using Transition-Metal Hydrides

7.2.2.1 General Aspects

While stoichiometric ionic hydrogenations using CF_3CO_2H and $HSiEt_3$ have enjoyed significant utility in organic synthetic reactions, they require stoichiometric quantities of both the hydrosilane and an acid. One of the principles of Green Chemistry [7] is that catalytic reactions are preferred over stoichiometric reagents. The development of a catalytic route to ionic hydrogenations would be environmentally attractive. In addition to the requirement of delivering a proton and a hydride to the substrate, catalytic methods additionally require a metal complex that is capable of reacting with H_2 to regenerate the proton and hydride sources. The following sections will demonstrate how kinetic and mechanistic studies separately documented the proton-transfer and hydride-transfer capabilities of transition-metal hydrides. These studies provided a firm mechanistic basis for the development and understanding of catalytic ionic hydrogenations.

7.2.2.2 Transition-Metal Hydrides as Proton Donors

The fact that metal hydrides can be acidic may seem paradoxical in view of the nomenclature that insists that all complexes with a M–H bond be referred to as "hydrides" regardless of whether their reactivity is hydridic or not. Not only can some metal hydrides donate a proton, but some can be remarkably acidic. Some cationic dihydrogen complexes are sufficiently acidic to protonate Et_2O [8], and some dicationic ruthenium complexes have an acidity comparable to or exceeding that of HOTf [9].

Systematic studies of the thermodynamic and kinetic acidity of metal hydrides in acetonitrile were carried out by Norton et al. [10, 11]. A review of the acidity of metal hydrides presents extensive tabulations of pK_a data [12]; only a few of the trends will be mentioned here. Metal hydrides span a wide range of pK_a values; considering only metal carbonyl hydrides shown in Table 7.1, the range exceeds 20 pK_a units. As expected, a substantial decrease in acidity is

Table 7.1 pK_a values of neutral metal carbonyl hydrides in CH_3CN.

Metal hydride	pK_a	Reference
$(CO)_4CoH$	8.3	11
$Cp(CO)_3MoH$	13.9	11
$(CO)_5MnH$	14.2	12
$(CO)_3(PPh_3)CoH$	15.4	11
$Cp(CO)_3WH$	16.1	11
$Cp^*(CO)_3MoH$	17.1	11
$(CO)_5ReH$	21.1	11
$Cp(CO)_2(PMe_3)WH$	26.6	11

found when an electron-accepting CO ligand is replaced by an electron-donating phosphine. The cobalt hydride [(CO)$_4$CoH] is quite acidic, being of comparable acidity in CH$_3$CN to that of HCl (pK_a = 8.9 in CH$_3$CN); substitution of one CO by PPh$_3$ to give [(CO)$_3$(PPh$_3$)CoH] reduces the acidity by about seven pK_a units. The stronger electron donor, PMe$_3$, has an even larger effect, as exemplified by the acidity of [Cp(CO)$_2$(PMe$_3$)WH] (Cp = η^5-C$_5$H$_5$) being less than that of [Cp(CO)$_3$WH] by about 10 pK_a units. Replacement of a Cp by Cp* [(Cp* = η^5-C$_5$Me$_5$)] also decreases the acidity, by about three pK_a units in the case of [Cp(CO)$_3$MoH] versus [Cp*(CO)$_3$MoH]. Third-row metals are less acidic than their first- or second-row analogues, as shown by [Cp(CO)$_3$MoH] being two pK_a units less acidic than [Cp(CO)$_3$WH]. A larger difference of about seven pK_a units was found between first-row [(CO)$_5$MnH] and third-row [(CO)$_5$ReH].

DuBois and coworkers have studied a wide range of metal hydrides, concentrating on those with two diphosphine ligands [13–19]. Measurements of pK_a values for a series of cobalt hydrides (Table 7.2, upper part) showed that the cationic dihydride [(H)$_2$Co(dppe)$_2$]$^+$ has a pK_a of 22.8, while the neutral cobalt hydride [HCo(dppe)$_2$] is far less acidic (pK_a = 38.1). Oxidation of this neutral hydride gives the paramagnetic Co(II) hydride, [HCo(dppe)$_2$]$^+$, which is much more acidic (pK_a = 23.6). The dicationic hydride [HCo(dppe)$_2$(NCCH$_3$)]$^{2+}$ (pK_a = 11.3) is by far the most acidic of this series. This remarkable series of complexes, all containing a Co(dppe)$_2$ core, span about 27 pK_a units as the oxidation states and formal charges are varied. The series of [HM(diphosphine)$_2$]$^+$ complexes in the lower part of Table 7.2 show that altering the electronic properties of substituents on the diphosphine ligand can have a profound effect on acidity. Complexes with two dmpe ligand (with methyl groups on the phosphorus) are six to ten pK_a units less acidic than corresponding complexes with a dppe ligand (with phenyl groups on the phosphorus). As was found for the metal carbonyl complexes discussed above, hydrides of the third row metal are

Table 7.2 pK_a values of metal bis(diphosphine) hydrides in CH$_3$CN.

Metal hydride [a)	pK_a	Reference
[(H)$_2$Co(dppe)$_2$]$^+$	22.8	16
[HCo(dppe)$_2$]$^+$	23.6	16
HCo(dppe)$_2$	38.1	16
[HCo(dppe)$_2$(NCCH$_3$)]$^{+2}$	11.3	16
[HNi(dppe)$_2$]$^+$	14.2	13
[HPt(dppe)$_2$]$^+$	22.0	13
[HNi(dmpe)$_2$]$^+$	24.3 [b)	13
[HPt(dmpe)$_2$]$^+$	31.1 [b)	15

a) dmpe = 1,2-bis(dimethylphosphino)ethane;
    dppe = 1,2-bis(diphenylphosphino)ethane.
b) Value determined in PhCN, though these values are usually
    similar to those found in CH$_3$CN, typically differing by less
    than 1 pK_a unit between the two solvents.

significantly less acidic than those of the first-row metal (Ni versus Pt in the examples shown in Table 7.2).

Morris et al. carried out extensive studies [20] of the acidity of metal hydrides in tetrahydrofuran (THF), including metal hydrides of very low acidity as well as dihydrogen complexes that are reactive with CH_3CN. The dielectric constant of THF is low compared to that of CH_3CN, so ion-pairing issues must be taken into account [21], though these measurements in THF provide useful comparisons to data in CH_3CN and other solvents.

7.2.3
Transition Metal Hydrides as Hydride Donors

An understanding of the factors that influence the propensity of metal hydrides to function as hydride donors is important in the use of such hydrides in ionic hydrogenations. Thermodynamic hydricities are immensely useful in considering the viability of potential hydride transfer reactions in stoichiometric or catalytic reactions. Since catalysis is a kinetic phenomenon, it is also necessary to understand the factors influencing the kinetics of hydride transfer reactions. Systematic studies that provided quantitative values for the thermodynamic and kinetic hydricity of a wide series of hydrides occurred after studies had been conducted that identified the factors influencing proton-transfer reactions of metal hydrides. A complicating factor in considering tests for hydricity is that removal of H^- from M–H produces a cationic species M^+ [22]. When the starting hydride M–H is an 18-electron complex, the M^+ that initially results will be an unsaturated 16-electron species. In most cases these 16-electron metal cations will not be stable but will voraciously seek an additional ligand. Since the overall reaction may involve ligand capture as well as the hydride transfer, comparisons of hydricity may be complicated by the subsequent reactivity of M^+ following the hydride transfer reaction.

It has long been recognized that the hydricity of a metal hydride can vary according to its position in the Periodic Table. Labinger and Komadina found evidence for this trend from an examination of the reactivity of a series of metallocene hydrides [23]. The Group 4 zirconium dihydride $[Cp_2ZrH_2]_n$ reacts quickly at room temperature with acetone to give $[Cp_2Zr(OCHMe_2)_2]$, which releases isopropyl alcohol upon hydrolysis. The Group 5 complex $[Cp_2NbH_3]$ reacts slowly with acetone, but quickly with the more electrophilic ketone $(CF_3)(CH_3)C=O$. The Group 6 dihydride, $[Cp_2MoH_2]$, did not react with acetone at $78\,^\circ C$, but did react slowly at $25\,^\circ C$ with $(CF_3)(CH_3)C=O$. The Group 7 hydride $[Cp_2ReH]$ did not react with either acetone or with $(CF_3)(CH_3)C=O$. This study provided evidence that hydricity is higher for metals to the left side of the Periodic Table, though if the ketone coordinates to the metal prior to hydride transfer, these measurements may reflect a combination of factors, rather than measuring only hydricity.

Darensbourg et al. have conducted extensive studies of the nucleophilic reactivity of a series of anionic metal carbonyl hydrides [24], which have been used for the reduction of alkyl halides [25], acyl chlorides [26], and ketones [27]. The

Table 7.3 Rate constants for reaction of anionic metal
hydrides with *n*-BuBr (THF solvent at 26 °C) [25].

Metal hydride	$10^3 \times k_H$ [M^{-1} s^{-1}]
[HW(CO)$_4$P(OCH$_3$)$_3$]$^-$	50
[HCr(CO)$_4$P(OCH$_3$)$_3$]$^-$	30
[HW(CO)$_5$]$^-$	3.3
[HV(CO)$_3$Cp]$^-$	2.2
[HCr(CO)$_5$]$^-$	1.8
[HRu(CO)$_5$]$^-$	1.0

kinetics of bromide displacement from *n*-BuBr by a series of anionic hydrides
(Eq. (6)) gave the results shown in Table 7.3.

$$MH^{\ominus} + Bu - Br \quad \rightarrow \quad MBr^{\ominus} + Bu - H \tag{6}$$

Although the range of rate constants observed is only about a factor of 50,
there is a clear trend indicating that the replacement of one CO by a phosphite
ligand increases the kinetic hydricity. The third-row tungsten hydrides are faster
hydride donors compared to the first-row chromium analogues. Whilst direct
displacement of the bromide by hydride is the prevalent mechanism in the reaction of this primary alkyl bromide, a radical chain mechanism involving hydrogen atom transfer from the metal hydride is also operative. This radical chain
mechanism (S$_H$2 pathway) is the predominant pathway in reactions of these
same anionic hydrides with sterically encumbered alkyl halides, where the S$_N$2
hydride displacement pathway is disfavored [28].

DuBois et al. carried out extensive studies on the thermodynamic hydricity of
a series of metal hydrides [13, 15–19]. The determination of thermodynamic hydricity generally requires several measurements (coupled with known thermochemical data) to constitute a complete thermochemical cycle. As with other
thermodynamic cycles, obtaining reliable values in an appropriate solvent can
be a difficult challenge, and this is sometimes coupled with problems in obtaining reversible electrochemical data. Scheme 7.2 illustrates an example in which
the hydricity of cationic monohydrides have been determined.

Thus, the thermodynamic hydricity shown in Eq. (10) is determined by evalu-
ating the values for Eqs. (7) to (9). Equation (7) is the pK_a of the hydride, and

$$L_nMH^+ \rightarrow L_nM + H^+ \tag{7}$$

$$L_nM \rightarrow L_nM^{2+} + 2e^- \tag{8}$$

$$\underline{H^+ + 2e^- \rightarrow H^-} \tag{9}$$

$$L_nMH^+ \rightarrow L_nM^{2+} + H^- \tag{10} \ (\Delta G^{\circ}_{H^-})$$

Scheme 7.2 Thermodynamic cycle for
determination of the hydricity of cationic metal hydrides.

$$L_nMH^+ + BH^+ \rightarrow L_nM^{2+} + H_2^+ + B \qquad (11)$$

$$B + H^+ \rightarrow BH^+ \qquad (12)$$

$$\underline{H_2 \rightarrow H^+ + H^- \qquad\qquad\qquad\qquad\qquad (13)}$$

$$L_nMH^+ \rightarrow L_nM^{2+} + H^- \qquad (\Delta G^\circ_{H^-}) \ (14)$$

Scheme 7.3 Thermodynamic cycle for determination of the hydricity using heterolytic cleavage of hydrogen.

Eq. (8) requires determination of the two-electron oxidation potential of L_nM by electrochemical methods. When combined with the two-electron reduction of protons in Eq. (9), the sum provides Eq. (10), the $\Delta G^0_{H^-}$ values of which can be compared for a series of metal hydrides. Another way to determine the $\Delta G^0_{H^-}$ entails the thermochemical cycle is shown in Scheme 7.3. This method requires measurement of the K_{eq} of Eq. (11) for a metal complex capable of heterolytic cleavage of H_2, using a base (B), where the pK_a of BH^+ must be known in the solvent in which the other measurements are conducted. In several cases, Du-Bois et al. were able to demonstrate that the two methods gave the same results. The thermodynamic hydricity data ($\Delta G^0_{H^-}$ in CH_3CN) for a series of metal hydrides are listed in Table 7.4. Transition metal hydrides exhibit a remarkably large range of thermodynamic hydricity, spanning some 30 kcal mol^{-1}.

Several trends are revealed by these data. For example, third-row metal hydrides are much more hydridic compared to their first-row analogues, so the trend of higher hydricity for third-row hydrides holds for both kinetic and thermodynamic hydricity. The hydricity of $[HPt(depe)_2]^+$ exceeds that of $[HNi(depe)_2]^+$ by 15 kcal mol^{-1}, and other Pt hydrides reflect the same trend, by 11–14 kcal mol^{-1} [13]. The second-row hydride $[HRh(dppb)_2]$ is a very powerful hydride donor, and its hydricity exceeds that of the first-row Co analogue by 14 kcal mol^{-1} [17]. Changing the ligands on a metal can also cause a profound change in thermodynamic hydricity. The chelating diphosphine ligand dmpe is much more electron-donating than dppe, with the Pt complex $[HPt(dmpe)_2]^+$ having a hydricity that exceeds that of $[HPt(dppe)_2]^+$ by 10 kcal mol^{-1} [13, 15]. The same conclusion is reached for comparisons of Ni complexes with different diphosphine ligands. Replacement of one of the electron-withdrawing CO ligands of $[HW(CO)_5]^-$ with PPh_3 leads to an increase of 4 kcal mol^{-1} in the hydricity of $[HW(CO)_4(PPh_3)]^-$ [18]. DuBois et al. found a dramatic dependence of hydricity on the bite angle of a series of Pd hydrides [19]. A variation of the bite angle of 33° results in tuning of the hydricity over a range of 27 kcal mol^{-1}, with smaller bite angles of the diphosphine leading to higher hydricity. Overall, DuBois et al. have concluded that, for isoelectronic complexes, the order of hydricity is second row > third row > first row [17]. The charge on a metal complex can affect its hydricity. For a series of cobalt complexes, the neutral complex $[HCo(dppe)_2]$ is a stronger hydride donor than the cationic, paramagnetic complex $[HCo(dppe)_2]^+$ [16]. While this follows the expected trend, it is clear from the data in Table 7.4 that the overall charge is just one factor influencing

Table 7.4 ΔG^0 for hydride transfer from metal hydrides in CH_3CN.

Hydride donor[a]	ΔG^0 [kcal mol^{-1}][b]	Reference
$HRh(dppb)_2$	34	17
$HW(CO)_4(PPh_3)^-$	36	18
$[HPt(dmpe)_2]^+$	42	15
$HW(CO)_5^-$	40	18
$[HPt(depe)_2]^+$	44	15
$[HPt(dmpp)_2]^+$	51	15
$HCo(dppe)_2$	49	16
$HCo(dppb)_2$	48	17
$[HNi(dmpe)_2]^+$	51	13
$[HPt(dppe)_2]^+$	52	15
$HMo(CO)_2(PMe_3)Cp$	55	91
$[HNi(depe)_2]^+$	56	13
$[HCo(dppe)_2]^+$	60	16
$[HNi(dmpp)_2]^+$	61	13
$[HNi(dppe)_2]^+$	63	13
$[(H)_2Co(dppe)_2]^+$	65	16
$[HPd(EtXantphos)_2]^+$	70	19
H_2	76	15
$HCPh_3$	99	92

a) dmpe and dppe as defined in Table 7.2; dmpp = 1,3-bis(dimethylphosphino)propane; depe = 1,2-bis(diethylphosphino)ethane; dppe = 1,2-bis(diphenylphosphino)ethane; depp = 1,2-bis(diethylphosphino)propane; EtXantphos = 9,9-dimethyl-4,5-bis(diethylphosphino)xanthene.
b) Estimated uncertainties on ΔG^0 values are typically ± 2 kcal mol^{-1}.

the hydricity, as many of the cationic hydrides are stronger hydride donors than neutral hydrides. For the series of $[HM(diphosphine)_2]^+$ complexes that have been the focus of much of the research by DuBois and colleagues, the high stability of the square-planar products, $[M(diphosphine)_2]^{+2}$, which result from hydride transfer, helps to explain the high hydricity of $[HM(diphosphine)_2]^+$.

The kinetics of hydride transfer from a series of neutral metal carbonyl hydrides have been determined by studying hydride transfer to $Ph_3C^+BF_4^-$ (Eq. (15)). In CH_2Cl_2 solvent, the M^+ cation resulting from hydride transfer from the metal hydride is captured by the BF_4^- anion, forming complexes with weakly bound FBF_3 ligands. A wide range of M–FBF$_3$ complexes have been studied by Beck and Sünkel [29]. The second-order rate constants for the hydride transfer reaction in Eq. (15) are listed in Table 7.5. The range of rate constants spans a factor of over 10^6, documenting a considerable range of kinetic hydricity.

$$M\text{-}H + Ph_3C^+BF_4^- \xrightarrow{k_{H^-}} M\text{-}F\text{-}BF_3 + Ph_3C\text{-}H \qquad (15)$$

Table 7.5 Rate constants for hydride transfer from metal
hydrides to Ph_3C^+ BF_4^- (CH_2Cl_2 solvent at 25 °C) [58, 93].

Metal hydride	k_H [M^{-1} s^{-1}]
$HRu(CO)_2Cp^*$	$>5 \times 10^6$
trans-$HMo(CO)_2(PMe_3)Cp$	4.6×10^6
$HFe(CO)_2Cp^*$	1.1×10^6
trans-$HMo(CO)_2(PPh_3)Cp$	5.7×10^5
trans-$HMo(CO)_2(PCy_3)Cp$	4.3×10^5
$HOs(CO)_2Cp^*$	3.2×10^5
$HW(NO)_2Cp$	1.9×10^4
cis-$HRe(CO)_4(PPh_3)$	1.2×10^4
$HMo(CO)_3Cp^*$	6.5×10^3
$HRe(CO)_5$	2.0×10^3
$HW(CO)_3(\eta^5\text{-indenyl})$	2.0×10^3
$HW(CO)_3Cp^*$	1.9×10^3
$HMo(CO)_3Cp$	3.8×10^2
$HW(CO)_3(C_5H_4Me)$	2.5×10^2
cis-$HMn(CO)_4(PPh_3)$	2.3×10^2
$HSiEt_3$	1.5×10^2
$HW(CO)_3Cp$	7.6×10^1
$HCr(CO)_3Cp^*$	5.7×10^1
$HMn(CO)_5$	5.0×10^1
$W(CO)_3(C_5H_4CO_2Me)$	7.2×10^{-1}

This systematic study reveals how changes in the metal and both steric and electronic changes of the ligand can alter the hydricity. Third-row metal hydrides are substantially more kinetically hydridic than their first-row analogues, with $[Cp^*(CO)_3WH]$ being about 33-fold faster at hydride transfer than $[Cp^*(CO)_3CrH]$, and a factor of about 40–50 being found for Re versus Mn hydrides. The second-row hydrides of Mo are about three- to five-fold faster than those of the third-row W hydrides. Changes in ligands cause an even more dramatic change in the kinetic hydricity. When a Cp ligand in $[CpM(CO)_3H]$ is replaced by Cp^*, the kinetic hydricity increases by a factor of 25 for M = W and a factor of 17 for M = Mo. Compared to Cp, the Cp^* ligand is much larger, so steric effects would predict a lower reactivity of the Cp^* compared to the Cp complex. Since Cp^* is a much better electron donor than Cp, the higher rate constant for the Cp^* complex makes it clear that electronic effects dominate over steric effects in these reactions. Even a single methyl group has a small, but measurable, effect on the hydricity, as indicated by the larger rate constant for $[(C_5H_4Me)(CO)_3WH]$ compared to $[Cp(CO)_3WH]$. An even more prominent effect upon substitution of a single substituent on the Cp ligand was found for the compound containing an electron-withdrawing group on the Cp ligand, with the hydricity of $[Cp(CO)_3WH]$ exceeding that of $[(C_5H_4CO_2Me)(CO)_3WH]$ by about a factor of 100. Replacement of one CO by a phosphine ligand has an even larger effect in enhancing the kinetic hydricity of metal hydrides. The

PMe$_3$-substituted Mo hydride *trans*-[Cp(CO)$_2$(PMe$_3$)MoH] is about 10^4 times as hydridic as [Cp(CO)$_3$MoH]. Some evidence for steric effects is apparent from the kinetic hydricity of *trans*-[Cp(CO)$_2$(PCy$_3$)MoH]. The PCy$_3$ is similar to PMe$_3$ electronically, but it is much more sterically demanding. The hydricity of *trans*-[Cp(CO)$_2$(PCy$_3$)MoH] is about 10 times less than that of *trans*-[Cp(CO)$_2$(PMe$_3$)-MoH], presumably due to steric effects. Yet even with this large ligand, the hydricity of *trans*-[Cp(CO)$_2$(PCy$_3$)MoH] exceeds that of Cp(CO)$_3$MoH by about three orders of magnitude, again providing strong evidence of the predominance of electronic over steric effects in these kinetic hydricity studies.

7.2.4
Stoichiometric Ionic Hydrogenation of Alkenes with Metal Hydrides as the Hydride Donor

As discussed earlier, Kursanov et al. showed that some alkenes can be hydrogenated using acids in conjunction with HSiEt$_3$ as the hydride donor. Compared with silanes, transition-metal hydrides as hydride donors offer substantial advantages, if their hydride donor capability can be coupled into a catalytic cycle that regenerates the M–H bond through reaction with H$_2$. Additionally, the versatile reactivity patterns of metal hydrides reveal another benefit. Whilst silanes generally react with acids to immediately evolve H$_2$, many transition-metal hydrides can be protonated (Scheme 7.4) to give dihydrogen complexes [30] or dihydrides. Protonation at the M–H to produce a dihydrogen complex is often kinetically preferred [31] over direct protonation at the metal to generate a dihydride; in many cases, the dihydrogen and dihydride forms are in equilibrium with each other. For the purposes of hydrogenation, it may not make much difference which form is predominant, as long as the complex is sufficiently acidic to transfer a proton to the substrate that is to be hydrogenated. Thus, the possibility of protonation of a metal hydride provides an alternate mechanistic pathway for stoichiometric ionic hydrogenations. Proton transfer from an external acid may occur to the substrate directly, or may involve protonation of a metal hydride, with subsequent delivery of the proton to the substrate. Deprotonation of a cationic dihydride or dihydrogen complex will generate a neutral metal hy-

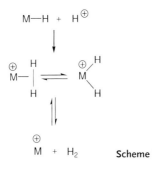

Scheme 7.4

dride, which can then serve as a hydride donor. Failures can occur if irreversible loss of H_2 from the metal occurs more quickly than proton transfer to the unsaturated substrate.

Ionic hydrogenation of certain C=C double bonds is readily accomplished through reaction with HOTf and transition-metal hydrides (Eq. (16)). These reactions proceed in high yield (>90%) in less than 5 minutes at 22 °C, and were shown in some cases to occur at temperatures as low as –50 °C. The limitations on the olefin substrate are the same as those encountered in the stoichiometric ionic hydrogenations using $HSiEt_3$ as the hydride donor – the alkene starting material must be able to be protonated to give a tertiary carbenium ion, which essentially limits the starting materials to 1,1-disubstituted, tri-substituted, and tetra-substituted alkenes. Styrene, stilbene and related phenyl-substituted C=C bonds were hydrogenated by this method also, but the yields were lower (46–57%).

$$\text{image: alkene + HOTf + M—H} \xrightarrow[\text{5 min}]{-50\,^{\circ}\text{C}}$$

$$\text{image: product + M—OTf} \tag{16}$$

Metal hydrides that were shown to be suitable hydride donors for this reaction included [Cp(CO)$_3$WH, Cp*(CO)$_3$WH], [Cp(CO)$_3$MoH], [(CO)$_5$MnH], [(CO)$_5$ReH] and [Cp*(CO)$_2$OsH]. As discussed above, a metal hydride may fail if it loses hydrogen too quickly after protonation (see Scheme 7.4). The failure of [Cp*(CO)$_2$FeH], [Cp*(CO)$_3$MoH], and [Cp(PPh$_3$)(CO)$_2$MoH] in the ionic hydrogenation of olefins is attributed to thermal decomposition occurring instead of proton transfer following protonation of these hydrides. Insufficient acidity of the protonated form (a cationic dihydride or dihydrogen complex) was found as the reason for failure of [Cp(PMe$_3$)(CO)$_2$WH] and [Cp(PMe$_3$)(CO)RuH]. These stoichiometric studies revealed the required characteristics of metal hydrides for them to be suitable for ionic hydrogenations. They must be able to function as hydride donors in the presence of acids. Protonation of the metal hydride is not required, but if it is protonated, it must be able to transfer the proton to the unsaturated substrate, and this requires both sufficient kinetic acidity and thermal stability to overcome alternate pathways that could thwart the desired reactivity.

Stoichiometric ionic hydrogenation of the C=C bond of a,β-unsaturated ketones by HOTf and [Cp(CO)$_3$WH] results in the formation of η^1-ketone complexes of tungsten [32]. As exemplified in Eq. (17), hydrogenation of methyl vinyl ketone gives a 2-butanone complex of tungsten. The bound ketone is displaced by the triflate counterion, giving the free ketone. Similar reactions were reported for hydrogenation of the C=C bond of a,β-unsaturated aldehydes.

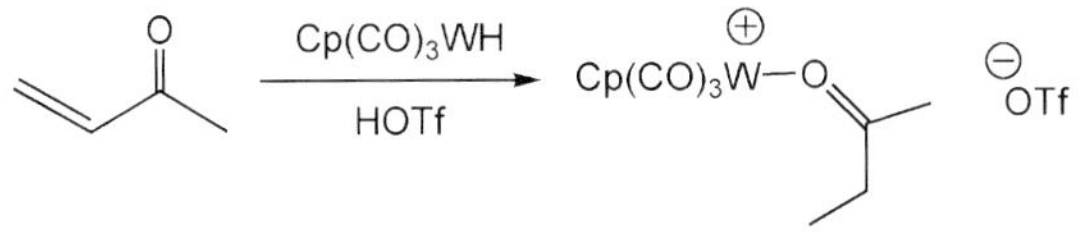

$$\text{(17)}$$

7.2.5
Stoichiometric Ionic Hydrogenation of Alkynes

Ionic hydrogenations of C≡C bonds of alkynes has only been studied in a few cases. Kursanov et al. reported that low yields were obtained upon attempted hydrogenation of aryl alkynes using CF_3CO_2H and $HSiEt_3$ [2]. In contrast, ionic hydrogenation of phenylacetylene by HOTf and [Cp(CO)$_3$WH] produced ethylbenzene as the product of double ionic hydrogenation of the C≡C bond [33]. In addition to the ethylbenzene that was promptly formed, the vinyl triflate and the geminal ditriflate shown in Scheme 7.5 were also observed; these organic triflates are formed by the addition of one or two equivalents of HOTf to the C≡C bond.

In the presence of HOTf and [Cp(CO)$_3$WH], these organic intermediates were slowly consumed, with more ethylbenzene being produced, the yield of which reached 92% after 28 h. Reaction of PhC≡CCH$_3$ with HOTf and [Cp(CO)$_3$WH] also led to the observation of vinyl triflates as intermediates. In addition to these organic intermediates, both *cis*- and *trans*-isomers of the β-methylstyrene complex [Cp(CO)$_3$W(η^2-PhHC=CHCH$_3$)]$^+$OTf$^-$ were observed, with this tungsten alkene complex reaching a maximum yield of 40% during the reaction. Ultimately, this complex as well as the vinyl triflates were converted to propylbenzene, which was observed in 91% yield. Ionic hydrogenation of BuC≡CH by HOTf and [Cp(CO)$_3$WH] led to vinyl triflate intermediates, but conversion to *n*-hexane was slow, requiring several days in the presence of excess HOTf. Since ionic hydrogenation of alkynes is so much slower than that of ionic hydrogenation of alkenes, the requirements of a suitable hydride donor are much more stringent. The ability of [Cp(CO)$_3$WH] to function as a hydride donor in the presence of acid is a key characteristic of this metal hydride that distinguishes it from HSiEt$_3$. Reaction of this tungsten hydride with HOTf leads to partial formation of the cationic dihydride [Cp(CO)$_3$W(H)$_2$]$^+$OTf$^-$ [34], but formation of H$_2$

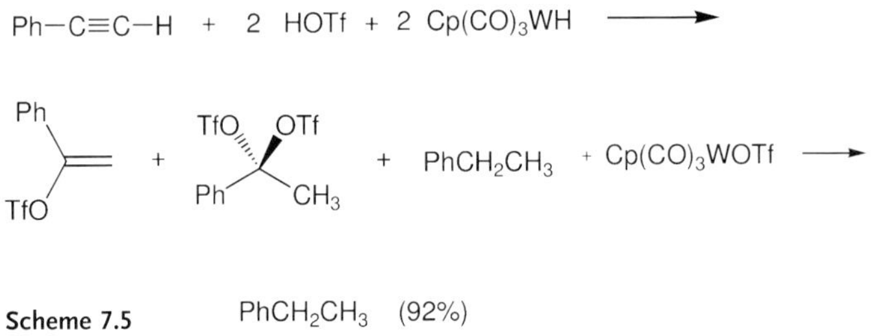

Ph−C≡C−H + 2 HOTf + 2 Cp(CO)$_3$WH ⟶

Scheme 7.5 PhCH$_2$CH$_3$ (92%)

from this dihydride is very slow, occurring over a time scale of weeks at room temperature.

7.2.6
Stoichiometric Ionic Hydrogenation of Ketones and Aldehydes using Metal Hydrides as Hydride Donors and Added Acids as the Proton Donor

Several systems have been reported involving stoichiometric hydrogenation of ketones or aldehydes by metal hydrides in the presence of acids. An ionic hydrogenation mechanism accounts for most of these hydrogenations, though in some examples alternative mechanisms involving the insertion of a ketone into a M–H bond are also plausible.

An early example came from the report in 1985 by Darensbourg et al. on the reactions of $[HCr(CO)_5]^-$ and $[HCr(CO)_4P(OMe_3)_3]^-$ with aldehydes and ketones, in the presence and absence of acids [27]. Paraformaldehyde reacts readily with $PPN^+[HCr(CO)_5]^-$ $\{PPN^+ = N(PPh_3)_2^+\}$ giving the alkoxide complex $[(CO)_5CrOCH_3]^-$ through insertion of formaldehyde into the Cr–H bond (Eq. (18)). The addition of HOAc produced methanol (Eq. (19)).

$$[(CO)_5CrH]^- + (CH_2O)_n \quad \rightarrow \quad [(CO)_5Cr\text{-}OCH_3]^- \tag{18}$$

$$[(CO)_5Cr\text{-}OCH_3]^- + HOAc \quad \rightarrow \quad CH_3OH + [(CO)_5Cr\text{-}OAc]^- \tag{19}$$

In contrast, only a sluggish reaction between propionaldehyde and $[HCr(CO)_5]^-$ was observed over several days, though addition of HOAc led to a 98% yield of *n*-propanol within 1 h. This striking change in reactivity between the two aldehydes suggests that propionaldehyde is hydrogenated to propanol not by an insertion mechanism, but rather through an ionic hydrogenation which protonation of the aldehyde activates it toward hydride transfer (Scheme 7.6). The phosphite-substituted anionic hydride $[HW(CO)_4P(OMe_3)_3]^-$ was more reactive with propionaldehyde in the absence of acids, providing evidence for a tungsten alkoxide complex that subsequently reacted with HOAc to produce a high yield of propanol. In the absence of acids, cyclohexanone showed little reactivity with any of the anionic hydrides $[HM(CO)_4L]^-$ (M = Cr or W, L = CO or $P(OMe_3)_3$).

As was found for aldehydes, however, the addition of HOAc led to the alcohol product. For less-reactive ketones, lower yields were found in some cases, and loss of some of the metal hydride occurs through formation of H_2 from reaction

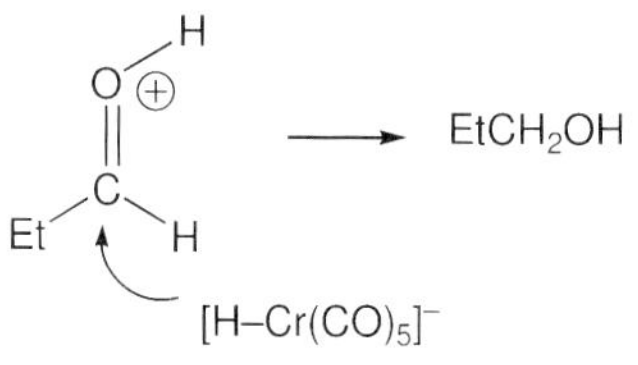

Scheme 7.6

of HOAc with [HCr(CO)$_5$]$^-$. In some cases, the weaker acid phenol could be used instead of HOAc.

In contrast to the lack of reactivity of ketones with PPN$^+$[HCr(CO)$_5$]$^-$, Brunet et al. reported different reactivity with K$^+$ rather than PPN$^+$ as the counterion. They found that K$^+$[HCr(CO)$_5$]$^-$ reacts with cyclohexanone in the absence of acid [35]. Hydrolysis with acid led to a 50% yield of cyclohexanol. These results suggest assistance from the K$^+$ cation; ion-pairing in metal anions has been studied in detail by Darensbourg [36].

Gibson and El-Omrani found that aldehydes were hydrogenated in refluxing THF by the bimetallic Mo hydride [(μ-H)Mo$_2$(CO)$_{10}$]$^-$ in the presence of HOAc [37]. These reactions most likely proceed through generation of the mononuclear hydride [HMo(CO)$_5$]$^-$, in analogy to the results discussed above for the Cr and W analogues.

Ito et al. found that hydrogenation of acetaldehyde, acetone, or cyclohexanone occurs at room temperature using [Cp$_2$MoH$_2$] and HOAc [38]. An ionic hydrogenation pathway was favored, in which protonation of the ketone or aldehyde was followed by hydride transfer from the metal, though a mechanism involving insertion of the C=O into the Mo–H bond was also considered possible. Both of the Mo–H bonds are active for this reaction. For example, in stoichiometric reactions using HOAc as the acid, the first hydride transfer occurs from [Cp$_2$MoH$_2$], which produces [Cp$_2$MoH(OAc)], and this complex functions as a hydride donor for the second equivalent. The reactivity of Cp$_2$MoH$_2$ is greater than that of [Cp$_2$MoH(OAc)], so that the first step is faster than the second. Using [Cp$_2$MoH(OTs)] (Ts = p-CH$_3$C$_6$H$_4$SO$_2$) as the hydride donor, a very high diastereoselectivity was found for hydrogenation of the C=O bond of 4-*tert*-butylcyclohexanone, which gave only the *cis* isomer of 4-*tert*-butylcyclohexanol. Hydrogenation of the C=N bond of imines is also accomplished using [Cp$_2$MoH$_2$] and HOAc; good yields of imines were obtained from reactions carried out at room temperature from 18 to 92 h. While most of these hydrogenations used protic acids, hydrogenation of N-cyclohexylidenecyclohexylamine was carried out using [Cp$_2$MoH$_2$] and the Lewis acid [Yb(OTf)$_3$], giving a 90% yield of the amine in 24 h at 50 °C (Eq. (20)).

$$\tag{20}$$

Hydride transfer from [(bipy)$_2$(CO)RuH]$^+$ occurs in the hydrogenation of acetone when the reaction is carried out in buffered aqueous solutions (Eq. (21)) [39]. The kinetics of the reaction showed that it was a first-order in [(bipy)$_2$(CO)RuH]$^+$ and also first-order in acetone. The reaction proceeds faster at lower pH. The proposed mechanism involved general acid catalysis, with a fast pre-equilibrium protonation of the ketone followed by hydride transfer from [(bipy)$_2$(CO)RuH]$^+$.

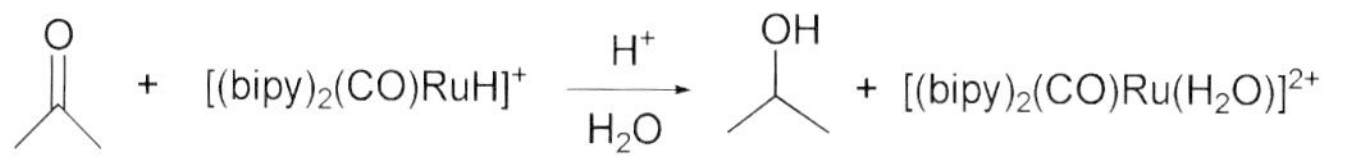

$$\text{(acetone)} + [(bipy)_2(CO)RuH]^+ \xrightarrow[H_2O]{H^+} \text{(isopropanol)} + [(bipy)_2(CO)Ru(H_2O)]^{2+} \qquad (21)$$

Harman and Taube found that the diamagnetic dihydrogen complex $[(NH_3)_5Os(\eta^2\text{-}H_2)]^{+2}$ does not react with acetone [40]. Oxidation gives the Os^{III} complex $[(NH_3)_5Os(\eta^2\text{-}H_2)]^{+3}$, which hydrogenates acetone to isopropyl alcohol. The reaction is slow, taking place over two days at room temperature. These results suggest that proton transfer to the ketone occurs from the acidic dihydrogen complex $[(NH_3)_5Os(\eta^2\text{-}H_2)]^{+3}$, and that hydride transfer from $[(NH_3)_5OsH]^{+2}$ to the protonated acetone generates the alcohol.

Extensive studies on stoichiometric hydrogenations of ketones have been carried out using HOTf as an acid, and metal carbonyl hydrides such as $[Cp(CO)_3WH]$ as the hydride donor [41, 42]. The addition of HOTf to a solution containing acetone and the tungsten hydride $[Cp(CO)_3WH]$ at 22 °C results in hydrogenation of the C=O bond to the alcohol, with the kinetically stabilized product having an alcohol ligand bound to the metal (Scheme 7.7) [41, 42]. Most previously reported alcohol complexes had been prepared by adding an alcohol to a metal complex with a weakly bound ligand, but in this case the alcohol ligand is formed in the reaction, without leaving the metal. The OH of the alcohol ligand is strongly hydrogen bonded to the triflate counterion, as shown by the short O···O distance of 2.63(1) Å found in the crystal structure of $[Cp(CO)_3\text{-}W(HO^iPr)]^+OTf^-$. Evidence that the hydrogen bonding is maintained in solution comes from the appearance of the OH of the bound alcohol ligand at δ 7.34 (d, J = 7.4 Hz) in the ^{1}H-NMR spectrum, a chemical shift substantially downfield of

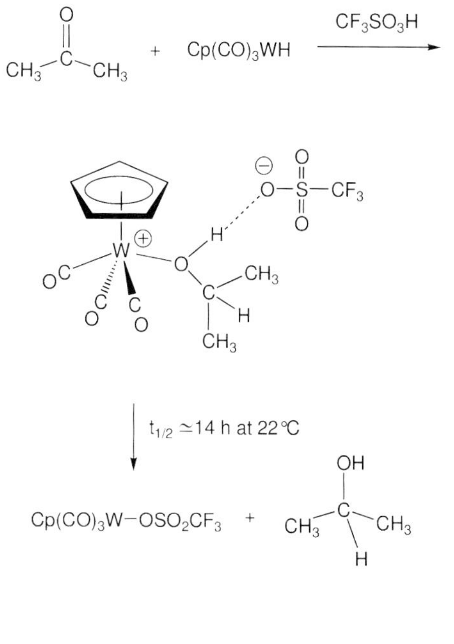

Scheme 7.7

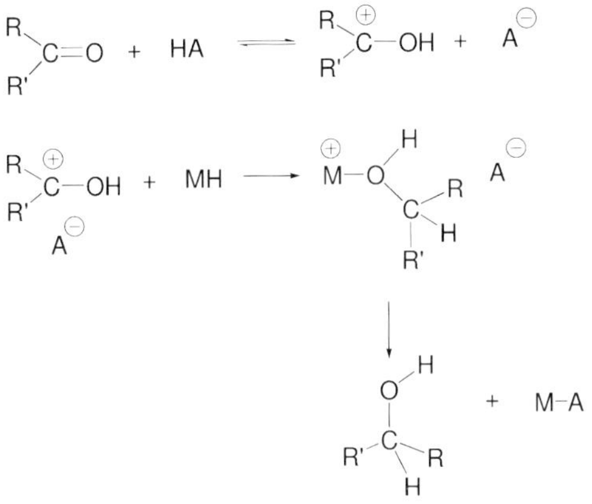

Scheme 7.8

that found for free alcohols. The bound alcohol is displaced by the triflate counterion, producing free alcohol and [Cp(CO)₃WOTf]. Other substrates that are readily hydrogenated by HOTf and [Cp(CO)₃WH] include propionaldehyde, cyclohexanone, 2-adamantanone; in all of these cases fully characterized W(alcohol) complexes were isolated.

The kinetics of the ionic hydrogenation of isobutyraldehyde were studied using [CpMo(CO)₃H] as the hydride and CF_3CO_2H as the acid [41]. The apparent rate decreases as the reaction proceeds, since the acid is consumed. However, when the acidity is held constant by a buffered solution in the presence of excess metal hydride, the reaction is first-order in acid. The reaction is also first-order in metal hydride concentration. A mechanism consistent with these kinetics results is shown in Scheme 7.8. Pre-equilibrium protonation of the aldehyde is followed by rate-determining hydride transfer.

Ionic hydrogenation of acetophenone by HOTf (1 equiv.) and [Cp(CO)₃WH] (1 equiv.) consumes only half of the ketone, and generates ethylbenzene (Eq. (22)) and [Cp(CO)₃WOTf] [42].

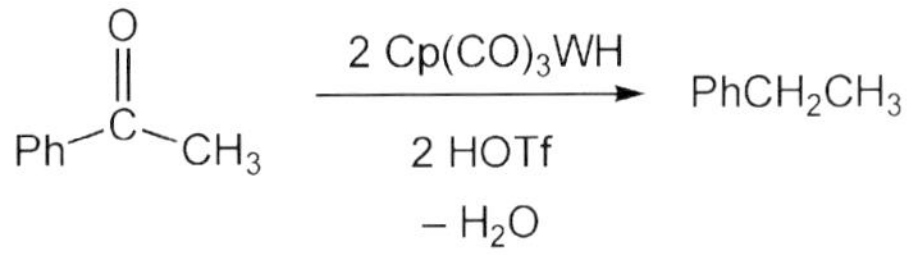

(22)

No intermediate tungsten complexes were observed in this reaction. The alcohol, *sec*-phenethylalcohol, is consumed at a rate which is much faster than that of its formation. It was shown separately to be converted to ethylbenzene (Eq. (23)) by HOTf and [Cp(CO)₃WH]. This reaction presumably proceeds through loss of water from the protonated alcohol, followed by hydride transfer from [Cp(CO)₃WH] to give ethylbenzene.

$$\underset{\underset{Ph}{\overset{H}{\diagdown}}\underset{CH_3}{\overset{OH}{\diagup}}C}{} \quad \xrightarrow[\substack{HOTf \\ -H_2O}]{Cp(CO)_3WH} \quad PhCH_2CH_3 \tag{23}$$

A competition between stoichiometric hydrogenation of acetone and acetophenone resulted in hydrogenation of the acetone [42]. Competitions of this type could be influenced by both the basicity of the ketone, as well as by the kinetics of hydride transfer to the protonated ketone. An intramolecular competition between an aliphatic and aromatic ketone resulted in preferential hydrogenation of the aliphatic ketone, with the product shown in Eq. (24) being isolated and fully characterized by spectroscopy and crystallography. Selective ionic hydrogenation of an aldehyde over a ketone was also found with HOTf and [Cp(CO)$_3$WH].

$$\tag{24}$$

Bakhmutov et al. reported the ionic hydrogenation of acetone or benzaldehyde by [ReH$_2$(CO)(NO)(PR$_3$)$_2$], (R=iPr, CH$_3$, OiPr) and CF$_3$CO$_2$H [43]. The resultant alcohol complexes were characterized by low-temperature NMR, and the OH protons had downfield chemical shifts. For example, the OH of the bound isopropyl alcohol in [ReH(CO)(NO)(PMe$_3$)$_2$(HO^{i}Pr)]$^+$CF$_3$CO$_2^-$ appears as a doublet at δ 8.17. The alcohol is subsequently released through displacement by the counterion, giving [ReH(CO)(NO)(PR$_3$)$_2$(O$_2$CCF$_3$)]. A significant kinetic preference was found for hydrogenation of benzaldehyde over acetone. Protonation of the dihydride [ReH$_2$(CO)(NO)(PR$_3$)$_2$] produces the cationic dihydrogen complexes [ReH(H$_2$)(CO)(NO)(PR$_3$)$_2$]$^+$, so protonation of the aldehyde or ketone can occur from these observable species, prior to hydride transfer to generate the alcohol. Whilst these hydrogenations produced alcohol complexes at low temperature, carrying out the reactions at room temperature gave mostly H$_2$ elimination, and only 10–15% yields of the alcohol as the hydrogenation product.

7.2.7
Stoichiometric Ionic Hydrogenation of Acyl Chlorides to Aldehydes with HOTf/Metal Hydrides

Conversion of acyl chlorides to aldehydes occurs upon reaction with HOTf and [Cp(CO)$_3$WH] [32]. The reaction of HOTf with benzoyl chloride and [Cp(CO)$_3$WH] led to the isolation of [Cp(CO)$_3$W(PhCHO)]$^+$OTf$^-$ (Eq. (25)), in which the aldehyde is bound to the metal [32]. The spectroscopic properties and crystal structure of this aldehyde complex revealed that it was bound through the oxygen, as an η^1-aldehyde. As was found in the case of the alcohol complexes, the aldehyde complex is kinetically stabilized. The triflate counterion displaces the bound aldehyde in a first-order process ($k = 3.6 \times 10^{-4}$ s^{-1}, $t_{1/2} \approx 33$ min at 25 °C), releasing the free aldehyde and generating [Cp(CO)$_3$WOTf].

(25)

An analogous reaction occurs when CH$_3$(C=O)Cl is reacted with HOTf and [Cp(CO)$_3$WH], with the acetaldehyde complex [Cp(CO)$_3$W(CH$_3$CHO)]$^+$OTf$^-$ being isolated [32]. Reaction with >1 equiv. each of HOTf and Cp(CO)$_3$WH led to subsequent hydrogenation of the CH$_3$CHO to ethanol, which was initially present as a bound ethanol ligand. It is possible that the mechanism of formation of the aldehyde complex involves protonation of the acyl chloride and hydride transfer from the metal, leading to a bound chlorohydrin complex, [Cp(CO)$_3$W(CH$_3$C(Cl)(H)OH)]$^+$OTf$^-$, which could expel HCl to produce [Cp(CO)$_3$W(CH$_3$CHO)]$^+$OTf$^-$. Since acyl chlorides are known to react with HOTf to produce acyl triflates, an alternative mechanism is formation of CH$_3$C(=O)OTf followed by reaction with [Cp(CO)$_3$WH]. The acyl triflate CH$_3$C(=O)OTf was prepared and shown to react with [Cp(CO)$_3$WH] to give [Cp(CO)$_3$W(CH$_3$CHO)]$^+$OTf$^-$, documenting the viability of this mechanistic pathway. It has not been established which of the two pathways is operative for these reactions.

Several anionic metal carbonyl hydrides stoichiometrically convert acyl chlorides to aldehydes. The anionic vanadium complex [Cp(CO)$_3$VH]$^-$ reacts quickly with acyl chlorides, converting them to aldehydes [44]. Although no further reduction of the aldehyde to alcohol was observed, the aldehydes reacted further under the reaction conditions in some cases, so a general procedure for isolation of the aldehydes was not developed.

Darensbourg et al. found that HCr(CO)$_5^-$ converts acyl chlorides to aldehydes rapidly at 25 °C (Eq. (26)) [26]. Yields >90% were detected by gas chromatography (GC) for preparation of CH$_3$CHO, n-BuCHO, PhCHO, and PhCH$_2$CHO. Since CH$_3$OD converts HCr(CO)$_5^-$ to DCr(CO)$_5^-$, the reaction of [HCr(CO)$_5$]$^-$ with PhCOCl in the presence of CH$_3$OD provided a convenient synthesis of the deuterated aldehyde, PhCDO.

$$R-\underset{\underset{Cl}{}}{\overset{\overset{O}{\|}}{C}} + [(CO)_5CrH]^- \longrightarrow R-\underset{\underset{H}{}}{\overset{\overset{O}{\|}}{C}} + [(CO)_5CrCl]^- \qquad (26)$$

As noted earlier, $[HCr(CO)_5]^-$ also converts alkyl halides to alkanes, but the reactivity of the acyl chloride is much higher, such that it was possible to selectively convert the acyl chloride to an aldehyde in one step, without interference from the alkyl bromide functionality. A second equivalent of $[HCr(CO)_5]^-$ further reduced the alkyl bromide (Eq. (27)).

$$(27)$$

7.2.8
Stoichiometric Ionic Hydrogenation of Ketones with Metal Dihydrides

Many examples of stoichiometric ionic hydrogenation discussed above involved hydride transfer from metal hydrides, following proton transfer from an *external* acid source. Achieving a catalytic system still requires the hydride transfer step, but will additionally require a source of protons from a metal complex. In many cases the proton source will be an acidic M–H bond, so examples of stoichiometric hydrogenation involving a metal-based proton and hydride source are an important step in documenting the viability of catalytic ionic hydrogenation methodology.

The cationic tantalum dihydride $Cp_2(CO)Ta(H)_2^+$ reacts at room temperature with acetone to generate the alcohol complex $[Cp_2(CO)Ta(HO^iPr)]^+$, which was isolated and characterized [45]. The mechanism appears to involve protonation of the ketone by the dihydride, followed by hydride transfer from the neutral hydride. The OH of the coordinated alcohol in the cationic tantalum alcohol complex can be deprotonated to produce the tantalum alkoxide complex $[Cp_2(CO)Ta(O^iPr)]$. Attempts to make the reaction catalytic by carrying out the reaction under H_2 at 60 °C were unsuccessful. The strong bond between oxygen and an early transition metal such as Ta appears to preclude catalytic reactivity in this example.

The cationic tungsten dihydride $[Cp(CO)_2(PMe_3)W(H)_2]^+$ hydrogenates the C=O bond of propionaldehyde within minutes at 22 °C, leading to the formation of *cis* and *trans* isomers of $[Cp(CO)_3W(HO^nPr)]^+OTf^-$ (Eq. (28)) [42]. The *cis* isomer of the alcohol complex released the free alcohol faster than the *trans* isomer. A similar stoichiometric ionic hydrogenation of acetone was also observed using $[Cp(CO)_2(PMe_3)W(H)_2]^+$.

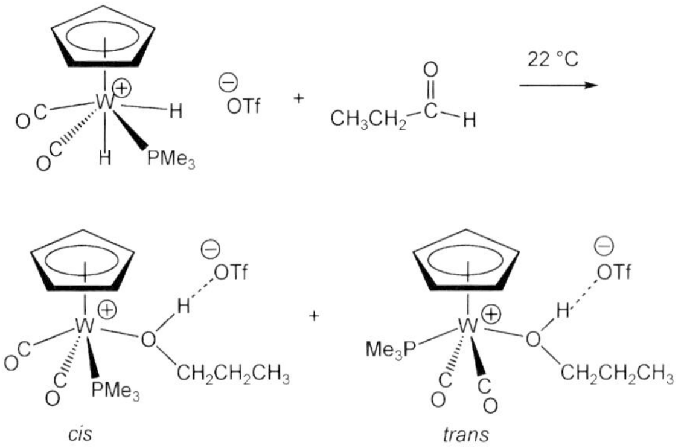

22 °C

OTf

CH$_3$CH$_2$—C—H

(28)

OTf

CH$_2$CH$_2$CH$_3$

Me$_3$P

OTf

CH$_2$CH$_2$CH$_3$

PMe$_3$

cis   *trans*

7.3
Catalytic Ionic Hydrogenation

7.3.1
Catalytic Ionic Hydrogenation of C=C Bonds

A series of cationic cobalt and rhodium complexes with a tetradentate chelating phosphine ligand have been reported [46]. These complexes were initially formulated as dihydrogen complexes, $[(PP_3)Co(H_2)]^+$ $\{PP_3 = P(CH_2CH_2PPh_2)_3\}$, but subsequent NMR studies conducted by Heinekey et al. showed that they were dihydride complexes $[(PP_3)Co(H)_2]^+$ [47]. Bianchini et al. found that, under an argon atmosphere, $[(PP_3)Co(H)_2]^+$ hydrogenates the C=C bond of dimethyl maleate in 3 h at room temperature [46]. When the reaction is conducted under H_2, catalytic hydrogenation of the C=C bond is observed, with a TOF of 1.5 h^{-1}. This reaction was suggested to proceed by an ionic mechanism, in which the cationic dihydride transfers a proton to the electron-deficient alkene, followed by hydride transfer from the neutral cobalt hydride complex. The acidity of Bianchini's $[(PP_3)Co(H)_2]^+$ would be expected to be roughly similar to that of $[(dppe)_2Co(H)_2]^+$, for which a pK_a of 22.8 was determined in CH_3CN (see Table 7.2). The low acidity of the dihydride raises questions about the likelihood of a proton-transfer mechanism for the initial step.

7.3.2
Catalytic Ionic Hydrogenation of Ketones by Anionic Cr, Mo, and W Complexes

Extensive studies on the hydride transfer reactivity of metal carbonyl anions such as $[HCr(CO)_5]^-$ presaged the development of anionic catalysts using Cr, Mo, and W. Darensbourg and coworkers found that 5 mol.% $[(CO)_5M(OAc)]^-$ (M = Cr, Mo, W) catalyzed the hydrogenation of cyclohexanone to cyclohexanol at 125 °C in THF using 36 bar H_2 (Eq. (29)) [48]. Under these conditions, TON determined after 24 h were 10 for W, 3.5 for Mo, and 18 for Cr.

$$\text{(cyclohexanone)} + H_2 \text{ (600 psi)} \xrightarrow[\substack{125\,°C \\ M = Cr,\ Mo,\ W}]{[(CO)_5MOAc]^-} \text{(cyclohexanol)} \qquad (29)$$

The organometallic products included recovered $[(CO)_5M(OAc)]^-$, along with $M(CO)_6$ and the bimetallic bridging hydride complex $[(\mu\text{-}H)M_2(CO)_{10}]^-$. It was proposed that, under the reaction conditions, $[(CO)_5MH]^-$ and HOAc were produced, and that insertion of the ketone into the M–H bond gave a metal alkoxide that reacted with HOAc to produce the alcohol.

Catalytic hydrogenations of cyclohexanone and benzaldehyde were also reported by Darensbourg et al. using 5% $[(\mu\text{-}H)M_2(CO)_{10}]^-$ (M = Cr, Mo, W) at 125 °C for 24 h; between four and 18 turnovers were observed under these conditions [48]. Related observations were made by Markó, who found that ketones (acetophenone, cyclohexanone, acetone, isobutyl methyl ketone) and aldehydes (benzaldehyde, butyraldehyde) could be catalytically hydrogenated at 160 °C under 100 bar of CO + H_2 [49]. Their experiments used 5 mol.% $[Cr(CO)_6]$ as the catalyst precursor, together with $NaOCH_3$ in methanol solution; under these conditions $[(CO)_5CrH]^-$ and $[(\mu\text{-}H)Cr_2(CO)_{10}]^-$ are formed. Similar hydrogenations were carried out starting with $[W(CO)_6]$ and with $[Mo(CO)_6]$. For the molybdenum example, milder conditions were used, with 91% hydrogenation of acetophenone being found after 3 h at 70 °C starting with 5 mol.% $Mo(CO)_6$ in methanol with added $NaOCH_3$. Fuchikami et al. found that $[(\mu\text{-}H)Cr_2(CO)_{10}]^-$ is much more active as a catalyst in dimethoxyethane (DME) than in THF [50]. Hydrogenation (50 bar H_2) of benzaldehyde at 100 °C using 1 mol.% $[(\mu\text{-}H)Cr_2(CO)_{10}]^-PPN^+$ produced 100 turnovers of benzyl alcohol after 13 h in DME, whereas using THF for 24 h at 125 °C gave only 14 turnovers.

Brunet et al. developed a transfer hydrogenation catalyst based on chromium, using 20% $K^+[(CO)_5CrH]^-$ as the catalyst precursor in THF solution, together with 5 equiv. each of HCO_2H and NEt_3 (Eq. (30)) [35, 51].

$$\text{(cyclohexanone)} + HCO_2H + NEt_3 \xrightarrow[\substack{\text{room temp.} \\ 24h,\ THF}]{20\ \%\ [(CO)_5CrH]^-K^+} \text{(cyclohexanol)} \qquad (30)$$

In reactions carried out for 24 h at room temperature, a 95% yield of cyclohexanol from cyclohexanone was obtained. Other ketones and aldehydes were also hydrogenated under identical conditions, but with slower rates (38% conversion for hydrogenation of 2-hexanone, 25% conversion of acetophenone, 45% for 3-methyl-2-butanone). Insertion of the C=O bond of the ketone or aldehyde into the Cr–H bond was proposed as the first step, producing a chromium alkoxide complex that reacts with acid to generate the alcohol product. The anionic chromium hydride $[(CO)_5CrH]^-$ is regenerated from the formate complex by decarboxylation (Eq. (31)). The role of the triethylamine is to moderate the strength of the formic acid, since formic acid alone is too strong of an acid, converting $[(CO)_5CrH]^-$ into $[(\mu\text{-}H)Cr_2(CO)_{10}]^-$.

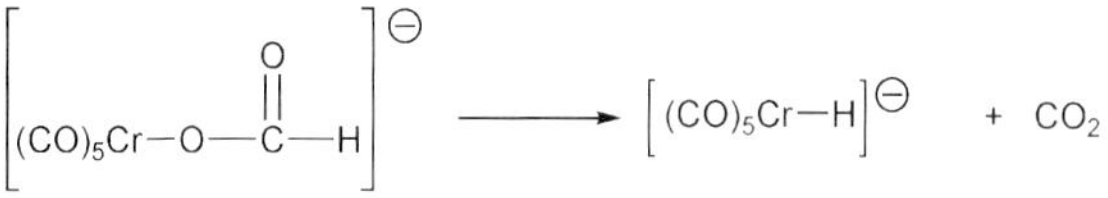

$$\left[(CO)_5Cr-O-\underset{\overset{\parallel}{O}}{C}-H \right]^{\ominus} \longrightarrow \left[(CO)_5Cr-H \right]^{\ominus} + CO_2 \tag{31}$$

7.3.3
Catalytic Ionic Hydrogenation of Ketones by Molybdenocene Complexes

Hydride transfer reactions from $[Cp_2MoH_2]$ were discussed above in studies by Ito et al. [38], where this molybdenum dihydride was used in conjunction with acids for stoichiometric ionic hydrogenations of ketones. Tyler and coworkers have extensively developed the chemistry of related molybdenocene complexes in aqueous solution [52–54]. The dimeric bis-hydroxide bridged dication dissolves in water to produce the monomeric complex shown in Eq. (32) [53]. In D_2O solution at 80 °C, this bimetallic complex catalyzes the H/D exchange of the α-protons of alcohols such as benzyl alcohol and ethanol [52, 54].

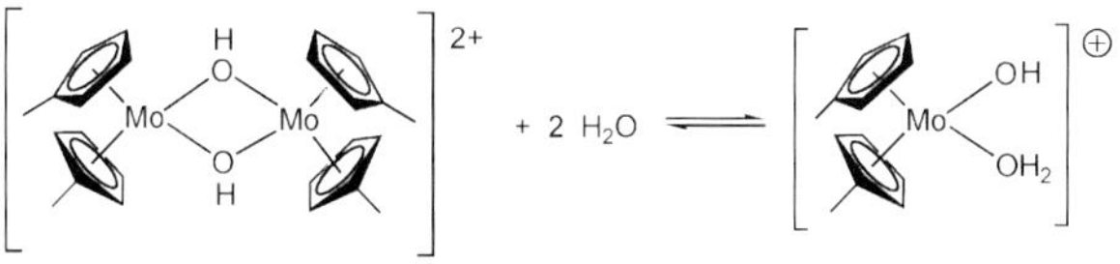

$$\left[\text{...} \right]^{2+} + 2\ H_2O \rightleftharpoons \left[\text{...} \right]^{\oplus} \tag{32}$$

The proposed mechanism for this H/D exchange is shown in Scheme 7.9. The formation of the alkoxide complex likely proceeds by displacement of the water ligand by the alcohol, forming an unobserved alcohol complex that transfers D^+ to the OD ligand, producing an OD_2 ligand.

The key step involves C–H bond activation, and produces a molybdenum complex with hydride and ketone ligands from the alkoxide ligand. Subsequent

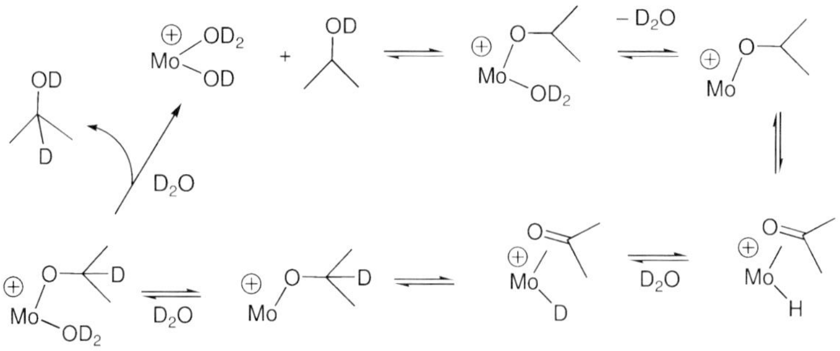

Scheme 7.9

H/D exchange leads to deuterium incorporation into the α-position of the alcohol.

Heating the bimetallic complex in D_2O solution also results in deuteration of the CH_3 sites on the cyclopentadienyl ring, through a mechanism involving oxidative addition of a C–H bond of the CH_3 group, followed by deuterium incorporation into the methyl group [54]. Heating of a mixture of the molybdenum complex with isopropyl alcohol and 2-butanone led to hydrogenation of the ketone, producing 2-butanol, and dehydrogenation of the isopropyl alcohol, generating acetone [52, 54]. Kuo et al. carried out further studies on these hydrogenations [55, 56]. Acetone is hydrogenated to isopropyl alcohol by [Cp$_2$MoH(OTf)] in water. The rates are faster in acidic solution than when the solution is buffered at pH 7, consistent with a general acid-catalyzed pathway in which the ketone is protonated prior to hydride transfer from the molybdenum hydride [55]. An alternative mechanism would involve insertion of the ketone into a Mo–H bond to give a metal alkoxide, which could then generate the alcohol by hydrolysis with water. The transfer hydrogenation of acetophenone by isopropyl alcohol in water is catalyzed by [Cp$_2$Mo(μ-OH)$_2$MoCp$_2$]$^{2+}$(OTs$^-$)$_2$ [56]. At 75 °C, the TOF is about 0.1 h^{-1}. Scheme 7.10 shows the proposed mechanism for this transfer hydrogenation; several of these steps have precedent in Tyler's H/D exchange reactions shown in Scheme 7.9 [52].

The catalytic ketone hydrogenation reaction is accelerated by addition of KOH. In the presence of 25 equiv. KOH, 1 mol.% of the molybdenum complex completely hydrogenated acetophenone overnight in refluxing 2-propanol

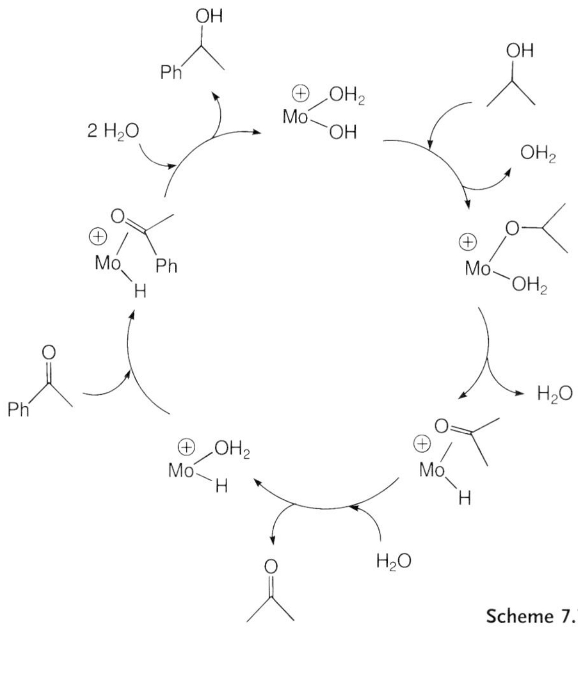

Scheme 7.10

(82 °C). The exact role of the base is not clear, but it may accelerate the formation of the molybdenum alkoxide complex from a bound alcohol ligand.

7.3.4
Catalytic Ionic Hydrogenation of Ketones by Cationic Mo and W Complexes

7.3.4.1 In Solution

Molybdenum and tungsten carbonyl hydride complexes were shown (Eqs. (16), (17), (22), (23), (24); see Schemes 7.5 and 7.7) to function as hydride donors in the presence of acids. Tungsten dihydrides are capable of carrying out stoichiometric ionic hydrogenation of aldehydes and ketones (Eq. (28)). These stoichiometric reactions provided evidence that the proton and hydride transfer steps necessary for a catalytic cycle were viable, but closing of the cycle requires that the metal hydride bonds be regenerated from reaction with H_2.

$$(33)$$

Tungsten and molybdenum ketone complexes, $[Cp(CO)_2(PR_3)M(O=CEt_2)]^+$-$BAr'_4$ [Ar' = 3,5-bis(trifluoromethyl)phenyl], could be isolated for $PR_3 = PMe_3$ and PPh_3 but were prepared *in situ* for the PCy_3 (Cy = cyclohexyl) complexes [57]. A series of experiments were carried out in CD_2Cl_2 solvent at 23 °C under 4 bar H_2, with about 10 equiv. $Et_2C=O$ (Eq. (33)). Formation of the alcohol (Et_2CH-OH) was accompanied at later times by small amounts of the ether, $(Et_2CH)_2O$, which arises from condensation of the alcohol. Under these reaction conditions, since the ketone substrate was not present in large excess, it was possible to monitor simultaneously the progress of the hydrogenation, as well as to detect the organometallic species present under catalytic hydrogenation conditions. As the reaction proceeds, the concentration of the ketone complex decreases, with concomitant formation of the alcohol complex. For example, in the case of the W complexes with a PPh_3 ligand, NMR evidence indicated the formation of *trans*-$[CpW(CO)_2(PPh_3)(Et_2CHOH)]^+$, with the concentration of this alcohol complex exceeding that of the ketone complex at later reaction times. As discussed earlier, alcohol complexes were previously found to be the kinetic product of stoichiometric ionic hydrogenation of ketones, so the observation under catalytic conditions indicates that the stoichiometric reactivity provides a good model in this case for the catalytic reactivity, even with some changes in ligands and counterions between the stoichiometric and catalytic reactions.

The proposed mechanism shown in Scheme 7.11 is supported by stoichiometric proton- and hydride-transfer reactions of metal hydrides that were dis-

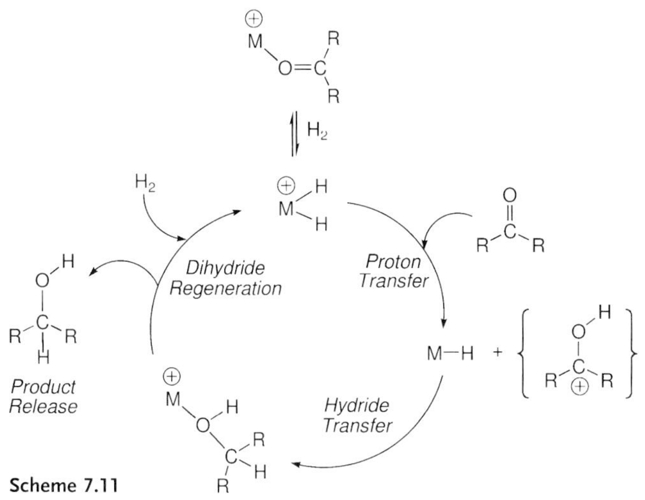

Scheme 7.11

cussed earlier. The key step to closure of the cycle, regeneration of the M–H bonds by H_2, is accomplished by reaction of the ketone complex with H_2. Tungsten dihydride complexes were sufficiently stable to be isolated and fully characterized [34], but molybdenum analogues were not directly observed. Evidence for the intermediacy of molybdenum dihydrides (or dihydrogen complexes) comes from heterolytic cleavage of H_2 by a molybdenum ketone complex in the presence of a hindered amine base (Eq. (34)). When H_2 is added to $[CpW(CO)_2(PPh_3)(Et_2C=O)]^+$ (with no added ketone), the dihydride $[CpW(CO)_2(PPh_3)(H)_2]^+$ is formed, providing further support for the operation of this step under catalytic conditions.

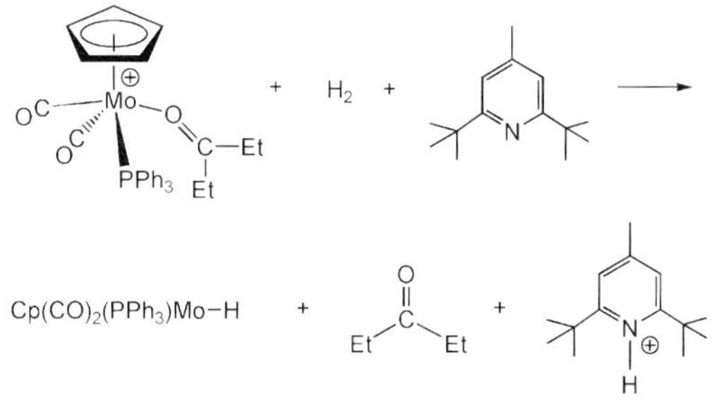

$$ (34) $$

Conversion of $[CpW(CO)_2(PPh_3)(Et_2C=O)]^+$ to *trans*-$[CpW(CO)_2(PPh_3)(Et_2CH-OH)]^+$ was observed when $Et_2C=O$ was hydrogenated under high pressure (65 bar) of H_2 for 17 h at 22 °C. Hydride transfer from $[CpW(CO)_2(PPh_3)H]$ to $Ph_3C^+BAr_4'^-$, followed by addition of the alcohol Et_2CHOH, led to the isolation of the *cis* isomer of $[CpW(CO)_2(PPh_3)(Et_2CHOH)]^+$. The studies of kinetic hydricity of metal hydrides had shown that hydride transfer from *trans*-$[Cp(CO)_2(PCy_3)MoH]$ to $Ph_3C^+BF_4^-$ occurs much faster than hydride transfer from the *cis* isomer of the hydride, *cis*-[Cp(CO)$_2$(PCy$_3$)MoH] [58]. Since the *trans* isomer of the alcohol complex is observed under catalytic conditions, the alcohol binds to the metal at the site from which hydride transfer took place. The studies of stoichiometric ionic hydrogenation of ketones had previously provided evidence that some amount of W–O bond formation is taking place in the transition state, before W–H bond cleavage is complete [41, 42].

Comparison of the rates of hydrogenation in these systematic studies of ionic hydrogenation of ketones by [Cp(CO)$_2$(PR$_3$)M(O=CEt$_2$)]$^+$ indicated the trends in metal and phosphine ligand [57]. For comparisons involving the same PR$_3$ ligand, the Mo complexes are invariably faster catalysts compared to the W analogue. The initial rate of hydrogenation is about eight-fold faster for Mo than for W for the PPh$_3$ complexes, and the difference is roughly two orders of magnitude for the PCy$_3$ complexes. For the Mo catalysts, the rate varies substantially with different phosphine ligands, in the order PCy$_3$ > PPh$_3$ > PMe$_3$. The approximate initial rate was about two turnovers per hour for the Mo–PCy$_3$ complex. For the W complexes, the same order was found, though the range of relative rates was smaller for W than for Mo. The phosphines PCy$_3$ and PMe$_3$ are similar electronically, so the much higher rate of catalysis found with the PCy$_3$ complexes makes it clear that steric effects predominate over electronic effects.

For the PPh$_3$ and PMe$_3$ complexes, the ketone or alcohol complexes were observed spectroscopically during the hydrogenation reaction, and those species are the resting states. Formation of the dihydride complexes under catalytic conditions is proposed to involve dissociation of the ketone or alcohol followed by addition of H$_2$. The higher rate of catalysis with the PCy$_3$ complexes suggests that the steric bulk of this ligand promotes ketone dissociation more than in the case of the PPh$_3$ or PMe$_3$ complexes. In contrast to the PPh$_3$ and PMe$_3$ complexes, the predominant tungsten complex observed during hydrogenation with the PCy$_3$ complex was [Cp(CO)$_2$(PCy$_3$)W(H)$_2$]$^+$. In this case, proton transfer from the metal to the ketone is slow and has become the turnover-limiting step of the catalytic cycle. Ketone binding to the metal is destabilized by steric factors for the PCy$_3$ complex, compared to analogues with PMe$_3$. In addition, the steric effects of the bulky PCy$_3$ ligand likely disfavor proton transfer from the metal to the free ketone. Norton et al. reported a pK_a of 5.6 in CH$_3$CN for [CpW(CO)$_2$(PMe$_3$)(H)$_2$]$^+$ [31]. The pK_a of protonated acetone in CH$_3$CN is about –0.1 [59, 60]. Presuming that there is not a large change in relative pK_a values in CH$_3$CN (in which the pK_a measurements were made) and CD$_2$Cl$_2$ (in which the hydrogenations were carried out), the thermodynamics of proton transfer from the dihydride to the ketone are uphill in all of these cases. The tungsten dihydride [CpW(CO)$_2$(PMe$_3$)(H)$_2$]$^+$ (for which pK_a data are available) is the one that leads to the slowest hydrogenation, so it is likely that this represents the least thermodynamically favorable example for proton transfer. Based on the trends in acidity identified above, [CpW(CO)$_2$(PPh$_3$)(H)$_2$]$^+$ is expected to be more acidic than [CpW(CO)$_2$(PMe$_3$)(H)$_2$]$^+$; similarly, Mo hydrides are more acidic than W hydrides. Rate constants for hydride transfer to protonated ketones have not been as extensively studied as those for hydride transfer to Ph$_3$C$^+$ (see Table 7.5), but hydride transfer from $Cp(CO)_2(PPh_3)MoH$ to protonated acetone occurs in CH_3CN with a reported rate constant of $1.2 \times 10^4\ M^{-1}\ s^{-1}$ at 25 °C [60]. Proton transfer from the dihydride to the ketone is thermodynamically uphill, but this does not prevent catalytic ionic hydrogenations with these W and Mo complexes from proceeding smoothly, since rapid hydride transfer from the neutral hydride follows the proton transfer.

Evidence for a major mode of catalyst deactivation in this system came from the observation of phosphonium cations (HPR_3^+) in the reaction mixture, which could form through the protonation of free PR_3 by the acidic dihydride complex. It is not known which species decomposes to release free PR_3, but the decomposition pathway is exacerbated by the subsequent reactivity in which protonation of phosphine removes a proton from the metal dihydride, effectively removing a second metal species from the cycle.

Knowledge of the pathway for catalyst deactivation suggested that catalysts with longer lifetimes might result if phosphine dissociation could be suppressed. A series of Mo complexes was prepared that had a two-carbon chain chelating the cyclopentadienyl ligand to the phosphine [61]. Hydride abstraction from $[HMo(CO)_2\{\eta^5:\eta^1\text{-}C_5H_4(CH_2)_2PR_2\}]$ (R = Cy, tBu, and Ph) using $Ph_3C^+BAr_4'^-$ in the presence of $Et_2C{=}O$ led to ketone hydrogenation catalysts (Eq. (35)) that exhibited several advantages over the unbridged complexes.

$$\text{(35)}$$

Comparisons of catalysis of these C_2-tethered P,C chelate complexes with the non-chelate analogues in CD_2Cl_2 solvent showed that the former were somewhat slower catalysts than the latter, but the compelling advantage of the chelate complexes is that they had much longer lifetimes.

7.3.4.2 Solvent-free

Environmental concerns have caused an intense emphasis on the development of chemical reactions that reduce waste. Solvents are used on a huge scale, with more than 15 billion kilograms of organic solvents being produced each year [62]. One of the Principles of Green Chemistry [7] indicates that the use of a solvent should be avoided whenever possible, and it has been found that the hydrides $[HMo(CO)_2\{\eta^5:\eta^1\text{-}C_5H_4(CH_2)_2PR_2\}]$ can be used as catalyst precursors for the solvent-free [62,63] hydrogenation of $Et_2C{=}O$. Hydrogenations can be carried out at higher temperatures, since the C_2–PR_2 complexes have significantly improved stability compared to the unbridged complexes. Another attractive feature of these bridged catalysts is that they can be used at low catalyst loading, less than 1 mol.% typically, and as low as 0.09 mol.% in one example. Solvent-free hydrogenation of $Et_2C=O$ using $[HMo(CO)_2\{\eta^5:\eta^1\text{-}C_5H_4(CH_2)_2PCy_2\}]$ (0.35 mol.%) as the catalyst precursor (activated by hydride removal using $Ph_3C^+BAr_4'^-$) was carried out under 4 bar H_2 at 50 °C for 10 days, and 132 turnovers were observed (Eq. (35)). Only 62 turnovers were found when the analogous C_2–PPh_2 complex was employed under identical conditions, indicating that the performance of the catalyst with R=Cy was superior to that found with R=Ph, the same trend that was found for the unbridged complexes. Although the steric bulk of the Cy group on the phosphines was recognized as an advantage, the use of the C_2–$PtBu_2$ complex (82 turnovers under identical conditions) was superior to that of the C_2–PPh_2 complex but not as high as the C_2–PCy_2 complex. Higher pressure of H_2 led to faster rates of catalysis – complete hydrogenation of $Et_2C=O$ was accomplished using $[HMo(CO)_2\{\eta^5:\eta^1\text{-}C_5H_4(CH_2)_2PCy_2\}]$ (0.35 mol.%) as the catalyst precursor at 50 °C under 55 bar H_2 for 8 days.

Most of the studies of these Mo catalysts were carried out with $BAr_4'^-$ as the counterion; catalysis is also observed using BF_4^- or PF_6^- as the counterion, albeit with lower turnover numbers [61]. Triflate is an attractive counterion; it offers advantages of lower cost compared to $BAr_4'^-$, but may be slightly different mechanistically since it coordinates to the metal. Protonation of $[HMo(CO)_2\{\eta^5:\eta^1\text{-}C_5H_4(CH_2)_2PCy_2\}]$ with HOTf leads to the formation of the triflate complex $[Mo(CO)_2\{\eta^5:\eta^1\text{-}C_5H_4(CH_2)_2PR_2\}OTf]$, presumably through an unobserved dihydride or dihydrogen complex. Solvent-free hydrogenation (4 bar H_2) of $Et_2C=O$ using 0.35 mol.% $[Mo(CO)_2\{\eta^5:\eta^1\text{-}C_5H_4(CH_2)_2PCy_2\}OTf]$ (50 °C, 10 days) gave 120 turnovers, only slightly less than the 132 turnovers found under identical conditions using the $BAr_4'^-$ counterion. Even lower catalyst loading was successfully carried out with $[Mo(CO)_2\{\eta^5:\eta^1\text{-}C_5H_4(CH_2)_2PCy_2\}OTf]$: 0.09 mol.% catalyst loading at 75 °C for 10 days provided 462 turnovers in the hydrogenation of $Et_2C=O$ under solvent-free conditions.

These Mo catalysts with a C_2-tether connecting the phosphine and cyclopentadienyl ligand provide an example of the use of mechanistic principles in the rational design of improved catalysts, in this case based on information about a decomposition pathway for the prior generation of catalysts. The new catalysts offer improved lifetimes, higher thermal stability, and low catalyst loading. The successful use of a triflate counterion and solvent-free conditions for the hydrogenation are additional features that move these catalysts closer to practical utility.

7.3.4.3 N-Heterocyclic Carbene Complexes

N-heterocyclic carbenes have recently been used as alternatives to phosphines in many catalytic reactions, owing in part to a decreased propensity to dissociate from the metal [64]. Hydride abstraction from the tungsten hydride $[Cp(CO)_2(IMes)WH]$ (IMes = the carbene ligand 1,3-bis(2,4,6-trimethylphenyl)-imidazol-2-ylidene) using $Ph_3C^+B(C_6F_5)_4^-$ leads to the formation of an unusual complex in which one C=C of a mesityl ring has a weak bonding interaction with the tungsten (Eq. (36)) [65].

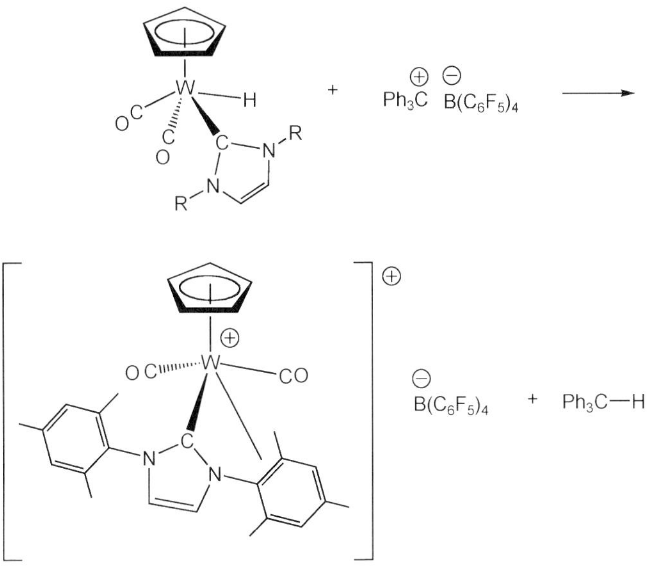

(36)

This complex can be used as a catalyst precursor for hydrogenation of ke-tones, though only two turnovers occurred in one day for solvent-free hydroge-nation of $Et_2C=O$ at 23 °C using 0.34 mol.% catalyst at 4 bar H_2. At the same catalyst loading, 61 turnovers occurred in 7 days at higher temperature (50 °C) and higher H_2 pressure (54 bar), though some decomposition of the catalyst is also observed. These catalysts are thought to operate by a mechanism analogous to that shown above (see Scheme 7.11) for the related phosphine complexes.

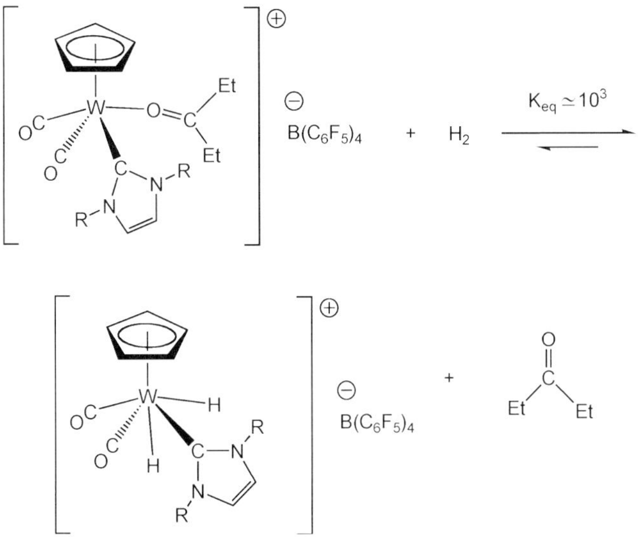

(37)

Displacement of the bound ketone by H_2 was directly observed by NMR (Eq. (37)), and an approximate equilibrium constant was determined. The cationic tungsten complex can also be used for catalytic hydrosilylation of ketones. In the case of catalytic hydrosilylation of aliphatic substrates using $HSiEt_3$, the catalyst precipitates at the end of the reaction, facilitating recycle and reuse [66].

7.3.5
Use of a Pd Hydride in Hydrogenation of C=C Bonds

DuBois et al. reported extensive studies on the thermodynamics of acidic and hydridic reactivity of a large series of complexes $[HM(diphosphine)_2]^+$. Aresta et al. found that protonation of the Pd complex $[Pd(dppe)_2]$ leads to the cationic hydride $[HPd(dppe)_2]^+$ [67]. The Pd–H bond can be cleaved as either a proton or as a hydride. Solutions of $[HPd(dppe)_2]^+$ decompose to give H_2, $[Pd(dppe)_2]$ and $[Pd(dppe)_2]^{2+}$. Reaction of $[HPd(dppe)_2]^+$ with methyl acrylate at 20 °C resulted in hydrogenation of the C=C bond, producing methyl propionate (Eq. (38)). In contrast to the previously discussed examples of ionic hydrogenation, two equivalents of the same palladium hydride complex are thought to furnish both the proton and the hydride in this case. Computations suggested that hydride transfer occurs first, producing a carbanionic intermediate that is then protonated by a second equivalent of the metal hydride. Catalytic hydrogenation of the C=C bond of cyclohexene-2-one was observed when $[HPd(dppe)_2]^+BF_4^-$ was used as the catalyst precursor under H_2 (4 MPa). A maximum TOF of about $16 \, h^{-1}$ was found at 50 °C. At higher temperatures (67 °C), higher TOF values were found, but decomposition of the Pd complex was observed, producing decomposition products (Pd black) that are also catalytically active.

$$2 \, [HPd(dppe)_2]^+ \quad + \quad \overset{O}{\overset{\|}{\diagdown}}-OCH_3 \quad \longrightarrow$$

(38)

$$[Pd(dppe)_2]^{2+} \quad + \quad [Pd(dppe)_2] \quad + \quad \overset{O}{\overset{\|}{\diagdown}}-OCH_3$$

7.3.6
Catalytic Hydrogenation of Iminium Cations by Ru Complexes

Norton and coworkers found that catalytic enantioselective hydrogenation of the C=N bond of iminium cations can be accomplished using a series of Ru complexes with chiral diphosphine ligands such as Chiraphos and Norphos [68]. Even tetra-alkyl-substituted iminium cations can be hydrogenated by this method. These reactions were carried out with 2 mol.% Ru catalyst and 3.4–3.8 bar H_2 at room temperature in CH_2Cl_2 solvent (Eq. (39)).

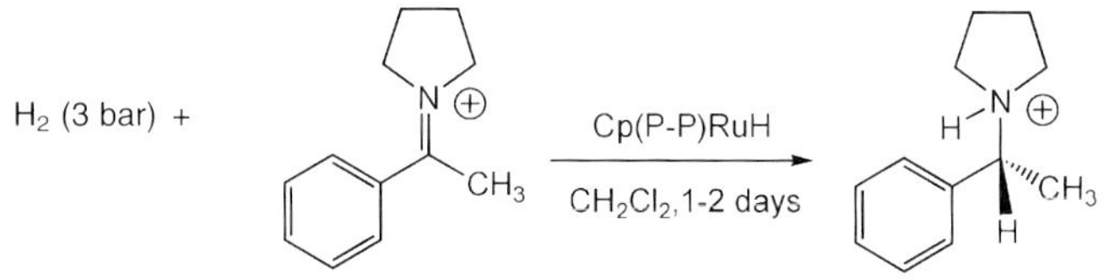

H$_2$ (3 bar) + [structure] $\xrightarrow[\text{CH}_2\text{Cl}_2, 1\text{-}2 \text{ days}]{\text{Cp(P-P)RuH}}$ [structure] (39)

The enantiomeric excess (ee) obtained under catalytic conditions was similar to that found when the hydride transfer was carried out in a stoichiometric reaction (Eq. (40)); these stoichiometric reactions were carried out in the presence of excess CH$_3$CN, which captures the 16-electron cationic Ru complex following hydride transfer.

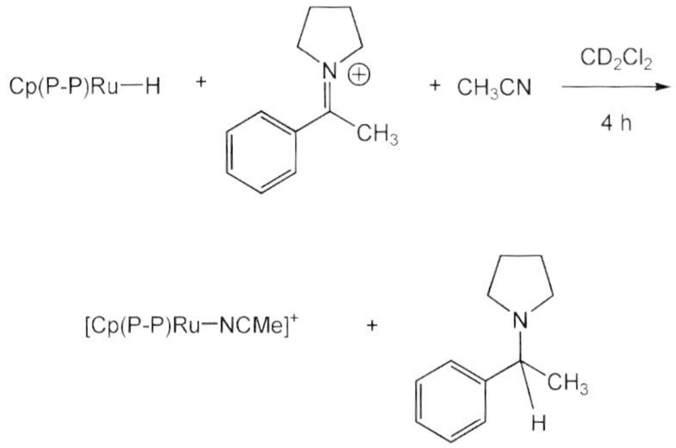

Cp(P-P)Ru—H   +   [structure]   + CH$_3$CN $\xrightarrow[\text{4 h}]{\text{CD}_2\text{Cl}_2}$

(40)

[Cp(P-P)Ru—NCMe]$^+$ +   [structure]

No change in the rate or ee of the catalytic reaction was observed when the pressure of H$_2$ was varied, indicating that H$_2$ does not play a role in the turnover-limiting step or in the determination of enantioselectivity. When the catalytic reaction was monitored by NMR under H$_2$ (5 bar), the neutral hydride was observed. All of these observations support the proposed mechanism shown in Scheme 7.12. This

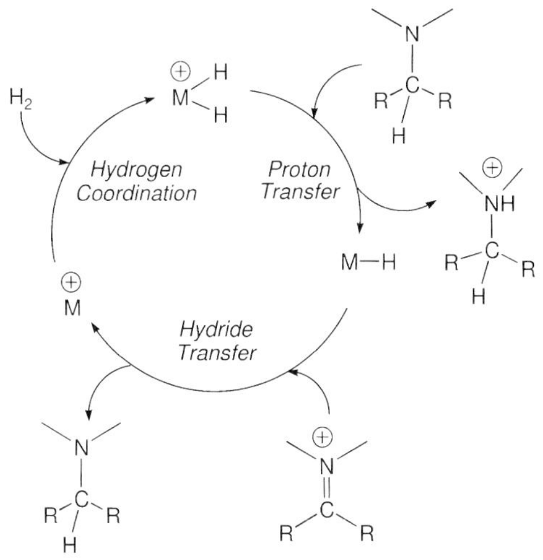

H$_2$

Hydrogen Coordination

Proton Transfer

M—H

Hydride Transfer

Scheme 7.12

mechanism is similar to the mechanism for ionic hydrogenation of ketones (Scheme 7.11), except that in the hydrogenation of iminium cations the hydride transfer occurs first, whereas proton transfer from cationic dihydrides to the ketone occurs first in Scheme 7.11. Hydride transfer to the iminium cation is the turnover-limiting and enantioselectivity-determining step of the mechanism.

The kinetics of hydride transfer (Eq. (40)) were determined for a series of chelating diphosphines. The rate constant of hydride transfer was found to be highly dependent on the bite angle of the diphosphine, increasing as the bite angle decreases [69]. The rate constant for [Cp(dppm)RuH] (dppm = $Ph_2PCH_2PPh_2$; bite angle 72°) was about 200 times higher than that for [Cp(dppp)RuH] (dppp = $Ph_2P(CH_2)_3PPh_2$; bite angle 92°). The rate constant for the Ru complex of the diphosphine with a C_2 bridge, [Cp(dppe)RuH] (dppe = $Ph_2P(CH_2)_2PPh_2$; bite angle 85°) was intermediate between the two. The increased *kinetic* hydricity resulting from a decreased bite angle parallels the observations of DuBois and colleagues, who found the same trend for *thermodynamic* hydricity [19].

7.4
Ruthenium Complexes Having an OH Proton Donor and a RuH as Hydride Donor

7.4.1
The Shvo System

Shvo and coworkers prepared a bimetallic complex in which the two metals are joined by a bridging hydride as well as by an O–H–O hydrogen bond joining the two substituted Cp ligands [70]. Shvo used this versatile catalyst precursor for hydrogenation of C=C and C=O bonds at 145 °C under 34 bar H_2 (Eq. (41)) [71].

Shvo Catalyst

$$+ \ H_2 \ (34 \ bar) \quad \xrightarrow[\text{hydrogenation of C=C and C=O}]{145\ °C} \quad (41)$$

Under these conditions, the bimetallic complex is cleaved into an 18-electron complex that performs the hydrogenations, and an unsaturated 16-electron complex. Addition of H_2 to the 16-electron complex produces the 18-electron complex that has an acidic OH and a hydridic RuH (Eq. (42)).

$$(42)$$

Solvent-free hydrogenations of 1-octene, 2-pentene, cyclohexene, and styrene were carried out with catalyst loadings as low as 0.05 mol.% of the dimer, in some cases with TOF values as high as 6000 h^{-1} [71]. Total turnover numbers of almost 2000 were obtained in most of these cases. Solvent-free hydrogenation of ketones such as Et$_2$C=O, cyclohexanone, and diisopropyl ketone were also reported at the same temperature and H$_2$ pressure, but with somewhat lower TOFs for the hydrogenation of C=O compared to C=C hydrogenations.

Kinetic and mechanistic studies by Casey et al. provided further insight into the mechanistic details of the hydrogenation of ketones and aldehydes, using a more soluble analogue of Shvo's catalyst (with *p*-tolyl groups instead of two of the Ph groups) [72]. The kinetics of hydrogenation of benzaldehyde by the Ru complex shown in Eq. (43) were first order in aldehyde and first order in the Ru complex; the

$$(43)$$

second-order rate constant at $-10\,^\circ C$ was determined to be $k = (3.0 \pm 0.2) \times 10^{-4}\ M^{-1}\ s^{-1}$. The activation enthalpy was $\Delta H^{\ddagger} = 12.0 \pm 1.5$ kcal mol^{-1}, and the very negative entropy of activation ($\Delta S^{\ddagger} = -28 \pm 5$ cal K^{-1} mol^{-1}) further supports a highly ordered transition state.

Isotope effects have been very useful in understanding the detailed mechanisms of many organometallic reactions [73], and have been used extensively in studies of Shvo complexes. Separate kinetic isotope effects were measured for hydrogenation of PhCHO at $0\,^\circ C$ (Eq. (43)); deuteration at the RuH(D) and OH(D) positions gave values of $k_{RuH}/k_{RuD} = 1.5 \pm 0.2$ and $k_{OH}/k_{OD} = 2.2 \pm 0.1$. Deuteration of both the RuH and OH sites gave a combined kinetic isotope effect of $k_{RuHOH}/k_{RuDOD} = 3.6 \pm 0.3$. The proposed mechanism involves concerted proton transfer from the OH site and hydride transfer from the Ru hydride, and is supported by the product of the two individual isotope effects ($1.5 \times 2.2 = 3.3$) being within experimental uncertainty of the combined isotope effect of 3.6. Since the actual hydrogenation step (Eq. (43)) occurs at low temperatures, the elevated temperature required for the catalytic reaction starting with the bimetallic complex as the catalyst precursor (see Eq. (41)) is necessary to generate the active mononuclear species. A pK_a of 17.5 was determined in CH$_3$CN for the OH of [[2,5-Ph$_2$-3,4-Tol$_2$(η^5-C$_4$COH)]Ru(CO)$_2$H], so this OH is significantly more acidic than either phenol ($pK_a = 26.6$ in CH$_3$CN) or benzoic acid ($pK_a = 20.7$ in CH$_3$CN). As shown in Table 7.5, [Cp*Ru(CO)$_2$H] has high kinetic hydricity, so this remarkable combination of acidity of an OH site and hydricity of the RuH combine to make the concerted proton- and hydride-transfer mechanism feasible in this type of complex.

A variety of ketones were hydrogenated using Shvo's catalyst at $100\,^\circ C$ using excess formic acid rather than H$_2$ as the source of hydrogen [74]. Excellent yields (>90%) of alcohols were generally obtained in 6 h or less, with total turnovers in the range of 6000–8000. The unsaturated 16-electron Ru complex that results after hydrogen is delivered to the substrate is proposed to react with for-

(44)

mic acid to produce a formate complex that expels CO_2, regenerating the metal hydride (Eq. (44)). Subsequent studies by Casey et al. showed that the formate complex is formed upon reaction of excess formic acid (HCO_2H) with Shvo's catalyst at low temperature [75]. This formate complex loses CO_2 above 0 °C, but the formate complex does not hydrogenate benzaldehyde directly.

Casey has suggested that the hydrogenation of alkenes by Shvo's catalyst may proceed by a mechanism involving loss of CO from the Ru–hydride complex, and coordination of the alkene. Insertion of the alkene into the Ru–H bond would give a ruthenium alkyl complex that can be cleaved by H_2 to produce the alkane [75]. If this is correct, it adds further to the remarkable chemistry of this series of Shvo complexes, if the same complex hydrogenates ketones by an ionic mechanism but hydrogenates alkenes by a conventional insertion pathway.

7.4.2
Hydrogenation of Imines by Shvo Complexes

Samec and Bäckvall found that the dinuclear Shvo complex catalyzes the transfer hydrogenation of imines using benzene as solvent and isopropanol as the hydrogen source (Eq. (45)) [76]. These catalytic hydrogenations were typically carried out at 70 °C, and gave >90% yields of the amine in 4 h or less.

$$(45)$$

Ketimines were hydrogenated faster than aldimines, and electron-donating groups accelerated the rate of hydrogenation. The OH and RuH bonds are regenerated by hydrogen transfer to the unsaturated 16-electron Ru complex from isopropanol, generating acetone (Scheme 7.13).

Kinetic studies were carried out by Bäckvall and coworkers at –54 °C on the hydrogenation of a ketimine, which produces a ruthenium complex with a bound amine (Eq. (46)) [77].

$$(46)$$

The crystal structure of an isopropylamine complex of Ru of this type has been reported [78]. Surprisingly, a negligible kinetic isotope effect ($k_{RuHOH}/k_{RuDOD} = 1.05 \pm 0.14$) was found when D labels on both the OH and RuH sites were used,

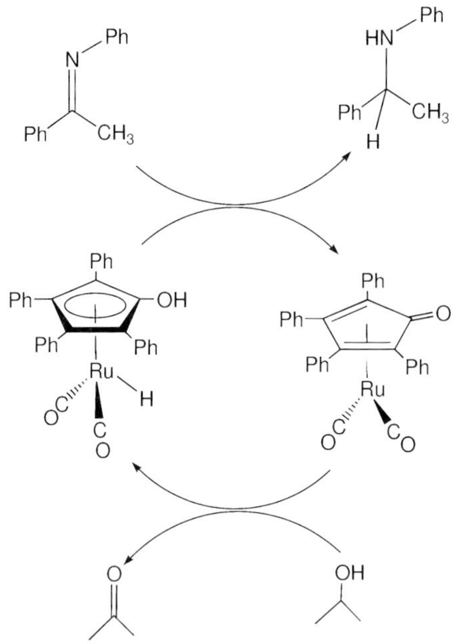

Scheme 7.13

indicating that the rate-determining step does not involve hydrogen transfer. The reaction is first order in amine and also first order in ruthenium. Bäckvall proposed a mechanism involving a ring slip ($\eta^5 \rightarrow \eta^3$) followed by coordination of the imine. Proton transfer from the OH concerted with hydride transfer from the RuH to an η^2-coordinated imine would give the coordinated amine, then rearrangement of the substituted cyclopentadiene ligand to η^4 was proposed to generate the observed product. An alternative mechanism consistent with the data was proposed by Casey and Johnson [79].

Casey and Johnson also reported kinetics and isotope effects for the hydrogenation of imines [79]. Hydrogenation of an electron-deficient imine, N-benzilidenepentafluoroaniline, gave the free amine as the organic product. Deuteration of the RuH site gave $k_{RuH}/k_{RuD} = 1.99 \pm 0.13$ at 11 °C, and deuteration of the acidic OH site gave a kinetic isotope effect of $k_{OH}/k_{OD} = 1.57 \pm 0.07$. The experimentally determined combined isotope effect ($k_{RuHOH}/k_{RuDOD} = 3.32 \pm 0.17$) is within experimental uncertainty of the product of the two individual isotope effects ($1.99 \times 1.57 = 3.12$). These observations are similar to those for hydrogenation of the C=O bond discussed above, and the data are consistent with the proposed concerted proton and hydride transfer for this imine hydrogenation.

Examination of a series of imines of differing electronic properties showed that a change in the rate-determining step of this stoichiometric C=N hydrogenation occurs as the imine becomes more electron-rich. Hydrogenation of N-isopropyl-(4-methyl)benzilidene amine led to an amine complex of ruthenium. In addition, the C=N hydrogenation was accompanied by isomerization of the imine to a ketimine (Eq. (47)).

(47)

When the hydrogenation was carried out with RuD labels, scrambling of D into the starting material and product was observed, indicating reversible hydrogen transfer. Hydrogenation of N-benzilidene-*tert*-butylamine was studied; this substrate has no β-hydrogens, so its reactions are not complicated by isomerization or exchange reactivity. Hydrogenation of this imine produces an amine complex (cf. Eq. (46)), and the kinetics at −48 °C were first order in imine and first order in Ru. In contrast to the normal kinetic isotope effects discussed above for the hydrogenation of aldehydes or other imines, inverse isotope effects were observed for the hydrogenation of this electron-rich *tert*-butyl imine. Deuteration of the RuH site resulted in $k_{RuH}/k_{RuD}=0.64\pm0.05$, and this is thought to be due to an inverse equilibrium isotope effect that favors deuterium on the carbon in the reversible transfer between ruthenium and carbon. Deuteration of the acidic OH site gave a kinetic isotope effect of $k_{OH}/k_{OD}=0.90\pm0.07$. The rate-determining step of the reaction is proposed to be coordination of the nitrogen of the amine to the ruthenium. Since the proton and hydride transfers occur prior to the amine coordination, the mechanistic information does not distinguish between concerted or stepwise proton and hydride transfer in this ionic mechanism.

7.4.3
Dehydrogenation of Imines and Alcohols by Shvo Complexes

Remarkably, the same Shvo complex can be used for the catalytic transfer *dehy*drogenation of aromatic amines to give imines (Scheme 7.14) [80]. This reaction produces high yields when carried out for 2–6 h in refluxing toluene with 2 mol.% catalyst. A quinone is used as the hydrogen acceptor, giving the corresponding hydroquinone.

The reaction can be made catalytic in 2,6-dimethoxy-1,4-benzoquinone (20 mol.%) by the addition of 1.5 equiv. MnO_2 to regenerate the quinone from the hydroquinone. Dehydrogenation is the slow step in this reaction; separate experiments had documented that conversion of the benzoquinone to the hydroquinone has a TOF of $> 4000\,h^{-1}$ [81]. Kinetic isotope effects showed that the rate-limiting step was cleavage of the C–H bond, and that the transfer of the two hydrogens is not concerted [82]. The proposed mechanism involved slow β-elimination from a coordinated amine followed by proton transfer to the oxygen of the cyclopentadienone ligand.

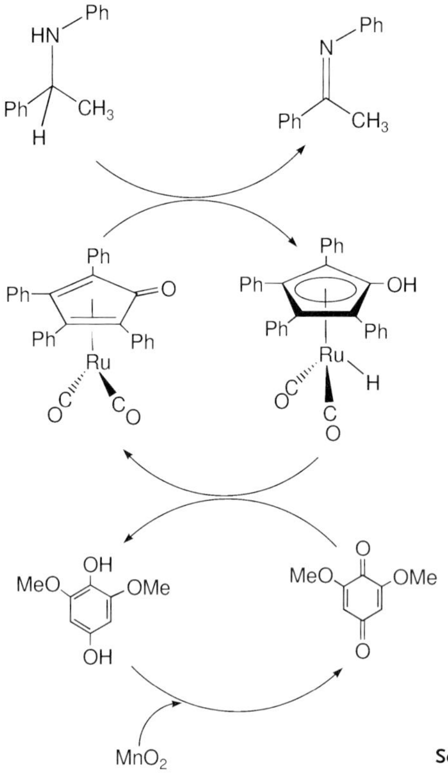

Scheme 7.14

Johnson and Bäckvall reported that the bimetallic Shvo catalyst can also catalyze the transfer dehydrogenation of alcohols (Eq. (48)) [83].

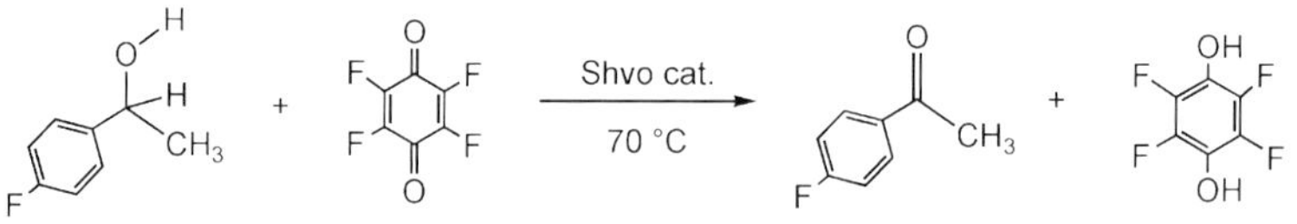

$$(48)$$

Using tetrafluorobenzoquinone as the hydrogen acceptor, the kinetics at 70 °C
showed that the reaction was first order in alcohol, zero order in the quinone,
and half-order in the Ru_2 catalyst. The half-order in bimetallic catalyst is expected in cases where a bimetallic species must dissociate into a monomeric active species. Proton and hydride transfer outside of the coordination sphere of
the ruthenium is a possible mechanism, and this is essentially the reverse of
the reaction shown in Eq. (42). An alternative mechanism favored by Bäckvall
and colleagues [81, 83] and by Menashe and Shvo [84] involves the formation of
an alcohol complex which undergoes proton transfer from the alcohol's OH to
the oxygen of the ligand, together with β-hydride elimination to form the RuH

bond. Kinetic isotope effects found with deuteration of the OH and CD sites of the alcohol were $k_{CHOH}/k_{CHOD}=1.87\pm0.17$ and $k_{CHOH}/k_{CDOH}=2.57\pm0.26$. This provides evidence for a concerted ionic mechanism, since the experimentally observed isotope effect with both sites labeled ($k_{CHOH}/k_{CDOD}=4.61\pm0.37$) is within experimental uncertainty of the product of the individual isotope effects ($1.87\times2.57=4.80$). The ability of the Shvo and other ruthenium catalysts to reversibly dehydrogenate alcohols has been used by Bäckvall and coworkers to accomplish the dynamic kinetic resolution of secondary alcohols, where the metal catalyst is used in conjunction with enzymes [85].

7.4.4
Catalytic Hydrogenations with Metal–Ligand Bifunctional Catalysis

The concerted delivery of protons from OH and hydride from RuH found in these Shvo systems is related to the proposed mechanism of hydrogenation of ketones (Scheme 7.15) by a series of ruthenium systems that operate by metal–ligand bifunctional catalysis [86]. A series of Ru complexes reported by Noyori, Ohkuma and coworkers exhibit extraordinary reactivity in the enantioselective hydrogenation of ketones. These systems are described in detail in Chapters 20 and 31, and mechanistic issues of these hydrogenations by ruthenium complexes have been reviewed [87].

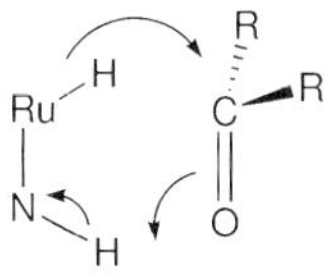

Scheme 7.15

7.5
Catalytic Hydrogenation of Ketones by Strong Bases

As documented throughout this handbook, the diversity of reaction patterns of transition-metal complexes leads to a remarkably rich chemistry, with a tremendous mechanistic diversity in the details of how H_2 is added to unsaturated substrates. Over forty years ago, Walling and Bollyky reported a catalytic hydrogenation of benzophenone that required no transition metal at all! They found that the C=O bond of benzophenone can be catalytically hydrogenated using KOtBu as a base [88], but harsh conditions (200 °C, 100 bar H_2) were used (Eq. (49)). Berkessel et al. recently examined details of this reaction and provided evidence that it was first order in ketone, first order in hydrogen, and first order in base [89].

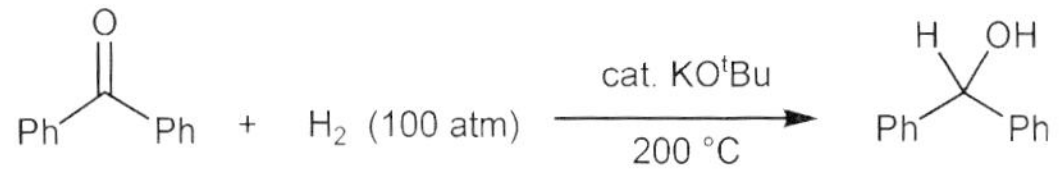

(49)

This ketone hydrogenation is limited to non-enolizable ketones; hydrogenation rates are similar for benzophenone and Ph(C=O)tBu, slower for (tBu)$_2$C=O, and still slower for 2,2,5,5-tetramethylcyclopentanone. The relative initial rates as the alkali metal was changed were the same for K$^+$ and Rb$^+$ (relative initial rate = 100), slightly faster for Cs$^+$ (relative initial rate = 105), slower for Na$^+$ (relative initial rate = 50), and much slower for Li$^+$ (relative initial rate = 7). The proposed six-membered cyclic transition state is shown in Scheme 7.16. Recent molecular orbital calculations on a simplified model of this system (hydrogenation of formaldehyde by NaOCH$_3$) suggest that the slow rates are due in large part to high entropic barriers to properly assemble and orient the highly ordered transition state required [90].

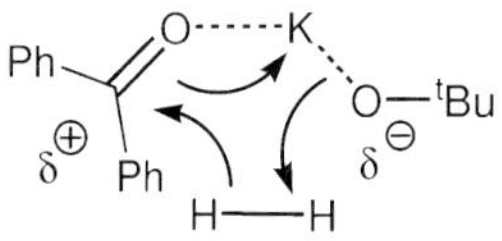

Scheme 7.16

7.6
Conclusion

Ionic hydrogenations are far less developed than hydrogenations that proceed by traditional insertion mechanisms. Despite this later historical development, recent developments have shown that ionic hydrogenations have become more widely recognized, and the prevalence of this area is expected to continue to expand. The development of ionic hydrogenations has benefited immensely from mechanistic experiments on model organometallic complexes. Fundamental studies of the acidity and hydricity of metal hydrides have – and will continue – to play an important role in helping the rational design of ionic hydrogenation catalysts. Both thermodynamic and kinetic studies of acidity and hydricity of metal hydrides are helpful in assessing the specific combinations of ligands and metals that may be suitable for consideration as catalysts. Despite being less extensively developed than other hydrogenation methods, there is burgeoning interest in ionic hydrogenations, and there are numerous promising avenues for future research into this area.

Acknowledgments

Research studies at the Brookhaven National Laboratory were carried out under contract DE-AC02-98CH10886 with the U.S. Department of Energy, and were supported by its Division of Chemical Sciences, Office of Basic Energy Sciences. The author thanks those dedicated postdoctorate scientists who have worked at Brookhaven on ionic hydrogenations, namely Drs. Jeong-Sup Song, Li Luan, Mark Voges, Marcel Schlaf, Prasenjit Ghosh, Barbara Kimmich, and Vladimir Dioumaev.

Abbreviations

DME   dimethoxyethane
HOTf   triflic acid
THF   tetrahydrofuran
TOF   turn over frequency
TON   turn over number

References

1   D. N. Kursanov, Z. N. Parnes, N. M.
    Loim, *Synthesis* **1974**, 633.
2   D. N. Kursanov, Z. N. Parnes, M. I. Kalin-
    kin, N. M. Loim, *Ionic Hydrogenation and
    Related Reactions*, Harwood Academic
    Publishers, New York, **1985**.
3   R. M. Bullock, J.-S. Song, *J. Am. Chem.
    Soc.* **1994**, *116*, 8602.
4   M. P. Doyle, C. C. McOsker, *J. Org.
    Chem.* **1978**, *43*, 693.
5   C. T. West, S. J. Donnelly, D. A. Kooistra,
    M. P. Doyle, *J. Org. Chem.* **1973**, *38*,
    2675.
6   M. P. Doyle, D. J. DeBruyn, S. J. Don-
    nelly, D. A. Kooistra, A. A. Odubela, C. T.
    West, S. M. Zonnebelt, *J. Org. Chem.*
    1974, *39*, 2740.
7   P. T. Anastas, M. M. Kirchhoff, *Acc.
    Chem. Res.* **2002**, *35*, 686.
8   (a) M. S. Chinn, D. M. Heinekey, N. G.
    Payne, C. D. Sofield, *Organometallics*
    1989, *8*, 1824–1826; (b) M. Schlaf, A. J.
    Lough, P. A. Maltby, R. H. Morris, *Orga-
    nometallics* **1996**, *15*, 2270.
9   T. P. Fong, C. E. Forde, A. J. Lough,
    R. H. Morris, P. Rigo, E. Rocchini, T. Ste-
    phan, *J. Chem. Soc., Dalton Trans.* **1999**,
    4475.
10 (a) R. F. Jordan, J. R. Norton, *J. Am.
    Chem. Soc.* **1982**, *104*, 1255;
    (b) R. T. Edidin, J. M. Sullivan, J. R. Nor-
    ton, *J. Am. Chem. Soc.* **1987**, *109*, 3945;
    (c) S. S. Kristjánsdóttir, A. E. Moody, R. T.
    Weberg, J. R. Norton, *Organometallics*
    1988, *7*, 1983.
11 E. J. Moore, J. M. Sullivan, J. R. Norton,
    J. Am. Chem. Soc. **1986**, *108*, 2257.
12 S. S. Kristjánsdóttir, J. R. Norton. In:
    Transition Metal Hydrides, A. Dedieu
    (Ed.), VCH, New York, **1991**, Chapter 9,
    p. 309.
13 D. E. Berning, B. C. Noll, D. L. DuBois,
    J. Am. Chem. Soc. **1999**, *121*, 11432.
14 (a) D. E. Berning, A. Miedaner,
    C. J. Curtis, B. C. Noll, M. C. Rakowski
    DuBois, D. L. DuBois, *Organometallics*
    2001, *20*, 1832; (b) A. Miedaner,
    J. W. Raebiger, C. J. Curtis, S. M. Miller,
    D. L. DuBois, *Organometallics* **2004**, *23*,
    2670.
15 C. J. Curtis, A. Miedaner, W. W. Ellis,
    D. L. DuBois, *J. Am. Chem. Soc.* **2002**,
    124, 1918.
16 R. Ciancanelli, B. C. Noll, D. L. DuBois,
    M. C. Rakowski DuBois, *J. Am. Chem.
    Soc.* **2002**, 2984.
17 A. J. Price, R. Ciancanelli, B. C. Noll, C. J.
    Curtis, D. L. DuBois, M. R. DuBois, *Orga-
    nometallics* **2002**, *21*, 4833.
18 W. W. Ellis, R. Ciancanelli, S. M. Miller,
    J. W. Raebiger, M. R. DuBois, D. L. Du-
    Bois, *J. Am. Chem. Soc.* **2003**, *125*,
    12230.
19 J. W. Raebiger, A. Miedaner, C. J. Curtis,
    S. M. Miller, O. P. Anderson,
    D. L. DuBois, *J. Am. Chem. Soc.* **2004**,
    126, 5502.
20 K. Abdur-Rashid, T. P. Fong, B. Greaves,
    D. G. Gusev, J. G. Hinman, S. E. Landau,
    A. J. Lough, R. H. Morris, *J. Am. Chem.
    Soc.* **2000**, *122*, 9155.
21 A. Streitwieser, Y.-J. Kim, *J. Am. Chem.
    Soc.* **2000**, *122*, 11783.
22 J. A. Labinger. In: *Transition Metal Hy-
    drides*, A. Dedieu (Ed.), VCH, New York,
    1991, Chapter 10, p. 361.
23 J. A. Labinger, K. H. Komadina, *J. Orga-
    nomet. Chem.* **1978**, *155*, C25.
24 M. Y. Darensbourg, C. E. Ash, *Adv. Orga-
    nomet. Chem.* **1987**, *27*, 1.
25 S. C. Kao, M. Y. Darensbourg, *Organome-
    tallics* **1984**, *3*, 646.

26 S.C. Kao, P.L. Gaus, K. Youngdahl, M.Y. Darensbourg, *Organometallics* **1984**, *3*, 1601.

27 P.L. Gaus, S.C. Kao, K. Youngdahl, M.Y. Darensbourg, *J. Am. Chem. Soc.* **1985**, *107*, 2428.

28 C.E. Ash, P.W. Hurd, M.Y. Darensbourg, M. Newcomb, *J. Am. Chem. Soc.* **1987**, *109*, 3313.

29 W. Beck, K. Sünkel, *Chem. Rev.* **1988**, *88*, 1405.

30 (a) G.J. Kubas, *Metal Dihydrogen and σ-Bond Complexes: Structure, Theory, and Reactivity*, Kluwer Academic/Plenum Publishers, New York, **2001**; (b) D.M. Heinekey, W.J. Oldham, Jr., *Chem. Rev.* **1993**, *93*, 913; (c) P.G. Jessop, R.H. Morris, *Coord. Chem. Rev.* **1992**, *121*, 155.

31 E.T. Papish, F.C. Rix, N. Spetseris, J.R. Norton, R.D. Williams, *J. Am. Chem. Soc.* **2000**, *122*, 12235.

32 J.-S. Song, D.J. Szalda, R.M. Bullock, *Inorg. Chim. Acta* **1997**, *259*, 161.

33 L. Luan, J.-S. Song, R.M. Bullock, *J. Org. Chem.* **1995**, *60*, 7170.

34 R.M. Bullock, J.-S. Song, D.J. Szalda, *Organometallics* **1996**, *15*, 2504.

35 J.-J. Brunet, R. Chauvin, P. Leglaye, *Eur. J. Inorg. Chem.* **1999**, 713.

36 M.Y. Darensbourg, *Prog. Inorg. Chem.* **1985**, *33*, 221.

37 D.H. Gibson, Y.S. El-Omrani, *Organometallics* **1985**, *4*, 1473.

38 (a) T. Ito, M. Koga, S. Kurishima, M. Natori, N. Sekizuka, K. Yoshioka, *J. Chem. Soc., Chem. Commun.* **1990**, 988; (b) M. Minato, Y. Fujiwara, M. Koga, N. Matsumoto, S. Kurishima, M. Natori, N. Sekizuka, K. Yoshioka, T. Ito, *J. Organomet. Chem.* **1998**, *569*, 139.

39 S.M. Geraty, P. Harkin, J.G. Vos, *Inorg. Chim. Acta* **1987**, *131*, 217.

40 W.D. Harman, H. Taube, *J. Am. Chem. Soc.* **1990**, *112*, 2261.

41 J.-S. Song, D.J. Szalda, R.M. Bullock, C.J.C. Lawrie, M.A. Rodkin, J.R. Norton, *Angew. Chem., Int. Ed. Engl.* **1992**, *31*, 1233.

42 J.-S. Song, D.J. Szalda, R.M. Bullock, *Organometallics* **2001**, *20*, 3337.

43 V.I. Bakhmutov, E.V. Vorontsov, D.Y. Antonov, *Inorg. Chim. Acta* **1998**, *278*, 122.

44 R.J. Kinney, W.D. Jones, R.G. Bergman, *J. Am. Chem. Soc.* **1978**, *100*, 7902.

45 J.-F. Reynoud, J.-F. Leboeuf, J.-C. Leblanc, C. Moïse, *Organometallics* **1986**, *5*, 1863.

46 C. Bianchini, C. Mealli, A. Meli, M. Peruzzini, F. Zanobini, *J. Am. Chem. Soc.* **1988**, *110*, 8725.

47 (a) D.M. Heinekey, A. Liegeois, M. van Roon, *J. Am. Chem. Soc.* **1994**, *116*, 8388; (b) D.M. Heinekey, M. van Roon, *J. Am. Chem. Soc.* **1996**, *118*, 12134.

48 P.A. Tooley, C. Ovalles, S.C. Kao, D.J. Darensbourg, M.Y. Darensbourg, *J. Am. Chem. Soc.* **1986**, *108*, 5465.

49 L. Markó, Z. Nagy-Magos, *J. Organomet. Chem.* **1985**, *285*, 193.

50 T. Fuchikami, Y. Ubukata, Y. Tanaka, *Tetrahedron Lett.* **1991**, *32*, 1199.

51 J.-J. Brunet, *Eur. J. Inorg. Chem.* **2000**, 1377.

52 C. Balzarek, T.J.R. Weakley, D.R. Tyler, *J. Am. Chem. Soc.* **2000**, *122*, 9427.

53 C. Balzarek, T.J.R. Weakley, L.Y. Kuo, D.R. Tyler, *Organometallics* **2000**, *19*, 2927.

54 C. Balzarek, D.R. Tyler, *Angew. Chem., Int. Ed.* **1999**, *38*, 2406.

55 L.Y. Kuo, T.J.R. Weakley, K. Awana, C. Hsia, *Organometallics* **2001**, *20*, 4969.

56 L.Y. Kuo, D.M. Finigan, N.N. Tadros, *Organometallics* **2003**, *22*, 2422.

57 (a) R.M. Bullock, M.H. Voges, *J. Am. Chem. Soc.* **2000**, *122*, 12594; (b) M.H. Voges, R.M. Bullock, *J. Chem. Soc., Dalton Trans.* **2002**, 759.

58 T.-Y. Cheng, B.S. Brunschwig, R.M. Bullock, *J. Am. Chem. Soc.* **1998**, *120*, 13121.

59 I.M. Kolthoff, M.K. Chantooni, Jr., *J. Am. Chem. Soc.* **1973**, *95*, 8539.

60 K.-T. Smith, J.R. Norton, M. Tilset, *Organometallics* **1996**, *15*, 4515.

61 (a) B.F.M. Kimmich, P.J. Fagan, E. Hauptman, R.M. Bullock, *Chem. Commun.* **2004**, 1014; (b) B.F.M. Kimmich, P.J. Fagan, E. Hauptman, W.J. Marshall, R.M. Bullock, *Organometallics* **2005**, *24*, 6220.

62 J.M. DeSimone, *Science* **2002**, *297*, 799.

63 (a) G.W.V. Cave, C.L. Raston, J.L. Scott, *J. Chem. Soc., Chem. Commun.* **2001**, 2159; (b) J.M. Thomas, R. Raja,

G. Sankar, B. F. G. Johnson, D. W. Lewis, *Chem. Eur. J.* **2001**, *7*, 2973.

64 (a) A. J. Arduengo, III, *Acc. Chem. Res.* **1999**, *32*, 913; (b) D. Bourissou, O. Guerret, F. P. Gabbaï, G. Bertrand, *Chem. Rev.* **2000**, *100*, 39; (c) W. A. Herrmann, *Angew. Chem., Int. Ed.* **2002**, *41*, 1290; (d) C. M. Crudden, D. P. Allen, *Coord. Chem. Rev.* **2004**, *248*, 2247.

65 V. K. Dioumaev, D. J. Szalda, J. Hanson, J. A. Franz, R. M. Bullock, *Chem. Commun.* **2003**, 1670.

66 V. K. Dioumaev, R. M. Bullock, *Nature* **2003**, *424*, 530.

67 M. Aresta, A. Dibenedetto, I. Pápai, G. Schubert, A. Macchioni, D. Zuccaccia, *Chem. Eur. J.* **2004**, *10*, 3708.

68 (a) M. P. Magee, J. R. Norton, *J. Am. Chem. Soc.* **2001**, *123*, 1778; (b) H. Guan, M. Imura, M. P. Magee, J. R. Norton, G. Zhu, *J. Am. Chem. Soc.* **2005**, *127*, 7805.

69 H. Guan, M. Iimura, M. P. Magee, J. R. Norton, K. E. Janak, *Organometallics* **2003**, 4084.

70 Y. Shvo, D. Czarkie, Y. Rahamim, D. F. Chodosh, *J. Am. Chem. Soc.* **1986**, *108*, 7400.

71 Y. Blum, D. Czarkie, Y. Rahamim, Y. Shvo, *Organometallics* **1985**, *4*, 1459.

72 C. P. Casey, S. W. Singer, D. R. Powell, R. K. Hayashi, M. Kavana, *J. Am. Chem. Soc.* **2001**, *123*, 1090.

73 (a) R. M. Bullock, B. R. Bender. In: *Encyclopedia of Catalysis*, I. Horváth (Ed.), Wiley, New York, **2002**; (b) R. M. Bullock. In: *Transition Metal Hydrides*, A. Dedieu (Ed.), VCH, New York, **1991**, Chapter 8, p. 263.

74 N. Menashe, E. Salant, Y. Shvo, *J. Organomet. Chem.* **1996**, *514*, 97.

75 C. P. Casey, S. W. Singer, D. R. Powell, *Can. J. Chem.* **2001**, *79*, 1002.

76 J. S. M. Samec, J.-E. Bäckvall, *Chem. Eur. J.* **2002**, *8*, 2955.

77 J. S. M. Samec, A. H. Éll, J.-E. Bäckvall, *Chem. Commun.* **2004**, 2748.

78 C. P. Casey, G. A. Bikzhanova, J.-E. Bäckvall, L. Johansson, J. Park, Y. H. Kim, *Organometallics* **2002**, *21*, 1955.

79 C. P. Casey, J. B. Johnson, *J. Am. Chem. Soc.* **2005**, *127*, 1883.

80 A. H. Éll, J. S. M. Samec, C. Brasse, J.-E. Bäckvall, *Chem. Commun.* **2002**, 1144.

81 G. Csjernyik, A. H. Éll, L. Fadini, B. Pugin, J.-E. Bäckvall, *J. Org. Chem.* **2002**, *67*, 1657.

82 A. H. Éll, J. B. Johnson, J.-E. Bäckvall, *Chem. Commun.* **2003**, 1652.

83 J. B. Johnson, J.-E. Bäckvall, *J. Org. Chem.* **2003**, *68*, 7681.

84 N. Menashe, Y. Shvo, *Organometallics* **1991**, *10*, 3885.

85 F. F. Huerta, A. B. E. Minidis, J.-E. Bäckvall, *Chem. Soc. Rev.* **2001**, *30*, 321.

86 R. Noyori, M. Yamakawa, S. Hashiguchi, *J. Org. Chem.* **2001**, *66*, 7931.

87 S. E. Clapham, A. Hadzovic, R. H. Morris, *Coord. Chem. Rev.* **2004**, *248*, 2201.

88 (a) C. Walling, L. Bollyky, *J. Am. Chem. Soc.* **1961**, *83*, 2968; (b) C. Walling, L. Bollyky, *J. Am. Chem. Soc.* **1964**, *86*, 3750.

89 A. Berkessel, T. J. S. Schubert, T. N. Müller, *J. Am. Chem. Soc.* **2002**, *124*, 8693.

90 B. Chan, L. Radom, *J. Am. Chem. Soc.* **2005**, *127*, 2443.

91 W. W. Ellis, J. W. Raebiger, C. J. Curtis, J. W. Bruno, D. L. DuBois, *J. Am. Chem. Soc.* **2004**, *126*, 2738.

92 X.-M. Zhang, J. W. Bruno, E. Enyinnaya, *J. Org. Chem.* **1998**, *63*, 4671.

93 T.-Y. Cheng, R. M. Bullock, *Organometallics* **2002**, *21*, 2325.

8
Homogeneous Hydrogenation by Defined Metal Clusters

Roberto A. Sánchez-Delgado

8.1
Introduction

The catalytic potential of well-defined transition-metal clusters in homogeneous reactions has attracted a great deal of attention over the years, as they represent a natural bridge between mononuclear complexes, metal nanoparticles, and metal-oxide, -sulfide and related surfaces used in heterogeneous catalysis. The molecular nature of metal clusters, together with their solubility properties, provides the advantages of classical mononuclear homogeneous catalysts (high activity, high selectivity, moderate operating conditions, possibility of catalyst design and modification), while the polynuclear framework can offer the possibility of multi-metallic cooperative effects often identified as a key element in the desirable properties of solid heterogeneous catalysts. Therefore, metal clusters can be expected, in principle, to combine the positive aspects of homogeneous and heterogeneous catalytic reactions and, perhaps more importantly, they may react through unique pathways associated with the cluster structures and thereby catalyze reactions not accessible by mononuclear or heterogeneous catalysts. Nevertheless, despite the impressive amount of work that has been devoted to develop these concepts over several decades, the great expectations first advanced during the mid-1970s have not yet been fully accomplished [1–6].

From a different perspective, well-defined metal clusters have served as useful models for discerning the complex mechanisms of heterogeneous catalytic systems. The structural trends for metal clusters are now well understood, and their reactions in solution can be studied in detail by relatively simple chemical and spectroscopic methods, thereby producing important information at the molecular level – something that is very difficult to achieve on solid catalysts. The knowledge thus gained from studying metal cluster chemistry can be extrapolated, with adequate caution, to heterogeneous reactions [1–6].

Another trend that has received considerable recent attention is the decomposition of metal clusters under controlled conditions on solid supports or on liquid suspensions, which generates small metallic particles of specific size, struc-

The Handbook of Homogeneous Hydrogenation.
Edited by J. G. de Vries and C. J. Elsevier
Copyright © 2007 WILEY-VCH Verlag GmbH & Co. KGaA, Weinheim
ISBN: 978-3-527-31161-3

ture or composition, displaying interesting catalytic features [7]. Although this is perhaps the area in which cluster chemistry has had the highest impact, such methods lead unambiguously to *heterogeneous* systems, and therefore it falls beyond the scope of this book.

Homogeneous hydrogenation has been one of the most frequently studied classes of reactions in an effort to demonstrate the principles of cluster catalysis and its links to heterogeneous catalysis. A good number of well-defined metal clusters have been claimed to promote hydrogen addition to C=C, C≡C, C=O bonds and aromatic rings, and a number of detailed mechanistic studies have been conducted. Many of these catalysts have later been shown to be mononuclear or heterogeneous in nature, but some others have proved to induce truly homogeneous cluster-catalyzed reactions. For the purpose of the discussion that follows, the classical definition of a cluster as a compound containing at least three metal atoms [8] will be adopted. Also, the concept of *cluster catalysis* is associated here to reaction mechanisms involving only polynuclear intermediates, regardless of whether the important interactions take place at only one or at several metal atoms.

8.1.1
Is a Cluster the Real Catalyst? Fragmentation and Aggregation Phenomena

One of the key points in discussing cluster catalysis is to determine whether a cluster is actually catalyzing the hydrogenation reaction, or if instead a mononuclear entity derived from cluster fragmentation, or metallic nanoparticles resulting from decomposition and aggregation are responsible for the catalytic transformation. An ideal model reaction would involve bonding of the substrate to more than one metal atom, so that if the cluster becomes degraded in the process, the catalytic activity would be lost. Nevertheless, sensible hydrogenation cycles have been proposed in which the cluster framework is maintained but the substrate does not need to bind to more than one metal atom in order to be transformed. In such cases it is difficult to ascertain which is the true catalytically active species. Several methods have been used to address the two questions that need to be answered:

- Is the catalyst truly homogeneous?
- If so, is the catalyst really a cluster?

Finke and coworkers have extensively addressed the question of homogeneous versus heterogeneous catalysis, and have provided a rather complex (but extremely reliable) set of experiments that allow the distinction of a molecular catalyst in solution from a suspended nanostructured metallic material [9]. The generation of metallic particles from metal complexes under a highly reducing hydrogen atmosphere is now recognized as a frequent phenomenon, particularly in arene hydrogenation studies [9, 10].

Once the homogeneity of a reaction has been established, it is never easy to determine the precise nuclearity of the active species, and a series of indicators or qualitative tests has been proposed [11]. Many publications provide as evidence for cluster catalysis simple statements such as "… the cluster was quantitatively recovered at the end of the reaction" or "… the cluster was the only species observed by IR or NMR spectroscopy". This type of assumption can be very misleading; it suffices to have 1% or less of the cluster transform into highly active mononuclear fragments or metallic particles in order to develop a high hydrogenation activity. Such small amounts of very reactive species easily go unnoticed in *in-situ* spectroscopic studies, and would certainly not be accounted for in a "quantitative" recovery of the cluster at the end of the reaction.

In order to establish the participation of a cluster, the best approach is to use a combination of experiments:

- Kinetic measurements, particularly the study of the rate-dependence on *cluster* concentration can be very informative; cluster-catalyzed reactions often display a first-order rate dependence on *cluster* concentration, whereas fractional or complex orders of reaction are associated with fragmentation processes.
- Reactions that are much faster than the analogous one catalyzed by a mononuclear complex, or that lead to different products or selectivities, are more likely to involve cluster intermediates.
- Heterobimetallic complexes that induce reactions at significantly faster rates than (or notably different product selectivities from) monometallic derivatives are probably genuine cluster catalysts.
- Edge- and face-capping chelating ligands have been proposed as a method to guarantee the stability of the cluster framework.
- Another interesting idea that has been explored without much success so far is the use of clusters with a chiral metal framework as catalysts for asymmetric hydrogenation, since only the intact cluster would induce enantioselectivity.
- NMR studies involving *para*-hydrogen has recently been introduced as a powerful tool to obtain direct evidence for cluster catalysis (*vide infra*).

This chapter reviews the literature involving well-defined molecular metal clusters as hydrogenation catalysts or catalyst precursors, with particular emphasis being placed on those systems that are likely to involve only or predominantly cluster intermediates throughout the hydrogenation cycle. The mechanisms in cases where cluster catalysis is strongly supported by experimental evidence are discussed in more detail.

8.2
Hydrogenation of C=C Bonds

Early investigations by Shapley [12], Basset [13], and Gladfelter [14] provided the first convincing examples of C=C bond hydrogenation cycles involving metal clusters; these are shown in Schemes 8.1–8.3. Shapley's mechanisms for $[Os_3H_2(CO)_{10}]$ was based on experiments performed under noncatalytic conditions, involving the isolation and/or NMR observation of all the species implicated in the cycle depicted in Scheme 8.1, as well as the pathways for their interconversion.

Basset's proposal for the silica-supported cluster $[Os_3(CO)_{10}(\mu\text{-}H)(\mu\text{-}OSi\equiv)]$ was made on the basis of surface IR spectroscopy studies, kinetic and gas uptake measurements, and reactions of the soluble analogue $[Os_3 (CO)_{10}(\mu\text{-}H)(\mu\text{-}OSi\text{-}Ph)]$; the supported catalyst hydrogenated ethylene at $90\,^{\circ}C$ and atmospheric pressure in a flow reactor at a TOF of $144\,h^{-1}$ for extended periods of time, achieving up to 24 000 turnovers overall. Gladfelter also used kinetic measurements and IR spectroscopy to deduce the mechanism of alkene hydrogenation by anionic clusters containing isocyanate ligands $[Ru_3(\mu\text{-}NCO) (CO)_{10}]^-$; this catalyst reduced 3,3-dimethylbutene at rates of about 300 to 360 turnovers h^{-1} under ambient conditions. The same group also characterized intermediates and individual reactions of the more stable, but catalytically less active, osmium analogue $[Os_3(\mu\text{-}NCO)(CO)_{10}]^-$ (eight turnovers after 24 h at $78\,^{\circ}C$ and 3.3 bar H_2 for 3,3-dimethylbutene hydrogenation). Although these are important pioneering cases from a fundamental point of view, they are of no practical use because catalytic activities were low and the scope of the reactions was limited.

Sánchez-Delgado et al. reported a comparative study of the hydrogenation of 1-hexene by use of $[Ru_3(CO)_{12}]$ in solution and supported on silica; IR evidence pointed to cluster catalysis in solution [turnover frequency (TOF) ca. $200\,h^{-1}$ at $90\,^{\circ}C$ and 40 bar H_2] and to the formation of mononuclear species on the silica surface (TOF ca. $600\,h^{-1}$ at $90\,^{\circ}C$ and 40 bar H_2) [15]. Another early proposal for a cluster-catalyzed reaction was provided by Doi et al. for $[H_4Ru_4(CO)_{12}]$ in the

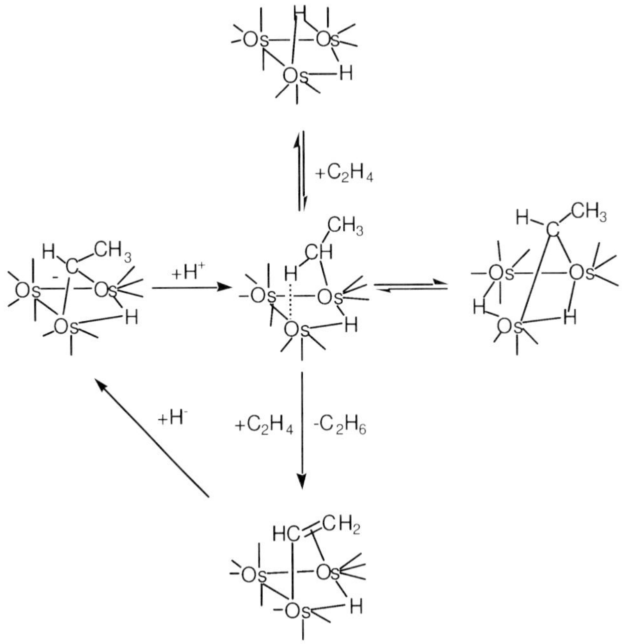

Scheme 8.1 Mechanism for the hydrogenation of alkenes catalyzed by $[Os_3H_2(CO)_{10}]$ (CO ligands omitted for clarity).

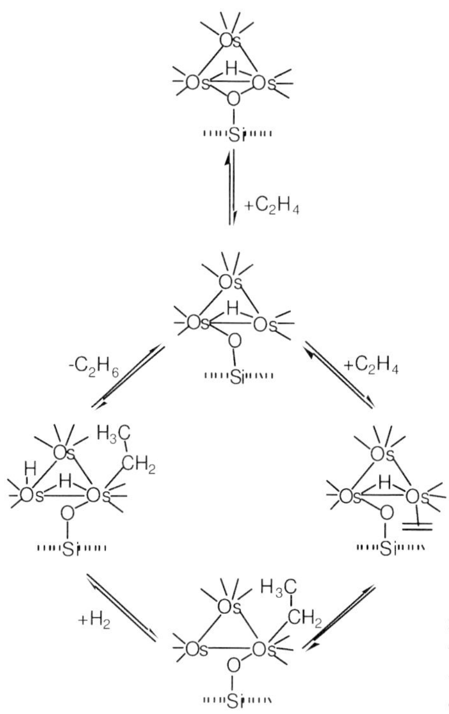

$+C_2H_4$

$-C_2H_6$

$+C_2H_4$

$+H_2$

Scheme 8.2 Mechanism for the hydrogenation of ethylene catalyzed by silica-supported osmium clusters (CO ligands omitted for clarity).

hydrogenation of ethylene (TOF 40 h^{-1} at 72 °C, 0.13 bar H$_2$, and 0.26 atm ethylene). Kinetic measurements were in agreement with a mechanism involving Eqs. (1)–(3); in particular, a first-order dependence of the reaction rate on cluster concentration was taken as evidence of cluster catalysis, although the hydrogenation cycle involved a single Ru atom [16].

$$H_4Ru_4(CO)_{11} + H_2 \rightleftharpoons H_6Ru_4(CO)_{11} \tag{1}$$
$$H_4Ru_4(CO)_{11} + C_2H_4 \rightleftharpoons H_3Ru_4(CO)_{11}(C_2H_5) \tag{2}$$
$$H_3Ru_4(CO)_{11}(C_2H_5) + H_2 \rightleftharpoons H_4Ru_4(CO)_{11} + C_2H_6 \tag{3}$$

Related studies by Sánchez-Delgado and coworkers on the kinetics of the hydrogenation of styrene (140 °C, 1.05 bar H$_2$) catalyzed by the tetranuclear Os clusters [H$_4$Os$_4$(CO)$_{12}$] (TOF 87 h^{-1}), [H$_3$Os$_4$(CO)$_{12}$]$^-$ (TOF 52 h^{-1}), [H$_3$Os$_4$(CO)$_{12}$I] (TOF 583 h^{-1}), and [H$_2$Os$_4$(CO)$_{12}$I]$^-$ (TOF 63 h^{-1}), pointed to cluster fragmentation as being responsible for the catalytic activity of these systems, which conclusion was based on a complex rate-dependence on cluster concentration and the similarity of the hydrogenation rate when a mononuclear Os complex was employed [17].

Phosphine-substituted complexes have shown promise for cluster catalysis, especially in the case of chelating ligands, because of the added stability that might help avoid cluster fragmentation. Bergounhou et al. reported a detailed study of the hy-

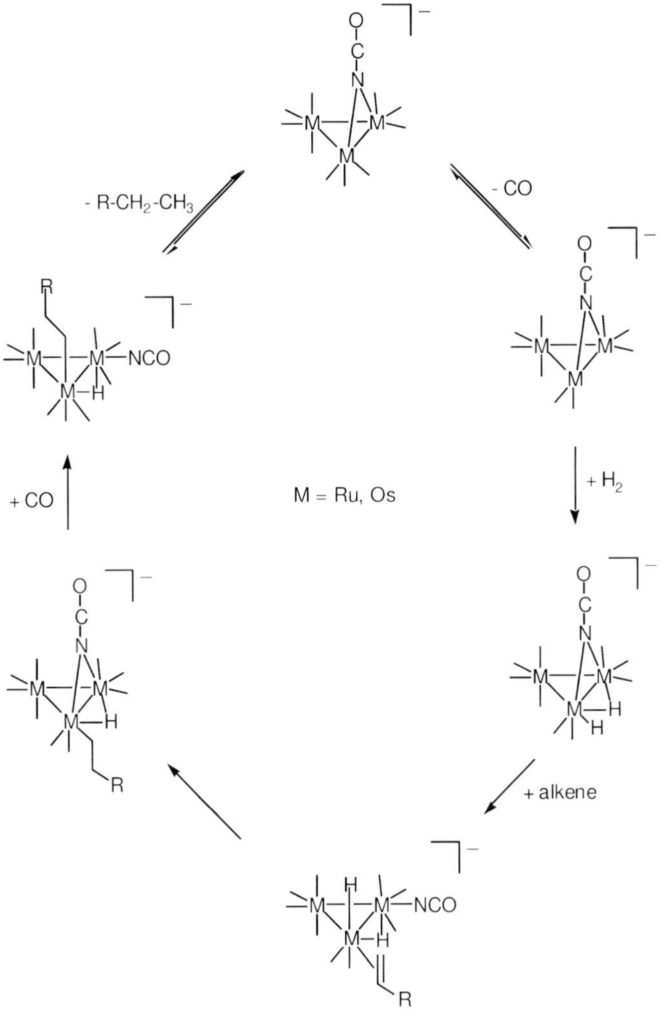

Scheme 8.3 Mechanism for the hydrogenation of alkenes catalyzed by anion-promoted osmium clusters (CO ligands omitted for clarity).

drogenation of 1-hexene by [Ru$_3$(μ-H)$_2$(μ^3-O)(CO)$_5$(dppm)$_2$] (TOF ca. 25 200 h^{-1}), and provided kinetic and spectroscopic evidence for the cycle depicted in Scheme 8.4, in which a Ru–Ru bond is broken but the cluster integrity is maintained by the oxo and diphosphine ligands [18]. Further catalytic studies with [Ru$_3$(CO)$_{12}$] substituted with chelating diphosphines were provided by Fontal et al. [19], and with PPh$_3$ by Dallmann and Buffon [20]. The suggestion was that the reactions proceed through cluster intermediates, although C=C bond hydrogenation was accompanied by extensive isomerization and no details of reaction mechanisms were provided. Clusters derived from [H$_4$Ru$_4$(CO)$_{12}$] by substitution with *chiral* dipho-

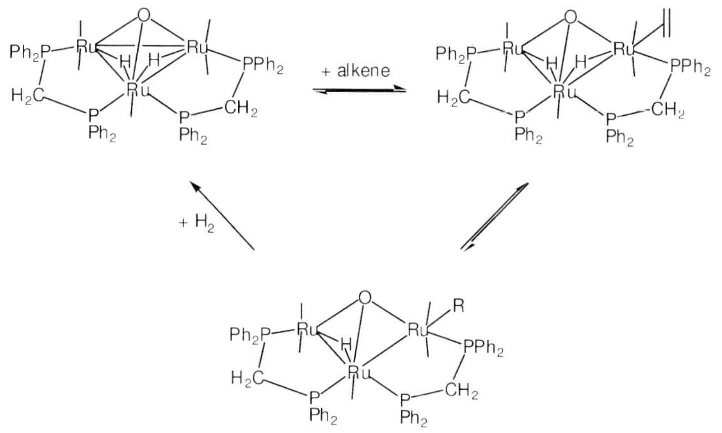

Scheme 8.4 Mechanism for the hydrogenation of alkenes catalyzed by ruthenium clusters stabilized by edge-bridging diphosphine ligands (CO ligands omitted for clarity).

sphines exhibited reasonable activities (TOF up to 60000 h^{-1}) and moderate enantioselectivities (6 to 46% ee) in the hydrogenation of α,β-unsaturated carboxylic acids. Although the clusters were generally "recovered intact" at the end of the reactions, the participation of mononuclear species cannot be ruled out [21, 22].

Moura et al. recently reported the first example of the use of Ir clusters in homogeneous diene hydrogenation [23]. [Ir$_4$(CO)$_{11}$(PPh$_2$H)], [Ir$_4$(CO)$_8$(μ^3-η^2-HCCPh)(μ-PPh$_2$)$_2$], [Ir$_4$(CO)$_9$(μ^3-η^3-Ph$_2$PC(H)CPh)(μ-PPh$_2$)], and [Ir$_4$(CO)$_{12}$] selectively reduce 1,5-cyclooctadiene to cyclooctene with high activities [average turnover number (TON) 2816], in contrast with mononuclear or metallic Ir catalysts, which quickly yielded the fully reduced product cyclooctane; this is strongly indicative of a cluster-catalyzed reaction. All the complexes lose the phosphine ligands during the course of the reactions to produce a common active species likely related to [Ir$_4$(CO)$_5$(C$_8$H$_{12}$)$_2$(C$_8$H$_{10}$)]; an "anchor-type" interaction between the two C=C bonds of the diene to one Ir atom allows the activation and hydrogenation of only one of those bonds by the cluster.

Much emphasis has been placed in recent times on easily recoverable liquid biphasic catalysts, including metal clusters in nonconventional solvents. For instance, aqueous solutions of the complexes [Ru$_3$(CO)$_{12-x}$(TPPTS)$_x$] (x = 1, 2, 3; TPPTS = triphenylphosphine-trisulfonate, P(m-C$_6$H$_4$SO$_3$Na)$_3$) catalyze the hydrogenation of simple alkenes (1-octene, cyclohexene, styrene) at 60 °C and 60 bar H$_2$ at TOF up to 500 h^{-1} [24], while [Ru$_3$(CO)$_9$(TPPMS)$_3$] (TPPMS = triphenylphosphine-monosulfonate, PPh$_2$(m-C$_6$H$_4$SO$_3$Na) is an efficient catalyst precursor for the aqueous hydrogenation of the C=C bond of acrylic acid (TOF 780 h^{-1} at 40 °C and 3 bar H$_2$) and other activated alkenes [25]. The same catalysts proved to be poorly active in room temperature ionic liquids such as [bmim][BF$_4$] (bmim = 1-butyl-3-methylimidazolium). No details about the active species involved are known at this point.

Well-known anionic clusters such as $[HFe_3(CO)_{11}]^-$, $[HWOs_3(CO)_{14}]^-$, $[H_3Os_4(CO)_{12}]^-$, and $[Ru_6C(CO)_{16}]^{2-}$ have also been tested as catalyst precursors in the hydrogenation of styrene in [bmim][BF$_4$], and in organic solvents such as octane and methanol [26]. The activity of the Fe cluster is the lowest of the series, and that for Ru is the highest, but the robust Ru$_6$ cluster was found to decompose under the reaction conditions to metallic particles, which are responsible for the catalytic activity. The WOs$_3$ and Os$_4$ clusters are much more active in the ionic liquid than in octane or methanol, and the improvement in activities (TOF 30 000 h^{-1}) was associated with increased stability of the clusters in the ionic medium, although no detailed mechanistic studies were conducted.

8.3
Hydrogenation of C≡C Bonds

The hydrogenation of alkynes is a very interesting reaction, since the selectivity toward the partially or the fully reduced product allows the *in-situ* comparison of the ability of a catalyst to reduce C≡C versus C=C bonds. This is perhaps the area in which cluster catalysis has been most extensively developed, as recently reviewed by Cabeza [27], Adams and Captain [4], and Dyson [28]. A good number of metal clusters have been employed as catalyst precursors in alkyne hydrogenation, the majority of them containing ruthenium.

Early studies by Valle and coworkers showed that $[Ru_3(CO)_{12}]$ [29], $[H_4Ru_4(CO)_{12}]$ [30], and the phosphine- and phosphite-substituted derivatives $[H_4Ru_4(CO)_{12-n}(PR_3)_n]$ (R=Bu^n, Ph, OEt, OPh) (n=1–3) [31] hydrogenate 1-pentyne and 2-pentyne efficiently at 80 °C and 1 atm H$_2$ to the corresponding alkenes; the internal C≡C bond is reduced more rapidly than the terminal one. Reaction rates increased with increasing number of P-donor ligands for the hydrogenation of 1-pentyne and decreased for 2-pentyne; rates were also increased with increasing basicity of the phosphine or phosphite. The complexes also promote C=C bond migration, and therefore 1-pentene, *cis*-2-pentene and *trans*-2-pentene are observed during the course of the reaction. Consecutive hydrogenation of the alkenes to *n*-pentane takes place only after the alkyne has been completely consumed. No attempt was made then to identify reaction intermediates or the catalytic mechanism, although common pathways were presumed for both complexes. The same clusters $[Ru_3(CO)_{12}]$, $[H_4Ru_4(CO)_{12}]$ [32], the related complex $[H_2Ru_4(CO)_{13}]$ [33, 34], and the diphenylphosphine and diphenylphosphido derivatives $[H_4Ru_4(CO)_{12-n}(PPh_2H)_n]$ (n=1–3), $[Ru_3(\mu\text{-}H)(\mu\text{-}PPh_2)_n(CO)_{11-n}]$ (n=1, 3), $[Ru_3(\mu\text{-}H)_{2-n}(\mu\text{-}PPh_2)_{2+n}(CO)_{8-n}]$ (n=0, 1), and $[Ru_4(\mu^3\text{-}PPh)(CO)_{13}]$ [33–38] were evaluated by the group of Sappa and found to hydrogenate *tert*-butylacetylene and diphenylacetylene at 120 °C and 1 atm H$_2$ (TOF 200–400 h^{-1}) to a mixture of *cis*- and *trans*-stilbene; complete hydrogenation to the alkane was not observed in this case, and the dihydride displayed the slowest rate. The direct participation of cluster structures in the catalytic cycle was suggested by the isolation of some intermediates that could be used as catalyst precursors with essentially the same activity.

Long-awaited direct evidence for cluster catalysis has recently been provided using *para* hydrogen induced polarization (PHIP) NMR techniques (see Chapter 12); hydride transfer to coordinated organic fragments and fluxional processes were shown to occur on metal clusters [39, 40]. When such methods were applied to the hydrogenation of alkenes and alkynes by $[Os_3(\mu\text{-}H)_2(CO)_{10}]$, $Ru_3(CO)_{10}L_2$ (L=PPh$_3$, PMe$_2$Ph, dppe), several active intermediates could be identified, including clusters and mononuclear species. It was further demonstrated that the catalytic route is dependent on the solvent; cluster catalysis is preferred in polar media and in that case, the active species are produced either by CO dissociation or, more slowly, by phosphine dissociation, which generates the vacant coordination site required for the alkene or alkyne to bind. In nonpolar media, fragmentation to a mononuclear complex was observed, and this complex actually competes with the clusters in the hydrogenation cycle (Scheme 8.5). Interestingly, for the phosphido-bridged cluster $[Ru_3(CO)_9(\mu\text{-}H)(\mu\text{-}PPh_2)]$, similar experiments show that, independently of the solvent used, only cluster catalysis takes place, according to Scheme 8.6 [41].

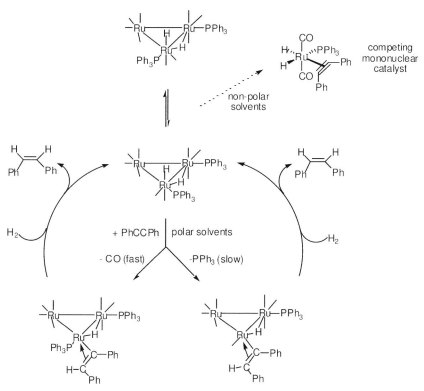

Scheme 8.5 Main species involved in the hydrogenation of diphenylacetylene catalyzed by ruthenium clusters, as determined by PHIP methods (CO ligands omitted for clarity).

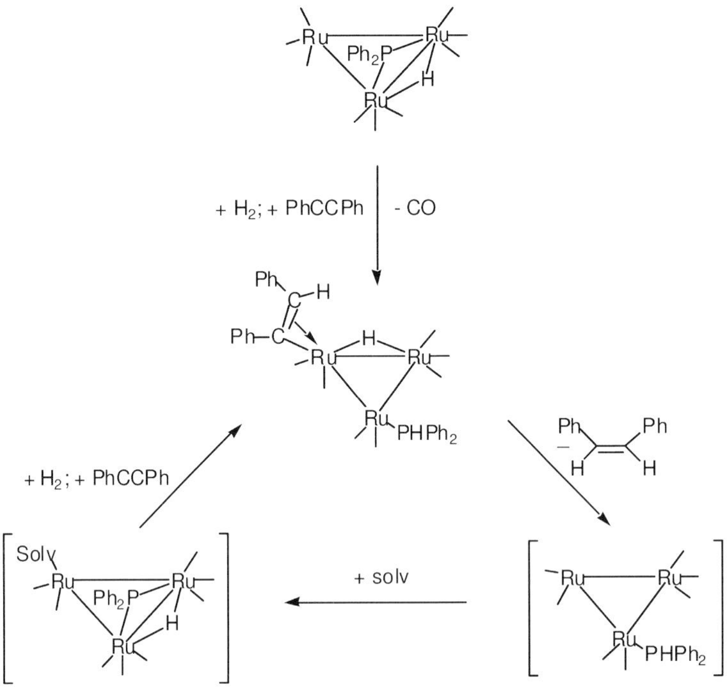

+ H₂; + PhCCPh | - CO

+ H₂; + PhCCPh

+ solv

Scheme 8.6 Mechanism for the hydrogenation of diphenyl-acetylene catalyzed by ruthenium clusters containing phosphido bridging ligands, as determined by PHIP methods (CO ligands omitted for clarity).

The Cp derivative [Ru₃Cp₂(μ^3-Ph₂C₂)(CO)₅] [42] also hydrogenates diphenyl-acetylene and 3-hexyne, but fragmentation probably takes place in this case to an important extent.

Cabeza and coworkers have extensively investigated the catalytic hydrogenation of alkynes by metal clusters substituted with N-donor ligands. Diphenylacetylene hydrogenation is induced by [Ru₃(μ-H)(μ-dmdab)(CO)₉] (Hdmdab = 3,5-dimethyl-1,2-diaminobenzene), but the catalytic activity is thought to be due to an unidentified mononuclear fragment [43]. This indicates that edge-bridging bidentate N-donor ligands do not stabilize the cluster structure sufficiently to avoid fragmentation. On the other hand, face-bridging N-donor ligands derived from 2-amino-6-methylpyridine (Hampy) do seem to stabilize clusters enough to maintain the polynuclear structure throughout a catalytic cycle. The complexes [Ru₃(μ-H)(μ_3-ampy)(CO)₉], [Ru₆(μ-H)₆(μ^3-ampy)(CO)₁₄], and their acetylene and phosphine derivatives [44–50], hydrogenate diphenylacetylene selectively to mixtures of E- and Z-stilbene readily at 80 °C and sub-atmospheric H₂ pressure (TOF 27 h⁻¹), as well as phenylacetylene to styrene under more forcing conditions (100 °C, 15 bar H₂). A related cluster containing the 2-anilinopyridine ligand has also been reported to hydrogenate phenyl-1-propyne [51]. Scheme 8.7

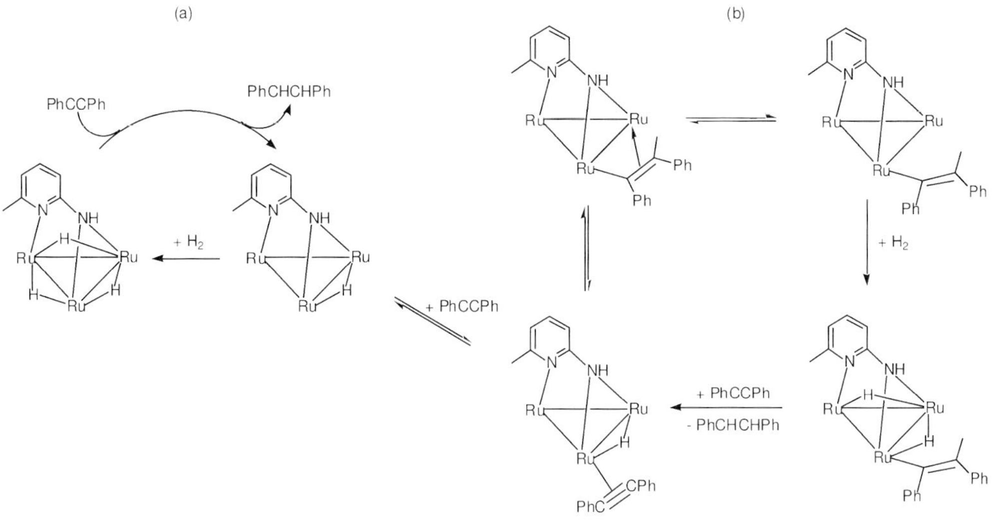

(a)

(b)

Scheme 8.7 Mechanism for the hydrogenation of diphenylacetylene catalyzed by ruthenium clusters containing face-
capping ampy ligands at: (a) low [substrate]:[cat] ratios; and
(b) high [substrate]:[cat] ratios (CO ligands omitted for clarity).

shows the main features of the diphenylacetylene hydrogenation mechanism
for $[Ru_3(\mu\text{-}H)(_{\mu_3}\text{-ampy})(CO)_9]$, where the active species is the alkenyl derivative
$[Ru_3(\mu_3\text{-ampy})(\mu\text{-PhC}=CHPh)(CO)_8]$, readily formed by reaction of the hydride
with the alkyne. If the latter complex is used as the catalyst precursor, the activity of diphenylacetylene hydrogenation is increased to TOF ca. 40 h^{-1} at 60 °C
and 0.8 bar H$_2$. The catalytic cycle in Scheme 8.7 is supported by detailed kinetic studies, together with isolation and identification of a number of intermediates, and the independent study of various elementary steps included in
the cycle. Phosphine-substituted ampy clusters also catalyze the hydrogenation
of alkynes, albeit at lower rates (TOF < 11 h^{-1} at 80 °C and 0.8 bar H$_2$), but probably through very similar mechanisms.

Heteronuclear clusters have also been used in homogeneous hydrogenation
with some success. Early studies by Ugo and Braunstein led to low-activity bimetallic catalysts [52]. $[RuFe_2(CO)_{12}]$, $[Ru_2Fe(CO)_{12}]$ and $[Ru_3FeH_2(CO)_{13}]$ were studied by Giordano and Sappa [32]; these promote the hydrogenation of diphenylacetylene at lower rates than the homonuclear Ru clusters, and the nature of the
active species was not established. Süss-Fink and coworkers reported the use of
$[IrRu_3(CO)_{13}(\mu\text{-}H)]$ in the hydrogenation of diphenylacetylene, curiously to E-stil-
bene (TOF 3900 h^{-1}), and proposed a cycle involving cluster intermediates [53].

More recently, Adams and coworkers have provided a very interesting case of
heteronuclear clusters that are very active for the hydrogenation of alkynes [4,
54, 55]. The high-nuclearity layer-segregated Pt–Ru complex $[Pt_3Ru_6(CO)_{21}(\mu^3\text{-}$
$H)(\mu\text{-}H)_3]$, consisting of three stacked triangular layers of metal atoms with an

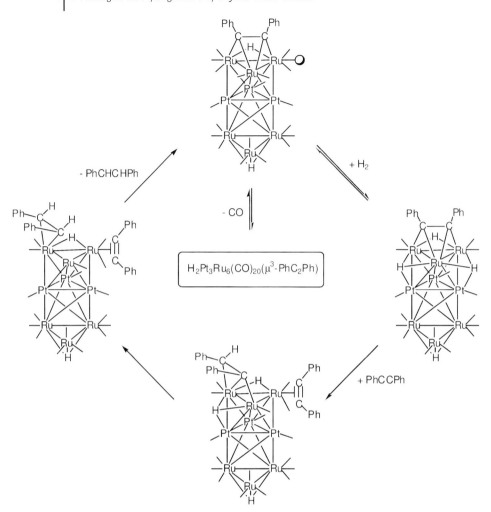

Scheme 8.8 Mechanism for the hydrogenation of diphenyl-acetylene catalyzed by layer-segregated Pt_3Ru_6 clusters (CO ligands omitted for clarity).

alternating arrangement $Ru_3Pt_3Ru_3$, readily reacts with diphenylacetylene to yield an alkyne complex, according to Eq. (4). In the process, two of the hydrides are transferred to a second diphenylacetylene molecule to yield 1 equiv. Z-stilbene, and one CO ligand is lost.

The resulting alkyne complex is capable of catalytically hydrogenating diphenylacetylene at $50\,^{\circ}C$ and 1 bar of H_2 with TOF close to $50\,h^{-1}$. The hydrogenation rate is first order in cluster concentration, indicating the participation of polynuclear species in the cycle, and it is also first order in substrate and hydrogen concentrations, while it is inhibited by CO. Labeling studies involving D_2

$$+ \ 2PhCCPh \longrightarrow \quad + \ PhCHCHPh + CO \quad (4)$$

and TolC≡CTol further pointed to cluster catalysis, according to the mechanism depicted in Scheme 8.8. The catalytic activity is high, but catalyst life is short, the cluster being degraded into various species after a few hundred turnovers. A related homonuclear cluster $[Ru_3(CO)_9(\mu^3\text{-}PhCCPh)(\mu\text{-}H)_2]$ was shown to hydrogenate the alkyne at considerably lower rates than those observed for the heteronuclear complex, showing that the presence of platinum in the vicinity of the "working" ruthenium triangle enhances the catalytic activity. Even though there is no evidence for any direct participation of platinum in the catalytic cycle, the activation of H_2 is thought to occur on Ru_2Pt triangular units (see Scheme 8.8). The same Pt_3Ru_6 cluster has also been found to catalyze the hydrosilylation of diphenylacetylene with triethylsilane [56].

8.4
Hydrogenation of Other Substrates

The anionic cluster $[Ru_4H_3(CO)_{12}]^-$ is a catalyst precursor for the transfer hydrogenation of simple and α,β-unsaturated ketones in boiling Pr^iOH, with reasonable rates (TOF up to 100 h^{-1}) and, in the latter case, with moderate diastereoselectivity (up to 77%). Although a detailed kinetic modeling was performed, the identity or nuclearity of the active species could not be ascertained, and a radical mechanism was proposed [57]. $[Ru_3(CO)_{12}]$, in combination with chiral tetradentate diimino phosphines (P_2N_2), catalyzes the transfer hydrogenation of prochiral ketones in boiling $Pr^iOH/KOPr^i$, with enantioselectivities up to 81% at 91% yield (TOF ca. 40 h^{-1}). Evidence pointing to a cluster-catalyzed reaction include the fact that an anionic cluster $[HRu_3(CO)_{12}(P_2N_2)]^-$ could be isolated at the end of the hydrogenation runs, and it was shown to catalyze the reaction in the absence of added base. The reaction rate was also first order in cluster concentration, and a related mononuclear complex containing the same ligand was inactive [58].

The ligand-stabilized cluster $[Ru_3(CO)_7(\mu^3, \eta^5 : \eta^5\text{-}4,6,8\text{-}trimethylazulene)]$ reacts with $PhMe_2Si\text{-}H$ to yield a new cluster containing a partially hydrogenated azulene ligand $[Ru_3(CO)_7(\mu^3, \eta^5 : \eta^5\text{-}4,5\text{-}dihydro\text{-}4,6,8\text{-}trimethylazulene)]$. Both of these complexes are efficient catalyst precursors for the hydrosilylation of acetophenone with moderate activities (TOF ca. 10 h^{-1}). Cluster participation is proposed on the basis of observed intermediates by NMR, and the fact that lower nuclearity-related complexes were not active in catalysis [59].

A series of tri- and tetra-nuclear Ru clusters previously reported by the groups of Süss-Fink [60] and of Dyson [61] as catalysts for the hydrogenation of benzene and other simple aromatics in biphasic media have later been shown to consist predominantly of active metallic particles [9, 10, 62].

8.5
Concluding Remarks

Despite the fact that the great expectations produced by cluster chemistry over two decades ago as a means of discovering novel catalysts with unique properties have not yet been realized, cluster catalysis continues to attract attention both as a conceptual issue and as a potential method to achieve unusual catalytic features. A number of catalysts originally thought to operate through cluster intermediates have subsequently been shown to owe their activity to the formation of mononuclear complexes or of metallic aggregates. Other systems do provide strong cases for catalytic cycles involving well-defined polynuclear intermediates, and a number of thorough kinetic and mechanistic studies have been performed which shed light on this fundamental question. The introduction of PHIP as a means of obtaining direct evidence for cluster catalysis is a most welcome development, and it is expected that further studies involving this technique will either corroborate or contradict the participation of polynuclear species in other catalytic systems. Although it is difficult to predict accurately whether fragmentation will occur when a given cluster is placed under catalytic hydrogenation conditions, some indicators are available to orient the search for polynuclear active species; in particular, the use of polydentate edge-bridging or face-capping ligands provides a good probability of obtaining cluster catalysis.

To date, there are no examples available of well-defined clusters that are of practical use, or that offer any advantages over mononuclear complexes in homogeneous hydrogenation reactions. The promising approach to enantioselective hydrogenation by use of chiral metal frameworks or of chiral polydentate ligands on clusters has been explored with limited success, but developments of possible utility are yet to be realized; continued efforts in that direction are certainly worthwhile. Heterobimetallic clusters appear as good candidates for promoting reactions not catalyzed by a single metal through synergistic enhancement of the properties of individual components; some interesting examples are now available where bimetallic high-nuclearity cluster catalysis has been demonstrated, and applications to unusual reactions are to be expected [4, 53–56]. The use of defined clusters as precursors of nanostructured materials is also of great interest, even if it falls outside the field of homogeneous hydrogenation. For example, applications to the synthesis of active catalysts for the important arene hydrogenation reaction are very appealing.

Until now, most studies on homogeneous hydrogenation by clusters have concentrated on alkenes and alkynes, though hopefully other substrates such as aldehydes, ketones, imines, and others will be further investigated, particularly using those systems that are now known to be genuine cluster catalysts.

Although the field of homogeneous hydrogenation by use of well-defined metal clusters has risen and fallen in popularity over the years, it has never been abandoned, most likely because the basic concept of a limited number of metal atoms in a well-defined structural and electronic molecular unit performing unique catalytic reactions still appears very seductive, and its realization poses exciting challenges in molecular design and synthetic chemistry. Hopefully, the expected breakthroughs toward distinctive catalytic properties in hydrogenation reactions by metal clusters will "see the light" before too long.

Abbreviations

bmim	1-butyl-3-methylimidazolium
ee	enantiomeric excess
IR	infra-red
NMR	nuclear magnetic resonance
PHIP	*para* hydrogen-induced polarization
TOF	turnover frequency
TON	turnover number
TPPMS	triphenylphosphine-monosulfonate
TPPTS	triphenylphosphine-trisulfonate

References

1 Adams, R. D., Cotton, F. A. (Eds.), *Catalysis by Di- and Poly-nuclear Metal Cluster Complexes*. Wiley-VCH: New York, **1998**.

2 Braunstein, P., Oro, L. A., Raithby, P. R. (Eds.), *Metal Clusters in Chemistry*. Wiley-VCH: New York, **1999**.

3 Dyson, P. J., *Coord. Chem. Rev.* **2004**, *248*, 2443.

4 Adams, R. D., Captain, B., *J. Organomet. Chem.* **2004**, *689*, 4521.

5 Dyson, P. J., McIndoe, J. S. (Eds.), *Transition Metal Carbonyl Chemistry*. Gordon and Breach: Amsterdam, **2000**.

6 Süss-Fink, G., Meister, G., *Adv. Organomet. Chem.* **1993**, *35*, 41.

7 Gates, B. C., *J. Mol. Catal.* **2000**, *163*, 55; Argo, A. M., Odzak, J. F., Lai, F. S., Gates, B. C., *Nature* **2002**, *415*, 623.

8 Johnson, B. F. G. (Ed.), *Transition Metal Clusters*. Wiley: Chichester, **1980**.

9 Widegreen, J. A., Finke, R. G., *J. Mol. Catal. A: Chem.* **2003**, *198*, 317.

10 Dyson, P. J., *Dalton Trans.* **2003**, 2964.

11 Rosenberg, E., Laine, R. In: Adams, R. D., Cotton, F. A. (Eds.), *Catalysis by Di- and Poly-nuclear Metal Cluster Complexes*. Wiley-VCH: New York, **1998**, p. 1.

12 Cree-Uchiyama, M., Shapley, J. R., St. George, G. M., *J. Am. Chem. Soc.* **1986**, *108*, 1316.

13 Choplin, A., Besson, B., D'Ornelas, L. D., Sánchez-Delgado, R., Basset, J. M., *J. Am. Chem. Soc.* **1988**, *100*, 2783.

14 Zuffa, J. L., Blohm, M. L., Gladfelter, W. L., *J. Am. Chem. Soc.* **1986**, *108*, 552;

Zuffa, J.L., Gladfelter, W.L., *J. Am. Chem. Soc.* **1986**, *108*, 4669.

15 Sánchez-Delgado, R.A., Durán, I., Monfort, J., Rodríguez, E., *J. Mol. Catal.*, **1981**, *11*, 193–203.

16 Doi, Y., Koshizuka, K., Keii, T., *Inorg. Chem.* **1982**, *21*, 2732.

17 Sánchez-Delgado, R.A., Andriollo, A., Puga, J., Martín, G., *Inorg. Chem.* **1987**, *26*, 1867.

18 Bregounhou, G., Fompegrine, P., Commenges, G., Bonnett, J.J., *J. Mol. Catal.* **1988**, *48*, 285.

19 Fontal, B., Reyes, M., Suárez, T., Bellandi, F., Díaz, J.C., *J. Mol. Catal. A: Chem.* **1999**, *149*, 75; Fontal, B., Reyes, M., Suárez, T., Bellandi, F., Ruiz, N., *J. Mol. Catal. A: Chem.* **1999**, *149*, 87.

20 Dallmann, K., Buffon, R., *J. Mol. Catal. A: Chem.* **2001**, *172*, 81.

21 Matteoli, U., Menchi, V., Frediani, P., Bianchi, M. Piacenti, F., *J. Organomet. Chem.* **1985**, *285*, 281; Matteoli, U., Beghetto, V., Scrivanti, A., *J. Mol. Catal. (A): Chem.* **1996**, *109*, 45; Salvini, A., Frediani, P. Bianchi, M., Piacenti, F., Pistolesi, L., Rosi, L., *J. Organomet. Chem.* **1999**, *582*, 218.

22 Homanen, P., Persson, R., Haukka, M., Pakkanen, T.A. Nordlander, E., *Organometallics* **2000**, *19*, 5568.

23 Moura, F.C.C., dos Santos, E.N., Lago, R.M., Vargas, M.D., Araujo, M.H., *J. Mol. Catal. A: Chem.* **2005**, *226*, 243.

24 Dyson, P.J., Ellis, D.J., Parker, D.G., Welton, T., *J. Mol. Catal. A: Chem.* **1999**, *150*, 71.

25 Gao, J.-X., Xu, P.-P., Yi, X.-D., Wan, H.-L., Tsai, K.-R., *J. Mol. Catal. A: Chem.* **1999**, *147*, 99.

26 Zhao, D., Dyson, P.J., Laurenczy, G., McIndoe, J.S., *J. Mol. Catal. A: Chem.* **2004**, *214*, 19.

27 Cabeza, J.A. In: Braunstein, P., Oro, L.A., Raithby, P.R. (Eds.), *Metal Clusters in Chemistry.* Wiley-VCH: New York, **1999**, Vol. 2, p. 715.

28 Dyson, P.J., *Coord. Chem. Rev.* **2004**, *248*, 2443.

29 Michelin-Lausarot, P., Vaglio, G.A., Valle, M., *J. Organomet. Chem.* **1984**, *275*, 233.

30 Michelin-Lausarot, P., Vaglio, G.A., Valle, M., *Inorg. Chim. Acta* **1977**, *25*, L107.

31 Michelin-Lausarot, P., Vaglio, G.A., Valle, M., *Inorg. Chim. Acta* **1979**, *36*, 213.

32 Giordano, R., Sappa, E., *J. Organomet. Chem.* **1993**, *448*, 157.

33 Cauzzi, D., Giordano, R., Sappa, E., Tiripicchio, A., Tiripicchio-Camellini, M., *J. Cluster Sci.* **1993**, *4*, 279.

34 Castiglioni, M., Giordano, R., Sappa, E., *J. Organomet. Chem.* **1983**, *258*, 217.

35 Castiglioni, M., Giordano, R., Sappa, E., *J. Organomet. Chem.* **1988**, *342*, 97.

36 Castiglioni, M., Giordano, R., Sappa, E., *J. Organomet. Chem.* **1989**, *362*, 399.

37 Castiglioni, M., Giordano, R., Sappa, E., *J. Organomet. Chem.* **1989**, *369*, 419.

38 Castiglioni, M., Giordano, R., Sappa, E., *J. Organomet. Chem.* **1991**, *407*, 377.

39 Aime, S., Gobetto, R., Canet, D., *J. Am. Chem. Soc.* **1998**, *120*, 6770; Aime, S., Dastru, W., Gobetto, R., Russo, A., Viale, D., Canet, D., *J. Phys. Chem.* **1999**, *103*, 9702; Bergman, B., Rosenberg, E., Gobetto, R., Aime, S., Milone, L., Reineri, F., *Organometallics* **2002**, *21*, 1508.

40 Gobetto, R., Milone, L., Reineri, F., Salassa, L., Viale, A., Rosenberg, E., *Organometallics* **2002**, *21*, 1919.

41 Blazina, D., Duckett, S.B., Dyson, P.J., Lohman, J.A.B., *Angew. Chem. Int. Ed. Engl.* **2001**, *40*, 3874; Blazina, D., Duckett, S.B., Dyson, P.J., Johnson, B.F.G., Lohman, J.A.B., Sleigh, C.J., *J. Am. Chem. Soc.* **2001**, *123*, 9760; Blazina, D., Duckett, S.B., Dyson, P.J., Lohman, J.A.B., *Chem. Eur. J.* **2003**, *9*, 1045; Blazina, D., Duckett, S.B., Dyson, P.J., Lohman, J.A.B., *Dalton Trans.* **2004**, 2108. Prestwich, T.G., Blazina, D., Ducket, S.B., Dyson, P.J., *Eur. J. Inorg. Chem.* **2004**, 4381.

42 Giordano, R., Sappa, E., Knox, S.A.R., *J. Cluster Sci.* **1996**, *7*, 179.

43 Cabeza, J.A., Fernández-Colinas J.M., Llamazares, A., Riera, V., *Organometallics* **1992**, *11*, 4355.

44 Cabeza, J.A., Fernández-Colinas, J.M., Llamazares, A., Riera, V., *J. Mol. Catal.* **1992**, *71*, L7.

45 Cabeza, J.A., Fernández-Colinas, J.M., Llamazares, A., Riera, V., *J. Organomet. Chem.* **1995**, *494*, 169.

46 Cabeza, J.A., Fernández-Colinas, J.M., Llamazares, A., Riera, V., García-Granda,

S., van der Maelen, J. F., *Organometallics*
1994, *13*, 4352; Cabeza, J. A., Fernández-
Colinas, J. M., Llamazares, A., Riera, V.,
García-Granda, S., van der Maelen, J. F.,
Organometallics **1994**, *13*, 3120.

47 Cabeza, J. A., del Rio, I., Fernández-Colinas J. M., Riera, V., *Organometallics* **1996**,
15, 449.

48 Cabeza, J. A., Fernández-Colinas J. M.,
Llamazares, A., Riera, V., *Organometallics*
1993, *12*, 4141.

49 Alvarez, S., Briard, P., Cabeza, J. A., del
Rio, I., Fernández-Colinas, J. M., Mulla,
F., Ouahab, L., Riera, V., *Organometallics*
1994, *13*, 4360.

50 Cabeza, J. A., Llamazares, A., Riera, V.,
Briard, P., Ouahab, L., *J. Organomet.
Chem.* **1994**, *480*, 205.

51 Lugan, N., Laurent, F., Lavigne, G.,
Newcomb, T. P., Liimatta, E. W.,
Bonnet, J. J., *J. Am. Chem. Soc.* **1990**,
112, 8607.

52 Fusi, A., Ugo, R., Psaro, R., Braunstein,
P., Dehand, J., *Philos. Trans. R. Soc. London A* **1982**, *308*, 125.

53 Ferrand, V., Süss-Fink, G., Neels, A.,
Stoeckli-Evans, H., *J. Chem. Soc. Dalton
Trans.* **1998**, 3825.

54 Adams, R. D., Barnard, T. S., Li, Z., Wu,
W., Yamamoto, J. H., *Organometallics*
1994, *13*, 2357.

55 Adams, R. D., Barnard, T. S., Li, Z., Wu,
W., Yamamoto, J. H., *J. Am. Chem. Soc.*
1994, *116*, 9103.

56 Adams, R. D., Barnard, T. S., *Organometallics* **1998**, *17*, 2567.

57 Bhaduri, S., Sharma, K., Mukesh, D.,
J. Chem. Soc. Dalton Trans. **1993**, 1191.

58 Zhang, H., Yang, Ch.-B., Li, Y.-Y., Donga,
Zh.-R., Gao, J.-X., Nakamura, H., Murata, K., Ikariya, T., *Chem. Commun.* **2003**,
142.

59 Matsubara, K., Ryu, K., Maki, T., Iura, T.,
Nagashima, H., *Organometallics* **2002**, *21*,
3023.

60 Plasseraud, L., Süss-Fink, G., *J. Organo-
met. Chem.* **1997**, *539*, 163; Süss-Fink,
G., Faure, M., Ward, T. R., *Angew. Chem.
Int. Ed.* **2002**, *41*, 99; Vielle-Petit, L.,
Therrien, B., Süss-Fink, G., Ward, T. R.,
J. Organomet. Chem. **2003**, *684*, 117;
Süss-Fink, G., Therrien, B., Vielle-Petit,
L., Tschan, M., Romack, V. B., Ward,
T. R., Dadras, M., Laurenczy, G., *J. Orga-
nomet. Chem.* **2004**, *689*, 1362.

61 Dyson, P. J., Ellis, D. J., Welton, T., Par-
ker, D. G., *Chem. Commun.* **1999**, 25;
Dyson, P. J., Russel, K., Welton, T., *Inorg.
Chem. Commun.* **2001**, *4*, 571.

62 Hagen, C. M., Vielle-Petit, L., Laurenczy,
G., Süss-Fink, G., Finke, R. G., *Organometallics* **2005**, *24*, 1819.

9
Homogeneous Hydrogenation:
Colloids – Hydrogenation with Noble Metal Nanoparticles

Alain Roucoux and Karine Philippot

9.1
Introduction

Today, metal nanoparticle science is a strategic research area in material development due to their particular physical and chemical properties. Catalysis is a traditional application of metal nanoparticles, but they also find application in diverse fields such as photochemistry, electronics, optics or magnetism [1, 2]. Metal nanocatalysts, defined as particles between 1 and 10 nm in size, can be obtained by a variety of methods according to the "organic" or "aqueous" nature of the media and the stabilizers used: polymers, ligands or surfactants [3]. During the past five years, the use of nanoparticles in this active research area has received increased attention since some homogeneous catalysts have been shown to be "nanoheterogeneous" [4–6]. Since that time, modern methods to distinguish the true nature of the catalysts have been described. From today onwards, soluble noble metal nanoparticles are to be considered as an unavoidable family of catalysts for hydrogenation under mild conditions at the border between homogeneous and heterogeneous chemistry. This chapter reviews recent progress in the hydrogenation of unsaturated compounds by noble metal nanoparticles in various liquid media.

9.2
Concepts

In materials chemistry, nanoparticles of noble metals are an original family of compounds. Well-defined in terms of their size, structure and composition, zerovalent transition-metal colloids provide considerable current interest in a variety of applications. Here, the main interest is their application in catalysis. Zerovalent nanocatalysts can be generated in various media (aqueous, organic, or mixture) from two strategic approaches according to the nature of the precursor, namely: (i) mild chemical reduction of transition-metal salt solutions; and (ii) metal atom

The Handbook of Homogeneous Hydrogenation.
Edited by J.G. de Vries and C.J. Elsevier
Copyright © 2007 WILEY-VCH Verlag GmbH & Co. KGaA, Weinheim
ISBN: 978-3-527-31161-3

extrusion starting from organometallic compounds able to decompose in solution under mild conditions [7]. To date, the key goal is the reproducible synthesis of nanoparticles in opposition to larger ones (nanopowders) and bulk materials. Consequently, nanostructured particles should have at least: (i) a specific size (1–10 nm); (ii) a well-defined surface composition; (iii) constant properties related to reproducible syntheses; and (iv) be isolable and redissolvable. In this respect, several synthetic methods have been described, including:

- chemical or electrochemical reduction;
- thermal, photochemical or sonochemical decomposition; and
- metal vapor synthesis.

Whichever method is followed, a protective agent able to induce a repulsive force opposed to the van der Waals forces is generally necessary to prevent agglomeration of the formed particles and their coalescence into bulk material. Since aggregation leads to the loss of the properties associated with the colloidal state, stabilization of metallic colloids – and therefore the means to preserve their finely dispersed state – is a crucial aspect for consideration during their synthesis.

The stabilization mechanisms of colloidal materials have been described in Derjaguin-Landau-Verway-Overbeek (DLVO) theory [8, 9]. Colloids stabilization is usually discussed in terms of two main categories, namely charge stabilization and steric stabilization.

9.2.1
Electrostatic Stabilization

Various anionic compounds such as halides, carboxylates or polyoxoanions, generally dissolved in aqueous solution, can establish electrostatic stabilization. Adsorption of these compounds onto the metallic surface and the associated countercations necessary for charge balance produces an electrical double-layer around the particles (Scheme 9.1). The result is a coulombic repulsion between the particles. At short interparticle distances, if the electric potential associated with the double layer is sufficiently high, repulsive forces opposed to the van der Waals forces will be significant to prevent particle aggregation.

Colloidal suspensions stabilized by electrostatic repulsion are highly sensitive to any phenomenon able to disrupt the double layer, such as ionic strength or thermal motion.

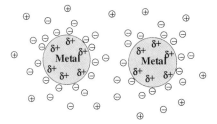

Scheme 9.1 Schematic representation of electrostatic stabilization: a coulombic repulsion between metal colloid particles.

9.2.2
Steric Stabilization

Nanoparticulate metal colloids can also be prevented from agglomeration by using protecting macromolecules such as polymers or oligomers [10, 11] or related stabilizers such as cyclodextrins [12] or cellulose derivatives [13]. The adsorption of these molecules at the surface of the particles provides a protective layer. In the interparticle space, the sterical environment of the adsorbed macromolecules reduces their mobility (Scheme 9.2). The result is an osmotic repulsion to restore the equilibrium by diluting the macromolecules, thereby separating the particles. By contrast with electrostatic stabilization, which is mainly used in aqueous media, steric stabilization can be used in either an organic or aqueous phase. Nevertheless, the length and/or nature of the adsorbed macromolecules can influence the thickness of the protective layer and thus modify the stability of the colloidal metal particles.

The electrostatic and steric effects can be combined to stabilize nanoparticles in solution. This type of stabilization is generally provided by means of ionic surfactants such as alkylammonium cations (Scheme 9.3). These compounds bear both a polar head group which is able to generate an electrical double layer, and a lipophilic side chain which is able to provide steric repulsion [14, 15].

Electrosteric stabilization can be also obtained from the couple ammonium (Bu_4N^+)/polyoxoanion $(P_2W_{15}Nb_3O_{62}^{9-})$. The significant steric repulsion of the bulky Bu_4N^+ countercation, when associated with the highly charged polyoxoanion (coulombic repulsion), provides efficient electrosterical stability towards agglomeration in solution of the resultant nanocatalysts [2, 5, 6].

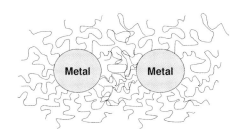

Scheme 9.2 Schematic representation of steric stabilization.

Steric stabilization   Electrostatic stabilization

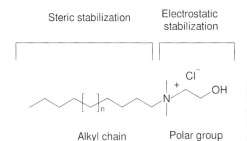

Alkyl chain   Polar group

Scheme 9.3 N-alkyl-N,N-dimethyl-N-(2-hydroxyethyl)-ammonium chloride salt, a typical cationic surfactant which combines electrostatic and steric stabilizations (electrosteric stabilization).

The term "steric stabilization" may also be used to describe protective transition-metal colloids with traditional ligands or solvents. This stabilization occurs by: (i) the strong coordination of various metal nanoparticles with ligands such as phosphines [16–18], thiols [19–22], amines [21, 23–26], oxazolines [27] or carbon monoxide [18]; or (ii) weak interactions with solvents such as tetrahydrofuran (THF) or various alcohols [18, 28–31].

Finally, the development of modified nanoparticles having better stability and a longer lifetime has involved interesting results in diverse catalytic reactions. Efficient activities are obtained with these transition-metal colloids used as catalysts for the hydrogenation of various unsaturated substrates. Consequently, several recent investigations in total, partial or selective hydrogenation have received significant attention.

9.3
Hydrogenation of Compounds with C=C Bonds

Alkene hydrogenation is a common field of catalytic application for metal nanoparticles. Various approaches have been utilized to obtain stable and active nanocatalysts in hydrogenation reactions. The main approaches are described in the following sections, and are classified according to the stabilizing mode retained for the nanoparticles.

9.3.1
Use of Polymers as Stabilizers

Organic polymers are very often used for the stabilization of metal nanoparticles by providing a steric stabilizing effect. Due to this embedding effect, it is generally considered that the diffusion of substrates through the polymer matrix can be limited. Nevertheless, some interesting results have been obtained.

Hirai and Toshima have published several reports on the synthesis of transition-metal nanoparticles by alcoholic reduction of metal salts in the presence of a polymer such as polyvinylalcohol (PVA) or polyvinylpyrrolidone (PVP). This simple and reproducible process can be applied for the preparation of monometallic [32, 33] or bimetallic [34–39] nanoparticles. In this series of articles, the nanoparticles are characterized by different techniques such as transmission electronic microscopy (TEM), UV-visible spectroscopy, electron diffraction (EDX), powder X-ray diffraction (XRD), X-ray photoelectron spectroscopy (XPS) or extended X-ray absorption fine structure (EXAFS, bimetallic systems). The great majority of the particles have a uniform size between 1 and 3 nm. These nanomaterials are efficient catalysts for olefin or diene hydrogenation under mild conditions (30 °C, $P_{H_2} = 1$ bar). In the case of bimetallic catalysts, the catalytic activity was seen to depend on their metal composition, and this may also have an influence on the selectivity of the partial hydrogenation of dienes.

Delmas et al. produced PVP-stabilized rhodium nanoparticles using the method reported by Hirai [32] to perform catalytic hydrogenation of oct-1-ene in a two-liquid-phase system [40]. These authors investigated the effect of various parameters on nanoparticle stability and activity under more or less severe conditions. It was also shown that PVP/Rh colloids could be reused twice or more, without any loss of activity.

9.3.2
Use of Non-Usual Polymers as Stabilizers

Several groups have developed the use of non-usual polymers for the stabilization of colloids for hydrogenation catalysis. Mayer reported the use of various nonionic polymers and cationic polyelectrolytes as stabilizers for the synthesis of metal nanoparticles [41]; poly(1-vinylpyrrolidone-*co*-acrylic) and poly(2-ethyl-2-oxazoline) can be cited as examples here. Stable palladium and platinum colloids were prepared by the reduction of noble metal salts by refluxing the alcoholic solutions containing the polymers [42, 43]. Depending on the polymer used, a range of particle sizes and narrow size distributions were obtained. It appeared that a hydrophobic backbone and hydrophilic side chains are required for the nonionic polymers to become well stabilized and to provide controlled-size particles. Cationic polyelectrolytes also provided interesting results. With regard to catalysis, cyclohexene hydrogenation in MeOH was performed with Pd and Pt colloids, and conversions of 100% could be obtained in many cases. These authors also compared the activities of the nanocatalysts for this reaction depending on their protective polymer, polyelectrolyte or nonionic polymers [44]. In the same context, the ability of amphiphilic block copolymers, namely poly(dimethylsiloxane)-*b*-poly(ethylene oxide) (PDMS-*b*-PEO), polystyrene-*b*-poly-(methacrylic acid) (PS-*b*-PMAA) and poly-styrene-*b*-poly(ethylene oxide) (PS-*b*-PEO), to stabilize colloidal palladium, platinum, silver and gold nanoparticles has been investigated [45, 46]. Transmission electron microscopy (TEM) revealed randomly distributed nanoparticles with narrow size distributions in the range of 1 to 10 nm. In the case of Pd and Pt, cyclohexene hydrogenation has been chosen as model reaction for evaluation of the nanocatalysts. The influence of the precursor type, the polymer nature and also the preparation conditions on the catalytic activities have also been studied [46].

Liu et al. prepared palladium nanoparticles in water-dispersible poly(acrylic acid) (PAA)-lined channels of diblock copolymer microspheres [47]. The diblock microspheres (mean diameter 0.5 μm) were prepared using an oil-in-water emulsion process. The diblock used was poly(*t*-butylacrylate)-*block*-poly(2-cinna-moyloxyethyl) methacrylate (P*t*BA-*b*-PCEMA). Synthesis of the nanoparticles inside the PAA-lined channels of the microspheres was achieved using hydrazine for the reduction of PdCl$_2$, and the nanoparticle formation was confirmed from TEM analysis and electron diffraction study (Fig. 9.1). The Pd-loaded microspheres catalyzed the hydrogenation of methylacrylate to methyl-propionate. The catalytic reactions were carried out in methanol as solvent under dihydro-

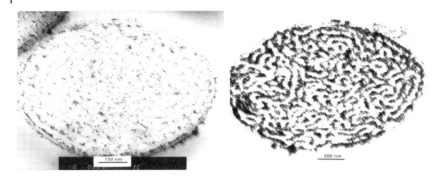

Fig. 9.1 Transmission electron microscopy images of Pd-loaded PCEMA-b-PAA microspheres containing 27% Pd (left) and 63% Pd (right). (Adapted from [47])

gen bubbling. The catalysts were recovered by centrifugation and methanol washing, and recovery efficiency was 100%.

Bronstein et al. also used block copolymer micelles for the preparation of noble-metal monometallic or bimetallic colloids (Pd, Pd/Au) [48]. These colloids were characterized by electron microscopy and wide-angle X-ray scattering (WAXS) and studied in the hydrogenation of cyclohexene, 1,3-cyclooctadiene, and 1,3-cyclohexadiene. The catalytic reactions were carried out in toluene at 30 °C with [catalyst]/[substrate] molar ratios of 1:250 and 1:500 in the case of cyclohexene, and 1:500 and 1:10000 for 1,3-cyclohexadiene. High catalytic activities were obtained in the hydrogenation reactions, which were found to depend strongly on a number of parameters such as colloid morphology and reducing agent residues at the colloid surface. In fact, the strength of the reducing agent determines the rate of nucleation and growth of the colloids inside the micelle core and, consequently, a number of different architectures can be tailor-made. The second effect of the reducing agent used is provided by its residues or reaction products after reduction of the metal colloids that can influence the catalytic properties. Finally, Pd colloids could selectively hydrogenate 1,3-cyclohexadiene in cyclohexene with rates of between 10 to 1230 (Table 9.1). Nevertheless, it was observed that the hydrogenation of 1,3-cyclohexadiene is accompanied by the disproportionation of 1,3-cyclohexadiene to cyclohexene and benzene as a side reaction.

Akashi and coworkers prepared small platinum nanoparticles by ethanol reduction of $PtCl_6^{2-}$ in the presence of various vinyl polymers with amide side chains [49]. These authors studied the effects of molecular weight and molar ratio [monomeric unit]/[Pt] on the particle sizes and size distributions by electron microscopy, and in some cases by the dispersion stability of the Pt colloids. The hydrogenation in aqueous phase of allyl alcohol was used as a model reaction to examine the change in catalytic activity of polymer-stabilized Pt colloids upon addition of Na_2SO_4 to the reaction solution. The catalytic tests were performed in water or in Na_2SO_4 aqueous solution at 25 °C under atmospheric pressure of

Table 9.1 Activity of block copolymer-stabilized Pd colloids in selective hydrogenation of 1,3-cyclohexadiene. (Adapted from [48])

Catalyst [a]	Reducing agent	Substrate/catalyst ratio [mol]	t [min]	substrate conversion [%]	Selectivity [%]			Rate [c]
					Cyclo-hexene	Cyclo-hexane	Benzene	
PS-1,2; Na₂PdCl₄	super-hydride [b]	500:1	10	100	71.3	0.1	28.6	40
PS-3,4; Pd(CH₃COO)₂	super-hydride	750:1	5	100	58.5	0.0	41.5	90
PS-3,4; Pd(CH₃COO)₂	NaBH₄	10000:1	5	100	65.9	0.0	34.1	1230
PS-3,4; Pd(CH₃COO)₂	no reduction	500:1	25	100	60.1	5.4	34.5	10
Pd on activated carbon (Aldrich, 1 wt% Pd)	–	5000:1	7	100	54.7	0.7	44.6	420

a) PS-1,2 and PS-3,4 are the two different chosen PS-*b*-P4VP block copolymers.
b) Superhydride = 1 M solution of $LiB(C_2H_5)_3H$ in THF.
c) Rate: mol cyclohexene min^{-1} g^{-1} atom metal as a total for both hydrogenation and disproportionation.

dihydrogen. 1-Propanol was the only product. The PNVF-Pt (PNVF = poly(*N*-vinylformamide)) colloid did not show any critical flocculation point, and was very stable in 0.8 M Na₂SO₄ solution. Moreover, it provided the same activity as that obtained in pure water, while the other polymer-stabilized colloids showed markedly different behavior. The same authors described the synthesis of poly(*N*-isopropylacrylamide)-protected Au/Pt nanoparticles (mean diameter = 1.9 nm), their characterization by UV-visible spectroscopy, electron microscopy and X-ray diffraction, and their temperature-dependent catalytic activity in the aqueous hydrogenation of allyl alcohol [50]. The Au/Pt bimetallic colloids were found to have an alloy structure, the two metals being fully mixed, and were more active than the Pt monometallic colloids for the hydrogenation of allyl alcohol in water at room temperature. The hydrogenation rate was also seen to depend on the molar ratio of the Au/Pt in the bimetallic nanoparticles.

Recently, Liew et al. reported the use of chitosan-stabilized Pt and Pd colloidal particles as catalysts for olefin hydrogenation [51]. The nanocatalysts with a diameter ca. 2 nm were produced from PdCl₂ and K₂PtCl₄ upon reduction with sodium borohydride in the presence of chitosan, a commercial biopolymer, under various molar ratios. These colloids were used for the hydrogenation of oct-1-ene and cyclooctene in methanol at atmospheric pressure and 30 °C. The catalytic activities in term of turnover frequency (TOF; mol. product mol. metal^{-1} h^{-1})

Table 9.2 Selective reduction of conjugated aromatic alkenes
catalyzed by "Pd-PMHS" nanocomposites. (Reprinted with the
permission of the American Chemical Society [52])

Substrate	Reaction conditions	Product	Yield [%]
Styrene	4 h/benzene/RT	Ethylbenzene	95[a]
4-Methoxystyrene	3 h/benzene/RT	1-Ethyl-4-methoxybenzene	95
2-Allylphenol	4 h/benzene/RT	2-Propylphenol	96
Cinnamonitrile	5 h/benzene/RT	3-Phenylpropionitrile	90
2-Chlorostyrene	6 h/benzene/RT	(2-Chloroethyl)benzene	85
1,2-Diphenylethene	4 h/benzene/RT	1,2-Diphenylethane	94
2-Vinylnaphthalene	4 h/benzene/RT	2-Ethylnaphthalene	96[b]
9-Vinylanthracene	6 h/benzene/RT	9-Ethylanthracene	92

a)   All reactions performed with 2:1 molar equivalents of col-
     loid:alkene and stirred under argon or nitrogen atmosphere.
     Reaction progress was monitored by ^{1}H-NMR and/or FT-IR
     spectroscopy.
b)   Isolated yields.
RT = room temperature.

ranged from 10^4 h^{-1} to 10^5 h^{-1}. Chitosan-stabilized Pt and Pd colloids had similar catalytic properties for the hydrogenation of cyclooctene, which led to cyclooctane in both cases. By contrast, they showed different properties for the hydrogenation of oct-1-ene, in which Pd isomerized octene, giving rise to oct-2-ene and oct-3-ene. The activity of the nanocatalysts decreased with increasing concentration of chitosan.

The group of Chauhan has reported the preparation of polysiloxane-encapsulated Pd nanoclusters [52]. These colloids were synthesized by the reduction of Pd(OAc)$_2$ with polymethylhydrosiloxane, which functions as a reducing agent as well as a capping material. Chemoselective hydrogenation of functional conjugated alkenes was achieved by these polysiloxane-stabilized particles under mild conditions in high yields (Table 9.2). These authors also investigated the nature of the true catalyst by performing several experiments such as UV-visible studies of the precursor transformation in nanoclusters, TEM studies during catalysis, quantitative poisoning studies, and recyclability and reproducibility studies of the nanoclusters. All of the results obtained confirmed Pd-nanoclusters as the active catalytic species.

As a final example of catalytic hydrogenation activity with polymer-stabilized colloids, the studies of Cohen et al. should be mentioned [53]. Palladium nanoclusters were synthesized within microphase-separated diblock copolymer films. The organometallic repeat-units contained in the polymer were reduced by exposing the films to hydrogen at 100 °C, leading to the formation of nearly monodisperse Pd nanoclusters that were active in the gas phase hydrogenation of butadiene.

9.3.3
Use of Dendrimers as Stabilizers

The use of dendrimers constitutes an attractive stabilization mode for the synthesis of metal nanoparticles for several reasons:
- Due to their specific structure, which contains cavities with functional groups, they can act both as template and stabilizer for the nanoparticles.
- Dendrimers can also act as selective gates controlling the access of small molecules to the encapsulated nanoparticles.
- Tailoring of their terminal groups can enhance their solubility.

Dendrimer interior functional groups and cavities can retain guest molecules selectively, depending on the nature of the guest and the dendritic endoreceptors, the cavity size, the structure, and the chemical composition of the peripheric groups. Two main methods are known for the synthesis of metal nanoparticles inside dendrimers. The first method consists of the direct reduction of dendrimer-encapsulated metal ions (Scheme 9.4); the second method corresponds to the displacement of less-noble metal clusters with more noble elements [54].

The team of Crooks is involved in the synthesis and the use of dendrimers and, more particularly, poly(amidoamine) dendrimers (PAMAM), for the preparation of dendrimer-encapsulated mono- or bimetallic nanoparticles of various metals (Pt, Pd, Cu, Au, Ag, Ni, etc.) [55, 56]. The dendrimers were used as nanocatalysts for the hydrogenation of allyl alcohol and *N*-isopropylacrylamide or other alkenes under different reaction conditions (water, organic solvents, biphasic fluorous/organic solvents or supercritical CO_2). The hydrogenation reaction rate is dependent on dendrimer generation, as higher-generation dendrimers are more sterically

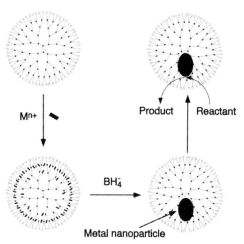

Scheme 9.4 Schematic of metal nanoparticles synthesis within dendrimer templates. (Reprinted from [54]; copyright 2002, Marcel Dekker.)

Table 9.3 Hydrogenation of alkenes using dendrimer-encapsulated Pd nanoparticles.[a] (Reprinted with permission of the American Chemical Society [59])

	Initial rate [mL H_2 min^{-1} × 10]			
Substrate	Pd(0)/G$_3$-TEBA	Pd(0)/G$_4$-TEBA	Pd(0)/G$_5$-TEBA	Pd/C
1,3-Cyclooctadiene	5.4	4.1	2.9	12.0
1,3-Cyclohexadiene	9.8	8.7	7.2	11.8
1,3-Cyclopentadiene	9.2	9.0	8.3	10.3
1-Heptene[b]	0.98	0.90	0.88	0.0
Cyclohexene	0.92	0.80	0.70	2.8
Methylacrylate	9.5	9.0	7.6	10.0
Tert-Butylacrylate	5.5	4.5	3.1	9.1
Allyl alcohol	12.1	11.5	11.2	11.6
2-Methyl-3-buten-2-ol	5.8	4.2	3.8	9.3

a) Reaction conditions: catalyst 5.0 μmol Pd, substrate 1.0 mmol, toluene 12.5 mL, H_2 1 atm, 30 °C.
b) After 15 min, the reaction mixture contained *n*-hexane, 1-hexene, 2-hexene and 3-hexene.

crowded on their periphery and thus less porous and less likely to admit substrates to interior metal nanoparticles than those of lower generations. Another advantage of dendrimer-encapsulated nanoparticles is that they can be recycled.

Rhee and coworkers published the synthesis of bimetallic Pt-Pd nanoparticles [57] or Pd-Rh nanoparticles [58] within dendrimers as nanoreactors. These nanocatalysts showed a promising catalytic activity in the partial hydrogenation of 1,3-cyclooctadiene. The reaction was carried out in an ethanol/water mixture at 20 °C under dihydrogen at atmospheric pressure. The dendrimer-encapsulated nanoclusters could be reused, without significant loss of activity.

Kaneda et al. reported substrate-specific hydrogenation of olefins using the tri-ethoxybenzamide-terminated poly(propylene imine) dendrimers (PPI) as nanoreactors encapsulating Pd nanoparticles (mean diameter 2–3 nm) [59]. The catalytic tests were performed in toluene at 30 °C under dihydrogen at atmospheric pressure (Table 9.3). The hydrogenation rates were seen to decrease with increasing generation of dendrimers, from G$_3$ to G$_5$.

9.3.4
Use of Surfactants as Stabilizers

Surfactants are well known as stabilizers in the preparation of metal nanoparticles for catalysis in water. Micelles constitute interesting nanoreactors for the synthesis of controlled-size nanoparticles from metal salts due to the confinement of the particles inside the micelle cores. Aqueous colloidal solutions are then obtained which can be easily used as catalysts.

Toshima et al. obtained colloidal dispersions of platinum by hydrogen- and photo-reduction of chloroplatinic acid in an aqueous solution in the presence of various types of surfactants such as dodecyltrimethylammonium (DTAC) and sodium dodecylsulfate (SDS) [60]. The nanoparticles produced by hydrogen reduction are bigger and more widely distributed in size than those resulting from the photo-irradiation method. Hydrogenation of vinylacetate was chosen as a catalytic reaction to test the activity of these surfactant-stabilized colloids. The reaction was performed in water under atmospheric pressure of hydrogen at 30 °C. The photo-reduced colloidal platinum catalysts proved to be best in terms of activity, a fact explained by their higher surface area as a consequence of their smaller size.

Larpent and coworkers were interested in biphasic liquid–liquid hydrogenation catalysis [61], and studied catalytic systems based on aqueous suspensions of metallic rhodium particles stabilized by highly water-soluble trisulfonated molecules as protective agent. These colloidal rhodium suspensions catalyzed octene hydrogenation in liquid–liquid medium with TOF values up to $78\,h^{-1}$. Moreover, it has been established that high activity and possible recycling of the catalyst could be achieved by control of the interfacial tension.

Gedanken et al. have reported the preparation of palladium nanoscale particles by sonochemical reduction of palladium acetate at room temperature in THF or methanol solution and in the presence of myristyltrimethylammonium bromide ($CH_3(CH_2)_{13}N(CH_3)_3Br$) as stabilizing agent [62]. XRD and TEM with selected area electron diffraction (SAED) techniques confirmed that the stabilized-Pd nanoclusters were nanocrystalline and composed of aggregates of spherical particles of size 10–20 nm. Cyclohexene hydrogenation reactions were carried out in diethyl ether under a hydrogen atmosphere (ca. 0.2 MPa) at room temperature. The formation of cyclohexane could be observed with a conversion of 64%, which was higher than that obtained using a commercial Pd/C catalyst.

9.3.5
Use of Polyoxoanions as Stabilizers

An original method for the stabilization of metal nanoparticles destined for catalytic applications was developed by Finke et al., who produced polyoxoanion- and tetrabutylammonium-stabilized iridium and rhodium nanoparticles (Scheme 9.5) [63, 64]. The synthesis method consists of the hydrogen reduction, in acetone, of the polyoxoanion-supported Ir(I) or Rh(I) complex [(n-$C_4H_9)_4N]_5$-$Na_3[(1,5$-COD$)Rh \cdot P_2W_{15}Nb_3O_{62}]$. The resultant Rh(0) nanoclusters are near-monodisperse with a mean size of 4 ± 0.6 nm, and can be isolated as a black powder and redispersed in non-aqueous solvents such as acetonitrile. The particles have been characterized by a variety of techniques, including TEM, energy-dispersive spectroscopy, electron diffraction, UV-visible spectroscopy and elemental analysis. Ion-exchange chromatography revealed that the Rh(0) nanoclusters are stabilized by the adsorption of the polyoxoanion onto their outer surface. These nanoclusters were active in cyclohexene hydrogenation at room

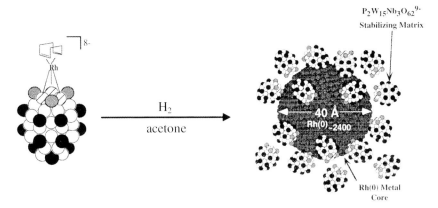

Scheme 9.5 Synthesis of Rh(0) nanoclusters from
(1,5-COD)RhP$_2$W$_{15}$Nb$_3$O$_{62}^{9-}$ polyoxoanion-supported nano-
cluster-forming precatalyst (space-filling representation).
(Adapted from [63].)

temperature and under dihydrogen pressure, with TOF of 3650 h^{-1}. In another
report, the same group studied the in-solution lifetimes of Rh(0) nanocatalysts;
these were found to approach those of a solid-oxide-supported Rh(0) catalyst,
with values higher than were previously reported [65].

9.3.6
Use of Ligands as Stabilizers

The use of ligands as protective agents for metal nanoparticle synthesis has become increasingly common during the past few years. The main advantage of
this stabilizing mode for nanocatalysts is the possibility of modulating the surface state of the particles by chemical influence of the ligand. Nevertheless, it is
necessary to identify ligands that give rise to stable, but active, nanocatalysts.
Until now, the ligands used have been generally simple organic ligands (thiols,
amines, carboxylic acids, phosphines), though on occasion more sophisticated ligands are used such as phenanthroline, β-cyclodextrins or ferritines.

As an initial example, Sastry et al. [66] described a single-step procedure for
the synthesis of catalytically active, hydrophobic platinum nanoparticles involved
in the spontaneous reduction of aqueous PtCl$_6^{2-}$ ions by hexadecylaniline (HDA)
at a liquid–liquid interface. HDA acts simultaneously as phase-transfer molecule, reducing agent and nanoparticle capping agent. The HDA chloroform solution of nanoparticles could be isolated as a black powder after solvent evaporation, while any uncoordinated HDA was removed by ethanol washing. The nanoparticles could then be redispersed in organic solvents (benzene, toluene,
hexane). UV-visible, TEM and XRD measurements confirmed the formation of
Pt nanoparticles of mean size 15.5±0.7 nm and with a face-centered cubic structure, while TGA analysis showed desorption of the HDA molecule (a weight

loss of 30% at 270 °C) and ^{1}H-NMR spectra proved close contact of the ligand with the metal surface. These Pt nanoparticles were investigated in styrene hydrogenation under H_2 pressure (14 bar) at 60 °C, and showed complete conversion into ethylbenzene with a TOF of 655 h^{-1}. They were also active in cyclohexene hydrogenation, allowing complete conversion into cyclohexane with 100% selectivity and a TOF of 2112 h^{-1}.

Vargaftik and Moiseev have obtained near-spherical palladium nanoclusters with an average size of 1.8 nm by reduction of palladium carboxylates $Pd_3(OCOR)_6$ (R = Me, Et, CHMe$_2$, CMe$_3$) with hydrogen in alcohol solutions containing 1,10-phenanthroline as ligand [67]. Based on elemental analysis, NMR, X-ray photoelectron spectroscopy and EXAFS investigations, these nanoclusters were described by the idealized formula $Pd_{147}phen_{32}O_{60}(OCOR)_{30}$. These nanocatalysts were used in the hydrogenation of alkynes and alkenes, the reduction of nitriles with formic acid, and the oxidation of aliphatic and benzylic alcohols. For alkene hydrogenation, styrene was used as substrate, the catalytic reaction being performed in alcohol at temperatures up to 70 °C. The results revealed that Pd-147 nanoclusters were much more active than the giant Pd-561 clusters or the traditional Pd/C catalyst. By comparison, ethylbenzene has been produced using all three catalysts, with respective TOF values of 242, 155, and 25 h^{-1}, respectively, at 20 °C.

Sahle-Demessie and Pillai studied the catalytic activity of phenanthroline-stabilized palladium nanoparticles in polyethylene glycol (PEG) as a recyclable catalyst system for the selective hydrogenation of olefins using molecular hydrogen under mild reaction conditions [68]. PEG acts not only as a reducing agent but also as a dispersing medium for the ligand-stabilized nanoparticles. The phenanthroline-stabilized Pd nanoparticles are well defined, with a mean size of 2–6 nm. Hydrogenation of various olefins (cyclohexene, then aliphatic, alicyclic and aromatic olefins) was performed in liquid phase under hydrogen at atmospheric pressure; some results are presented in Table 9.4. The catalyst system was found to be active and selective for the hydrogenation of a variety of olefins, with good to excellent conversion. In the case of 1,5-cyclooctadiene, cyclooctene is formed, but the selectivity of the di-hydrogenated product was enhanced with an increase in reaction time. The hydrogenation of unsaturated alcohols, aldehydes and ketones, such as cinnamyl alcohol and citral, formed the C=C hydrogenated product selectively without affecting the C=O group. However, the selectivity of the di-hydrogenated product also increased with reaction time. The catalyst system was easily separable from the reaction mixture by extraction of the products with diethyl ether, and could be reused several times (six cycles) without any loss of activity or selectivity. These authors demonstrated that phenanthroline stabilizes the palladium nanoparticles in PEG, which may act as a mobile supporting phase, thereby achieving high stability and high activity for the catalyst system.

Kaifer and coworkers showed interest in the modification of metal nanoparticles with organic monolayers prepared with suitable molecular hosts. They reported the preparation of water-soluble platinum and palladium nanoparticles modified with thiolated β-cyclodextrin (β-CD) [69]. Nanoparticle synthesis was

Table 9.4 Phenanthroline-stabilized palladium nanoparticles in PEG for the hydrogenation of alkenes.[a] (Adapted from [68])

Substrate	Product	Temp. [°C]	Time [h]	Conversion [%]	Selectivity [%]	TON
Cyclopentene	Cyclopentane	30	20	100	100	448
1-Hexene	*n*-Hexane	50	8	100	100	448
Cyclohexene	Cyclohexane	30	8	100	100	448
1-Methyl Cyclohexene	Methyl cyclohexane	50	20	54	100	242
1-Phenyl Cyclohexene	Phenyl cyclohexane	50	20	100	94[b]	448
1-Octene	*n*-Octane	50	8	100	100	448
2-Octene	*n*-Octane	50	20	100	100	448
Cyclooctene	Cyclooctane	50	8	100	100	448
1,5-Cyclooctadiene	Cyclooctene	40	20	100	38[c]	448
Styrene	Ethyl benzene	30	20	100	100	448
Stilbene	Dibenzyl	50	4	100	100	448
Norbornylene	Norborane	40	4	100	100	448
Cinnamyl alcohol	3-Phenyl propanol	40	20	100	100	448
Citral	Citronellal	50	20	100	15[d]	448
3-Methyl-2-butenal	3-Methyl butyraldehyde	40	8	94	100	422
Mesityl oxide	4-Methyl-2-pentanone	40	8	94	100	422
Cyclohexenone	Cyclohexanone	40	8	51	100	229
3-Methyl cyclohexene-1-ol	3-Methyl cyclohexanol	40	20	100	100	448

a) Reaction conditions: 10 mmol substrate, 5 mg Pd acetate, 1.5–3.0 mg 1,10-phenanthroline, 4 g PEG (400), stir, H$_2$ balloon.
b) Remaining biphenyl.
c) Remaining cyclooctane.
d) Remaining hydrocitronellal.
TON: calculated as the number of moles product formed per mol palladium.

realized by the reduction of a DMSO-water solution of Na$_2$PtCl$_4$ (or Na$_2$PdCl$_4$) by NaBH$_4$ in the presence of HS-β-cyclodextrin at room temperature. The nanoparticles were isolated as a dark precipitate after the addition of ethanol. Centrifugation and washing with DMSO and ethanol, followed by drying at 60 °C, provided dried powders containing Pt or Pd nanoparticles with a respective mean size of 14.1±2.2 and 15.6±1.3 nm. These Pt and Pd nanoparticles could be redispersed in water, and were tested for catalytic allylamine hydrogenation under atmospheric pressure of H$_2$ and room temperature (Table 9.5). Both types of CD-modified metal nanoparticles were soluble in the reaction media and could be easily recovered at the end of the reaction by precipitation with ethanol.

Table 9.5 Percentage conversion from allylamine (1.8 mmol) to propylamine obtained with cyclodextrin (CD)-modified Pt and Pd nanoparticles under 1.0 atm $H_2(g)$ at room temperature in D_2O solution (2.0 mL). (Reprinted with permission of the Royal Society of Chemistry [69])

Catalyst	Quantity [mg]	Time [h]	Conversion [%]
CD-mod. Pt	10	6	>95
CD-mod. Pd	10	6	100
None	–	6	0
CD-mod. Pt	5	1	10
CD-mod. Pd	5	1	30

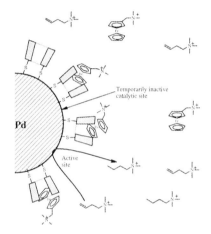

Scheme 9.6 Deactivation of active catalytic sites by binding ferrocene derivatives to the CD hosts on the Pd nanoparticles. (Reprinted with permission of the American Chemical Society [70].)

Following the same procedure, these authors also reported [70] water-soluble Pd nanoparticles (diameter 3.5±1.0 nm) which were modified with covalently attached cyclodextrin receptors and obtained by $NaBH_4$ reduction of $PdCl_4^{2-}$ in dimethylformamide solution containing perthiolated β-CD. After characterization by usual methods (TEM, UV-visible, NMR), hydrogenation of 1-butenyl(trimethyl)ammonium bromide under H_2 (1 atm) in D_2O at 25 °C gave a TOF of 320 h^{-1}. Since these Pd nanoparticles behave as active catalyst for the hydrogenation of alkenes in aqueous media, the group investigated whether this catalytic activity could be modulated through binding of guests to the surface-immobilized CD hosts. Since it is well known that ferrocene derivatives form stable inclusion complexes with β-CD, ferrocene compounds were added to the reaction mixture (Scheme 9.6). Under these circumstances, a substantial reduction in the rate of hydrogenation, dependent upon the ferrocene concentration, was noted. In addition, the catalytic ability of such a system for the hydrogenation of an olefin bearing a ferrocenyl group was compared. The inhibitory effect was lower in this case, probably because of the affinity of the olefin through its ferrocene moiety for the

CD-modified nanoparticles sites. These studies afford an interesting example of "tunable catalyst" design at the molecular level. Manipulation of the surface of catalytically active metal nanoparticles seems possible, and can be used to modulate the catalytic activity on demand.

9.3.7
Biomaterial as a Protective Matrix

Ueno et al. have published results relating to olefin hydrogenation by a Pd nanocluster confined in an apo-ferritin cage [71]. Ferritin, an iron-storage protein, comprises 24 subunits that assemble to form a hollow cage-like structure with a diameter of 12 nm and an internal cavity of 8 nm. These authors used ferritin as a stabilizing agent for the synthesis of Pd nanoparticles and their use in hydrogenation catalysis. The aqueous synthesis was realized by reduction of K_2PdCl_4 with $NaBH_4$ in the presence of apo-ferritin, giving rise to a clear brown solution containing monodispersed spherical Pd nanoparticles with a mean size of

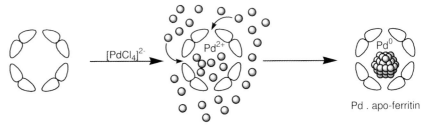

Scheme 9.7 Schematic synthesis of apo-ferritin stabilized-Pd nanoparticles. (Reprinted with permission of Wiley [71].)

Table 9.6 Hydrogenation activity of Pd-apo-ferritin nanoparticles in water. (Reprinted with permission of Wiley [71].)

Olefin	TOF [h^{-1}] for Pd apo-ferritin[a), b)]	TOF [h^{-1}] for Pd particles[b), c)]
CH$_2$=CHCONH$_2$ (1)	72 ± 0.7	58 ± 5.9
CH$_2$=CHCOOH (2)	6.3 ± 1.1	12 ± 2.6
CH$_2$=CHCONH-*i*Pr (3)	51 ± 6.5	15 ± 0.3
CH$_2$=CHCONH-*t*Bu (4)	31 ± 5.9	6.1 ± 0.6
CH$_2$=CHCO-Gly-OMe (5)	6.3 ± 3.8	28 ± 2.6
CH$_2$=CHCO-D,L-Ala-OMe (6)	Not detected	23 ± 0.3

a) Hydrogenation reactions catalyzed by Pd apo-ferritin carried out at 7 °C (pH 7.5) with 30 μM Pd.
b) TOF = [product(mol)] per atom Pd per hour.
c) The same conditions were used, but apo-ferritin was omitted.

2.0±0.3 nm (Scheme 9.7). The catalytic hydrogenation of olefins (acrylamide derivatives) by Pd apo-ferritin nanoparticles was evaluated in aqueous medium (Table 9.6). Since the observed TOFs are dependent upon substrate size, it could be concluded that Pd-apo-ferritin particles induce size-selective olefin hydrogenation.

9.3.8
Ionic Liquids used as Templates for the Stabilization of Metal Nanoparticles

More efficient catalytic systems that might combine the advantages of both homogeneous (catalyst modulation) and heterogeneous catalysis (catalyst recycling) are the subjects of great attention by the scientific community working on catalysis. For such purpose, ionic liquids are interesting systems as they can provide simple product separation and catalyst recycling.

The group of Dupont has developed the synthesis of metal nanoparticles in ionic liquids for use as catalysts. The group has published several reports relating to ionic liquid-stabilized nanoparticles of various metals, including iridium [72], platinum [73] and ruthenium [74]. Mostly, the nanoparticles are synthesized from organometallic precursors such as [Ir COD Cl]$_2$, Pt$_2$(dba)$_3$ or Ru COD COT (COD = cycloocta-1,5-diene; COT = cycloocta-1,3,5-triene; dba = dibenzylydeneacetone). The decomposition of these complexes is accomplished under molecular hydrogen (4 atm) at 75 °C in the chosen ionic liquid such as 1-*n*-butyl-3-methylimidazolium hexafluorophosphate (BMI PF$_6$) or 1-*n*-butyl-3-methylimidazolium tetrafluoroborate (BMI BF$_4$). These reaction conditions lead to well-dispersed nano-

Table 9.7 Catalytic performance of Pt(0) nanoparticles in solventless, homogeneous, and biphasic conditions.[a]
(Reprinted with permission of the American Chemical Society [73].)

Medium	Substrate	Product	Time [h]	Conversion [%][b]	TOF [h^{-1}][c]
Solventless	Hex-1-ene	Hexane	0.25	100	1000
Acetone	Hex-1-ene	Hexane	0.25	100	1000
BMI PF$_6$	Hex-1-ene	Hexane	0.4	100	625
BMI PF$_6$	Cyclohexene	Cyclohexane	1.6	100	156
Solventless	Cyclohexene	Cyclohexane	0.3	100	833
Acetone	Cyclohexene	Cyclohexane	0.3	100	833
BMI PF$_6$	2,3-Dimethyl-1-butene	2,3-Dimethyl-butane	3	82	68
Solventless	2,3-Dimethyl-1-butene	2,3-Dimethyl-butane	0.6	100	417
Solventless	1,3-Cyclohexadiene	Cyclohexane	0.3	100	833

a) Reaction conditions: [substrate]/[Pt] = 250 at 75 °C and under 4 atm H$_2$ (constant pressure).
b) Substrate conversion.
c) [mol product] [mol Pt]$^{-1}$ [h].

Table 9.8 Hydrogenation of alkenes by Ru(0) nanoparticles
under multiphase and solventless conditions (75 °C and
constant pressure of 4 atm, substrate/Ru = 500).
(Adapted from [74])

Medium	Substrate	Time [h]	Conversion [%]	TON [a]	TOF [h^{-1}] [b]
Solventless	1-hexene	0.7	>99	500	714
BMI BF$_4$	1-hexene	0.6	>99	500	833
BMI PF$_6$	1-hexene	0.5	>99	500	1000
Solventless	cyclohexene	0.5	>99	500	1000
BMI BF$_4$	cyclohexene	5.0	>99	500	100
BMI PF$_6$	cyclohexene	8.0	>99	500	62
Solventless	2,3-dimethyl-2-butene	1.2	76	380	316

a) Turnover number (TON) = mol hydrogenated product mol^{-1} Ru.
b) Turnover frequency (TOF) = TON h^{-1}.

metric particles (d$_m$ = 2–3 nm) in the ionic liquid. The thus-obtained nanoparticles
are stable and isolable by centrifugation and acetone washing, and were characterized using a variety of techniques such as TEM, XRD, EDX, or XPS. The isolated
nanoparticles can be redispersed in the ionic liquid, acetone, or used in solventless
conditions for respectively, liquid–liquid biphasic, homogeneous or heterogeneous
hydrogenation of alkenes and arenes under mild conditions (75 °C, 4 atm). Various
olefins were used as substrates for the catalytic experiments, and results obtained
with the Pt and Ru nanoparticles are listed in Tables 9.7 and 9.8, respectively. In
general, the best results were obtained in solventless conditions. The recovered catalysts could be reused as solids or redispersed in the ionic liquid several times,
without significant loss in catalytic activity.

In order to avoid the problem of aggregation, some research groups have
studied the addition of a stabilizer (ligand or polymer) to increase stability of
the nanocatalysts in ionic liquid. Han and coworkers reported the use of ligand-
stabilized palladium nanoparticles for the hydrogenation of olefins in an ionic liquid [75]. These authors prepared phenanthroline-stabilized Pd nanoparticles (2–
5 nm) in BMIM PF$_6$ and tested them directly as catalyst for olefin hydrogenation
at 40 °C. The catalytic system showed high activity, with TOFs up to 234 h^{-1}, and
the catalyst could be recycled (Table 9.9). The authors proposed that the ligand protected the Pd nanoparticles while the ionic liquid acted as a mobile support for the
nanocatalyst and enhanced their stability.

Similarly, Kou et al. published the synthesis of PVP-stabilized noble-metal nanoparticles in ionic liquids BMI PF$_6$ at room temperature [76]. The metal nanoparticles (Pt, Pd, Rh) were produced by reduction of the corresponding metal halide
salts in the presence of PVP into a refluxing ethanol-water solution. After evaporation to dryness the residue was redissolved in methanol and the solution added to
the ionic liquid. The methanol was then removed by evaporation to give the ionic
liquid-immobilized nanoparticles. These nanoparticles were very stable. TEM ob-

Table 9.9 Hydrogenation of olefins catalyzed by Phen-protected Pd nanoparticles in [BMIM][PF$_6$]. [a] (Adapted from [75])

Olefin	Olefin/Pd [mol mol^{-1}]	Temperature [°C]	Time [h]	Conversion [%]
Cyclohexene	2000	40	2.0	35
Cyclohexene	500	40	5.0	100
Cyclohexene	500	30	7.0	100
Cyclohexene	500	50	4.0	100
Cyclohexene	500	60	3.5	100
1-Hexene	500	20	3.0	100
1-Hexene	500	40	1.5	100
1,3-Cyclohexadiene	500	40	2.0	95 [b]
1,3-Cyclohexadiene	500	40	7.0	100 [c]

a) 1 bar H$_2$ (constant pressure).
b) Product is cyclohexene; no cyclohexane was detected.
c) Product is cyclohexane.

servations indicated that the distribution of particles size in the range 2–5 nm were similar before and after their immobilization in the ionic liquid. The catalytic performance was evaluated in the hydrogenation of olefins at 40 °C under hydrogen pressure (1 atm) in biphasic conditions (Table 9.10). The results showed that nanoparticles were highly active catalysts for the hydrogenation of olefins under very mild conditions. The ionic liquid-immobilized nanoparticles were easily separated from the product mixture by simple decantation or reduced pressure distillation, and could be reused several times, without loss of activity. TEM analysis carried out after several catalytic experiments showed that the particles did not aggregate; thus, the combination of PVP and ionic liquid appeared to be successful in inhibiting particle aggregation.

Table 9.10 Hydrogenation of alkenes catalyzed by PVP-stabilized noble-metal nanoparticles in [BMI][PF$_6$]. (Adapted from [76])

Substrate	Metal	Substrate/metal [mol mol^{-1}]	Time [h]	Conversion [%] [a]	TOF [h^{-1}] [b]
Cyclohexene	Pt	2000	16	100	125
1-Hexene	Pt	1000	1	100	1000
1-Dodecene	Pt	1000	1	100	1000
Cyclohexene	Pd	250	1	100	250
Cyclohexene	Rh	250	2	100	125

a) Substrate conversion.
b) Turnover frequency (TOF) = [mol product] [mol metal]$^{-1}$ h^{-1}.

Table 9.11 Hydrogenation reactions with Pd nanocatalysts in methanol, toluene or [BMIM][PF$_6$].[a] (Adapted from [77])

Substrate	Time [h]	Product	Yield [%]
(benzyl acrylate)	20	(benzyl propanoate)	97
(2-allyl benzyl phenyl ether)	22	(2-propyl benzyl phenyl ether)	98
Ph⌇⌇O⌇Ph (cinnamyl benzyl ether)	20	Ph⌇⌇O⌇Ph	82
		Ph⌇⌇O⌇Ph	17
⌇⌇(⌇)$_8$O⌇Ph	40	⌇⌇(⌇)$_8$O⌇Ph	92

a) Reactions carried out at room temperature using a balloon filled with H$_2$ and 0.001 equiv. Pd$_{OAc}$ as catalyst in [BMIM][PF$_6$] as solvent. Substrate/metal ratio = 100.

Recently, tetrabutylammonium bromide-stabilized Pd nanoparticles have been described for the hydrogenation of carbon-carbon double bonds bearing benzyloxy groups in [BMIM][PF$_6$] under 1 bar of hydrogen and at room temperature [77]. Some results are summarized in Table 9.11. The nanoparticles were synthesized from PdCl$_2$ or Pd(OAc)$_2$ precursors. These metal salts were mixed with tetrabutylammonium bromide and heated under vacuum at 120 °C before addition of tri-n-butylamine (n-Bu$_3$N) and an additional treatment at 120 °C for 3 h. The black powders formed could be isolated and characterized by a variety of techniques (IR, NMR, elemental analysis, TEM). TEM analysis revealed dispersed nanoparticles (d$_m$=4.1±1.0 nm) in the case of Pd(OAc)$_2$ as precursor, but agglomerated particles (d$_m$=7.5±1.7 nm) with PdCl$_2$. It appeared that the hydrogenation of double bonds was chemoselective. Recycling of the system [BMIM][PF$_6$]/palladium nanoparticles has been carried out, without noticeable modification of the chemoselectivity and yield.

9.3.9
Supercritical Microemulsions Used as Templates for the Stabilization of Metal Nanoparticles

In recent years, supercritical fluids such as scCO$_2$ were considered to be modern "green" solvents: they were non-toxic, readily available, inexpensive, and environmentally benign. They are studied as a reaction medium for catalytic applications because of their interest in product separation and catalyst recovery, and several reports have described the use of metal nanoparticles in supercritical fluids.

Wai et al. used water-in-CO_2 microemulsions as a medium for synthesizing metallic Pd nanoparticles [78]. The water-in-CO_2 microemulsion was prepared by mixing an aqueous $PdCl_2$ solution with a mixture of bis(2-ethylhexyl)sulfosuccinate (AOT) as surfactant and perfluoropolyetherphosphate (PFPE-PO_4) as co-surfactant into CO_2 at 80 atm to ensure the formation of an optically transparent microemulsion. Injection of dihydrogen into the microemulsion allowed the formation of Pd nanoparticles in the size range 5–10 nm, as confirmed by UV-visible spectroscopy and TEM analysis. The hydrogenation of 4-methoxy-cinnamic acid to 4-methoxyhydrocinnamic acid was performed first in liquid CO_2 and then in supercritical CO_2 at 35 °C and 50 °C. The hydrogenation process is much faster in the scCO_2 phase (35 and 50 °C) compared with that in the liquid CO_2 phase (20 °C), probably due to a better diffusion of the reactant from bulk CO_2 to the Pd nanoparticles surface, the diffusion coefficient of CO_2 dramatically changing at the critical point (31 °C). *Trans*-stilbene and maleic acid hydrogenations were also performed, undergoing the production of 1,2-diphenylethane and succinic acid respectively, as determined by NMR investigation.

The same authors also reported the dispersion of palladium nanoparticles in a water/AOT/*n*-hexane microemulsion by hydrogen gas reduction of $PdCl_4^{2-}$ and its efficiency for hydrogenation of alkenes in organic solvents [79]. UV-visible spectroscopy and TEM analysis revealed the formation of Pd nanoparticles with diameters in the range of 4 to 10 nm. Three olefins (1-phenyl-1-cyclohexene, methyl *trans*-cinnamate, and *trans*-stilbene) were used as substrates for the catalytic hydrogenation experiments under 1 atm of H_2 (Table 9.12). All of the start-

Table 9.12 Catalytic hydrogenation of olefins with Pd nanoparticles in a water-in-hexane microemulsion. (Reprinted with the permission of the American Chemical Society [79])

Olefin	Catalyst	Reaction time [min]	Conversion [%]	Product
	Pd/ME [a]	5	>97	
	Pd/C [b]	20	50	
	Pd/C [b]	40	>97	
	Pd/ME [a]	6	>97	
	Pd/C [b]	25	60	
	Pd/C [b]	45	>97	
	Pd/ME [a]	7	>97	
	Pd/C [b]	40	68	
	Pd/C [b]	60	>97	

a) Molar ratio of alkene/AOT/Na_2PdCl_4 = 1/1/0.01 and 10 mL of *n*-hexane used as solvent (W = 15) at 30 °C.
b) 1.7 mg Pd/C (0.17 mg or 1.6 µmol Pd) in 10 mL *n*-hexane.

ing alkenes were converted into the saturated hydrocarbons within 6–7 min, while the Pd nanoparticles dispersed in the microemulsion proved to be more efficient catalysts than the Pd/C catalyst tested for comparison.

9.3.10
Conclusion

A variety of approaches has been explored for the synthesis of metal nanoparticles, with the objective of using them as catalysts for alkenes hydrogenation. It appears, clearly, that whichever the chosen stabilizing mode, it is possible to obtain active nanoparticles for olefin hydrogenation. Nevertheless, the activities obtained are difficult to compare as the experimental conditions are, by necessity, different. Even if the tested catalytic reaction is for most of the time the hydrogenation of simple olefins, other colloidal systems have permitted the hydrogenation of more interesting alkenes. Recycling of the nanocatalysts is also of major interest, as colloidal catalysts may be considered to be pseudo-homogeneous catalysts (soluble nanocatalysts) with clear advantages over their heterogeneous counterparts. Indeed, colloidal nanoparticles represent highly interesting systems for the future development of catalysis in terms of both activity and selectivity.

9.4
Hydrogenation of Compounds with C≡C Bonds

The use of dispersed or immobilized transition metals as catalysts for partial hydrogenation reactions of alkynes has been widely studied. Traditionally, alkyne hydrogenations for the preparation of fine chemicals and biologically active compounds were only performed with heterogeneous catalysts [80–82]. Palladium is the most selective metal catalyst for the semihydrogenation of monosubstituted acetylenes and for the transformation of alkynes to *cis*-alkenes. Commonly, such selectivity is due to stronger chemisorption of the triple bond on the active center.

The liquid-phase hydrogenation of various terminal and internal alkynes under mild conditions was largely described with metal nanoparticles deposited/incorporated in inorganic materials [83, 84], although several examples of selective reduction achieved by stabilized palladium, platinum or rhodium colloids have been reported in the literature.

The selective hydrogenation of hex-2-yne into *cis*-hex-2-ene with Pd colloids stabilized by 1,10-phenanthroline and derivatives has been reported by Schmid. Selectivity in alkenes up to 99% was obtained [25]. The use of PVP-stabilized Pt colloids with an average particle size of 1.4 nm dispersed in a propanol mixture prepared from $Pt_2(dba)_3$ provided 81% and 62% selectivity to *cis*-hexene at 50% and 90% hex-2-yne conversion, respectively. Bradley has shown that selectivity up to 89% in *cis*-hex-2-ene could be obtained with colloids supported in an amorphous microporous mixed oxides [85]. Following the adapted procedure, Bönnemann performed the hydrogenation of hex-3-yn-1-ol into *cis*-hex-3-en-1-ol with previously synthesized Pd colloids and immobilized on CaCO$_3$; the obtained selectivity was about 98% [86].

Partial hydrogenation of acetylenic compounds bearing a functional group such as a double bond has also been studied in relation to the preparation of important vitamins and fragrances. For example, selective hydrogenation of the triple bond of acetylenic alcohols and the double bond of olefin alcohols (linalol, isophytol) was performed with Pd colloids, as well as with bimetallic nanoparticles Pd/Au, Pd/Pt or Pd/Zn stabilized by a block copolymer (polystyrene-poly-4-vinylpyridine) (Scheme 9.8). The best activity (TOF 49.2 s^{-1}) and selectivity (>99.5%) were obtained in toluene with Pd/Pt bimetallic catalyst due to the influence of the modifying metal [87, 88].

Recently, Chaudhari compared the activity of dispersed nanosized metal particles prepared by chemical or radiolytic reduction and stabilized by various polymers (PVP, PVA or poly(methylvinyl ether)) with the one of conventional supported metal catalysts in the partial hydrogenation of 2-butyne-1,4-diol. Several transition metals (e.g., Pd, Pt, Rh, Ru, Ni) were prepared according to conventional methods and subsequently investigated [89]. In general, the catalysts prepared by chemical reduction methods were more active than those prepared by radiolysis, and in all cases aqueous colloids showed a higher catalytic activity (up to 40-fold) in comparison with corresponding conventional catalysts. The best results were obtained with cubic Pd nanosized particles obtained by chemical reduction (Table 9.13).

Catalytic studies and kinetic investigations of rhodium nanoparticles embedded in PVP in the hydrogenation of phenylacetylene were performed by Choukroun and Chaudret [90]. Nanoparticles of rhodium were used as heterogeneous catalysts (solventless conditions) at 60 °C under a hydrogen pressure of 7 bar with a [catalyst]/[substrate] ratio of 3800. Total hydrogenation to ethylbenzene was observed after 6 h of reaction, giving rise to a TOF of 630 h^{-1}. The kinetics of the hydrogenation was found to be zero-order with respect to the alkyne compound, while the reduction of styrene to ethylbenzene depended on the concentration of phenylacetylene still present in solution. Additional experi-

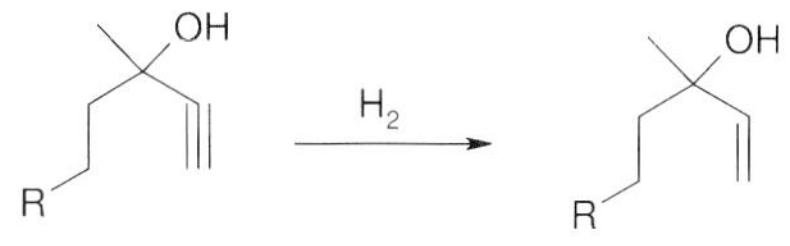

| R is | (CH$_3$)$_2$C=CH- | Dehydrolinalol | Linalol |
| | (CH$_3$)$_2$CH((CH$_2$)$_3$CH(CH$_3$)-)$_2$CH$_2$ - | Dehydroisophytol | Isophytol |

Scheme 9.8 Semihydrogenation of olefin alcohols with Pd colloids stabilized by a block copolymer polystyrene-poly-4-vinylpyridine.

Table 9.13 Comparison of colloidal and heterogeneous catalysts in the hydrogenation of 2-butyne-1,4-diol. (Adapted from [89])

Catalyst	Size [nm]	Selectivity [%]	TOF [$\times 10^{-5}$ h^{-1}]	TOF$_{Mt-PVP}$/ TOF$_{Mt-CaCO_3}$
Pt/PVP	5.1	96	5.5	
Pt/CaCO$_3$	–	83	0.2	27
Rh/PVP	5	96	4.2	
Rh/CaCO$_3$	–	85	0.1	42
Ru/PVP	4.8	95.2	5.1	
Ru/CaCO$_3$	–	75	0.14	36
Ni/PVP	9.4	99	0.1	
Ni/C	–	65	0.01	10
Pd/PVA	5.7	99	5	–
Pd/PMVE	5.4	89.2	5.2	–
Pd/PVP	5	91.1	5.7	
Pd/CaCO$_3$	–	83	0.15	38

ments conducted in the presence of phosphine showed that the amount of styrene increased while the formation of ethylbenzene versus styrene decreased.

Colloidal catalysts in alkyne hydrogenation are widely used in conventional solvents, but their reactivity and high efficiency were very attractive for application in scCO$_2$. This method, which is based on colloidal catalyst dispersed in scCO$_2$, yields products of high purity at very high reactions rates. Bimetallic Pd/Au nanoparticles (Pd exclusively at the surface, while Au forms the cores) embedded in block copolymer micelles of polystyrene-block-poly-4-vinylpyridine

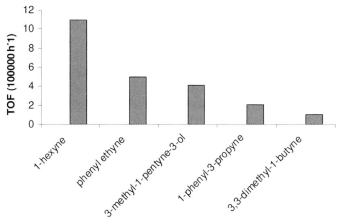

Fig. 9.2 TOF-values calculated at 50% conversion in the hydrogenation of various alkynes with a [substrate]/[Pd] ratio near 6500. (Adapted from [91])

were reported in highly efficient single-phase hydrogenation of alkynes in
scCO$_2$ [91]. Several substrates, including 1-hexyne, 3-methyl-1-pentyne-3-ol, 3,3-
dimethyl-1-butyne, phenyl ethyne and 1-phenyl-3-propyne, were investigated at
a hydrogen pressure of 15 bar, 50 °C and a CO$_2$ pressure of 150 bar (Fig. 9.2).

According to the hydrogen pressure and substrate/Pd ratio, a TOF up to
4×10^6 h^{-1} was observed for the hydrogenation of 1-hexyne, this being the highest TOF ever reported for alkyne hydrogenation.

9.5
Arene Hydrogenation

In most cases, the industrial hydrogenation of benzene and derivative compounds is performed using heterogeneous catalysis [92, 93]. In many cases (if
not the majority) these systems require drastic conditions (high hydrogen pressure and/or temperature), but the use of nanoparticles under mild conditions
has received an increasing amount of attention since some homogeneous catalysts were shown to be micro- or nano-heterogeneous [4–6]. Recent progress
[94, 95] in the complete catalytic hydrogenation of monocyclic aromatic compounds by noble metal (0) nanoparticles such as Pt, Rh, Ru or Ir in various liquid media has been described in the literature (Scheme 9.9).

In this strategic research area, significant results based on a critical combination of various parameters were obtained by several catalytic systems [94]. Three
parameters seem to be highly important:
- the stabilizing agent, which may be either one of PVP, polyoxoanion, surfactant or ionic liquids;
- the nature of the precursor: metal salts or organometallic compounds; and
- the type of the catalytic system: monophasic (organic) or biphasic liquid–liquid (organic/organic and water/organic) media.

Ammonium salts are commonly used to stabilize aqueous colloidal suspensions
of nanoparticles. The first such example was reported in 1983–84 by Januszkiewicz and Alper [96, 97], who described the hydrogenation of several benzene derivatives under 1 bar H$_2$ and biphasic conditions starting with [RhCl(1,5-hexa-
diene)]$_2$ as the metal source and with tetraalkylammonium bromide as a stabilizing agent. Some ten years later, Lemaire and coworkers investigated the *cis/*

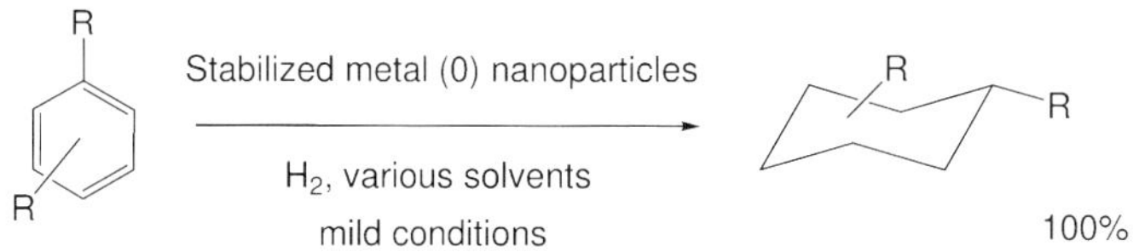

Scheme 9.9 Total hydrogenation of monocyclic arene
compounds by various zerovalent noble-metal nanoparticles.

trans selectivity in the hydrogenation of methylanisole and cresol derivatives. Rhodium colloids in the 2- to 3-nm size range were obtained from rhodium trichloride in the presence of tricaprylylmethylammonium chloride salt or trioctylamine. A Total Turn Over (TTO) of 40 in 24 h was reported for the hydrogenation of 2-methylanisole, and the *cis* compound was formed with selectivities exceeding 97% [98]. These authors also observed a partial hydrogenolysis of the methoxy group (10%). Similarly, James and coworkers used tetrabutylammonium salts to stabilize rhodium and ruthenium nanoparticles. Several substrates containing the 4-propylphenol fragment were hydrogenated in biphasic media and under various conditions (20–100 °C, 1–50 bar H_2). The best results were obtained for the hydrogenation of 2-methoxy-4-propylphenol by ruthenium nanoparticles with a TTO of 300 in 24 h [99–101].

In 1999, the group of Roucoux succeeded in the synthesis of aqueous suspensions of rhodium(0) colloids by reducing $RhCl_3 \cdot 3H_2O$ with $NaBH_4$ in the presence of surfactants which provide an electrosterical stabilization. Nanoparticles stabilized by *N,N*-dimethyl-*N*-cetyl-*N*-(2-hydroxyethyl)ammonium salts (counteranion: Br, Cl, I, CH_3SO_3, BF_4) are well-defined with a mean size of 2–3 nm. Various mono-, di-substituted and/or functionalized arene derivatives are hydrogenated, with TOFs up to 200 h^{-1}, in pure biphasic liquid–liquid (water/substrate) media at 20 °C and 1 bar H_2. The hydrogenation of anisole is observed with a TOF of 4000 h^{-1} under 40 bar H_2. The nanocatalyst could be separated by simple decantation and reused in successive hydrogenation reactions. Significant results have been obtained in the hydrogenation of anisole, with 2000 TTO in 37 h [14,15]. In the same manner, the efficient hydrogenation of various benzene compounds in aqueous media at room temperature and under 40 bar H_2 has also been described by using similar reusable surfactant-protected iridium(0) nanoparticles [102]. In all cases, the conversion is complete after a few hours. A TTO of 3000 was obtained for anisole hydrogenation in 45 h. TEM observations showed the particles to be monodispersed, with an average diameter of 1.9±0.7 nm (Scheme 9.10). Selective hydrogenation of di-substituted benzenes such as xylene, methylanisole, and cresol was also observed with these aqueous suspensions of rhodium(0) and iridium(0) nanoparticles. In all cases, the *cis*-compound is the major product (>80%). The *cis/trans* ratio decreases with the position of the substituents *ortho > meta > para*, but the nature of the metal does not seem important with this surfactant-stabilized system [14, 102].

Similar surfactant-stabilized colloidal systems have been reported by Albach and Jautelat, who prepared aqueous suspensions of Ru, Rh, Pd, Ni nanoparticles and bimetallic mixtures stabilized by dodecyldimethylammonium propanesulfonate [103]. Benzene, cumene and isopropylbenzene were reduced in biphasic conditions under various conditions at 100–150 °C and 60 bar H_2, and TTO up to 250 were obtained.

The immobilization of metal nanoparticles with a water-soluble polymeric material such as PVP has also been described. The groups of Choukroun and Chaudret have described the hydrogenation of benzene in a biphasic mixture with PVP-protected native Rh nanoparticles synthesized from the organometal-

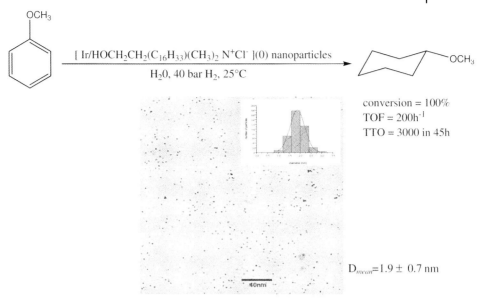

$$[Ir/HOCH_2CH_2(C_{16}H_{33})(CH_3)_2 \ N^+Cl^-](0) \ \text{nanoparticles}$$

$$H_2O, \ 40 \ \text{bar} \ H_2, \ 25°C$$

conversion = 100%
TOF = 200h^{-1}
TTO = 3000 in 45h

D_{mean}=1.9 ± 0.7 nm

Scheme 9.10 Hydrogenation of anisole by reusable surfactant-stabilized Ir(0) colloids in water. TEM micrograph and
size distribution of Ir(0). (Adapted from [102].)

lic complex [RhCl(C$_2$H$_4$)$_2$]$_2$ [90]. The presence of soluble Rh nanoparticles with
sizes in the 2- to 3-nm range was confirmed by TEM. In water/benzene biphasic conditions at 30 °C and under 7 bar H$_2$, complete benzene hydrogenation is
observed at a substrate:catalyst ratio of 2000 after 8 h, giving rise to a TOF of
675 h^{-1} (related to H$_2$ consumed).

Details of a similar polymer-stabilized colloidal system were published by
James and coworkers [104]. Rhodium colloids are produced by reducing
RhCl$_3$·3H$_2$O with ethanol in the presence of PVP. The monophasic hydrogenation of various substrates such as benzyl acetone and 4-propylphenol and benzene derivatives was performed under mild conditions (25 °C, 1 bar H$_2$). The nanoparticles are poorly characterized and benzyl acetone is reduced, with 50 TTO
in 43 h.

Recently, Dupont and coworkers described the use of room-temperature imidazolium ionic liquids for the formation and stabilization of transition-metal nanoparticles. The potential interest in the use of ionic liquids is to promote a biphasic organic–organic catalytic system for a recycling process. The mixture
forms a two-phase system consisting of a lower phase which contains the nanocatalyst in the ionic liquid, and an upper phase which contains the organic
products. Rhodium and iridium [105], platinum [73] or ruthenium [74] nanoparticles were prepared from various salts or organometallic precursors in dry 1-bu-
tyl-3-methylimidazolium hexafluorophosphate (BMI PF$_6$) ionic liquid under hydrogen pressure (4 bar) at 75 °C. Nanoparticles with a mean diameter of 2–3 nm

were isolated by centrifugation. The isolated colloids could be used as solids (heterogeneous catalyst), in acetone solution (homogeneous catalyst), or re-dispersed in BMI PF_6 (biphasic system) for benzene hydrogenation studies. Iridium and rhodium nanoparticles have also been studied in the hydrogenation of various aromatic compounds [105]. In all cases, total conversions were not observed in BMI PF_6. A comparison between Ir(0) and Rh(0) nanoparticles shows that iridium colloids are much more active for the benzene hydrogenation in biphasic conditions, with TOFs of 50 h^{-1} and 11 h^{-1}, respectively, and 24 h^{-1} and 5 h^{-1} for *p*-xylene reduction at 75 °C and under 4 bar H_2. The best results were obtained with platinum nanoparticles prepared from simple decomposition of $Pt_2(dba)_3$ in heterogeneous (solventless) conditions with a TOF of 28 h^{-1} for 100% conversion. The authors reported that the TOF dramatically decreased in biphasic liquid–liquid condition (BMI PF_6) to 11 h^{-1} at 46% conversion, justifying the absence of recycling studies with this substrate [73]. Finally, Dupont and coworkers have described the preparation of Ru(0) nanoparticles by H_2 reduction of the organometallic precursor Ru(COD)(COT) in BMI PF_6, a room-temperature ionic liquid, with hydrogen pressure [74]. These nanoparticles are efficient catalysts for the complete hydrogenation of benzene (TOF = 125 h^{-1}) under solventless conditions (heterogeneous catalyst). In a biphasic system, the authors observed a partial conversion in BMI PF_6 with a modest TOF of 20 h^{-1} at 73% conversion in the benzene hydrogenation.

More recently, Dupont and coworkers studied the impact of the steric effect in the hydrogenation of monoalkylbenzenes by zerovalent nanoparticles (Ir, Rh, Ru) in the ionic liquid BMI PF_6. The results, when compared with those obtained with the classical supported heterogeneous catalysts, showed a relationship between the reaction constants and the steric factors [106].

Finally, these particles generated in ionic liquids are efficient nanocatalysts for the hydrogenation of arenes, although the best performances were not obtained in biphasic liquid–liquid conditions. The main importance of this system should be seen in terms of product separation and catalyst recycling. An interesting alternative is proposed by Kou and coworkers [107], who described the synthesis of a rhodium colloidal suspension in BMI BF_4 in the presence of the ionic copolymer poly[(*N*-vinyl-2-pyrrolidone)-*co*-(1-vinyl-3-butylimidazolium chloride)] as protective agent. The authors reported nanoparticles with a mean diameter of ca. 2.9 nm and a TOF of 250 h^{-1} in the hydrogenation of benzene at 75 °C and under 40 bar H_2. An impressive TTO of 20 000 is claimed after five total recycles.

An alternative approach to stabilize nanoparticles is to use polyoxoanions (see Scheme 9.5). Finke and coworkers described polyoxoanion- and ammonium-stabilized rhodium zerovalent nanoclusters for the hydrogenation of classical benzene compounds [95, 108]. This organometallic approach allows reproducible preparation of stable nanoparticles starting from a well-defined complex in terms of composition and structure (see Section 9.3.5).

The polyoxoanion-stabilized Rh(0) nanoclusters were investigated in anisole hydrogenation [6,95]. The catalytic reactions were carried out in a single phase using a propylene carbonate solution under mild conditions (22 °C, 3.7 bar H_2). Under these standard conditions, anisole hydrogenation with a substrate:catalyst (S:Rh) ratio of 2600 was performed in 120 h, giving rise to a TTO of 1500 ± 100. The authors observed that the addition of proton donors such as HBF_4 Et_2O or H_2O increased the catalytic activity, and reported 2600 TTO for complete hydrogenation in 144 h at 22 °C and 3.7 bar H_2 with a ratio HBF_4 Et_2O:Rh of 10. A black precipitate of bulk Rh(0) is visible at the end of the reaction as a result of the destabilization of nanoclusters due to the interaction of H^+ or H_2O with the basic $P_2W_{15}Nb_3O_{62}^{9-}$ polyoxoanion.

During the past decade, a variety of stabilized systems based on transition-metal nanoparticles has been seriously investigated, and better lifetimes and activities for the total hydrogenation of monocyclic arene derivatives under mild conditions have been reported. New catalytic systems have been described in various media such as supercritical fluids [109], and these represent a promising area of future research. Several recent investigations have shown modest, but promising, results in partial or selective arene hydrogenation with well-defined colloids [94]. In the future, the partial hydrogenation of arene derivatives into cyclohexene or cyclohexadiene compounds should be highlighted as they are key intermediates in organic synthesis.

The process developed by Asahi Chemical Industry in Japan [110], and performed in a tetraphasic system combining gas, oil, water and ruthenium particles with an average diameter of 20 nm, is a significant milestone in this area. The selectivity is very high and a yield of 60% in cyclohexene is obtained with this "bulk" ruthenium catalyst in the presence of zinc as co-catalyst at 150 °C and under 50.4 bar of H_2. The cyclohexene produced by this process is used as a feedstock for caprolactam.

At present, the efficient partial hydrogenation of benzene and its derivatives has been rarely described with well-defined soluble nanoparticles catalysts. Nonetheless, this remains an interesting area for research, with promising future applications.

9.6
Hydrogenation of Compounds with C=O Bonds

Several reports have been made of the hydrogenation of compounds bearing C=O bonds by colloidal catalytic systems.

In 1996, Liu et al. reported the selective hydrogenation of cinnamaldehyde, an *α,β*-unsaturated aldehyde, to cinnamyl alcohol, an *α,β*-unsaturated alcohol, by means of PVP-protected Pt/Co bimetallic colloids prepared by the polyol process [111]. The colloids were obtained as a dark-brown homogeneous dispersion in a mixture of ethylene glycol and diethylene glycol, and characterized by TEM and XRD. These authors prepared different samples of nanoparticles with Pt:Co ratios of 3:1 and 1:1, the mean diameters of which measured 1.7 and 2.2 nm, respectively. These colloidal systems were also compared with the single metal-based colloids Pt/PVP and Co/PVP. The catalytic tests were carried out under a H$_2$ pressure of 4 MPa at 333 K in EtOH. While Pt/PVP and Co/PVP colloids gave low activity and selectivity, bimetallic colloids exhibited interesting behavior, with the Pt:Co/PVP (3:1) system being the most interesting. Selectivity in cinnamyl alcohol up to 99.8% was indeed obtained with colloid Pt:Co (3:1) with very good conversions (up to 96.2%). It was observed that activity and selectivity could be affected by the presence of added water or NaOH in the reaction mixture. These results are of interest as it is a difficult task to reduce only the C=O bond when it is conjugated to a C=C double bond, with almost all metal catalysts readily reducing the C=C double bond. These results also showed that the colloids were stable enough for catalytic hydrogenation reaction at elevated temperature and pressure.

Another example from Liu's team in this field concerns the selective hydrogenation of citronellal to citronellol by using a Ru/PVP colloid obtained by NaBH$_4$ reduction method [112]. This colloid contains relatively small particles with a narrow size distribution (1.3 to 1.8 nm by TEM), whereas the metallic state of Ru was confirmed by XPS investigation. This colloid exhibited a selectivity to citronellol of 95.2% with a yield of 84.2% (total conversion 88.4%), which represented a good result for a monometallic catalyst.

Gin and coworkers have presented an original strategy for the synthesis of Pd nanoparticles with both good stability and catalytic activity in benzaldehyde hydrogenation by using a cross-linked lyotropic liquid crystal (LLC) assembly as an organic template (Fig. 9.3) [113]. Incorporation of Pd atoms has been performed by ion exchange on the cross-linked inverted hexagonal phase of the sodium salt with an acetonitrile solution of dichloro(1,5-cyclooctadiene)palladium II complex. The Pd-(II)-LLC composite was treated with dihydrogen to afford spherical Pd nanoparticles of 4–7 nm mean size dispersed in the polymer ma-

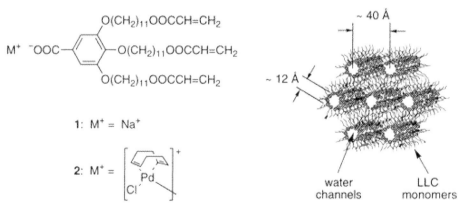

Fig. 9.3 The structure of LLC monomers, 1 and 2, and the inverted hexagonal phase. (Reprinted with the permission of the American Chemical Society [113])

trix. The benzaldehyde hydrogenation was carried out under 1 bar H_2 and 60 °C, and a 98% yield of benzylalcohol was observed after 3 h of reaction.

Pertici et al. described the synthesis of ruthenium nanoparticles on polyorganophosphazenes (PDMP) as stabilizing polymers [114]. The synthesis method consisted of the decomposition of a THF solution of the organometallic $Ru(\eta^4\text{-}COD)(\eta^6\text{-}COT)$ complex at 45 °C under dihydrogen atmosphere in the presence of various polyorganophosphazenes. This procedure gave rise to new materials in which small ruthenium clusters were bound to the arene groups of the polymers. High-resolution TEM analysis revealed well-dispersed and very small nanoparticles of 1.55±0.5 nm mean size in the polymer matrix. These materials could be purified as fine powders containing 5 wt.% of Ru, the dispersion of which could be easily obtained in organic solvent such as ethanol or THF, or in water, allowing their use as homogeneous catalysts. The catalytic experiments were carried out under mild conditions (25 °C, 1 bar H_2) in ethanol, THF or water as solvent. Olefin and ketone hydrogenations were performed. A wide range of carbonyl compounds has been tested such as cyclohexanone, ethyl

Table 9.14 Hydrogenation of ketones with Ru/PDMP at atmospheric hydrogen pressure and 25 °C. (Reprinted with the permission of Elsevier [114])

Substrate[a]	Solvent	Phase[b]	Time [h]	Product[c] [%]	TOF[d]
Cyclohexanone	Ethanol	Hom	10	Cyclohexanol (100)	10
Ethyl acetoacetate	Ethanol	Hom	24	Ethyl 3-hydroxybutyrate (100)	4.2
Ethyl pyruvate	Ethanol	Hom	6	Ethyl lactate (100)	16.6
Pyruvic acid	Ethanol	Hom	8	Lactic acid (100)[e]	12.5
Acetophenone	Ethanol	Hom	96	1-Phenylethanol (100)	1
Pyruvic acid	Water	Hom	7	Lactic acid (100)[e]	14.3
Cyclohexanone[f]	Ethanol	Hom	30	Cyclohexanol (100)	10
Ethyl acetoacetate[f]	Ethanol	Hom	6	Ethyl 3-hydroxybutyrate (100)	3.3
Ethyl pyruvate[f]	Ethanol	Hom	10	Ethyl lactate (100)	16.6
Pyruvic acid[f]	Ethanol	Hom	8	Lactic acid (100)[e]	12.5
Acetophenone[f]	Ethanol	Hom	96	1-Phenylethanol (100)	1
Pyruvic acid[f]	Water	Hom	7	Lactic acid (100)[e]	14.3
Cyclohexanone	THF	Het	10	Cyclohexanol (30)	3
Ethyl acetoacetate	THF	Het	30	Ethyl 3-hydroxybutyrate (20)	0.7
Ethyl pyruvate	THF	Het	6	Ethyl lactate (20)	3
Acetophenone	THF	Het	96	1-Phenylethanol (20)	0.2

a) Substrate (13 mmol); catalyst Ru on PDMP (5 wt.% Ru), 0.26 g (0.13 mg atoms Ru); solvent, 15 mL.
b) Hom: homogeneous phase; Het: heterogeneous phase.
c) Composition determined by GLC analysis.
d) Moles converted substrate per gram-atom ruthenium h^{-1}.
e) Composition determined by ^1H-NMR spectroscopy.
f) Catalyst recovered from runs 1–6, respectively, and reused.

acetoacetate, ethyl pyruvate and pyruvic acid as examples of aliphatic carbonyl compounds, and acetophenone as an example of ketone bearing an aryl group. All carbonyl compounds were quantitatively reduced to the corresponding alcohols (Table 9.14). The catalysts could be reused after precipitation with a non-solvent.

A HRTEM study showed these catalysts to be more resistant towards agglomeration, as only a low degree of aggregation was noted after dissolution and precipitation of the catalysts. However, this slight agglomeration did not lead to catalyst deactivation. Finally, a catalytic study of a Ru/poly[bis(aryloxy)]phosphazene system indicated that the structure of the side chain of the support had considerable influence on the activity of the deposited metal, providing the possibility of modifying the catalytic activity by changing the support structure [114].

The group of Dupont has described the use of ionic liquids for the formation and stabilization of various metal nanoparticles and their application in hydrogenation reactions. In a report concerning the use of Ir(0) nanoparticles (mean diameter 2.1 ± 0.3 nm) prepared by reduction with molecular hydrogen (4 bar) of $[Ir(COD)Cl]_2$ (COD = 1,5-cyclooctadiene) dissolved in BMI PF$_6$ ionic liquid at 75 °C, it was mentioned that the solvent, acetone, used for the "homogeneous" hydrogenation reaction of benzene (75 °C, 4 bar H$_2$) is also hydrogenated, even in the early stages of the reaction [115]. These nanoparticles could be isolated by centrifugation and characterized by TEM and XRD. The authors also reported the hydrogenation of acetophenone. Under "solventless" conditions, they observed the formation not only of 1-cyclohexylethanol but also of ethylcyclohexane, the hydrogenolysis product in a high yield (up to 42%). It was concluded that this result suggested a "heterogeneous" behavior of the Ir(0) nanoparticles in terms of active sites. On pursuing these investigations, it was found that such Ir(0) nanoparticles constitute an efficient and recyclable catalyst for the "solventless" or biphasic hydrogenation of various cyclic and acyclic ketones un-

Table 9.15 Hydrogenation of various carbonyl compounds by Ir(0) nanoparticles in solventless conditions ([substrate]/[Ir] ratio = 250, 75 °C, 4 bar). (Adapted from [116])

Substrate	Product	Time [h]	Yield [%][a]	TOF[b] [h^{-1}]
Benzaldehyde	Phenylmethanol	15	100	17
Cyclopentanone	Cyclopentanol	4	100[c]	62.5
2-Pentanone	2-Pentanol	2.5	96	96
4-Methyl-2-pentanone	4-Methyl-2-pentanol	2.5	96	96
3-Pentanone	3-Pentanol	3.7	100	68
Ethylpyruvate	Ethyl-lactate	2.5	98	98
Acetone	2-Propanol	2.0	95	119

a) Conversions determined by GLC.
b) Mol ketone mol^{-1} iridium h^{-1}.
c) Products obtained were cyclopentanol and bicyclopentyl ether (12%).

der mild conditions (Table 9.15) after optimization of the reaction conditions with cyclohexanone [116]. In the case of ketones containing aromatic cycles, a tendency for the selective hydrogenation of the aromatic ring was observed. The method was also applied to the hydrogenation of benzaldehyde.

9.7
Enantioselective Hydrogenation

The enantioselective hydrogenation of prochiral substances bearing an activated group, such as an ester, an acid or an amide, is often an important step in the industrial synthesis of fine and pharmaceutical products. In addition to the hydrogenation of β-ketoesters into optically pure products with Raney nickel modified by tartaric acid [117], the asymmetric reduction of α-ketoesters on heterogeneous platinum catalysts modified by cinchona alkaloids (cinchonidine and cinchonine) was reported for the first time by Orito and coworkers [118–121]. Asymmetric catalysis on solid surfaces remains a very important research area for a better mechanistic understanding of the interaction between the substrate, the modifier and the catalyst [122–125], although excellent results in terms of enantiomeric excesses (up to 97%) have been obtained in the reduction of ethyl pyruvate under optimum reaction conditions with these Pt/cinchona systems [126–128].

Supported palladium and platinum modified by chiral compounds are largely used as pure heterogeneous hydrogenation catalysts. However, recent studies have been performed starting with catalysts of colloidal nature and particles with dimensions of only a few nanometers. Their development continues to attract substantial interest for three main reasons:

- The elimination of the support such as Al_2O_3, SiO_2, TiO_2, zeoliths and of its influences.
- The possibility of obtaining size- and shape-controlled nanoparticles, thereby giving efficient activities.
- The possibility of adapting chiral molecules as inducer or stabilizer (form and amount) for better selectivities.

The concept of using colloids stabilized with chiral ligands was first applied by Bönnemann to hydrogenate ethyl pyruvate to ethyl lactate with Pt colloids. The nanoparticles were stabilized by the addition of dihydrocinchonidine salt (DHCin, HX) and were used in the liquid phase or adsorbed onto activated charcoal and silica [129, 130]. The molar ratio of platinum to dihydrocinchonidine, which ranged from 0.5 to 3.5 during the synthesis, determines the particle size from 1.5 to 4 nm and contributes to a slight decrease in activity (TOF = 1 s^{-1}). In an acetic acid/MeOH mixture and under a hydrogen pressure up to 100 bar, the (R)-ethyl lactate was obtained with optical yields of 75–80% (Scheme 9.11).

Several mechanistic investigations concerning the hydrogenation of pyruvate derivatives (ethyl or methyl esters) were performed with platinum, rhodium and

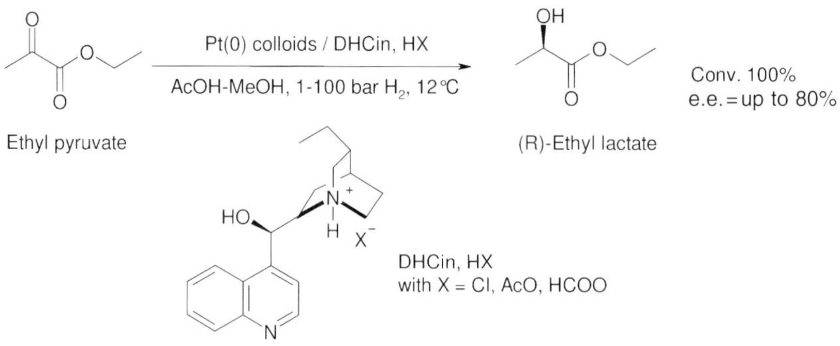

Scheme 9.11 Enantioselective hydrogenation of ethyl pyruvate with platinum colloids stabilized by protonated-dihydrocinchonidine.

iridium nanoparticles stabilized by PVP. In all cases, nanoparticles of size range 2 to 4 nm were modified by cinchonidine or quinine [127, 131–135]. The Pt/PVP catalyst hydrogenated ethyl pyruvate, with ee-values up to 92%, and methyl pyruvate with ee-values up to 98%, depending upon the nanoparticle size. The best results were obtained in acetic acid. The asymmetric hydrogenation of trifluoroacetophenone and α-diketones such as 2,3-butanedione and 3,4-hexanedione was also investigated using finely dispersed PVP-stabilized Pt nanoparticles modified with cinchonidine, and ee-values of up to 30% were reported, according to the solvent [136, 137]. Similar results (ee 30%) were described with solvent-stabilized Pt and Pd nanoparticles prepared by the metal vapor synthesis route for the enantioselective hydrogenation of ethyl pyruvate [138].

Recently, platinum nanoparticles protected by N,N-dimethyl-N-cetyl-N-(2-hydroxyethyl)ammonium chloride salt and modified with cinchonidine were investigated in the enantiomeric hydrogenation of ethyl pyruvate in pure biphasic liquid–liquid (water/substrate) media at room temperature [139]. For the first time, the aqueous phase containing Pt(0) nanocatalysts with an average size of 2.5 nm could be reused for successive hydrogenations, and with a total conversion of activity and enantioselectivity in (R)-(+)-ethyl lactate up to 55% (Scheme 9.12).

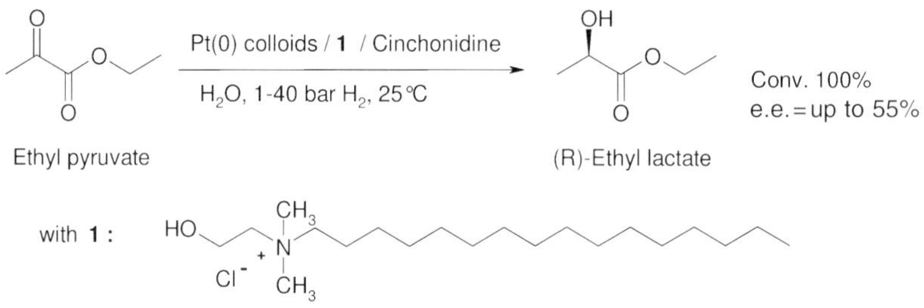

Scheme 9.12 Reusable aqueous suspension of Pt nanoparticles for enantioselective hydrogenation of ethyl pyruvate.

Table 9.16 Comparison of stabilized nanoparticle systems modified with cinchonidine for the hydrogenation of various α-ketoesters and α-diketones.

Catalyst	Substrate	Solvent	Conversion [%]	Temp. [°C]	$PH_{2\ [bar]}$	ee [%] (config.)	Refer- ence(s)
Pt-PVP	Methyl pyruvate [a]	AcOH	100	25	40	97.6 (R)	127, 135
Pt-PVP	Ethyl pyruvate [b]	AcOH	100	25	40	93.8 (R)	127, 135
Pt-PVP	n-propyl pyruvate [b]	AcOH	100	25	40	95.6 (R)	135
Pt-PVP	Iso-propyl pyruvate [b]	AcOH	58.5	25	40	77.1 (R)	135
Pt-PVP	n-butyl pyruvate [b]	AcOH	86.7	25	40	90.5 (R)	135
Pt-PVP	Iso-butyl pyruvate [b]	AcOH	95.5	25	40	93.1 (R)	135
Pt-PVP	Methyl pyruvate [c]	EtOH	61.3	25	40	84.7 (R)	127
Pt-DHCin,HX	Ethyl pyruvate [d]	AcOH/MeOH	100	19	1–100	75–80 (R)	129, 130
Pt-MEK	Ethyl pyruvate [e]	MEK	100	25	70	25 (R)	138
Pt-MMK	Ethyl pyruvate [f]	MMK-H_2O	100	25	70	36 (S)	138
Pt-HEA16Cl	Ethyl pyruvate [g]	H_2O	100	25	40	55 (R)	139
Pt/PVP	2,3-Butanedione [h]	EtOH	95.5	25	40	27.9 (R)	136
Pt/PVP	2,3-Butanedione [h]	AcOH	92.9	25	40	22.6 (R)	136
Ir-PVP	2,3-Butanedione [h]	AcOH	80.5	25	40	20.4 (R)	136
Pt/PVP	3,4-Hexanedione [h]	EtOH	94.9	25	40	18.0 (R)	136
Pt/PVP	3,4-Hexanedione [h]	AcOH	87.8	25	40	20.6 (R)	136
Pd-MEK-KD1	Ethyl pyruvate [i]	MEK	13	25	70	29 (R)	138
Rh-PVP	Ethyl pyruvate [j]	EtOH/THF	100	25	50	42.2 (R)	133
Ir-PVP	Methyl pyruvate [k]	EtOH	85	20	25	17.0 (R)	134

a) TOF = 1.21 s^{-1}.
b) [substrate]/[Pt] ratio = 1600.
c) TOF = 0.76 s^{-1}.
d) TOF $\approx$ 1 s^{-1}.
e) Initial rate = 1025 mmol s^{-1} mol$_{metal}^{-1}$.
f) Cinchonine is used as modifier, initial rate = 2253 mmol s^{-1} mol$_{metal}^{-1}$.
g) [substrate]/[Pt] ratio = 400.
h) [substrate]/[Pt] ratio = 1765.
i) Initial rate = 47 mmol s^{-1} mol$_{metal}^{-1}$.
j) TOF = 941 h^{-1}.
k) Average rate = 838 mmol h^{-1} g$_{metal}^{-1}$.

Although several noble-metal nanoparticles have been investigated for the enantiomeric catalysis of prochiral substrates, platinum colloids remain the most
widely studied. PVP-stabilized platinum modified with cinchonidine showed ee-
values >95%. Several stabilizers have been also investigated such as surfactants,
cinchonidinium salts and solvents, and promising ee-values have been observed.
Details of a comparison of various catalytic systems are listed in Table 9.16; in
one case, the colloid suspension was reused without any loss in enantioselectivity. Clearly, the development of convenient two-phase liquid–liquid systems for
the recycling of chiral colloids remains a future challenge.

9.8
Conclusion

This chapter provides a non-exhaustive overview of the hydrogenation of carbon-carbon double or triple bonds, carbonyl groups and aromatic compounds with colloids as soluble catalysts. The subject, while of crucial importance, is generally not covered in detail in books on homogeneous catalysis, despite several efficient molecular homogeneous complexes having been shown to be precursors of "nanoheterogeneous" catalysts. Such materials, when correctly characterized, constitute an interesting class of both homogeneous and/or heterogeneous catalysts, and the use of colloids is generally seen as being compatible with a variety of reaction media according to the organic- or water-soluble nature of the stabilizers. Colloids can also be adapted for use in biphasic conditions, thereby allowing recovery of the nanocatalysts by simple decantation/filtration, and subsequent recycling. Although stabilized colloids are neither "traditional" nor "routine" catalysts, their performances in some cases – and their potential role as catalysts – is now clearly recognized by the scientific community, and nanoparticle systems represent an interesting compromise between homogeneous and heterogeneous catalytic systems, both in terms of activity and selectivity. Finally, the number of reports related to the use of colloids in catalysis has increased significantly, with interest in colloidal systems limited not only to current hydrogenation reactions but also being indicative of future processes.

Abbreviations

β-CD	*β*-cyclodextrin
BMI PF$_6$	1-*n*-butyl-3-methylimidazolium hexafluorophosphate
COD	cycloocta-1,5-diene
COT	cycloocta-1,3,5-triene
DLVO	Derjaguin-Landau-Verway-Overbeek
DTAC	dodecyltrimethylammonium
EDX	electron diffraction X-ray
EXAFS	extended X-ray absorption fine structure
HDA	hexadecylaniline
HRTEM	high-resolution TEM
LLC	lyotropic liquid crystal
NMR	nuclear magnetic resonance
PAA	poly(acrylic acid)
PAMAM	poly(amidoamine)
PDMS-*b*-PEO	poly(dimethylsiloxane)-*b*-poly(ethylene oxide)
PEG	polyethylene glycol
PPI	poly(propylene imine)
PS-*b*-PEO	polystyrene-*b*-poly(ethylene oxide)
PS-*b*-PMAA	polystyrene-*b*-poly(methacrylic acid)

PtBA-*b*-PCEMA	poly(*t*-butyl acrylate)-*block*-poly(2-cinnamoyloxyethyl) methacrylate
PVA	polyvinylalcohol
PVP	polyvinylpyrrolidone
SAED	selected area electron diffraction
SDS	sodium dodecylsulfate
TEM	transmission electronic microscopy
TGA	thermogravimetric analysis
TOF	turnover frequency
THF	tetrahydrofuran
TTO	total turnover
WAXS	wide-angle X-ray scattering
XPS	X-ray photoelectron spectroscopy
XRD	X-ray diffraction

References

1 G. Schmid (Ed.), *Nanoparticles. From theory to application.* Wiley-VCH, Weinheim, **2004**.

2 D.L. Feldheim, C.A. Foss, Jr. (Eds.), *Metal Nanoparticles: Synthesis, Characterization and Applications.* Marcel Dekker, New York, **2002**.

3 A. Roucoux, J. Schulz, H. Patin, *Chem. Rev.* **2002**, *102*, 3757.

4 P.J. Dyson, *Dalton Trans.* **2003**, 2964.

5 J.A. Widegren, R.G. Finke, *J. Mol. Catal. A Chem.* **2003**, *198*, 317.

6 M.C. Hagen, L. Vieille-Petit, G. Laurenczy, G. Suss-Fink, R. Finke, *Organometallics* **2005**, *24*, 1819.

7 K. Philippot, B. Chaudret, *C. R. Chimie* **2003**, *6*, 1019.

8 J.T.G. Overbeek, in: Goodwin, J.W. (Ed.), *Colloidal Dispersions.* Royal Society of Chemistry, London **1981**, pp. 1–23.

9 D.F. Evans, H. Wennerström, in: *The Colloidal Domain*, 2nd edn. Wiley-VCH, New York, **1999**.

10 R.J. Hunter, in: *Foundations of Colloid Science.* Oxford University Press, New York **1987**, vol. 1, pp. 316.

11 D.H. Napper, in: *Polymeric Stabilization of Colloidal Dispersions.* Academic Press, London, **1983**.

12 M. Komiyama, H. Hirai, *Bull. Chem. Soc. Mater.* **1993**, *56*, 2833.

13 A. Duteil, R. Queau, B. Chaudret, C. Roucau, J.S. Bradley, *Chem. Mater.* **1993**, *5*, 341.

14 J. Schulz, A. Roucoux, H. Patin, *Chem. Eur. J.* **2000**, *6*, 618.

15 A. Roucoux, J. Schulz, H. Patin, *Adv. Synth. Catal.* **2002**, *345*, 222.

16 L. Manna, S.C. Scher, A.P. Alivisatos, *J. Am. Chem. Soc.* **2000**, *122*, 12700.

17 A. Duteil, G. Schmid, W. Meyer-Zaika, *J. Chem. Soc. Chem. Commun.* **1995**, 31.

18 A. Rodriguez, C. Amiens, B. Chaudret, M.J. Casanove, P. Lecante, J.S. Bradley, *Chem. Mater.* **1996**, *8*, 1978.

19 F. Dassenoy, K. Philippot, T. Ould Ely, C. Amiens, P. Lecante, E. Snoeck, A. Mosset, M.J. Casanove, B. Chaudret, *New J. Chem.* **1998**, *22*, 703.

20 S. Chen, K. Kimura, *J. Phys. Chem. B* **2001**, *105*, 5397.

21 C. Pan, K. Pelzer, K. Philippot, B. Chaudret, F. Dassenoy, P. Lecante, M.J. Casanove, *J. Am. Chem. Soc.* **2001**, *123*, 7584.

22 C.J. Kiely, J. Fink, M. Brust, D. Bethell, D.J. Schiffrin, *Nature* **1998**, *396*, 444.

23 G. Schmid, V. Maihack, F. Lantermann, S. Peschel, *J. Chem. Soc. Dalton Trans.* **1996**, 589.

24 N. Cordente, M. Respaud, F. Senocq, M.J. Casanove, C. Amiens, B. Chaudret, *Nano Lett.* **2001**, *1*, 565.

25 G. Schmid, S. Emde, V. Maihack, W. Meyer-Zaika, S. Peschel, *J. Mol. Catal. A Chem.* **1996**, *107*, 95.

26 K. Soulantica, A. Maisonnat, M.C. Fromen, M.J. Casanove, P. Lecante, B. Chaudret, *Angew. Chem. Int. Ed.* **2001**, *38*, 3736.

27 M. Gomez, K. Philippot, V. Collière, P. Lecante, G. Muller, B. Chaudret, *New J. Chem.* **2003**, *27*, 114.

28 O. Vidoni, K. Philippot, C. Amiens, B. Chaudret, O. Balmes, J.O. Malm, J.aO. Bovin, F. Senocq, M.J. Casanove, *Angew. Chem. Int. Ed.* **1999**, *38*, 3736.

29 K. Pelzer, K. Philippot, B. Chaudret, *Z. Phys. Chem.* **2003**, *217*, 1539.

30 K. Pelzer, O. Vidoni, K. Philippot, B. Chaudret, V. Colliere, *Adv. Funct. Mater.* **2003**, *13*, 118.

31 Y. Wang, J. Ren, K. Deng, L. Gui, Y. Tang, *Chem. Mater.* **2000**, *12*, 1622.

32 H. Hirai, *J. Macromol. Sci. Chem.* **1979**, *A13*(5), 633.

33 Y. Shiraishi, M. Nakayama, E. Takagi, T. Tominaga, N. Toshima, *Inorg. Chim. Acta* **2000**, *300–302*, 964.

34 N. Toshima, K. Kushihashi, T. Yonezawa, H. Hirai, *Chem. Lett.* **1989**, 1769.

35 N. Toshima, T. Yonezawa, M. Harada, K. Asakura, Y. Iwasawa, *Chem. Lett.* **1990**, 815.

36 N. Toshima, T. Yonezawa, K. Kushihashi, *J. Chem. Soc. Faraday Trans.* **1993**, *89*(14), 2537.

37 M. Harada, K. Asakura, N. Toshima, *J. Phys. Chem.* **1993**, *97*, 5103.

38 N. Toshima, Y. Wang, *Langmuir* **1994**, *10*, 4574.

39 N. Toshima, *Fine Particles Science and Technology* **1996**, 371.

40 A. Borsla, A.M. Wilhelm, H. Delmas, *Catal. Today* **2001**, *66*, 389.

41 A.B.R. Mayer, J.E. Mark, *Polym. Mater. Sci. Eng.* **1995**, *73*, 220.

42 A.B.R. Mayer, J.E. Mark, *Polymer Bulletin* **1996**, *37*, 683.

43 A.B.R. Mayer, J.E. Mark, *Macromol. Rep.* **1996**, *A33*(suppl. 7/8), 451.

44 A.B.R. Mayer, J.E. Mark, *J. Polym. Sci. Part A: Polym. Chem.* **1997**, *35*(15), 3151.

45 A.B.R. Mayer, J.E. Mark, *Polymers Preprints* **1996**, *37*(1), 459.

46 A.B.R. Mayer, J.E. Mark, R.E. Morris, *Polym. J.* **1998**, *30*(3), 197.

47 Z. Lu, G. Liu, H. Phillips, J.M. Hill, J. Chang, R.A. Kydd, *Nano Lett.* **2001**, *1*(12), 683.

48 M.V. Seregina, L.M. Bronstein, O.A. Platonova, D.M. Chernyshov, P.M. Valetsky, *Chem. Mater.* **1997**, *9*, 923.

49 C.-W. Chen, D. Tano, M. Akashi, *J. Colloid Interface Sci.* **2000**, *225*, 349.

50 C.-W. Chen, M. Akashi, *Polym. Adv. Technol.* **1999**, *10*, 127.

51 M. Adlim, M.A. Bakar, K.Y. Liew, J. Ismail, *J. Mol. Catal. A: Chem.* **2004**, *212*, 141.

52 B.P.S. Chauhan, J.S. Rathore, T. Bandoo, *J. Am. Chem. Soc.* **2004**, *126*, 8493.

53 J.F. Cieben, R.E. Cohen, A. Duran, *Mater. Sci. Eng.* **1999**, *C7*, 45.

54 R.M. Crooks, V. Chechik, B.I. Lemon, III, L. Sun, L.K. Yeung, M. Zhao, in: D.L. Feldheim, C.A. Foss, Jr. (Eds.), *Metal Nanoparticles: Synthesis, Characterization and Applications*. Marcel Dekker, New York, **2002**, pp. 262.

55 R.M. Crooks, M. Zhao, L. Sun, V. Chechik, L.K. Yeung, *Acc. Chem. Res.* **2001**, *34*(3), 181.

56 R.M. Crooks, Y. Niu, *C.R. Chimie* **2003**, *6*, 1049.

57 Y.-M. Chung, H.-K. Rhee, *Catal. Lett.* **2003**, *85*(3/4), 159.

58 Y.-M. Chung, H.-K. Rhee, *J. Mol. Catal. A: Chem.* **2003**, *206*, 291.

59 M. Ooe, M. Murata, T. Mizugaki, K. Ebitani, K. Kaneda, *NanoLett.* **2002**, *2*(9), 999.

60 N. Toshima, T. Takahashi, H. Hirai, *Chemistry Lett.* **1985**, 1245.

61 C. Larpent, E. Bernard, F. Brisse-le-Menn, H. Patin, *J. Mol. Catal. A: Chem.* **1997**, *116*, 277.

62 N.A. Dhas, A. Gedanken, *J. Mater. Chem.* **1999**, *8*(2), 445.

63 J.D. Aiken, III, R.G. Finke, *J. Mol. Catal. A: Chem.* **1996**, *114*, 29.

64 J.D. Aiken, III, R.G. Finke, *Chem. Mater.* **1999**, *11*, 1035.

65 J.D. Aiken, III, R.G. Finke, *J. Am. Chem. Soc.* **1999**, *121*, 8803.

66 S. Mandal, P.R. Selvakannan, D. Roy, R.V. Chaudhari, M. Sastry, *Chem. Commun.* **2002**, 3002.

67 I. P. Stoolyarov, Y. V. Gaugash, G. N. Kryukova, M. N. Vargaftik, I. I. Moiseev, *Russ. Chem. Bull., Int. Ed.* **2004**, *53*(6), 1194.

68 U. R. Pillai, E. Sahle-Demessie, *J. Mol. Catal. A: Chem.* **2004**, *222*, 153.

69 J. Alvarez, J. Liu, E. Román, A. E. Kaifer, *Chem. Commun.* **2000**, 1151.

70 J. Liu, J. Alvarez, W. Ong, E. Román, A. E. Kaifer, *Langmuir* **2001**, *17*, 6762.

71 T. Ueno, M. Suzuki, T. Goto, T. Matsumoto, K. Nagayama, Y. Watanabe, *Angew. Chem. Int. Ed.* **2004**, *43*, 2527.

72 J. Dupont, G. S. Fonseca, A. P. Umpierre, P. F. P. Fichtner, S. R. Teixera, *J. Am. Chem. Soc.* **2002**, *124*, 4228.

73 C. W. Scheeren, G. Machado, J. Dupont, P. F. P. Fichtner, S. R. Texeira, *Inorg. Chem.* **2003**, *42*, 4738.

74 E. T. Silveira, A. P. Umpierre, L. M. Rossi, G. Machado, J. Morais, G. V. Soares, I. J. R. Baumvol, S. R. Teixeira, P. F. P. Fichtner, J. Dupont, *Chem. Eur. J.* **2004**, *10*, 3734.

75 J. Huang, T. Jiang, B. Han, H. Gao, Y. Chang, G. Zhao, W. Wu, *Chem. Commun.* **2003**, 1654.

76 X.-D. Mu, D. G. Evans, Y. Kou, *Catal. Lett.* **2004**, *97*(3-4), 151.

77 J. Le Bras, D. K. Mukherjee, S. González, M. Tristany, B. Ganchegui, M. Moreno-Maas, R. Pleixats, F. Hénin, J. Muzart, *New J. Chem.* **2004**, *28*, 1550.

78 H. Ohde, C. M. Wai, H. Kim, J. Kim, M. Ohde, *J. Am. Chem. Soc.* **2002**, *124*, 4540.

79 B. Yoon, H. Kim, C. M. Wai, *Chem. Commun.* **2003**, *9*, 1040.

80 H. Lindlar, *Helv. Chim. Acta* **1952**, *35*, 446.

81 M. Bartok, J. Czombos, K. Felfoldi, L. Gera, G. Göndos, A. Molnar, F. Notheisz, I. Palinko, G. Wittmann, A. G. Zsigmond, *Stereochemistry of Heterogeneous Metal Catalysis.* John Wiley & Sons, New York, **1985**.

82 S. Bailey, F. King, in: R. A. Sheldon, H. van Bekkum (Eds.), *Fine Chemicals through Heterogeneous Catalysis.* Wiley, New York, **2001**, p. 351.

83 For recent examples, see A. Mastalir, Z. Kiraly, *J. Catal.* **2003**, *220*, 372.

84 A. Mastalir, Z. Kiraly, Gy. Szöllosi, M. Bartok, *Appl. Catal. A* **2001**, *213*, 133.

85 C. Lange, D. De Caro, A. Gamez, S. Stork, J. S. Bradley, W. F. Maier, *Langmuir* **1999**, *15*, 5333.

86 H. Bönnemann, W. Brijoux, K. Siepen, J. Hormes, R. Franke, J. Pollmann, J. Rothe, *Appl. Organomet. Chem.* **1997**, *11*, 783.

87 L. M. Bronstein, D. M. Chernyshov, I. O. Volkov, M. G. Ezernitskaya, P. M. Valetsky, V. G. Matveeva, E. M. Sulman, *J. Catal.* **2000**, *196*, 302.

88 E. Sulman, V. Matveeva, A. Usanov, Y. Kosivtov, G. Demidenko, L. Bronstein, D. Chernyshov, P. Valetsky, *J. Mol. Catal. A: Chem.* **1999**, *146*, 265.

89 M. M. Telkar, C. V. Rode, R. V. Chaudhari, S. S. Joshi, A. M. Nalawade, *Appl. Catal. A* **2004**, *273*, 11.

90 J. L. Pellagatta, C. Blandy, V. Collière, R. Choukroun, B. Chaudret, P. Cheng, K. Philippot, *J. Mol. Catal. A: Chem.,* **2002**, *178*, 55.

91 H. G. Niessen, A. Eichhorn, K. Woelk, J. Bargon, *J. Mol. Catal. A: Chem.* **2002**, *182-183*, 463.

92 K. Weissermel, H. J. Arpe, *Industrial Organic Chemistry*, 2nd edn. VCH, New York, **1993**, p. 343.

93 J. A. Moulijn, P. W. N. M. van Leeuwen, R. A. van Santen (Eds.), *An Integrated Approach to Homogeneous, Heterogeneous and Industrial Catalysis.* Elsevier, Amsterdam, **1995**.

94 A. Roucoux, Stabilized noble metal nanoparticles: An unavoidable family of catalysts for arene derivatives hydrogenation, in: C. Copéret, B. Chaudret (Eds.), *Topics in Organometallic Chemistry.* Springer, **2005**, Vol. 16, p. 261.

95 J. A. Widegren, R. G. Finke, *J. Mol. Catal. A: Chemical* **2003**, *187*, 207.

96 K. R. Januszkiewicz, H. Alper, *Organometallics* **1983**, *2*, 1055.

97 K. R. Januszkiewicz, H. Alper, *Can. J. Chem.* **1984**, *62*, 1031.

98 K. Nasar, F. Fache, M. Lemaire, J. C. Beziat, M. Besson, P. Gallezot, *J. Mol. Catal.* **1993**, *78*, 257.

99 T. Q. Hu, B. R. James, S. J. Rettig, C. L. Lee, *Can. J. Chem.* **1997**, *75*, 1234.

100 T. Q. Hu, B. R. James, S. J. Rettig, C. L. Lee, *J. Pulp. Pap. Sci.* **1997**, *23*, 153.

101 B. R. James, Y. Wang, C. S. Alexander, T. Q. Hu, *Chem. Ind.* **1998**, *75*, 233.

102 V. Mévellec, A. Roucoux, E. Ramirez, K. Philippot, B. Chaudret, *Adv. Synth. Catal.* **2004**, *346*, 72.

103 R. W. Albach, M. Jautelat, German Patent DE 19807995, Bayer AG **1999**.

104 T. Q. Hu, B. R. James, C. L Lee, *J. Pulp. Pap. Sci.* **1997**, *23*, 200.

105 G. S. Fonseca, A. P. Umpierre, P. F. P. Fichtner, S. R. Teixeira, J. Dupont, *Chem. Eur. J.* **2003**, *9*, 3263.

106 G. S. Fonseca, E. T. Silveira, M. A. Gelesky, J. Dupont, *Adv. Synth. Catal.* **2005**, *347*, 847.

107 X. D. Mu, J. Q. Meng, Z. C. Li, Y. Kou, *J. Am. Chem. Soc.* **2005**, *27*, 127.

108 R. G. Finke, in: D. L. Feldheim, C. A. Foss, Jr. (Eds.), *Metal Nanoparticles: Synthesis, Characterization and Applications.* Marcel Dekker, New York, **2002**, Chapter 2, pp. 17.

109 R. J. Bonilla, P. G. Jessop, B. R. James, *Chem. Commun.* **2000**, 941.

110 H. Nagahara, M. Ono, M. Konishi, Fukuoka, *Appl. Surf. Sci.* **1997**, *121/122*, 448.

111 W. Yu, Y. Wang, H. Liu, W. Zheng, *J. Mol. Catal. A: Chemical* **1996**, *112*, 105.

112 W. Yu, M. Liu, H. Liu, X. Ma, Z. Liu, *J. Colloid Interface Sci.* **1998**, *238*, 439.

113 H. D. Ding, D. L. Gin, *Chem. Mater.* **2000**, *12*, 22.

114 A. Spitaleri, P. Pertici, N. Scalera, G. Vitulli, M. Hoang, T. W. Turney, M. Gleria, *Inorg. Chim. Acta* **2003**, 61.

115 G. S. Fonseca, A. P. Umpierre, P. F. P. Fichtner, S. R. Teixeira, J. Dupont, *Chem. Eur.* **2003**, *9*, 3263.

116 G. S. Fonseca, J. D. Scholten, J. Dupont, *Synlett* **2004**, *9*, 1525.

117 Y. Izumi, *Adv. Catal.* **1983**, *32*, 215.

118 Y. Orito, S. Imai, S. Niwa, G.-H. Nguyen, *J. Synth. Org. Chem. Jpn.* **1979**, *37*, 173.

119 Y. Orito, S. Imai, S. Niwa, *J. Chem. Soc. Jpn.* **1979**, 1118.

120 Y. Orito, S. Imai, S. Niwa, *J. Chem. Soc. Jpn.* **1980**, 670.

121 S. Niwa, S. Imai, Y. Orito, *J. Chem. Soc. Jpn.* **1982**, 137.

122 A. Baiker, in: D. E. De Vos, I. F. J. Vankelecom, P. A. Jacobs (Eds.), *Chiral Catalyst Immobilization and Recycling.* Wiley-VCH, Weinheim, **2000**, pp. 155.

123 P. B. Wells, R. P. K. Wells, in: D. E. De Vos, I. F. J. Vankelecom, P. A. Jacobs (Eds.), *Chiral Catalyst Immobilization and Recycling.* Wiley-VCH, Weinheim, **2000**, pp. 123.

124 M. Studer, H.-U. Blaser, C. Exner, *Adv. Synth. Catal.* **2003**, *345*, 45 and references cited therein.

125 P. B. Wells, K. E. Simons, J. A. Slipszenko, S. P. Griffiths, D. F. Ewing, *J. Mol. Catal.* **1999**, *146*, 159.

126 H. U. Blaser, H. P. Jallet, *J. Mol. Catal.* **1991**, *68*, 215.

127 X. Zuo, H. Liu, M. Liu, *Tetrahedron Lett.* **1998**, *39*, 1941.

128 B. Torok, K. Felfoldi, G. Szakonyi, K. Balazsik, M. Bartok, *Catal. Lett.* **1998**, *52*, 81.

129 H. Bönnemann, G. A. Braun, *Angew. Chem. Int. Ed. Engl.* **1996**, *35*, 1992.

130 H. Bönnemann, G. A. Braun, *Chem. Eur. J.* **1997**, *3*, 1200.

131 J. U. Köhler, J. S. Bradley, *Catal. Lett.* **1997**, *45*, 203.

132 J. U. Köhler, J. S. Bradley, *Langmuir* **1998**, *14*, 2730.

133 Y. Huang, J. Chen, H. Chen, R. Li, Y. Li, L. Min, X. Li, *J. Mol. Catal.* **2001**, *170*, 143.

134 X. Zuo, H. Liu, C. Yue, *J. Mol. Catal.* **1999**, *147*, 63.

135 X. Zuo, H. Liu, D. Guo, X. Yang, *Tetrahedron* **1999**, *55*, 7787.

136 X. Zuo, H. Liu, J. Tian, *J. Mol. Catal.* **2000**, *157*, 217.

137 J. Zhang, X. Yan, H. Liu, *J. Mol. Catal.* **2001**, *175*, 125.

138 P. J. Collier, J. A. Iggo, R. Whyman, *J. Mol. Catal.* **1999**, *146*, 149.

139 V. Mévellec, C. Mattioda, J. Schulz, J. P. Rolland, A. Roucoux, *J. Catal.* **2004**, *225*, 1.

10
Kinetics of Homogeneous Hydrogenations:
Measurement and Interpretation

Hans-Joachim Drexler, Angelika Preetz, Thomas Schmidt, and Detlef Heller

10.1
Introduction

Recently, the results of kinetic measurements have been summarized for homogeneous hydrogenations with transition metal complexes in a review [1]. Essential new results of kinetic investigations leading to the completion of hitherto existing ideas regarding the reaction mechanism of particular catalyses are represented in the respective chapters of this book, and shall not be repeated here. Rather, this chapter will introduce the kinetic treatment of reaction sequences with pre-equilibria typical for catalyses, together with the analysis and interpretation of Michaelis-Menten kinetics, the monitoring of hydrogenations, and a discussion of possible problems, with selected examples.

Kinetic investigations deliver quantitative correlations regarding the concentration–time dependence of the participating reactants, and therefore serve as the major methodical approach in the elucidation of reaction mechanisms. A knowledge of funded mechanistic ideas opens the possibility of an aimed manipulation of activity and selectivity, respectively, which are important parameters of catalyses. As "operating values", pressure and temperature – as well as the concentration of particular reaction partners – are available. However, when scaling-up from a laboratory standard to an industrial application, kinetic results are indispensable. Moreover, kinetics provides essential indications about the nature of the actual catalyst and the distinction between homogeneous and heterogeneous catalysis. This objective has been investigated more intensively during the past few years, partly with surprising results, and naturally plays an important role when transition-metal complexes meet with hydrogen as reducing agent [2]. Thereby, the problem does not lie in the kinetics as the method. ("In the kinetic approach no frontiers exist today between homogeneous, enzymatic, and heterogeneous catalysis. There is a consistent science which permits the definition of useful and efficient rate laws describing sequences of elementary steps." [3])

In spite of these capabilities of kinetics it is necessary to emphasize here that, in principle, it is not possible to *prove* that a reaction mechanism occurs only by

The Handbook of Homogeneous Hydrogenation.
Edited by J. G. de Vries and C. J. Elsevier
Copyright © 2007 WILEY-VCH Verlag GmbH & Co. KGaA, Weinheim
ISBN: 978-3-527-31161-3

using kinetic investigations! Rather, it is the nature of kinetics to describe quantitative dependences between reaction partners and thus to exclude specific reaction sequences. This model discrimination, however, does not principally allow the favoring of one reaction mechanism among a few remaining possibilities [4]. Furthermore, it is possible that formal-kinetically equivalent reaction sequences are chemically different and hence are not to be distinguished by merely applying kinetic methods [5]. Only additional findings such as the detection (or rather the isolation) of intermediates, the interpretation of isotope labeling studies, as well as computational chemistry, allow descriptions to be made of experimental results which are consistent in the form of a closed catalytic cycle – the reaction mechanism most probable on the basis of the existing indications.

There is, however, no doubt about the significance of kinetics for catalysis as the following statements indicate:

- "Kinetic measurements are essential for the elucidation of any catalytic mechanism since catalysis, by definition and significance, is purely a kinetic phenomenon" [6].
- "Asymmetric catalysis is four-dimensional chemistry. Simple stereochemical scrutiny of the substrate or reagent is not enough. The high efficiency that the reactions provide can only be achieved through a combination of both an ideal three-dimensional structure (x,y,z) and suitable kinetics (t)" [7].
- "Carefully determined conversion–time diagrams, *in-situ* spectroscopic studies and, if possible, kinetic time laws belong to the fundamentals of catalysis research and are prerequisites for a mechanistic understanding" [8].

Although the outstanding relevance of kinetics is clear, very few publications relate to in-depth kinetic analyses. The reasons for this are complex, and some of these are detailed below:

- The field of homogeneous catalysis deals primarily with the organometallic complex catalysis, besides organocatalysis, which is at present experiencing a renaissance [9]. One problem of most of the transition-metal complexes used today is a need for anaerobic reaction conditions, and this is why many conventional possibilities of kinetic investigations are restricted in their application.
- A further problem results from the catalysis itself. Only the permanent repetition of a catalytic cycle demonstrates clearly the advantage of catalysis over a simple stoichiometric reaction. A good catalyst must be very effective, leading to a desired product with a high turnover number (TON, defined as moles of substrate per mole of catalyst) and turnover frequency (TOF, defined as TON per unit time) [10–12]. Because of the large substrate:catalyst ratio, however, detailed kinetic investigations are complicated as the interesting intermediates of the catalytic cycle must be detected and quantified, beside large quantities of substrate and/or product. In addition, in a catalytic cycle, the amount of transition-metal complex is shared by several intermediates. In the case of stereoselective catalyses, the number of relevant intermediates might also easily be multiplied [13]. Furthermore, one condition of catalytic reaction se-

quences with transition-metal complexes must not be neglected, namely that intermediates can relatively easily be transposed into one another, mostly reversibly. Due to disadvantageous equilibrium positions, the intermediates might not be detectable, even under stationary catalytic conditions [14].

- In almost every case differential equations for the quantitative description of the time dependence of particular species resulting from a catalytic cycle cannot be solved directly. This requires approximate solutions to be made, such as the equilibrium approximation [15], the Bodenstein principle [16], or the more generally valid steady-state approach [17]. A discussion of differences and similarities of different approximations can be found in [18].

- Another problem arises from the fact that good kinetic studies in the field of homogeneous catalysis require not only complex-chemical and methodical experience, but also a solid knowledge of physical chemistry. Yet, this additional requirement is seldom requested at a time when financial pressure on research is steadily growing [19].

10.2
The Basics of Michaelis-Menten Kinetics

Most catalytic cycles are characterized by the fact that, prior to the rate-determining step [18], intermediates are coupled by equilibria in the catalytic cycle. For that reason Michaelis-Menten kinetics, which originally were published in the field of enzyme catalysis at the start of the last century, are of fundamental importance for homogeneous catalysis. As shown in the reaction sequence of Scheme 10.1, the active catalyst first reacts with the substrate in a pre-equilibrium to give the catalyst–substrate complex [20]. In the rate-determining step, this complex finally reacts to form the product, releasing the catalyst.

Under isobaric conditions ($k_2 = k'_2 \cdot [H_2]$), many hydrogenations exactly follow this model. The classical example is the asymmetric hydrogenation of prochiral dehydroamino acid derivatives with Rh or Ru catalysts [21].

The so-called Michaelis-Menten equation (Eq. 1) [22] follows independently of the approximation chosen to solve the differential equation resulting from Scheme 10.1. Its derivation in detail can, for example, be found in [23].

$$V = \frac{dP}{dt} = \frac{k_2 \cdot [E]_0 \cdot [S]}{K_M + [S]} = \frac{V_{sat} \cdot [S]}{K_M + [S]} \qquad (1)$$

$$\text{with (a) } K_M = \frac{k_{-1}}{k_1} = \frac{[E] \cdot [S]}{[ES]}, \text{ (b) } K_M = \frac{k_{-1} + k_2}{k_1} = \frac{[E] \cdot [S]}{[ES]},$$

$$\text{(c) } K_M = \frac{k_2}{k_1} = \frac{[E] \cdot [S]}{[ES]}$$

where (a) is the equilibrium approximation; (b) is the steady-state approach; and (c) is the irreversible formation of the substrate complex ($k_{-1} = 0$).

$$E + S \underset{k_{-1}}{\overset{k_1}{\rightleftharpoons}} ES \xrightarrow{k_2} P + (E)$$

Scheme 10.1 Reaction sequence of the simplest case of Michaelis-Menten kinetics. E = catalyst; S = substrate; ES = catalyst–substrate complex; P = product; k_i = rate constants.

The Michaelis-Menten equation is characterized by two constants:
- the rate constant for the reaction of the catalyst–substrate complex to the product (k_2); and
- the Michaelis constant (K_M).

A more detailed examination shows that, in case of equilibrium approximation, the value of K_M corresponds to the inverse stability constant of the catalyst–substrate complex, whereas in the case of the steady-state approach the rate constant of the (irreversible) product formation is additionally included. As one cannot at first decide whether or not the equilibrium approximation is reasonable for a concrete system, care should be taken in interpreting K_M-values as inverse stability constants. At best, the reciprocal of K_M represents a lower limit of a "stability constant"! In other words, the stability constant quantifying the pre-equilibrium can never be smaller than the reciprocal of the Michaelis constant, but can well be significantly higher.

The Michaelis constant has the dimension of a concentration and characterizes – independently of the method of approximation – the substrate concentration at which the ratio of free catalyst to catalyst–substrate complex equals unity. At this point, exactly one-half of the catalyst is complexed by the substrate. Likewise, one finds that at a value of $[S] = 10\,K_M$, the ratio of $[E]/[ES]$

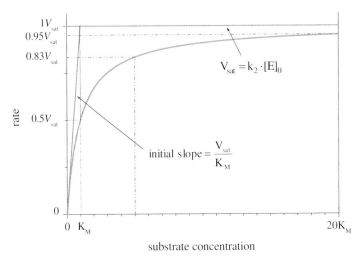

Fig. 10.1 Product formation rate as a function of substrate concentration (Eq. (1)).

equals 0.1, which means that virtually 91% of the initial catalyst ($[E_0]$) is present as substrate complex. The product formation rate is shown schematically as a function of substrate concentration in Figure 10.1.

Because of the complexity of biological systems, Eq. (1) as the differential form of Michaelis-Menten kinetics is often analyzed using the initial rate method. Due to the restriction of the initial range of conversion, unwanted influences such as reversible product formation, effects due to enzyme inhibition, or side reactions are reduced to a minimum. The major disadvantage of this procedure is that a relatively large number of experiments must be conducted in order to determine the desired rate constants.

An analysis of the product formation is, in principle, not limited to the initial range of rates, however. Laidler investigated the problem of the validity range of the Michaelis-Menten equation as a function of time under the assumption of steady-state conditions for the catalyst–substrate complex [24]. As long as either condition shown in Eq. (2) is fulfilled – by choice of experimental conditions it is usually $[S]_0 \gg [E]_0$ – Eq. (3) applies up to high conversions for hydrogenations, which corresponds to Eq. (1) [23]. In fact, each point of a hydrogenation curve can be understood as an "initial rate experiment". By analyzing a hydrogenation over a wide range of conversion, a large number of initial rate experiments can be omitted. "Reaction progress kinetic analysis" as a powerful methodology was very recently described by Blackmond in a highly recommended review [25].

$$[S]_0 \gg [E]_0 \quad [E]_0 \gg [S]_0 \quad k_{-1} + k_2 \gg k_1 \cdot [E]_0 \quad k_{-1} + k_2 \gg k_1 \cdot [S]_0 \qquad (2)$$

$$\frac{d[P]}{dt} = \frac{k_2 \cdot [E]_0 \cdot ([S]_0 - n_{H_2})}{K_M + ([S]_0 - n_{H_2})} \quad (n_{H_2} = \text{hydrogen consumption}) \qquad (3)$$

There are two limiting cases of Michaelis-Menten kinetics. Beginning from Eq. (1) at high substrate excesses (or very small Michaelis constants) Eq. (4a) results. This corresponds to a zero-order reaction with respect to the substrate, the rate of product formation being independent of the substrate concentration. In contrast, very low substrate concentrations [26] (or large Michaelis constants) give the limiting case of first-order reactions with respect to the substrate, Eq. (4b):

$$\text{(a)} \ V = \frac{d[P]}{dt} = k_2 \cdot [E]_0 = V_{sat} \ [27] \quad \text{(b)} \ V = \frac{d[P]}{dt} = \frac{k_2 \cdot [E]_0}{K_M} \cdot [S] = k_{obs} \cdot [S] \quad (4)$$

In Figure 10.1, it can be seen that even with substrate excesses of $[S] = 20 \ K_M$, the saturation range is not yet reached. Conversely, the data in Figure 10.2 indicate that even for very small substrate concentrations ($[S] = 0.05 \ K_M$) the limiting case for the first-order reaction – when the rate is directly proportional to the substrate concentration – is not identical with the values from Eq. (1).

Since methods to analyze Michaelis-Menten kinetics have been sufficiently described in the literature [23, 28], this problem is discussed only briefly at this point.

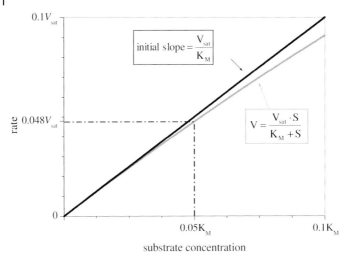

Fig. 10.2 Comparison of Eq. (1) (upper line) with the limiting case of a first-order reaction Eq. (4b) (lower line) for very low substrate concentrations.

In principle, the differential form (Eq. (1)), as well as the integrated form (Eq. (8)), can be used. The differential form of the Michaelis-Menten equation is applied in many cases, since differential values (e.g., flow meter or heat flow data) are often available; by contrast, time-dependent substrate or product concentrations (or proportional quantities) can easily be differentiated numerically.

Initial values for a non-linear fit of Eq. (1) can be achieved by linearizations. Most conventional linearizations result from the transformation of the Michaelis-Menten equation, and are plotted according to:

$$\text{Lineweaver-Burk [29]:} \quad \frac{1}{V} = \frac{K_M}{V_{sat}} \cdot \frac{1}{[S]} + \frac{1}{V_{sat}} \quad \text{plot}: 1/V \text{ versus } 1/[S] \tag{5}$$

$$\text{Eadie-Hofstee [30]:} \quad V = V_{sat} - \frac{V}{[S]} \cdot K_M \quad \text{plot}: V \text{ versus } V/[S] \tag{6}$$

$$\text{Hanes [31]:} \quad \frac{[S]}{V} = \frac{K_M}{V_{sat}} + \frac{1}{V_{sat}} \cdot [S] \quad \text{plot}: [S]/V \text{ versus } [S] \tag{7}$$

An analysis of the influence of errors shows clearly that the double-reciprocal plot according to Lineweaver-Burk [32] is the least suitable. "Although it is by far the most widely used plot in enzyme kinetics, it cannot be recommended, because it gives a grossly misleading impression of the experimental error: for small values of v small errors in v lead to enormous errors in $1/v$, but for large values of v the same small errors in v lead to barely noticeable errors in $1/v$" [23]. Due to the error distribution, that is much more uniform, the plot according to Hanes (Eq. (7)), is the most favored.

The integrated form of the simple Michaelis-Menten kinetics (Eq. (8)), is most suitable to analyze the time-dependent progressive substrate conversion or the corresponding product formation.

$$\frac{1}{t} \cdot \ln \frac{[S]_0}{[S]_t} = -\frac{[P]_t}{K_M \cdot t} + \frac{V_{sat}}{K_M} \tag{8}$$

A more detailed discussion of further possibilities for the analysis of Eq. (1) can be found in [23].

In homogeneous catalysis, the quantification of catalyst activities is commonly carried out by way of TOF or half-life. From a kinetic point of view, the comparison of different catalyst systems is only reasonable if, by giving a TOF, the reaction is zero order or, by giving a half-time, it is a first-order reaction. Only in those cases is the quantification of activity independent of the substrate concentration utilized!

As derived above, there are two limiting cases of Michaelis-Menten kinetics, which is often the basis of homogeneous catalysis. Depending upon the substrate concentration, a reaction of either first or zero order is possible as a limiting case. For hydrogenations of various substrates involving pre-equilibria, reaction orders of unity or zero have been reported for the substrate. Data relating to the kinetics of homogeneous hydrogenations with transition metal complexes before the year 2000 can be found in reference [1], and more recent examples in reference [33] (olefins: [21c, 33a–h]; ketones: [33i–k]; imines: [33l]; alkynes: [33m]; nitro groups: [33m]; N-hetero aromatic compounds: [33n, o]; CO_2: [33p]).

If a reaction that must be investigated follows a reaction sequence as in Scheme 10.1, and if the reaction order for the substrate equals unity, it means that (with reference to Eq. (4b)), the observed rate constant (k_{obs}) is a complex term. Without further information, a conclusion about the single constants k_2 and K_M is not possible. Conversely, from the limiting case of a zero-order reaction, the Michaelis constant cannot be determined for the substrate. For particular questions such as the reliable comparison of activity of various catalytic systems, however, both parameters are necessary. If they are not known, the comparison of catalyst activities for given experimental conditions can produce totally false results. This problem is described in more detail for an example of asymmetric hydrogenation (see below).

10.3
Hydrogenation From a Kinetic Viewpoint

10.3.1
Measurement of Concentration–Time Data and Possible Problems

There exists a multitude of possibilities to monitor hydrogenations in various pressure ranges. In principle, isochoric and isobaric techniques are feasible. In the latter case, the kinetics allows simplification because the concentration of the reaction partner, hydrogen, is constant. The classical method for measuring concentration–time data is to take samples from the reaction vessel during the hydrogenation, and then to analyze those samples via common methods such as gas chromatography (GC), high-performance liquid chromatography (HPLC), and nuclear magnetic resonance (NMR). In so doing, the sampling over various temperature and pressure ranges can be automated, as can the analysis. The advantage of this method is that any eventually occurring intermediates are detected individually as a function of time, and thus are accessible for kinetic interpretation. The disadvantage, however, is the major analytical effort required. For rapid reactions this method is also hardly appropriate. Moreover, it is sometimes difficult to stop the reaction immediately after sampling, this being a problem which is often underestimated.

A significantly more elegant solution is an *in-situ* monitoring of hydrogenations, as this advantageously provides a large amount of data available for analyses.

Both integrally and differentially measured values can be detected *in situ*. In the first case, substrate- or product-specific signals, or directly proportional quantities, are suitable. Hence, Noyori describes the monitoring of a ketone hydrogenation via the intensity of the infra-red (IR) carbonyl stretching band at $1750\,\text{cm}^{-1}$ [21c]. To register the hydrogen consumption of a hydrogenation, a product-proportional concentration as a function of time is monitored. However, the measurement of rates – for example using flow meters or via a heat flow with a calorimeter – represents a typical differential method.

In those cases where concentrations are not measured directly, the problem of "calibration" of the *in-situ* technique becomes apparent. An assurance must be made that no additional effects are registered as systematic errors. Thus, for an isothermal reaction, calorimetry as a tool for kinetic analysis, heat of mixing and/or heat of phase transfer can systematically falsify the measurement. A detailed discussion of the method and possible error sources can be found in [34].

High-throughput methods for catalyst screening and optimization, as described in the literature even for hydrogenations [35], are not suitable for kinetic analyses in most cases.

10.3.1.1 Monitoring of Hydrogenations via Hydrogen Consumption

One method, which is still used frequently to follow hydrogenations *in situ*, is the registration of hydrogen consumption. There is a multitude of solutions that can be simply subdivided into normal-pressure and high-pressure measurements. Due to common isobaric reaction conditions the hydrogen concentration is constant, which simplifies kinetics. An isochoric mode of operation is not advisable because of the complexity of the measurement. In fact, the decreasing hydrogen concentration in solution during the course of the hydrogenation must also be taken into account.

For hydrogen, deviations from the ideal gas law must be considered only at higher pressures [36]. Nonetheless, the virial equation allows the amount of hydrogen to be calculated, for example in a reservoir of known volume, by applying Eq. (9). By using mass balances – based on the initial pressure or cumulatively on the previous value – hydrogen consumption can be determined with accuracy [37]. The problem of such measurements rather lies in a possible temperature gradient between the reservoir and the reaction vessel.

$$n_{H_2} = \frac{\text{reservoir volume}}{\text{real molar volume}} \quad \text{real molar volume} = \frac{R \cdot T}{p} + B + \left(\frac{C - B^2}{R \cdot T}\right) \quad (9)$$

where R=gas constant, T=temperature, p=pressure, and B and C=virial coefficients.

For flow rate measurements the volume or, more conveniently, the mass flow is suitable. In the first case a pressure- and temperature-dependent calibration is necessary if the gas does not show ideal behavior. This also applies for heat conductivity as the measured quantity often used in flow meters. Currently, real pressure- and temperature-independent measurement of a hydrogen mass flow of a hydrogenation remains problematic on the laboratory scale, at least for low substrate concentrations.

By contrast, the measurement of the hydrogen consumption under normal pressure is relatively simple. The elementary structure of many such measuring devices is similar, and is based principally on the fact that the pressure drop is balanced by reduction in the reaction volume or by supply of the consumed gas, thus ensuring isobaric conditions. An appropriate device for monitoring major gas consumptions is described in [38].

For hydrogenations under normal pressure and isobaric conditions, we use a device which registers gas consumption automatically (Fig. 10.3). Possible error sources resulting from such gas consumption measurements and possibilities of their minimization will be discussed.

The basic principle to realize isobaric conditions for the hydrogenation apparatus shown in Figure 10.3 is to change the volume of the closed reaction space via a (not commercially available) gas-tight syringe in order to ensure a permanent atmospheric pressure as the reference. For this purpose, a sensible pressure sensor registers the pressure drop caused by hydrogen consumption in the closed reaction system. Using a processor-controlled stepping motor axis, the piston of the syringe is depressed until the initial pressure is reached. At this point the position of the piston is registered as a function of time and finally visualized as the hydrogenation curve. (The same arrangement also allows the automatic registration of gas formation.)

This method, although being used analogously in other devices, incorporates a number of principal error sources. These result substantially from transport phenomena, vapor pressure of the solvent, gas solubility, and tempering problems. Particular points, together with possible means of their minimization, will be discussed in the following section.

One problem encountered when monitoring gas-consuming reactions is the influence of transport phenomena. The reaction partner hydrogen must be transported to the catalyst, and thereby it should penetrate the gas–liquid inter-

Fig. 10.3 Normal pressure hydrogenation device for the automatic registration of hydrogen consumption under isobaric conditions.

face at a distinctly higher rate than it is consumed by the hydrogenation. Only if such a regime holds it can be guaranteed that the detected effect can be interpreted as being exclusively kinetic.

Blackmond et al. investigated the influence of gas-liquid mass transfer on the selectivity of various hydrogenations [39]. It could be shown – somewhat impressively – that even the pressure-dependence of enantioselectivity of the asymmetric hydrogenation of a-dehydroamino acid derivatives with Rh-catalysts (as described elsewhere [21 b]) can be simulated under conditions of varying influence of diffusion! These results demonstrate the importance of knowing the role of transport phenomena while monitoring hydrogenations.

Several possibilities exist to determine the influence of transport phenomena. The measurement of gas consumption in dependence on the interfacial area, the physical absorption coefficient, the rate of a chemical reaction following the absorption, and the concentration gradient (as the driving force of the absorption) allows decisions to be made on which regime is, in fact, in existence [40].

For the rate of physical absorption of a gas into a liquid without subsequent chemical reaction, Eq. (10) is valid.

$$\frac{d[C]}{dt} = k_L \cdot a \cdot ([C^*] - [C]) \quad \text{or} \quad [C] = [C^*] \cdot (1 - e^{-k_L \cdot a \cdot t}) \tag{10}$$

where k_L = physical mass-transfer coefficient (liquid side), a = interfacial area, $[C^*]$ = gas concentration in solution at time $t\infty$ (gas solubility), and $[C]$ = gas concentration in solution at time t.

The analysis of gas absorption proceeding exponentially under experimental conditions provides the gas solubility $[C^*]$ and the value of $k_L \cdot a$. As a rule of thumb, this value should be approximately ten-fold larger than the rate of a subsequent chemical reaction in order to eliminate diffusion influences on the latter reaction [41].

The often-applied method of determining the dependence of initial rate on stirring speed must be treated with caution, for two reasons. On the one hand, the initial rate can be lower than at higher conversions due to induction periods [42], and on the other hand an increase in stirring speed does not enforce a proportionally higher interfacial area.

The following procedure has been approved as being straightforward (also see [43]). A zero-order dependence is achieved by monitoring a reaction in the range of diffusion control. The rate is determined only by the constant concentration gradient in the interfacial area. The systematic investigation of whether diffusion influences hydrogenations is appropriate only if they also follow zero order, but in the range of kinetic control. An example of this is the catalytic hydrogenation of dienes as COD (cycloocta-1,5-diene) or NBD (norborna-2,5-diene) with cationic rhodium(I) chelates. Up to high conversions this reaction proceeds in the saturation range of Michaelis-Menten kinetics, and hence as a zero-order reaction. The pseudo-rate constant $k_{obs} = k'_2 \cdot [H_2] \cdot [E]_0$ is a linear function of the initial catalyst concentration. A continuous increase of the employed catalyst concentration ($[E]_0$) under given experimental conditions (reactor geometry, stirring speed, stirrer size) leads to a straight line, and the hydrogen consumption is independent of the predominating regime (kinetics versus diffusion). Plotting the slopes of the straight lines as a function of the catalyst concentration provides information about the limitations of the regime, which is exclusively controlled by kinetics. Figure 10.4 illustrates the hydrogenation curves of the catalytic hydrogenation of NBD with [Rh(Ph-β-glup-OH)NBD]BF$_4$ (Ph-β-glup-OH = phenyl 2,3-bis(O-diphenylphosphino)-β-D-glyco-pyranoside).

The plot of measured rates as a function of the initial catalyst concentration is shown in Figure 10.5. For the range from 0 to ca. 12 mL min^{-1} the straight line passing through the origin proves the direct proportionality between rate and catalyst concentration. In other words, the hydrogen concentration in solution (gas solubility) for the mentioned range of rates is constant, and it is measured in the kinetically controlled range. As the figure indicates, rates of hydrogen consumption of 30 mL min^{-1} are indeed nonproblematic with regard to the registration. However, for rates greater than 12 mL min^{-1}, gas consumption under the given experimental conditions is increasingly determined by transport phenomena. Because of the rising influence of diffusion, the bulk concentration of hydrogen in solution continuously decreases below the value of the hydrogen solubility. The hydrogenations are slower than would be expected for the kinetically controlled range.

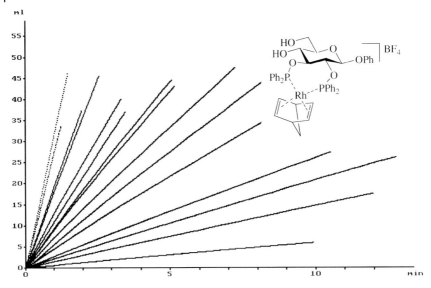

Fig. 10.4 Hydrogenation curves of catalytic NBD hydrogenations (each at least 2.0 mmol) with [Rh(Ph-β-glup-OH)NBD]BF$_4$ at varying catalyst concentrations (0.0025, 0.005, 0.0075, 0.01, 0.015, 0.02, 0.025, 0.03, 0.035, 0.04, 0.05, 0.08, 0.1, 0.15, and 0.2 mmol) each in 15.0 mL MeOH at 25.0 °C and 1.013 bar total pressure.

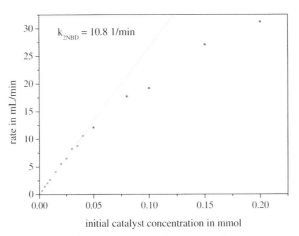

Fig. 10.5 Rate of gas consumption from Figure 10.4 as a function of initial catalyst concentration [E]$_0$.

One further source of error is that of vapor pressure of the solvent. Whilst this plays only a minor role at higher hydrogen pressures, its neglect for hydrogenations under normal pressure is a problem that is often underestimated. Figure 10.6 illustrates the vapor pressure of various solvents often used in hydrogenations as a function of temperature [44].

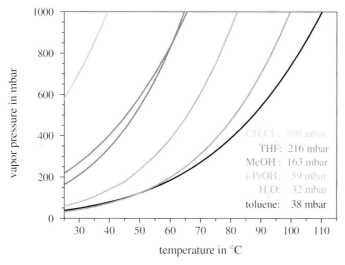

Fig. 10.6 Vapor pressure of CH_2Cl_2, THF, MeOH, i-PrOH, H_2O and toluene as a function of temperature. The concrete values refer to 25.0 °C.

Although in the case of methylene chloride under normal pressure more than one-half of the gas phase consists of solvent vapor (57%), in the case of toluene and water this share amounts to only ca. 3–4% of the total pressure. In order to compare activities in various solvents at the same hydrogen pressure above the reaction solution, besides a different gas solubility for the solvents (i.e., the hydrogen concentration in solution), a different partial pressure of hydrogen must be taken into account.

Another problem results from high vapor pressures of relatively low-boiling solvents. With regard to the dependence on reactor geometry, it can take some time for the vapor pressure of the solvent to become established in the gas phase of the closed system. As this equilibration of vapor pressure provides a positive volume contribution (the pressure above the reaction solution increases in a closed system), measured gas consumptions can be considerably falsified not only as a function of time but also in respect of the overall balance! One way to avoid this problem is to separate the gas phase above the reaction solution from the gas in the measuring burette by using a tempered bubble counter (cf. Fig. 10.3).

A further problem is constituted by the different solubilities of hydrogen in the conventional solvents used for hydrogenations. In Table 10.1 (column 2), data are listed of gas solubility (expressed as mole fraction) $[x_{H_2} = n_{H_2}/(n_{solvent} + n_{H_2})]$ of various solvents at 25.0 °C and 1.013 bar H_2 partial pressure [45].

Because very small mole fraction solubilities correspond in practice to the molar ratio [45 a], the values can (considering the molar volume and density of the solvent) be easily transformed into hydrogen concentrations (see Table 10.1,

Table 10.1 Hydrogen solubilities in various solvents at 25.0 °C [46].

Solvent	H$_2$-solubility [mole fraction x at 1.013 bar]	Mol H$_2$ in 1 L solvent [1.013 bar H$_2$]	Mol H$_2$ "effective" in 1 L solvent [1.013 bar total pressure]
THF	0.000270	3.3291×10^{-3}	2.62×10^{-3}
MeOH	0.000161	3.9752×10^{-3}	3.33×10^{-3}
i-PrOH	0.000266	3.4742×10^{-3}	3.27×10^{-3}
H$_2$O	0.0000141	7.80576×10^{-4}	7.56×10^{-4}
Toluene	0.000317	2.9576×10^{-3}	2.85×10^{-3}

column 3). In hydrogenations under normal pressure, however, the different vapor pressure of the solvent, by which the relevant hydrogen partial pressure is reduced, must be taken into account (Fig. 10.6). Consideration of the vapor pressure of the solvent at a total pressure of 1.013 bar above the reaction solution leads to an "effective" gas solubility – that is the actually interesting hydrogen concentration in solution (see Table 10.1, column 4). The results show that under equal conditions (25.0 °C and 1.013 bar total pressure above the reaction solution), the hydrogen concentration in solution differs markedly for different solvents. The solvents THF and *i*-PrOH indeed show a similar hydrogen solubility (Table 10.1, column 2), despite differing in the "effective" hydrogen concentration (Table 10.1, column 4) by ca. 20%. In contrast, the solvents H$_2$O and toluene exhibit approximately the same vapor pressure above the reaction solution (Fig. 10.6), yet the "effective" hydrogen concentration in solution differs by a factor of 3.7. In the first case, variation in vapor pressure is the cause for such behavior, but in the second case it is the variation in gas solubility.

For meaningful comparisons of the activity of catalysts in various solvents under seemingly equal conditions these factors must, of course, be considered.

In order to determine the reaction order in hydrogen of a homogeneously catalyzed hydrogenation under isobaric conditions, the variation of partial pressure is an essential precondition. Commonly, hydrogen/inert gas mixtures are used, yet the change in composition of the gas mixture (the share of H$_2$ is reduced due to consumption in the hydrogenation) is generally neglected. However, this may lead to a dependence on the volume of the gas phase and, potentially, to a major systematic error. By contrast, the method described in the following section permits the use of isobaric conditions by varying the partial pressure.

While the gas phase above the reaction solution contains the reactive gas at a chosen concentration (obtained by dissolution with an inert gas such as argon; H$_2$/Ar gas mixtures are available commercially), pure hydrogen is arranged in the gas burette. Mixing of the gas phases, each of which has a different hydrogen concentration, is prevented by a bubble counter. After beginning the hydrogenation, hydrogen is delivered exclusively from the gas burette in order to obtain pressure equalization. The main problem with this type of measurement is that a concentration gradient (caused by the higher concentration of gas streaming into the

Fig. 10.7 An apparatus used to monitor hydrogenations at different hydrogen partial pressures.

gas space above the reaction solution) must be avoided. In addition to thorough mixing of the gas phase above the reaction solution, this problem could be solved by including an arrangement whereby the bubble counter between the gas volume above the reaction solution and the gas burette is located directly above the reaction solution in the gas phase of the hydrogenation vessel (cf. Fig. 10.7).

The result of the described methodical solution to monitor gas-consuming reactions at reduced partial pressure under isobaric conditions is shown in Figure 10.8 for the catalytic hydrogenation of COD with a cationic Rh-complex. The slope of the measured straight lines corresponds to the maximally obtainable rate ($V_{sat} = k_2 \cdot [E]_0 = k_2' \cdot [H_2] \cdot [E]_0$) [42 b], which is directly proportional to the hydrogen concentration in solution and at validity of Henry's law to the hydrogen partial pressure above the reaction solution. The experiments prove that the "dilution factor" of the gas phase can adequately be found in the rate constant. (Further examples can be found in [47].)

Besides the above-mentioned errors, further difficulties may arise in the *in-situ* monitoring of hydrogenations. For example, in order to start a hydrogenation it is necessary to exchange inert gas and hydrogen by evacuation. In fact, this procedure leads to a cooling of the solution caused by the evaporation enthalpy of the solvent. The time taken to reach the equilibrium value of the vapor pressure of the solvent above the reaction solution must also be taken into account. In order to avoid such problems, it has been proven of value to seal the catalyst (or the substrate) in a glass ampoule under argon (cf. Fig. 10.3), and to start the hydrogena-

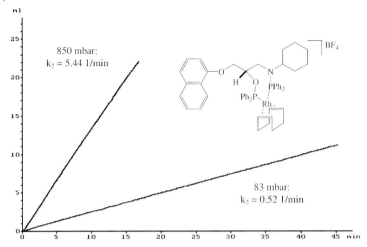

Fig. 10.8 Rate constant k_2 for the catalytic COD-hydrogenation with [Rh(cyclohexyl-PROPRAPHOS)COD]BF$_4$ as catalyst at various hydrogen partial pressures (normal pressure and commercial argon/hydrogen mixtures (AGA) which contain 9.71% H$_2$). Reaction mixture: 15.0 mL MeOH; 0.01 mmol catalyst; 1.0 mmol COD.

tion by destroying the ampoule only when the thermal equilibria have been established. It must be borne in mind, however, that it is extremely difficult to exclude all error sources, and at best a minimization of the problem is possible for a concrete case. However, it is important to assess – and at least report – the expected relative importance of those errors that have been neglected.

10.3.1.2 Monitoring of Hydrogenations by NMR and UV/Visible Spectroscopy

The details of a series of *in-situ* methods and appropriate investigations have been described concerning NMR spectroscopic monitoring of catalytic reactions with gases in various pressure ranges [42 e, 48]. However, disadvantages might include the reactive gas not being supplied, that isobaric conditions during the gas consumption are not possible, that thorough mixing of the reaction solution is insufficient (diffusion problems), or that special NMR probe heads are necessary. A very interesting solution has been described by Iggo et al. [48d], in which the NMR flow cell for the *in-situ* study of homogeneous catalysis allows measurements up to 190 bar (!), but requires the use of a standard wide-bore NMR probe. Details of state-of-the-art methods for the *in-situ* monitoring of reactions using NMR spectroscopy can be found in [49].

An improvement of the possibility for experiments under normal pressure (as described in [42e]) is shown in Figure 10.9 [50]. During registration of the spectrum, the reactive gas is continuously bubbled into the reaction solution below the NMR-active sample volume; thus, diffusion problems can be excluded for mod-

(a)

(b)

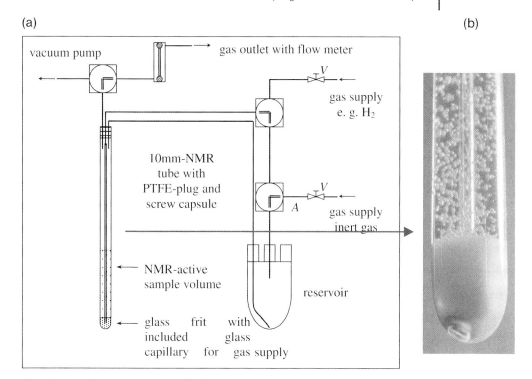

Fig. 10.9 (a) Schematic arrangement for the NMR-spectro-
scopic monitoring of gas-consuming reactions under catalytic
conditions according to [50]. (b) Gas flow during the measurement (a – argon, b – hydrogen).

erately fast reactions. In spite of the introduction of gas during the measurement
(cf. Fig. 10.9b) and hence a deterioration in the homogeneity of the magnetic field,
sufficiently good spectra (^{1}H, ^{13}C, ^{31}P) can be obtained under *in-situ* catalytic conditions using the non-rotating NMR tube. One disadvantage of this arrangement is
that the gas excess is withdrawn from the device and disappears. For this reason, it
is not economical to employ expensive, isotopically labeled gaseous reaction partners such as ^{2}H$_2$ and ^{13}CO. Moreover, because of the permanent loss of gas – especially in long-term measurements – the solvent is also discharged.

An application of the arrangement shown in Figure 10.9 is depicted in Figure
10.10. For the hydrogenation of (Z)-N-acetylamino methyl cinnamate (AMe) with
[Rh(DIPAMP)(solvent)$_2$]anion (DIPAMP = 1,2-bis-(o-methoxy-phenyl-phenyl phos-
phino)ethane) in isopropyl alcohol at 25 °C and under normal pressure, the ^{31}P-
NMR spectrum shown in Figure 10.10b was measured during hydrogenation under steady-state conditions. The comparison with the spectrum taken under argon
(Fig. 10.10a) proves that the ratio major/minor catalyst–substrate complex is higher during the hydrogenation than under thermodynamic conditions (argon). The
measurement of a similar spectrum after termination of hydrogen supply and in-

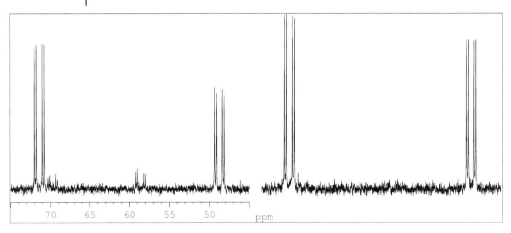

Fig. 10.10 ^{31}P-NMR spectra of [Rh(DIPAMP)(MeOH)$_2$]$^+$ and
1.0 mmol AMe in 5.0 mL iso-propyl alcohol-d_8 at 25 °C with
the arrangement shown in Figure 10.9. Spectrum (a) is registered under argon; spectrum (b) is accumulated during a reaction time of 30 min (1200 pulses).

troduction of argon virtually congruently leads to the initial spectrum. Thus, it
could clearly be proven that the change in the phosphorus spectra should be exclusively attributed to the reaction with hydrogen.

Recently, a new (and now commercially available) methodology was reported for
measuring *in-situ* high pressure NMR spectra up to 50 bar under stationary conditions. The instrument uses a modified sapphire NMR tube, and gas saturation of
the sample solution and exact pressure control is guaranteed throughout the overall measurement, even at variable temperatures. For this purpose, a special gas cy-
cling system is positioned outside the magnet in the routine NMR laboratory [51].

Today, stirring inside UV/visible cells, cell tempering, the use of flow-through
cells, and the detection of smallest amounts of samples in microcells are all
possible, without problems. However, a complete gas exchange (e.g., argon for
hydrogen) is still difficult. Moreover, because of the disadvantageous geometry
of a cell in terms of the ratio of surface to volume, it is generally only possible
to trace relatively slow reactions with gases in the kinetically controlled regime.
After all, the realization of isobaric conditions for the gaseous reaction partner
in case of a cell represents high requirements to the pressure adjustment since
the gas consumptions are relatively small. For such problems, the application of
immersion probes (known also as "fiber-optical probes") represents a good alternative. Due to the arbitrary dimensioning of the reaction vessel and the immersion probe (an example for analyses under normal pressure is shown in
Fig. 10.11), hydrogenations can be monitored *in situ* over a variety of pressure
and temperature ranges, and in an elegant manner.

The "external" measurement of UV/visible spectra principally allows the trac-
ing of other quantities at the same time, such as conductivity, pH, and gas con-

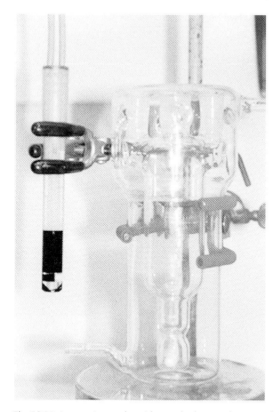

Fig. 10.11 Immersion probe with standard ground joint and reaction vessel for UV/visible spectroscopic analyses under normal pressure.

sumption under catalytic conditions. Monitoring of the time-dependent change of extinction at the maximum (441 nm) of $[Rh(DIOP)(COD)]BF_4$ (DIOP = 4,5-bis(diphenyl-phosphino-methyl)-2,2-dimethyl-1,3-dioxolane)] for the catalytic hydrogenation of COD and the simultaneous registration of hydrogen consumption is shown graphically in Figure 10.12.

The rate constant can be obtained directly from the slope of the graph [42f]. Furthermore, it can clearly be seen that the concentration of $[Rh(DIOP)(COD)]BF_4$, both under argon and during the hydrogenation, is the same until depletion of the substrate COD. This confirms that hydrogenation proceeds in the range of saturation kinetics of the underlying Michaelis-Menten kinetics. Thus, the experimental procedure provides information regarding the catalyst via UV/visible spectroscopy; subsequently, the rate of product formation can be quantified from the hydrogen consumption of the very same reaction solution [52].

In addition to the above-mentioned possibilities for the *in-situ* monitoring of hydrogenations, there are, of course, also techniques involving calorimetry and IR spectroscopy [34, 35c, 39, 41, 53, 54].

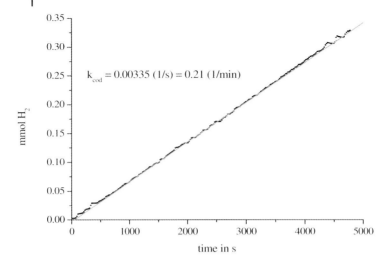

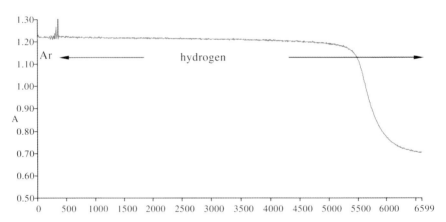

Fig. 10.12 Simultaneous monitoring of the time-dependent UV/ visible spectrum at 441 nm (maximum of the catalyst extinction) and hydrogen consumption for the hydrogenation of COD with [Rh(DIOP)COD]BF$_4$. Conditions: 0.02 mmol precatalyst; 0.33 mmol COD; 20.0 mL methanol; 25.0 C; 1.013 bar total pressure.

10.3.2
Gross-Kinetic Measurements

10.3.2.1 Derivation of Michaelis-Menten Kinetics with Various Catalyst-Substrate Complexes

The catalytic asymmetric hydrogenation with cationic Rh(I)-complexes is one of the best-understood selection processes, the reaction sequence having been elucidated by Halpern, Landis and colleagues [21 a, b], as well as by Brown et al. [55]. Diastereomeric substrate complexes are formed in pre-equilibria from the solvent complex, as the active species, and the prochiral olefin. They react in a series of elementary steps – oxidative addition of hydrogen, insertion, and reductive elimination – to yield the enantiomeric products (cf. Scheme 10.2) [56].

The rate law for two diastereomeric catalyst–substrate complexes (C_2-symmetric ligands) resulting from Michaelis-Menten kinetics (Eq. (11)) has already been utilized by Halpern et al. for the kinetic analysis of hydrogenations according to Scheme 10.2, and corresponds to Eq. (3) of this study.

$$\frac{d[H_2]}{dt} = \frac{d[R]}{dt} + \frac{d[S]}{dt} = \frac{\left(\dfrac{(k_{2min} \cdot K_{ESmin}) + (k_{2maj} \cdot K_{ESmaj})}{K_{ESmin} + K_{ESmaj}}\right) \cdot [Rh]_0 \cdot [\text{olefine}]}{\left(\dfrac{1}{K_{ESmin} + K_{ESmaj}}\right) + [\text{olefine}]}$$

with

$$K_{ESmin} = \frac{k_{1min}}{k_{-1min} + (k_{2min} \cdot [H_2])} \qquad K_{ESmaj} = \frac{k_{1maj}}{k_{-1maj} + (k_{2maj} \cdot [H_2])} \qquad (11)$$

In answer to the question, "why are there not much more kinetic analyses of selection processes in analogy to these classic works?", it should be realized that particular prerequisites are necessary. In the concrete case, such prerequisites included a major stability of the substrate complexes, a convenient ratio of the diastereomeric substrate complexes, and a pressure-dependence of the enantioselectivities.

The following section deals with kinetic equations for the simple Michaelis-Menten kinetics with more than two intermediates; subsequently, their application for the interpretation of hydrogenations in practical examples is discussed.

If C_1-symmetric ligands are employed in asymmetric hydrogenation instead of the corresponding C_2-symmetric ligands, there coexist principally four stereoisomeric substrate complexes, namely two pairs of each diastereomeric substrate complex. Furthermore, it has been shown that, for particular catalytic systems, intramolecular exchange processes between the diastereomeric substrate complexes should in principle be taken into account [57]. Finally, the possibility of non-established pre-equilibria must be considered [58]. The consideration of four intermediates, with possible intramolecular equilibria and disturbed pre-equilibria, results in the reaction sequence shown in Scheme 10.3. This is an example of the asymmetric hydrogenation of dimethyl itaconate with a Rh-complex, which contains a C_1-symmetrical aminophosphine phosphinite as the chiral ligand.

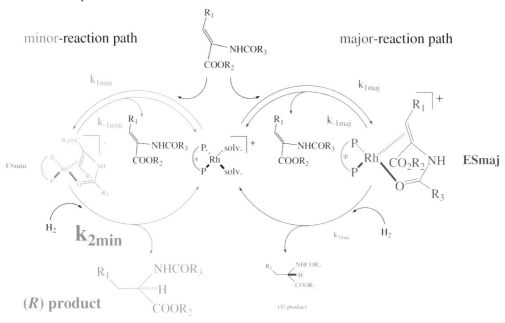

Scheme 10.2 Selection model of the Rh(I)-complex-catalyzed asymmetric hydrogenation to the (R)-amino acid derivative (according to [21a, b, 55]). ES$_{maj}$ and ES$_{min}$ correspond to the diastereomeric catalyst–substrate complexes.

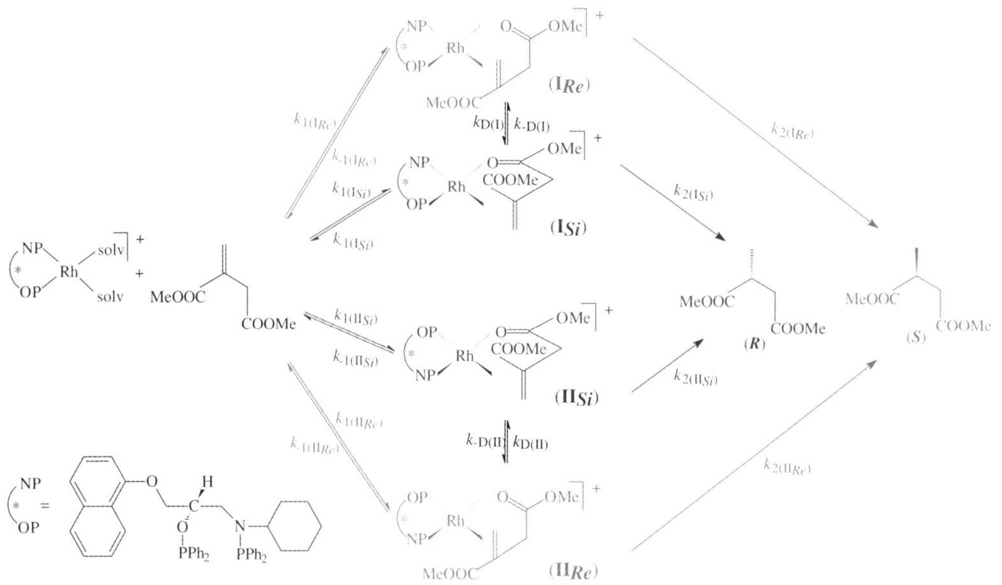

Scheme 10.3 Reaction sequence for the asymmetric hydrogenation of dimethyl itaconate with a C$_1$-symmetric ligand under consideration of intramolecular exchange processes between the intermediates and of disturbed pre-equilibria.

For the time-dependent change of the particular concentrations, Eqs. (12 a) and (12 b) result.

$$\text{(a)} \quad \frac{d[S]}{dt} = k_{2(I_{Re})} \cdot [I_{Re}] + k_{2(II_{Re})} \cdot [II_{Re}] \qquad \text{(b)} \quad \frac{d[R]}{dt} = k_{2(I_{Si})} \cdot [I_{Si}] + k_{2(II_{Si})} \cdot [II_{Si}]$$

$$(12)$$

The common further treatment of the approach – assumption of steady-state conditions for the intermediate substrate complexes, consideration of the catalyst balance ($[\text{catalyst}]_0 = [\text{solvent complex}] + [I_{Re}] + [I_{Si}] + [II_{Re}] + [II_{Si}]$) and of the stoichiometry of the hydrogenation – provides the rate of hydrogen consumption under isobaric conditions (Eq. (13)) [57 f]. A more general derivation can be found in [59].

$$
\begin{aligned}
-\frac{d[H_2]}{dt} &= \frac{d[S]}{dt} + \frac{d[R]}{dt} \\
&= \frac{\dfrac{(k_{2(I_{Re})} \cdot K_{I_{Re}} + k_{2(II_{Re})} \cdot K_{II_{Re}} + k_{2(I_{Si})} \cdot K_{I_{Si}} + k_{2(II_{Si})} \cdot K_{II_{Si}})}{(K_{I_{Re}} + K_{I_{Si}} + K_{II_{Re}} + K_{II_{Si}})} \cdot [\text{cat}]_0 \cdot [S]}{\dfrac{1}{(K_{I_{Re}} + K_{I_{Si}} + K_{II_{Re}} + K_{II_{Si}})} + [S]} \\
&= \frac{k_{obs} \cdot [\text{cat}]_0 \cdot ([S]_0 - n_{H_2})}{K_M + ([S]_0 - n_{H_2})}
\end{aligned}
$$

$$(13)$$

This relationship corresponds to the simplest Michaelis-Menten kinetics (Eq. (3)). In addition to the equation derived earlier by Halpern et al. for the simplest model case of a C_2-symmetric ligand without intramolecular exchange [21b], every other possibility of reaction sequence corresponding to Scheme 10.3 can be reduced to Eq. (13). Only the physical content of the values of k_{obs} and K_M, which must be determined macroscopically, differs depending upon the approach (see [59] for details). Nonetheless, the constants k_{obs} and K_M allow conclusions to be made about the catalyses:

- The value $1/K_M$ corresponds to the ratio of concentrations of the sum of all catalyst–substrate complexes to the product $\{[\text{solvent complex}] \cdot [\text{substrate}]\}$, and thus is a measure of how much catalyst–substrate complex is present [60].
- The k_{obs}-values are all to be interpreted as the sum of all rate constants for the oxidative addition of hydrogen, each multiplied by the mole fraction of the corresponding catalyst–substrate complex. Hence this "gross-rate constant" is dependent only on the ratio of intermediates, and not on their absolute concentrations.

Clearly, a comprehensive description of catalytic systems is not possible from the hydrogen consumption alone. The reaction sequence represented in Scheme 10.3 already contains 16 rate constants. However, valuable data regarding the catalysis can be obtained from an analysis of the gross hydrogen consumption on the basis of Eq. (13), for various catalytic systems. Some practical examples of this are described in the following section.

10.3.2.2 Data from Gross Kinetic Measurements

The hydrogen consumption and enantioselectivities for the asymmetric hydrogenation of dimethyl itaconate with various substituted catalysts of the basic type [Rh(PROPRAPHOS)COD]BF$_4$ are illustrated in Figure 10.13 [61]. The systems are especially suitable for kinetic measurements because of the rapid hydrogenation of COD in the precatalyst. There are, in practice, no disturbances due to the occurrence of induction periods.

NMR-analyses suggest that the hydrogenation runs corresponding to Scheme 10.3. Three of the four possible catalyst–substrate complexes are detectable in the ^{31}P-NMR-spectrum [57f].

A comparison of the activities for various catalyst derivatives shown in Figure 10.13 seems to prove that the ligand with the cyclohexyl residue leads to the most active catalyst for the hydrogenation of dimethyl itaconate. The catalyst containing the methyl derivative apparently exhibits the lowest activity.

A more detailed analysis, however, shows that such comparisons of activity can be completely misleading, because Michaelis-Menten kinetics are principally described by two constants. The Michaelis constant contains information regarding the pre-equilibria, the rate constants quantify the product formation from the intermediates.

An analysis of the hydrogenation curves shown in Figure 10.13 indicates, for those precatalysts with R=2-propyl, 3-pentyl and cyclopentyl, that they can be described quantitatively as first-order reactions. The comparison between experimental and calculated data (the latter being determined by least-squares regres-

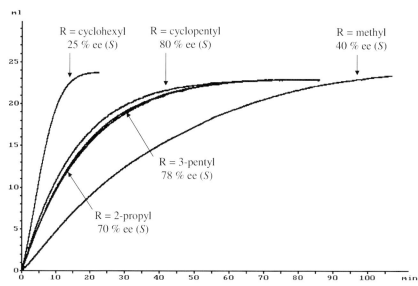

Fig. 10.13 Catalytic asymmetric hydrogenation of 1.0 mmol dimethyl itaconate with 0.01 mmol [Rh(PROPRAPHOS)COD]BF$_4$- precatalysts ((S)-PROPRAPHOS: R=2-propyl) in 15.0 mL MeOH at 1.013 bar total pressure and 25 °C.

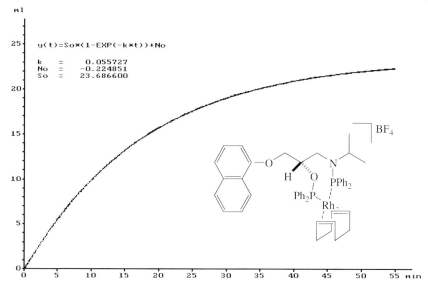

ml

$y(t)=So*(1-EXP(-k*t))+No$

k   =   0.055727
No =   -0.224851
So =   23.686600

BF$_4$

Fig. 10.14 Comparison of the experimental hydrogenation curve and first-order calculated values for the asymmetric hydro-genation of the PROPRAPHOS-precatalyst with dimethyl itaconate.

sion analysis) is shown in Figure 10.14. Due to the excellent conformity, these curves lie on top of each other.

If the ligands containing R = methyl and R = cyclohexyl are employed, the hydrogenations describe not only the initial range of the Michaelis-Menten equation, but also a range which cannot be assigned to the limiting case of the first-order reaction (cf. Figs. 10.1 and 10.2). Determination of the sought constants is carried out using nonlinear regression, with the initial values determined by linearization of Eq. (7). The comparison between experimental and calculated values corresponding to Eq. (13) for the ligand containing the methyl residue is shown in Figure 10.15. The results prove a good correspondence between the experiment and the model. For the initial quantity of substrate (1.0 mmol), the range of half-saturation concentration is reached almost at the start of the hydrogenation (Fig. 10.15).

The results of the kinetic analysis for the investigated systems are summarized in Table 10.2, the substrate concentration used being the same for all trials. In the case of methyl- and cyclohexyl-substituted ligands the Michaelis constant is smaller than the initial substrate concentration of $[S]_0 = 0.06666$ mol L^{-1} (Table 10.2). However, a description of the hydrogenations with other catalyst ligands as first-order reactions shows that in each of these cases the Michaelis constant must be much greater than the experimentally chosen substrate concentration.

Even at a rather higher substrate concentration, the limiting value of a concentration-independent rate is not reached in these cases. This is illustrated for the example of the PROPRAPHOS-type catalyst in Figure 10.16. It is, further-

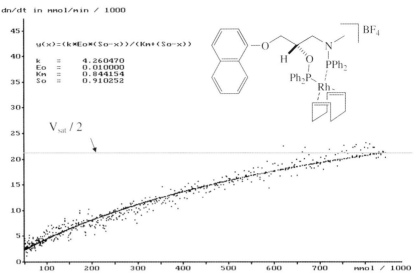

dn/dt in mmol/min / 1000

$y(x)=(k*Eo*(So-x))/(Km+(So-x))$

k	=	4.260470
Eo	=	0.010000
Km	=	0.844154
So	=	0.910252

$V_{sat} / 2$

mmol / 1000

Fig. 10.15 Comparison of experimental and calculated values for the asymmetric hydrogenation of dimethyl itaconate with the methyl substituted ligand of the PROPRAPHOS precatalyst. Specifications of concentration refer to 15.0 mL of solvent.

Table 10.2 Kinetic analysis of the asymmetric hydrogenation of dimethyl itaconate with derivatives of [Rh(PROPRAPHOS)-COD]BF$_4$ (see Fig. 10.13).

Ligand	% ee	k (1st order) [1 s^{-1}]	K_M [mol L^{-1}]	$1/K_M$ [L mol^{-1}]	k_{obs} (Eq. (13)) [1 s^{-1}]	V_{sat} [mol L^{-1} s]
R = cyclohexyl	25	7.27×10^{-3}	0.0286	35	3.12×10^{-1}	2.08×10^{-4}
R = methyl	40	8.42×10^{-4}	0.0562	18	7.10×10^{-2}	4.73×10^{-5}
R = 2-propyl	70	9.28×10^{-4}	–	–	–	–
R = 3-pentyl	78	8.50×10^{-4}	–	–	–	–
R = cyclopentyl	80	1.07×10^{-3}	–	–	–	–

more, remarkable that at an increase in the substrate/catalyst ratio from 100 (standard conditions) to 1000 the PROPRAPHOS catalyst already shows an initial rate which is more than twice the maximum reachable rate with the cyclohexyl derivative (4.2×10^{-4} mol L^{-1} s versus 2.1×10^{-4} mol L^{-1} s). Indeed, PROPRAPHOS is still not used optimally with regard to its activity (see Fig. 10.16).

Interpretation of the reciprocals of the Michaelis constants allows the following conclusions to be made regarding hydrogenations under specified experimental conditions. In the case of the methyl and cyclohexyl ligand, the prevailing form of the catalyst in solution is the catalyst–substrate complex. However, for the other examples of first-order reactions, large Michaelis constants (or very

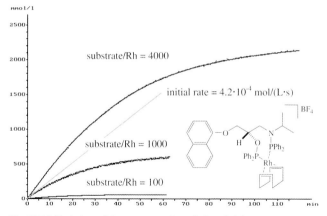

Fig. 10.16 Variation of the concentration of dimethyl itaconate for the hydrogenation with [Rh((S)-PROPRAPHOS)COD]BF$_4$.

small reciprocals of the same) prove that in these cases the equilibrium between solvent complex and the diastereomeric catalyst–substrate complexes is shifted to the side of the solvent complex.

These results, obtained from the gross-hydrogen consumption under normal conditions on the basis of the model developed above, make it clear that even catalysts of the same basic type can give rise to considerably different pre-equilibria. As a consequence, comparison of activities of various catalytic systems under "standard conditions" can provide the wrong picture. Hence, the cyclohexyl precatalyst with dimethyl itaconate seems to be the most active one (by reference to Fig. 10.13). Nonetheless, an increase in the initial substrate concentration by a factor of ten already leads to a different order in activity.

In addition to comparisons of activity of various catalysts, the choice of an appropriate solvent represents yet another problem in catalysis. The choice is usually made by direct comparison of the activity of a catalyst in various solvents. Nonetheless, analogous problems as mentioned above must be considered. Variable substrate concentrations can lead to seemingly different orders in the activity of solvents. The reason for this is based on the fact that macroscopic activity is caused by different amounts of catalyst–substrate complex.

These results underline the fact that "gross-activities" based on TOFs or half-lives only are not appropriate to compare catalytic systems that are characterized by pre-equilibria. Rather, only an analysis of gross-kinetics on the basis of suitable models can provide detailed information concerning the catalysis.

As explained earlier, the pre-equilibria are characterized by the limiting values of Michaelis-Menten kinetics. In the case of first-order reactions with respect to the substrate, we have: $K_M \gg [S]_0$. Since the pre-equilibria are shifted to the side of educts during hydrogenation, only the solvent complex is detectable. In contrast, in the case of zero-order reactions only catalyst–substrate complexes are expected under stationary hydrogenation conditions in solution. These consequences resulting from Michaelis-Menten kinetics can easily be proven by var-

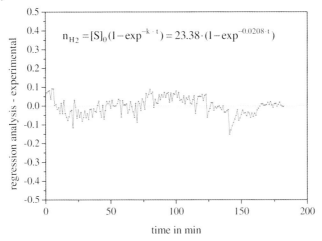

Fig. 10.17 Asymmetric hydrogenation of dimethyl itaconate with [Rh(Ph-β-glup-OH)(MeOH)$_2$]BF$_4$; comparison between first-order fit (x-axis) and experimental values. Conditions: 0.01 mmol catalyst; 1.0 mmol substrate; 15.0 mL MeOH; 1.013 bar total pressure.

ious methods such as NMR- and UV/visible spectroscopy, and this is demonstrated by some examples in the following section.

The asymmetric hydrogenation of dimethyl itaconate with [Rh(Ph-β-glup-OH-MeOH)$_2$]BF$_4$ runs as a first-order reaction. The deviation of experimental values from the hydrogen consumption calculated from parameters of nonlinear regression analysis (x-axis) is shown in Figure 10.17. In solution, only the solvent complex should be detectable during hydrogenation. In order to monitor hydrogenation via UV/visible spectroscopy, a 100-fold excess of the prochiral olefin is added to the solvent complex. The exchange of argon for hydrogen starts the hydrogenation, which is then monitored by cyclic measurement of the spectra (Fig. 10.18).

Although the substrate complexes absorb in the range of measurement (see Fig. 10.18, inset), the spectra observed during hydrogenation do not differ from the spectrum of the pure solvent complex. On completion of the reaction, gas chromatographic analysis proves that hydrogenation has occurred and that the usual values for conversion and selectivity have resulted. Thus, only solvent complex is present during the hydrogenation, and this corresponds to expectations from kinetic interpretations of the hydrogen consumption curve.

NMR spectroscopy provides analogue results. Inspection of hydrogen consumption curves following the hydrogenation of *Z*- or *E*-methyl 3-acetamidobutenoate with [Rh(Et-DuPHOS)(MeOH)$_2$]BF$_4$ (Et-DuPHOS = 1,2-bis(2,5-diethylphospholanyl)benzene)) showed the reaction to exhibit first-order kinetics (Fig. 10.19).

For both cases in these examples only the solvent complex is detectable, besides traces of non-hydrogenated COD-precatalyst (cf. Fig. 10.20).

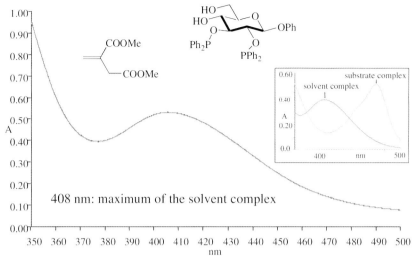

Fig. 10.18 UV/visible spectrum of 0.02 mmol [Rh(Ph-β-glup-OH)(MeOH)$_2$]BF$_4$ in 35.0 mL MeOH under Ar, and five spectra (cyclic, monitored at 6.0 min intervals) after addition of a 100-fold excess of dimethyl itaco-nate and exchange of Ar for H$_2$. Inset: the spectrum of the catalyst–substrate complex. GC analysis after 30 min hydrogenation: 45% conversion, 78% ee (*R*), which agrees well with corresponding hydrogenations.

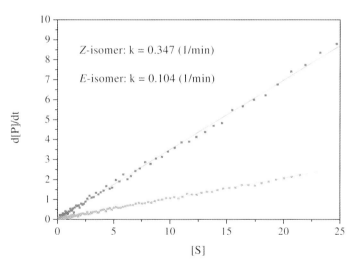

Fig. 10.19 Asymmetric hydrogenation of *E*- and *Z*-methyl 3-acetamidobutenoate with [Rh(Et-DuHOS)MeOH$_2$)BF$_4$ as first-order reactions; d[P]/d*t* versus [S] (Eq. (4 b)).

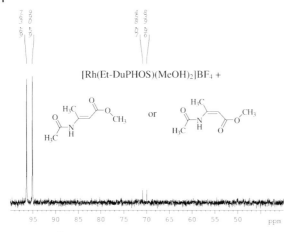

Fig. 10.20 ^{31}P-NMR spectrum of a solution of 0.01 mM [Rh(Et-DuPHOS)(MeOH)$_2$]BF$_4$ and 0.1 mM *E*- or *Z*-methyl 3-acetamidobutenoate.

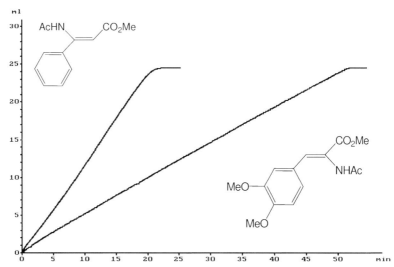

Fig. 10.21 Hydrogen consumption for the hydrogenation of (*Z*)-3-*N*-acetylamino-3-(phenyl)-methyl propenoate with [Rh((*R*,*R*)-Et-DuPHOS)(MeOH)$_2$]BF$_4$ in *i*-PrOH (59% ee) and (*Z*)-2-benzoylamino-3-(3,4-dimethoxy phenyl)-methyl acrylate with [Rh((*S*,*S*)-DI-PAMP)(MeOH)$_2$]BF$_4$ in MeOH (98% ee). Conditions: 0.01 mmol catalyst; 1.0 mmol substrate; 15.0 mL solvent; 1.013 bar total pressure.

Nonetheless, if zero-order reactions are analyzed in terms of the validity of Michaelis-Menten kinetics, all of the catalyst is present in solution as catalyst–substrate complex up to high conversions. The hydrogenation rate is independent of the substrate concentration; two such examples are provided in Figure 10.21.

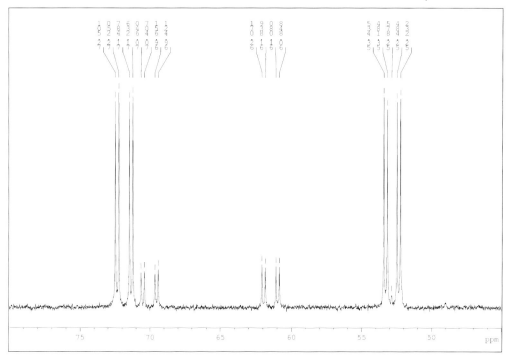

Fig. 10.22 ^{31}P-NMR spectrum of a solution of 0.02 mM [Rh((S,S)-DIPAMP)(MeOH)$_2$]BF$_4$ and 0.1 mM (Z)-2-benzoyl-amino-3-(3,4-dimethoxyphenyl)-methyl acrylate.

In these cases, the ^{31}P-NMR spectrum exhibits only signals of substrate complexes; there is almost no solvent complex visible. This is illustrated for (Z)-2-ben-zoylamino-3-(3,4-dimethoxyphenyl)-methyl acrylate with [Rh((S,S)-DIPAMP) (MeOH)$_2$]BF$_4$ in Figure 10.22.

Thus, if information is being sought about intermediates for this type of catalysis, it does not make sense to analyze systems that lead to first-order reactions! Rather, systems in which the hydrogenation rate is independent of the substrate concentration would be more appropriate. Indeed, for both catalytic systems shown in Figure 10.21, in each case one of the catalyst–substrate complexes could be isolated and characterized by crystal structure analysis (Fig. 10.23).

In the case of the α-dehydroamino acid (Fig. 10.23, right), it could be shown by using low-temperature NMR spectroscopy that the isolated crystals correspond to the major substrate complex in solution. However, according to the major–minor concept (see Scheme 10.2), it does not lead to the main enantiomer [63]. On the contrary, it could be proven unequivocally for various substrate complexes with β-dehydroamino acids that the isolated substrate complexes are major-substrate complexes. Surprisingly, they also gave the main enantiomer of the asymmetric hydrogenation, which would not be expected on the basis of

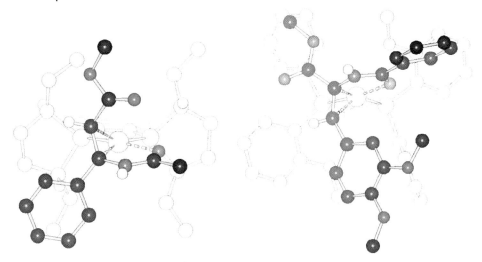

Fig. 10.23 X-ray structure of [Rh((*R,R*)-Et-DuPHOS)((*Z*)-3-*N*-acetylamino-3-(phenyl)-methyl propenoate)]$^+$ and of [Rh((*S,S*)-DIPAMP)((*Z*)-2-benzoylamino-3-(3,4-dimethoxyphenyl)-methyl acrylate)]$^+$ [62].

classical ideas [62a]. In these cases the major-substrate complex determines the selectivity, in analogy to the well known lock-and-key concept of enzyme catalysis proposed by Emil Fischer. This result could only be gained by quantitative monitoring of the hydrogenation and the subsequent interpretation of kinetic findings within the frame of Michaelis-Menten kinetics.

Abbreviations

COD cycloocta-1,5-diene
NBD norborna-2,5-diene
TOF turnover frequency
TON turnover number

References

1 R. A. Sanchez-Delgado, M. Rosales, *Coord. Chem. Rev.* **2000**, *196*, 249.
2 J. A. Widegren, R. G. Finke, *J. Mol. Cat. A: Chemical* **2003**, *198*, 317.
3 G. Djéga-Mariadassou, M. Boudart, *J. Catal.* **2003**, *216*, 89.
4 The number of reaction possibilities that fulfill particular limiting conditions can be surprisingly high. This is shown by mechanisms of various catalyzed reactions generated by computer programs such as ChemNet or MECHEM. (a) R. E. Valdes-Perez, A. V. Zeigarnik, *J. Chem. Inf. Comput. Sci.* **2000**, *40*, 833; (b) L. G. Bruk, S. N. Gorodskii, A. V. Zeigarnik, R. E. Valdes-Perez, O. N. Temkin, *J. Mol. Catal. A: Chem.* **1998**, *130*, 29; (c) A. V. Zeigarnik, R. E. Valdes-Perez, O. N. Tem-

kin, L. G. Bruk, S. I. Shalgunov, *Organometallics* **1997**, *16*, 3114.

5 According to Chen et al., alkali cation co-catalysis kinetics cannot be distinguished from classic ideas (proton instead of alkali) for the asymmetric hydrogenation of acetophenone with the Noyori-catalyst (*trans*-RuCl$_2$[(*S*)-BINAP][(*S,S*)DPEN]) (see [43b]).

6 J. Halpern, *Science* **1982**, *217*, 401.

7 R. Noyori, *Asymmetric catalysis in organic synthesis.* John Wiley & Sons, Inc., New York, **1994**.

8 W. A. Herrmann, B. Cornils, *Angew. Chem. Int. Ed.* **1997**, *36*, 1049.

9 (a) P. I. Dalko, L. Moisan, *Angew. Chem. Int. Ed.* **2004**, *43*, 5138; (b) Special edition *Accounts Chem. Res.* **2004**, *37*, 8, 487.

10 We point out that in enzyme kinetics TON is understood as TOF! "It is also sometimes called the turnover *number*, because it is a reciprocal time and defines the number of catalytic cycles (or "turnovers") that the enzyme can undergo in unit time, or the number of molecules of substrate that one molecule of enzyme can convert into products in one unit of time." Quotation from [23].

11 For the hydrogenation of acetophenone with the Ru-BINAP-DPEN system by Noyori, for example, in the course of 48 h 2.4 million conversions at the catalyst are reached (H. Doucet, T. Ohkuma, K. Murata, T. Yokozawa, M. Kozawa, E. Katayama, A. F. England, T. Ikariya, R. Noyori, *Angew. Chem. Int. Ed.* **1998**, *37*, 1703). Rautenstrauch could – under slightly changed conditions – reach a TOF of 333 000 h^{-1} (V. Rautenstrauch, X. Hoang-Cong, R. Churlaud, K. Abdur-Rashid, R. H. Morris, *Chem. Eur. J.* **2003**, *9*, 4954). The best results as concerns the activity of transfer hydrogenations were reported by Mathey, according to who cyclohexanone can, during the course of 15 h, be completely converted with a substrate/catalyst ratio of 20 million by a catalyst containing a P/N-ligand (C. Thoumazet, M. Melaimi, L. Ricard, F. Mathey, P. Le Floch, *Organometallics* **2003**, *22*, 1580).

12 The following TON and TOF specifications refer to minimum quantities for technically relevant processes: H. U. Blaser, B. Pugin, F. Spindler, *Enantioselective Synthesis, Applied Homogeneous Catalysis by Organometallic Complexes*, 2nd edn, Volume 3, B. Cornils, W. A. Herrmann (Eds.), Wiley-CH, Weinheim, **2002**, pp. 1131. "Catalyst productivity, given as substrate/catalyst ratio (*s/c*) or turnover number (TON), determines catalyst costs. These *s/c* ratios ought to be >1000 for small-scale, high-value products and >50 000 for large-scale or less-expensive products (catalyst re-use increases the productivity)" and "Catalyst activity, given as turnover frequency for >95% conversion (TOF$_{av}$, h^{-1}), determines the production capacity. TOF$_{av}$ ought to be >500 h^{-1} for small-scale and >10 000 h^{-1} for large-scale products" (p. 1133).

13 Due to *re*- and *si*-coordination of prochiral substrates at a catalyst with C_2-symmetric chiral ligands two diastereomeric catalyst-substrate complexes emerge. In the case of C_1-symmetric ligands already four stereoisomer intermediates result.

14 The so-called "minor-substrate complexes" of asymmetric hydrogenations are not detectable in many cases, even though they determine the stereochemical course of the hydrogenation due to their high reactivity according to classical ideas.

15 L. Michaelis, M. L. Menten, *Biochem. Z.* **1913**, *49*, 333.

16 M. Bodenstein, *Z. Phys. Chem.* **1913**, *85*, 329.

17 G. E. Briggs, J. B. S. Haldane, *Biochem. J.* **1925**, *19*, 338.

18 (a) F. G. Helfferich, *Kinetics of Homogeneous Multistep Reactions, Comprehensive Chemical Kinetics*, Volume 38, Elsevier, Amsterdam, **2001**; (b) K. A. Connors, *Chemical Kinetics. The Study of Reaction Rates in Solution*, VCH Publishers Inc., New York, **1990**.

19 "Quite often, kinetic studies are not performed because it appears time-consuming, expensive (with respect to the price or availability of most of chiral ligands) and not rewarding") (see [35a]).

20 An irreversible formation of the catalyst–substrate complex is described in: D. D.

van Slyke, G. E. Cullen, *J. Biol. Chem.* **1914**, *19*, 141.

21 (a) A. S. C. Chan, J. J. Pluth, J. Halpern, *J. Am. Chem. Soc.* **1980**, *102*, 5952; (b) C. R. Landis, J. Halpern, *J. Am. Chem. Soc.* **1987**, *109*, 1746; (c) M. Kitamura, M. Tsukamoto, Y. Bessho, M. Yoshimura, U. Kobs, M. Widhalm, R. Noyori, *J. Am. Chem. Soc.* **2002**, *124*, 6649.

22 Today, this synonym is used for the more common steady-state approach by Briggs and Haldane (see [17]).

23 A. Cornish-Bowden, *Fundamentals of Enzyme Kinetics*, 3rd edn, Portland Press Ltd., London, **2004**.

24 K. J. Laidler, *Can. J. Chem.* **1955**, *33*, 1614.

25 D. G. Blackmond, *Angew. Chem. Int. Ed.* **2005**, *44*, 4302.

26 At the end of a hydrogenation the substrate concentration is naturally also low.

27 The term for the saturation rate (V_{sat}) in earlier works commonly employed as "maximum rate" (V_{max}) should not – according to the recommendation of the International Union of Biochemistry – be used any more because it does not describe a real maximum, but a limit.

28 (a) A. G. Marongoni, *Enzyme Kinetics – A Modern Approach*, Wiley-Interscience, **2003**; (b) R. A. Copeland, *Enzymes – A Practical Introduction to Structure, Mechanism, and Data Analysis*, VCH Publishers, Inc., New York, **1996**; (c) H. Bisswanger, *Enzymkinetik*, VCH Verlagsgesellschaft mbH, Weinheim, **1994**; (d) A. Cornish-Bowden, *Principles of Enzyme Kinetics*, Butterworth & Co. Ltd., London, **1976**; (e) H. J. Fromm, *Initial Rate Enzyme Kinetics*, Springer, Berlin, **1975**; (f) I. H. Segel, *Enzyme kinetics*, Wiley & Sons, New York, **1975**.

29 H. Lineweaver, D. Burk, *J. Am. Chem. Soc.* **1934**, *56*, 658.

30 (a) G. S. Eady, *J. Biol. Chem.* **1942**, *146*, 85; (b) B. H. J. Hofstee, *J. Biol. Chem.* **1952**, *199*, 357; (c) B. H. J. Hofstee, *Nature* **1959**, *184*, 1296.

31 C. S. Hanes, *Biochem. J.* **1932**, *26*, 1406.

32 The original work of Lineweaver and Burk [29] is the most cited one in *J. Am. Chem. Soc.* (according to *Chem. Eng. News* **2003**, *81*, 27).

33 (a) Y. Fu, X.-X. Guo, S.-F. Zhu, A.-G. Hu, J.-H. Xie, Q.-L. Zhou, *J. Org. Chem.* **2004**, *69*, 4648; (b) H.-J. Drexler, J. You, S. Zhang, C. Fischer, W. Baumann, A. Spannenberg, D. Heller, *Org. Process Res. Dev.* **2003**, *7*, 355; (c) D. Heller, H.-J. Drexler, J. You, W. Baumann, K. Drauz, H.-P. Krimmer, A. Börner, *Chemistry – A European Journal* **2002**, *8*, 5196; (d) D. Heller, H.-J. Drexler, A. Spannenberg, B. Heller, J. You, W. Baumann, *Angew. Chem. Int. Ed.* **2002**, *41*, 777; (e) O. Pàmies, M. Diéguez, G. Net, A. Ruiz, C. Claver, *J. Org. Chem.* **2001**, *66*, 8364; (f) I. M. Angulo, E. Bouman, *J. Mol. Catal. A: Chemical* **2001**, *175*, 65; (g) N. Tanchoux, C. de Bellefon, *Eur. J. Inorg. Chem.* **2000**, 1495; (h) O. Pàmies, G. Net, A. Ruiz, C. Claver, *Eur. J. Inorg. Chem.* **2000**, 1287; (i) M. Rosales, A. González, M. Mora, N. Nader, J. Navarro, L. Sánchez, H. Soscín, *Trans. Met. Chem.* **2004**, *29*, 205; (j) C. A. Sandoval, T. Ohkuma, K. Muniz, R. Noyori, *J. Am. Chem. Soc.* **2003**, *125*, 13490; (k) A. Salvini, P. Frediani, S. Gallerini, *Appl. Organometal. Chem.* **2000**, *14*, 570; (l) V. Herrera, B. Munoz, V. Landaeta, N. Canudas, *J. Mol. Catal. A: Chemical* **2001**, *174*, 141; (m) P. K. Santra, P. Sagar, *J. Mol. Catal. A: Chemical* **2003**, *197*, 37; (n) M. Rosales, J. Castillo, A. González, L. González, K. Molina, J. Navarro, I. Pacheco, H. Pérez, *Trans. Met. Chem.* **2004**, *29*, 221; (o) M. Rosales, F. Arrieta, J. Castillo, A. Gonzàlez, J. Navarro, R. Vallejo, *Stud. Surf. Sci. Catal.* **2000**, *130*, 3357; (p) C. A. Thomas, R. J. Bonilla, Y. Huang, P. G. Jessop, *Can. J. Chem.* **2001**, *79*, 719.

34 A. Zogg, F. Stoessel, U. Fischer, K. Hungerbühler, *Thermochim. Acta* **2004**, *419*, 1.

35 (a) C. deBellefon, N. Pestre, T. Lamouille, P. Grenouillet, V. Hessel, *Adv. Synth. Catal.* **2003**, *345*, 190; (b) C. deBellefon, R. Abdallah, T. Lamouille, N. Pestre, S. Caravieilhes, P. Grenouillet, *Chimia* **2002**, 621; (c) J. L. Bars, T. Häußner, J. Lang, A. Pfaltz, D. G. Blackmond, *Adv. Synth. Catal.* **2001**, *343*, 207; (d) C. deBellefon, N. Tanchoux, S. Caravieilhes, P. Grenouillet, V. Hessel, *Angew. Chem. Int. Ed.* **2000**, *39*, 3442.

36 The real molar volume of hydrogen amounts to 24.48 cm^3 mol^{-1} at 25.0 °C and 1.013 bar (ideal gas: 24.46 cm^3 mol^{-1}). Due to non-linear compressibility, pV/p-isotherms at 100 bar deviate from ideal gas behavior by a few per cent.

37 Virial coefficients of hydrogen can, for example, be found in: J.H. Dymond, E.B. Smith, *The Virial Coefficients of Pure Gases and Mixtures*, Clarendon Press, Oxford, **1980**. By interpolation (e.g., with cubic spline functions), virial coefficients can be determined for any temperature.

38 B. Bogdanovic, B. Spliethhoff, *Chem.-Ing.-Tech.* **1983**, *55*, 156.

39 (a) Y. Sun, J. Wang, C. Le Blond, R.N. Landau, J. Laquidara, J.R. Sowa, Jr., D.G. Blackmond, *J. Mol. Catal. A: Chemical* **1997**, *115*, 495; (b) Y. Sun, R.N. Landau, J. Wang, C. LeBlond, D.G. Blackmond, *J. Am. Chem. Soc.* **1996**, *118*, 1348.

40 (a) J.-C. Charpentier, Mass-transfer rates in gas-liquid absorbers and reactors. In: *Advances in Chemical Engineering*, T.B. Drew, G.R. Cokeleit, J.W. Hoopes, Jr., T. Vermeulen (Eds.), Academic Press Inc., New York, **1981**, Volume 11, pp. 1; (b) G. Astarita, *Mass transfer with chemical reaction*, Elsevier Publishing Company, Amsterdam, **1967**.

41 Y. Sun, J. Wang, C. Le Blond, R.A. Reamer, J. Laquidara, J.R. Sowa, Jr., D.G. Blackmond, *J. Organomet. Chem.* **1997**, *548*, 65.

42 Such induction periods can, for example, result from transferring a precatalyst into the active species. For the asymmetric hydrogenation this is described in detail in: (a) W. Braun, A. Salzer, H.-J. Drexler, A. Spannenberg, D. Heller, *Dalton Trans.* **2003**, 1606; (b) H.-J. Drexler, W. Baumann, A. Spannenberg, C. Fischer, D. Heller, *J. Organomet. Chem.* **2001**, *621*, 89; (c) C.J. Cobley, I.C. Lennon, R. McCague, J.A. Ramsden, A. Zanotti-Gerosa, *Tetrahedron Lett.* **2001**, *42*, 7481; (d) A. Börner, D. Heller, *Tetrahedron Lett.* **2001**, *42*, 223; (e) W. Baumann, S. Mansel, D. Heller, S. Borns, *Magn. Res. Chem.* **1997**, *35*, 701; (f) D. Heller, S. Borns, W. Baumann, R. Selke, *Chem. Ber.* **1996**, *129*, 85.

43 (a) L. Greiner, M.B. Ternbach, *Adv. Synth. Catal.* **2004**, *346*, 1392 (supplementary information); (b) R. Hartmann, P. Chen, *Adv. Synth. Catal.* **2003**, *345*, 1353.

44 (a) A.G. Osborn, D.R. Douslin, *J. Chem. Eng. Data* **1974**, *19*, 114; (b) T. Boublik, K. Aim, *Collect. Czech. Chem. Comm.* **1972**, *11*, 3513; (c) D.W. Scott, *J. Chem. Thermodynamics* **1970**, *2*, 833; (d) T.E. Jordan, *Vapor pressure of Organic Compounds*, Interscience Publishers, Inc. New York, **1954**.

45 (a) P.G.T. Fogg, W. Gerrard, *Solubility of Gases in Liquids*, John Wiley & Sons, Chichester, **1991**; (b) E. Brunner, *J. Chem. Eng. Data* **1985**, *30*, 269.

46 To the best of our knowledge there are no experimental data available for CH_2Cl_2.

47 D. Heller, J. Holz, S. Borns, A. Spannenberg, R. Kempe, U. Schmidt, A. Börner, *Tetrahedron: Asymmetry* **1997**, *8*, 213.

48 (a) A. Aghmiz, A. Orejón, M. Dieguez, M.D. Miquel-Serrano, C. Claver, A.M. Masdeu-Bultó, D. Sinou, G. Laurenczy, *J. Mol. Catal. A: Chemical* **2003**, *195*, 113; (b) H.G. Niessen, P. Trautner, S. Wiemann, J. Bargon, K. Woelk, *Rev. Sci. Instrum.* **2002**, *73*, 1259; (c) S. Gaemers, H. Luyten, J.M. Ernsting, C.J. Elsevier, *Magn. Res. Chem.* **1999**, *37*, 25; (d) J.A. Iggo, D. Shirley, N.D. Tong, *New J. Chem.* **1998**, 1043; (e) A. Cusanelli, U. Frey, D. Marek, A.E. Merbach, *Spectr. Eur.* **1997**, *9*, 22; (f) S. Mansel, D. Thomas, C. Lefeber, D. Heller, R. Kempe, W. Baumann, U. Rosenthal, *Organometallics* **1997**, *16*, 2886; (g) W. Baumann, S. Mansel, D. Heller, S. Borns, *Magn. Reson. Chem.* **1997**, *35*, 701; (h) P.M. Kating, P.J. Krusic, D.C. Roe, B.E. Smart, *J. Am. Chem. Soc.* **1996**, *118*, 10000; (i) M. Haake, J. Natterer, J. Bargon, *J. Am. Chem. Soc.* **1996**, *118*, 8688; (j) K. Woelk, J. Bargon, *Rev. Sci. Instrum.* **1992**, *63*, 3307; (k) D.C. Roe, *J. Magn. Res.* **1985**, *63*, 388.

49 (a) L. Damoense, M. Datt, M. Green, C. Steenkamp, *Coord. Chem. Rev.* **2004**, *248*, 2393; (b) E.M. Vincente, P.S. Pregosin, D. Schott, *Mechanism in Homogeneous Catalysis – A Spectroscopic Approach*. B. Heaton (Ed.), Wiley-VCH,

2005, Chapter 1, pp. 1–80; (c) G. Laurenczy, L. Helm, *Mechanism in Homogeneous Catalysis – A Spectroscopic Approach*. B. Heaton (Ed.), Wiley-VCH, **2005**, Chapter 2, pp. 81.

50 D. Heller, W. Baumann, DE 102 02 173 C2, **2003**.

51 (a) D. Selent, W. Baumann, A. Börner, DE 10333143.A1 (3.3.**2005**); (b) D. Selent, W. Baumann, K.-D. Wiese, D. Ortmann, A. Börner, 14th International Symposium on Homogeneous Catalysis, Munich, 5.–9.7. **2004**, poster no. 0044.

52 Analogue trials in a cell are not practicable. To monitor hydrogen consumption accurately (ca. 8 mL in the example), an absolute quantity of COD is necessary. With equal catalyst/substrate ratios (as in the example), the resultant intracellular high catalyst concentration could only be compensated by a very small cell thickness, which would in turn considerably restrict intracellular mixing.

53 (a) S. Richards, M. Ropic, D. Blackmond, A. Walmsley, *Analytica Chim. Acta* **2004**, *519*, 1; (b) C. LeBlond, J. Wang, R. Larsen, C. Orella, Y.-K. Sun, *Topics Catal.* **1998**, 149; (c) C. LeBlond, J. Wang, R.D. Larsen, C.J. Orella, A.L. Forman, R.N. Landau, J. Laquidara, J.R. Sowa, D.G. Blackmond, Y.-K. Sun, *Thermochim. Acta* **1996**, *289*, 189.

54 A. Haynes, *Mechanism in Homogeneous Catalysis – A Spectroscopic Approach*. B. Heaton (Ed.), Wiley-VCH, **2005**, Chapter 3, pp. 107.

55 (a) J.M. Brown, P.A. Chaloner, *J. Chem. Soc., Chem. Commun.* **1980**, 344; (b) J.M. Brown, P.A. Chaloner, *Homogeneous Catalysis with Metal Phosphine Complexes*. L.H. Pignolet (Ed.), Plenum Press, New York, **1983**, pp. 137; (c) J.M. Brown, *Chem. Soc. Rev.* **1993**, *22*, 25.

56 A review regarding experimental findings, which seemingly speak for the alternatively discussed dihydride mechanism, can be found in: I.D. Gridnev, T. Imamoto, *Acc. Chem. Res.* **2004**, *37*, 633. However, it must be stressed that verified results such as the pressure dependence of enantioselectivity cannot be described by this model. Models related to the dihydride mechanism and developed

substantially on the basis of low-temperature NMR spectroscopy have not yet been investigated in terms of kinetic consequences.

57 (a) J.M. Brown, P.A. Chaloner, G.A. Morris, *J. Chem. Soc., Chem. Commun.* **1983**, 664; (b) J.M. Brown, P.A. Chaloner, G.A. Morris, *J. Chem. Soc., Perkin Trans. II* **1987**, 1583; (c) H. Bircher, B.R. Bender, W. von Philipsborn, *Magn. Reson. Chem.* **1993**, *31*, 293; (d) R. Kadyrov, T. Freier, D. Heller, M. Michalik, R. Selke, *J. Chem. Soc., Chem. Commun.* **1995**, 1745; (e) J.A. Ramsden, T.D.W. Claridge, J.M. Brown, *J. Chem. Soc., Chem. Commun.* **1995**, 2469; (f) D. Heller, R. Kadyrov, M. Michalik, T. Freier, U. Schmidt, H.W. Krause, *Tetrahedron: Asymmetry* **1996**, *7*, 3025; (g) A. Kless, A. Börner, D. Heller, R. Selke, *Organometallics* **1997**, *16*, 2096.

58 In case of the asymmetric hydrogenation with Rh complexes this disturbance in the equilibrium establishment is shown in pressure-dependent e.e. values (see [21b]). Djega-Mariadassou and Boudart [3] describe this phenomenon as "kinetic coupling"; see also G. Djega-Mariadassou, *Catal. Lett.* **1994**, 7. In this context, we point out that under "kinetic coupling" conditions it is principally not possible experimentally to determine a partial order of 1 with respect to hydrogen.

59 D. Heller, R. Thede, D. Haberland, *J. Mol. Cat. A: Chemical* **1997**, *115*, 273.

60 This statement also applies if intermolecular pre-equilibria are not established. In case of established intermolecular pre-equilibria the value of $1/K_M$ corresponds to the sum of all stability constants.

61 (a) H.W. Krause, H. Foken, H. Pracejus, *New J. Chem.* **1989**, *13*, 615; (b) H.W. Krause, U. Schmidt, S. Taudien, B. Costisella, M. Michalik, *J. Mol. Catal. A: Chemical* **1995**, *104*, 147; (c) C. Döbler, H.-J. Kreuzfeld, M. Michalik, H.W. Krause, *Tetrahedron: Asymmetry* **1996**, *7*, 117; (d) H.J. Kreuzfeld, C. Döbler, U. Schmidt, H.W. Krause, *Amino Acids* **1996**, *11*, 269; (e) U. Schmidt, H.W. Krause, G. Oehme, M. Michalik, C. Fischer, *Chirality* **1998**, *10*, 564;

(f) C. Döbler, H.-J. Kreuzfeld, C. Fischer, M. Michalik, *Amino Acids* **1999**, *16*, 391; (g) T. Dwars, U. Schmidt, C. Fischer, I. Grassert, H.W. Krause, M. Michallik, G. Oehme, *Phosphorus, Sulfur, and Silicon* **2000**, *158*, 209.

62 (a) H.-J. Drexler, W. Baumann, T. Schmidt, S. Zhang, A. Sun, A. Spannenberg, C. Fischer, H. Buschmann, D. Hel-

ler, *Angew. Chem. Int. Ed.* **2005**, *44*, 1184; (b) H.-J. Drexler, S. Zhang, A. Sun, A. Spannenberg, A. Arrieta, A. Preetz, D. Heller, *Tetrahedron: Asymmetry* **2004**, *15*, 2139.

63 T. Schmidt, W. Baumann, H.-J. Drexler, A. Arrieta, D. Heller, H. Buschmann, *Organometallics* **2005**, *24*, 3842.

Part II
Spectroscopic Methods in Homogeneous Hydrogenation

The Handbook of Homogeneous Hydrogenation.
Edited by J. G. de Vries and C. J. Elsevier
Copyright © 2007 WILEY-VCH Verlag GmbH & Co. KGaA, Weinheim
ISBN: 978-3-527-31161-3

11
Nuclear Magnetic Resonance Spectroscopy in Homogeneous Hydrogenation Research

N. Koen de Vries

11.1
Introduction

11.1.1
Nuclear Magnetic Resonance (NMR)

NMR spectroscopy is a very powerful tool for structural characterization purposes. Following the discovery of the technique, organic chemists were quick to recognize its potential and as soon as commercial instruments came on the market they began to use NMR to elucidate the structures of their synthetic efforts. During the late 1960s and early 1970s, some very important developments took place that made the technique what it is today. The first development was that of the so-called Fourier transform (or FT-NMR) measurement [1]. Until then, the magnetic field was varied, and as it swept through the whole frequency range a recorder noted the moment when a nucleus was on resonance. In the FT-mode, all frequencies are excited at the same time by a short pulse, and the response in time of all the nuclei together is recorded as they relax back to equilibrium. The Fourier transformation converts this time-dependent signal to a frequency-dependent signal, which is the NMR-spectrum. As in all FT-techniques, the major advantage is that one can acquire more than one scan. This improves the signal-to-noise ratio because it increases with the square root of the number of scans. From then on, insensitive nuclei such as ^{13}C-NMR could also be measured routinely. This nucleus has a much larger chemical shift range and thus allows a better resolution.

The second development that has revolutionized the practice of NMR was the introduction of multidimensional spectroscopy. This was initialized by Jeener [2], who showed that, by introducing a second pulse and varying the time between them, a second "time-axis" could be constructed. A double Fourier transformation yields the familiar two-dimensional spectrum, nowadays known by everyone as the COSY spectrum. Ernst, already involved in the development of FT-NMR, showed that the concept was more generally applicable [3], and paved

The Handbook of Homogeneous Hydrogenation.
Edited by J.G. de Vries and C.J. Elsevier
Copyright © 2007 WILEY-VCH Verlag GmbH & Co. KGaA, Weinheim
ISBN: 978-3-527-31161-3

the way for the whole variety (hundreds, if not thousands) of multidimensional experiments that exist today. A typical and well-known example of what can be achieved is the complete three-dimensional structure elucidation of peptides consisting of hundreds of amino acids.

NMR continued to develop during the 1980s and 1990s with techniques such as inverse NMR [4] and gradient pulses [5]. It would also be valuable to mention here the various instrumental improvements such as cryo-probes and the ever-increasing magnetic-field strength (with commercial 1000-MHz spectrometers almost in reach), all aimed at improving resolution and sensitivity.

It is this relative insensitivity that is usually considered as the major drawback of NMR spectroscopy. However, the flexibility of the NMR technique, with the ability to obtain structural information, quantitative data (e.g. kinetic parameters), as well as an indication of molecular volume, using pulsed gradient spin echo (PGSE) NMR diffusion methods [6], makes NMR a most valuable tool.

11.1.2
NMR in Homogeneous Hydrogenation Research

On the subject of NMR spectroscopy in homogeneous hydrogenation research, we can recognize that most developments have found their way into this research area, albeit with some delay. For example, the PGSE methods had been used for over twenty years to determine diffusion coefficients of organic molecules before their potential in organometallic chemistry was investigated [7]. Pioneering studies were also conducted by von Philipsborn who, during the early 1980s, was already performing transition metal NMR spectroscopy in order to probe structures of organometallic compounds and their reactivity. For further information, the reader is referred to an excellent review on this topic [8].

In this chapter, we will provide an introduction into the application of NMR in the area of homogeneous hydrogenation research, and suggest further reading for those who wish to obtain more information on the subject. The principles of NMR spectroscopy have been covered thoroughly by Ernst [9], and numerous books and review articles have appeared on all aspects of the application of NMR in homogeneous hydrogenation research. Typical reviews include "NMR and homogeneous catalysis" in general [10], "NMR at elevated gas pressures and its application to homogeneous catalysis" [11], the measurement of transition metals by NMR [12] and, more recently, "^{103}Rh NMR spectroscopy and its application to rhodium chemistry" [13].

Other important topics, such as the use of *para*-hydrogen-induced polarization (PHIP) NMR, are discussed in more detail elsewhere in this book. Basically, this approach enhances the NMR signal one thousandfold, thus allowing the detection of intermediates that go unnoticed when using "classical" NMR techniques. PHIP is particularly suited for homogeneous hydrogenation research because a prerequisite of the method is that both former *para*-hydrogen nuclei must be present (and J-coupled) in the molecule of interest.

In summary, NMR in homogeneous hydrogenation research is used for:

- Structure elucidation of various species present in the reaction mixtures; typical tools used for this purpose include chemical shifts, coupling constants, PHIP-NMR, and 2D-NMR.
- Determination of reaction mechanisms by combining the observed intermediates in a catalytic cycle. To do this, it is often necessary to measure under different conditions – that is, variable temperature NMR. The use of high-pressure NMR cells is crucial in order to measure under the real catalytic conditions. The EXSY experiment helps to unravel exchange pathways, both intra- and intermolecular.
- Determination of the reaction kinetics if the reaction is slow enough to record a series of NMR spectra. This can be done with standard 1D-NMR measurements if the concentrations are high enough, and if not, by making use of sensitivity enhancement through *para*-hydrogen. NMR experiments designed for kinetic investigations in combination with PHIP techniques include ROCHESTER (Rates Of Catalytic Hydrogenation Estimated Spectroscopically Through Enhanced Resonances) [14] and DYPAS. An example of the investigation of kinetics of homogeneous hydrogenation reactions using both experiments can be found in [15].

11.2
NMR Methods

11.2.1
General

In order to perform the various tasks mentioned in Section 11.1.2, it is necessary to use one or several methods to gather information by NMR spectroscopy. Typically, chemical shift and coupling constant information, 2D-NMR measurements, variable temperature or pressure studies are used. If appropriate, specific examples of the particular topic as applied in homogeneous hydrogenation research are detailed below.

The most popular nucleus for NMR measurements is the proton, and this is obviously related to its high sensitivity. As a rotating charged particle, a proton will generate a magnetic moment. Because, for a proton, the spin quantum number $I = 1/2$, there are two possible spin states. In the absence of a field these states have the same energy, but in a strong magnetic field they are no longer equivalent. The population of the lower level is slightly higher than that of the upper level, according to the Boltzmann distribution. This results in a small excess magnetization vector along the z-axis. By applying a radiofrequency field for a short time, this vector can be moved and ends up in the x-y-plane (90° pulse), where it will rotate. This time-response is registered by a radiofrequency coil aligned along the x-axis. This oscillatory decay (or FID) is transformed to a frequency response spectrum by the FT calculation.

11.2.2
Chemical Shift

11.2.2.1 General

The chemical shift is caused by shielding of the proton due to a local field, generated by circulations of electrons. It is this local effect that makes it possible to use NMR as a structural characterization tool. For ^{1}H-NMR there is a relatively straightforward correlation between chemical shift and electron density, because only diamagnetic contributions play a role, but for most other nuclei a paramagnetic term also contributes to the shielding. This makes it difficult to relate, for example, the geometry around a metal with its chemical shift [13]. For qualitative discussions of chemical shifts, the following simplified expression can be used:

$$\delta = -A + B \times (k^2 \langle r^{-3} \rangle)(\Delta E)^{-1}$$

where r is the valence shell radius and ΔE the ligand field-splitting energy.

If either one of the terms dominates, it is sometimes possible to observe correlations. For example, a linear dependence of ^{195}Pt chemical shift with UV-visible data for a series of octahedral platinum complexes has been identified [16].

Another example is the linear correlation of the ^{59}Co chemical shifts of the catalyst with the regioselectivity of a trimerization reaction of acetylenes [15].

Despite the difficulties in explaining the metal NMR shifts, it is still worthwhile measuring them because the huge chemical shift range makes it easy to observe the different species present, for example diastereoisomers [18].

Bender et al. [19] used ^{103}Rh-NMR chemical shift differences to demonstrate an electronic difference at rhodium between two diastereoisomers, and suggested that this might influence the crucial hydrogen addition step.

Another popular NMR nucleus is ^{31}P; this is because it is quite sensitive, it already has a large chemical shift range (as compared to ^{1}H-NMR), the phosphorus atom is usually coupled directly to the metal atom, and phosphorus is often present in ligands of homogeneous hydrogenation complexes. Verkade [20] has reviewed aspects of ^{31}P-NMR including chemical shifts, coupling constants and 2D-NMR experiments involving ^{31}P.

11.2.2.2 Chemical Shifts in Homogeneous Hydrogenation Research

In the context of ^{1}H chemical shifts and determination of the reaction mechanism of homogeneous hydrogenation catalysts, one usually tries to observe hydride-intermediates that typically resonate at high field (–5 to –30 ppm). Agostic bonds (see Fig. 11.1) also tend to have a hydride-like proton chemical shift.

Similarly, bridging hydrides have negative chemical shifts (e.g., at –11.18 ppm for the bridging hydride in Figure 11.2 [21]).

An excellent example of the use of ^{31}P chemical shifts, in combination with ^{1}H-NMR and ^{103}Rh-NMR data as well as coupling constants information, can be found in the report by Duckett et al. [21] on the activation of

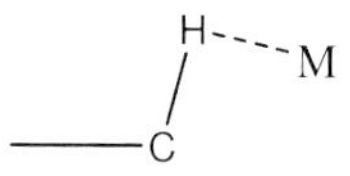

Fig. 11.1 An agostic bond.

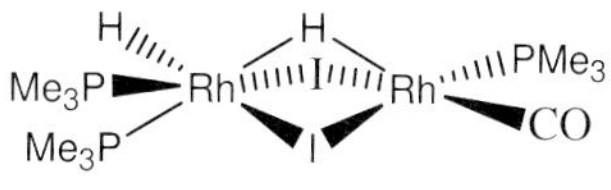

Fig. 11.2 A bridging hydride.

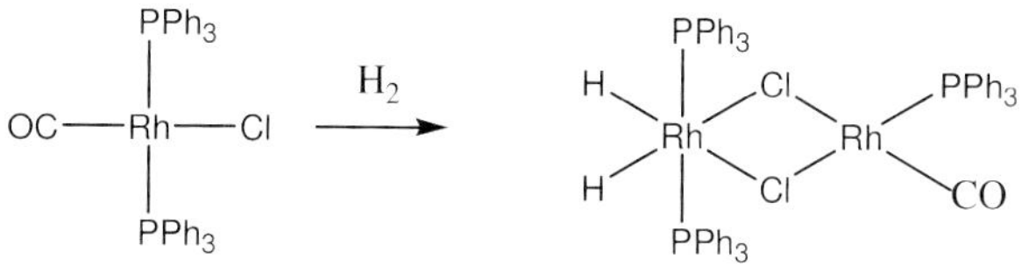

Fig. 11.3 Activation of RhCl(CO)(PPh$_3$)$_2$.

RhX(CO)(PPh$_3$)$_2$ [X=halogen] complexes with hydrogen. The reaction for X=Cl is shown in Figure 11.3.

The resulting product showed a single ^{31}P chemical shift at 40 ppm with a ^{1}J coupling to rhodium of 118 Hz, indicative of a rhodium (III) center. As this rhodium atom was shown to couple to two phosphorus atoms, it could be concluded that the phosphorus atoms are in a symmetrical position, as shown in Figure 11.3. The other Rh atom couples only to a phosphorus atom with a ^{1}J$_{RhP}$ of 196 Hz, indicative of a rhodium (I) center.

When the smaller P(Me)$_3$ ligand was used, not only the halogens but also a hydride were found in the bridging position, as was concluded from the fact that the phosphorus atom connected to the rhodium (I) center was now coupled to a hydride.

11.2.3
Coupling Constant

A second aspect that is important for the structural characterization, as mentioned above, is that of the coupling constant. Both the magnitude of the coupling constant and the multiplicity (which reflects the number of (equivalent) nuclei that couple) are important. For example, the magnitude of the ^{1}J$_{RhH}$ coupling constant provides information about the degree of s-character of the bond [13].

In general, in square-planar and octahedral complexes ^{1}J$_{ML}$ depends on the *trans*-ligand: stronger donors reduce the ^{1}J$_{ML}$-value [10]. Similarly, ^{2}J$_{L1,M,L2}$-values are usually greater for *trans*-interactions than for *cis*-interactions [10], which is clearly an important tool for structural characterization (see Fig. 11.4).

The binuclear Rh-complex in Figure 11.2 contains two hydride resonances at −11.18 and −14.75 ppm [21]. The former has two smaller phosphorus couplings

Fig. 11.4 *Trans* (left) and *cis* (right) interactions.

($^2J_{PH}$ = 21 Hz and 16 Hz) and one large coupling ($^2J_{PH}$ = 99.5 Hz), indicative of a *trans* arrangement, as mentioned above. In addition, two essentially equal couplings to the two rhodium centers were found, which is consistent with the bridge position.

Gridnev et al. showed in their study of the asymmetric hydrogenation of enamides by Rh-catalysts another useful application of coupling constant patterns. By selectively labeling certain atoms, for example with ^{13}C or ^2D, additional couplings appear (as compared to the non-labeled product) and this will provide information about the exact structure [22].

11.2.4
2D-NMR

11.2.4.1 General

As mentioned above, 2D-NMR (or more generally multidimensional NMR) is based on the transfer of magnetization during the evolution/mixing period.

The transfer of magnetization is not restricted to protons as in the $^1H-^1H$ COSY experiment, but can also be applied between other nuclei (i.e., $^1H-^{13}C$-correlation or $^{31}P-^{31}P$ correlation experiments).

A major advance in the field has been the reverse or inverse detected NMR experiment [4]. Originally, signals were enhanced by transferring magnetization from protons to insensitive nuclei that were consequently detected, as in the INEPT sequence. By reversing this process and detecting the more abundant and receptive nucleus (mostly ^1H, but it can also be ^{19}F or ^{31}P), the sensitivity of the experiment was greatly enhanced. An additional advantage is that the relaxation delay between the pulses is now governed by the ^1H relaxation time, which is usually relatively short compared to relaxation time of other nuclei.

Technically, the inverse experiment used to be very demanding because the excess of protons not coupled to the nucleus of interest (e.g., protons coupled to the almost hundred-fold excess of ^{12}C instead of ^{13}C) needed to be suppressed. Originally, this was achieved by the use of elaborate phase-cycling schemes, but today the coherence pathway selection by gradient pulses facilitates this process.

11.2.4.2 2D-NMR in Homogeneous Hydrogenation Research

A variety of examples of 2D-NMR experiments is provided in reference [21]. The structure elucidation of the di-rhodium compound shown in Figure 11.3 was mostly carried out in this way. For example, 2D $^1H-^{31}P$ heteronuclear multiple quantum correlation (HMQC) experiments were used to show that two rhodium-coupled hydride resonances are connected to a single type of ^{31}P nucleus.

A 2D ^{1}H–^{103}Rh HMQC experiment was used to show that both the hydrides are connected to a Rh center that resonates at 925 ppm and is coupled to two ^{31}P-nuclei. A ^{1}H–^{1}H COSY experiment shows that the two hydrides are coupled.

An example of a ^{103}Rh–^{31}P correlation experiment of Rh(MonoPhos)$_2$-(COD)BF$_4$ is shown in Figure 11.5. Here also, two inequivalent ^{31}P atoms are connected to the same Rh center [23].

A second class of multidimensional experiments uses through-space interactions; perhaps the best-known example is the NOESY sequence, which is as follows:

90° (+x) – evolution time – 90° (–x) – mixing time – 90° (+x) – acquisition

As with the COSY experiment, the sequence starts with a pulse followed by an evolution period, but now the mechanism that couples the two spins (which must be in close proximity, typically <6 Å) is the Nuclear Overhauser Effect (NOE). The second pulse converts magnetization into population disturbances, and cross-relaxation is allowed during the mixing time. Finally, the third pulse transfers the spins back to the x-y-plane, where detection takes place. The spectrum will resemble a COSY spectrum, but the off-diagonal peaks now indicate through-space rather than through-bond interactions.

Just as in the COSY type of experiments this cross-relaxation effect is not restricted to protons, but can also involve heteronuclei; the acronym HOESY (heteronuclear Overhauser effect) is used in these cases. This can be used, for example, to show that an anion such as BF$_4^-$ is in close proximity to the ligands of the organometallic compound, as was carried out by Macchioni et al. with a ^{19}F-^{1}H HOESY experiment [24].

Another important experiment here is that of EXSY (exchange spectroscopy). Here, the pulse sequence is identical to the NOESY sequence, but during the mixing time the spins physically migrate to another site due to slow chemical exchange [25]. This exchange can be both intra- and intermolecular. An example of a ^{31}P–^{31}P EXSY spectrum of the hydrogenation catalyst Rh(MonoPhos)$_2$-(COD)BF$_4$ is shown in Figure 11.6.

This molecule exists of isomers of the complex shown in Figure 11.7 that inter-convert rapidly, even at temperatures below 200 K, as shown by the off-diagonal peaks between the double doublets of one structure and the broad doublet of the other structure [23]. The isomeric structures are caused by the relative positions of the two ligands with respect to each other (i.e., parallel or anti-parallel orientation of the NMe$_2$ groups).

By recording EXSY spectra at different temperatures, information can be obtained about exchange rates and activation parameters. Duckett et al. [21] used 2D-EXSY experiments to show that, for the dihydride shown in Figure 11.3, the hydride exchange process is intramolecular. The only observable off-diagonal peaks are between the two hydride resonances. Because of a negative entropy of activation of –61 J K^{-1} mol^{-1} it was concluded that a step with a degree of ordering is required: an additional rotation step around a Rh–halogen bond (see Fig. 11.8).

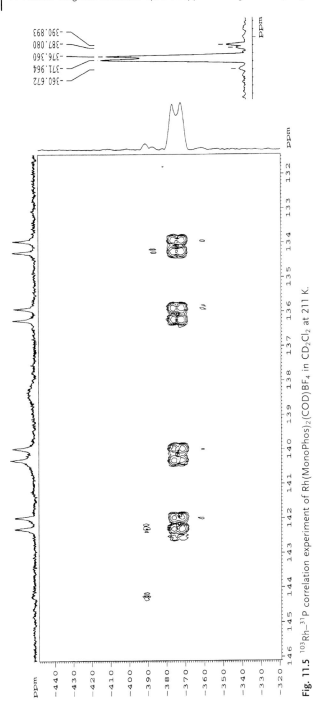

Fig. 11.5 ^{103}Rh–^{31}P correlation experiment of Rh(MonoPhos)$_2$(COD)BF$_4$ in CD$_2$Cl$_2$ at 211 K.

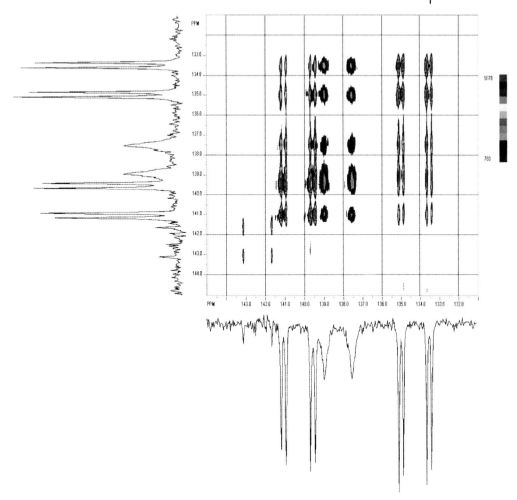

Fig. 11.6 ^{31}P–^{31}P EXSY spectrum of Rh(MonoPhos)$_2$(COD)BF$_4$ in C$_2$D$_2$Cl$_2$.

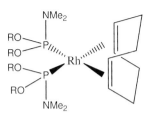

Fig. 11.7 Rh(MonoPhos)$_2$(COD)BF$_4$.

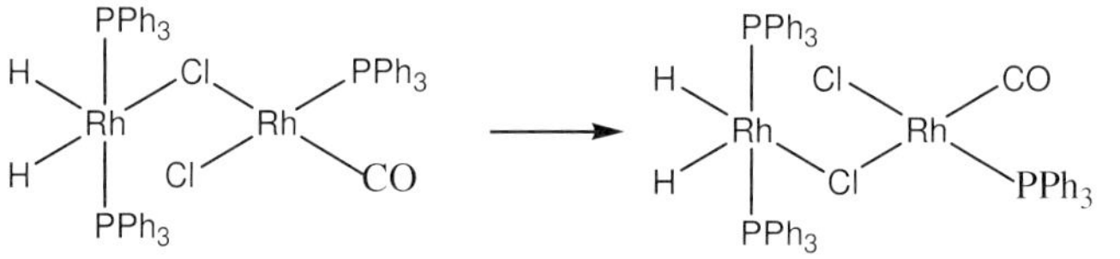

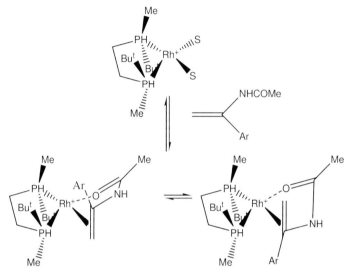

Fig. 11.8 Rotation about the Rh–Cl bond.

Fig. 11.9 Intermolecular exchange of Rh–*t*-Bu-BisP* complex at 348 K.

Gridnev et al. studied the mechanism of the enantioselective hydrogenation of enamides with Rh–BisP* and Rh–MiniPHOS catalysts [22].

By using EXSY measurements, it was shown that at 323 K there is only intramolecular exchange, whereas at 348 K intermolecular exchange also occurs via complete dissociation of the complex, producing a free substrate and a solvate complex (Fig. 11.9).

One example of the use of 2D-NMR experiments in conformational analysis is the study of molecular interactions between cinchonidine and acetic acid [26]. These alkaloids are used as chiral auxiliaries in enantioselective hydrogenations, and the enantiomeric excess is dependent on solvent polarity, acetic acid being a good solvent. This suggests that protonation and a preferred conformation play a role in achieving high enantioselectivities. With a combination of COSY-experiments, ^3J coupling constants and NOESY experiments, it was shown that one conformer is preferred in acidic solutions.

11.2.5
Variable Temperature and Variable Pressure Studies

11.2.5.1 General

Both temperature and pressure are important parameters/variables in NMR measurements of homogeneous hydrogenation catalysts. Usually, a certain hydrogen pressure is needed to form the active catalyst. The temperature controls the rate of reactions. Sometimes, temperatures above room temperature are needed; for example, the reaction shown in Figure 11.3 occurs at a hydrogen pressure of 3 atmos and temperatures above 318 K. In other cases, intermediates can only be observed at temperatures below room temperature. Modern NMR instruments routinely allow measurements to be made in the range of, for example 170 to 410 K, but this range can easily be extended by the use of special NMR probes.

As mentioned above, variable-temperature measurements are also used to obtain exchange rates and activation parameters. In the slow exchange regime, the average lifetime of a spin is long enough to observe individual lines for the various states. By increasing the exchange rate (e.g., by raising the temperature), the individual lines broaden and shift to each other until they merge into one single line. This is termed the "coalescence phenomenon" (for background information, see [27]).

11.2.5.2 Variable-Temperature Studies in Homogeneous Hydrogenation Research

Figure 11.10 illustrates an example of a ^{31}P-NMR variable-temperature measurement, in this case of the isomeric complexes of $Rh(MonoPhos)_2(COD)BF_4$ (see

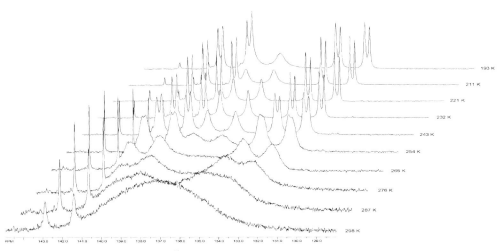

Fig. 11.10 Temperature-dependency of ^{31}P-NMR spectra of $Rh(MonoPhos)_2(COD)BF_4$ in CD_2Cl_2.

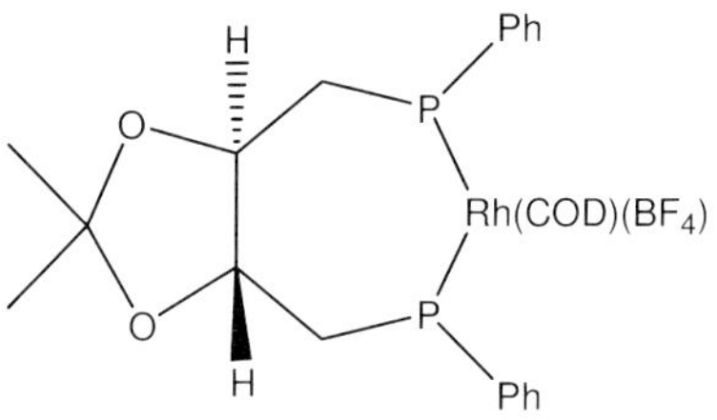

Fig. 11.11 Hydrogenation catalyst showing temperature-dependent ^{31}P-NMR spectra.

Fig. 11.7) [23]. This measurement shows clearly the broadening and merging of the peaks as the temperature is increased.

Another example is the temperature-dependent study of the ^{31}P-NMR spectra of the hydrogenation catalyst shown in Figure 11.11 [28]. At 240 K the ^{31}P-NMR signals are equivalent, whereas at 123 K two sets of signals are observed which are attributed to a seven-membered ring in a twist-chair and a distorted boat conformation.

11.2.5.3 **Variable-Pressure Studies in Homogeneous Hydrogenation Research**

The use of high-pressure cells and probes has been reviewed recently [29]. For homogeneous hydrogenation reactions, there is no requirement for very high pressures, and in these cases the popular sapphire tubes designed by Roe [30] will do well. In fact, the present author's group has used this type of tube in the laboratory to carry out hydrogenations, carbonylations and hydroformylations, and they have been found to be very convenient to work with. They can be pressurized in a safety hood and transported to the NMR in a transport cylinder [11]. Moreover, since the sapphire tube is transparent, color changes can also be observed.

When the tube has been placed inside the NMR spectrometer, the procedure used is exactly as for a standard NMR tube.

An example of these pressure studies is provided by the studies of Elsevier et al. [31], who investigated the dependence of the hydrogenation rate of 4-octyne by a Pd-catalyst on the dihydrogen pressure, which was varied between 0 and 40 bar. The hydrogenation rate was shown to depend linearly on the dihydrogen pressure. In order to elucidate the reaction mechanism, the dependence of the reaction rate on substrate and catalyst concentration, and on the temperature, was also measured. NMR experiments with deuterium gas as well as PHIP-experiments were also carried out.

Another interesting application of high-pressure tubes is the *in-situ* investigation of reactions in supercritical solvents such as carbon dioxide. For example, the iridium-catalyzed enantioselective hydrogenation of imines was investigated in a sapphire tube at 313 K [32].

Despite the advantages and ease of use of sapphire tubes, great care must be taken. It is very important that protective measurements are put in place when a pressurized tube is handled outside the spectrometer, because the tube's ability to cope with high pressure can be diminished if the sapphire surface is scratched. Another potentially weak point is the glue that is used to fix the titanium head to the sapphire tube. In the author's laboratory, on one occasion the head separated from the tube and was launched to the ceiling when operating close to the tube's design limits (for both temperature and pressure). Consequently, it is advised that the transport cylinder be left on top of the magnet when the tube is in the spectrometer.

11.2.6
PGSE NMR Diffusion Methods

Pulsed-field gradient NMR methods are relatively new in homogeneous hydrogenation research. They allow the investigation of the diffusion of a molecule in solution by applying a defocusing and refocusing gradient pulse and measuring the decrease in intensity of the NMR signals belonging to this molecule. By carrying out a series of measurements with increasing gradient strengths, a diffusion coefficient can be calculated. If the technique is applied to a mixture of compounds and the resulting diffusion coefficients are plotted against the chemical shifts in a "2D-like fashion", the experiment is named DOSY (diffusion-ordered spectroscopy).

An example of the use of PGSE NMR spectroscopy can be found in the studies of Selke et al. [33], who investigated the dependence of enantioselectivity on the distribution of a chiral hydrogenation catalyst between aqueous and micellar phases. When a compound is incorporated into a micelle, its mobility is much lower compared to its mobility in solution. This effect is exactly what is probed with PGSE NMR. The calculated diffusion coefficient is a time-averaged value of the lower diffusion coefficient of the catalyst incorporated into the micelles, and of the diffusion coefficient of the free catalyst. An increased amount of micelle-embedded catalyst was found to lead to an increased enantioselectivity.

11.3
Outlook

The use of NMR continues to improve existing methods, and to develop new concepts. By cleverly combining existing pulse-sequences, new sequences are formed with improved properties. An example is the combination of the COSY and DOSY sequence to a new 3D-NMR COSY-IDOSY sequence with improved sensitivity, a 32-fold decrease in experiment time, and an improved resolution resulting in better data analysis [34].

Other developments are based on a completely new concept, for example the possibility of recording multi-dimensional NMR spectra with a single scan which, in theory, will allow the elucidation of structures to be made in seconds [35].

In time, all of these developments will be used in future chemical research studies, making NMR an even more useful tool than it is today.

Abbreviations

DOSY	diffusion-ordered spectroscopy
EXSY	exchange spectroscopy
FT	Fourier transform
HMQC	heteronuclear multiple quantum correlation
HOESY	heteronuclear Overhauser effect
NMR	nuclear magnetic resonance
NOE	nuclear Overhauser effect
PGSE	pulsed-gradient spin echo
PHIP	*para*-hydrogen-induced polarization
ROCHESTER	Rates Of Catalytic Hydrogenation Estimated Spectroscopically Through Enhanced Resonances

References

1 R. R. Ernst, W. A. Anderson, *Rev. Sci. Instr.* **1966**, *37*, 93.

2 J. Jeener, Ampere International Summer School, Basko Polje, Yugoslavia, **1971**.

3 W. P. Aue, E. Bartholdi, R. R. Ernst, *J. Chem. Phys.* **1976**, *64*, 2229.

4 L. Müller, *J. Am. Chem. Soc.* **1979**, *101*, 4481.

5 (a) R. E. Hurd, *J. Magn. Reson.* **1990**, *87*, 422; (b) A. Bax, P. G. de Jong, A. F. Mehlkopf, J. Schmidt, *Chem. Phys. Lett.* **1978**, *55*, 9.

6 E. O. Stejskal, J. E. Tanner, *J. Chem. Phys.* **1965**, *42*, 288.

7 M. Valentini, P. S. Pregosin, H. Ruegger, *Organometallics* **2000**, *19*, 2551.

8 W. von Philipsborn, *Chem. Soc. Rev.* **1999**, *28*, 95.

9 R. R. Ernst, G. Bodenhausen, A. Wokaun, *Principles of Nuclear Magnetic Resonance in One and Two Dimensions*, Oxford University Press, Oxford, **1987**.

10 E. M. Viviente, P. S. Pregosin, D. Scott. In: B. Heaton (Ed.), *Mechanisms in Homogeneous Catalysis. A Spectroscopic Approach.* Wiley-VCH, Weinheim, **2005**.

11 C. J. Elsevier, *J. Mol. Catal.* **1994**, *92*, 285.

12 P. S. Pregosin (Ed.), *Transition Metal Nuclear Magnetic Resonance*, Elsevier, Amsterdam, **1991**.

13 J. M. Ernsting, S. Gaemers, C. J. Elsevier, *Magn. Reson. Chem.* **2004**, *42*, 721.

14 M. S. Chinn, R. J. Eisenberg, *J. Am. Chem. Soc.* **1992**, *114*, 1908.

15 P. Hübler, R. Giernoth, G. Kümmerle, J. Bargon, *J. Am. Chem. Soc.* **1999**, *121*, 5311.

16 W. Juranic, *J. Chem. Soc., Dalton Trans.* **1984**, 1537.

17 W. von Philipsborn, *Pure Appl. Chem.* **1986**, *58*, 513.

18 C. Weidemann, W. Priebsch, D. Rehder, *Chem. Ber.* **1989**, *122*, 235.

19 B. R. Bender, M. Koller, D. Nanz, W. von Philipsborn, *J. Am. Chem. Soc.* **1993**, *115*, 5889.

20 L. D. Quin, J. G. Verkade (Eds.), *Phosphorus-31 NMR Spectral Properties in Compound Characterization and Structural Analysis*, VCH Publishers, New York, **1994**.

21 P. D. Morran, S. B. Duckett, P. R. Howe, J. E. McGrady, S. A. Colebrooke, R. Eisenberg, M. G. Partridge, J. A. B. Lohman, *J. Chem. Soc., Dalton Trans.* **1999**, 3949.

22 I. D. Gridnev, M. Yasutake, N. Higashi, T. Imamoto, *J. Am. Chem. Soc.* **2001**, *123*, 5268.

23 M. van den Berg, Dissertation, University of Groningen, The Netherlands, **2006**.

24 B. Binotti, C. Carfagna, E. Foresti, A. Macchioni, P. Sabatino, C. Zuccaccia, D. Zuccaccia, *J. Organomet. Chem.* **2004**, *689*, 647.

25 S. Schaeublin, A. Hoehener, R. R. Ernst, *J. Magn. Reson.* **1974**, *13*, 196.

26 D. Ferri, T. Bürgi, A. Baiker, *J. Chem. Soc., Perkin Trans. 2,* **1999**, 1305.

27 J. J. Delpuech (Ed.), *Dynamics of Solutions and Fluid Mixtures by NMR*, John Wiley & Sons Ltd, Chichester, **1995**.

28 R. Kadyrov, A. Borner, R. Selke, *Eur. J. Inorg. Chem.* **1999**, 705.

29 G. Laurenczy, L. Helm. In: B. Heaton (Ed.), *Mechanisms in Homogeneous Catalysis. A Spectroscopic Approach*, Wiley-VCH, Weinheim, **2005**.

30 D. C. Roe, *J. Magn. Reson.* **1985**, *63*, 388.

31 A. M. Kluwer, T. S. Koblenz, Th. Jonischkeit, K. Woelk, C. J. Elsevier, *J. Am. Chem. Soc.* **2005**, *127*, 15470.

32 S. Kainz, A. Brinkmann, W. Leitner, A. Pfaltz, *J. Am. Chem. Soc.* **1999**, *121*, 6421.

33 M. Ludwig, R. Kadyrov, H. Fiedler, K. Haage, R. Selke, *Chem. Eur. J.* **2001**, *7*(15), 3298.

34 M. Nilsson, A. M. Gil, I. Degadillo, G. Morris, *Chem. Commun.* **2005**, 1737.

35 L. Frydman, L. Lupulescu, *Proc. Natl. Acad. Sci. USA* **2002**, *99*, 15858.

12
Parahydrogen-Induced Polarization: Applications
to Detect Intermediates of Catalytic Hydrogenations

Joachim Bargon

12.1
In-Situ Spectroscopy

The level of understanding the mechanisms of homogeneously catalyzed reactions correlates with information about reactive intermediates. Therefore, *in-situ* methods – that is, investigations based upon physical techniques conducted *during* the reactions – are highly desirable and important. For this purpose, a considerable scope of *in-situ* techniques based upon time-proven classical types of spectroscopy has been developed. Since many of the industrially important catalyzed reactions require the use of high pressures and high temperatures for optimum performance, ideally, these techniques should accommodate such conditions. Contemporary *in-situ* methods include specialized forms of optical or magnetic resonance spectroscopy, in particular vibration and rotation spectroscopy, such as infrared or Raman spectroscopy, or nuclear magnetic resonance (NMR) [1, 2]. This chapter will focus on the latter technique.

12.1.1
In-Situ NMR Spectroscopy

For the elucidation of chemical reaction mechanisms, *in-situ* NMR spectroscopy is an established technique. For investigations at high pressure either sample tubes from sapphire [3] or metallic reactors [4] permitting high pressures and elevated temperatures are used. The latter represent autoclaves, typically machined from copper-beryllium or titanium-aluminum alloys. An earlier version thereof employs separate torus-shaped coils that are imbedded into these reactors permitting *in-situ* probing of the reactions within their interior. However, in this case certain drawbacks of this concept limit the filling factor of such NMR probes; consequently, their sensitivity is relatively low, and so is their resolution. As a superior alternative, the metallic reactor itself may function as the resonator of the NMR probe, in which case no additional coils are required. In this way gas/liquid reactions or reactions within supercritical fluids can be studied

The Handbook of Homogeneous Hydrogenation.
Edited by J.G. de Vries and C.J. Elsevier
Copyright © 2007 WILEY-VCH Verlag GmbH & Co. KGaA, Weinheim
ISBN: 978-3-527-31161-3

conveniently at high pressure and elevated temperatures with much higher sensitivity [4].

12.1.2
In-Situ PHIP-NMR Spectroscopy

A number of years ago, an additional *in-situ* method to investigate hydrogenation reactions via NMR spectroscopy was introduced, which gives rise to a significant signal enhancement primarily of the proton NMR spectra; consequently, the intermediates and minor reaction products of homogeneous hydrogenations can be identified much more readily. This signal enhancement is due to proton spin polarization originally derived from separating the spin isomers of dihydrogen, which can readily be achieved at low temperatures. Using only one of these spin isomers for the hydrogenation – namely parahydrogen – NMR spectra recorded during homogeneous hydrogenations of various substrates in the presence of suitable organometallic catalysts exhibit considerable signal enhancement in the form of intense emission or absorption lines. In principle, either one of the two spin isomers – that is, ortho- or parahydrogen – may be used instead of ordinary dihydrogen (H_2), though typically parahydrogen is employed. This phenomenon was originally observed experimentally by chance [5], but it has later been independently thought of theoretically, and has been attributed to symmetry breaking during the hydrogenation reaction [6].

This concept has originally been named PASADENA (*P*arahydrogen *A*nd *S*ynthesis *A*llow *D*ramatically *E*nhanced *N*uclear *A*lignment) [6], but the spectroscopic method based on this phenomenon has subsequently also been called PHIP (*P*ara-*H*ydrogen *I*nduced *P*olarization) [7]. In this chapter the abbreviation PHIP will be used throughout.

The signal enhancement due to this approach can, in principle, be as high as 10^5-fold – that is, equal to the reciprocal Boltzmann factor; however, the experimentally achievable enhancement factors typically range between 10 and 10^3. Thanks to this increase in sensitivity, the PHIP phenomenon, therefore, provides for a powerful tool to investigate the fate of the dihydrogen, the catalysts, and of the substrates during hydrogenation reactions.

The observed polarization is primarily associated with the former parahydrogen protons. However, other protons may also experience a drastic signal enhancement due to nuclear spin polarization transferred to these nuclei via the nuclear Overhauser effect (NOE) or similar processes, both in the final reaction products as well as in their precursor intermediates.

Likewise, the original proton polarization can be transferred to other magnetically active heteronuclei, notably to ^{13}C, ^{15}N, ^{19}F, ^{29}Si, ^{31}P or appropriate isotopes of various transition metals. This is especially attractive because of the frequently low sensitivity of many heteronuclei, in particular of those with low magnetic moments [8].

Furthermore, in favorable cases, (para-)hydrogenation-derived spin polarization may be carried on to subsequent follow-up reactions, where it can serve to boost the notoriously low sensitivity of NMR spectroscopy in general. For this purpose, the follow-up reactions have to be reasonably fast, such as the bromination of alkenes, for example [51].

The polarization patterns are dependent upon the strength of the magnetic field, in which the reactions are carried out. If the reactions are carried out at high fields (i.e., notably within the NMR spectrometer), the resonances appear in "antiphase" – that is, there is an equal number of absorption and emission lines and no net polarization. At low field however (i.e., when the reaction is carried out at zero or a very low field and then transferred into the high field of the NMR spectrometer for subsequent investigation), the resonances display net polarization, as has been outlined by Pravica and Weitekamp [9].

Initially in this chapter, the various features of the PHIP phenomenon, of the apparatus to enrich parahydrogen and orthodeuterium, and of the computer-based analysis or simulations of the PHIP spectra to be observed under specific assumptions will be outlined. In the following sections, comparisons of the experimentally obtained and of the simulated spectra reveal interesting details and mechanistic information about the hydrogenation reactions and their products.

12.2
Ortho- and Parahydrogen

H_2 consists of two nuclear spin isomers, namely p-H_2 with antiparallel proton spins, and o-H_2 with parallel arrangement. Whereas p-H_2 is diamagnetic (i.e., it represents a singlet state (S)), o-H_2 shows nuclear paramagnetism with a resulting nuclear spin of one (Fig. 12.1). Its three allowed alignments relative to an external magnetic field – namely with, against, or orthogonal to the field – are labeled T_{+1}, T_{-1} or T_0, respectively (Fig. 12.2). Accordingly, in the absence of a magnetic field, o-H_2 is degenerate threefold, and in thermal equilibrium at room temperature, n-H_2 consists of 25.1% p-H_2 and of 74.9% o-H_2; that is, the ratio is approximately 1:3 [10].

Due to symmetry requirements, only specific quantum states of the spin and of the rotational states of H_2 are mutually compatible. Therefore, the energeti-

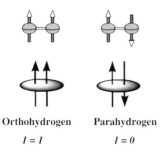

Orthohydrogen   Parahydrogen

$I = 1$ $I = 0$

magnetically active   *magnetically inactive*

Fig. 12.1 The spin isomers of molecular hydrogen ("dihydrogen").

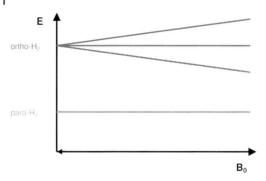

Figure 12.2 Magnetic field dependence of the energy levels of ortho- and para-H_2. Parahydrogen (p-H_2) is a singlet that is unaffected by the magnetic field, whereas orthohydrogen (o-H_2) is a triplet. Its energy levels split, showing the Zeeman effect.

cally lowest state of H_2 corresponds to p-H_2, which prevails at cryogenic temperatures. This fact is used to enrich fractions in p-H_2. Even though both spin isomers, p-H_2 and o-H_2, can even be separated completely, the $\sim$50% enrichment of p-H_2 achieved at 77 K using liquid nitrogen as a coolant is sufficient for conducting parahydrogen labeling experiments [11].

12.2.1
Magnetic Field Dependence of the PHIP-Phenomenon: PASADENA and ALTADENA Conditions

According to Bowers and Weitekamp [6], ^{1}H spin polarization occurs in the hydrogenation products of p-H_2, if a chemical reaction breaks its initial symmetry. Pure p-H_2 itself is NMR-inactive (i.e., it cannot be detected via ^{1}H-NMR). Correspondingly, since hydrogenation of an asymmetrically substituted acetylene yields an olefin with chemically inequivalent protons, the use of p-H_2 give rise to two resonances, typically consisting of doublets with one component in emission and one in absorption each ("antiphase doublets") [6]. Bowers and Weitekamp named this phenomenon PASADENA, which applies to reactions carried out within the high magnetic field of a NMR spectrometer.

Figure 12.3 outlines the essential features of the PASADENA/PHIP concept for a two-spin system. If the symmetry of the p-H_2 protons is broken, the reaction product exhibits a PHIP spectrum (Fig. 12.3, lower). If the reaction is carried out within the high magnetic field of the NMR spectrometer, the PHIP spectrum of the product consists of an alternating sequence of enhanced absorption and emission lines of equal intensity. This is also true for an AB spin system due to a compensating balance between the individual transition probabilities and the population rates of the corresponding energy levels under PHIP conditions. The NMR spectrum after the product has achieved thermal equilibrium exhibits intensities much lower than that of the intermediate PHIP spectrum.

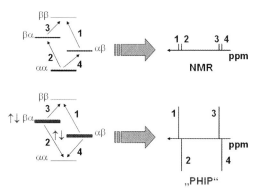

Fig. 12.3 Regular NMR (upper) and high-field PHIP-NMR spectrum (lower) of a two-spin AX system.

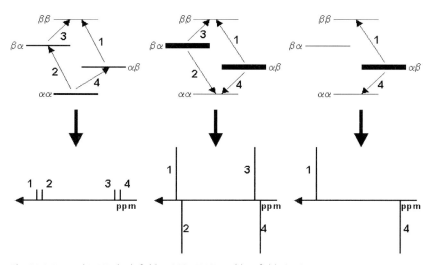

Fig. 12.4 Normal NMR, high-field (PASADENA) and low-field (ALTADENA) PHIP.

A related phenomenon occurs if the reactions are carried out at low field outside, but where the samples are transferred immediately thereafter into the spectrometer for subsequent NMR analysis. This variety has been termed ALTA-DENA (*Adiabatic Longitudinal Transport After Dissociation Engenders Net Alignment*) [9]. Other authors [7] have since used the acronym PHIP as an alternative for the same phenomenon.

An essential requirement for the occurrence of PHIP is that the rates of the reactions compete favorably with the relaxation of the nuclear spins in the products; therefore, the reaction rates must be faster than the relaxation rates, and the polarization must be detected during the reactions – that is, *in situ*.

Today, many examples have been reported which demonstrate the potential of the PHIP technique as a powerful analytical tool to investigate the reaction mechanisms of homogeneously catalyzed hydrogenations [12].

For more complex spin systems, a computer program PHIP$^+$ has been developed [13, 45] which allows the expected PHIP spectra to be calculated from the chemical shifts and coupling constants of the products. Depending upon which proton pair in the product molecule stems from p-H$_2$, different – but characteristic – polarization patterns result [14]. The patterns also depend on the sign of the coupling constants. Simple "sign rules" governing the relative sequence of the emission and absorption lines in the PHIP spectra (i.e., their "phase") can be formulated in similar manner to the "Kaptein Rules" of chemically induced dynamic nuclear polarization (CIDNP) [15].

12.2.2
PHIP, CIDNP, and Radical Mechanisms

In appearance, the PHIP phenomenon closely resembles those due to CIDNP [16], another phenomenon, which also gives rise to emission and enhanced absorption lines in NMR spectra. However, CIDNP is the consequence of the occurrence of free radicals, and previously has frequently been considered unequivocal proof for free radical reactions.

The fact that two entirely different phenomena can both yield nuclear spin polarization may cause confusion; therefore, the appearance of intense emission and absorption lines during *in-situ* NMR investigations of hydrogenation reactions is not necessarily proof for free radical intermediates, and examples of erroneous conclusions do exist [5].

For the mechanisms of homogeneous catalysis this may seem irrelevant, but as early as 1977 – long before the discovery of PHIP – Halpern [16e] had used "…the observation of CIDNP…" to "…establish a radical reaction path…" during the hydrogenation of styrene with an organometallic catalyst. Given the fact that even at room temperature (i.e., without any cooling of dihydrogen) there is a slight excess of parahydrogen, which is well capable of giving rise to "CIDNP-like phenomena" in the ^{1}H-NMR spectra, observations of emission and enhanced absorption lines alone are no longer a reliable proof of the occurrence of free radicals, if H$_2$ is involved. Therefore, caution is recommended, especially when exploiting the results of earlier investigations. Today, it is easy to err on the safe side, since a reliable discrimination exists between the alternatives CIDNP and PHIP, using both enriched ortho- and parahydrogen in independent runs [17]. In this way it should be possible to ascertain, whether a "…radical reaction pathway in homogeneous hydrogenation" [16e] really exists by using both para- and orthohydrogen in subsequent experiments.

12.2.3
Preparation of Parahydrogen

Molecular hydrogen or dihydrogen (H_2) occurs in two isomeric forms, namely with its two proton spins aligned either parallel (orthohydrogen) or antiparallel (parahydrogen) (see Fig. 12.1). Parahydrogen was first prepared during the 1920s by Bonhoeffer and Harteck [18]. The separation of both spin isomers [10] has also been possible for many years, even though orthohydrogen – the energetically less favorable form – is occasionally still described in the literature as "obtainable only in theory" [20a], especially in older text books [20b].

12.2.3.1 Parahydrogen Enrichment

As stated earlier, in the state of thermal equilibrium at room temperature, dihydrogen (H_2) contains 25.1% parahydrogen (nuclear singlet state) and 74.9% orthohydrogen (nuclear triplet state) [19]. This behavior reflects the three-fold degeneracy of the triplet state and the almost equal population of the energy levels, as demanded by the Maxwell-Boltzmann distribution. At lower temperatures, different ratios prevail (Fig. 12.5) due to the different symmetry of the singlet and the triplet state [19].

Since interconversions between different states of symmetry (i.e., between ortho- and parahydrogen) are forbidden, the adjustment of the relative ratios of the two spin isomers to the values corresponding to the thermal equilibrium at an arbitrary temperature is normally very slow and, therefore, must be catalyzed. In the absence of a catalyst, dihydrogen samples retain their once achieved ratio and, accordingly, they can be stored in their enriched or separated forms for rather long periods (a few weeks or even a few years in favorable cases).

In spite of the fact that ortho- or parahydrogen, once separated, last for a long time, they are normally not available commercially; therefore, they must be prepared as needed. A time-proven process for the enrichment of parahydrogen follows a procedure first described by Bonhoeffer and Harteck [18]. This is based upon the fact that, at low temperature, the energetically more favorable isomer

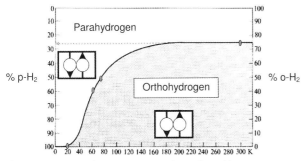

Fig. 12.5 Fraction of spin isomers at thermal equilibrium between o-H_2 and p-H_2.

(i.e., parahydrogen) becomes enriched, provided that a suitable catalyst renders the conversion possible. In this respect, activated charcoal functions as a convenient catalyst.

12.2.3.2 High-Pressure Apparatus for Parahydrogen Enrichment

The apparatus used (Fig. 12.6) follows the Bonhoeffer and Harteck concept [18]. Using liquid nitrogen as a coolant, it produces a constant flow of a mixture of 51% parahydrogen and 49% orthohydrogen (i.e., 51% enriched parahydrogen), at pressures up to 20 bar. As such it is especially suited for NMR investigations at elevated pressures using recently developed NMR probes [4]. The U-shaped charcoal reactor A shown in Figure 12.7 is manufactured from brass tubing, 4 cm in diameter and 30 cm in height, and capable of withstanding pressures of 20 bar and more. Copper tubing (3 mm) is used throughout to connect the reactor with the pressure relief valve (P) and the other components of the apparatus. The reactor is filled up to two-thirds of its height with coarse-grained charcoal, which is topped off with glass wool to constrain the filling. It is not advantageous to use powdered charcoal, as this tends to restrict the flow of the hydrogen through the reactor. Furthermore, tiny particles of charcoal are then carried along by the gas and drift into other parts of the apparatus.

In order to prime the apparatus, the charcoal-filled reactor is initially heated to 400 °C and maintained at this temperature for 10 h while operating the vacuum pump. This procedure must be repeated subsequently at regular intervals in order to regenerate the charcoal filling. The so-activated charcoal removes contaminants from the hydrogen stream, and thereby also serves as a gas cleaner.

Upon appropriate priming, the activated charcoal reactor A must be evacuated and is subsequently chilled in a liquid nitrogen bath. After an induction period, which depends on the operating pressure, 51% enriched parahydrogen is avail-

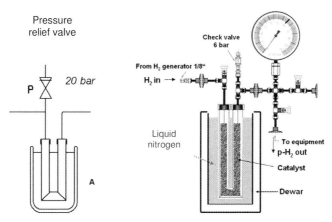

Fig. 12.6 Apparatus for the enrichment of parahydrogen at high pressure; principle and details.

able and can be supplied continuously. Typical induction periods are 30 min at 1 bar hydrogen pressure and 60 min at 10 bar. Flow rates of 50 mL min^{-1} have been achieved and tested in this mode, without any loss of parahydrogen enrichment during a prolonged flow period.

Upon termination of the experiment, all parts of the apparatus should be evacuated once again, before the liquid nitrogen bath is removed. The same precaution should also be maintained during removal of the bath, in order to safeguard against pressure surges caused by thermal expansion of the residual hydrogen or due to sudden desorption from the charcoal.

Integrated thermal conductivity cells (see Fig. 12.8) allow a quantitative determination of the corresponding ortho/para ratios of the dihydrogen. The enriched parahydrogen is well-suited for *in-situ* NMR studies of hydrogenation reactions that yield nuclear spin polarization due to symmetry breaking during the reaction. The same apparatus has also been used successfully to enrich ortho- and paradeuterium mixtures.

12.2.3.3 Enrichment of Parahydrogen using Closed-Circuit Cryorefrigeration

Virtually pure parahydrogen can conveniently be obtained upon cooling molecular hydrogen to temperatures between 20 and 30 K. For this purpose, commercially manufactured systems are available, and details have been described elsewhere [52]. It is advantageous (for thermodynamic considerations) and preferable (for safety reasons) to maintain the cooling temperature above the boiling point of dihydrogen at the corresponding pressure; therefore, the cryo-system should be operated at ca. 30 K. The level of enrichment for parahydrogen at 1.5 bar and flow rates of 10 to 30 mL min^{-1} for the two systems operating at 30 or 77 K, respectively, is illustrated in Figure 12.7.

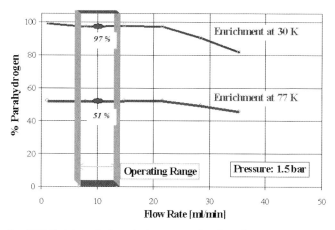

Fig. 12.7 Enrichment of parahydrogen as a function of temperature and flow rate.

12.2.4
Preparation of Orthohydrogen

During the 1950s, the enrichment or isolation of orthohydrogen was occasionally assumed to be impossible [19], until Sandler [10] described a procedure based on the preferential adsorption of the ortho form of dihydrogen on diamagnetic surfaces at low temperatures. The energy difference of orthohydrogen between its free and adsorbed state can be explained using the model of the hindered rotator [53]. Cunningham et al. [54] have presented an apparatus for the enrichment of orthohydrogen up to 99% using carefully selected γ- alumina in a cascade of selective desorption at 20.4 K. The present author's group have used single-stage separation of orthohydrogen on γ-alumina at 77 K (liquid nitrogen), and routinely obtained a mixture of 15% parahydrogen and 85% orthohydrogen (i.e., 85% enriched orthohydrogen), which is sufficient for most purposes [17]. Since the enrichment of orthohydrogen is carried out at ambient pressure, it is superfluous to design this apparatus for higher pressures. The alumina adsorption cell (B in Fig. 12.8) forms the core of the low-pressure apparatus. This cell is designed as a glass tube helix, and is fabricated from 150 cm of 20-mm diameter glass tubing. The helix provides for a rapid and homogeneous temperature change using a Dewar flask containing liquid nitrogen. This helix is charged with 260 g of γ-alumina that must not contain paramagnetic impurities (E. Merck, Darmstadt, Germany, Art. # 1095) [55]. The helix is terminated with fritted glass, through which it is connected to the remainder of the all-glass apparatus. Prior to the first orthohydrogen enrichment the γ-alumina filling is activated and degassed at $400\,^\circ$C for 10 h, similar to the process described for the activated charcoal filling. Batches of orthohydrogen enriched to 85% can be prepared in this low-pressure apparatus, and the degree of enrichment can be determined accurately using thermal conductivity cells.

12.2.5
Thermal Conductivity Cells for Ortho/Para Determination

A pair of thermal conductivity cells (C in Fig. 12.8) according to Grilly [56] is used to determine the ortho/para ratio. Each cell is constructed from 120 mm of 10-mm diameter glass tubing, whereby two VACON pins lead into the interior carrying a tungsten filament, which is aligned along the center axis of the cell. Electrically, the filaments of the two cells form a bridge circuit together with appropriate resistors, which can be balanced externally. For maximum sensitivity reasons their resistance should be of the same order of magnitude as the filament resistance. The bridge voltage (here 25 V) is chosen such that the filaments heat up to 250 K, since under these conditions the resulting temperature gradient between the filaments and the walls of the liquid nitrogen-chilled cells is appropriately within the range of optimum sensitivity and performance.

The thermal conductivity cells are connected to their respective valves V1 and V2, using U-shaped glass tubing. This design has been found advantageous, as it minimizes any unavoidable convection of hydrogen within the cells.

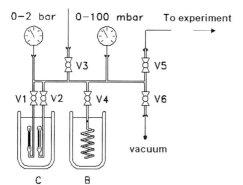

Fig. 12.8 Apparatus for orthohydrogen enrichment at low pressure, and integrated thermal conductivity cells.

12.2.6
Determination of the Ortho/Para Ratio

The specific heats of ortho- and parahydrogen differ one from another, whereby the difference reaches a maximum at temperatures in between 140 and 170 K. Within this temperature range the difference between heat capacities is sufficient and suitable to determine the ortho/para ratio of H_2 using thermal conductivity cells. Typically, the setup according to Grilly is used; this consists of two conductivity cells, whereby one functions as a reference to eliminate long-term thermal drift [56]. The resistance of the tungsten filament in these cells is almost proportional to the orthohydrogen concentration [57]; therefore, it can be used as a measure for the ortho/para ratio.

12.2.7
Enrichment of Ortho- or Paradeuterium

Using the same apparatus, molecular deuterium (D_2) can also be enriched in its ortho- or para-forms. It must be noted, however, that in the case of D_2 the ortho-form is the energetically more favorable and hence easier to prepare [19]. Therefore, it is orthodeuterium which becomes enriched when using an appropriately cooled charcoal cell. The achievable degree of enrichment at the temperature of liquid nitrogen (77 K) is considerably lower than in the case of H_2, but nonetheless the apparatus has been used successfully for the enrichment of the D_2 spin isomers, albeit with appropriate modifications of the procedure. Essentially, in this case the Dewar flask to be filled with liquid nitrogen should tolerate a partial vacuum (underpressure) because, by lowering the pressure of the boiling nitrogen, the temperature falls, and eventually the nitrogen solidifies somewhere at around 60 K. This temperature suffices to achieve an adequate enrichment to observe orthodeuterium-induced polarization (ODIP) [46].

Far superior results have been obtained, however, using closed-circuit cryorefrigeration for the required cooling of the D_2, as this allows the temperature to be lowered to 20–30 K. In this case, when determining the enrichment the resistors of the bridge circuit and the operating temperature of the thermal conductivity cells must be optimized for D_2, due to the different physical parameters of its individual spin isomers [52].

12.3
Applications of PHIP-NMR Spectroscopy

12.3.1
In-Situ PHIP-NMR Spectroscopy of Homogeneous Hydrogenations

12.3.1.1 Activation of Dihydrogen

Dihydrogen shows a weak tendency to undergo chemical reactions, unless it is activated by certain types of transition-metal compounds. Buntkowsky et al. [21] have investigated the early stages of activation of H_2 using parahydrogen, and the results and conclusions derived thereof have been reported.

In general, the activation of dihydrogen by transition-metal complexes has been investigated intensely since the 1960s, when Wilkinson and colleagues discovered the first successful homogeneous hydrogenation catalyst $RhCl(PPh_3)_3$ [22]. Herewith, terminal alkenes and alkynes can be readily hydrogenated at 25 °C at a hydrogen pressure of 1 bar. The mechanism and kinetics thereof have since been extensively studied. Some intermediate dihydrides such as $Rh(H)_2Cl(PPh_3)_3$ and $Rh(H)_2Cl_2(PPh_3)_4$ have been previously observed and characterized using NMR spectroscopy [23], and even PHIP-NMR spectroscopy [24].

The data available in the literature, however, are typically restricted to those containing PPh_3 as the phosphine ligands. In general, the spectroscopic observation of such dihydride species is rather difficult. Due to their intermediate character they have a short lifetime, and hence they typically occur only at rather low concentrations. Whereas conventional NMR spectroscopy frequently fails to identify their existence, PHIP has allowed the detection of several new dihydride products, even at these low concentrations. If hydrogenation with parahydrogen is carried out *in situ* using a spectrometer-activated apparatus, the kinetic constants of the reactions of the formation and disappearance of intermediates may be determined; details of this method are outlined in the following section.

12.3.1.2 Concepts of Reaction Mechanisms

Time-proven concepts for the reaction mechanisms of homogeneous hydrogenations follow two approaches which, according to Halpern's step-wise analysis of hydrogenations using "Wilkinson's catalyst" [25] and the cationic catalyst DI-PHOS [26], respectively, can be grouped into the so-called "dihydride" or "unsaturate" routes [27] (Fig. 12.9).

Due to these alternatives, the detection of intermediates is of considerable interest, as this would allow differentiation to be made between these two principal al-

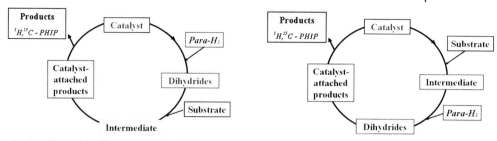

Fig. 12.9 Alternative approaches to the sequence of events during the catalytic activation of dihydrogen.

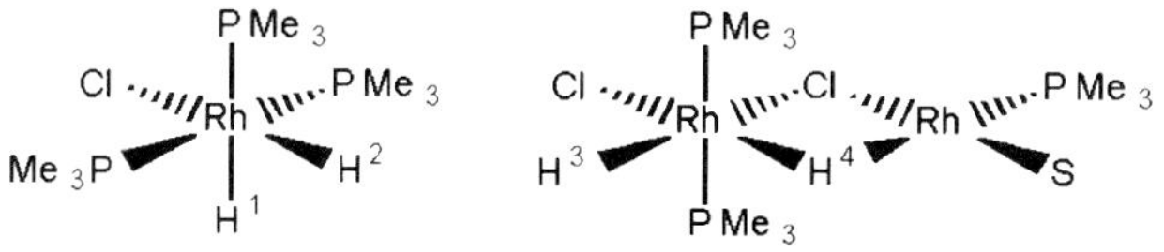

Fig. 12.10 Mono- and binuclear rhodium dihydride complexes.

ternatives. Evidence in favor of one or the other route has been accumulated and associated with the nature of the catalysts – that is, whether they are cationic or neutral – and extensive accounts of this have been published. Within the scope of this chapter, a number of special cases will be discussed that provide evidence for the existence of different intermediates, including a number of dihydrides.

12.3.2
In-Situ PHIP-NMR Observation of Mono- and Binuclear Rhodium Dihydride Complexes

PHIP can also be used to identify and characterize mono- and binuclear rhodium dihydride complexes such as $[RhH_2ClL_3]$ (L=PMe_3, PMe_2Ph, $PMePh_2$, PEt_3, PEt_2Ph, $PEtPh_2$, or $P(n\text{-butyl})_3$), $[(H)(Cl)Rh(PMe_3)_2(\mu\text{-}Cl)(\mu\text{-}H)Rh(PMe_3)]$, and $[(H)(Cl)Rh(PMe_2Ph)_2(\mu\text{-}Cl)(\mu\text{-}H)Rh(PMe_2Ph)]$ (Fig. 12.10) obtained from the binuclear complex $[RhCl(2,5\text{-norbornadiene})]_2$ when treated with the corresponding phosphine and parahydrogen. By substituting chloride with trifluoroacetate, the complexes $[RhH_2(CF_3COO)L_3]$ (L=PPh_3, PEt_2Ph, PEt_3, and $P(n\text{-butyl})_3$) are analogously generated [58].

12.3.2.1 Reactions of [RhCl(NBD)]₂ with Parahydrogen in the Presence of Tertiary Phosphines

The binuclear precursor (di-μ-chloro-bis-[η^4-2,5-norbornadiene]-rhodium(I))= $[(Rh(NBD)Cl]_2$ is well suited for the *in-situ* preparation of a variety of homogeneous hydrogenation catalysts, if tertiary phosphines (here: PMe_3, PMe_2Ph,

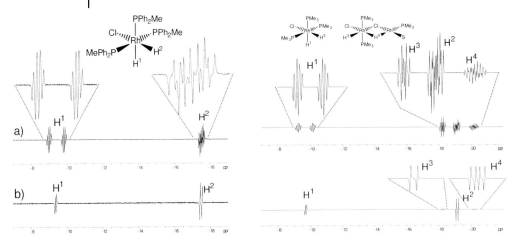

Fig. 12.11 PHIP-NMR spectra of some mono- (left) and binuclear Rh-complexes (right).

PMePh$_2$, PEt$_3$, PEt$_2$Ph, PEtPh$_2$, or P(n-butyl)$_3$) are added. Upon the addition of dihydrogen to solutions of these mixtures, NBD is hydrogenated off, and mononuclear dihydride species [Rh(H)$_2$ClL$_3$] are generated, most likely via the complexes [RhClL$_3$] as intermediates. These dihydride complexes play a key role as intermediates in any subsequent catalytic hydrogenation.

As characteristic examples, Figure 12.11 (left) shows the results obtained upon addition of parahydrogen to a solution of 10 mg [Rh(NBD)Cl]$_2$ and 19 µL PMePh$_2$ in acetone-d$_6$ (Rh:P=1:3). Strongly enhanced resonances of the dihydride protons of the complex [Rh(H)$_2$Cl(PMePh$_2$)$_3$] are observed in the ^{1}H-NMR spectrum, whereby the hydride *trans* to a PMePh$_2$ ligand occurs at $\delta_H=-9.4$ ppm, whereas the hydride *trans* to the chloride has a higher chemical shift and appears at $\delta_{H'}=17.6$ ppm. The latter is characteristic for hydride protons in the *trans* position to such an electronegative ligand. The hydride resonance at lower field shows a large coupling to one *trans* phosphorus (J$_{HP(trans)}$=178.6 Hz), an additional coupling to two equivalent *cis* phosphorus nuclei (J$_{HP(cis)}$=14.1 Hz), a coupling to the central rhodium (J$_{HRh}$=13 Hz), and a coupling to the upfield hydride proton (J$_{HH'}$=−7.7 Hz).

The coupling between the two former parahydrogen protons causes the antiphase character of this resonance, exhibiting emission and absorption maxima, accordingly. The second hydride resonance at higher field is a complex multiplet with couplings to rhodium (J$_{HRh}$=22 Hz), one equatorial and two axial phosphorus nuclei (J$_{HP(ax.)}$=16.5 Hz, J$_{HP(eq.)}$=11 Hz), and the lower field hydride proton (J$_{HH'}$ =−7.7 Hz). These data have been tested by simulating the resulting multiplet and its polarization pattern using the computer program PHIP^{++}, upon which very good agreement has been obtained [28].

If the ^{1}H-NMR spectrum is recorded with complete ^{31}P decoupling, both hydride resonance at −9.4 ppm and −17.6 ppm collapse into a doublet of antiphase doublets. The remaining 13-Hz and 22-Hz couplings correspond to J$_{HRh}$ and J$_{HRh'}$,

Table 12.1 ^{1}H-NMR data of the observed hydrides.

No.	Complex	^{1}H chemical shifts δ [ppm] in acetone-d$_6$ and coupling constants J [Hz]
1	[Rh(H)$_2$Cl(PMe$_3$)$_3$]	-9.7 (^{1}J$_{HRh}$=15, ^{2}J$_{HP(trans)}$=178.6, ^{2}J$_{HP(cis)}$=20, ^{2}J$_{HH}$=-7.5)
		-19.0 (^{1}J$_{HRh}$=27, ^{2}J$_{HP(ax.)}$=19, ^{2}J$_{HP(eq.)}$=13, ^{2}J$_{HH}$=-7.5)
2	[Rh(H)$_2$Cl(PMe$_2$Ph)$_3$]	-9.6 (^{1}J$_{HRh}$=18, ^{2}J$_{HP(trans)}$=172.5, ^{2}J$_{HP(cis)}$=18, ^{2}J$_{HH}$=-7.2)
		-18.2 (^{1}J$_{HRh}$=24.7, ^{2}J$_{HP(ax.)}$=16.5, ^{2}J$_{HP(eq.)}$=11, ^{2}J$_{HH}$=-7.2)
3	[Rh(H)$_2$Cl(PMePh$_2$)$_3$]	-9.4 (^{1}J$_{HRh}$=13, ^{2}J$_{HP(trans)}$=164, ^{2}J$_{HP(cis)}$=14.1, ^{2}J$_{HH}$=-7.7)
		-17.6 (^{1}J$_{HRh}$=22, ^{2}J$_{HP(ax.)}$=15, ^{2}J$_{HP(eq.)}$=10, ^{2}J$_{HH}$=-7.7)
4	[Rh(H)$_2$Cl(PEt$_3$)$_3$]	-10.7 (^{1}J$_{HRh}$=14.7, ^{2}J$_{HP(trans)}$=161.7, ^{2}J$_{HP(cis)}$=16, ^{2}J$_{HH}$=-7.9)
		-19.8 (^{1}J$_{HRh}$=24.1, ^{2}J$_{HP(ax.)}$=16, ^{2}J$_{HP(eq.)}$=13, ^{2}J$_{HH}$=-7.9)
5	[Rh(H)$_2$Cl(PEt$_2$Ph)$_3$]	-10.1 (^{1}J$_{HRh}$=15.4, ^{2}J$_{HP(trans)}$=162, ^{2}J$_{HP(cis)}$=16.3, ^{2}J$_{HH}$=-7.3)
		-19.3 (^{1}J$_{HRh}$=24.6, ^{2}J$_{HP(ax.)}$=17, ^{2}J$_{HP(eq.)}$=15.1, ^{2}J$_{HH}$=-7.3)
6	[Rh(H)$_2$Cl(PEtPh$_2$)$_3$]	-9.9 (^{1}J$_{HRh}$=12.5, ^{2}J$_{HP(trans)}$=155, ^{2}J$_{HP(cis)}$=13.3, ^{2}J$_{HH}$=-7.9)
		-18.3 (^{1}J$_{HRh}$=20, ^{2}J$_{HP(ax.)}$=12.5, ^{2}J$_{HP(eq.)}$=13.5, ^{2}J$_{HH}$=-7.9)
7	[Rh(H)$_2$Cl(P(n-butyl)$_3$)$_3$]	-10.7 (^{1}J$_{HRh}$=13.5, ^{2}J$_{HP(trans)}$=163, ^{2}J$_{HP(cis)}$=16.7, ^{2}J$_{HH}$=-7.8)
		-19.8 (^{1}J$_{HRh}$=23, ^{2}J$_{HP(ax.)}$=15.8, ^{2}J$_{HP(eq.)}$=13.1, ^{2}J$_{HH}$=-7.8)
8	[(H)(Cl)Rh(PMe$_3$)$_2$(μ-Cl)(μ-H)-Rh(PMe$_3$)]	-18.2 (^{1}J$_{HRh}$=25, ^{2}J$_{HP(cis)}$=17.3, ^{2}J$_{HH}$=-4.2)
		-20.2 (^{1}J$_{HRh}$=30.3, 20, ^{2}J$_{HP(tans)}$=29, ^{2}J$_{HP(cis)}$=14.6, ^{2}J$_{HH}$=-4.2)
9	[(H)(Cl)Rh(PMe$_2$Ph)$_2$(μ-Cl)-(μ-H)Rh(PMe$_2$Ph)]	-17.3 (^{1}J$_{HRh}$=23, ^{2}J$_{HP(cis)}$=15.4, ^{2}J$_{HH}$=-4.2)
		-20.0 (^{1}J$_{HRh}$=30, 21.4, ^{2}J$_{HP(trans)}$=29, ^{2}J$_{HP(cis)}$=15.5 ^{2}J$_{HH}$=-4.2)
10	[Rh(H)$_2$(CF$_3$COO)(PPh$_3$)$_3$]	-8.9 (^{1}J$_{HRh}$=6, ^{2}J$_{HP(trans)}$=161, ^{2}J$_{HP(cis)}$=12, ^{2}J$_{HH}$=-9.5)
		-19.1 (^{1}J$_{HRh}$=18, ^{2}J$_{HP(ax.)}$=17, ^{2}J$_{HP(eq.)}$=17, ^{2}J$_{HH}$=-9.5)
11	[Rh(H)$_2$(CF$_3$COO)(PEt$_3$)$_3$]	-10.1 (^{1}J$_{HRh}$=14.8, ^{2}J$_{HP(trans)}$=160, ^{2}J$_{HP(cis)}$=16, ^{2}J$_{HH}$=-8.7)
		-22.5 (^{1}J$_{HRh}$=13, ^{2}J$_{HP(ax.)}$=17, ^{2}J$_{HP(eq.)}$=17, ^{2}J$_{HH}$=-8.7)
12	[Rh(H)$_2$(CF$_3$COO)(PEt$_2$Ph)$_3$]	-9.5 (^{1}J$_{HRh}$=15.5, ^{2}J$_{HP(trans)}$=161, ^{2}J$_{HP(cis)}$=16.8, ^{2}J$_{HH}$=-8.6)
		-21.9 (^{1}J$_{HRh}$=26.8, ^{2}J$_{HP(ax.)}$=16.7, ^{2}J$_{HP(eq.)}$=16.7, ^{2}J$_{HH}$=-8.6)
13	[Rh(H)$_2$(CF$_3$COO)(P(n-butyl)$_3$)$_3$]	-9.5 (^{1}J$_{HRh}$=15.5, ^{2}J$_{HP(trans)}$=161.5, ^{2}J$_{HP(cis)}$=16.7, ^{2}J$_{HH}$=-7.1)
		-21.9 (^{1}J$_{HRh}$=28.8, ^{2}J$_{HP(ax.)}$=17.5, ^{2}J$_{HP(eq.)}$=17.5, ^{2}J$_{HH}$=-7.1)

respectively (Fig. 12.11). Analogous complexes with other phosphine ligands have also been observed (Table 12.1). If the temperature is elevated, the intensities of the polarized hydride resonances increase significantly. The spectrum displayed in Figure 12.11 was recorded at 315 K. Experiments in acetone-d$_6$ can be carried out up to about 330 K, since above this value, boiling and associated evaporation of the solvent becomes so significant that this badly interferes with the quality of the spectra. Furthermore, the resolution of the corresponding hydride resonances deteriorates with increasing temperature. Therefore, a lower temperature is advantageous for better resolution, since the rate of phosphine dissociation slows down with decreasing temperature, upon which the resonances sharpen.

The rate of the observed exchange reaction of the phosphine ligands in the dihydrides increases in the above-listed series of phosphines from PMe$_3$ to P(n-butyl)$_3$ (Table 12.1), which in turn correlates with the activity of the corresponding complexes as hydrogenation catalysts.

Furthermore, the intensities of the polarized hydride resonances increase with temperature. Since these intensities correlate with the rate of the oxidative addition of parahydrogen to the corresponding Rh complexes and with the subsequent decay of the so-formed dihydrides, both the rates of their formation as well as the rates of their decay seem to increase with temperature.

12.3.2.2 Formation of the Binuclear Complexes [(H)(Cl)Rh(PMe$_3$)$_2$(μ-Cl)-(μ-H)Rh(PMe$_3$)] and [(H)(Cl)Rh(PMe$_2$Ph)$_2$(μ-Cl)(μ-H)Rh(PMe$_2$Ph)]

Upon the addition of parahydrogen to a solution of [RhCl(NBD)]$_2$ and PMe$_3$ (ratio of Rh:P = 1:3) in acetone-d$_6$ at 315 K, intense polarization signals of the dihydride complex [Rh(H)$_2$Cl(PMe$_3$)$_3$] (1) can be observed, together with strongly polarized resonances in the aliphatic region due to hydrogenation of the NBD ligand. This complex 1 has also been investigated by others using PHIP-NMR spectroscopy [29]. In the later stages of the reaction, however, two new resonances emerge at –18.2 ppm and –20.2 ppm, respectively (Fig. 12.11, right), which can be assigned to the hydride protons of the complex 8 (Scheme 12.1). These signals also have antiphase character due to coupling between the two parahydrogen protons ($J_{HH'}$ = -4.2 Hz). Thereby, the hydride resonance at –18.2 ppm exhibits a coupling to Rh (J_{HRh} = 25 Hz) and the *cis* phosphorus ($J_{HP(cis)}$ = 17.3 Hz). The second hydride resonance consists of a complex multiplet with couplings to the two inequivalent Rh atoms (J_{HRh} = 30.3 Hz and $J_{Rh'H}$ = 20 Hz, respectively), as well as to the *trans* ($J_{HP(trans)}$ = 29 Hz) and the two *cis* phosphorus nuclei ($J_{HP(cis)}$ = 14.6 Hz). This assignment is confirmed upon ^{31}P decoupling of the hydride protons: In the ^{1}H{^{31}P} NMR spectrum (see Fig. 12.11) the signal at –18.2 ppm has collapsed into a doublet of antiphase doublets with the 25-Hz coupling corresponding to J_{HRh}. The second hydride resonance has simplified to a doublet of doublet of antiphase doublets, with couplings of 30.3 Hz and 20 Hz corresponding to J_{HRh} and $J_{HRh'}$, respectively.

Complexes of the type [(H)(Cl)Rh(PMe$_3$)$_2$(μ-Cl)(μ-H)Rh(CO)(PMe$_3$)] and [(H)(Cl)Rh(PMe$_3$)$_2$(μ-I)(μ-H)Rh(CO)(PMe$_3$)], with NMR data corresponding to our results, have also been observed before by Duckett, Eisenberg, and coworkers, using PHIP-NMR spectroscopy [30, 31].

In these earlier studies the phosphine ligand at the rhodium(I) center has been shown to be *trans* to the μ-hydride with a phosphorus coupling of $J_{HP(trans)}$ = 30 Hz and 32 Hz, respectively. Therefore, we assign to complex 8 the structure as outlined in Scheme 12.1.

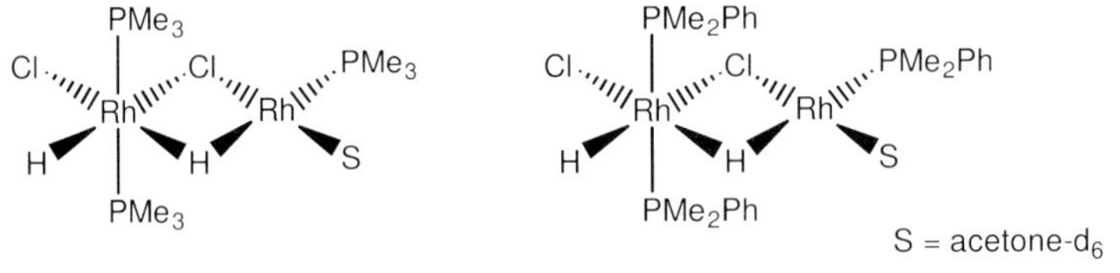

8

9

S = acetone-d$_6$

Scheme 12.1

In their study of Wilkinson's catalyst, Duckett and Eisenberg postulated binuclear complexes containing an alkene at the rhodium(I) center [24].

Additional dihydride complexes can be obtained using PEt_3, PEt_2Ph, or $P(n$-butyl)$_3$ as the phosphine ligands. However, in the case of PEt_2Ph or $P(n$-butyl)$_3$ the resonances of the dihydrides are broadened due to a rapid exchange of these phosphine ligands, which blurs the resonances of the hydride protons. Upon cooling the sample, the rate of this exchange process can be slowed down resulting in improved resolution.

Upon using either acetic acid or tetrafluoroboronic acid instead of trifluoroacetic acid, however, no analogous rhodium-containing dihydrides could be observed. A few similar Rh-containing complexes have been described before by other authors [32].

12.3.2.3 General Procedure for the Generation of the Complexes $[Rh(H)_2ClL_3]$ (L = Phosphine)

In a typical experiment, the rhodium complex $[RhCl(NBD)]_2$ (10 mg) and the corresponding amount of phosphine ligand (Rh:P=1:3) are placed into an NMR tube together with 700 µL of degassed acetone-d_6 (Scheme 12.2). p-H_2 is then bubbled *in situ* through the solution within the magnetic field of the spectrometer, using a thin capillary that can be lowered into the spinning NMR tube. This lowering is synchronized by the NMR spectrometer, whereby the spectra are recorded not until the capillary is raised again and the p-H_2 bubbles have vanished.

Likewise, the complex $[Rh(NBD)(acac)]$ (10 mg), together with trifluoroacetic acid (1 µL) and a corresponding amount of phosphine ligand (Rh:P=1:3) is dissolved in 700 µL acetone-d_6 and treated as described above to obtain the complexes $[Rh(H)_2(CF_3COO)L_3]$ (L=phosphine).

12.3.3.3 Intermediate Dihydrides of Cationic Rh Catalysts

NMR evidence for intermediate dihydrides of cationic Rh catalysts remained elusive for a long time, ever since the first demonstrations [33] of effective enantioselective catalysis, for example in the homogeneous hydrogenation of dehydroamino acid derivatives for the synthesis of L-DOPA.

Para-enriched hydrogen offers considerable advantages for the NMR identification of transient intermediates [12d, 34]. PHIP experiments carried out *in situ* under PASADENA conditions are especially powerful in this regard. The PHANEPHOS [MM]-derived Rh catalyst is unusually reactive, with turnover possible even at −40 °C. This high reactivity, coupled with good enantioselectivity, provides an ideal case for characterizing the elusive Rh dihydrides.

Upon displacement of the NBD ligand with parahydrogen according to Scheme 12.3, the ^{1}H-PHIP-NMR spectra displayed in Figure 12.12 were observed, whereby the details of their parameters depended on the type and polarity of the solvent.

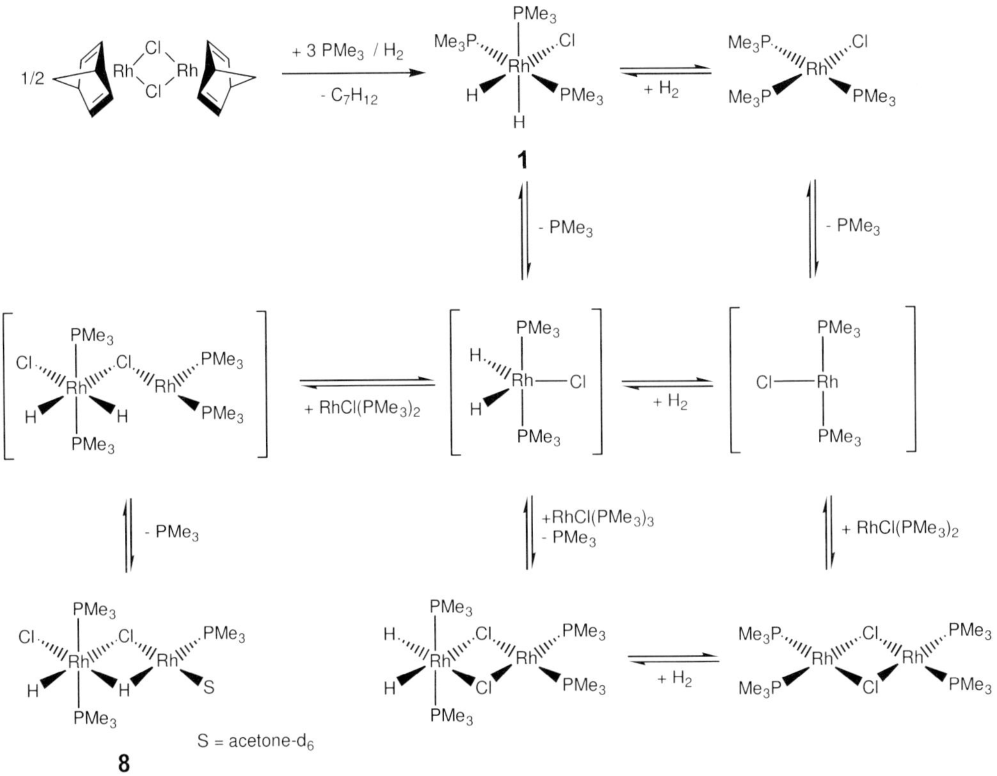

Scheme 12.2 The sequence of reactions leading to the observed intermediates.

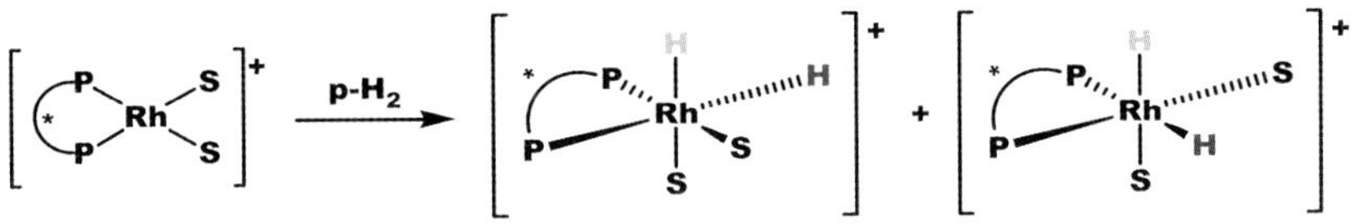

Scheme 12.3 Formation of dihydride intermediates of a cationic Rh complex via displacement of the NMD ligand in the DIPHOS-derived catalyst (S = solvent).

Upon addition of dehydroamino acid esters, the Rh complexes from the above-illustrated NBD-precursor formed *in situ* with displacement of the solvents even at $-40\,°C$. Using differently substituted substrates and mixtures thereof, the spectra shown in Figure 12.13 were observed.

The relative shift of the resonances of the dihydride nuclei listed in Table 12.2 follow a free energy correlation, as is outlined in the Hammett plots shown in Figure 12.14.

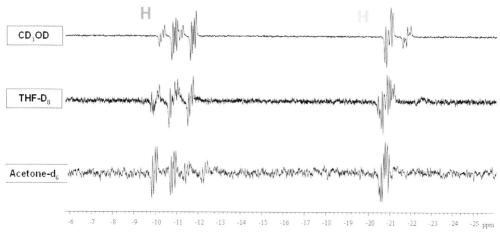

Fig. 12.12 Cationic Rh-dihydrides derived from [Rh(NBD)(DIPHOS)]⁺.

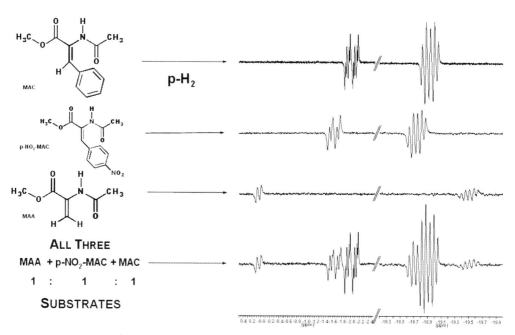

Fig. 12.13 Simultaneous ¹H-PHIP-NMR spectra of dihydrides
from enamide complexes of the DIPHOS-containing Rh-catalyst.

Table 12.2 NMR data and substituent effects of the observed dihydride intermediates.

R	H	![structure]—NO₂	![structure]—F	![structure]—H
δ_H	+0.07 ppm	−1.60 ppm	−1.93 ppm	−1.97 ppm
$^2J_{HH}$	−1.0 Hz	−3.0 Hz	−4.3 Hz	−4.3 Hz
$^1J_{HRh}$	1.0 Hz	14.2 Hz	15 Hz	13.4 Hz
$^2J_{HP}$	17.5 Hz	32.8 Hz	36.0 Hz	34.0 Hz
	1.0 Hz	5.25 Hz	6.5 Hz	5.8 Hz

$^1J_{H^{13}C} = 85.8$ Hz

Substituent	σ^0	$\Delta\delta_{LF}$ [ppm]	$\Delta\delta_{HF}$ [ppm]
p-OMe	-0.14	+0.052	+0.061
p-F	+0.15	-0.038	-0.028
p-NO₂	+0.81	-0.366	-0.190
	–		

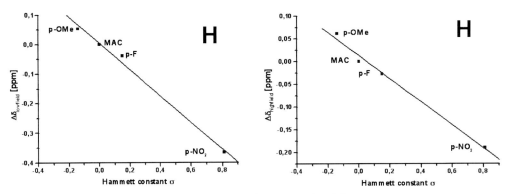

Fig. 12.14 Free energy correlation and Hammett plot of chemical shift data of the intermediate dihydrides.

12.3.3.4 Obtaining Structural Information using ^{13}C-Labeled Substrates

By conducting experiments with the β-^{13}C-labeled ester outlined in Figure 12.15, an additional ^{13}C coupling of 86 Hz to the low-field Rh hydride can be detected. In addition to the expected polarization transfer to the ^{13}CH$_2$Ph of the product at 38 ppm, a strong reactant signal appears in the ^{13}C INEPT($+\pi/4$) spectrum at 135 ppm, implying reversibility of enamide complexation in the observed transient. With the α-^{13}C-labeled enamide, however, weak ^{13}C coupling (ca. 3 Hz) to the low-field hydride is observed [35].

The experimental result and the simulated spectrum are shown in Figure 12.16. The spectroscopic data of this intermediate allows the structure and geo-

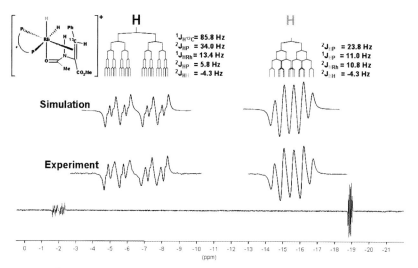

Fig. 12.15 β-^{13}C-labeled esters used as substrates.

R = Me, ^{13}C-β
R = Me, ^{13}C-α

$^1J_{H^{13}C}$= 85.8 Hz
$^2J_{HP}$ = 34.0 Hz
$^1J_{HRh}$= 13.4 Hz
$^2J_{HP}$ = 5.8 Hz
$^2J_{HH}$ = -4.3 Hz

$^2J_{HP}$ = 23.8 Hz
$^1J_{HP}$ = 11.0 Hz
$^2J_{HRh}$ = 10.8 Hz
$^2J_{HH}$ = -4.3 Hz

Simulation

Experiment

Fig. 12.16 ^{1}H-NMR spectrum of the dihydride of the enamide Rh complex.

metric details of this intermediate to be predicted. These postulates agree well with the results of high-level density functional theory (DFT) calculations. According to such calculations on the simple [Rh(PH$_3$)$_2$] complex [36], the reaction proceeds through a η^2-dihydrogen complex to a classical dihydride. The thermodynamically favored Rh diastereomer of this dihydride has a low energy pathway to an agostic species closely resembling the agostic dihydride intermediate outlined in Figure 12.17, although this is not on the computationally preferred pathway of hydrogenation. More specific DFT calculations [36] on the model dehydroamino ester indicate that the structure outlined in Figure 12.17 represents a significant minimum, with a computed Rh–H bond length of 1.76 Å [35, 36].

Whereas, from all of these informative ^{1}H-PHIP-NMR spectra, the structure of the dihydride intermediate (including geometric details about peculiar bonding therein) can be determined rather exactly and reliably, a degree of uncertainty remains as to whether this intermediate represents the "major" or the "minor" diastereomer according to the nomenclature of Halpern [27]. This is the consequence of different kinetic constants associated with the two alternative cycles with different stereochemistry, and which accounts for the "major" and "minor" reaction product (Fig. 12.18). In fact, it is the difference in the rate

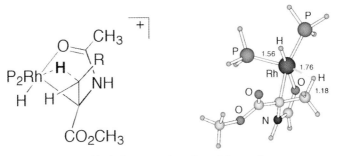

Fig. 12.17 Agostic dihydride intermediate derived from a dehydroamino ester substrate.

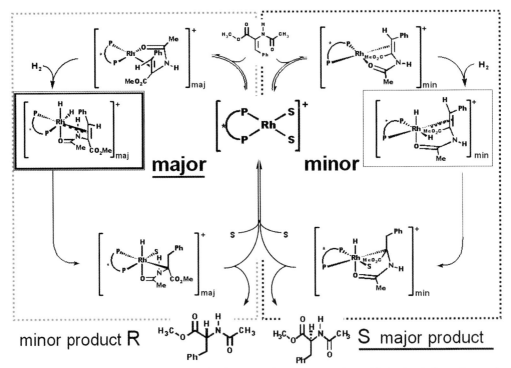

Fig. 12.18 The two equivalent but stereochemically different reaction cycles of homogeneous hydrogenation correlating the "minor" intermediate with the "major" reaction product, and vice-versa (according to [27]).

of generation and the rate of decay of the two respective intermediates that accounts for their respective lifetimes and concentrations.

In consequence, the "major" intermediate (which, in principle, is easier to detect because it should occur at a somewhat higher concentration) correlates with the "minor" reaction product, which is the irrelevant one for most synthetic purposes. The opposite is true for the "minor" intermediates, which correlate with the "major" reaction product. Although of course the latter two are of more importance, it remains unclear as to which intermediate is seen here – the "minor", more difficult to detect, or (more likely) the "major" intermediate [35].

12.4
Catalyst-Attached Products as Observable Intermediates

During the homogeneous hydrogenation of a variety of substrates – and in particular of those containing aryl groups – "satellites" appear in addition to the "expected" PHIP-NMR spectra of the usual para-hydrogenated products, typically shifted upfield relative to those of the authentic parent compounds [37].

The shift correlates in magnitude with the separation of each particular group distance-wise from the aromatic moiety of the substrate or product; this points to the formation of an intermediate π-complex, for which the rate of formation and the rate of decay can be determined. The ^{1}H-PHIP-NMR spectrum, as well as the anticipated intermediate product–catalyst-π-complex observed during the hydrogenation of styrene, is outlined in Figure 12.19.

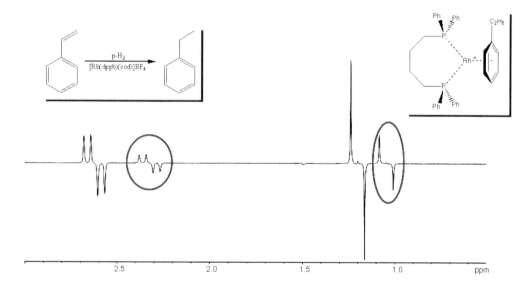

Fig. 12.19 The ^{1}H-PHIP-NMR spectrum and the anticipated intermediate product–catalyst-π-complex observed during the hydrogenation of styrene.

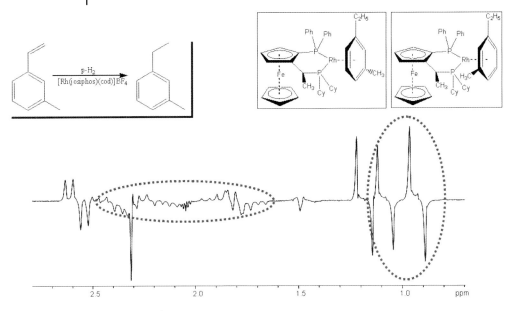

Fig. 12.20 The ^{1}H-PHIP-NMR spectrum and the anticipated two-intermediate product–catalyst–complexes observed during the hydrogenation of *meta*-methylstyrene.

12.4.1
Enantioselective Substrates

If the substrate is enantioselective, as for example *meta*-methylstyrene, the "satellites" split into (at least) two separate resonances, respectively, for the CH_2- and the CH_3-resonances of the resulting product *meta*-methyl-ethylbenzene (Fig. 12.20).

12.4.2
Chiral Catalysts

As opposed to the *meso* forms of catalyst, their chiral counterparts give rise to more complex ^{1}H-PHIP-NMR spectra of the catalyst-attached intermediates. This fact is outlined for the selection of precatalysts and catalysts derived from those listed in Figure 12.21.

Figure 12.22 shows the CH_2-resonance of the catalyst-attached intermediates observed during the hydrogenation of styrene using an achiral catalyst, whilst Figure 12.23 depicts the same region of the spectrum of the intermediate when employing a chiral catalyst. In the latter case the more complex splitting pattern is clear. Finally, Figure 12.24 depicts the CH_2- and CH_3-resonances observed in the ^{1}H-PHIP-NMR spectrum of the catalyst-attached intermediates observed during the hydrogenation of styrene, using the catalysts listed in Figure 12.21.

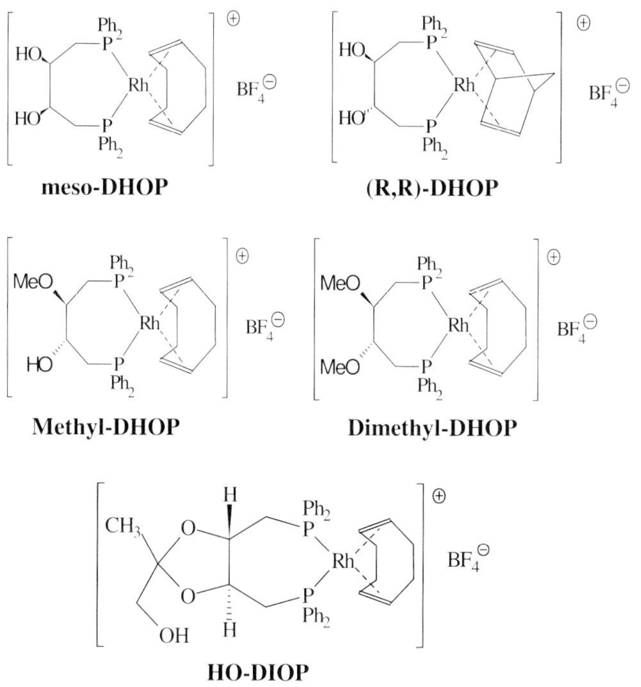

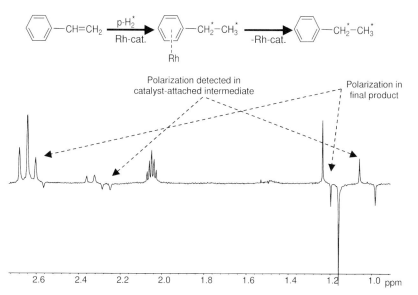

meso-DHOP

(R,R)-DHOP

Methyl-DHOP

Dimethyl-DHOP

HO-DIOP

Fig. 12.21 Chiral and achiral Rh-catalysts employed for the hydrogenation of styrene.

Polarization detected in catalyst-attached intermediate

Polarization in final product

Fig. 12.22 CH_2- and CH_3-resonances observed in the ^{1}H-PHIP-NMR spectrum of the intermediate attached to an achiral catalyst during the hydrogenation of styrene.

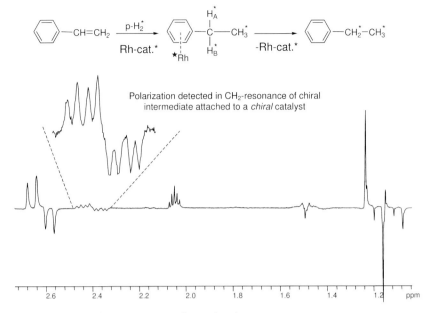

Fig. 12.23 CH$_2$- and CH$_3$-resonances observed in the ^{1}H-PHIP-NMR spectrum of the intermediate attached to a chiral catalyst during the hydrogenation of styrene.

12.4.3
Determination of Kinetic Constants

The formation and decay of these product–catalyst-π-complexes are expected to occur according to the sequence of reactions as outlined in Scheme 12.4. The kinetic constants associated with the occurrence of k_{HYD} and the decay of k_{OFF}, respectively, can both be determined by PHIP-NMR using a process termed dynamic PASADENA (DYPAS) spectroscopy, as has been outlined previously [37]. For this purpose the addition of parahydrogen to the reaction is synchronized with the pulse sequences of the NMR spectrometer, whereby the time for acquiring the NMR spectra is delayed by variable amounts. The results thereof are listed in Table 12.3. A variety of kinetic constants can be determined, and the method is reasonably accurate; the margins of error are also indicated in Table 12.3 [37].

The data reveal that electron-donating groups such as amino- or alkoxy-groups increase the rate of formation of k_{HYD} according to their donor strength; for the rate of decay, such a correlation is more likely opposite – that is, they decrease the rate of decay of k_{OFF}. Similar observations have been made for electron-withdrawing substituents, which decrease the rate of formation of k_{HYD} according to their acceptor strength, but increase the rate of decay of k_{OFF}. Again, a free-energy correlation appears to be possible (unpublished results). One clear consequence of this correlation is a variable relative intensity of "satellite" resonances

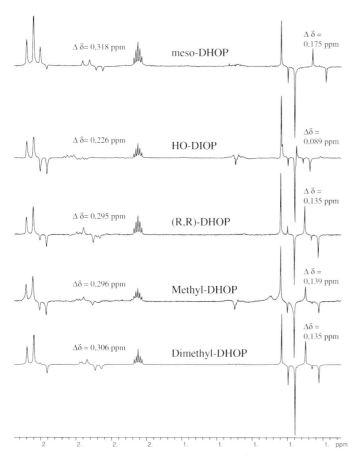

Fig. 12.24 CH$_2$- and CH$_3$-resonances observed in the ^{1}H-PHIP-NMR spectrum of the intermediate attached to the selection of various catalysts during the hydrogenation of styrene.

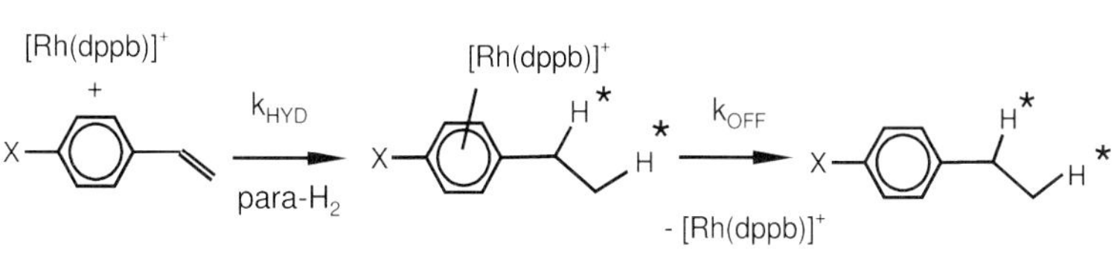

Scheme 12.4

and those of the final hydrogenation products. In cases where the rate of formation k_{HYD} is high, but the rate of decay of k_{OFF} is low, the intensity of the satellites can exceed those of the final hydrogenation products.

The principle of the DYPAS experiment is outlined in Figure 12.25. The results derived from the DYPAS experiments (see Table 12.3) of reactions following Scheme 12.4 may be represented pictorially as the time-dependence of the

Table 12.3 Rates of formation and of decay of the interim product–catalyst-π-complexes [44].

X-	K_{HYD} [s^{-1}]	k_{OFF} [s^{-1}]
-H	0.111 ± 0.009	0.35 ± 0.03
-OMe	0.138 ± 0.010	0.45 ± 0.04
-OEt	(0.40)	0.273 ± 0.024
-NH$_2$	(0.84)	0.113 ± 0.036

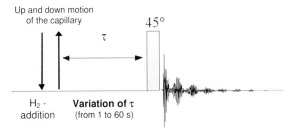

Fig. 12.25 The principle of the DYPAS experiment.

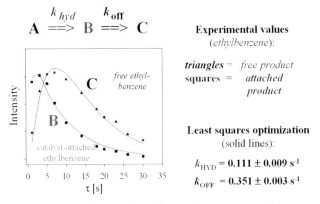

Fig. 12.26 Plots of the time-dependence of concentrations of the intermediates and products of the hydrogenation of styrene.

intermediate and the final product of two consecutive reactions (Fig. 12.26). Unlike the standard appearance of a plot of time-dependence of final product for two consecutive reactions, which typically shows asymptotic saturation behavior, the curve for the signal intensity of C also decays, due to relaxation of the PHIP-derived signal enhancement of this product. The DYPAS method is not restricted to the determination of the kinetic constants listed in Table 12.3, but may also be used to determine other rates of formation and decay (Table 12.4).

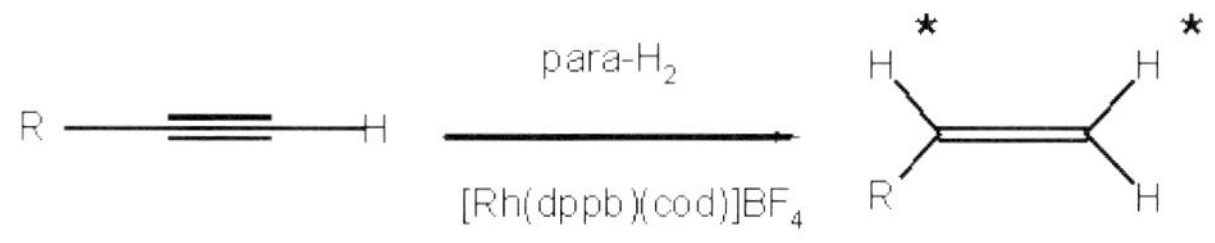

R ———≡≡≡——— H $\xrightarrow[\text{[Rh(dppb)(cod)]BF}_4]{\text{para-H}_2}$ (alkene product with *★* labels)

Scheme 12.5 Hydrogenation of differently substituted ethynyl-benzenes.

Table 12.4 Rates of hydrogenation of differently substituted ethynylbenzenes.

Substrate	R-	K_{HYD} [s^{-1}]
1	Ph-	0.140 ± 0.005
2	$(CH_3)_3C$-	0.085 ± 0.002
3	$(CH_3)Si$-	0.064 ± 0.004
4	$(Ph)_3Si$-	0.085 ± 0.003

As a further example, the hydrogenation of differently substituted ethynylben-zenes have been investigated using the catalyst [Rh(dppb)(cod)]BF$_4$. The reaction is outlined in Scheme 12.5 and the results are listed in Table 12.4.

12.4.4
Computer-Assisted Prediction and Analysis of the Polarization Patterns: DYPAS2

For *in-situ* studies of reaction mechanisms using parahydrogen it is desirable to compare experimentally recorded NMR spectra with those expected theoretically. Likewise, it is advantageous to know, how the individual intensities of the inter-mediates and reaction products depend on time. For this purpose a computer simulation program DYPAS2 [45] has been developed, which is based on the density matrix formalism using superoperators, implemented under the C^{++} class library GAMMA.

DYPAS2 [45] offers a large variety of simulation modes and allows the calcu-lation of polarization patterns of the expected NMR spectra derived either from parahydrogen or orthodeuterium. Furthermore, the individual boundary condi-tions can be taken into account, under which the hydrogenations take place. Ac-cordingly, the magnetic field dependence of the resulting polarization patterns can be predicted, for example in spectra based on the PASADENA or ALTADE-NA effect, respectively.

The kinetics of hydrogenation transfer is covered by the use of an exchange superoperator assuming a pseudo first-order reaction. Thereby, competing hy-drogenations of the substrate to more than one product can also be accommo-dated. In addition, the consequences of relaxation effects or NOEs can be in-cluded into the simulations if desired. Furthermore, it is possible to simulate the consequences of different types of pulse sequences, such as PH-INEPT or INEPT+, which have previously been developed for the transfer of polarization from the parahydrogen-derived protons to heteronuclei such as ^{13}C or ^{15}N. The individual delays required in these pulse sequences are critical parameters for the associated magnitude of the transferred polarization, but it is not trivial to estimate their optimal values. Therefore – and in particular for large spin systems – it is desirable to obtain access to intensity plots, which display the calculated intensities of the polarization-enhanced NMR spectra of the heteronuclei as a function of the individual delay times. DYPAS2 contains this option and provides an even greater variety of other possibilities [45].

12.5
Colloidal Catalysts

12.5.1
In-Situ PHIP-NMR Investigation of the Hydrogenation of Ethynylbenzene by Pd$_x$[N(octyl)$_4$Cl]$_y$

It is recognized that some colloidal catalysts show "homogeneous"-like behavior, though for the most part it is unclear which parameters define the borderline between heterogeneous- (i.e., surface reactions) and homogeneous-type catalysis of colloidal systems. The reaction of ethynylbenzene (phenylacetylene) with parahydrogen using the colloidal palladium catalyst Pd$_x$[N(octyl)$_4$Cl]$_y$ leads to nuclear spin polarization in the hydrogenation products [38]. Heterogeneous catalysts are not expected to give rise to the PHIP effect, as the spin correlation in the adsorbed p-H$_2$ is considered to be lost once the dihydrogen molecules interact with a catalytically active surface. Therefore, most likely hydrogen atoms transferred in such fashion would stem from different p-H$_2$ molecules. However, it has been shown that the colloidal transition-metal catalyst system Pd$_x$[N(octyl)$_4$Cl]$_y$ gives rise to the PHIP phenomenon, thereby implying a homogeneous reaction pathway and proving that the two transferred hydrogen atoms stem from the same dihydrogen molecule (Scheme 12.6).

For this purpose, standard 5-mm NMR tubes were charged with 100 μL ethynylbenzene, 6 mg of the catalyst Pd$_x$[N(octyl)$_4$Cl]$_y$, and 0.7 mL acetone-d$_6$ and placed into a 200-MHz spectrometer. Charges of 51%-enriched p-H$_2$ were prepared as previously outlined via catalytic equilibration over charcoal at 77 K and injected repeatedly in synchronization with the pulsed NMR-experiment via an electromechanically lowered glass capillary mechanism.

The PHIP-NMR spectra shown in Figure 12.27 were obtained during the hydrogenation of 4-chlorostyrene with p-H$_2$ parahydrogen [38]. In order to elimi-

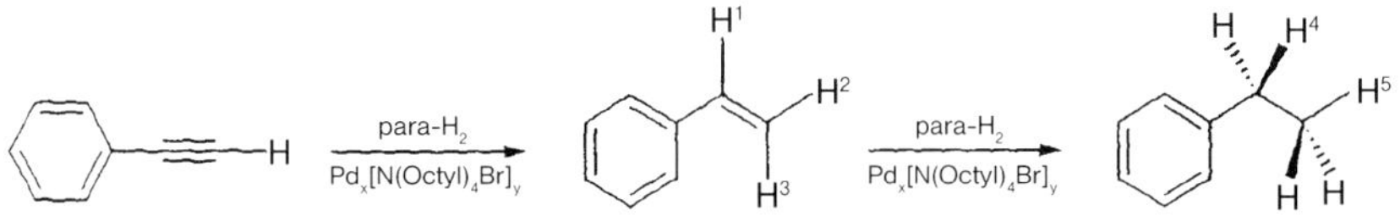

Scheme 12.6 Hydrogenation of phenylacetylene using a colloidal Pd-catalyst system.

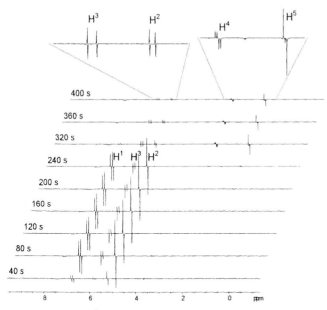

Fig. 12.27 200 MHz ^{1}H-PHIP-NMR spectra recorded during the hydrogenation of phenylacetylene using a colloidal Pd catalyst.

nate interfering signals originating from spins in thermal equilibrium (i.e., "non-PHIP signals"), the NMR spectra were acquired after a hydrogenation time of 10 s accumulating the responses to different pulse angles.

The maximum signal intensity for ^{1}H-PHIP signals occurs at a flip angle of 45°, in contrast to the usual "90° pulse" used for substrates in thermal equilibrium. Therefore, by alternately adding and subtracting scans acquired using pulse angles of –45° and 135°, respectively, it is possible to minimize or even suppress the signal intensity of the "unpolarized", so-called "thermal" proton spins, whereas the signals of the proton spins in polarization then display the maximum of their signal intensity. Less than perfect elimination of the thermal signals may occur, nevertheless, in part because of a slight drift of the external magnetic field (B_0), or because of changes of the homogeneity. Each spectrum shown in Figure 12.27 is the result of four accumulated scans and 40 s of hydrogenation time each.

Cis hydrogenation of ethynylbenzene leads to polarization signals in the respective positions H^1 and H^2 of styrene (Scheme 12.6), as is detected. Furthermore, polarization also occurs in position H^3. This signal shows an anti-phase coupling of 1 Hz with its geminal hydrogen. In order to distinguish whether this is a result of geminal transfer of p-H$_2$ parahydrogen to the positions H^2 and H^3, or the consequence of a NOE, the use of deuterated ethynylbenzene is required.

With proceeding hydrogenation, the concentration of ethynylbenzene decreases; therefore, the positions H^1 and H^2 show increasingly less polarization.

Subsequently, the polarization signals H^4 and H^5 of ethylbenzene (i.e., the hydrogenation product of styrene) appear. Accordingly, ethylbenzene is also formed via homogeneous catalysis as well – that is, two more p-H$_2$ protons are transferred simultaneously.

The hydrogenation of ethynylbenzene to styrene (as well as that of styrene to ethylbenzene) mediated by the colloidal catalyst Pd$_x$[N(octyl)$_4$Cl]$_y$ occurs in a "pairwise" fashion – that is, the two hydrogen atoms of dihydrogen are transferred simultaneously, which is otherwise characteristic of a homogeneous hydrogenation. This *in-situ* NMR experiment convincingly demonstrates the unique power of PHIP for the investigation of hydrogenations mediated by colloidal systems.

12.6
Transfer of Proton Polarization to Heteronuclei

12.6.1
General Aspects

Homogeneously catalyzed hydrogenations of unsaturated substrates with parahydrogen leading to strong polarization signals in ^{1}H-NMR spectra also give rise to strong heteronuclear polarization, especially if the hydrogenations are carried out at low magnetic fields. The mechanisms underlying the process of PHIP transfer to heteronuclei, however, are still not fully understood, and different mechanisms have been discussed. The polarization transfer from protons to carbon nuclei during the hydrogenation of alkynes may serve as a typical example, which has been investigated for several substrates. It could be shown that in systems containing easily accessible triple bonds (Table 12.5) (e.g., ethynylbenzene or 2,2-dimethylbutyne), a polarization transfer takes place to all carbon nuclei in the molecule. Accordingly, all ^{13}C resonances can be observed in NMR spectra recorded *in situ* with good to excellent signal-to-noise ratios (SNRs) using only a single transient [42]. This technique has made it possible to identify a Ru-based catalyst system, which yields *E*-alkenes via homogeneous hydrogenations of alkynes with central triple bonds, as opposed to the more conventional *Z*-alkenes that result from hydrogenation with Rh-based catalysts [47].

PHIP studies together with polarization transfer permit elucidation of the structure (i.e., the carbon skeleton of compounds) in a fashion which is far superior to the conventional ^{1}H-PHIP technique outlined above.

Furthermore, the qualitative influence of substituents on the symmetry and electronic structure of the substrate and its hydrogenation product on the efficiency of the transfer of polarization to the ^{13}C-nuclei have been discussed, as well as the feasibility of a polarization transfer to other heteronuclei. Evidence in the form of a shift of the aromatic ^{13}C resonances has been found for an initial attachment of hydrogenation products containing aromatic segments to the metal center of the cationic hydrogenation catalyst – probably in the form of a π-complex.

Table 12.5 Reactions studied using ^{13}C-PHIP-NMR spectroscopy.

$H_3C-C(CH_3)(CH_3)-C\equiv CH$	$\xrightarrow[\text{Cat. / Acetone-}d_6]{p\text{-}H_2}$	$H_3C-C(CH_3)(CH_3)-C(H)=CH_2$ (2, 1)
$C_6H_5-C\equiv CH$ (phenyl)	$\xrightarrow[\text{Cat. / Acetone-}d_6]{p\text{-}H_2}$	$C_6H_5-C(H)=CH_2$ (ring positions 3, 2, 4, 5, 6, 1; vinyl 7, 8)
$HC\equiv C-(CH_2)_3-CH_3$	$\xrightarrow[\text{Cat. / Acetone-}d_6]{p\text{-}H_2}$	$H_2C=C(H)-(CH_2)_3-CH_3$ (1, 2)
$H_3C-C\equiv C-(CH_2)_2-CH_3$	$\xrightarrow[\text{Cat. / Acetone-}d_6]{p\text{-}H_2}$	$H_3C-C(H)=C(H)-(CH_2)_2-CH_3$ (2, 3)
$H_3C-CH_2-C\equiv C-CH_2-CH_3$	$\xrightarrow[\text{Cat. / Acetone-}d_6]{p\text{-}H_2}$	$H_3C-CH_2-C(H)=C(H)-CH_2-CH_3$ (3, 4)
$H_3C-(H_2C)_2-C\equiv C-(CH_2)_2-CH_3$	$\xrightarrow[\text{Cat. / Acetone-}d_6]{p\text{-}H_2}$	$H_3C-(H_2C)_2-C(H)=C(H)-(CH_2)_2-CH_3$ (4, 5)

It has been postulated [39] and demonstrated previously that "Hetero-PHIP" for nuclei such as ^{13}C, ^{29}Si, and ^{31}P can result in a signal enhancement (SE) >10 [8, 40, 41], particularly if the reactions are carried out at low magnetic fields (i.e., under ALTADENA conditions). ^{31}P INEPT($+\pi/4$) experiments have also been carried out to transfer the initial proton polarization to ^{31}P [8].

Barkemeyer et al. [8a] showed previously that high enhancement can also be achieved at high magnetic fields when hydrogenating symmetric systems, where the breakdown of symmetry is caused by the naturally abundant ^{13}C nuclei occurring individually in the two other equivalent carbon atoms of the double bond of the substrate (see Scheme 12.8) [8a].

Although this phenomenon provides a powerful tool for the NMR investigation of low-γ nuclei, the full potential of this sizeable polarization transfer from parahydrogen to other nuclei has not been applied very much as compared to ^{1}H-PHIP-NMR spectroscopy. In particular, the use of the PHIP effect to enhance and correspondingly simplify the detection of carbon nuclei is a powerful tool for investigating the structure of organic molecules.

12.6.2
Polarization Transfer to ^{13}C

The PHIP phenomenon may be used to label specifically individual, naturally occurring ^{13}C nuclei magnetically via specific enhancement of their corresponding resonances. This frequently simplifies the assignment, thereby facilitating the interpretation of ^{13}C-NMR spectra of more complex molecules. The polarization phenomena occurring in the ^{13}C-PHIP-NMR spectra during the *in-situ* hydrogenation of various alkynes using the catalyst [Rh(cod)(dppb)]$^+$ in CDCl$_3$ at 5 bar of parahydrogen may serve as a characteristic example [42].

The spectra depicted in Figure 12.28 were recorded during the hydrogenation of 3,3-dimethylbut-1-yne to 3,3-dimethylbut-1-ene, and occur according to Scheme 12.7. As is apparent from Figure 12.28, signals from all carbon atoms are visible and nearly all expected couplings can be determined accordingly.

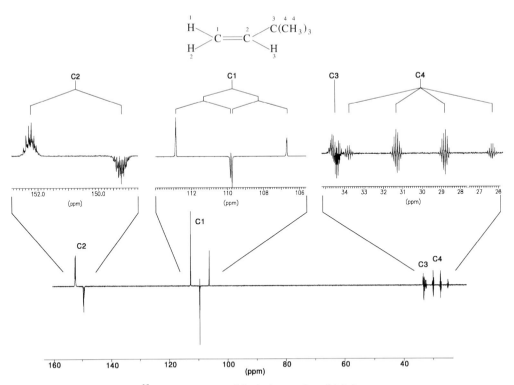

Fig. 12.28 ^{13}C-PHIP spectrum of the hydrogenation of 3,3-dimethylbut-1-yne to 3,3-dimethylbut-1-ene.

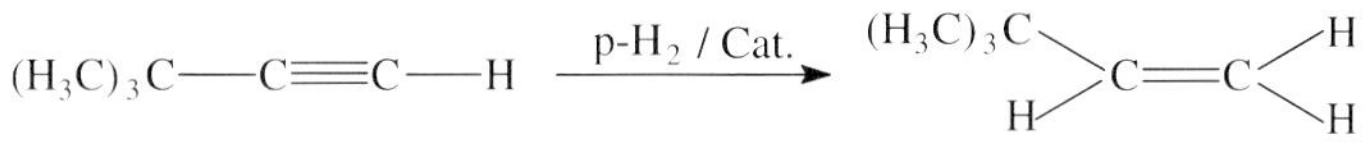

$$(H_3C)_3C\text{---}C\equiv C\text{---}H \quad \xrightarrow{\text{p-H}_2\,/\,\text{Cat.}} \quad \begin{array}{c}(H_3C)_3C\\[2pt]H\end{array}\!\!\!\Big\rangle C\!\!=\!\!C\Big\langle\!\!\!\begin{array}{c}H\\[2pt]H\end{array}$$

Scheme 12.7 Hydrogenation of 3,3-dimethylbut-1-yne to 3,3-dimethylbut-1-ene.

Other alkynes investigated include ethynylbenzene, 1-hexyne, 2-hexyne, 3-hexyne, and 4-octyne (Table 12.5). These substrates have in common that the triple bond is rather accessible. The substituents cover a wide range of different electronic structures and steric requirements, and in addition, they provide for symmetric and asymmetric substrates. Within this selection of hexynes the electronic structure and accessibility of the individual triple bonds are almost the same, but the symmetry of the bond is different. Furthermore, 1-hexyne, 3,3-dimethylbutyne, and ethynylbenzene all have terminal triple bonds, but the electronic structure and accessibility of the π-system of these triple bonds are influenced differently by the catalyst, depending on the group adjacent to the triple bond. Within this series it could be shown how +M and +I effects or steric hindrance influence the hydrogenation and polarization transfer from the protons to the carbon atoms [42]. The six reactions investigated are outlined in Table 12.5 – that is, the starting materials and the corresponding reaction products, together with the numeration of the individual carbon atoms following IUPAC nomenclature.

The corresponding signal enhancements, as well as the magnitude of the polarization transfer from the former parahydrogen atoms to the directly adjacent former acetylenic carbon, and also to other carbon atoms in the product molecules, have been compared, whereby the polarization spectra were measured under ALTADENA conditions. This means that the hydrogenation was conducted within the Earth's magnetic field, upon which the reaction solution was transferred into the high magnetic field of the superconducting magnet of the NMR spectrometer (corresponding to a proton resonance frequency of 200 MHz or ~50 MHz for the ^{13}C nuclei, respectively) to detect the PHIP spectra. By means of this procedure, polarization is transferred to all magnetically active nuclei in the following way. The resonance frequencies of all nuclei in this very low magnetic field of the Earth (or less) are all virtually the same. Therefore, a highly coupled spin system of a high order is obtained – that is, the differences of resonance frequencies between the carbons (^{13}C) and the protons are small compared to their coupling constants. This is essential for the polarization transfer from protons to a large number of carbons.

By contrast, in the high field of a superconducting magnet, polarization transfer to heteronuclei is less efficient, because now the difference in resonance frequency of ^{1}H and ^{13}C is significant and exceeds the magnitude of the coupling constants between the carbons and the protons. In this latter case, polarization transfer can be achieved most effectively by appropriate pulse sequences (e.g., via cross-polarization).

For these experiments, standard 5-mm NMR tubes equipped with a screw cap and a septum were charged with 200 µL of the substrate, 6 mg of the catalyst [Rh(cod)(dppb)]BF$_4$ (Cat.), 0.7 mL of degassed chloroform-d$_1$, and 5 bar of 50%-

enriched parahydrogen. This degree of enrichment was achieved via catalytic equilibration over charcoal at 77 K. Higher levels of enrichment, namely >97%, have also been achieved using a closed-cycle cooler cryostat.

In this fashion, ^{13}C-PHIP-NMR spectra were obtained using the six substrates 3,3-dimethylbut-1-yne, ethynylbenzene, 1-hexyne, 2-hexyne, 3-hexyne, and 4-oc-tyne, respectively. The relevant spectroscopic data for these substrates were obtained from a database [43]. All ^{13}C spectra could be recorded using only a single transient, as they exhibited a higher SNR than expected. The enhanced SNR is due to the PHIP effect, and results from a transfer of the initial proton polarization of the parahydrogen to the carbon atoms. Because of the strong signal enhancement caused by the PHIP effect, no signals were visible for the carbon atoms of the starting material. In general, it is not possible to obtain carbon spectra using just a single transient, at least not with such a high SNR.

Normally in ^{13}C-NMR spectra, the solvent gives rise to the strongest signal of the spectrum; therefore, it frequently must be suppressed via an appropriate pulse sequence. In ^{13}C-PHIP-NMR spectroscopy as outlined here, however, this is not necessary; rather, the signal due to the solvent CDCl$_3$ (normally occurring as a triplet at $\delta = 77.7$ ppm) is not even visible, at least not in the more favorable cases. In all reactions investigated, the signals of all carbon atoms in the products could be detected exploiting only a single transient, which yields a good to excellent SNR.

Although a polarization transfer to all carbon atoms in the corresponding substrates is observed universally, the associated SNR is quite different for the individual substrates. By contrast to the ^{1}H-PHIP-NMR spectra, namely, the ^{13}C-PHIP-NMR spectra observed exhibit much more significant differences depending on the individual substrates. In general, in the ^{1}H-NMR the SNR is much higher; therefore, small differences cannot be distinguished so easily, whereas for ^{13}C-NMR and accordingly for ^{13}C-PHIP-NMR this is much easier. In addition, in ^{1}H-PHIP-NMR the polarization is only transferred to the adjacent protons not further away than approximately corresponding to a ^{3}J coupling. For ^{13}C, however, transfers of the polarization to carbon nuclei as far away as six bonds have been detected.

It is worthwhile pointing out that it is desirable to acquire all spectra under identical conditions, such as hydrogen pressure, elapsed time prior to the acquisition of the spectrum, temperature, and amount of catalyst used. For this reason it is desirable to use a set-up which permits the spectra to be recorded in totally standardized fashion, which does not depend on any individual "human factor". Such a system would allow the reaction to be conducted at a low magnetic field and would thereafter transfer the solution automatically (notably quickly and "adiabatically") into the NMR spectrometer for subsequent analysis.

In addition to the transfer of some degree of polarization to all carbon atoms in the product molecule, a slight low field shift of the ^{13}C resonances in the aromatic region of the PHIP-NMR spectrum of the product recorded during the hydrogenation of ethynylbenzene can be observed. A similar effect has been observed before in the ^{1}H-PHIP-NMR spectra recorded during the hydrogenation of styrene derivatives using cationic rhodium catalysts. This phenomenon has been termed "product attachment" (see Section 12.4), implying that the hydrogenation product binds to the catalyst's metal center via the phenyl ring in the form of a π-complex. The strength of this attachment depends on the electronic structure of the product and the catalyst, and it can either lead to a shift to high or low frequencies in the ^{1}H-NMR spectrum. By contrast to the ^{1}H-PHIP-NMR investigations of this product attachment (as reported earlier), the effect as observed here in ^{13}C-PHIP-NMR spectra is more pronounced. For ^{1}H-NMR the magnitude of this shift is about 0.5 ppm, whereas for ^{13}C-NMR a shift of about 4 ppm was observed. However, considering the wider range of ^{13}C-shifts compared to those of ^{1}H-nuclei, this is to be expected.

These ^{13}C-PHIP-NMR investigations show clearly that it is possible to effectively transfer polarization from parahydrogen to ^{13}C atoms, especially when conducting the hydrogenations at low magnetic fields. This polarization transfer clearly depends on the electronic and steric features of the groups adjacent to the triple bond. The strongest ^{13}C-PHIP-NMR signals have been observed using either 3,3-dimethylbutyne or ethynylbenzene as the substrate, whereby a transfer of the polarization has been observed for all carbon atoms in the molecule, associated with an excellent SNR. Furthermore, the hydrogenation of 1-hexyne also shows polarization and signal enhancement for all carbon nuclei, but the SNR is not as good as it is for 3,3-dimethylbutyne or ethynylbenzene. With 2-hexyne as the substrate, polarization transfer is observed for all carbon nuclei likewise. Finally, during the hydrogenation of 3-hexyne to 3-hexene, and of 4-octyne to 4-octene, respectively, polarization signals are also observed, with good intensity for the two carbons of the olefinic group and of all aliphatic carbon nuclei. This is consistent with previous studies of other symmetrically substituted alkynes showing strong ^{13}C-PHIP-NMR signals, and it is especially true for symmetrically substituted alkynes carrying electron-poor substituents (e.g., acetylene dicarboxylic acid and its esters). This finding has now been extended to electron-rich alkyne substrates.

When acquiring NMR spectra under ALTADENA conditions, polarization is transferred to all magnetically active nuclei in the substrate. This means that the initial polarization of the parahydrogen is distributed over the number of atoms present in the substrate. This "sharing of the polarization" permits an explanation of the observed decrease in intensity from 3,3-dimethylbut-1-yne or ethynylbenzene to the hexynes, and 4-octyne. In the latter two cases there are more nuclei participating in this sharing. Additionally, the good accessibility of the terminal triple bond of 3,3-dimethylbut-1-yne, ethynylbenzene, and 1-hexyne implies an easy hydrogenation of these substrates, yielding a higher SNR in contrast to the SNR of 2- and 3-hexyne carrying a more central and hence less-accessible triple bond. Finally, the electron-richness of 3,3-dimethylbut-1-yne or ethynylbenzene facilitates the coordination of these substrates to the metal center of the catalyst, thereby resulting in a slightly better SNR, than is the case for 1-hexyne.

Because of the low natural abundance of ^{13}C nuclei (1.1%), practically all observed product molecules contain only one ^{13}C nucleus. Accordingly, the polarization signals in the ^{13}C NMR spectrum clearly do not originate from one and the same product molecule, but rather stem from different, singly labeled molecules. Since the fraction of the product molecules that contain two or even more ^{13}C nuclei are practically negligible, the observed ^{13}C-NMR spectra are the result of a superimposition of the spectra stemming from product molecules which contain only a single ^{13}C nucleus at the respective position. Therefore, a transfer of polarization from the former parahydrogen atoms to an individual ^{13}C nucleus cannot occur via transfer along the backbone (i.e., not along the carbon chain of the molecules), but it is caused by direct or indirect coupling, either scalar or dipolar, between the former parahydrogen nuclei, other protons, and the corresponding ^{13}C nucleus.

In these experiments it was shown possible to transfer polarization, induced by the PHIP effect, to almost any given carbon in the hydrogenation product of an alkyne, and that certain asymmetrical and electronically activated substrates give rise to the strongest polarization transfer to ^{13}C nuclei. In the cases of 3,3-dimethylbutyne or ethynylbenzene, all carbon atoms were detected using only a single transient, and thereby yielding an excellent SNR [42]. Even with substrates that yield only a weaker polarization transfer (i.e., 1-, 2-, 3-hexyne, and 4-octyne) the SNR was significantly better than was obtained using conventional ^{13}C-NMR without accumulation. In favorable cases, the normally dominating ^{13}C signals of the solvent (e.g., CDCl$_3$) are barely visible in the PHIP-enhanced spectra.

Because of large coupling constants between ^{13}C and ^{1}H (as compared to the coupling constants between ^{1}H nuclei), polarization transfer to ^{13}C nuclei is more effective than among protons. It appears that the maximum distance for transfer of polarization between protons (i.e., in ^{1}H-PHIP-NMR) is about three bonds, whereas it is five for ^{13}C-PHIP-NMR.

In addition to the observed polarization transfer, attachment of the hydrogenated product to the catalyst – most likely in the form of a π-complex between the aromatic portion of a product and the cationic catalyst – has also been observed in the ^{13}C-PHIP-NMR spectra. The associated larger shift range of the affected ^{13}C will make it possible to characterize the nature of this attachment as well as the associated binding energies of the hydrogenation product to the catalyst's metal center more precisely and effectively.

Figure 12.29 illustrates the ^{13}C-PHIP spectrum obtained when hydrogenating 4-fluoro-1-ethynylbenzene under ALTADENA conditions in the Earth's magnetic field. The reaction proceeds according to Scheme 12.5 and the catalyst used is outlined there. This system and the other isomers thereof, namely the hydrogenation of 2-, 3-, and 4-fluoro-1-ethynylbenzene, have also been studied using ^{1}H- and ^{19}F-PHIP-NMR, and the results of the latter will be outlined in the following section (Fig. 12.30).

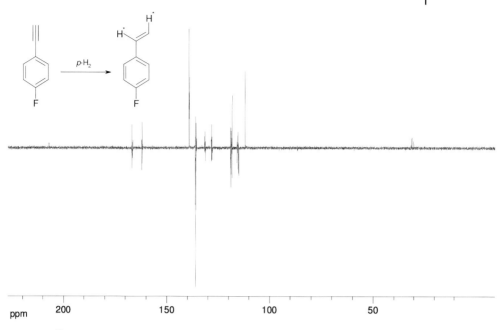

Fig. 12.29 ^{13}C-PHIP spectrum of the hydrogenation of 4-fluoro-1-ethynylbenzene.

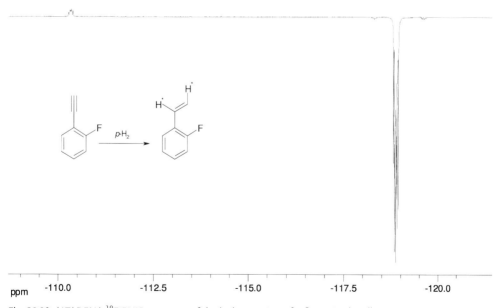

Fig. 12.30 ALTADENA-^{19}F-PHIP-spectrum of the hydrogenation of 2-fluoro-1-ethynylbenzene.

12.6.3
Polarization Transfer to ^{19}F

Transfer of the initial proton polarization is not confined to other protons or ^{13}C, but the signals of other heteronuclei (^{2}H, ^{15}N, ^{29}Si, ^{31}P) in the hydrogenation products can also become substantially enhanced, thereby also increasing their receptivity. Accordingly, the transfer of the PHIP-derived high spin order to ^{19}F has been accomplished using a set of chemically similar fluorinated styrene and ethynylbenzene derivatives.

Additionally, the transfer to ^{19}F occurs not only when the hydrogenation is initiated in the Earth's magnetic field (ALTADENA condition) but also when the whole reaction is carried out in the presence of the strong field of the NMR spectrometer (PASADENA). Both through-bond and through-space interactions are responsible for this process, termed parahydrogen-aided resonance transfer (PART). The high-field and low-field PHIP transfer mechanisms must be strictly distinguished, because they give rise to different phenomena [48].

The best results are obtained when using substrates associated with high hydrogenation rates and long spin-lattice relaxation time for all nuclei of interest, and if the reactions are carried out in the absence of the strong field of the NMR spectrometer. Therefore, in order to study the consequences of polarization transfer to ^{19}F, the hydrogenations of ^{19}F-containing ethynylbenzenes and

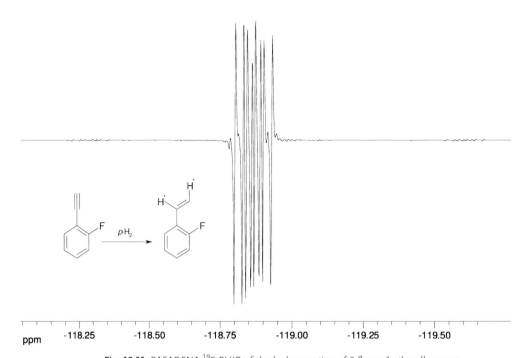

Fig. 12.31 PASADENA-^{19}F-PHIP of the hydrogenation of 2-fluoro-1-ethynylbenzene.

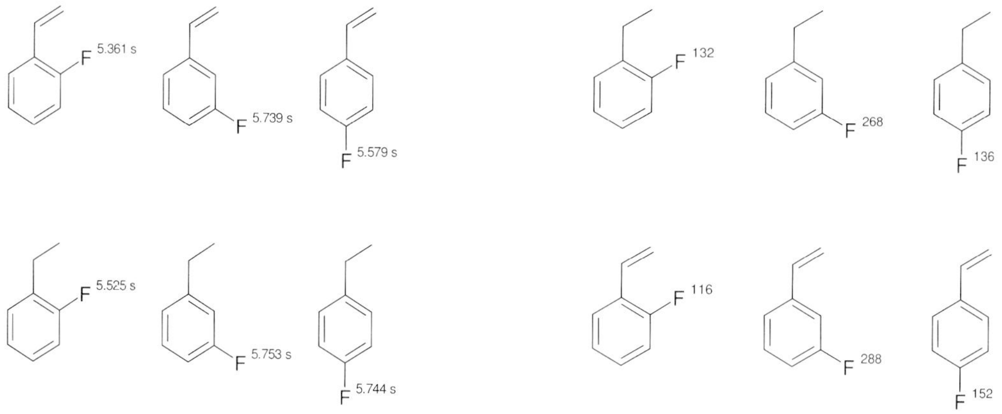

Fig. 12.32 ^{19}F-NMR T_1-spin relaxation times (left) and ^{19}F-PHIP ALTADENA enhancement factors (right) for the hydrogenation products shown.

styrenes have been investigated. It might be expected that if the chemical reaction occurred at high field (i.e., within a MRI or NMR magnet), the signal enhancement would be confined to certain protons only. Instead, if the reaction occurs at low field, the polarization might be transferred to heteronuclei such as ^{31}P, ^{13}C, or ^{19}F, without requiring any special pulse sequences. It transpires, however, that in these systems the ^{19}F nuclei become spin polarized both at low field (Fig. 12.30) and at high field (Fig. 12.31), which opens up new possibilities.

The pentafluorophenyl derivatives are particularly revealing, since the polarization transferred to the individual ^{19}F positions can be compared and correlated with different concepts of their interactions with the primarily spin polarized protons [48]. The enhancement factors achieved in the hydrogenation products of monosubstituted compounds at low field under ALTADENA conditions are given in Figure 12.32.

Likewise, the initial proton polarization may be transferred to other magnetically active heteronuclei, most attractively to those associated with a low γ-value of their nucleus (i.e., to ^{15}N, ^{29}Si), and similarly "difficult" ones, using heteroatom PHIP at low magnetic fields [8, 45].

12.6.4
Parahydrogen-Assisted Signal Enhancement for Magnetic Resonance Imaging

The initial observation of a significant PHIP-derived enhancement factor of >2500 for ^{13}C has rendered this approach attractive for heteronuclear magnetic resonance imaging (MRI) [49]. Using the system as outlined in Scheme 12.8 that is associated with an enhancement factor for ^{13}C of 2500 [8 a], Golman et al. [49] have demonstrated the feasibility of heteronuclear MRI originally in the form of ^{13}C angiography investigations.

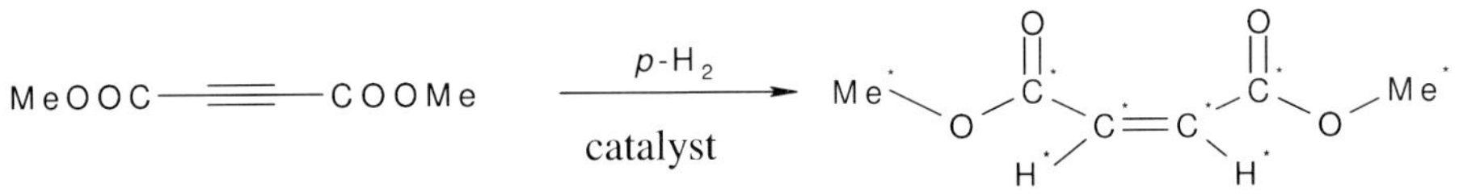

MeOOC——≡——COOMe $\xrightarrow[\text{catalyst}]{p\text{-}H_2}$

Scheme 12.8 Hydrogenation of acetylene dicarboxylic acid dimethylester to the corresponding maleate [8 a].

This initial system, though not optimized for this particular application in medicine, has since been superseded by more appropriate systems with other choices of the components [48]. Finally, possible applications for using selectively hyperpolarized fluorine-containing molecules as contrast agents for medical imaging techniques appear especially attractive; however, this subject matter is discussed elsewhere [50].

12.7
Catalysts Containing other Transition Metals

Over the years, a considerable number of systems have been investigated, comprising catalysts that contain a variety of transition metals, including Co [5], Pd [59, 60], Pt [60, 61], Ru [47, 62], and Ta [63], although the most detailed studies have been conducted on systems containing either Rh or Ir (including some rather exotic ones [64]) as the transition metals. Even investigations on the surfaces of ZnO have been conducted successfully [59]. Further details have also been reported in earlier reviews [12 d, 34 c, 65].

12.8
Summary and Conclusions

PHIP-NMR spectroscopy is an *in-situ* method, which is associated with a significant signal enhancement of both the protons as well as of various heteronuclei. On the basis of an associated increase in sensitivity, this approach and method allows the identification of unstable intermediates and previously elusive minor reaction products of homogeneous hydrogenations and other reactions. It brings about strong signal enhancement (factor $\cong 1000$ for protons and even higher for heteronuclei such as ^{13}C). This increase in sensitivity renders it possible to detect and analyze crucial reaction intermediates. Therefore, structural and geometric conclusions can be made, and even the rate of formation and of the decay of the intermediates can be determined. The initial proton polarization can be transferred to heteronuclei and to yield enhancement factors for ^{13}C nuclei of more than 2500. The method has interesting applications in MRI, especially in medicinal imaging and molecular spectroscopy.

The method yields unique mechanistic information, it permits the determination of kinetic parameters, it allows the determination of the degree of reversibility of various homogeneous hydrogenations, it yields information about simultaneous (i.e., "pairwise" hydrogen transfer versus asynchronous transfer of individual hydrogens in subsequent events), and it also promises attractive applications to boost the normally low sensitivity of ^{1}H-MRI and, in particular, of heteronuclear MRI.

By using PHIP-NMR studies, various intermediates such as the previously elusive dihydrides of neutral and cationic hydrogenation catalysts, as well as hydrogenation product/catalyst complexes, have already been detected during the hydrogenation of styrene derivatives using cationic Rh catalysts. Information about the substituent effect on chemical shifts and kinetic constants has been obtained via time-resolved PASADENA NMR spectroscopy (DYPAS).

Acknowledgment

The financial support of the Deutsche Forschungsgemeinschaft (DFG), Bonn, Germany, and of the Fonds der Chemischen Industrie, Frankfurt/M., Germany, is gratefully acknowledged.

Abbreviations

ALTADENA	Adiabatic Longitudinal Transport After Dissociation Engenders Net Alignment
CIDNP	chemically induced dynamic nuclear polarization
DFT	density functional theory
MRI	magnetic resonance imaging
NOE	nuclear Overhauser effect
ODIP	orthodeuterium-induced polarization
PART	parahydrogen-aided resonance transfer
PASADENA	Parahydrogen And Synthesis Allow Dramatically Enhanced Nuclear Alignment
PHIP	Para-Hydrogen Induced Polarization
SE	signal enhancement
SNR	signal-to-noise ratio

References

1 Moser, W.R., Reaction Monitoring by High-Pressure Cylindrical Internal-Reflectance and Optical-Fiber Coupled Reactors. In: Moser, W.R., Slocum, D.W. (Eds.), *Homogeneous Transition Metal-Catalyzed Reactions*, American Chemical Society, Washington DC, **1992**, Advanced Chemistry Series 230, p. 5.

2 Whyman, R., *In Situ Spectroscopic Studies in Homogeneous Catalysis*. In: Moser, W.R., Slocum, D.W. (Eds.), *Homogeneous Transition Metal-Catalyzed Reactions*, American Chemical Society, Washington DC, **1992**, Advanced Chemistry Series 230, p. 19 and references therein.

3 (a) Roe, D.C., *Adv. Chem. Ser.* **1992**, *230*, 33; (b) Elsevier, C.J., *J. Mol. Catal.* **1994**, *92*, 258.

4 (a) Rathke, J.W., Klingler, R.J., Chen, M.J., Woelk, K., *Trends Organomet. Chem.* **1994**, *1*, 117; (b) Woelk, K., Bargon, J., *Rev. Sci. Instrum.* **1992**, *63*, 33070; (c) Trautner, P., Woelk, K., Bargon, J., et al., *J. Magn. Reson.* **2001**, *151*, 284; (d) Niessen, H.P., Trautner, P., Wiemann, S., Bargon, J., Woelk, K., *Rev. Sci. Inst.* **2002**, *73* Part 1, 1259.

5 (a) Bryndza, H.E., Thesis, Berkeley, **1981**; (b) Seidler, P.F., Bryndza, H.E., Frommer, J.E., Stuhl, L.S., Bergman, R.G., *Organometallics* **1983**, *2*, 1701.

6 (a) Bowers, C.R., Weitekamp, D.P., *Phys. Rev. Lett.* **1986**, *57*, 2645; (b) Bowers, C.R., Weitekamp, D.P., *J. Am. Chem. Soc.* **1987**, *109*, 5541.

7 Eisenschmid, T.C., Kirrs, R.U., Deutsch, P.P., Hommeltoft, S.I., Eisenberg, R., Bargon, J., Lawler, R.G., Balch, A.L., *J. Am. Chem. Soc.* **1987**, *109*, 8089.

8 (a) Barkemeyer, J., Haake, M., Bargon, J., *J. Am. Chem. Soc.* **1995**, *117*, 2927; (b) Haake, M., Natterer, J., Bargon, J., *J. Am. Chem. Soc.* **1996**, *118*, 8688, and references therein.

9 Pravica, M.G., Weitekamp, D.P., *Chem. Phys. Lett.* **1988**, *145*, 255.

10 (a) Sandler, Y.L., *J. Phys. Chem.* **1954**, *58*, 54–57; (b) Sandler, Y.L., *J. Chem. Phys.* **1958**, *29*, 97.

11 Depatie, D.A., Mills, R.L., *Rev. Sci. Instr.* **1968**, *39*, 105.

12 (a) Bowers, C.R., Jones, D.H., Kurur, N.D., Labinger, J.A., Pravica, M.G., Weitekamp, D.P., *Adv. Mag. Res.* **1990**, *14*, 269; (b) Eisenberg, R., *Acc. Chem. Res.* **1991**, *24*, 110; (c) Bargon, J., Kandels, J., Woelk, K., *Z. Phys. Chem.* **1993**, *180*, 65.

13 Greve, T., Bargon, J., unpublished results.

14 Bargon, J., Kandels, J., Kating, P., *J. Chem. Phys.* **1993**, *98*, 6150.

15 Kaptein, R., *J. Chem. Soc. Chem. Commun.* **1972**, 872.

16 (a) Lepley, A.R., Closs, G.L., *Chemically Induced Magnetic Polarization*, J. Wiley and Sons, New York, **1973**; (b) Pine, S.H., *J. Chem. Edu.* **1972**, 49, 664, and references therein; (c) Carey, F.A., Sundberg, R.J., *Advanced Organic Chemistry*, Plenum Press, New York, **1984**, 2nd edn, Part A, p. 623; (d) Buchachenko, A.L., Frankevich, E.F., *Chemical Generation and Reception of Radio- and Microwaves*, VCH Publishers, New York, **1994**, p. 33; (e) Sweany, R.L., Halpern, J., *J. Am. Chem. Soc.* **1977**, 99, 8335.

17 Bargon, J., Kandels, J., Woelk, K., *Angew. Chem. Int. Ed.* **1990**, *29*, 58; *Angew. Chemie* **1990**, *102*, 70.

18 Bonhoeffer, K.F., Harteck, P., *Z. Phys. Chem.* **1929**, *B4*, 113.

19 (a) Sandler, Y.S., *J. Phys. Chem.* **1954**, *58*, 58; (b) Sandler, Y.S., *J. Chem. Phys.* **1958**, *29*, 97.

20 (a) Steinfeld, J.I., *Molecules and Radiation*, The MIT Press, Cambridge, Massachusetts, London, Third Printing, **1981**, p. 98; (b) Sidgwick, N.V., *The Chemical Elements and Their Compounds*, Vol. I, Oxford University Press, Oxford, **1951**; (c) Driesen, A., van der Poll, E., Silvera, I.F., *Phys. Rev.* **1984**, *B30*, 2317.

21 Buntkowsky, G., Bargon, J., Limbach, H.H., *J. Am. Chem. Soc.* **1996**, *118*, 8677.

22 (a) Young, J.F., Osborn, J.A., Jardine, F.H., Wilkinson, G., *J. Chem. Soc., Chem. Commun.* **1965**, 131; (b) Osborn, J.A., Jardine, F.H., Young, J.F., Wilkinson, G., *J. Chem. Soc. A* **1966**, 1711

23 Tolman, C.A., Meakin, P.Z., Lindner, D.L., Jesson, J.P., *J. Am. Chem. Soc.* **1974**, *96*, 2762.

24 Duckett, S. B., Newell, C. L., Eisenberg, R., *J. Am. Chem. Soc.* **1994**, *116*, 10548.

25 (a) Halpern, J., Okamoto, T., Zakhariev, A., *J. Mol. Catal.* **1976**, *2*, 65; (b) Dawans, F., Morel, D., *J. Mol. Catal.* **1977–78**, *3*, 403.

26 Schrock, R. R., Osborn, J. A., *J. Am. Chem. Soc.* **1976**, *98*, 2134.

27 (a) Halpern, J., Riley, D. P., Chan, A. S. C., Pluth, J. J., *J. Am. Chem. Soc.* **1977**, *99*, 8055; (b) Chan, A. S. C., Halpern, J., *J. Am. Chem. Soc.* **1980**, *102*, 838; (c) Halpern, J., *Inorg. Chim. Acta.* **1981**, *50*, 11; (d) Halpern, J., *Science* **1982**, *217*, 401; (e) Collman, J. P., Hegedus, L. S., Norton, J. R., Finke, R. G., *Principles and Applications of Organotransition Metal Chemistry*, University Science Books, Mill Valley, California, **1987**, p. 529.

28 Greve, T., PhD Thesis, Faculty of Sciences/Physics, University of Bonn, **1996**.

29 Duckett, S. B., Eisenberg, R., Goldman, A. S., *J. Chem. Soc., Chem. Commun.* **1993**, 1185.

30 Duckett, S. B., Eisenberg, R., *J. Am. Chem. Soc.* **1993**, *115*, 5292.

31 D., Colebrooke, S. A., Duckett, S. B., Lohman, J. A. B., Eisenberg, R., *J. Chem. Soc., Dalton Trans.* **1998**, 3363.

32 Esteruelas, M. A., Lahuerta, O., Modrego, J., Nürnberg, O., Oro, L. A., Rodriguez, L., Sola, E., Werner, H., *Organometallics* **1993**, *12*, 266.

33 (a) Dang, T. P., Kagan, H. B., *Chem. Commun.* **1971**, 481; (b) Knowles, W. S., Sabacky, M. J., Vineyard, B. D., *Chem. Commun.* **1972**, 10.

34 (a) Harthun, A., Selke, R., Bargon, J., *Angew. Chem. Int. Ed.* **1996**, *35*, 2505; (b) Harthun, A., Barkemeyer, J., Selke, R., Bargon, J., *Tetrahedron Lett.* **1995**, *36*, 7423; (c) Natterer, J., Bargon, J., *Progr. NMR Spectr.* **1997**, *31*, 293; (d) Duckett, S. B., Sleigh, C. J., *Prog. NMR Spect.* **1999**, *34*, 71.

35 (a) Heinrich, H., Giernoth, R., Bargon, J., Brown, J. M., *Chem. Commun.* **2001**, 1296; (b) Giernoth, R., Heinrich, H., Adams, N. J., Bargon, J., Brown, J. M., *J. Am. Chem. Soc.* **2000**, *122*, 12381; (c) Giernoth, R., Hydrogenation. In:

Mechanisms in Homogeneous Catalysis: A Spectroscopic Approach, B. Heaton (Ed.), John Wiley & Sons Inc., **2005**, p. 359.

36 (a) Landis, C. R., Feldgus, S., *Angew. Chem. Int. Ed.* **2000**, *39*, 2863; (b) Landis, C. R., Hilfenhaus, P., Feldgus, S., *J. Am. Chem. Soc.* **1999**, *121*, 8741; (c) Kless, A., Börner, A., Heller, D., Selke, R., *Organometallics* **1997**, *16*, 2096; (d) Bray, M. R., Deeth, R. J., Paget, V. J., Sheen, P. D., *Int. J. Quant. Chem.* **1997**, *61*, 85.

37 Giernoth, R., Hübler, P., Bargon, J., *Angew. Chem. Int. Edit.* **1998**, *37*, 2473.

38 Eichhorn, A., Koch, A., Bargon, J., *J. Mol. Catal. A – Chem.* **2001**, *174*, 293.

39 Bowers, C. R., Jones, D. H., Kurur, N. D., Labinger, J. A., Pravica, M. G., Weitekamp, D. P., *Adv. Magn. Reson.* **1990**, *14*, 269 and references therein.

40 Eisenberg, R., Eisenschmid, T. C., Chinn, M. S., Kirss, R. U., *Homogenous Transition Metal Catalyzed Reactions*. Moser, W.R., Slocum, D.W. (Eds.), Advances in Chemistry 230, Washington DC, **1992**, p. 45 and references therein.

41 Duckett, S. B., Newell, C. L., Eisenberg, R., *J. Am. Chem. Soc.* **1993**, *11*, 1156.

42 Stephan, M., Kohlmann, O., Niessen, H. G., Eichhorn, A., Bargon, J., *Magn. Reson. Chem.* **2002**, *40*, 157.

43 Aldrich/ACD Library of FT NMR SpectrA Pro, V 1.7, **1998**.

44 Wildschütz, S., Hübler, P., Bargon, J., *Chem. Phys. Chem.* **2001**, *2*, 328.

45 (a) Schmidt, T., Bargon, J., unpublished results; (b) Schmidt, T., PhD Thesis, Bonn University, **2003**, http://hss.ulb.uni-bonn.de/diss_online/math_nat_fak/2003/schmidt_thorsten/index.htm.

46 (a) Limbacher, A., Jonischkeit, T., Bargon, J., to be published; (b) Limbacher, A., PhD Thesis, Bonn University, **2004**. http://hss.ulb.uni-bonn.de/diss_online/math_-nat_fak/2004/limbacher_arndt/index.htm; (c) Jonischkeit, T., PhD Thesis, Bonn University, **2004**. http://hss.ulb.uni-bonn.de/diss_online/math_nat_fak/2004/jonischkeit_thorsten/index.htm.

47 (a) Niessen, H. G., Schleyer, D., Wiemann, S., Bargon, J., et al., *Magn. Reson. Chem.* **2000**, *38*, 747; (b) Schleyer, D., Niessen, H. G., Bargon, J., *New. J. Chem.*

2001, *25*, 423; (c) Schleyer, D., PhD Thesis, Bonn University, **2000**. http://hss.ulb.uni-bonn.de/diss_online/math_nat_fak/2000/schleyer_dana/index.htm.

48 (a) Kuhn, L.T., Fligg, R., Bargon, J., unpublished results; (b) Kuhn, L.T., Bommerich, U., Bargon, J., submitted; (c) Bommerich, U., PhD Thesis, Bonn University, **2005**. http://hss.ulb.uni-bonn.de/diss_online/math_nat_fak/2005/bommerich_ute/index.htm.

49 (a) Golman, K., Axelsson, O., Jóhannesson, H., Månsson, S., Olofsson, C., Petersson, J.S., *Magn. Reson. Med.* **2001**, *46*, 1; (b) Golman, K., Ardenkjaer-Larsen, J.H., Svensson, J., Axelsson, O., Hansson, G., Hansson, L., Jóhannesson, H., Leunbach, I., Månsson, S., Petersson, J.S., et al., *Acad. Radiol.* **2002**, *9* (Suppl. 2), S507; (c) Ardenkjær-Larsen, J.H., Fridlund, B., Gram, A., Hansson, G., Hansson, L., Lerche, M.H., Servin, R., Thaning, M., Golman, K., *Proc. Natl. Acad. Sci. USA* **2003**, *100*, 10158; (d) Mansson, S., Johansson, E., Magnusson, P., Chai, C.M., Hansson, G., Petersson, J.S., Stahlberg, F., Golman, K., *Eur. Radiol.* **2005**, June 14 [Epub ahead of print]. *http://www.ncbi.nlm.nih.gov/entrez/query.fcgi?cmd=Retrieve&db=PubMed&list_uids=15954020 &dopt=Citation* (November 15, 2005).

50 Bargon, J., to be published.

51 Koch, A., Bargon, J., *Magn. Reson. Chem.* **2000**, *38*, 216.

52 Fligg, R., Bargon, J., Enrichment of parahydrogen using the Oxford closed cycle cooler cryostat. *Research Matters,* **2000**, 12, Oxford Instruments, Oxford, UK.

53 (a) Evett, A.A., *J. Chem. Phys.* **1959**, *31*, 565; (b) White, D., Lassettre, E.N., *J. Chem. Phys.* **1960**, *32*, 72; (c) Freeman, M.P., Hagyard, M.J., *J. Chem. Phys.* **1969**, *49*, 4020.

54 Cunningham, C.M., Chapin, D.S., Johnston, H.L., *J. Am. Chem. Soc.* **1958**, *80*, 2382.

55 Dosiere, M., *J. Chem. Edu.* **1985**, *62*, 891.

56 Grilly, E.R., *Rev. Sci. Instrum.* **1953**, *24*, 72.

57 Bradshaw, T.W., Norris, J.O.W., *Rev. Sci. Instr.* **1987**, *58*, 83.

58 Koch, A., Bargon, J., *Inorg. Chem.* **2001**, *40*, 533.

59 Carson, P.J., Bowers, C.R., Weitekamp, D.P., *J. Am. Chem. Soc.* **2001**, *123*, 11821.

60 Sulman, E., Deibele, C., Bargon, J., *React. Kinet. Catal.* **1999**, *L67* (1), 117.

61 Jang, M., Duckett, S.B., Eisenberg, R., *Organometallics* **1996**, *15*, 2863.

62 Duckett, S.B., Mawby, R.J., Partridge, M.G., *Chem. Commun.* **1996**, 383.

63 Millar, S.P., Zubris, D.L., Bercaw, J.E., Eisenberg, R., *J. Am. Chem. Soc.* **1998**, *120*, 5329.

64 Suardi, G., Cleary, B.P., Duckett, S.B., Sleigh, C., Rau, M., Reed, E.W., Lohman, J.A.B., Eisenberg, R., *J. Am. Chem. Soc.* **1997**, *119*, 7716.

65 Bowers, C.R., Sensitivity enhancement utilizing parahydrogen. In: Grant, D.M., Harris, R.K. (Eds.), *Encyclopedia of Nuclear Magnetic Resonance*, Volume 9, **2002**, p. 1.

13

A Tour Guide to Mass Spectrometric Studies of Hydrogenation Mechanisms

Corbin K. Ralph, Robin J. Hamilton, and Steven H. Bergens

13.1
Introduction

The ability of electrospray ionization (ESI) to generate intact complex organometallic ions in the gas phase has revolutionized the application of mass spectrometry (MS) to the study of reactions catalyzed by organometallic compounds [1]. To date, the catalytic reactions studied using ESI-MS include oxy transfer reactions with manganese-oxo-salen complexes [2], Ziegler-Natta polymerizations of alkenes [3], alkene metathesis using ruthenium-alkylidene catalysts [4], C–H activation by iridium complexes [5], Suzuki couplings [6], nickel-catalyzed coupling reactions [7], chiral catalysts for allylations [8], and catalytic hydrogenations [9]. The objectives of these studies have ranged from understanding elementary reactions between substrates and coordinatively unsaturated catalytic intermediates in the gas phase, to rapid screening of one-pot, complex catalyst mixtures for activity towards various reactions. Although MS studies of alkene hydrogenations have been carried out using other ionization techniques [10], for example to characterize product isotopomers or organometallic species present in solution, the focus of this chapter will be on ESI-MS studies.

The reported ESI-MS studies of catalytic reactions can roughly be divided into two categories: 1) characterization of species present in catalytically active solutions; and 2) application of tandem ESI-MS systems to study the chemistry of complex organometallic ions in the gas phase [1]. It is the study of reactivity of ions in the gas phase that has provided the most mechanistic information about catalytic hydrogenations using MS. A paramount concern which attends such studies from a solution-phase chemist's point of view is the relevance of gas-phase ion chemistry to a reaction of interest carried out in solution under a given set of conditions (temperature, pressure, concentrations, solvent, etc.). A number of factors come into play when making such comparisons. Ionic species in the gas phase can induce polarizations in neutral reagent molecules, for example, causing an electrostatic attraction between the reactants, and thereby resulting in substantially larger Arrhenius frequency factors and different activa-

The Handbook of Homogeneous Hydrogenation.
Edited by J. G. de Vries and C. J. Elsevier
Copyright © 2007 WILEY-VCH Verlag GmbH & Co. KGaA, Weinheim
ISBN: 978-3-527-31161-3

tion energies in the gas phase than in solution [3a, 4a, 11]. Further, the concentrations and contact times between reagents are smaller in the gas phase than in solution. Also, a large component of activation energies for reactions carried out in solution can involve solvent effects [12]. Various steps in reactions catalyzed by transition-metal complexes in solution involve changes in coordination number, and thereby are accompanied by coordination or dissociation of solvent molecules. These and other factors are involved when comparing the reactivity of organometallic ions in solution to the gas phase, and no rigorous mechanistic theories or studies currently exist that specifically address these concerns [13]. The approach adopted for the present chapter is to provide a tour guide-level description and an understanding of the methods used to carry out such experiments. The studies of catalytic hydrogenation are then discussed in some detail, with the discussions being accompanied by relevant information and perspective from the other reactions and systems studied by ESI-MS. The objective of this tour guide is to provide chemists studying catalytic hydrogenations with sufficient perspective and information to determine if ESI-MS can provide experimental data relevant to a catalytic hydrogenation of interest. As with all tour guides, more detailed and insightful information can be obtained from the locals. Therefore, the interested hydrogenation tourist is encouraged to "step off the bus", to study the source literature in more detail, and perhaps to contact the researchers involved to discuss a reaction of interest.

13.2
A General Description of ESI-MS

Perhaps of high interest to the hydrogenation chemist are the pertinent details of how gas-phase ions of organometallic catalysts are prepared. Of specific interest are the types of fragmentation events that can occur during the preparation of the gas phase ions, and what forms coordinatively unsaturated organometallic ions adopt in the gas phase.

ESI was first developed for MS by Fenn et al. in the mid-1980s [14]. ESI is the mildest method to generate ions in the gas phase from species dissolved in solution. Species with high molecular masses, such as synthetic polymers, DNA fragments, and proteins can be transferred to the gas phase without fragmentation. ESI has been applied to a wide variety of organometallic species [11]. The application of ESI has allowed the production of various organometallic ions in the gas phase with control over the number of solvent molecules associated with the ion, and control over gas-phase ligand dissociation reactions that generate coordinatively unsaturated ions for study [15]. Of note, however, is that ESI does not, in the strictest sense, generate charged molecules. Rather, it separates the charged species from each other, and from the neutral species in solution. In fact, ESI cannot easily be applied to neutral molecules unless they have, for example, basic groups that can associate with charged species also present in solution (e.g., protons or sodium ions). An interesting illustration of the ability of ESI-MS to detect and separate minute amounts of charged species in solution is the study by Chen et al. of a mixture containing a variety of neutral alkene metathesis catalyst precursors with the general formula [RuCl$_2$(alkylidene)(diphosphane)] [4c]. When dissolved in solution, these neutral catalyst precursors are in equilibrium with small amounts of the cationic catalysts, [RuCl(alkylidene)(diphosphane)]$^+$, formed by chloride dissociation. The application of ESI-MS to these mixtures allowed the detection of trace amounts of cationic catalysts, which could also be separated in the gas phase on the basis of their m/z ratio. The separated catalysts were then individually screened for activity towards the ring-opening olefin metathesis of norbornene in the gas phase. It was found that the cationic catalysts with the highest activities towards norbornene in the gas phase also had the highest activities in solution.

ESI operates in stages (Fig. 13.1) [1, 14]. A typical sequence is as follows. The organometallic ion is introduced as a salt dissolved in a sprayable solvent through a capillary into the electrospray chamber. Polar solvents such as dichloromethane, acetonitrile, and alcohols are easily sprayed, whilst nonpolar solvents can be sprayed as mixtures with polar solvents. The concentration of the organometallic complex typically ranges from 0.001 to 10 mM, which is similar to the concentration range used in homogeneous catalysts. The electrospray chamber often contains a nitrogen "bath" gas at or near atmospheric pressure when used to transfer organometallic ions to the gas phase. The capillary contains an electrode which is in contact with the sample solution. A large potential difference (typically 4–6 kV) is applied between this electrode and the walls of the chamber. The electric field is intensified at the tip of the capillary, causing ion separation and formation of charged droplets in the atmosphere of the chamber. The droplets contain an excess of either negative or positive ions, depending on whether the capillary tip is negative relative to the chamber walls, or positive. Solvent evaporation occurs as the charged droplets containing the ions are accelerated towards the chamber walls by the potential difference. Rapid solvent evaporation leads to a decrease in the droplet diameter. The droplet diameter decreases until the coloumbic repulsions between the ions near the surface exceed the surface tension of the solvent. The droplet will then undergo fission into smaller droplets containing fewer ions than the parent, and/or expulsion of the ions from the surface of the droplet into the gas phase. This sequence repeats itself until all the ions in the gas phase are free from solvent. In practice, desolvation occurs over several stages of the process (*vide infra*). The charged ions or solvated ions are driven by a pressure differential through a heated capillary (held typically at temperatures ranging from 150 to 200 C) into the first vacuum chamber that is typically maintained at 0.5–1 Torr. The ions are essentially pulled through the capillary by the viscous drag forces of the nitrogen gas escaping into the first vacuum chamber to emerge as a supersonic stream. The outlet of the capillary is sheathed in a metal tube, and a potential is applied between this tube and a skimmer at the other end of the chamber. This "skimmer-" or "tube potential" accelerates the ions in the expanding stream towards the skimmer. The skimmer potential typically ranges from 35 to 100 V. The first

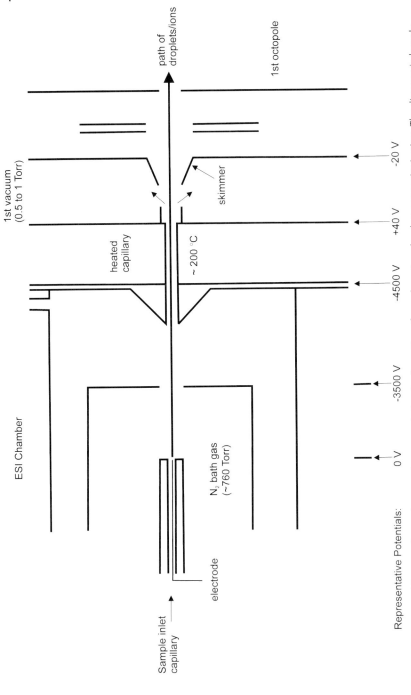

Fig. 13.1 A schematic simplification of typical events that occur during the formation of gas-phase ions by electrospray ionization. The diagram is loosely based upon those found elsewhere [1b, c, 12].

pumping chamber typically serves three functions. First, it completes desolvation of the ions. Second, it imparts translational kinetic energy to the ions via the skimmer potential before the ions pass into the next chamber (either an octopole or quadrupole). Third, collisions occur between the background gas in the chamber and the ions as they are accelerated towards the skimmer. The energy of these collisions is in part controlled by the magnitude of the skimmer potential. At higher voltages, the collisions with the remaining background gas can induce ligand dissociation from the parent organometallic ion. The skimmer potential is thereby used as an adjustable parameter to control the translational kinetic energy of the ions before they pass into the next chamber, and if desired, to control collision-induced dissociation (CID) of ligands from the parent organometallic ions before they leave the skimmer chamber.

A point of interest at this stop in our tour is that fragmentation of organometallic ions in ESI-MS often proceeds via ligand dissociation (e.g., phosphane loss) to generate coordinatively unsaturated organometallic ions [1–9]. One of the strengths of this technique is that such unsaturated ions are typically proposed as reactive intermediates in catalytic reactions carried out in solution (*vide infra*), allowing ESI-tandem-MS systems to study directly the gas-phase reactivity of such species.

For the published ESI-MS mechanistic hydrogenation studies, the ions pass through the skimmer into the first octopole [9]. The longitudinal kinetic energy of the ions entering the first octopole is determined in part by the skimmer potential. The octopole serves several functions. The first is as an ion guide in which any neutral molecules remaining in the ion stream are removed by vacuum. A radiofrequency is applied to the poles that functions, in effect, to contain the ions in the region near the radial center of the octopole without imparting significant kinetic energy upon them as they move longitudinally through the unit. The second function of the octopole is to act as a gas-phase reactor. The octopole is typically held at moderate temperatures (e.g., 70 °C) and it contains a neutral inert or reagent gas at pressures up to ~ 0.1 Torr. The organometallic ions interact with the neutral gas molecules as they move through the octopole via collisions and/or reactions. One experimental concern is to keep the energy distribution of the gas molecules as near as possible to thermal equilibrium inside the octopole. The process is called "thermalization", and it occurs via collisions between the ions that have been accelerated by the skimmer potential and the neutral gas present in the octopole. One incentive for establishing and maintaining thermal equilibrium among the ions in the octopoles is that the system then better represents ions in solution. Another incentive is that Maxwell-Boltzmann statistics can then be applied to model the distribution of vibrational, rotational, and kinetic energies among the molecules in the gas phase. Models have been developed that incorporate Maxwell–Boltzmann statistics as well as certain experimental parameters (e.g., temperature, pressure, and longitudinal velocity) [1 b, c]. These models provide information such as the number of collisions between an ion and the neutral gas molecules as it travels through the octopole, or information about the kinetic energies of the collisions, or in some cases, bond dissociation energies can be obtained through CID studies. Reaction probabilities can be obtained by comparing product yields to the number of calculated collisions between an organometallic ion and the reagent gas in the octopole. Apart from thermalization, collisions between the ions and the background gas can induce either CID (at high skimmer potentials) or reactions with reagent gases such as hydrogen [1b,c, 16].

The resulting mixture of organometallic ions leaves the octopole and is then separated on the basis of m/z using the first quadrupole, typically at 10^{-6} Torr. The quadrupole allows ions of only one m/z ratio to pass through. The separated organometallic ions are characterized as much as possible by subsequent CID and gas-phase reactivity studies, and by isotopic labeling (*vide infra*). After leaving the first quadrupole, the separated organometallic ions are then driven into a second octopole containing a neutral or reagent gas *via* another potential difference. The magnitude of this potential difference determines the kinetic energy of the organometallic ions entering the second octopole. The product ions leaving the second octopole (or gas-phase reactor) are analyzed with a second quadrupole and detector. The experimental strategy typically applied to mechanistic studies with ESI-MS is thereby to vary the operating parameters of the ESI, the first octopole, and quadrapole to generate an organometallic ion of interest. The ion of interest is fragmented or reacted with a reagent molecule in the second octopole, and the products characterized in the last quadrupole and detector. It is the gas-phase reactions of the organometallic ions in the octopoles that are used to model the solution chemistry of the reaction. A point of interest at this part of our tour is that one-pot catalyst mixtures have been screened with ESI-MS systems by separating out the ions using the first octopole-quadrupole and then studying their individual gas-phase reactivity in the second octopole [1c, 4c, 8].

13.3
Mechanistic Hydrogenation Studies

Although a number of reactions catalyzed by organometallic complexes have been studied using ESI-MS (*vide supra*), the number of reports detailing ESI-MS mechanistic studies of alkene hydrogenation is small [9]. The first was carried out by Chen et al. using cationic rhodium(I)-phosphine complexes as catalyst systems. A number of cationic Rh(I) precursors were sprayed, including $[Rh(P(CD_3)_3)_4(H)_2]^+$ (**1**), $[Rh(P(CH_3)_3)_3(nbd)]^+$ (**2**, nbd=norbornadiene), $[Rh(PPh_3)_2(nbd)]^+$ (**3**), $[Rh((S, S)$-chiraphos)(nbd)]^+$ (**4**, (S, S)-chiraphos=$(2S, 3S)$-(–)-*bis*(diphenylphosphino)butane), and $[Rh((R)$-BINAP)(MeOH)_2]^+$ (**5**, BINAP=2,2'-*bis*(diphenylphosphino)-1,1'-binaphthyl). As discussed in the previous section, the tube lens potential was adjusted to control successively the amount of desolvation to produce gas phase ions of the parent rhodium compound, or to induce ligand and/or hydrogen loss by CID. The mixture of ions leaving the first octopole were separated by the first quadrupole, and then studied by CID or reactions with reagent gases in the second octopole.

Spraying dichloromethane solutions of **1** and using N_2 gas in the first octopole allowed mass selection at the first quadrupole of either $[Rh(P(CD_3)_3)_4]^+$ (**6**), from loss of hydrogen from **1**; $[Rh(P(CD_3)_3)_3]^+$ (**7**), from subsequent loss of $P(CD_3)_3$; and $[Rh(P(CD_3)_3)_2]^+$ (**8**), from further loss of $P(CD_3)_3$ (Scheme 13.1). Compound **8** is a gas-phase mimic of catalyst species that have been observed in solution such as *trans*-$[Rh(PPh_3)_2(MeOH)_2]^+$ (**9**), that contain extremely labile solvento ligands, and that are proposed to be catalytic intermediates in hydrogenations of alkenes using cationic rhodium-$(PR_3)_2$ complexes as catalysts [17]. One immediate difference between **9** in methanol solvent, and **8** in the gas phase is the absence of coordinating solvent molecules, making **8** an extremely reactive, coordinatively-unsaturated species. This difference is not unique to this system and warrents further discussion here. Most fundamental steps of catalytic cycles such as oxidative additions, reductive eliminations, and insertions involve changes in the ligands coordinating and changes in coordination number, and are thereby influenced by solvent molecules that act as ligands. The empirical observation can be made that organometallic ions in the gas phase tend to relieve coordination unsaturation by intramolecular bonding or activation. For example, gas-phase reactivity, CID, and deuterium-labeling experiments indicate that the formally 12-electron species **8** exists in equilibrium with the mono- and di-C–D activated species **10** and **11** (Scheme 13.1).

As another example, studies of the catalytic activity of the gas-phase ions [Ru-Cl(alkylidene)(diphosphane)]$^+$ toward ring-opening olefin metathesis of norbornene show that an alkene group in the growing polymer chain reaches back to oc-

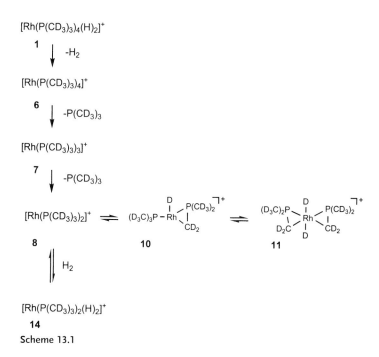

Scheme 13.1

cupy a vacant coordination site on ruthenium [4d]. In some cases, this tendency towards intramolecular bonding or activation has led to new discoveries that have borne true in the corresponding solution-phase chemistry. For example, an ESI-MS study of the gas-phase chemistry of the unsaturated, formally 16-electron cation $[(\eta^5\text{-}C_5Me_5)Ir(CH_3)(P(CH_3)_3)]^+$ (**12**) showed that it undergoes C–H activation down a P–CH$_3$ group, followed by elimination of methane to generate $[(\eta^5\text{-}C_5Me_5)Ir(CH_3)(\eta^2\text{-}CH_2P(CH_3)_2)]^+$ (**13**) [5]. Further study in the gas phase showed that **13** is substantially more active towards intermolecular C–H activation than **12**. Subsequent solution-phase chemistry by another group confirmed that $[(\eta^5\text{-}C_5Me_5)Ir(CH_3)(\eta^2\text{-}CH_2P(CH_3)_2)(OTf)]$ (**13′**, OTf=triflate, a weakly coordinating anion) is substantially more active towards intermolecular C–H activation in solution than is $[(C_5Me_5)Ir(CH_3)(P(CH_3)_3)(OTf)]$ (**12′**) [18].

Organometallic ions that undergo reversible intramolecular activation or bonding are more likely to provide gas-phase analogues to reactions of catalytic intermediates in solution containing labile solvento ligands. Returning to hydrogenation for an example, the mass-selected compound **8**, which likely exists as a mixture with **10** and **11**, reacts with H$_2$ gas in the second octopole to give the dihydride $[Rh(P(CD_3)_3)_2(H)_2]^+$ (**14**) (see Scheme 13.1). The reaction can be reversed without H–D exchange with the CD$_3$ groups, showing that it was **8**, not **10** or **11**, that reacted with the H$_2$ gas. The reversible activation of the methyl groups to form **10** and **11** thereby functions partially to relieve the coordination unsaturation of **8**, as do the methanol solvento ligands in solutions of **9**. In this sense, reaction of **8** with H$_2$ gas to generate the dihydride **14** provides a gas-phase analogue to the known oxidative addition of H$_2$ to **9** in methanol solution to generate *trans*-$[Rh(H)_2(PPh_3)_2(MeOH)_2]^+$ (**15**) [17b].

Intramolecular bonding or activation to alleviate coordination unsaturation in the gas phase does fail, at times, to mimic the behavior of labile solvento complexes in solution. For example, the gas-phase ion $[Rh(PPh_3)_2]^+$ (**16**), which is the solvent-free, direct analogue of **9**, failed to add H$_2$ or D$_2$ in the second octopole. The authors attributed this lack of analogous reactivity for **16** in the gas phase to the proposed formation of an intramolecular η^6-arene complex with one of the phosphines. The formation of such an 18-electron arene complex is presumably irreversible on the timescale that the ion resides in the second octopole. Consistent with this behavior, the gas-phase ions $[Rh((S,\ S)\text{-chiraphos})]^+$ (**17**) and $[Rh((R)\text{-BINAP})]^+$ (**18**) also failed to react with H$_2$ in the second octopole. Worthy of mention, however, is that the analogues of complexes **17** and **18** in methanol solution react reversibly, but only to a small extent with hydrogen to generate traces of $[Rh(diphosphane)(H)_2(MeOH)_2]^+$ [19]. One therefore cannot rule out, on the bases of the data in the literature, that the gas-phase ions **17** and **18** did not mimic the behavior of their solvento analogues in methanol solution. The authors also reported that **17** reacts, at very low collision energies, with methyl acrylate to generate the adduct $[Rh((S,\ S)\text{-chiraphos})(methyl\ acrylate)]^+$ (**19**). Although the authors did not pursue this avenue further, formation of the adduct **19** does provide an analogue to the known reactions between $[Rh((S,\ S)\text{-chiraphos})(CH_3OH)_2]^+$ and $[Rh((R)\text{-BINAP})(MeOH)_2]^+$ and various

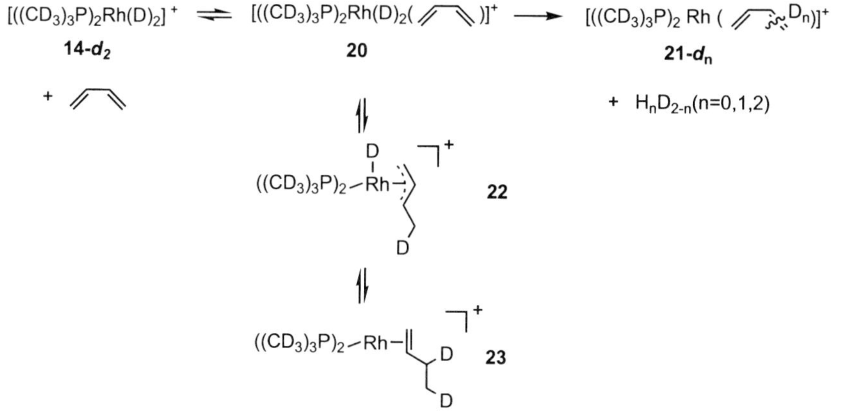

$[((CD_3)_3P)_2Rh(D)_2]^+ \rightleftharpoons [((CD_3)_3P)_2Rh(D)_2(\diagup\diagdown)]^+ \longrightarrow [((CD_3)_3P)_2 Rh (\diagup\diagdown D_n)]^+$

14-d_2   **20**   **21-d_n**

$+ \diagup\diagdown$ $+ \; H_nD_{2-n}(n=0,1,2)$

$((CD_3)_3P)_2-Rh$ ... D ... D   **22**

$((CD_3)_3P)_2-Rh$ ... D, D   **23**

Scheme 13.2

unsaturated organic carbonyl compounds to form adducts similar to **19** in methanol solution [19].

Further studies using butadiene and isotopic labeling established the reaction pathway shown in Scheme 13.2. Complex **14-d_2** was shown to react with butadiene to generate the observed diene adduct **20**. Compound **20** then reacted further to lose H_nD_{2-n} (n=0, 1, 2) and generate the observed butadiene compound **21-d_n** (n=0, 1, 2). The observation of H–D exchange between the deuterides and buta-diene-hydrogen atoms in **20** is evidence for the reversible formation of the allyl-deuteride **22** and perhaps for the butene complex **23** (both not observed). This path-way thereby provides a gas-phase analogue for the following plausible sequence of steps in a catalytic alkene hydrogenation: oxidative addition of H_2, alkene coordi-nation, and alkene-hydride insertion. In support of this sequence in the gas phase, reaction of $[Rh(P(CD_3)_3)_2]^+$ (**8**) with 1-butene leads to the butadiene-dihydride iso-topomer of **20**, presumably via the protio isotopomers of **22** and **23**.

Noyori et al. recently used ESI-MS to characterize species present in catalytically active solutions during the hydrogenation of aryl-alkyl ketones using their base-free catalyst precursors *trans*-[Ru((R)-tol-BINAP)((R, R)-dpen)(H)(η^1-BH$_4$)] (tol-BI-NAP = 2,2′-bis(ditolylphosphino)-1,1′-binaphthyl; dpen = 1,2-diphenylethylenedia-mine) in 2-propanol [9b]. Based upon ESI-MS observations, deuterium-labeling studies, kinetics, NMR observations, and other results, the authors proposed that the cationic dihydrogen complex *trans*-[Ru((R)-tol-BINAP)((R, R)-dpen)(H)(η^2-H$_2$)]$^+$ is an intermediate in hydrogenations carried out in the absence of base.

The mechanism for hydrogenation of styrene using the iridium-phosphanylox-azoline compound [(PHOX)Ir(COD)]$^+$ (**24**, Scheme 13.3) as catalyst precursor has also been studied using ESI-MS [9c]. Catalytically active solutions were prepared by reacting the catalyst precursor and substrate dissolved in methylene chloride with hydrogen gas until all the cyclooctadiene-containing iridium species were consumed. To minimize loss of H_2 during spraying, the mixture was forced into the spray chamber using the pressure of hydrogen in the reactor (6 bar H_2). Three

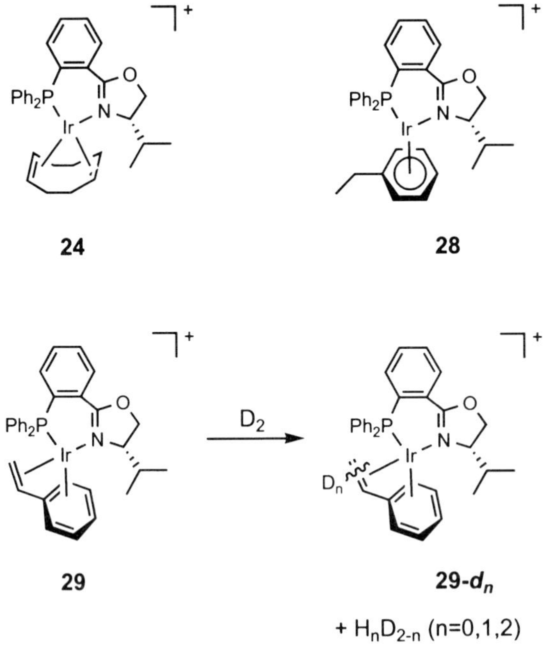

Scheme 13.3

species were observed at low tube potentials: one with a m/z ratio corresponding to the sum of $[(PHOX)Ir]^+$, styrene, and H_2 (**25**); the second with a m/z ratio corresponding to the sum of $[(PHOX)Ir]^+$, styrene, and $2 \times H_2$ (**26**); and the third with a m/z ratio corresponding to the sum of $[(PHOX)Ir]^+$ and styrene (**27**). Further, the gas-phase reaction of $[(PHOX)Ir(H)_2]^+$ and ethylbenzene formed a mixture from which a species $[(PHOX)Ir(ethylbenzene)]^+$ (**28**) was separated in the gas phase by its m/z ratio at the first quadrupole. Although precise structural determinations are not possible with MS data, the authors quite reasonably proposed that **28** is an 18-electron compound with ethylbenzene acting as η^6-arene ligand (Scheme 13.3). Based upon this proposal, it is possible that **25**, the species isolated from the active catalytic solution, and **28** are the same compound.

The ion **28** loses H_2 by CID with argon to form $[(PHOX)Ir(styrene)]^+$ (**29**). Compound **29** then undergoes H–D exchange with D_2 gas to form the mixture of isotopomers **29**, **29**-d_1, and **29**-d_2 (Scheme 13.3). When combined, these observations show that the oxidative addition of H_2 to **29** is followed by alkene hydride insertion, and that both these steps occur rapidly and reversibly in the gas phase. These results thereby provide gas-phase analogues for catalytic elementary steps that are proposed to occur in solution. Support for this proposed sequence of steps was obtained from a solution-phase catalytic deuteration of styrene. Analysis showed no deuterium incorporation in the unreacted styrene at various conversions, and clean formation of dideuterio ethylbenzene as sole product.

Based upon their data and upon results in the literature, the authors concluded that hydrogenations using **24** or related species as catalyst precursor proceed in solution by mechanisms involving iridium(I)/(III) formal oxidation states. During the course of their discussions, the authors made the interesting observation that the rate of gas-phase collisions between the thermalized iridium organometallic ions and D_2 under their experimental conditions in the octopole were similar to the rate of diffusion-controlled encounters between iridium species and D_2 in solution.

13.4
Conclusions

ESI represents a powerful method by which to transfer organometallic ions from catalytically active solutions into the gas phase. ESI-MS systems allow the characterization of the gas-phase ions using CID, reactivity, and isotope-labeling studies. The application of ESI-tandem-MS systems allows gas-phase preparations and isolation of desired organometallic ions in the first ESI-octopole-quadrupole, followed by characterization or reactivity studies in the second octopole-quadrupole.

ESI-MS study of gas-phase chemistry is a relative young area of interest in organometallic chemistry. ESI-MS mechanistic investigations have shown that organometallic reactions occurring as part of catalytic reactions in solution – including oxidative additions, insertions, metathesis, and eliminations – also occur in the gas phase. The extent to which such gas-phase analogues are relevant to reactions that occur in solution has not, to date, been addressed by a rigorous theoretical and experimental investigation. However, we end this tour by pointing out that a number of examples now exist where the gas-phase organometallic reactivity has been shown experimentally to be relevant to solution-phase catalytic reactions. In some cases, the gas-phase organometallic reactions have provided successful new leads to solution-phase studies. We predict that with careful gas-phase experimentation, combined with confirmation by experiments carried out in solution, ESI-MS studies will continue to discover and explore hydrogenation systems for which the gas-phase organometallic chemistry is relevant to the analogous reactions carried out in solution. Finally, there are a number of hydrogenations catalyzed by cationic catalyst systems that have not yet been studied using ESI-MS [20]. Further, hydrogenations using non-ionic catalyst systems can, in principle, be modified for ESI-MS studies by substituting neutral for ionic groups in the ancillary ligands. Recent advances in the use of ESI-tandem-MS systems to evaluate chiral catalysts for enantioselective allylations can perhaps also be modified for enantioselective catalytic hydrogenations [8].

Acknowledgments

The production of this chapter was supported by the University of Alberta. The authors are very grateful to Professor Paul Kebarle for many useful discussions, and for his kind patience.

Abbreviations

CID   collision-induced dissociation
ESI   electrospray ionization
MS   mass spectrometry

References

1 Reviews: (a) R. Colton, A. D'Agostino, J.C. Traeger, *Mass Spectrom. Rev.* **1995**, *14*, 79; (b) D.A. Plattner, *Int. J. Mass. Spec.* **2001**, *207*, 125; (c) P. Chen, *Angew. Chem. Int. Ed.* **2003**, *42*, 2832.

2 (a) D. Feichtinger, D.A. Plattner, *Angew. Chem. Int. Ed.* **1997**, *36*, 1718; (b) D.A. Plattner, D. Feichtinger, J. El-Bahraoui, O. Wiest, *Int. J. Mass. Spec.* **2000**, *195/196*, 351; (c) D. Feichtinger, D.A. Plattner, *J. Chem. Soc., Perkin Trans. 2* **2000**, 1023; (d) D. Feichtinger, D.A. Plattner, *Chem. Eur. J.* **2001**, *7*, 591.

3 (a) D. Feichtinger, D.A. Plattner, P. Chen, *J. Am. Chem. Soc.* **1998**, *120*, 7125; (b) C. Hinderling, P. Chen, *Int. J. Mass. Spec.* **2000**, *195/196*, 377.

4 (a) C. Hinderling, C. Adlhart, P. Chen, *Angew. Chem. Int. Ed.* **1998**, *37*, 2685; (b) C. Adlhart, C. Hinderling, H. Baumann, P. Chen, *J. Am. Chem. Soc.* **2000**, *122*, 8204; (c) C. Adlhart, P. Chen, *Helv. Chim. Acta* **2000**, *83*, 2192; (d) M.A.O. Volland, C. Adlhart, C.A. Kiener, P. Chen, P. Hofmann, *Chem. Eur. J.* **2001**, *7*, 4621.

5 (a) C. Hinderling, D.A. Plattner, P. Chen, *Angew. Chem. Int. Ed.* **1997**, *36*, 243; (b) C. Hinderling, D. Feichtinger, D.A. Plattner, P. Chen, *J. Am. Chem. Soc.* **1997**, *119*, 10793.

6 O. Aliprantis, J.W. Canary, *J. Am. Chem. Soc.* **1994**, *116*, 6985.

7 S.R. Wilson, Y. Wu, *Organometallics* **1993**, *12*, 1478.

8 C. Markert, A. Pfaltz, *Angew. Chem. Int. Ed.* **2004**, *43*, 2498.

9 (a) Y.-M. Kim, P. Chen, *Int. J. Mass. Spec.* **1998**, *185/186/187*, 871; (b) C.A. Sandoval, T. Ohkuma, K. Muniz, R. Noyori, *J. Am. Chem. Soc.* **2003**, *125*, 13490; (c) R. Dietiker, P. Chen, *Angew. Chem. Int. Ed.* **2004**, *43*, 5513.

10 Examples: (a) J.R. Morandi, H.B. Jensen, *J. Org. Chem.* **1969**, *34*, 1889; (b) M. Castiglioni, R. Giordano, E. Sappa, *J. Organomet. Chem.* **1995**, *491*, 111; (c) X.-Y. Wang, Z.-Y. Hou, W.-M. Lu, F. Chen, X.-M. Zheng, *J. Chromatogr. A* **1999**, *855*, 341.

11 For example, see: (a) W.E. Farneth, J.I. Brauman, *J. Am. Chem. Soc.* **1976**, *98*, 7891; (b) T.B. McMahon, T. Heinis, G. Nicol, J.K. Hovey, P. Kebarle, *J. Am. Chem. Soc.* **1988**, *110*, 7591.

12 R.B. Jordan, *Reaction Mechanisms of Inorganic and Organometallic Systems*, 2nd edn. Oxford University Press, New York, **1998**.

13 See [1 b,c] and references cited therein for discussions of these factors.

14 (a) M. Yamashita, J.B. Fenn, *J. Phys. Chem.* **1984**, *88*, 4451; (b) C.M. Whitehouse, R.N. Dreyer, M. Yamashita, J.B. Fenn, *Anal. Chem.* **1985**, *57*, 675.

15 The first example, V. Katta, S.K. Chowdhury, B.T. Chait, *J. Am. Chem. Soc.* **1990**, *112*, 5348.

16 Y.-M. Kim, P. Chen, *Int. J. Mass. Spec.* **2000**, *202*, 1.

17 (a) R. R. Schrock, J. A. Osborn, *J. Am. Chem. Soc.* **1976**, *98*, 2134; (b) J. Halpern, D. P. Riley, A. S. C. Chan, J. J. Pluth, *J. Am. Chem. Soc.* **1977**, *99*, 8055; (c) J. M. Brown, P. A. Chaloner, P. N. Nicholson, *J. Chem. Soc., Chem. Comm.* **1978**, 646.

18 H. F. Luecke, R. G. Bergman, *J. Am. Chem. Soc.* **1997**, *119*, 11538.

19 For examples, see: (a) J. M. Brown, P. A. Chaloner, *J. Chem. Soc., Chem. Commun.* **1978**, 321; (b) A. S. C. Chan, J. Halpern, *J. Am. Chem. Soc.* **1980**, *102*, 838; (c) A. S. C. Chan, J. J. Pluth, J. Halpern, *J. Am. Chem. Soc.* **1980**, *102*, 5952; J. Halpern, *Science* **1982**, *217*, 401; (d) A. Miyashita, H. Takaya, T. Souchi, R. Noyori, *Tetrahedron* **1984**, *40*, 1245; reference [15 b]; and references cited therein.

20 For examples, see: (a) J. A. Wiles, S. H. Bergens, *Organometallics* **1999**, *18*, 3709; (b) D. A. Dobbs, K. P. M. Vanhessche, E. Brazi, V. Rautenstrauch, J.-V. Lenoir, J.-P. Genêt, J. Wiles, S. H. Bergens, *Angew. Chem. Int. Ed.* **2000**, *39*, 1992.

Part III
Homogeneous Hydrogenation by Functional Groups

The Handbook of Homogeneous Hydrogenation.
Edited by J. G. de Vries and C. J. Elsevier
Copyright © 2007 WILEY-VCH Verlag GmbH & Co. KGaA, Weinheim
ISBN: 978-3-527-31161-3

14
Homogeneous Hydrogenation of Alkynes and Dienes

Alexander M. Kluwer and Cornelis J. Elsevier

14.1
Stereoselective Homogeneous Hydrogenation of Alkynes to Alkenes

14.1.1
Introduction

The reduction of carbon-carbon double and triple bonds is a very important transformation in synthetic organic chemistry. In this context, the conversion of alkynes into alkenes (i.e., semihydrogenation) is particularly useful, especially the stereoselective addition of one molar equivalent of hydrogen to the triple bond. This allows for the selective preparation of the corresponding (*E*)- or (*Z*)-alkenes depending on the choice of reaction conditions during the reduction. The classical catalytic hydrogenation using a heterogeneous transition-metal catalyst and molecular hydrogen constitutes the most general method for the selective reduction of carbon-carbon triple bonds. However, other pathways involving for instance organoaluminum and organoboron intermediates or hydride-transfer reagents in combination with metal salts have also been successfully applied [1, 2].

The most widely used catalytic procedures for the catalytic hydrogenation of alkynes to afford (*Z*)-alkenes generally employ palladium or nickel as the catalytically active transition metal. The Lindlar catalyst (lead-poisoned Pd on $CaCO_3$) and the P2-Ni catalyst are among the most prominent members of this group [1–3]. These systems show considerable selectivity for a variety of alkynes; however, substrates with triple bonds conjugated to other unsaturated moieties or electron-poor alkynes display low selectivity due to overreduction or other side reactions [1]. The complex nature of the surface of the Lindlar catalyst, containing different domains each of which contributes to the product distribution, makes the outcome of the stereoselective hydrogenation unpredictable. Thus, the yields and selectivity will generally vary, even for identical compounds under identical conditions [2, 4].

A large number of homogeneous transition-metal complexes have been reported as catalysts for the stereoselective hydrogenation of alkynes, although the

The Handbook of Homogeneous Hydrogenation.
Edited by J. G. de Vries and C. J. Elsevier
Copyright © 2007 WILEY-VCH Verlag GmbH & Co. KGaA, Weinheim
ISBN: 978-3-527-31161-3

details of this apparently simple reaction remain mostly obscure. Only a few homogeneous catalysts have been investigated in more detail, and some of these show a remarkable selectivity towards a variety of alkynes containing various functional groups. The origin of the stereoselectivity of these catalysts in the semihydrogenation of alkynes can often be ascribed to kinetic factors and sometimes to the lack of interaction of the catalyst with the product alkene, which is consequently not further reduced. The exhaustive reduction of alkynes to alkanes, which can in synthetic terms be useful, is very similar to alkene hydrogenation and will, therefore, not be treated here. This chapter will mainly focus on the homogeneous catalytic semihydrogenation of alkynes using molecular hydrogen. Supported catalysts and cluster-catalyzed hydrogenations will not be treated here. The performance of the catalysts has been the major criterion for selecting the homogeneous catalytic systems discussed in this chapter. Special attention will be given to the mechanistic details of selected systems.

14.1.2
Chromium Catalysts

One of the most selective semihydrogenation catalysts reported concerns a class of chromium tricarbonyl compounds with the generic formula [Cr(CO)$_3$(arene)] (**1**) [5]. This catalyst is able to hydrogenate a wide variety of polyunsaturated compounds, and has been successfully applied in, for example, the 1,4-hydrogenation of conjugated dienes and α,β-unsaturated carbonyl compounds [6]. The outstanding performance of this catalyst in alkyne hydrogenation has been attributed to its complete inactivity towards compounds containing isolated carbon-carbon double bonds (i.e., neither isomerization nor over-reduction is observed). The (Z)-alkene is the sole product, which is isolated in very high yields (87–100%; see Table 14.1). Generally, the hydrogenation only proceeds under rather forcing reaction conditions, although, less strongly coordinated arenes (e.g., naphthalene or methyl benzoate) allow for milder conditions while maintaining the high stereoselectivity (see Table 14.1). Details about the reaction mechanism have not been revealed, but the similarity with the [Cr(CO)$_3$]-catalyzed 1,4-hydrogenation of conjugated dienes suggests that [Cr(CO)$_3$(S)$_3$] (S = solvent) species is the catalytically active complex.

The applicability of the [Cr(CO)$_3$(arene)] 1,2-*syn*-hydrogenation has been demonstrated in the syntheses of pheromones where the hydrogenation of the alkyne to the (Z)-isomer is a key step in the synthetic scheme [5, 7]. For such compounds, obtaining the correct the stereo- and regio-isomer is essential for its biological activity. In these cases, selectivities up to 100% have been reported. Special attention has been given to conjugated alkyne-diene systems of which the stereo- and regio-chemical outcome can be precisely predicted based on the specific chemistry of the [Cr(CO)$_3$(arene)] catalyst towards alkynes and conjugated dienes.

Table 14.1 Hydrogenation of alkynes using [Cr(CO)$_3$(arene)] (**1**) [5].

Entry	Alkyne	Catalyst	Time [h]	H$_2$ pressure [bar]	(Z) alkene [%][c]
1	(phenyl–C≡C–CH$_3$ structure)	[Cr(CO)$_3$(mbz)][a]	23	69	92
2	(phenyl–C≡CH structure)	[Cr(CO)$_3$(nap)][b]	24	20	92
3	C$_6$H$_{13}$—≡—C$_6$H$_{13}$	[Cr(CO)$_3$(nap)][b]	15	69	100
4	(alkyne chain with OH)	[Cr(CO)$_3$(mbz)][a]	8	69	95
5	(alkyne chain with OH)	[Cr(CO)$_3$(nap)][b]	8	49	87

a) [Cr(CO)$_3$(mbz)] = [Cr(CO)$_3$(methyl benzoate)] (**1a**). Reaction conditions: solvent acetone, reaction temp. 120 °C, substrate : catalyst ratio = 5 : 1.

b) [Cr(CO)$_3$(nap)] = [Cr(CO)$_3$(naphthalene)] (**1b**). Reaction conditions: solvent THF, reaction temp. 45 °C, substrate : catalyst ratio = 5 : 1.

c) Determined by GC analysis relative to internal standard.

14.1.3
Iron Catalysts

Terminal alkynes are selectively hydrogenated to alkenes by the iron(II) catalyst precursors [(PP$_3$)FeH(N$_2$)]BPh$_4$ (**2**) and [(PP$_3$)FeH(H$_2$)]BPh$_4$ (**3**) (PP$_3$ = P(CH$_2$CH$_2$PPh$_2$)$_3$) in tetrahydrofuran (THF) under mild conditions (1 bar H$_2$) [8]. The hydrogenation rates when using complexes **2** and **3** were found to be low at ambient temperature, but increased with increasing reaction temperature. The turnover frequency (TOF) ranges from 8 to 20 mol mol^{-1} h^{-1} at 66 °C for various substrates. Under these conditions, alkynes are converted chemoselectively into alkenes, regardless of the reaction temperature. The only exception is ethynyltrimethylsilane HC≡CSiMe$_3$, which mainly produces the dimeric product 1,4-bis(trimethylsilyl)butadiene by a reductive coupling reaction [9]. A thorough kinetic study of the hydrogenation of phenyl acetylene employing **2** has provided more insight into the mechanistic details of this reaction.

The mechanism is dominated by the remarkable stability of the Fe(η^2-H$_2$) bond, which is one of the most stable η^2-H$_2$ complexes reported in the literature [8, 10]. Remarkably, the free coordination site for the incoming alkyne is provided by the reversible dissociation of one of the phosphine moieties of the PP$_3$ ligand rather than dissociation of the dihydrogen ligand (see Scheme 14.1). The coordinated alkyne subsequently inserts into the Fe–H bond and the emerging Fe–vinyl bond is

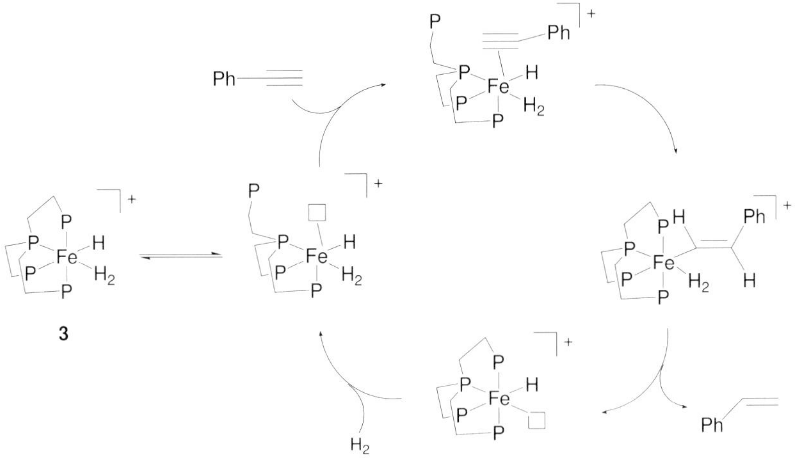

Scheme 14.1 Proposed mechanism for the hydrogenation of phenyl acetylene catalyzed by [(PP₃)FeH(H₂)]⁺BPh₄⁻ (**3**); the anion is BPh₄⁻ throughout in this scheme.

then cleaved *via* an intramolecular protonolysis (i.e., heterolytic splitting of the dihydrogen molecule occurs). The binding of the alkyne to the metal center appears to be the rate-determining step as it is supported by the zero-order in dihydrogen gas and a first-order in phenyl acetylene. The activation parameters for this reaction are $\Delta H^{\ddagger} = 11 \pm 1$ kJ mol^{-1}, $\Delta S^{\ddagger} = -27 \pm 3$ J K^{-1}, and $\Delta G^{\ddagger} = 19 \pm 4$ kJ mol^{-1}.

The cationic complex $[\text{FeH}(\eta^2\text{-H}_2)(\text{PP}_3)]^+$ (**3**; $\text{PP}_3 = \text{P}(\text{CH}_2\text{CH}_2\text{PPh}_2)_3$) displays an identical octahedral structure as the comparable ruthenium complex $[\text{RuH}(\eta^2\text{-H}_2)(\text{PP}_3)]^+$ (**4**; $\text{PP}_3 = \text{P}(\text{CH}_2\text{CH}_2\text{PPh}_2)_3$) [11]. Both complexes have been found to be selective catalysts for the hydrogenation of alkynes. Despite their structural similarities, the chemistry of compounds **3** and **4** in the semihydrogenation of alkynes is dominated by the difference in metal–dihydrogen bond strength, which decreases in the order Fe > Ru and has been attributed to a stronger back-donation from iron into the antibonding $\sigma^*(\text{H}_2)$ orbital compared to ruthenium [12].

14.1.4
Ruthenium Catalysts

Numerous compounds containing ruthenium are found to be active catalysts for the semihydrogenation of alkynes. Much attention has been devoted to the class of ruthenium carbonyl clusters, and it has been demonstrated that both internal and terminal alkynes can be effectively hydrogenated using such compounds, affording the corresponding alkenes [13]. However, the activities of these catalysts are generally lower than those of the mononuclear complexes and their stereoselectivity is also significantly lower due to extensive isomerization of the (primary) reaction products.

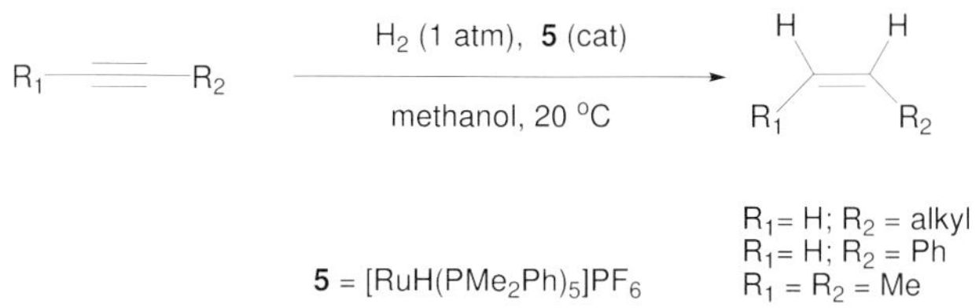

$$R_1 \underset{}{=\!\!=\!\!=} R_2 \quad \xrightarrow[\text{methanol, 20 °C}]{\text{H}_2\ (1\ \text{atm}),\ \textbf{5}\ (\text{cat})} \quad \begin{array}{c} \text{H} \qquad \text{H} \\ \diagdown \quad / \\ R_1 \qquad R_2 \end{array}$$

R_1 = H; R_2 = alkyl
R_1 = H; R_2 = Ph
R_1 = R_2 = Me

5 = [RuH(PMe$_2$Ph)$_5$]PF$_6$

Scheme 14.2 Hydrogenation of terminal and internal alkynes by [RuH(PMe$_2$Ph)$_5$]PF$_6$ (**5**) [14].

Mononuclear ruthenium complexes, particularly ruthenium–phosphine complexes, are among the best-characterized catalytic systems. A particular active catalyst for this reaction is based on the sterically crowded complex [RuH(PMe$_2$Ph)$_5$]PF$_6$ (**5**) which, in the presence of 25 equiv. PMe$_2$Ph in methanol, has been shown to be an extremely active catalyst system for the selective hydrogenation of alkynes to alkenes [14] (Scheme 14.2). The rates of hydrogenation were limited by mass transfer processes in this case – that is, by diffusion of hydrogen into the solution. Thus, under these conditions, the TOF reported amounts to 130 mol mol^{-1} h^{-1}. The reaction proceeds affording the corresponding (Z)-alkenes without subsequent isomerization or further hydrogenation of the products. In addition, at high phosphine concentrations, the catalyst is completely inactive in the hydrogenation of alkenes to alkanes. In order to prevent catalyst deactivation, a large excess of PMe$_2$Ph is needed and the reaction temperature should be kept below 20 °C.

A related cationic ruthenium catalyst precursor, [RuH(cod)(PMe$_2$Ph)$_3$]PF$_6$ (**6**; cod = cyclooctadiene), was studied in the semihydrogenation of alkynes, and the results demonstrate a distinctly different catalyst behavior and chemoselectivity compared to [RuH(PMe$_2$Ph)$_5$]PF$_6$ (**5**) [15]. Under 1 bar H$_2$, **6** reacts to produce a complex, regarded as [RuH(PMe$_2$Ph)$_3$(solvent)$_2$]$^+$, which is a very active hydrogenation catalyst for alkynes as well as alkenes. The catalyst reveals an unusual order of reactivity – that is, the rate of hydrogenation to *cis*-alkenes increased in the order 1-hexyne < 2-hexyne < 3-hexyne, which has been attributed to the lower tendency of 3-hexyne to form the alleged catalytically inactive ruthenium-bis(alkyne) complexes. The activity of the catalyst for alkene hydrogenation could be significantly suppressed by the addition of 1 equiv. PMe$_2$Ph, and **6** becomes completely inactive towards alkenes by the addition of 2 equiv. PMe$_2$Ph. This addition will effectively lead to the more saturated, less-active ruthenium complex ions [RuH(PMe$_2$Ph)$_4$]$^+$ (**8**) and [RuH(PMe$_2$Ph)$_5$]$^+$ (**5**), respectively.

The two ruthenium complexes [RuH(PMe$_2$Ph)$_5$]PF$_6$ (**5**) and [RuH(cod)PMe$_2$Ph)$_3$]PF$_6$ (**6**) are considered to be related by Scheme 14.3. Since the cationic catalyst precursors **5** and **6** are both 18-electron species, hydrogenation of cyclooctadiene in **6** and phosphine dissociation from **5** will produce the catalytically active intermediate ions [RuHL$_4$]$^+$ (**8**) and [RuHL$_3$]$^+$ (**7**) (L = PMe$_2$Ph), respectively. It has thus been proposed that **7** is active for the hydrogenation of alkenes as well as alkynes, whereas **8** only hydrogenates alkynes. This scheme

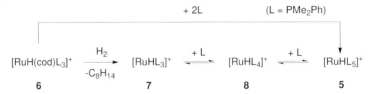

Scheme 14.3 Proposed relationship between different ruthenium–phosphine complexes. The anion has been omitted and is PF_6^+ throughout.

can be further extended to include the ruthenium complex $[Ru(H)(H_2) (PMe_2Ph)_4]^+$ (not shown in Scheme 14.3), which is formed by a reaction of **8** with hydrogen [16]. In the absence of additional phosphine, **8** shows rapid deactivation in the hydrogenation of 1-alkynes and in the isomerization of internal alkenes. However, under similar phosphine-deficient conditions, $[RuH (PMe_2Ph)_5]PF_6$ (**5**) remains active as a hydrogenation catalyst, but it displays a lower rate and a higher selectivity towards (Z)-alkenes.

The related cationic complex $[RuH(\eta^2\text{-}H_2)(PP_3)]^+$ (**4**; $PP_3 = P(CH_2CH_2PPh_2)_3$) is a non-classical trihydride complex where the metal is coordinated by the four phosphorus atoms, by a dihydrogen ligand, and a terminal hydride ligand. Although this complex is generally less active than, for instance, $[RuH (PMe_2Ph)_5]^+$ (**5**), it has a better defined chemistry [11]. The kinetic study of the selective hydrogenation of phenyl acetylene employing **4** reveals a first-order dependence on the catalyst concentration and second-order in hydrogen. The order in substrate changes from apparently first-order to zero-order with increasing substrate concentration. This catalytic behavior has been attributed to the formation of $[Ru(\eta^3\text{-}1,4\text{-diphenylbutenynyl})(PP_3)]$ (**9**) complexes by coupling of two substrate molecules. Complex **9** was found to be equally active in the hydrogenation of phenyl acetylene as compared to complex **4** (Scheme 14.4).

The proposed reaction mechanism follows a sequence of reaction steps usually adopted for a monohydride-metal hydrogenation catalyst, although other, more complex ruthenium intermediates contribute to the overall hydrogenation [11, 16]. Here, the η^2-coordinated dihydrogen serves merely as a stabilizing ligand and the formally unsaturated species $[RuH(PP_3)]^+$ (**10**) is the actual, reactive intermediate. Possibly, such a species must be thought of as labile, transient solvento species. Coordination of phenyl acetylene, insertion into the Ru–H bond followed by reaction of dihydrogen will eventually generate styrene and the ruthenium-monohydride species.

A particularly interesting hydrogenation catalyst for the semihydrogenation of alkynes concerns the ruthenium complex $[Cp^*Ru(\eta^4\text{-}CH_3CH=CHCH= CHCO_2H)][CF_3SO_3]$ (**11**) [17]. This ruthenium complex represents one of the few homogeneous catalysts that allow for the direct *trans*-hydrogenation of internal alkynes, yielding the (E)-alkene stereoselectively. Generally, internal alkynes can be transformed selectively into E-alkenes using different homogeneous and heterogeneous methods (e.g., dissolving metal reduction). However, under these

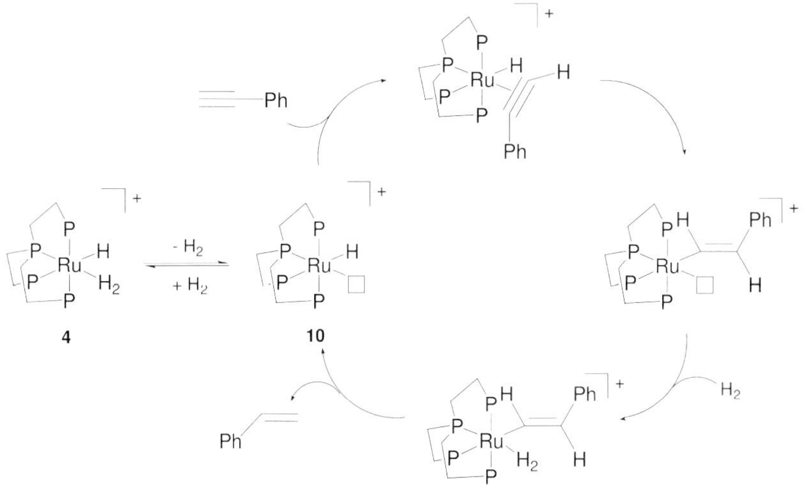

Scheme 14.4 Proposed mechanism for the hydrogenation of phenyl acetylene catalyzed by [RuH(η^2-H$_2$)(PP$_3$)]$^+$ (**4**) [11].

reaction conditions other functional groups will be affected and, hence, low chemoselectivities are obtained in these cases.

The [Cp*Ru]$^+$-catalyzed semihydrogenation of alkynes has been studied in more detail using PHIP-NMR (PHIP = *Para*-Hydrogen Induced Polarization; see Chapter 12). With this method the initially formed hydrogenation products can be identified and characterized, even at very low concentrations and low conversions. Different types of alkynes were probed, and it was found that the reaction is not influenced greatly by the functional groups present in the substrate. However, internal alkynes readily produced polarized signals in the hydrogenation product (i.e., the *E*-isomer), even at low temperature. This finding indicates that the hydrogen is transferred in a *pair-wise* manner and it confirms that the *E*-isomer emerges as the primary reaction product. No polarized signals were detected for terminal alkynes, which has been ascribed to an allegedly inactive vinylidene complex that has been formed *via* a 1,2-hydrogen shift [17].

The remarkable stereoselectivity of this catalyst – that is, the *trans*-addition of the two hydrogen atoms – has been explained by the involvement of a dimeric ruthenium complex. In the proposed reaction mechanism (Scheme 14.5) the intermediates are assumed to be alkyne-bridged dinuclear ruthenium complexes (**13**, **14** and **15**). It has been suggested that the hydrogenation should be fast and without total separation of the two *para*hydrogen atoms – that is, without loss of the spin correlation between the two hydrogen atoms. Only under these conditions can *para*hydrogen polarized signals be observed. The observation of these polarized signals reveals that the hydrogen atoms transferred to the alkyne stem from the same hydrogen molecule and supports the mechanism depicted in Scheme 14.5.

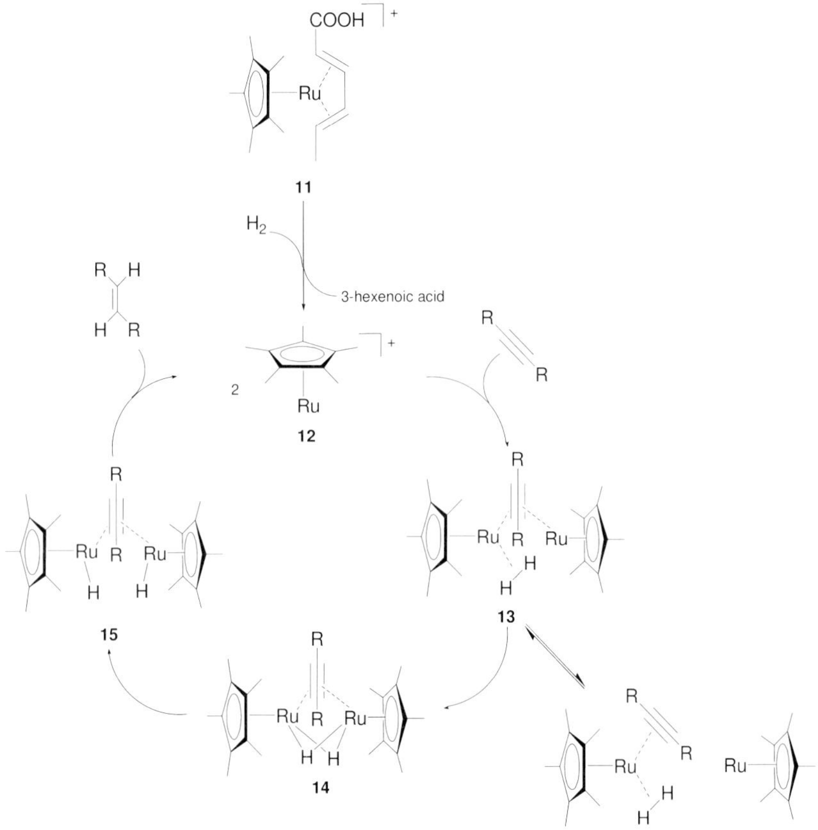

Scheme 14.5 Stereoselective hydrogenation of internal alkynes by complex $[Cp^*Ru(\eta^4\text{-}CH_3CH=CHCH=CHCOOH)][CF_3SO_3]$ (**11**); the anion is CF_3SO3^- throughout [17].

14.1.5
Osmium Catalysts

An interesting example of the alkyne hydrogenation by homogeneous transition-metal catalysts is provided by the complexes $[OsH(Cl)(CO)(PR_3)_2]$ (**16**; $PR_3 = PMeBu_2^t$, PPr_3^i), which catalyze the hydrogenation of phenyl acetylene at 1 bar H_2 in 2-propanol solution at 60 °C [18]. These complexes react rapidly with phenyl acetylene to produce the respective stable 16-electron alkenyl-osmium compounds $[Os((E)\text{-}CH=CHPh)Cl(CO)(PR_3)_2]$ (**17**) almost quantitatively. Subsequent reaction with dihydrogen produces styrene, ethylbenzene and the dihydrogen complex $[OsH(Cl)(\eta^2\text{-}H_2)(CO)(PR_3)_2]$ (**18**). These reactions constitute a catalytic cycle for the reduction of alkynes affording the corresponding alkenes with selectivities close to 100%.

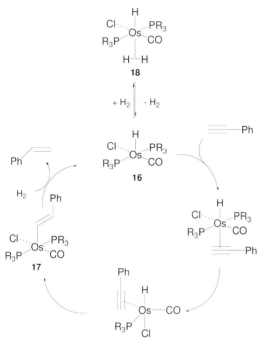

Scheme 14.6 Proposed mechanism for the hydrogenation of phenyl acetylene catalyzed by [OsHCl(CO)(PR$_3$)$_2$] (**16**).

The kinetic investigation of this reaction reveals the reaction is first-order in substrate, catalyst and hydrogen concentration, and thus yields the rate law: $r = k_{cat}$[Os][alkyne][H$_2$]. The proposed mechanism as given in Scheme 14.6 is based on the rate law and the coordination chemistry observed with these osmium complexes.

In the absence of phenyl acetylene, these Os-complexes catalyze the hydrogenation of styrene to ethyl benzene at rates about ten-fold faster than those observed for the C≡C bond reduction. The authors concluded that the styrene hydrogenation is kinetically favored, so the observed selectivity must originate from a thermodynamic difference. This difference is established by the formation of the very stable vinyl-osmium intermediates, which forms a thermodynamic sink that causes all the osmium present to be tied up in this form; consequently the kinetically unfavorable pathway becomes essentially the only one available in the presence of alkyne. Furthermore, it has been shown that the hydrogenation of phenyl acetylene to styrene by [OsHCl(CO)(PPr$_3^i$)$_2$] can also be performed under hydrogen-transfer conditions using 2-propanol as the hydrogen donor [19, 20].

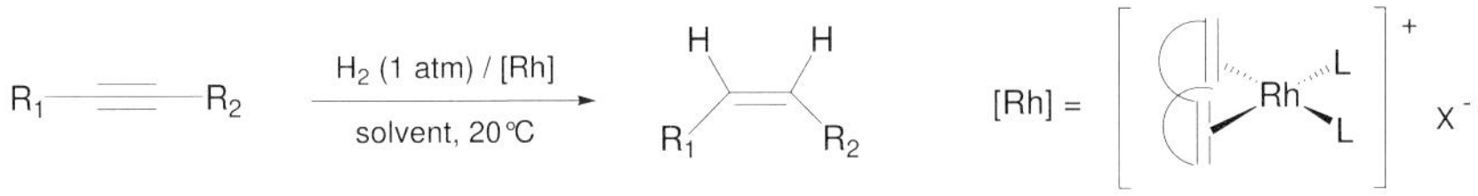

R_1	R_2	Selectivity
Ph	H	>95%
Ph	COOEt	95%
CH_3	C_3H_7	99%
$Me_2C(OH)$	CH_3	n.d.

Scheme 14.7 Hydrogenation of various alkynes using complex $[Rh(nbd)L_n]^+X^-$ (**19**) [21].

14.1.6
Rhodium Catalysts

The Schrock/Osborn cationic Rh-catalyst with the general formula $[Rh(nbd)L_n]^+X^-$ (**19**; wherein nbd is norbornadiene, L is a phosphine ligand and X^- is a – weakly coordinating – anion) is counted among the most successful and most widely applied catalysts for the semihydrogenation of alkynes [21]. The reaction proceeds well in coordinating solvents (such as acetone, ethanol, or 2-methoxyethanol), and under these conditions internal alkynes are reduced efficiently affording the corresponding (Z)-alkene in yields as high as 99% (see Scheme 14.7). This catalyst system is a moderately active hydrogenation catalyst with TOF (determined at 50% conversion) ranging between 20 and 80 mol $mol^{-1} h^{-1}$ [21–23]. However, the reduction of terminal alkynes by these systems is less successful since the fairly acidic character of the alkyne destroys the active species (such as $[RhH(L_n)S_y]$) by formation of an alkynyl-rhodium derivative [21].

The mechanism of the semihydrogenation of alkynes in coordinating solvents is assumed to be analogous to the alkene hydrogenation – that is, a catalytic scheme consisting of a mono-hydride route (*via* $RhH(L)_n(solvent)_y$) and a dihydride route (*via* $Rh(H)_2(L)_n(solvent)_y$) (see Scheme 14.8) [21, 24]. The active rhodium(solvento) species are formed by a reaction of **19** with molecular hydrogen. Generally, the cationic rhodium mono-hydride $RhH(L)_n(solvent)_y$ is an extremely active isomerization and hydrogenation catalyst, whereas the rhodium dihydride $[Rh(H)_2(L)_n(solvent)_y]^+$ is a less active hydrogenation catalyst but displays very little isomerization activity [21]. Under catalytic conditions, the mono-hydride and the dihydrido-rhodium(solvento) species are in equilibrium; the amount of each species – and thus the contribution of each route to the product formation – can be shifted by the addition of acid or base. Therefore, performing the semihydrogenation under acidic conditions will generate a highly selective hydrogenation system for the reduction of alkynes to (Z)-alkenes. Another method of generating related rhodium species involves protonation of the acetate of a neutral Rh(I) or Rh(II) complex in the presence of a phosphine ligand (such as $[Rh(OAc)(PPh_3)_3]$ or $[Rh_2(OAc)_4/PR_3]$) [20, 25]. Catalytic hydrogenation systems prepared by this method behave in many respects similarly to the systems based on $[Rh(diene)L_2]^+A^-$.

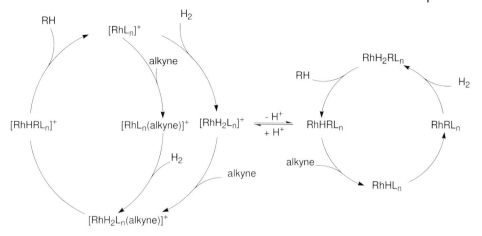

Scheme 14.8 General hydrogenation mechanism catalyzed by cationic $[Rh(nbd)L_n]^+$ species (**19**) [21].

Recovery of the precious Rh-catalyst has been successfully developed by using various strategies. One of the approaches concerns the use of highly fluorous Rh-catalysts for selective hydrogenation of terminal and internal alkynes under fluorous biphasic conditions yielding catalyst retentions of >99.92% (for FC-75/hexanes, 1:3 v/v) [23]. Furthermore, the immobilization of the rhodium catalyst on a solid support has been reported as a feasible approach, which shows only little leaching of the rhodium, while maintaining a high selectivity and activity [26].

In less-coordinating solvents such as dichloromethane or benzene, most of the cationic rhodium catalysts $[Rh(nbd)(PR_3)_n]^+A^-$ (**19**) are less effective as alkyne hydrogenation catalysts [21, 27]. However, in such solvents, a few related cationic and neutral rhodium complexes can efficiently hydrogenate 1-alkynes to the corresponding alkene [27–29]. A kinetic study revealed that a different mechanism operates in dichloromethane, since the rate law for the hydrogenation of phenyl acetylene by $[Rh(nbd)(PPh_3)_2]^+BF_4^-$ is given by: $r = k[catalyst][alkyne][pH_2]^2$ [29].

$$[Rh(nbd)(PPh_3)_2]BF_4 + H_2 \rightleftharpoons [RhH_2(nbd)(PPh_3)_2]BF_4 \qquad (1)$$

$$[RhH_2(nbd)(PPh_3)_2]BF_4 \rightleftharpoons [RhH_2(nbd)(PPh_3)]BF_4 + PPh_3 \qquad (2)$$

$$[RhH_2(nbd)(PPh_3)]BF_4 + PhC\equiv CH \rightleftharpoons [RhH(CH=CHPh)(nbd)(PPh_3)]BF_4 \qquad (3)$$

$$[RhH(CH=CHPh)(nbd)(PPh_3)]BF_4 + H_2 \rightarrow [RhH_2(nbd)(PPh_3)]BF_4 + CH_2=CHPh \qquad (4)$$

$$[RhH_2(nbd)(PPh_3)]BF_4 + PPh_3 \rightarrow [RhH_2(nbd)(PPh_3)_2]BF_4 \qquad (5)$$

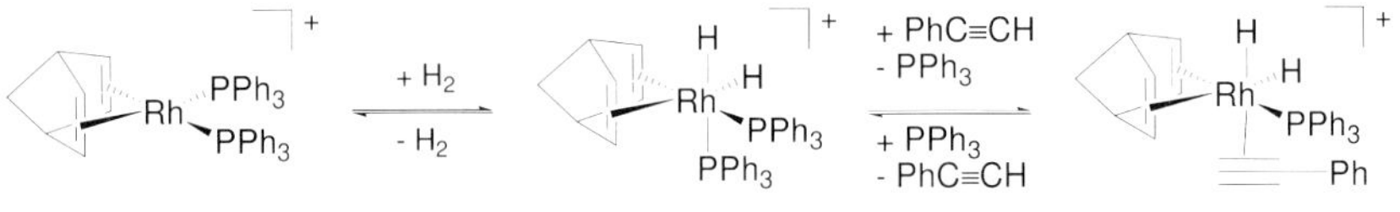

Scheme 14.9 Reaction of $[Rh(nbd)(PPh_3)_2]^+BF_4^-$ (**19**) with H_2 to give *cis,cis*-$[RhH_2(nbd)(PPh_3)_2]^+BF_4^-$ (**20**) and the subsequent formation of $[RhH_2(nbd)(PPh_3)(alkyne)]^+BF_4^-$ (**21**).

The proposed mechanism for this case is summarized by Eqs (1) to (5). Note that norbornadiene is *not* hydrogenated under the reaction conditions and remains coordinated to the metal center during the hydrogenation process. This feature has been observed for both rhodium and iridium systems of the general formula $[M(diene)(L)_2]^+X^-$ [28–31]. In the first step of the catalytic sequence, $[Rh(nbd)(PPh_3)_2]^+BF_4^-$ (**19**) reacts with dihydrogen to form the coordinatively saturated intermediate *cis,cis*-$[RhH_2(nbd)(PPh_3)_2]^+BF_4^-$ (**20**), which must then dissociate a phosphine ligand in order to create a vacant coordination site for the incoming alkyne, to give the corresponding alkyne complex $[RhH_2(nbd)(PPh_3)(PhC_2H)]^+$ BF_4^- (**21**) (see Scheme 14.9). Most likely, the phosphine *trans* to the hydride would be most labile due to the large *trans*-influence of the hydride ligand. The alkyne would then arrive *cis* to one hydride but *trans* to the other, and migratory insertion will thus lead to the *trans*-hydrido-vinyl species. If the metal center in **21** is stereochemically integer under the hydrogenation conditions, this complex would not facilitate the reductive elimination. Therefore, a subsequent reaction with dihydrogen is required for the formation of the alkene, rationalizing the second-order in hydrogen.

14.1.7
Iridium Catalysts

The iridium complex $[Ir(cod)(\eta^{2-i}PrPCH_2CH_2OMe)]^+BF_4^-$ (**22**) in dichloromethane at 25 °C at 1 bar H_2 is a particularly active catalyst for the hydrogenation of phenyl acetylene to styrene [29]. In a typical experiment, an average TOF of 50 mol mol^{-1} h^{-1} was obtained (calculated from a turnover number, TON, of 125) with a selectivity close to 100%. The mechanism of this reaction has been elucidated by a combination of kinetic, chemical and spectroscopic data (Scheme 14.10).

The main species in solution has been identified to be the hydrido-alkynyl complex $[IrH(C_2Ph)(cod)(\eta^{2-i}PrPCH_2CH_2OMe)]^+BF4^-$ (**23**). This is, however, only a sink that results from direct reaction of **22** with the 1-alkyne, draining the active catalyst from the system. The catalysis proceeds *via* the dihydrido-diene intermediate $[IrH_2(cod)(\eta^{2-i}PrPCH_2CH_2OMe)]^+$ BF_4^- (**24**), which reacts reversibly with the alkyne to yield the hydrido-iridium-styryl complex **25**, followed by a rate-determining reaction of this hydrido-vinyl species with hydrogen to re-

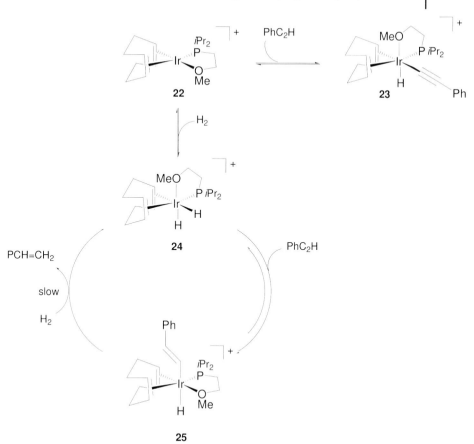

Scheme 14.10 Proposed mechanism for the hydrogenation of phenyl acetylene catalyzed by [Ir(cod)($\eta^{2\text{-}i}$PrPCH$_2$CH$_2$O-Me)]$^+$BF$_4^-$ (**22**); the anion is BF$_4^-$ throughout [29].

generate the dihydrido-diene complex **24** and liberate the alkene. This mechanism is in agreement with the observed rate law $r = k[\text{Ir}][\text{pH}_2]$. As in the case of the hydrogenation of phenyl acetylene by [Rh(nbd)(PPh$_3$)$_2$]$^+$BF$_4^-$ in dichloromethane, the diene is not hydrogenated under these reaction conditions and remains coordinated to the iridium. Furthermore, the origin of the selectivity has been ascribed to kinetic reasons – that is, the alkyne competes more effectively for the position liberated by displacement of the methoxy group of the etherphosphine ligand than the alkene [29].

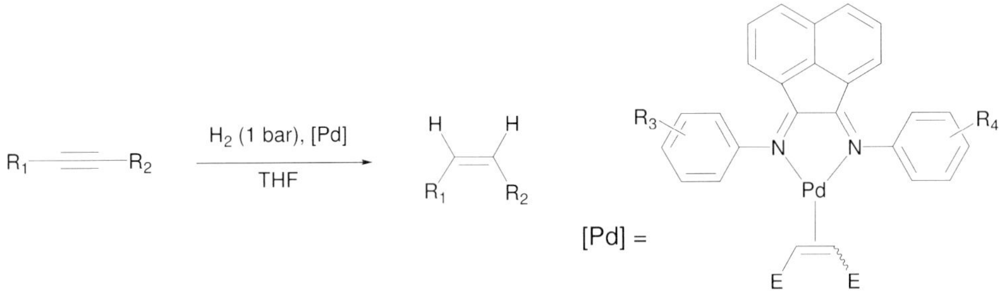

26a; $R_3 = R_4 = 4\text{-OCH}_3$
26b; $R_3 = R_4 = 3,5\text{-}(CF_3)_2$

Scheme 14.11 Stereoselective hydrogenation of alkynes to alkenes by [Pd(Ar-bian)(alkene)] (**26**). In compound **26a** the electron-poor alkene is dimethyl fumarate, $E = CO_2Me$; in **26b** the electron-poor alkene is maleic anhydride, $E = C(O)OC(O)$.

14.1.8
Palladium Catalysts

As mentioned earlier, the Lindlar catalyst has been the most widely used heterogeneous hydrogenation catalyst for the semihydrogenation of alkynes. However, the surface of these catalysts is a rather complex assembly of various domains, each of which contributes to the product distribution in its own way, often resulting in unpredictable overall results. In order to circumvent the presence of different palladium species, much attention has been devoted to the application of homogeneous mononuclear palladium complexes; either in the form of a Pd(II) complex [32, 33] or a Pd(0) compound [34–36]. In most of the reported cases, these mononuclear palladium complexes display a higher chemo- and (sometimes) higher stereoselectivity than the heterogeneous counterparts.

An important advantage of a heterogeneous catalyst over a homogeneous one concerns the ease of catalyst separation from the products. Therefore, much research effort in this field has been directed towards immobilized palladium complexes, where the palladium is coordinated to a ligand attached to a polymer, a clay, or another inorganic support [37–43]. Assuming that a single type of palladium species is formed, these heterogenized palladium catalysts are potentially very important and might compete with the Lindlar catalysts in certain applications. For the sake of completeness, it should be noted that palladium-clusters have been found to be active, with some showing extremely high activities and very good selectivities in the hydrogenation of alkynes to alkenes [44–46]. These Pd-clusters will not be treated here, however.

Elsevier et al. have reported the stereoselective hydrogenation of alkynes by zerovalent palladium catalysts bearing a bidentate nitrogen ligand, which are able homogeneously to hydrogenate a wide variety of alkynes to the corresponding (Z)-alkenes (see Scheme 14.11). The semihydrogenation occurs under very mild conditions (25 °C, 1 bar H_2), and the observed selectivity towards the (Z)-alkene for the various alkynes is very high indeed [35]. The precatalysts employed are isolable [Pd(Ar-bian)(alkene)] compounds (**26**), which have previously been used in the homogeneous hydrogenation of electron-poor alkenes [47] and in carbon–element bond-formation reactions [48]. With respect to most other diimine ligands, Ar-bian derivatives are rigid, which imposes the correct geometry for coordination and imparts a high chemical stability. The ease of modifying the electronic as well as the steric properties of these ligands make them ideal to study their complexes in a variety of catalytic reactions.

Using [Pd(p-MeO-C_6H_4-bian)(dmfu)] (**26a**; dmfu = dimethyl fumarate), the observed selectivity towards semihydrogenation to the (Z)-isomer for the various alkynes is very high (>99%) at 25 °C and 1 bar H_2 [35]. Typically, TOFs of 100 to 200 mol $mol^{-1} h^{-1}$ have been obtained with this catalyst. The high selectivity, which is in many cases superior to that obtained with Lindlar catalyst, is maintained until full conversion of the alkyne; the selectivity has been determined at >99.5% conversion. As can be seen from Table 14.2, internal as well as terminal alkynes are reduced to the corresponding (Z)-alkenes with great ease and selectivity. Besides the high stereoselectivity, the chemoselectivity is also remarkably high, as was demonstrated by the presence of other reducible functional groups in various substrates, such as carboalkoxy, nitro-groups, or even alkene moieties in conjugated enynes.

The complex [Pd{(m,m'-$(CF_3)_2C_6H_3$)bian}(ma)] (**26b**; ma = maleic anhydride) displays an unprecedented high reaction rate, while maintaining the high stereo- and chemoselectivity, which is typical for the [Pd(Ar-bian)] systems. In a detailed kinetic study [49] the catalyst behavior was investigated in the hydrogenation of 4-octyne. Under the reaction conditions (7 bar H_2, 21 °C) an average TOF of 16 000 (calculated from TON of 1600) can be reached, though hydrogen diffusion limitations prevent higher rates. The reaction rate is given by the equation: $r = k[Pd][H_2][alkyne]^{0.65}$. Thus, the reaction is first-order in palladium and dihydrogen, confirming that mononuclear species are involved in the catalysis. The broken reaction order for the substrate suggests that an equilibrium between palladium complexes containing the substrate and the reaction product is operative under the reaction conditions. High-level density functional theory (DFT) calculations and *para*hydrogen induced polarization (PHIP) NMR measurements support a mechanism consisting of the following consecutive steps: alkyne coordination, heterolytic dihydrogen activation (hydrogenolysis of one Pd–N bond). Subsequently, hydropalladation of the alkyne, followed by addition of N–H to palladium, reductive coupling of the vinyl and hydride, and finally, substitution of the product alkene by the alkyne substrate (see Scheme 14.12) [49].

At high substrate, or low hydrogen concentration, the semihydrogenation of 4-octyne is inhibited by the formation of catalytically inactive palladacycle species. These species are formed by oxidative coupling of two substrate molecules. In addition, careful kinetic measurements and product analysis has revealed that the activation of the catalyst precursor **26b** during the induction period occurs by hydrogenation of the coordinated maleic anhydride to succinic anhy-

Table 14.2 Product distribution in the hydrogenation of alkynes using [Pd(p-MeO-C$_6$H$_4$-bian)(dmfu)] (**26a**) [35].[a]

Entry	Substrate	Product distribution [%][b]		
		(Z)-alkene	(E)-alkene	alkane
1	R━━━R R = alkyl, CO$_2$Me	>98	–	–
2	(structure with OH)	>99	–	–
3	(phenyl−C≡C−R) R = H, Me, n-Bu, CO$_2$H	>99 (R=H)[c] 91 (R=Me) 92 (R=n-Bu) 95 (R=COOMe)	– 2 6 5	– 6 3 –
4	(macrocyclic diester structure)	95	5	–
5	(biphenyl structure) R = H, NO$_2$	87 (R=H) 97 (R=NO$_2$)	– 3	13 –
6	(CH$_2$)$_n$ (cyclic alkyne structure)	n=4, ethenylcyclohexene exclusively[c] n=6, ethenylcyclooctene exclusively		

a) Reaction conditions: 80 mM substrate and 0.8 mM using [Pd(p-MeO Ar-bian)(dmfu)] (**26a**) in THF at 20 °C and 1 bar H$_2$.
b) Product distribution determined by GC and ^{1}H-NMR at >99.5% conversion of alkyne.
c) Determined by reaction with D$_2$.

dride, concurrently producing the catalytically active [Pd(Ar-bian)(alkyne)] complex [49].

It was concluded that the high selectivity observed in the hydrogenation experiments using **26b** is explained by the relatively strong coordination of the alkyne to the palladium center, which only allows for the presence of small amounts of alkene complexes. Only the latter are responsible for the observed minor amounts of (E)-alkene, which was shown to be a secondary reaction product formed by a subsequent palladium-catalyzed, hydrogen-assisted isomerization reaction. Since no n-octane was detected in the reaction mixture, only a tiny

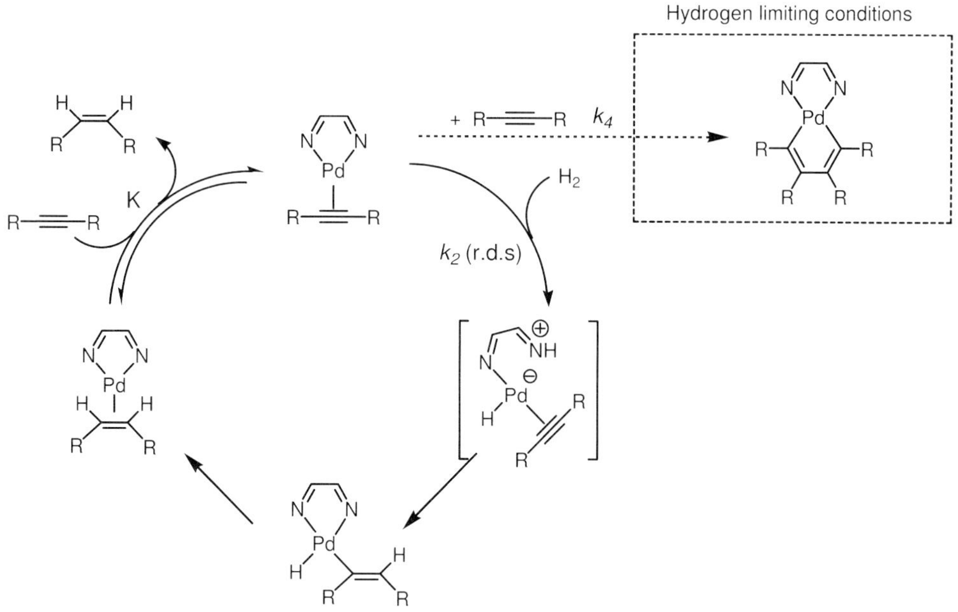

Scheme 14.12 Mechanism for the selective hydrogenation of alkynes by [Pd{(m,m'-(CF$_3$)$_2$C$_6$H$_3$)bian}(ma)] (**26b**) [49].

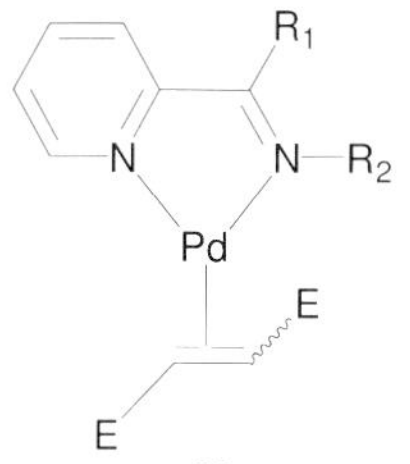

Fig. 14.1 Zerovalent [Pd(pyca)(alkene)] (**27**) precatalyst complex.

amount of the intermediate Pd(hydrido)(alkyl) species will reductively eliminate to form the alkane.

Zerovalent [palladium(alkene)] complexes **27** containing pyca ligands (pyca = pyridine-2-carbaldimine; Fig. 14.1) have been investigated in the stereoselective hydrogenation reaction of 1-phenyl-1-propyne in THF [34]. Most of the obtained inherently high *cis*-selectivities in the semihydrogenation of 1-phenyl-1-propyne are comparable to each other and to the results of the [Pd(Ar-bian)] system. However, the stability of the [Pd(pyca)] systems is generally lower and decomposition occurs in several cases before full consumption of the alkyne has taken place. It was concluded that the nature of the substituents on the imine nitrogen atom seems to be the most important factor determining the stability of the various catalysts precursors under hydrogenation conditions. It appeared that more donating capacity of the N-substituent results in a more stable catalyst, whereas increased steric bulk reduces its stability.

Members of the Parma group have shown that a large variety of divalent palladium complexes, with the generic formula [Pd(L)X] (X=OAc⁻ or Cl⁻) bearing a hydrazinato-based tridentate ligand L (containing different donor atoms), are able to hydrogenate alkynes to the corresponding alkenes with different degrees of success [32, 50–52]. Considerable mechanistic details have been reported for this reaction employing the catalysts [Pd(thiosemicarbazonato)X] (**28**) and [Pd(thiobenzoylhydrazonato)X] (**29**) (see Fig. 14.2) [32]. The hemilability of these ligands plays an intricate role in the catalysis with such compounds. Under the reaction conditions used (1 bar H₂, 30 °C, substrate:catalyst ratio=100), the chloro-analogue [Pd(NNS)Cl] (**28 a**) exhibits a full conversion in 24 h with 92% selectivity for styrene. The overall reaction rate is low, and TOFs around 4 mol · mol⁻¹ h⁻¹ have been reported. Although these catalysts are completely unreactive towards styrene present in solution, the ethylbenzene formed during the reaction was proposed to be the result from a second, consecutive hydrogenation step of the primary reaction product – that is, styrene which remains coordinated to the palladium (see Scheme 14.13). The stereoselectivity of the reaction could be tailored by changing the counter ion of the Pd(II) complex.

A kinetic investigation was performed using the [Pd(methyl-2-pyridylketone-thiosemicarbazonato)Cl] (**29 a**) complex at 50 °C, and this yielded the experimentally obtained rate law; $r = k$[catalyst][alkyne][pH₂] [32]. Based on this rate law and the activity data, a catalytic cycle as depicted in Scheme 14.13 was postulated. The first step of the mechanism is the activation of molecular hydrogen *via* heterolytic hydrogen splitting to generate a palladium hydride species (**30**) and accommodating the proton on one of the basic sites of the ligand (the hydrazone nitrogen). It is assumed that the pyridine moiety acts as a hemilabile ligand, creating the coordination site for the hydride. After hydrogen activation, the incoming alkyne replaces the counterion X⁻ from the first coordination sphere and generates the [Pd(NNS)(alkyne)] species (**32**). This step appears to be the predominant factor in driving the chemoselectivity of the hydrogenation reaction. It was shown that the weakly bonded acetate anion is easily replaced by both phenyl acetylene and the reaction product styrene, hence displaying low chemoselectivity, while tightly associated anions such as iodide show no hydrogenation activity under the reaction conditions used. The chloride ion appears to take an intermediate position, allowing it to be replaced by phenyl acetylene but not by the reaction product styrene; thus it forms a basis for discriminating between these potential substrates. Subsequently, the hydride is transferred to the coordinated phenyl acetylene forming a [Pd(NNS)(alkenyl)]⁺X⁻ species (**32**) which then reacts with molecular hydrogen, leading to the [Pd(NNS)(H)(styrene)]⁺X⁻ complex (**33**). Either the chloride anion or the phenyl acetylene then displaces the coordinated styrene resulting in complex **30** or **31**, respectively.

In addition, the related divalent palladium complexes [Pd(PNO)X] (**35**) (PNO = N′-(2-(diphenylphosphino)benzylidene)acetohydrazonato and related ligands) [50, 51] and [Pd(NNN)X] (**36**) (NNN = N-pyridin-2-yl-N′-pyridin-2-ylmethy-

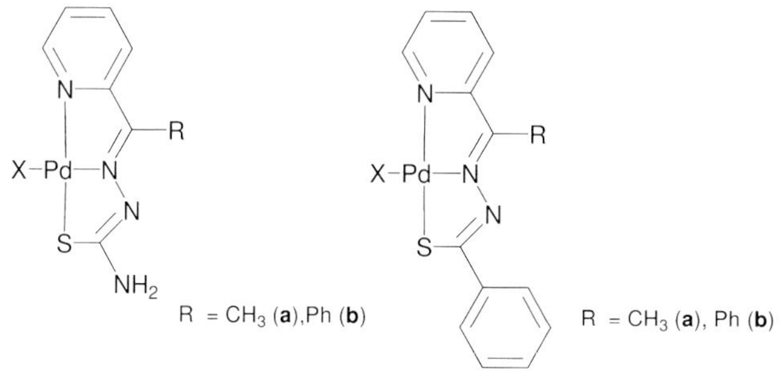

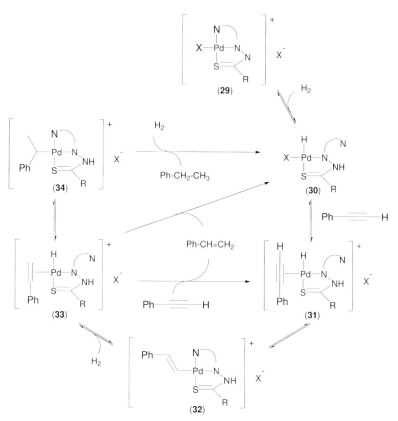

Fig. 14.2 Structure of the precursor catalysts [Pd(thiosemicar-bazonato)X] (**28**) and [Pd(thiobenzoylhydrazonato)X] (**29**).

Scheme 14.13 Proposed mechanism for the hydrogenation of phenyl acetylene by [Pd(methyl 2-pyridyl ketone thiosemicarba-zonato)Cl] (**29a**).

lene hydrazinato) (X = Cl⁻ or OAc⁻) [52] have been reported to be able to hydrogenate alkynes to the corresponding alkenes. For these systems, a heterolytic hydrogen activation similar to catalysts **28** and **29** is suggested. The hemilability of the ligand and the nature of the anion is important feature of the catalytic activity. The rate of phenyl acetylene hydrogenation employing **36** appears similar to the rate reported for **28** and **29**, while **35** displays a distinctively lower reaction rate. Furthermore, no further mechanistic details concerning the chemo- or stereoselectivity of these catalytic systems have been reported.

14.1.9
Conclusions

The stereoselective hydrogenation of alkynes to alkenes can be effected by a wide variety of homogeneous catalysts. The appropriate choice of catalyst and reaction conditions allows the selective formation of either the (*Z*)- or the (*E*)-alkene. Most of the catalysts display a very high chemoselectivity, as they are not reactive towards reducible functional groups such as carbonyl, ester, and double bonds. Many of the details related to catalyst behavior and intricate mechanistic details concerning semihydrogenation of alkynes have often not been unraveled, and will remain a topic of research for the coming years.

14.2
Homogeneous Hydrogenation of Dienes to Monoenes

14.2.1
Introduction

The hydrogenation of polyenes is a research topic that has attracted substantial attention over the past three decades. Especially, the removal of diene constituents from (light) hydrocarbon fractions, polydiene rubbers and fatty acids has been a major focus in the research effort. Although the importance of this reduction is generally recognized, there are only a small number of homogeneous catalysts that have been reported to be active in the diene hydrogenation reaction. Details about the catalytic behavior and mechanism are often very scarce, or these have not been explored at all. Frequently, a distinction is made between the hydrogenation of conjugated and non-conjugated dienes. The latter type is often considered as consisting of two single, isolated double bonds, which are thus fairly easily reduced by most hydrogenation catalysts. Substrates containing conjugated double bonds (either present in the substrate or formed by prior isomerization reactions of non-conjugated double bonds) often pose more difficulty in a hydrogenation reaction. Especially, the formation of stable diene- or allyl complexes often inhibits the reaction and, hence, the selective hydrogenation of conjugated dienes to monoenes forms a particular challenge.

Selectivity in the hydrogenation reaction of dienes to monoenes can be achieved by two types of catalytic system: (i) those which are completely inert with respect to the hydrogenation of the resulting monoenes; and (ii) those for which the selectivity is due to the discrimination based on thermodynamic and/or kinetic factors that suppress the rate of formation of the saturated hydrocarbon. The latter approach is the most common way of achieving selectivity for these hydrogenations.

In this section, an overview will be provided of reported catalysts that can perform the hydrogenation of conjugated dienes to monoenes with a considerable degree of selectivity using molecular hydrogen as a hydrogen source. Most attention will be given to those catalytic systems that show remarkable (high) selectivities, high reaction rates, or which have demonstrated their value in organic synthesis. Furthermore, special attention will be given to the mechanistic details of these homogeneous hydrogenation systems.

14.2.2
Zirconium Catalysts

The oligomeric hydrido-zirconium complexes $[Cp_2Zr(H)(CH_2PPh_2)]_n$ (**37**) constitute selective systems for the hydrogenation of dienes [53]. The reaction is performed at 80 °C at hydrogen pressures between 10 and 40 bar; under these conditions the linear dienes (e.g., 1,3-pentadiene) are hydrogenated consecutively to the corresponding alkanes. The monoene *E*-2-pentene is detected only as intermediate in the reaction mixture. Remarkably, cyclic olefins such as cycloheptatriene, 1,3- and 1,5-cyclooctadiene are selectively hydrogenated to the cycloheptene and cyclooctene, respectively (>98% selectivity). In contrast to linear dienes, cyclic dienes are not fully hydrogenated to the alkane and the hydrogenation stops at the monoene stage.

In order to understand the unusual reactivity of $[Cp_2Zr(H)(CH_2PPh_2)]_n$ towards cyclic dienes, the reaction mechanism was studied using 1,3-cyclooctadiene and butadiene as substrate. It appears that the first step, the activation of the initial zirconium precursor catalyst **37**, is achieved by a sequence of decomposition steps consisting of the reductive elimination of the phosphine PPh_2Me and the simultaneous formation of the zirconocene intermediate $[Cp_2Zr]$. This intermediate will then be trapped, either by the diene, producing $[Cp_2Zr(diene)]$ (**44**) or by **37** to give the homometallic hydride bridged complex $[Cp_2Zr(\mu\text{-}H)(\mu\text{-}CH_2PPh_2)ZrCp_2]$ (**38**) (see Scheme 14.14) [54]. The complexes **44** and **38** are related species. From detailed stoichiometric reactions, using the less sterically demanding butadiene as substrate, it has been shown that the reaction of intermediate **44** with starting material **37** and the reaction of compound **38** with butadiene are viable and produce a common intermediate. This common intermediate has been identified as a dinuclear Zr(IV)–Zr(II) complex in which the butenyl fragment bridges between the two zirconium centers in a $\mu^1\text{-}\eta^1:1,2\text{-}\eta^2$-fashion, structurally similar to compound **39** (Scheme 14.14) [55].

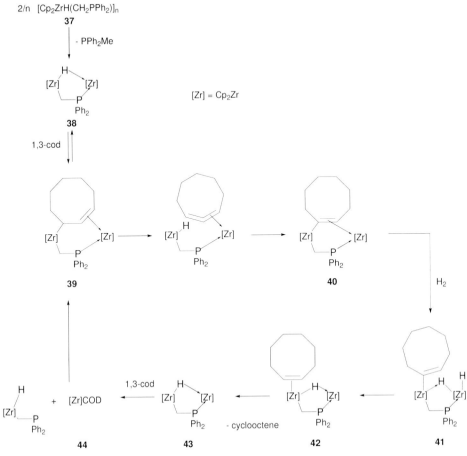

Scheme 14.14 Proposed mechanism for the hydrogenation of
1,3-cyclooctadiene catalyzed by [Cp$_2$Zr(H)(CH$_2$PPh$_2$)]$_n$ (**37**) [53].

Kinetic studies revealed that the order of reaction with respect to the zirconium is a half, reflecting the dissociation of active catalytic species [53]. Since the order in substrate is zero and one in hydrogen, the addition of H$_2$ is the rate-limiting step in the reaction mechanism. The rate of the reaction, expressed in average TOFs, ranges between 180 and 750 mol mol^{-1} h^{-1}, and the activation energy of the hydrogenation of 1,3-cyclooctadiene was determined to be 29 kJ mol^{-1}. The proposed catalytic cycle, as is depicted in Scheme 14.14, consists of the formation of the initial Zr(IV)–Zr(II) bimetallic complex **39** in which a cyclooctenyl ligand is bridging between the two zirconium centers. Compound **39** then isomerizes to the μ-alkenyl zirconium(IV)/zirconium(II) intermediate **40**, which is considered the key intermediate in the reaction mechanism. Formation of complex **40** is proposed to explain the effective protection of the remaining C=C double bond in the catalytic cycle. This species then reacts with molecular hydrogen in an oxidative addition manner to form the dinuclear Zr(IV)–Zr(IV) dihydride complex (**41**), having one bridging and one terminal hydride. Reductive elimination of the alkenyl and one of the hydrides yields the dinuclear-monohydride bridged complex (**42**) which then loses the cyclooctene. The unusual selectivity is ascribed to the cooperative effect of two zirconium centers, stabilized and held together by the bridging CH_2PPh_2-ligand, to combine the action of a zirconium(IV) and a zirconium(II) centre, bringing about rapid mono-hydrogenation and at the same time effective protection of the remaining C=C double bond.

14.2.3
Chromium Catalysts

Group VI metal carbonyls and their derivatives, especially chromium carbonyl compounds, are able to hydrogenate a variety of substrates (e.g., alkynes, enones) with high stereospecificity [6]. The regio- and stereospecific 1,4-hydrogenation of conjugated dienes to (Z)-monoenes catalyzed by $[Cr(CO)_3(arene)]$ complexes (**45**) was first reported in 1968 by the groups of Cais et al. [56] and of Frankel et al. [57]. The catalysts $[Cr(CO)_3(benzene)]$, and similar ones containing a benzene derivative that were initially explored, required rigorous reaction conditions (150–160 °C and 50 bar H_2) in order for the stereoselective 1,4-hydrogenation to proceed. However, the activity depends on the substituents attached to the benzene ligand – that is, the presence of more electron-withdrawing substituents and more coordinating solvents allowed milder reaction conditions (Table 14.3).

Spectroscopic studies have revealed that $[Cr(CO)_3(S)_3]$ (S = solvent) is the active species for the hydrogenation, and the formation of these species is promoted either by less strongly coordinated ligands (electron-poor benzene or polyaromatic compounds) or by more strongly coordinating solvents. The use of polyaromatic compounds such as naphthalene, anthracene and phenanthrene as ligands combined with polar coordinating solvents (THF or acetone) has proven to be particular effective for constructing a highly active catalytic system under mild conditions [58]. Whereas several mechanistic schemes have been suggested for the chromium-catalyzed diene hydrogenation, they all share a common feature: the active metal fragment $[Cr(CO)_3]$ is generated *in situ* by thermal or photochemical activation of a coordinatively saturated $[Cr(CO)_3(L)_3]$ complex ((L)$_3$ = arene; L = CH_3CN or CO) [58–60].

At ambient reaction temperatures and hydrogen pressures, kinetic studies employing $[Cr(CO)_3(nap)]$ (nap = naphthalene) (**45a**) as catalyst revealed that two reaction mechanisms operate simultaneously, namely a hydride route and a diene route (Scheme 14.15) [61]. After the initial step, in which $[Cr(CO)_3(nap)]$ reacts with the solvent to generate the active catalyst $[Cr(CO)_3(S)_3]$ (**46**), this solvent complex can then subsequently react either with hydrogen or the conjugated diene. In the hydride route, $[Cr(CO)_3(S)_3]$ undergoes oxidative addition of dihydrogen to give complex **48**, which still has two labile coordination sites nec-

Table 14.3 Hydrogenation of methyl sorbate to 3-hexenoic acid methyl ester with [Cr(CO)$_3$(arene)] (**45**) catalysts.

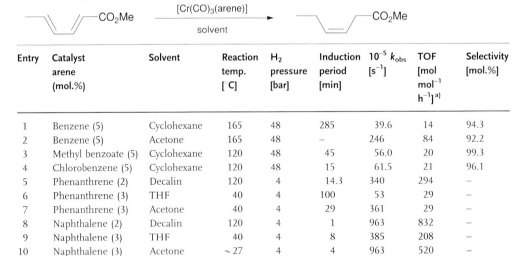

Entry	Catalyst arene (mol.%)	Solvent	Reaction temp. [°C]	H$_2$ pressure [bar]	Induction period [min]	$10^{-5} k_{obs}$ [s^{-1}]	TOF [mol mol^{-1} h^{-1}][a]	Selectivity [mol.%]
1	Benzene (5)	Cyclohexane	165	48	285	39.6	14	94.3
2	Benzene (5)	Acetone	165	48	–	246	84	92.2
3	Methyl benzoate (5)	Cyclohexane	120	48	45	56.0	20	99.3
4	Chlorobenzene (5)	Cyclohexane	120	48	15	61.5	21	96.1
5	Phenanthrene (2)	Decalin	120	4	14.3	340	294	–
6	Phenanthrene (3)	THF	40	4	100	53	29	–
7	Phenanthrene (3)	Acetone	40	4	29	361	29	–
8	Naphthalene (2)	Decalin	120	4	1	963	832	–
9	Naphthalene (3)	THF	40	4	8	385	208	–
10	Naphthalene (3)	Acetone	~27	4	4	963	520	–

a) TOF determined at 50% conversion using first-order kinetics.
TOF = 0.5 k_{obs}[substrate]/[catalyst].

essary to accommodate the diene in a η^4-fashion. After coordination of the conjugated diene producing complex **49**, the two hydrides are rapidly transferred, producing a (Z)-monoene. The diene route is established by initial coordination of the diene to the solvento complex **46** to produce [Cr(CO)$_3$(diene)(S)] **47** prior to the hydrogen activation. Although both hydrogenation routes are active under the typical hydrogenation conditions, additional kinetic studies performed at high temperatures and elevated hydrogen pressures employing [Cr(CO)$_3$(arene)] (arene=substituted benzene derivative) show that, under these conditions, only the hydride route is operative. The oxidative addition of molecular hydrogen is the rate-determining step for this hydrogenation route [62, 63].

The [Cr(CO)$_3$(arene)]-catalyzed 1,4-hydrogenation of conjugated dienes has become an established route for the stereocontrolled synthesis of alkenes in organic synthesis [6, 7]. The potential of this method has been clearly demonstrated in the synthesis of olfactory compounds (fragrances and insect pheromones), where the 1,4-hydrogenation of dienes was employed in a key step of the synthesis. For such compounds, the stereo- and regio control of the double-bond geometry is extremely important for maintaining its biological activity. Furthermore, the [Cr(CO)$_3$(arene)]-catalyzed hydrogenation displays a remarkably high chemoselectivity, and the outcome of the reaction is not affected by the presence of other functional groups such as non-conjugated carbon-carbon double bonds, esters, ketones, carboxylic acids, epoxides, phosphonate esters, sulfonamides, or even cyano groups [6].

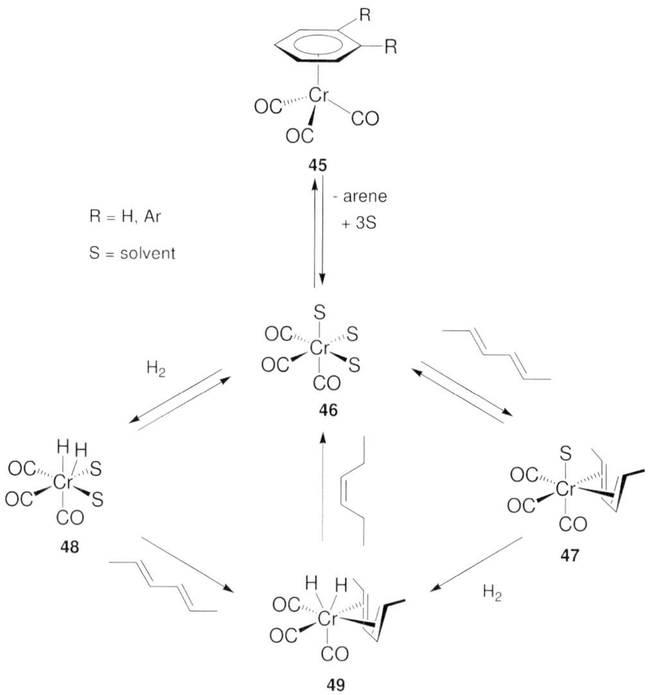

Scheme 14.15 Proposed hydrogenation mechanism for the
1,4-hydrogenation of dienes by [Cr(CO)$_3$(arene)] (**45**).

An interesting feature of the [Cr(CO)$_3$]-catalyzed 1,4-hydrogenation is the pre-
determined outcome of the stereocontrolled reaction; the diene must coordinate
in a η^4-s-*cis* fashion to the chromium, allowing only one conformationally rigid
and predisposed intermediate [6]. The presence of only one accessible catalyst–
substrate intermediate structure invokes the 1,4-hydrogenation to proceed with
complete regio- and stereocontrol, regardless of the thermodynamic stability of
the hydrogenation products. This was demonstrated in the stereocontrolled synthesis of carbacyclin analogues for the production of the exclusively (*E*)-exocyclic
isomer (Scheme 14.16). The chromium–diene intermediate, in conjunction with
the 1,4-addition, dictates the formation of the exocyclic isomer, and since the
chromium catalyst is inactive in the isolated double-bond isomerization reaction, the (*E*)-isomer is obtained solely as the reaction product.

Apart from Cr(0) carbonyl complexes, similar Mo, W and Co complexes also
catalyze 1,4-*cis*-hydrogenation of dienes, though the selectivity of these catalysts
is relatively low [63].

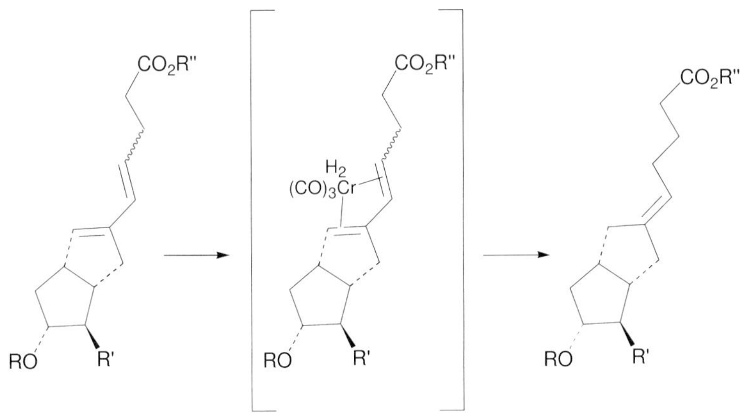

Scheme 14.16 Stereocontrolled 1,4-hydrogenation of a carba-cyclin analogue using [Cr(CO)$_3$(arene)] (**45**) complexes.

14.2.4
Ruthenium Catalysts

Ruthenium complexes are active hydrogenation catalysts for the reduction of dienes to monoenes. Both zerovalent and divalent ruthenium complexes containing various (alkene, diene and phosphine) ligands have been employed as catalysts that have met with different degrees of success.

Ruthenium(0) complexes containing cyclic polyenes such as Ru(cod)(η^6-triene) (**50**) (cod = 1,5-cyclooctadiene; triene = 1,3,5-cyclooctatriene or 1,3,5-cyclohepta-triene) have proven to be selective hydrogenation catalysts for the reduction of cycloheptatriene and cyclooctadiene to the corresponding cyclic monoenes [64, 65]. The [Ru(cod)] fragment is maintained as the catalytic unit throughout the hydrogenation reaction and the η^6-coordinated triene (e.g., cyclooctatriene) is hydrogenated to the monoene during the induction period.

For instance, cycloheptatriene has been selectively hydrogenated at 1 bar H$_2$ pressure at 20 °C, yielding cycloheptene. The selectivity depended largely on the solvent used, ranging from 100% when *n*-hexane was used, or 99.5% in THF, to very low values when ethanol was employed. The conversion is quantitative in THF and ethanol, but in *n*-hexane it did not exceed 65%; consequently, the authors concluded that THF gives the best combination of selectivity and conversion. In this case, the formation of cycloheptane was observed only after the substrate cycloheptatriene had completely been consumed.

Cyclooctadiene isomers (i.e., 1,5-cod or 1,3-cod) are selectively hydrogenated by [Ru(η^4-cod)(η^6-C$_8$H$_{10}$)] (**51**) to produce exclusively cyclooctene in THF, under ambient temperature (20 °C) and 1 bar H$_2$ pressure [64]. Again, cyclooctane is only detected when the diene substrate is completely transformed to the monoene. The rate of hydrogenation is higher in case of the conjugated 1,3-cyclooctadiene substrate, whereas isomerization of the non-conjugated 1,5-cyclooctadiene

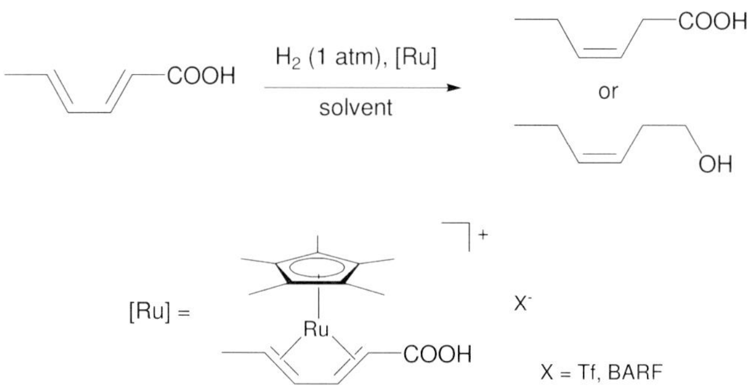

Scheme 14.17 Hydrogenation of sorbic acid by [Cp*Ru(η^4-sorbic acid)]$^+$X$^-$ (**52**). X$^-$ = CF$_3$SO$_3^-$ or BARF$^-$.

to 1,3-cyclooctadiene has been proposed. This isomerization appeared to be necessary for the catalytic hydrogenation of 1,5-cod.

Cationic ruthenium complexes containing the fragment [Cp*Ru] (**52a**) can stereoselectively hydrogenate sorbic acid or sorbic alcohol to *cis*-3-hexenoic acid or *cis*-3-hexen-1-ol, respectively (Scheme 14.17). The highest rate and stereoselectivity have been obtained with the "naked" [Cp*Ru] – that is, a monocyclopentadiene–ruthenium complex without any additional, inhibiting ligands and [Cp*Ru(η^4-sorbic acid)]$^+$X$^-$ (**52**) (sorbic acid = (2*E*,4*E*)-MeCH=CHCH=CHCO$_2$H and X = CF$_3$SO$_3^-$ or BARF$^-$) displays the best results. These ionic ruthenium catalysts have been used successfully in a single phase as well as liquid two-phase systems (such as liquid–liquid, ionic liquid–liquid solvent systems) [66, 67]. The activity of these catalysts depends strongly on the solvent or solvent mixtures used, with the rate of hydrogenation of sorbic acid (in TOF) ranging from 92 to 1057 mol mol^{-1} h^{-1}. The highest rate and selectivity have been reported for the solvent methyl-*tert*-butyl ether (TOF = 1057, *cis:trans* ratio 96 : 3). In general, the hydrogenation of sorbic alcohol proceeds with higher activity and almost complete selectivity when using **52** as the catalyst.

Mechanistic studies employing the PHIP phenomenon showed that the homogeneous hydrogenation of sorbic acid (in acetone under 1 bar H$_2$) proceeds by concerted 1,4-hydrogenation of the diene moiety [68]. Both atoms of the same dihydrogen molecule are transferred to the substrate in a synchronous fashion, yielding *cis*-3-hexenoic acid as the primary reaction product. Furthermore, it was found that the *trans*-isomer is formed by subsequent rearrangement of the *cis*-3-hexenoic acid, and is not the result of direct hydrogenation of sorbic acid.

14.2.5
Cobalt Catalysts

In spite of the number of disadvantages, considerable interest has been given to the hydrogenation of dienes by the water-soluble catalyst $K_3[Co(CN)_5H]$ (**53**) [69, 70]. The catalyst shows a high chemoselectivity towards conjugated carbon-carbon double bonds; that is, isolated carbon-carbon double bonds or carbonyl functional groups will not be affected by the catalyst under the hydrogenation conditions. However, the hydrogenation reaction suffers from short catalyst lifetime, substrate inhibition, low hydrogenation rate (generally TOF ≤2) and low regioselectivity of the monoene products [71]. The addition of phase-transfer reagents (e.g., ammonium salts [71], β-cyclodextrin [72]) or surfactants [73, 74] largely overcomes these complications, but the rate of the reaction remains low. In general, the product ratios are dominated by the overall 1,4-addition of hydrogen under the phase-transfer conditions, although in some cases 1,2-addition has been observed [73, 75]. Several modified versions of this catalyst are known where one or more cyanides are replaced by diamines such as ethylenediamine, bipy or phen [75].

14.2.6
Rhodium Catalysts

The well-known cationic Osborn catalyst $[Rh(nbd)L_n]^+A^-$ (**54**), which has been an extremely useful catalyst for the reduction of alkenes and alkynes, also facilitates the rapid and selective hydrogenation of dienes in polar solvents, affording the corresponding monoenes [76]. In order to probe this reaction, various substrates such as norbornadiene, substituted butadienes and conjugated and non-conjugated cyclic dienes have been tested. The TOF depends strongly on the reaction conditions (i.e., solvent), the catalyst employed, and on the structure of the substrate. Hence, the rate of hydrogenation typically ranges from 140 to 250 mol mol^{-1} h^{-1}. More sterically encumbered substrates such as 2,5-dimethyl-hexa-2,4-diene show very low hydrogenation rates (~6 mol mol^{-1} h^{-1}), while 1,3-cyclooctadiene is rapidly hydrogenated to the monoene (330 mol mol^{-1} h^{-1}). The product monoenes, which are formed almost quantitatively (up to 99% yield), result from overall 1,2- and 1,4-addition of H_2 onto the diene moiety [25, 76, 77]. The ratio of the two addition modes depends strongly on the structure of the substrate and on the nature of the ligand, L. For the diene reduction, chelating diphosphines or diarsines are preferred as stabilizing ligands, as catalysts derived from monodentate ligands tend to become easily deactivated due to the formation of an unsaturated $[Rh(diene)_2L]^+$ fragment (**55**).

The proposed hydrogenation mechanism for (conjugated) dienes by the cationic rhodium catalyst $[P_2Rh(nbd)]^+X^-$ (**54**) is depicted in Scheme 14.18. Activation of the precursor catalyst **54** has been studied in considerable detail, and reveals that the hydrogenation of the initially coordinated norbornadiene is fast and complete for most investigated systems (contrary to the often-used $[P_2Rh(cod)]^+X^-$ analogues) [78–80]. In contrast to the reported hydrogenation mechanisms for monoenes and alkynes, the hydrogenation of dienes does not depend on the addition of acid and, therefore, it is assumed to proceed by a different catalytic route (Scheme 14.18) [21, 76, 81]. The proposal of an "unsaturated" mechanism, in which the diene coordinates to form [P$_2$Rh(diene)]$^+$ (57), prior to hydrogen activation, can be rationalized by the high coordination constant of the *cis-s*-coordinated (chelating) diene to the rhodium. The hydrogenation mechanism shows a zero order in diene, and it has been suggested that the rate-determining step involves the reaction of H$_2$ with the cationic [P$_2$Rh(diene)]$^+$ species (57). Detailed *in-situ* NMR studies employing deuterium and *para*-hydrogen-enriched molecular hydrogen has allowed the proposal of the structure of the [P$_2$Rh(diene)(H)$_2$] (58) intermediate, which was revealed by the cross-relaxation transfer, originating from the enhanced magnetization inflicted by the *para*-hydrogen adducts [78]. In addition, these studies demonstrated that the hydrogenation of norbornadiene and 1,4-cyclooctadiene occurs successively and confirms the mechanism presented in Scheme 14.18 [82]. To date, many details of the hydrogenation of 1,3-dienes have not been resolved, such as the true nature of the [P$_2$Rh(R)(H)] intermediate between complex 58 and 59 (not shown in Scheme 14.18), possibly being either a [Rh(alkenyl)] or a [Rh(allyl)] species. The first species would lead to 1,2-addition, and the latter one to either 1,2- or 1,4-addition.

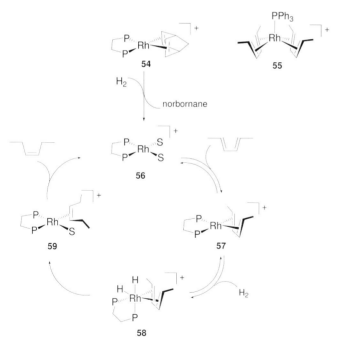

Scheme 14.18 Proposed 1,4-hydrogenation mechanism for the hydrogenation of dienes by cationic rhodium complexes [P$_2$Rh(diene)]$^+$A$^-$ (54).

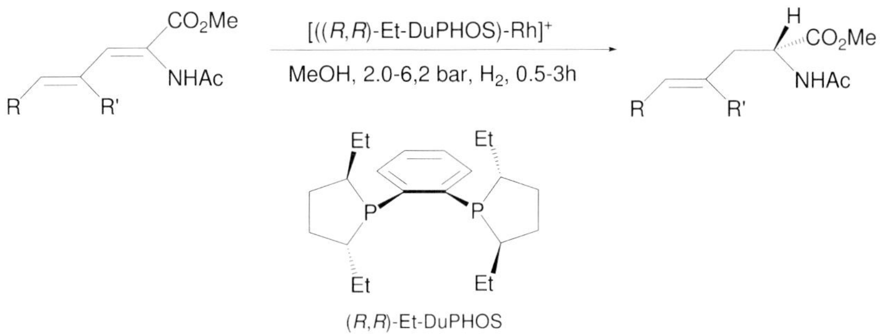

CO$_2$Me

[((*R*,*R*)-Et-DuPHOS)-Rh]$^+$

NHAc

MeOH, 2.0-6,2 bar, H$_2$, 0.5-3h

R   R'

H

CO$_2$Me

NHAc

R   R'

Et   Et

P   P

Et   Et

(*R*,*R*)-Et-DuPHOS

Scheme 14.19 Hydrogenation of conjugated α,γ-dienamide esters by [((*R*,*R*)-Et-DuPHOS)-Rh(cod)]OTf (**54b**) (OTf = triflate) [83].

The characteristics of the hydrogenation of norbornadiene, substituted butadienes and conjugated and cyclic dienes are all very similar. In the case of conjugated dienes, there appears to be hardly any isomerization activity, while in the case of 1,4-dienes an isomerization step to form the corresponding 1,3-diene is assumed prior to hydrogenation. The catalyst behavior changes after the diene has been completely converted to the monoene, whereupon the rhodium catalyst resumes its "normal" monoene hydrogenation behavior.

An interesting example of the use of [Rh(nbd)L$_n$]$^+$A$^-$ (**54**) in organic synthesis is provided by the cationic rhodium complex [((*R*,*R*)-Et-DuPHOS)-Rh(cod)]OTf (**54b**), which appears to be particularly effective in the enantiomeric hydrogenation of conjugated α,γ-dienamide esters [83]. Full conversion to the corresponding γ,δ-unsaturated amino acids could be obtained using the Et-DuPHOS-Rh catalyst system, yielding very high regio- and enantioselectivity (>95% and 99% ee, respectively; see Scheme 14.19) with over-reductions as low as <0.5%. The rate of the reaction (expressed as average TOFs, determined at full conversion) for different substrates ranged from 160 to 1000 mol mol^{-1} h^{-1}. Either enantiomer of the γ,δ-unsaturated amino acid could be obtained by the use of (*R*,*R*)- or (*S*,*S*)-Et-DuPHOS. Interestingly, the reaction proceeds by hydrogenation of the enamide C=C double bond, with complete regioselectivity over the distal C=C double bond. Effectively, only 1,2-addition is observed, which can be explained by the chelating effect of the enamide group to the metal center and preventing over-reduction, even after the enamide double bond has been completely converted.

The selective diene hydrogenation of monoterpenes such as myrcene, which contain both isolated monoene and diene moieties, forms a particular challenge [84]. The catalyst [RhH(CO)(PPh$_3$)$_3$] (**60**) has been reported to perform remarkably well for such hydrogenation reactions, and the diene moiety was shown to be selectively reduced to the monoene, while the isolated double bond remained unaffected under the reaction conditions used (Scheme 14.20). The rates of reaction expressed as average TOF (determined at ca. 80% conversion) ranged from ca. 640 (in benzene, 20 atm H$_2$ at 100 °C) to 7600 mol mol^{-1} h^{-1} (in cyclohexane, 20 atm H$_2$ at 80 °C). The hydrogenation in benzene solution resulted in

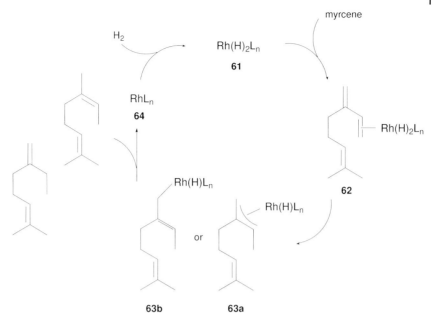

Scheme 14.20 Proposed mechanism for the hydrogenation of myrcene by [RhH(CO)(PPh$_3$)$_3$] (**60**) and the formation of the major reaction products [84].

the highest chemoselectivity of the reduction of the conjugated diene to the corresponding monoene (98% selectivity). The products arise from overall 1,2- and 1,4-addition onto the diene moiety.

The proposed reaction mechanism differs from the mechanism reported for the cationic rhodium complexes [Rh(nbd)L$_n$]$^+$A$^-$ (**54**); that is, in the present case a "dihydride route" is favored by the authors, implying that first molecular hydrogen is activated to yield the rhodium dihydride species [Rh(H)$_2$L$_n$] (**61**). This species subsequently reacts with the substrate's conjugated diene fragment. However, formation of the proposed rhodium(dihydride) species from the precursor catalyst [RhH(CO)(PPh$_3$)$_3$] **60** has not been disclosed. Based on the product distribution, the authors suggest that the diene initially coordinates in an η^2-fashion (complex **62**), showing preference for the least-substituted double bond (Scheme 14.20). The presence of [Rh(η^4-diene)] complexes has not been ruled out as possible intermediates in the product formation, and could be the bias for the observed chemoselectivity. After transfer of the first hydride to the substrate, either a [Rh(η^3-allyl)] complex (**63a**) or the [Rh(η^1-allyl)] complex (**63b**) is formed; the latter species leads to overall 1,2-addition, while [Rh(η^3-allyl)] leads to both 1,2- as well as 1,4-addition. Although the dihydride route is the favored mechanism, a catalytic cycle based on a rhodium-monohydride has not been excluded.

Finally, other rhodium catalysts for the selective diene hydrogenation worth mentioning include [RhCl(PPh$_3$)$_3$] (**64**), [RhCl(nbd)]$_2$ (**65**), and the catalytic sys-

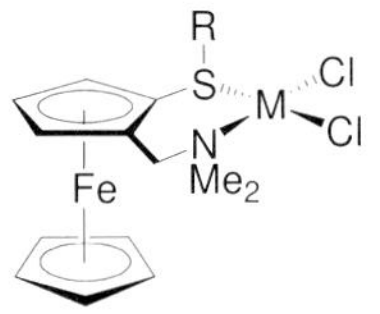

68 M = Pd, Pt
        R = alkyl, aryl

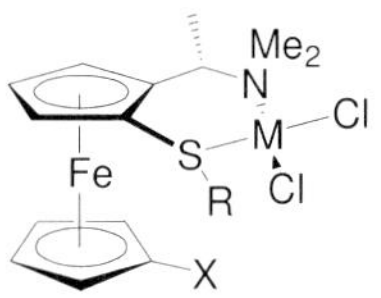

69 M = Pd, Pt
        R = alkyl, aryl
        X = H, SR

Fig. 14.3 General structures of compounds **68** and **69**.

tems resulting from protonation of $[Rh(CO_2Me)_2(PPh_3)_2]$ (**66**) or $[Rh(CO_2Me)(PPh_3)_3]$ (**67**) precursors [25, 77]. Generally, these complexes show chemoselectivity that is similar to or lower than that of the cationic rhodium complexes $[Rh(nbd)L_n]^+A^-$ (**54**), with comparable reaction rates. The observed selectivity for these systems is attributed principally to the much higher coordinating power of dienes compared with monoenes. Catalysts **64**, **65**, **66** and **67** reduce conjugated dienes by 1,2- as well as 1,4-addition of hydrogen, yielding mixtures of monoenes.

14.2.7
Palladium and Platinum Catalysts

Palladium has been frequently used to reduce conjugated dienes to monoenes. Most of the catalysts based on palladium consist of palladium species – that is, mononuclear complexes and Pd nanoparticles that are heterogenized on a solid support. Both mineral and organic supports have been successfully employed in these systems and, in addition, various hydrogen donors can be used, namely dihydrogen, Group XIII or XIV metal hydrides, and various formate salts. The number of homogeneous palladium complexes able to hydrogenate conjugated dienes to monoenes using molecular hydrogen is very limited. The majority of the accounts reported in the open literature concerning these homogeneous palladium catalysts merely state the observed activity and, hence, the mechanistic details mostly remain obscure.

A large variety of palladium complexes of the general type **68** and **69** (Fig. 14.3), containing ferrocenyl-amine-sulfide or -selenide ligands, are active catalysts for the selective hydrogenation of conjugated dienes to monoenes [85, 86]. The best results (at H_2 pressure of 4–7 bar and room temperature) have been obtained with cyclic substrates. 1,3-Cyclooctadiene is converted into cyclooctene, with the most efficient catalyst reported for this transformation being $[\eta^2\text{-}\{FeCpC_5H_3(CH_2NMe_2)(S(t\text{-}Bu)\}PdCl_2]$, with a TOF of 345 mol mol^{-1} h^{-1} and a selectivity of 97.2% [87]. Higher activities are accompanied by lower selectivities due to a higher degree of over-reduction to cyclooctane. Acyclic dienes yielded less satisfactory results, although these substrates are selectively hydro-

Table 14.4 Selective hydrogenation of 1,3-cyclooctadiene to cyclooctene with various palladium complexes.

	Catalyst	Induction time [h]	Conversion [%]	TOF [mol mol^{-1} h^{-1}]	Selectivity [%]	Reference
1	[{FeCpC$_5$H$_3$(CH$_2$NMe$_2$)(SMe)}PdCl$_2$][a]	49.7	100	14	94.1	87
2	[{FeCpC$_5$H$_3$(CH$_2$NMe$_2$)(SEt)}PdCl$_2$][a]	42.3	100	15	98.4	87
3	[{FeCpC$_5$H$_3$(CH$_2$NMe$_2$)(S(n-Pr))}PdCl$_2$][a]	37.5	100	16	90.9	87
4	[{FeCpC$_5$H$_3$(CH$_2$NMe$_2$)(S(t-Bu))}PdCl$_2$][a]	0.0	100	345	97.2	87
5	[{FeCpC$_5$H$_3$(CH$_2$NMe$_2$)(S(4-tolyl))}PdCl$_2$][a]	0.0	100	691	78.5	87
6	[{FeCpC$_5$H$_3$(CHMeNMe$_2$)(SMe)}PdCl$_2$]	42.5	100	17	96.7	85
7	[{FeCpC$_5$H$_3$(CHMeNMe$_2$)(S(4-tolyl))}PdCl$_2$]	0	98.6	465	91.9	85
8	[{FeCpC$_5$H$_3$(CHMeNMe$_2$)(S(4-ClPh))}PdCl$_2$]	0	100	684	96.8	85

a) 9.0 mL acetone, 2.0×10^{-5} mol, 7.45×10^{-3} mol substrate, room temperature, 4 bar H$_2$.
b) 9.0 mL acetone, 2.0×10^{-5} mol, 7.45×10^{-3} mol substrate, room temperature, 7 bar H$_2$.

genated by 1,4-addition to give the monoenes. A subsequent isomerization reaction leads to product mixtures and, hence, lower stereoselectivities.

The testing of a wide variety of substituted palladium complexes with ferrocene-based diphosphines has revealed several trends. One of the most apparent trends is that increasing the steric bulk on both the amine and the sulfide substituents increases the rate of the reaction, while the monoene selectivity is retained (Table 14.4). The hydrogenation reaction is strongly dependent on the solvent used, and the best results are obtained in acetone. It has been suggested that the solvent is involved in dissociation of the thioether ligand to create a free coordination site, to accommodate the diene [87]. In line with this observation, it appears that neither the platinum aminothioether complexes nor the palladium or platinum complexes of the homologous amino-selenide ligands are active hydrogenation catalysts. Apparently, the Pt–S, Pt–Se and the Pd–Se bonds are too strong, which prevents the required ligand dissociation needed to form an active catalyst. In addition, some catalysts display an induction period before the active catalyst is formed. No further mechanistic details about these catalytic systems have been reported.

Palladium-allyl complexes in combination with stabilizing ligands (e.g., phosphine, halide, allyl, and dienes) are able to hydrogenate conjugated dienes and non-conjugated dienes to the corresponding monoenes under relatively mild reaction conditions [88]. One of the most effective catalytic systems concerns [Pd(L)(PPh$_3$)Cl] (**70a**, L=allyl; **70b**, L=1-Me-allyl) in *N,N*-dimethylformamide. Hydrogenation reactions performed at 55 °C and 1 bar H$_2$ using **70b** show a moderate rate (∼10 mol mol^{-1} h^{-1} determined at 50% conversion) in the hydrogenation of 1,5- and 1,3-cyclooctadiene. However, these catalysts appear to be completely inactive towards the hydrogenation of monoenes (cyclooctene, cyclohexene and 1-octene), thereby corroborating the proposal that the catalysis is

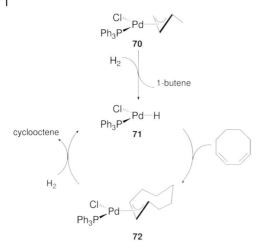

performed by the mononuclear palladium species. In case of hydrogenation of non-conjugated dienes, the substrate is isomerized to the corresponding conjugated diene prior to hydrogenation.

The influence of hydrogen pressure, substrate and catalyst concentration has briefly been mentioned. The reaction rate is dependent upon the catalyst concentration and hydrogen pressure, but appears to be independent of substrate concentration. The mechanism is proposed to involve the activation of the parent [Pd(allyl)] species producing an unstable hydrido-Pd(II) species (**71**), ensued by a fast reaction with the diene to restore the [Pd(allyl)] moiety (**72**) (Scheme 14.21). The observation that most of the starting material is isolated after the reaction suggests that only a small portion of the catalyst is active under the reaction conditions. Although a complete selectivity for the monoene is observed (even after full conversion), the presence of catalytically active colloidal palladium has not been completely excluded.

14.2.8
Conclusions

The homogeneous hydrogenation of dienes, especially of conjugated dienes remains a challenge, and only a few catalysts have been reported that show good chemo- and stereoselectivity. Until now, the most widely and best explored catalyst for this reaction is constituted by the group of [Cr(CO)₃(arene)] (**45**) and [Rh(nbd)Lₙ]⁺A⁻ (**54**) complexes. Clearly, this is a research area that has received relatively little attention in the past, and several of the catalytic systems discussed seem to be promising candidates for future investigations.

Abbreviations

DFT   density functional theory
PHIP   *para*hydrogen induced polarization
THF   tetrahydrofuran
TOF   turnover frequency
TON   turnover number

References

1 Hutchins, R.O., Hutchins, M.G.K., in: Patai, S., Rappoport, Z. (Eds.), *Reduction of Triple-Bonded Groups, The Chemistry of Functional Groups*. John Wiley & Sons Ltd, New York, **1983**; Vol. 1, pp. 571.

2 Schlögl, R., Noack, K., Zbinden, H., Reller, A., *Helv. Chim. Acta* **1987**, *70*, 627.

3 Molnár, A., Sárkány, A., Varga, M., *J. Mol. Catal. A* **2001**, *173*, 185.

4 Ulan, J.G., Maier, W.F., Smith, D.A., *J. Org. Chem.* **1987**, *52*, 3132.

5 Sodeoka, M., Shibasaki, M., *J. Org. Chem.* **1985**, *50*, 1147.

6 Sodeoka, M., Shibasaki, M., *Synthesis* **1993**, 643.

7 Vasil'ev, A.A., Serebryakov, E.P., *Russ. Chem. Bull.* **2002**, *51*, 1341.

8 Bianchini, C., Meli, A., Peruzzini, M., Frediani, P., Bohanna, C., Esteruelas, M.A., Oro, L.A., *Organometallics* **1992**, *11*, 138.

9 Bianchini, C., Meli, A., Peruzzini, M., Vizza, F., Zanobini, F., *Organometallics* **1989**, *8*, 2080.

10 Esteruelas, M.A., Oro, L.A., *Chem. Rev.* **1998**, *98*, 577.

11 Bianchini, C., Bohanna, C., Esteruelas, M.A., Frediani, P., Meli, A., Oro, L.A., Peruzzini, M., *Organometallics* **1992**, *11*, 3837.

12 Eckert, J., Albinati, A., White, R.P., Bianchini, C., Peruzzini, M., *Inorg. Chem.* **1992**, *31*, 4241.

13 Cabeza, J.A., in: Braunstein, P., Oro, L.A., Raithby, P.R. (Eds.), *Homogeneous Catalysis with Ruthenium Carbonyl Cluster Complexes: Metal Clusters in Chemistry*, Vol. 2. Wiley-VCH GmbH, Weinheim, **1999**, pp. 715.

14 Albers, M.O., Singleton, E., Viney, M.M., *J. Mol. Cat.* **1985**, *30*, 213.

15 Nkosi, B.S., Coville, N.J., Albers, M.O., Singleton, E., *J. Mol. Cat.* **1987**, *39*, 313.

16 Lough, A.J., Morris, R.H., Ricciuto, L., Schleis, T., *Inorg. Chim. Acta* **1998**, *270*, 238.

17 Schleyer, D., Niessen, H.G., Bargon, J., *New J. Chem.* **2001**, *25*, 423.

18 Andriollo, A., Esteruelas, M.A., Meyer, U., Oro, L.A., Sanchez-Delgado, R.A., Sola, E., Valero, C., Werner, H., *J. Am. Chem. Soc.* **1989**, *111*, 7431.

19 Werner, H., Meyer, U., Esteruelas, M.A., Sola, E., Oro, L.A., *J. Organomet. Chem.* **1989**, *366*, 187.

20 Espuelas, J., Esteruelas, M.A., Lahoz, F.J., Oro, L.A., Valero, C., *Organometallics* **1993**, *12*, 663.

21 Schrock, R.R., Osborn, J.A., *J. Am. Chem. Soc.* **1976**, *98*, 2143.

22 Dickson, R.D., in: *Homogeneous Catalysis with Compounds of Rhodium and Iridium, Catalysis by Metal Complexes*. D. Reidel Publishing Co., Dordrecht, **1985**, pp. 40.

23 de Wolf, E., Spek, A.L., Kuipers, B.W.M., Philipse, A.P., Meeldijk, J.D., Bomans, P.H.H., Frederik, P.M., Deelman, B.-J., van Koten, G., *Tetrahedron* **2002**, *58*, 3911.

24 Chaloner, P.A., Esteruelas, M.A., Joó, F., Oro, L.A., in: *Homogeneous Hydrogenation*. Kluwer Academic Publishers, Dordrecht, **1993**.

25 Spencer, A., *J. Organomet. Chem.* **1975**, *93*, 389.

26 Burk, M.J., Gerlach, A., Semmeril, D., *J. Org. Chem.* **2000**, *65*, 8933.

27 Crabtree, R. H., Gautier, A., Giordano, G., Khan, T., *J. Organomet. Chem.* **1977**, *141*, 113.

28 Usón, R., Oro, L. A., Sariego, R., Valderrama, M., Rebullida, C., *J. Organomet. Chem.* **1980**, *197*, 87.

29 Esteruelas, M. A., González, I., Herrero, J., Oro, L. A., *J. Organomet. Chem.* **1998**, *551*, 49.

30 Esteruelas, M. A., López, A. M., Oro, L. A., Pérez, A., Schulz, M., Werner, H., *Organometallics* **1993**, *12*, 1823.

31 Chen, W., Esteruelas, M. A., Herrero, J., Lahoz, F. J., Martín, M., Oñate, E., Oro, L. A., *Organometallics* **1997**, *16*, 6010.

32 Pelagatti, P., Venturini, A., Leporati, A., Carcelli, M., Costa, M., Bacchi, A., Pelizzi, G., Pelizzi, C., *J. Chem. Soc., Dalton Trans.* **1998**, 2715.

33 Costa, M., Pelagatti, P., Pelizzi, C., Rogolino, D., *J. Mol. Catal. A* **2002**, *178*, 21.

34 van Laren, M. W., Duin, M. A., Klerk, C., Naglia, M., Rogolino, D., Pelagatti, P., Bacchi, A., Pelizzi, C., Elsevier, C. J., *Organometallics* **2002**, *21*, 1546.

35 van Laren, M. W., Elsevier, C. J., *Angew. Chem., Int. Ed.* **1999**, *38*, 3715.

36 Sulman, E., Deibele, C., Bargon, J., *React. Kinet. Catal. Lett.* **1999**, *67*, 117.

37 Choudary, B. M., Sharma, G. V. M., Bharathi, P., *Angew. Chem.* **1989**, *101*, 506.

38 Elman, B., Moberg, C., *J. Organomet. Chem.* **1985**, *294*, 117.

39 Ferrari, C., Predieri, G., Tiripicchio, A., Costa, M., *Chem. Mater.* **1992**, *4*, 243.

40 Holy, N. L., Shelton, S. R., *Tetrahedron* **1981**, *37*, 25.

41 Islam, M., Bose, A., Mal, D., Saha, C. R., *J. Chem. Res., Synop.* **1998**, 44.

42 Moberg, C., Rakos, L., *J. Organomet. Chem.* **1987**, *335*, 125.

43 Sobczak, J. W., Lesiak, B., Jablonski, A., Kosinski, A., Palczewska, W., *Pol. J. Chem.* **1995**, *69*, 1732.

44 Stern, E. W., Maples, P. K., *J. Catal.* **1972**, *27*, 120.

45 Evrard, D., Groison, K., Mugnier, Y., Harvey, P. D., *Inorg. Chem.* **2004**, *43*, 790.

46 Niessen, H. G., Eichhorn, A., Woelk, K., Bargon, J., *J. Mol. Catal. A* **2002**, *182*, 463.

47 van Asselt, R., Elsevier, C. J., *J. Mol Catal. A* **1991**, *65*, L13.

48 Elsevier, C. J., *Coord. Chem. Rev.* **1999**, *185*, 809.

49 Kluwer, A. M., Koblenz, T. S., Jonischkeit, T., Woelk, K., Elsevier, C. J., *J. Am. Chem. Soc.* **2005**, *127*, 15470.

50 Bacchi, A., Carcelli, M., Costa, M., Leporati, A., Leporati, E., Pelagatti, P., Pelizzi, C., Pelizzi, G., *J. Organomet. Chem.* **1997**, *535*, 107.

51 Pelagatti, P., Bacchi, A., Carcelli, M., Costa, M., Fochi, A., Ghidini, P., Leporati, E., Masi, M., Pelizzi, C., Pelizzi, G., *J. Organomet. Chem.* **1999**, *583*, 94.

52 Costa, M., Pelagatti, P., Pelizzi, C., Rogolino, D., *J. Mol. Catal. A* **2002**, *178*, 21.

53 Raoult, Y., Choukroun, R., Basso-Bert, M., Gervais, D., *J. Mol. Cat.* **1992**, *72*, 47.

54 Raoult, Y., Choukroun, R., Blandy, C., *Organometallics* **1992**, *11*, 2443.

55 Raoult, Y., Choukroun, R., Gervais, D., *J. Organomet. Chem.* **1990**, *399*, C1.

56 Cais, M., Frankel, E. N., Rejoan, A., *Tetrahedron Lett.* **1968**, *9*, 1919.

57 Frankel, E. N., Selke, E., Grass, C. A., *J. Am. Chem. Soc.* **1968**, *90*, 2446.

58 Yagupsky, G., Cais, M., *Inorg. Chim. Acta* **1975**, *12*, L27.

59 Wrighton, M. S., Schroeder, M. A., *J. Am. Chem. Soc.* **1973**, *95*, 5764.

60 Schroeder, M. A., Wrighton, M. S., *J. Organomet. Chem.* **1974**, *74*, C29.

61 Chandiran, T., Vancheesan, S., *J. Mol. Cat.* **1992**, *71*, 291.

62 Chandiran, T., Vancheesan, S., *J. Mol Cat.* **1994**, *88*, 31.

63 Frankel, E. N., Butterfield, R. O., *J. Org. Chem.* **1969**, *34*, 3930.

64 Airoldi, M., Deganello, G., Dia, G., Gennaro, G., *Inorg. Chem. Acta* **1983**, *68*, 179.

65 Airoldi, M., Deganello, G., Dia, G., Gennaro, G., *J. Organomet. Chem.* **1980**, *187*, 391.

66 Steines, S., Engelert, U., Drießen-Hölscher, B., *Chem. Commun.* **2000**, 217.

67 Steines, S., Wasserscheid, P., Drießen-Hölscher, B., *J. Prakt. Chem.* **2000**, *342*, 348.

68 Niessen, H. G., Schleyer, D., Wiemann, S., Bargon, J., Steines, S., Drießen-Hölscher, B., *Magn. Reson. Chem.* **2000**, *38*, 747.

69 Funabiki, T., Matsumoto, M., Tarama, K., *Bull. Chem. Soc. Jpn.* **1972**, *45*, 2723.

70 Funabiki, T., Kasaoka, S., Matsumoto, M., Tarama, K., *J. Chem. Soc., Dalton Trans.* **1974**, 2043.

71 Reger, D. L., Habib, M. M., Fauth, D. J., *J. Org. Chem.* **1980**, *45*, 3860.

72 Lee, J. T., Alper, H., *J. Org. Chem.* **1990**, *55*, 1854.

73 Reger, D. L., Habib, M. M., *J. Mol. Cat.* **1978**, *4*, 315.

74 Reger, D. L., Habib, M. M., *J. Mol. Cat.* **1980**, *7*, 365.

75 Reger, D. L., Gabrielli, A., *J. Mol. Cat.* **1981**, *12*, 173.

76 Schrock, R. R., Osborn, J. A., *J. Am. Chem. Soc.* **1976**, *98*, 4450.

77 Heldal, J. A., Frankel, E. N., *J. Am. Oil Chem. Soc.* **1985**, *62*, 1117.

78 Aimé, S., Canet, D., Dastrù, W., Gobetto, R., Reineri, F., Viale, A., *J. Phys. Chem. A* **2001**, *105*, 6305.

79 Drexler, H. J., Baumann, W., Spannenberg, A., Fischer, C., Heller, D., *J. Organomet. Chem.* **2001**, *621*, 89.

80 Cobley, C. J., Lennon, I. C., McCague, R., Ramsden, J. A., Zanotti-Gerosa, A., in: H. U. Blaser, E. Schmidt (Eds.), *Asymmetric Catalysis on Industrial Scale.* Wiley-VCH, Weinheim, Germany, **2004**, p. 269.

81 Schrock, R. R., Osborn, J. A., *J. Am. Chem. Soc.* **1976**, *98*, 2134.

82 Bargon, J., Kandels, J., Kating, P., Thomas, A., Woelk, K., *Tetrahedron Lett.* **1990**, *31*, 5721.

83 Burk, M. J., Allen, J. G., Kiesman, W. F., *J. Am. Chem. Soc.* **1998**, *120*, 657.

84 Speziali, M. G., Moura, F. C. C., Robles-Dutenhefner, P. A., Araujo, M. H., Gusevskaya, E. V., dos Santos, E. N., *J. Mol. Cat. A* **2005**, *239*, 10.

85 Naiini, A. A., Lai, C. K., Ward, D. L., Brubaker, C. H., Jr., *J. Organomet. Chem.* **1990**, *390*, 73.

86 Lai, C. K., Naiini, A. A., Brubaker, C. H., Jr., *Inorg. Chim. Acta* **1989**, *164*, 205.

87 Okoroafor, M. O., Shen, L. H., Honeychuck, R. V., Brubaker, C. H., Jr., *Organometallics* **1988**, *7*, 1297.

88 Strukul, G., Carturan, G., *Inorg. Chim. Acta* **1979**, *35*, 99.

15

Homogeneous Hydrogenation of Aldehydes, Ketones, Imines and Carboxylic Acid Derivatives: Chemoselectivity and Catalytic Activity

Matthew L. Clarke and Geoffrey J. Roff

15.1
Introduction

The reduction of aldehydes, ketones, esters, acids, and anhydrides to alcohols is one of the most fundamental and widely employed reactions in synthetic chemistry. Sodium borohydride, lithium aluminum hydride and other stoichiometric reducing agents are often perfectly adequate reagents for laboratory-scale syntheses. In an industrial setting, however, the increased demands for atom economy, cleaner synthesis and straightforward work-up procedures make the use of these reagents disadvantageous. Reduction procedures that make use of molecular hydrogen show better ecology, are more cost-effective, and are potentially easier to operate than those that require the clean-up of boron or aluminum waste at the end of the reaction. The hydrogenation of C=O (and C=N) bonds is therefore the preferred method for their reduction.

Heterogeneous catalysts such as Pd/C and Pt/C are widely used for this purpose, and often represent the most economical method to carry out these reductions. However, in cases where milder conditions, functional group tolerance and chemoselectivity are required, heterogeneous catalysts can be unsuitable for the task. There has therefore been a substantial research effort aimed towards developing homogeneous catalysts for this purpose.

This chapter aims to provide an overview of the current state of the art in homogeneous catalytic hydrogenation of C=O and C=N bonds. Diastereoselective or enantioselective processes are discussed elsewhere. The chapter is divided into sections detailing the hydrogenation of aldehydes, the hydrogenation of ketones, domino-hydroformylation-reduction, reductive amination, domino hydroformylation-reductive amination, and ester, acid and anhydride hydrogenation.

The Handbook of Homogeneous Hydrogenation.
Edited by J. G. de Vries and C. J. Elsevier
Copyright © 2007 WILEY-VCH Verlag GmbH & Co. KGaA, Weinheim
ISBN: 978-3-527-31161-3

15.2
Hydrogenation of Aldehydes

15.2.1
Iridium Catalysts

The first report of a catalytic system for the effective homogeneous hydrogenation of an aldehyde to an alcohol was published during the late 1960s [1]. Coffey reported that the use of a catalyst prepared *in situ* by the reaction of $[Ir(H)_3(PPh_3)_3]$ with acetic acid was effective for the hydrogenation of n-butyraldehyde to n-butanol at 50 °C and at 1 bar (Scheme 15.1). The reaction was found to be first order in both substrate and catalyst concentration, and to be highly dependent upon the solvent. No hydrogenation occurred in undiluted aldehyde or in toluene, but the addition of acetic acid initiated gas uptake. The active catalytic species was thought to be $[Ir(H)_2(CH_3COO)(PPh_3)_3]$.

This catalytic system was further studied by Strohmeier and Steigerwald, who performed reactions at 10 bar without solvent to achieve hydrogenation of a series of aldehydes (Table 15.1) [2]. Turnover numbers (TON) of up to 8000 were achieved in the case of the hydrogenation of benzaldehyde. The chemoselectivity of this catalyst towards carbonyl hydrogenation over alkene hydrogenation was

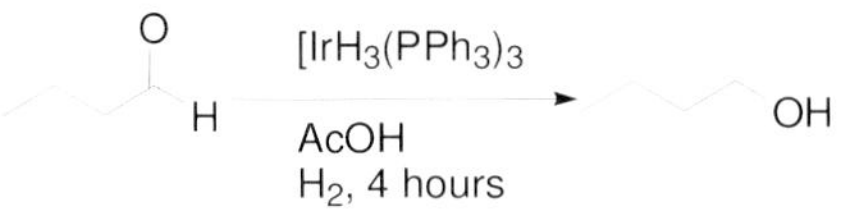

Scheme 15.1 Hydrogenation of n-butyraldehyde.

Table 15.1 Hydrogenation of aldehydes with $[IrH_3(PPh_3)_3]$ in acetic acid.

Substrate	Catalyst [mol.%]	Temperature [°C]	Yield [%]	TON	TOF [h^{-1}]
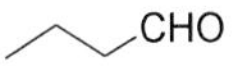 CHO	0.022	80	73	3280	492
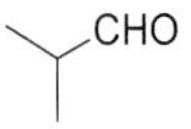 CHO	0.023	110	82	3540	177
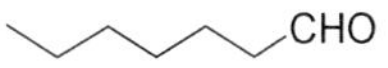 CHO	0.032	90	64	2000	89
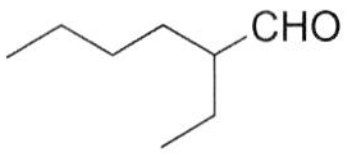 CHO	0.039	110	80	2030	81
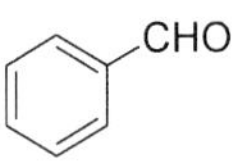 CHO	0.013	110	98	7780	259

examined for α,β-unsaturated aldehydes (Scheme 15.2). Using the $[IrH_3(PPh_3)_3]$ complex in acetic acid for the hydrogenation of crotonaldehyde resulted in the formation of the saturated alcohol (Scheme 15.3). It was also noted that this catalyst did not allow for ketone hydrogenation at 10 bar.

Other attempts to use iridium PPh_3 complexes such as $[IrCl(PPh_3)_3]$, $[IrCl(CO)(PPh_3)_2]$, $[Ir(ClO_4)(CO)(PPh_3)_2]$ and $[Ir(CO)(PPh_3)_3]ClO_4$ to hydrogenate unsaturated aldehydes did not yield great results [3], mainly because these catalysts suffered from low activity and selectivity.

The catalytic system of $[Ir(COD)Cl]_2$ with an excess of the bulky phosphine $P(o\text{-MeOPh})_3$ under transfer hydrogenation conditions of propan-2-ol and KOH was used successfully in the selective hydrogenation of cinnamaldehyde (Scheme 15.4) [4]. Selectivity and activity were found to increase with increasing P/Ir ratios, and complete conversion was achieved in as little as 5 minutes (turnover frequency (TOF) $\sim 6000\ h^{-1}$).

Scheme 15.2 Distribution of products in the hydrogenation of α,β-unsaturated aldehydes.

$$\text{[IrH}_3\text{(PPh}_3\text{)}_3\text{], AcOH, 110°C, 27 hours, H}_2\text{ (10 atm)}$$

82% + 4% + 4%

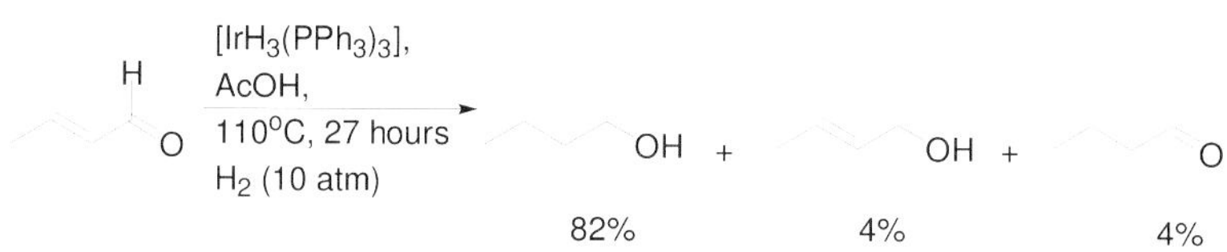

Scheme 15.3 Hydrogenation of crotonaldehyde with $[IrH_3(PPh_3)_3]$ in acetic acid.

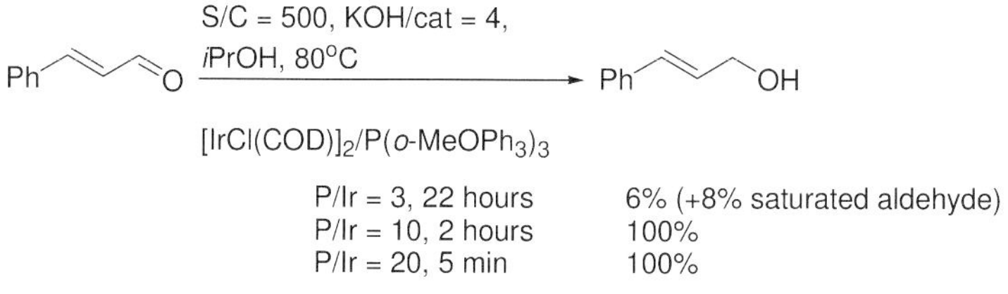

S/C = 500, KOH/cat = 4,
iPrOH, 80°C

$[IrCl(COD)]_2/P(o\text{-MeOPh}_3)_3$

P/Ir = 3, 22 hours	6% (+8% saturated aldehyde)
P/Ir = 10, 2 hours	100%
P/Ir = 20, 5 min	100%

Scheme 15.4 Transfer hydrogenation of cinnamaldehyde.

Using molecular hydrogen as the reducing agent, $[Ir(COD)(OCH_3)]_2$ with an excess of tertiary phosphine was better than $[Ir(COD)Cl]_2$ for the selective hydrogenation of cinnamaldehyde [5]. In these studies, a great dependence on solvent and ligand was reported. A variety of different phosphines, which were markedly different in their steric and electronic properties, were examined in this reaction. In propan-2-ol the most effective phosphine was PCy_2Ph which gave 94% yield (TOF 235 h^{-1}) of the unsaturated alcohol in a 2 h reaction under 30 bar H_2 at 100 °C. Phosphines such as $PCyPh_2$, $PPhPr^i_2$, PPh_2Pr^i and $PEtPh_2$ were also effective in giving over 95% selectivity. The less-effective phosphines were PEt_2Ph, $PMePh_2$, PBu^i_3 and PMe_2Ph. Reactions that were performed in toluene were generally less effective.

More recent advances in iridium-catalyzed aldehyde hydrogenation have been through the use of bidentate ligands [6]. In the hydrogenation of citral and cinnamaldehyde, replacing two triphenylphosphines in $[IrH(CO)(PPh_3)_3]$ with bidentate phosphines BDNA, BDPX, BPPB, BISBI and PCP (Fig. 15.1) led to an increase in catalytic activity.

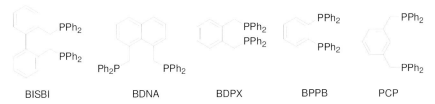

Fig. 15.1 Bidentate ligands employed in Ir-catalyzed hydrogenation.

Table 15.2 Bidentate ligands used for the hydrogenation of citral and cinnamaldehyde under 50 bar H_2 at 100 °C.

Complex	Substrate	Conversion [%]	Selectivity [%][a]	TOF [h^{-1}][b]
$[IrH(CO)(PPh_3)_3]$	Citral	3.1	70.5	7
	Cinnamaldehyde	11.4	35.0	61
$[IrH(CO)(PPh_3)(BPPB)]$	Citral	7.7	61.2	18
	Cinnamaldehyde	27.3	19.3	146
$[IrH(CO)(PPh_3)(BISBI)]$	Citral	11.6	92.2	28
	Cinnamaldehyde	44.6	13.1	238
$[IrH(CO)(PPh_3)(BDNA)]$	Citral	19.3	95.8	46
	Cinnamaldehyde	20.6	77.4	110
$[IrH(CO)(PPh_3)(BDPX)]$	Citral	58.8	96.4	141
	Cinnamaldehyde	58.1	9.0	310
$[IrHCl(CO)(PCP)]$	Citral	13.9	42.5	33
	Cinnamaldehyde	52.0	1.3	277

a) Selectivity of allylic alcohol formed as a percentage of total hydrogenation products.
b) TOF (h^{-1}) expressed for conversion of starting material.

Fig. 15.2 Citral.

Even though the activity was better than the PPh$_3$ analogue (Table 15.2), the conversion was still low. Of the ligands tested, BDPX showed the greatest promise in selectivity for citral (Fig. 15.2).

The selectivities in forming cinnamyl alcohol from cinnamaldehyde using these catalysts were poor, and generally resulted in the formation of the saturated aldehyde. This could be overcome by the use of a large excess of phosphine, though at the expense of yield. The same group have demonstrated that ruthenium analogues of the BDNA complex are more active and selective [7].

15.2.2
Rhodium Catalysts

Wilkinson's catalyst [RhCl(PPh$_3$)$_3$] [8], a convenient catalyst for the hydrogenation of olefins, was found to be deactivated by aldehydes to give the catalytically non-active complex [RhCl(CO)(PPh$_3$)$_2$] as a result of the competing decarbonylation reaction. Despite the lack of activity of this catalyst, extensive investigations have been made into rhodium catalysis for aldehyde hydrogenation, and these have led to the development of some highly efficient catalysts.

15.2.2.1 Rh-amine Catalysts
The first report of rhodium catalysts for aldehyde reduction came from Marko who reported the use of RhCl$_3$ · 3H$_2$O under hydroformylation conditions [9]. It was suggested that the active species were rhodium carbonyls, and the catalyst system was successfully utilized in the hydrogenation of ethanal, propanal, and benzaldehyde.

In the presence of strongly basic amines, RhCl$_3$ · 3H$_2$O was effective in catalyzing the hydrogenation of cinnamaldehyde [10]. In the absence of carbon monoxide or triethylamine, however, only small amounts of hydrogenated products were obtained. Under hydroformylation conditions with increasing concentrations of triethylamine, catalytic activity and selectivity to cinnamyl alcohol were increased. The effect of the amines was found to be very important. Primary or secondary amines were ineffective in producing hydrogenation products. Strongly basic tertiary amines such as triethylamine and *N*-methylpyrolidine were more effective for activity and selectivity. The addition of triphenylphosphine increased hydrogenation of the carbon–carbon double bond, giving dihydrocinnamaldehyde. The activity of RhCl$_3$ · 3H$_2$O at lower temperatures can be increased by the pretreatment with CO giving RhCl$_2$(CO)$_4$ and allowing hydrogenations to occur at 60 °C with up to 94% yield and 85% selectivity in a 1-h reaction (TOF = 289 h^{-1}).

The rhodium carbonyl cluster [Rh$_6$(CO)$_{16}$], in combination with the diamine *N,N,N',N*-tetramethyl-1,3-propanediamine is an effective catalytic system for the hydrogenation of saturated and unsaturated aldehydes in water under a pressure of carbon monoxide and hydrogen [11]. In reactions lasting only a few hours, and using a substrate catalyst ratio of 300, simple aldehydes are converted in quantitative yields. The unsaturated aldehydes take longer to react, but the selectivity favors the formation of unsaturated alcohols in high yields.

15.2.2.2 Cationic Rhodium Phosphine Catalysts

The effect of the phosphines has been further studied by the hydrogenation of aldehydes and ketones in the presence of the cationic species $[Rh(nbd)(PR_3)_2]ClO_4$ [12]. Both, triethylphosphine and trimethylphosphine complexes showed the greatest activity (triethylphosphine being preferred), whereas triphenylphosphine-based catalysts showed little or no activity and the diphosphine (dppe) complex inhibited the reaction completely. At 30 °C and under 1 bar H_2, the triethylphosphine catalyst could complete the hydrogenation of benzaldehyde in 24 h, whereas under the same conditions it could only manage 80% and 41% hydrogenation for phenylacetaldehyde and n-butyraldehyde, respectively. The presence of propane and propene in the reaction mixture of n-butyraldehyde hydrogenation suggests the occurrence of a certain degree of decarbonylation, which leads to deactivation of the catalyst.

An alternative air-stable cationic rhodium complex $[(COD)Rh(DiPFc)]OTf$ is an efficient catalyst precursor for the hydrogenation of aldehydes and ketones [13]. This is the first useful diphosphine-based catalyst, possibly due to backbone rigidity and strong electron-donating alkyl-substituted phosphorus atoms. Using this commercially available catalytic system, benzaldehyde can be converted to benzyl alcohol under mild conditions. Using a substrate:catalyst ratio (SCR) of 500:1, a quantitative yield was obtained in 3 h at 25 °C and under only 4 bar hydrogen (TOF ~165 h^{-1}). Unlike some alternative catalysts, there was no deactivation of the catalyst through decarbonylation, and a range of saturated aldehydes have been successfully hydrogenated in the presence of this catalyst (Fig. 15.3).

The hydrogenations were active either with the isolated catalyst or *in situ* generation of the catalyst by the reaction of DiPFc with $[(COD)_2Rh]OTf$ in methanol (Scheme 15.5). Both, the components and the isolated catalysts are available from Strem Chemicals.

The solid catalyst is stable to oxygen and moisture, showing no loss of activity when exposed to the atmosphere for several days. However, the catalyst reacts fairly rapidly with oxygen when in solution and this leads to catalyst deactivation, a problem which is easily overcome by simply degassing the reaction solvent.

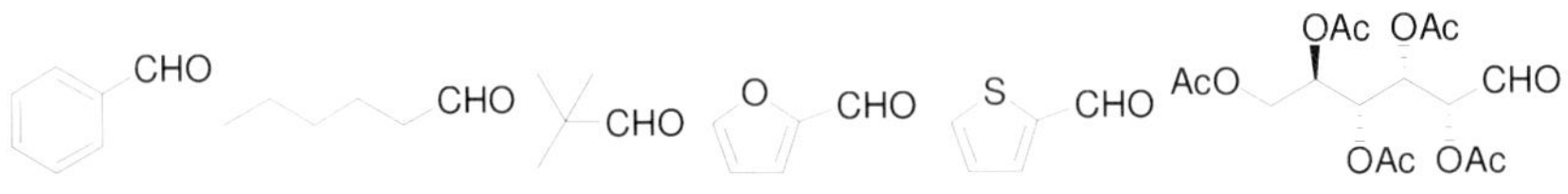

Fig. 15.3 Substrates hydrogenated using $[Rh(dippf)(COD)]^+OTf^-$.

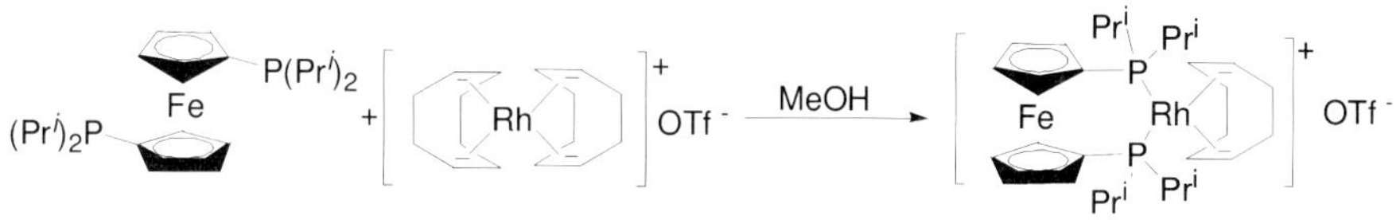

Scheme 15.5 Preparation of cationic DiPFc catalyst.

The Rh-DiPFc catalyst has recently been immobilized on modified alumina essentially to provide an immobilized homogeneous catalyst [14]. Although in this form it is strictly a heterogeneous catalyst, it provides the best of both worlds. These catalysts may be superior to the traditional heterogeneous catalysts in terms of reactivity and selectivity, but benefit over homogeneous catalysts in their ease of removal and re-use. Similar immobilizations have involved the binding of rhodium carbonyl clusters to polymers [15, 16]. Generally high selectivity was observed in the hydrogenation of a series of unsaturated aldehydes either under hydrogen and carbon monoxide or formic acid transfer hydrogenations.

15.2.2.3 Water-Soluble Rh Catalysts

Water-soluble complexes constitute an important class of rhodium catalysts as they permit hydrogenation using either molecular hydrogen or transfer hydrogenation with formic acid or propan-2-ol. The advantages of these catalysts are that they combine high reactivity and selectivity with an ability to perform the reactions in a biphasic system. This allows the product to be kept separate from the catalyst and allows for an ease of work-up and cost-effective catalyst recycling. The water-soluble Rh-TPPTS catalysts can easily be prepared *in situ* from the reaction of [RhCl(COD)]$_2$ with the sulfonated phosphine (Fig. 15.4) in water [17].

In the reduction of benzaldehyde performed in water in the presence of Na$_2$CO$_3$ and *i*-PrOH, the yields were generally very high. This system was highly effective in 2 h with complete conversion when H$_2$ was used as the hydrogen donor. Using *i*-PrOH as the hydrogen donor, yields were well above 90% even after recycling of the catalyst several times. Sodium formate could also provide efficient hydrogenation, with over 90% yield. Using iridium analogues resulted in very poor yields. Several other aldehydes were reduced with good yields using the transfer hydrogenation protocol (Fig. 15.5).

This catalyst is chemoselective in the reduction of α,β-unsaturated aldehydes, without any decarbonylation [18]. However, the resulting product was the saturated aldehyde. Generally, at pressures <20 bar H$_2$ and temperatures between 30 and 80 °C, selectivities exceeding 95% can be achieved in 1 h. Recycling posed no problem with successive runs, showing the same selectivity and activity.

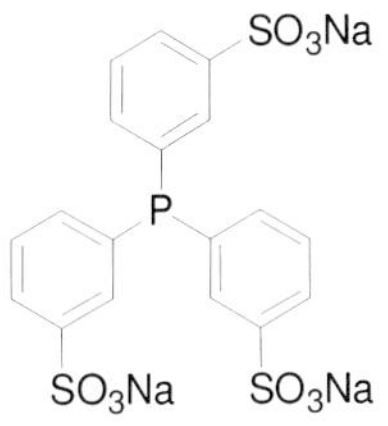

Fig. 15.4 TPPTS.

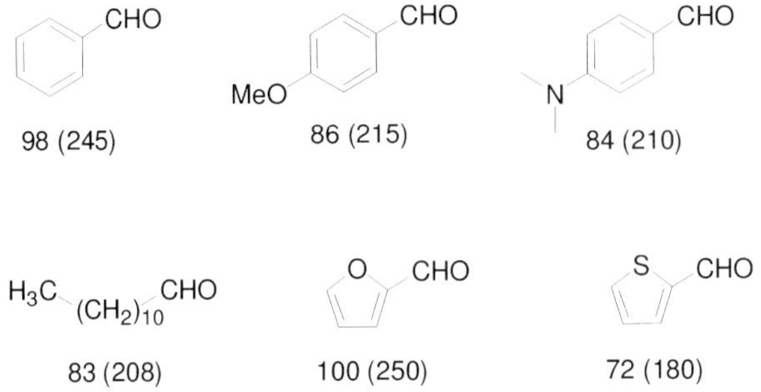

98 (245) 86 (215) 84 (210)

83 (208) 100 (250) 72 (180)

Fig. 15.5 Transfer hydrogenation of aldehydes using [Rh(COD)Cl]$_2$/TPPTS at 80 °C using *i*-PrOH as hydrogen donor. Values shown are yields (TOF, h^{-1}).

15.2.3
Ruthenium Catalysts

15.2.3.1 **Ru-PPh$_3$ Catalysts**

The most common carbonyl hydrogenation catalysts are derived from ruthenium species. In early studies, these were generally based on the phosphine-coordinated ruthenium carbonyls that are more commonly used for hydroformylation reactions. Thus, the hydroformylation catalyst [Ru(CO)$_3$(PPh$_3$)$_2$] was shown to be effective in the hydrogenation of propionaldehyde under 20 bar H$_2$ and at 120 °C [19]. Increasing the temperature and pressure led to an increase in reaction time. Tsuji and Suzuki used the complex [RuCl$_2$(PPh$_3$)$_3$] to hydrogenate a series of aliphatic and aromatic aldehydes [20]. Under 10 bar H$_2$ the reactions were found to be sluggish at room temperatures, but proceeded smoothly above 70 °C. Hydrogenation of aldehydes in the presence of ketones showed selectivity exclusively for the aldehydes. Benzaldehyde was also exclusively reduced in the presence of nitrobenzene, a substrate which is known to be reduced to aniline under harsh conditions by this catalyst [21]. Strohmeier and Weigelt used the catalyst [RuCl$_2$(CO)$_2$(PPh$_3$)$_2$] to hydrogenate a series of aldehydes at 15 bar H$_2$ and at 160–180 °C, with generally high yield and turnover numbers (Table 15.3) [22]. Although these are amongst the highest turnover numbers reported for aldehyde hydrogenation, the reactions were carried out at relatively high temperatures.

Table 15.3 Hydrogenation of simple aldehydes using RuCl$_2$(CO)$_2$(PPh$_3$)$_2$.

Substrate	Catalyst [mol%]	Temperature [C]	Time [h]	Yield [%]	TON
(linear aldehyde, CHO)	0.0083	180	4	90	10 800
(branched aldehyde, CHO)	0.0033	160	11	98	29 400
(branched aldehyde, CHO)	0.0017	160	12	99	59 400
(benzaldehyde, CHO)	0.0017	180	14	93	56 000

Sanchez-Delgado and De Ochoa achieved excellent conversion of linear aldehydes by introducing chloride ligands [23]. The catalyst precursors [RuHCl(CO)(PPh$_3$)$_3$], [RuHCl(PPh$_3$)$_3$] and [RuCl$_2$(PPh$_3$)$_3$] were used to reduce both aliphatic and aromatic aldehydes, although benzaldehyde reduction was less efficient than with the previously mentioned [RuCl$_2$(CO)$_2$(PPh$_3$)$_2$] catalyst. Although [RuHCl(PPh$_3$)$_3$] was found to be the more active catalyst, it required inert conditions and promoted decarbonylation of the aldehyde. Evidence of this comes from the presence of metal carbonyl species at the end of reaction. Having carbonyl ligands appears to solve this problem, and [RuHCl(CO)(PPh$_3$)$_3$] was found to be the most convenient catalyst. Using propionaldehyde with a SCR of 50 000, turnover numbers of up to 32 000 were achieved after 50 h of continuous reaction at 140 °C under 30 bar H$_2$ [24]. Using this same catalyst in the reduction of crotonaldehyde, the favored product was the fully saturated alcohol [25].

Hotta, using [RuHCl(PPh$_3$)$_3$] and HCl, allowed for highly selective reduction of citral [26]. Using 2.5 mol% [RuHCl(PPh$_3$)$_3$] in toluene under 50 bar H$_2$ at 30 °C, the selectivity achieved was 66%. The addition of 12.5% HCl, and performing the reaction in toluene:ethanol (27:3) further increased selectivity to 98%, with 99% conversion. The desirable mild conditions were offset by the relatively low turnover numbers.

15.2.3.2 Polydentate Ru Catalysts

The use of polydentate ligands is rare for aldehyde hydrogenation. The ruthenium complex [RuCl$_2$(TRIPHOS)] (TRIPHOS = PhP(CH$_2$CH$_2$PPh$_2$)$_2$ catalyzes the hydrogenation and isomerization of alkenes, as well as the hydrogenation of aldehydes, ketones, and nitriles [27]. For simple aldehydes such as n-propanol, n-butanol, and n-hexanol, reasonable conversions can be achieved in 2 h under 34 bar H$_2$ at 100 °C, with turnover numbers of around 1000. In the hydrogenation of crotonaldehyde and cinnamaldehyde, it is the olefinic bond that is reduced favorably, although some unsaturated alcohol is also produced.

Table 15.4 Ru-BDNA-catalyzed hydrogenations of citral and cinnamaldehyde.

Catalyst	Substrate	Conversion [%][a]	Selectivity [%][b]	TOF [h^{-1}][c]
[RuCl$_2$(PPh$_3$)$_3$]	Citral	35	36	141
	Cinnamaldehyde	40	62	212
[RuHCl(CO)(PPh$_3$)(BDNA)]	Citral	93	96	371
	Cinnamaldehyde	86	94	461
[RuH$_2$(CO)(PPh$_3$)(BDNA)]	Citral	64	>99	255
	Cinnamaldehyde	62	95	328

a) 50 bar H$_2$, 70–80 °C, toluene, SCR 1200 (citral) or 1600 (cinnamaldehyde).
b) Selectivity of allylic alcohol formed as a percentage of total hydrogenation products.
c) TOF expressed for conversion of starting material.

The bidentate ligand BDNA shows good conversions and selectivity in the hydrogenation of citral and cinnamaldehyde [7]. These crystalline complexes are easily prepared by the replacement of triphenylphosphines in several Ru–PPh$_3$ complexes with BDNA by refluxing for several hours in toluene. In comparison with [RuCl$_2$(PPh$_3$)$_3$], the most promising complexes were [RuHCl(CO)(PPh$_3$)(BDNA)] and [RuH$_2$(CO)(PPh$_3$)(BDNA)] (Table 15.4).

Another bidentate ligand which shows improvements over triphenylphosphine is that of BISBI [28]. The complex [RuCl$_2$(PPh$_3$)(BISBI)] shows selectivities over 80%, but the yields are only 40–50% [28]. The analogous iridium complexes are less active, but show similar selectivity [6].

15.2.3.3 Diamine-Modified Ru Catalysts

RuCl$_2$(PPh$_3$)$_3$ has been shown to catalyze the reduction of several aldehydes, but does not have widespread scope. This catalyst is not chemoselective and, in the presence of alkenes, would favor olefin reduction over that of the aldehyde. Noyori and coworkers showed that chemoselectivity is easily introduced by the addition of ethylene-diamine as a ligand (Scheme 15.6) [29, 30]. This system requires the presence of co-catalytic KOH/*i*-PrOH as an activator.

Using an easily prepared stock solution of [RuCl$_2$(PPh$_3$)$_3$]/NH$_2$(CH$_2$)$_2$NH$_2$ and KOH in *i*-PrOH, unsaturated aldehydes are quantitatively reduced exclusively to unsaturated alcohols (Scheme 15.7).

Direct comparisons of the diamine system against the parent complex led to the conclusion that the effect of the diamine and KOH/*i*-PrOH activator decelerate olefin hydrogenation and in turn accelerate carbonyl hydrogenation. In the published report, there were no attempts to optimize turnover numbers or TOF for aldehyde hydrogenation. However, the catalyst has been shown to hydrogenate ketones with a SCR of 10000 at room temperature, which suggests that these catalysts represent the current state of the art in terms of activity and selectivity.

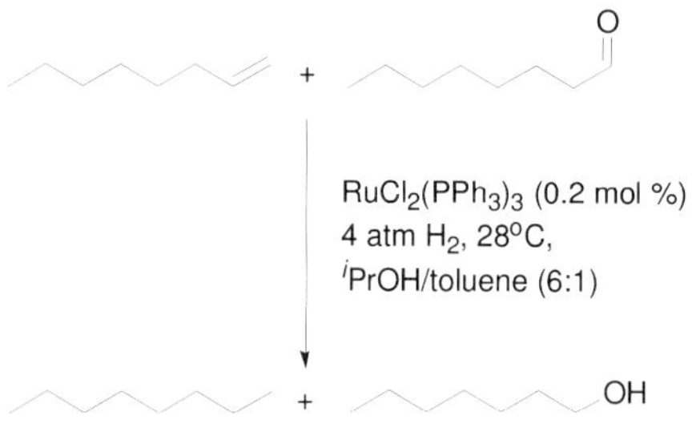

Scheme 15.6 Direct comparison of aldehyde and alkene hydrogenation.

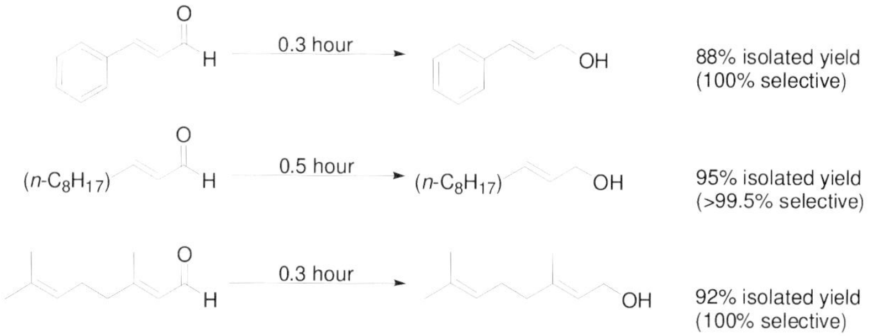

No additive, 150 mins	250	1
H$_2$N(CH$_2$)$_2$NH$_2$ + KOH, 10 Mins (Ru:en:KOH = 1:1:2)	1	1500

Conditions: 28°C in 6:1 propanol-toluene under 4 atm H$_2$; Substrate:Ru:en:KOH = 500:1:1:2

Scheme 15.7 Hydrogenation of unsaturated aldehydes using Noyori's system.

15.2.3.4 Ru-TPPMS/TPPTS Catalysts

It has been shown previously how water-soluble rhodium Rh-TPPTS catalysts allow for efficient aldehyde reduction, although chemoselectivity favors the olefinic bond in the case of unsaturated aldehydes [17]. The analogous ruthenium complex shows selectivity towards the unsaturated alcohol in the case of crotonaldehyde and cinnamaldehyde [31].

The biphasic reduction of 3-methyl-2-butenal under a pressure of hydrogen demonstrated a non-dependence on solvent (Table 15.5) [18]. Good conversions and selectivities were achieved in a selection of immiscible solvents in just over an hour, using a SCR of 200:1. No phase-transfer agents were needed as the slight solubility of the substrate in water ensured a rapid reaction. Under the same conditions, the catalyst was recycled three times with no loss of activity or selectivity: in fact, the reactions were faster than the initial reaction [18]. This was due to the initial run requiring an induction period for formation of the active catalyst. The analogous rhodium catalysts could cleanly reduce unsaturated aldehydes, but the high selectivity was towards the saturated aldehyde.

Table 15.5 Biphasic reduction of 3-methyl-2-butenal.

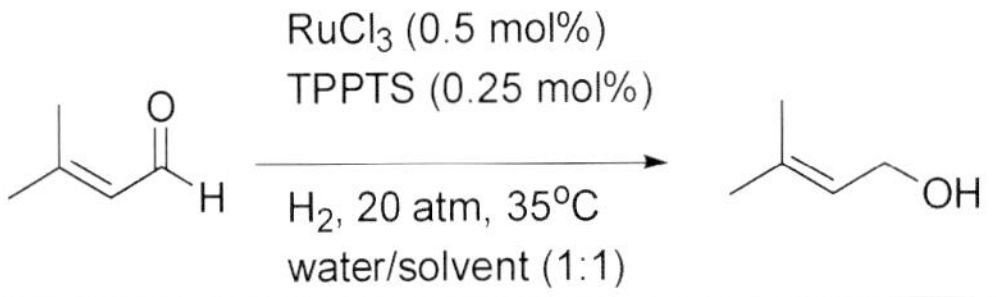

RuCl$_3$ (0.5 mol%)
TPPTS (0.25 mol%)
H$_2$, 20 atm, 35°C
water/solvent (1:1)

Solvent	Time [min]	Conversion [%]	Selectivity [%]
Cyclohexane	80	99	92
Chloroform	70	84	96
Ethyl acetate	75	93	96
Toluene (fresh catalyst)	60	100	96
Toluene (1st recycle)	30	99	97
Toluene (2nd recycle)	30	99	97

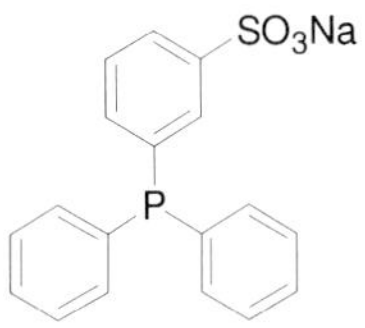

SO$_3$Na
P

Fig. 15.6 TPPMS.

Table 15.6 Hydrogenation of α,β-unsaturated aldehydes using [(PPh$_3$)CuH]$_6$ (5 mol% Cu) and PhPMe$_2$ (30 mol%) at room temperature.

Substrate	Pressure [bar]	Conversion [%]	Selectivity [%]	TOF [h^{-1}]
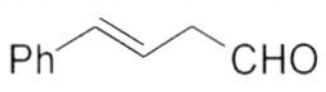 Ph⌒⌒CHO	5	94	97	5
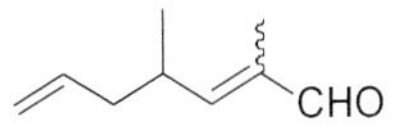 ⌒⌒⌒CHO	28	91	94	<1
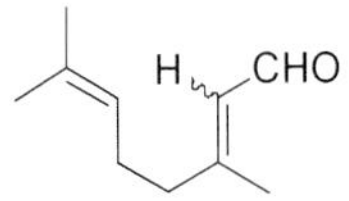 H⌒CHO	34	90	92	1
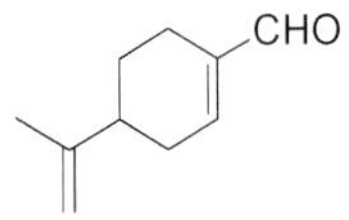 ⌒CHO	34	95	97	1

The ruthenium complex of the mono-sulfonated TPPMS (Fig. 15.6) is not only good for the transfer hydrogenation of simple substituted benzaldehydes with yields over 90% and with over 98% selectivity [32], but it is also chemo-selective in the transfer hydrogenation of α,β-unsaturated aldehydes without the aid of phase-transfer agents [33]. The RuCl$_2$/(TPPMS) catalyst was far more effective than either rhodium or iridium TPPMS catalysts [32]. The solution of the catalyst is air-stable in the presence of HCOO$^-$, and the reactions and work-ups are very simple. In a direct comparison of homogeneous and biphasic reductions of cinnamaldehyde using Ru-PPh$_3$ catalysts against Ru-TPPMS/TPPTS, the homogeneous Ru-PPh$_3$ systems were found to favor complete reduction of both the carbonyl and the olefinic bond. In contrast, if aqueous biphasic systems were employed, selectivity was restricted to the carbonyl bond [34].

15.2.4
Other Metal Catalysts

15.2.4.1 Copper
Phenyldimethylphosphine-stabilized copper(I) hydrides catalyze a highly chemo-selective hydrogenation of unsaturated aldehydes and ketones [35]. The reaction tolerates the use of either benzene or tetrahydrofuran (THF) as solvent, but requires a high concentration of *tert*-butanol as co-solvent to ensure high turnover and reaction homogeneity. Although high pressures are not required, they must exceed 1 bar in order to obtain complete conversion. In the reduction of α,β-unsaturated aldehydes using [(PPh$_3$)CuH]$_6$ and PhPMe$_2$, chemoselectivity was high, in most cases giving greater than 90% yields although the TOF was very low in all cases (Table 15.6). The minor byproducts were the saturated alcohols that arise from complete reduction.

As allylic alcohols are unaffected by use of this catalyst it is proposed that the complete reduction occurs through competitive conjugate reduction, followed by subsequent reduction of the carbonyl. Although this catalyst is slower in action and results in low turnover numbers compared to some catalysts, it is inexpensive and provides good selectivity at room temperature.

15.2.4.2 Osmium
The Osmium cluster Os$_3$(CO)$_{12}$ and clusters in the presence of various phosphines and triphenylphosphite have been utilized for the hydrogenation of cinnamaldehyde and crotonaldehyde (Table 15.7) [36]. The results show that good yields of unsaturated alcohols can be obtained by using a large excess of phosphine at elevated hydrogenation temperatures.

In such reactions, a temperature exceeding 130 °C has a dramatic effect on the catalytic activity. The pressure of hydrogen has a similar effect, with a large increase in activity above 30 bar. These catalysts did not exhibit the same selectivity for ketones. Osmium triphenylphosphine systems have been briefly exam-

Table 15.7 Hydrogenation of crotonaldehyde and cinnamaldehyde under 45 bar H_2 at 140 °C for 9 h.

Catalytic system	Substrate	Conversion [%]	Unsaturated alcohol [%]	Saturated aldehyde [%]	Saturated alcohol [%]
$[Os_3(CO)_{12}]$	Crotonaldehyde	18	13	5	0
	Cinnamaldehyde	15	7	6	2
$[Os_3(CO)_{12}]/P^nBu_3$	Crotonaldehyde	93	89	0	4
(15:1)	Cinnamaldehyde	97	86	0	11
$[Os_3(CO)_{12}]/PPh_3$	Crotonaldehyde	47	35	4	8
(15:1)	Cinnamaldehyde	98	91	1	6
$[Os_3(CO)_{12}]/P(OPh)_3$	Crotonaldehyde	28	9	0	19
(15:1)	Cinnamaldehyde	81	79	1	1

ined as potential catalysts for hydrogenation. However, in the reduction of crotonaldehyde, it is generally the unsaturated aldehyde which is produced [25].

The use of water-soluble ligands was referred to previously for both ruthenium and rhodium complexes. As in the case of ruthenium complexes, the use of an aqueous biphasic system leads to a clear enhancement of selectivity towards the unsaturated alcohol [34]. Among the series of systems tested, the most convenient catalysts were obtained from mixtures of $OsCl_3 \cdot 3H_2O$ with TPPMS (or better still TPPTS) as they are easily prepared and provide reasonable activities and modest selectivities. As with their ruthenium and rhodium analogues, the main advantage is the ease of catalyst recycling with no loss of activity or selectivity. However, the ruthenium-based catalysts are far superior.

15.3
Hydrogenation of Ketones

15.3.1
Iridium Catalysts

The cyclometallated iridium complex $[Ir(H)_2(P,C\text{-}Ph_2PC_6H_4N(Me)CH_2)(P,N\text{-}Ph_2PC_6H_4NMe_2)]$ (Fig. 15.7) is formed from the reaction of $[Ir(COD)(OMe)]_2$ with o-(diphenylphosphino)-N,N-dimethylaniline [37].

The product, although sensitive to light and air, was an effective catalyst for the transfer hydrogenation of several ketones in propan-2-ol. Unsaturated ketones were used with a SCR of 500:1, and mostly gave high selectivity and modest yields (Table 15.8).

This was the first example of catalytic chemoselective reduction of a,β-unsaturated ketones to allylic alcohols by hydrogen transfer and, unusually, did not require the use of a basic co-catalyst.

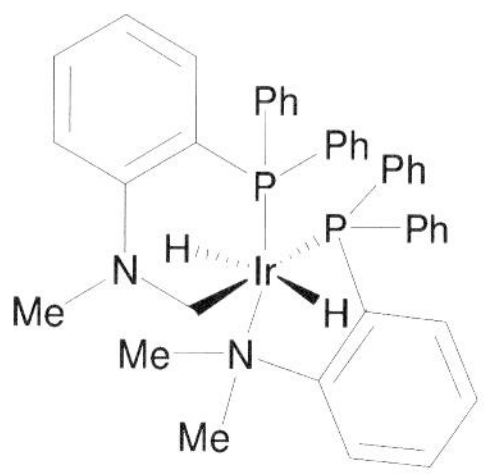

Fig. 15.7 [Ir(H)$_2$(P,C-Ph$_2$PC$_6$H$_4$N(Me)CH$_2$)(P,N-Ph$_2$PC$_6$H$_4$NMe$_2$)].

Table 15.8 Hydrogenation of unsaturated ketones in propan-2-ol at 83 °C (SCR = 500).

Substrate	Conversion [%]	Saturated ketone [%]	Saturated alcohol [%]	Unsaturated alcohol [%]	Selectivity [%]
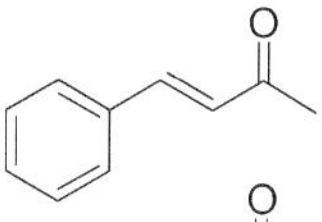	99 (1 h)	1	6	92	93
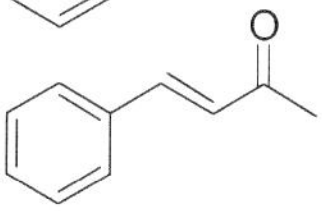	94 (1 h)	13	14	67	71
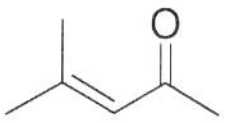	65 (7 h)	2	1	62	95
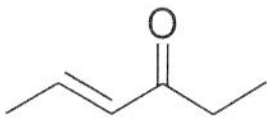	35 (7 h)	23	0	12	34

In the selective hydrogenation of benzylideneacetone (PhCH=CHCOMe) using an iridium phosphine system generated *in situ* from [Ir(COD)(OMe)]$_2$ and the appropriate phosphine [38], a heavy dependence was found on the nature and amount of phosphine used. Both of these factors are important in the activity and selectivity of the catalyst. Using PMe$_2$Ph as the model phosphine in a two-fold excess, the C=C bond was hydrogenated and the saturated ketone further hydrogenated to the saturated alcohol. However, increasing the excess of phosphine led to a switch in selectivity towards the carbonyl, although a loss of catalytic activity was reported. The cone angle of the phosphine is also important. Regardless of the solvent used, selectivity is raised above 90% when the cone angle is between 135 and 150°. The selectivity falls to zero at cone angles above and below this range.

These results suggest that, depending on the cone angle and relative concentration of the phosphines, different catalytic species are formed, and only catalysts formed from a large excess of relatively small phosphines are selective.

Generally, the selective reactions were complete in less than 24 h (SCR = 500, 30 bar H_2, 100 °C; TOF ~ 20 h^{-1}).

The mixed donor polydentate ligands Pr^n-$N(CH_2CH_2PPh_2)_2$ (PNP) and $Et_2NCH_2CH_2N(CH_2CH_2PPh_2)_2$ (P_2N_2) have been reacted with $[Ir(COD)(OMe)]_2$ to produce complexes that were active in the reduction of PhCH=CHCOMe [39]. Conversions of 90% with modest selectivity were achieved in 2–4 h in propan-2-ol at 83 °C. At 140 °C in cyclopentanol, similar results are obtained in less than 30 min.

15.3.2
Rhodium Catalysts

15.3.2.1 Rh-Phosphine Catalysts
The cationic species $[RhH_2(PPh_3)_2L_2]^+$ (L = solvent) has been used by Schrock and coworkers to catalyze the hydrogenation of alkenes, dienes and allynes [40]. These authors discovered that when the triphenylphosphine groups are replaced with more basic phosphines, ketones were reduced under mild conditions [41]. Using the $[RhH_2(PPhMe_2)_2L_2]^+X^-$ ($X^- = PF_6^-$ or ClO_4^-), acetone was reduced under atmospheric pressure of H_2 at 25 °C in the presence of 1% water. Under identical conditions, cyclohexanone, acetophenone and butan-2-one were also successfully reduced. Benzophenone was not hydrogenated, and it is thought that it may have formed a stable Rh complex. The addition of water was vital for activity, with the maximum rate achieved when 1% water is used. This addition of water also inhibited the reduction of alkenes. When the same catalysts were used for aldehyde reduction they proved to be effective initially, but their activity fell rapidly.

Rossi and coworkers successfully hydrogenated a series of simple ketones, with over 96% yields, using the complex $[Rh_2H_2Cl_2(COD)(PPh_2)_3]$ in the presence of a strong base [42].

$[Rh(bpy)_2]^+$, obtained by the *in-situ* reduction of $[Rh(bpy)_2Cl_2]Cl$ with hydrogen in methanolic sodium hydroxide [43], can reduce a series of simple ketones under 1 bar H_2 and at 30 °C [44]. Yields of over 98% were obtained in all cases with a SCR of up to 680:1. When a mixture of ketones and aldehydes was placed under such conditions, the ketones were found to be reduced preferentially, although unsaturated ketones were generally reduced to saturated ketones.

Although the complex $[RhCl(PPh_3)_3]$ is inactive towards the hydrogenation of ketones, the addition of triethylamine dramatically increases the rate. Yields were increased from only 0.5% to over 98% for the reduction of acetophenone at 50 °C under 71 bar H_2 in a 1:1 mixture of methanol and benzene [45]. Several other ketones have been reduced this manner, including benzophenone, which has proved difficult (see above; Fig. 15.8).

The catalyst derived from $[Rh(NBD)Cl]_2$ and PPh_3 showed the same enhancement with triethylamine [45]. Later studies [46] showed that increasing the amount of methanol increased the rate, although some benzene must be retained to dissolve the catalyst. The presence of triethylamine as co-catalyst must be at least 5 equivalents relative to the rhodium in order to obtain a maximum

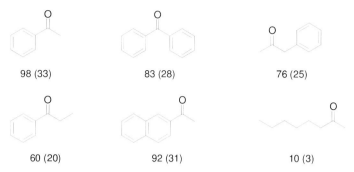

98 (33)   83 (28)   76 (25)

60 (20)   92 (31)   10 (3)

Fig. 15.8 Hydrogenation of ketones using RhCl(PPh$_3$)$_3$ + 5Et$_3$N. Values shown are yields (TOF, h^{-1}).

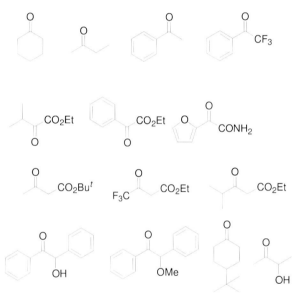

Fig. 15.9 The range of ketones hydrogenated using [Rh(DiPFc)(COD)]OTf.

rate. Coupled with this, an increase of triphenylphosphine from 2 to 4 equivalents also increases the activity. Combining all of these factors provides an idealized catalytic system of [Rh(NBD)Cl]$_2$ (2.5 mol%) with PPh$_3$ (8 mol%) and NEt$_3$ (12.5 mol%). With a SCR of 40:1, this system was used to reduce a vast range of ketones in benzene:methanol (30:70) at 50 °C under 1 bar H$_2$, with yields that were still below 80%.

The cationic complex [Rh(DiPFc)(COD)]OTf was discussed earlier as being an excellent catalyst for the hydrogenation of aldehydes under mild conditions. Under similarly mild conditions (25 °C, 4 bar H$_2$, SCR 450, 4 h, TOF ∼110 h^{-1}), a range of ketones was hydrogenated quantitatively (Fig. 15.9) [13].

In optimizing the conditions for such a reduction, protic solvents such as methanol and ethanol are required over dichloromethane (DCM), ethyl acetate (EtOAc), and THF which deactivate the catalyst. High substrate concentrations were also required, presumably due to dimerization of the catalyst that can occur in the absence of ketone or olefinic substrates. Finally, increasing the hydrogen pressure also gave an increase in yield. When this catalyst is used for the hydrogenation of unsaturated ketones, the C=C bond is first reduced very rapidly to give the saturated ketone. A slower reduction of the carbonyl group then occurs to yield the saturated alcohol.

15.3.2.2 Water-Soluble Rh Catalysts

The water-soluble ligand (TPPTS) was discussed earlier with regard to aldehyde reduction [17]. Similarly, in ketone transfer hydrogenation, high yields are obtained for a variety of substrates with the ability for efficient catalyst recycling at no expense of activity or selectivity (Fig. 15.10).

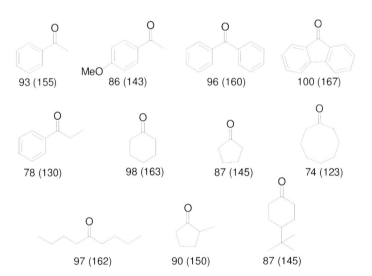

Fig. 15.10 Transfer hydrogenation of ketones at 80 °C catalyzed by [RhCl(COD)]$_2$/TPPTS. Values shown in brackets are yields (TOF, h^{-1}).

15.3.3
Ruthenium Catalysts

15.3.3.1 Ruthenium Carbonyl Clusters

Early efforts into ruthenium-catalyzed ketone hydrogenation experiments were performed using ruthenium-carbonyl clusters [47]. With cyclohexanone as a substrate and $[H_4Ru_4(CO)_{12}]$ as the catalyst, a range of solvents was tested for applicability. The greater reaction rates were achieved using alcohols, although the use of primary or secondary alcohols led to a decrease in selectivity due to the formation of ethers. The catalyst could be recovered at the end of the reaction. Partial displacement of the carbonyls with phosphines led to a decrease in activity, but further replacement of carbonyls with phosphines increased activity. By modifying such complexes with chiral bidentate phosphines, the first example of enantioselective transfer hydrogenation using $[H_4Ru_4(CO)_8[(–)\text{-}DIOP)]_2]$ was realized, although optical yields were less than 10% [48].

15.3.3.2 Ru–PPh₃ Complexes

Mononuclear ruthenium complexes were found to be superior to carbonyl clusters during a comprehensive comparison of a variety of catalysts in the reduction of acetone [49]. Without solvent, most catalysts were highly selective, although the activity was quite low. The addition of water to the system vastly increased yields, in agreement with Schrock and Osborn's observations into rhodium-catalyzed hydrogenations (Table 15.9) [41].

The addition of aqueous NaOH or acetic acid resulted in an increase in rate, but the selectivity was reduced – perhaps due to the formation of aldol condensation products. This is in contrast to the findings of Rossi with rhodium systems [42] and Strohmeier, who claimed that the addition of acid or base also increased selectivity when using $RuCl_2(PPh_3)_3$ and $Ru(CF_3CO_2)_2(CO)(PPh_3)_2$ as catalyst [50]. It was found that catalysts possessing carbonyl ligands or nitrosyl ligands were

Table 15.9 Hydrogenation of acetone. Conditions: SCR = 1300, 150 °C, 69 bar, 4 h.

Complex	Conversion (2.5% H_2O) [%]	Selectivity (2.5% H_2O) [%]	Conversion (dry) [%]	Selectivity (dry) [%]
$[RuHCl(CO)(PPh_3)_3]$	95	95	25	93
$[RuH(NO)(PPh_3)_3]$	97	95	22	95
$[RuCl_2(CO)_2(PPh_3)_2]$	90	92	26	87
$[Ru(H)_2(CO)(PPh_3)_3]$	69	94	67	100
$[Ru(H)_2(PPh_3)_4]$	39	82	56	98
$[RuCl_2(PPh_3)_3]$	33	83	18	82
$[RuH_4(PPh_3)_3]$	30	86	78	100
$[RuHCl(PPh_3)_3]$	13	70	6	70
$[Ru_3(CO)_{12}]$	3	41	6	69

higher in activity and selectivity. This was attributed to the complexes of general formula $[RuX_2(PPh_3)_n]$ (X=H, Cl; n=3, 4) having a competing decarbonylation reaction, as demonstrated by the presence of metal carbonyl complexes in the reaction mixture after completion. In the hydrogenation of acetone under 69 bar H_2 at 150 °C with a SCR of 100000, turnover numbers of up to 15000 h^{-1} could be achieved over three days, using $[RuHCl(CO)(PPh_3)_3]$ and a little water. These were similar findings to the hydrogenation of aldehydes under the same conditions.

Table 15.10 Hydrogenation of unsaturated ketone at 28 °C and 4 bar H_2 (ketone : $RuCl_2(PPh_3)_3$: $H_2N(CH_2)_2NH_2$: KOH = 500 : 1 : 1 : 2).

Substrate	Yield [%]	Unsaturated alcohol [%]	Saturated alcohol [%]	TON	TOF [h^{-1}]
	100	98.2	1.8	491	714
	99.5	100	0	498	332
	100	99.9	0	10000	555[a]
	98.2	99.6	0.4	489	327
	100	70	30	350	500
	99.8	99.9	0.1	499	333[b]
	100	92.8	7.2	464	71
	99	100	0	495	495

Yields of saturated ketone are <0.1% in all cases.
a) Reaction performed with ketone : Ru = 10000 : 1.
b) H_2 pressure of 8 bar used.

The [RuCl$_2$(PPh$_3$)$_3$] catalysts can be used more effectively to hydrogenate ketones using formic acid as the hydrogen source [51]. In solvent-free reactions, the formic acid completely decomposes and the products are easily obtained from the reaction mixture. Thus, in reactions carried out at 125 °C with a SCR of 800:1, simple ketones and aldehydes are reduced with excellent yields. Applying formic acid as a hydrogen source to the Ru cluster catalysts and other Ru phosphine catalysts gave less favorable results.

15.3.3.3 Diamine-Modified Ru Catalysts

Noyori and coworkers discovered that the activity of [RuCl$_2$(PPh$_3$)$_3$] could be enhanced by the addition of ethylenediamine (en) and KOH/i-PrOH [52]. Using the system for acetophenone hydrogenation (Ru:en:KOH, 1:1:20, SCR 5000 at 28 °C under 3 bar H$_2$), TOFs of 6700 h^{-1} were obtained. The pressure of hydrogen is important, as demonstrated by a TOF of 880 h^{-1} under 1 bar H$_2$ (SCR = 500). By increasing the pressure to 50 bar and using a SCR of 10 000, TOFs in excess of 23 000 were obtained. The reaction was even shown to work at −20 °C, showing just how mild the conditions employed can be. In order for the catalytic system to work, both the organic and inorganic bases are required with at least one primary amine end to the diamine. Applying this catalytic system to unsaturated ketones shows a remarkable selectivity towards the unsaturated alcohol (Table 15.10) [29]. Reaction times vary from substrate to substrate between 1 and 18 h, with yields and selectivities of over 99% easily achieved. The catalyst will even reduce the acetylenic ketones without the alkyne group being affected. The catalyst shows great scope and, with ligand modification, a highly enantioselective catalyst can be produced. The mechanism of this unique catalyst is described in Chapters 20 and 32.

An alternative variation to this catalyst, *trans*-[RuCl$_2$[P(C$_6$H$_4$-4-CH$_3$)$_3$]$_2$ (H$_2$NCH$_2$CH$_2$NH$_2$)] and KOtBu in isopropanol, is excellent for the selective hydrogenation of benzophenones (Scheme 15.8) [53].

The products of such reactions can be useful intermediates in the synthesis of commercial drugs. The nature of the substituents within the benzophenones has an effect on rate, with electron-withdrawing groups favoring the reaction more than electron-donating groups. For example, kinetic studies showed that *p*-trifluoromethylbenzophenone was hydrogenated 11-fold faster than the *p*-

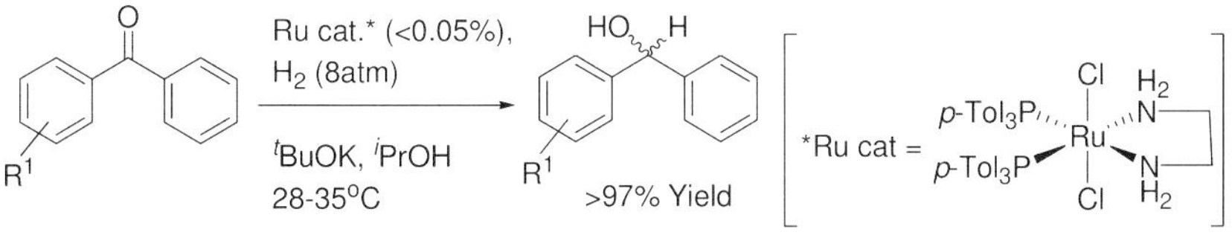

R^1 = H, *o*-CH$_3$, *o*-Cl, *m*-Cl, *p*-C$_6$H$_5$, *p*-CH$_3$O, *p*-F, *p*-Cl, *p*-CF$_3$

Scheme 15.8 Hydrogenation of benzophenones.

Table 15.11 Hydrogenation of benzophenones with *trans*-[RuCl$_2${P(C$_6$H$_4$-4-CH$_3$)$_3$}$_2$(H$_2$NCH$_2$CH$_2$NH$_2$)] and KO*t*Bu in *i*PrOH under 8 bar H$_2$ at 28–35 °C.

R^1	SCR	Concentration [M]	Yield [%]	TON	TOF [h^{-1}]
H	20 000	2.7	99	19 800	413
o-CH$_3$	3000	1.5	99	2970	165
p-CH$_3$O	3000	1.5	99	2970	165
p-Cl	3000	1.3	100	3000	375
p-CF$_3$	2000	0.4	99	1980	1980

methoxy derivative. However, a range of benzophenones was reduced smoothly at 30 °C (Table 15.11). In an optimized experiment demonstrating the practicability of the method, a slurry of 200 g benzophenone in 200 mL *i*-PrOH was hydrogenated with an SCR of 20 000 within 48 h at 30 °C.

Recently, several catalysts based on ligands containing an NH$_2$ or NH grouping within the phosphine ligand, such as Ph$_2$PCH$_2$CH$_2$NH$_2$, have been shown to have considerable activity and chemoselectivity for ketone hydrogenation [54–56].

15.3.3.4 Other Ru Catalysts

As for some of the monodentate phosphine-based catalysts, *cis*-[Ru(6,6'-Cl$_2$bpy)$_2$(OH$_2$)$_2$][CF$_3$SO$_3$]$_2$ was found to require water for the best catalytic activity in the reduction of aldehydes and ketones [57]. Aldehydes and ketones were found to be hydrogenated, with reasonable yields. Unsaturated aldehydes were reduced with selectivity towards the unsaturated alcohol, whereas unsaturated ketones showed selectivity towards the saturated ketones.

The water-soluble ruthenium TPPTS system which functioned well for saturated and unsaturated aldehydes has also been tested for the hydrogenation of ketones [31]. Although good yields for simple ketones could be obtained depending on the substrate, the selectivity when used for unsaturated ketones was in favor of the C=C bond. The polyphosphine catalysts RuHCl(CO)(PPh$_3$)(dppe) (dppe = Ph$_2$PCH$_2$CH$_2$PPh$_2$) and RuHCl(CO)(tdpme) (tdpme = CH$_3$C(CH$_2$PPh$_2$)$_3$) show greater activity than RuHCl(CO)(PPh$_3$)$_3$ in the hydrogenation of cyclohexanone [58]. Turnover numbers of 450 and 625 are achieved, respectively, for the polydentate complexes, compared to 82 for the triphenylphosphine complex.

Highly efficient transfer hydrogenation of ketones can be achieved by the use of the transfer hydrogenation catalyst *trans,cis,cis*-[RuX$_2$(CNR)$_2$(dppf)] (X = Cl or Br; R = CH$_2$Ph, cy, *t*-Bu, 2,6-C$_6$H$_3$Me$_2$) [59]. These are the first examples of isocyanide–ruthenium species being used for the transfer hydrogenation of ketones. The complexes are prepared by the reaction of bis(allyl)-ruthenium(II) complex [Ru(η^3-2-C$_3$H$_4$Me)$_2$(dppf)] with HX acid in the presence of the isocyanide. All the catalysts were effective in the hydrogenation of acetophenone, giving quanti-

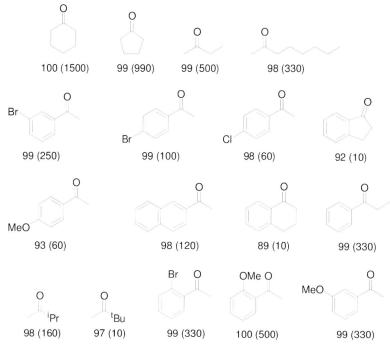

Fig. 15.11 *trans,cis,cis*-[RuCl$_2$(CNCH$_2$Ph)$_2$(dppf)] catalysed hydrogenation of ketones. Values shown are yields (TOF, h^{-1}).

tative yields between 0.5 and 8 h. *trans,cis,cis*-[RuCl$_2$(CNCH$_2$Ph)$_2$(dppf)] proved to be the most active, and was further utilized in the transfer hydrogenation of a series of ketones at 82 °C using a ketone:Ru:NaOH ratio of 250:1:24 (Fig. 15.11).

15.3.4
Other Metal Catalysts

15.3.4.1 Copper

The use of phosphine-stabilized copper complexes as hydrogenation catalyst was discussed previously for aldehydes. The same catalysts have been used in the hydrogenation of simple ketones, with high yields achieved in reactions lasting for up to 48 h [35]. Several unsaturated ketones were hydrogenated, with high chemoselectivity, to the unsaturated alcohol. These catalysts are sensitive to the structure of the phosphine ligand. In the hydrogenation of 4-phenyl-3-butan-2-one, it is possible to obtain any of the three possible products by varying the phosphine (Fig. 15.12) [60].

No phosphine, 70 atm H$_2$	91	:	9	:	0	89% (3)
PPh$_3$, 117 atm H$_2$	0	:	92	:	8	95% (1.5)
Me$_2$PPh, 34 atm H$_2$	0	:	8	:	92	91% (6)

Fig. 15.12 Effect of phosphine on selectivity. Values shown are yields (TOF, h^{-1}).

15.3.4.2 Metal Carbonyls

The metal carbonyls $Cr(CO)_6$, $Mo(CO)_6$, $W(CO)_6$ and $Fe(CO)_5$ have all been tested in the hydrogenation of acetophenone in the presence of a strong base [61, 62]. In reactions performed in either triethylamine of sodium methoxide in methanol using 5 mol% of catalyst, the Mo and Cr complexes proved to be superior. The different bases had an effect on the yield that was further demonstrated when $Cr(CO)_6$ was used in the hydrogenation of a series of ketones under the same conditions. In most cases, the reactions were found to be better in the methoxide system, with over 98% yields obtained in reactions lasting 3 h at 120 °C.

15.4
Domino-Hydroformylation-Reduction Reactions

15.4.1
Cobalt Catalysts

Cobalt-catalyzed hydroformylation of terminal alkenes using $[Co(H)(CO)_4]$ as catalyst delivers mixtures of branched and linear aldehydes under elevated pressures and high temperatures (160–200 °C). In 1968, it was found that adding a trialkylphosphine to the cobalt catalyst reduces activity, but stabilizes the catalyst for use under 100 bar syngas pressure [63]. The use of phosphine ligands increases the hydrogenation activity such that the aldehydes are directly hydrogenated to alcohols as the only oxygenated products isolated. This is a desirable process, since linear alcohols are often the target products from many hydroformylation processes. Tributylphosphine can serve as a ligand for this purpose, but the ligands which provide the best catalyst stability are those that have a bicyclic structure such as the "phobane" ligand [64] and, more recently, the limonene-derived phosphines shown below [65]. Recent studies of the hydroformylation of dodecene at 170 °C, 85 bar syngas pressure using 1000 ppm $[Co(H)(CO)_4]$ show that 70% linear alcohols can be formed, with relatively small amounts of branched alcohol ($n:iso=4.9$) and alkane (6%) as the side products. Under these typical conditions, aldehyde hydrogenation appears to be the most facile step in the process, as aldehydes are not observed.

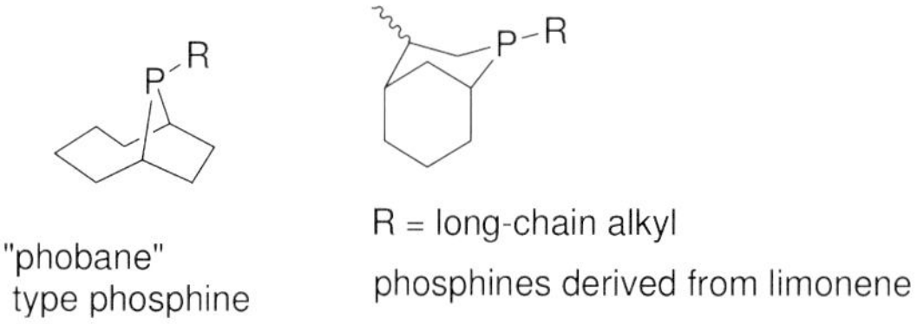

"phobane"
type phosphine

R = long-chain alkyl

phosphines derived from limonene

Fig. 15.13 Bicyclic phosphines used in cobalt-catalyzed hydroformylation.

15.4.2
Rhodium Catalysts

Rhodium-catalyzed hydroformylation can be carried out under much milder conditions (5–50 bar H_2/CO, $T = 20$–$120\,^{\circ}$C), shows higher TOFs, fewer alkane byproducts, and can be manipulated to give very high selectivity towards the linear aldehydes [66]. Given that linear alcohols are frequently the desired products, several investigations have been made on the use of Rh catalysts to hydrogenate aldehydes under the reaction conditions. This has indeed been observed in several cases, using strongly electron-donating phosphines [67–69] or amines [70, 71] as ligands. The most detailed studies on this topic have been carried out by Cole-Hamilton and coworkers, who used $[Rh_2(OAc)_4]$/PEt_3 as a catalyst [72–74]. In the hydroformylation of hex-1-ene in aprotic solvents, hydrogenation of the initially formed heptanal and 2-methylhexanal products to the corresponding alcohols occurs as the reactions proceeds. High TOFs were observed at $120\,^{\circ}$C (40 bar syngas) with modest linear-to-branched regioselectivity: low linear selectivity is often observed using alkyl phosphine ligands in hydroformylation. Pure heptan-1-al is also readily hydrogenated under similar reaction conditions using the same catalysts. However, when the reactions were carried out in alcoholic solvents, mechanistic investigations established that alcohols are actually the initial reaction products with no aldehyde intermediates being formed.

More recently, during research aimed at supporting the highly linear selective hydroformylation catalyst $[Rh(H)(Xantphos)(CO)_2]$ onto a silica support, the presence of a cationic rhodium precursor in equilibrium with the desired rhodium hydride hydroformylation catalyst was observed. The presence of this complex gave the resulting catalyst considerable hydrogenation activity such that high yields of linear nonanol could be obtained from oct-1-ene by domino hydroformylation-reduction reaction [75].

15.5
Reductive Amination of Ketones and Aldehydes

Although imine hydrogenation is discussed in greater detail in Chapter 34, it seems appropriate at this point to describe one-pot reductive amination of aldehydes and ketones. The reductive amination of aldehydes and ketones using sodium borohydride or sodium cyanoborohydride is well established. However, a more environmentally benign and economical method to carry out this reaction is to use molecular hydrogen. Several heterogeneous catalysts have been shown to be effective in this transformation, but interest has been expressed in the use of more controllable homogeneous catalysts for this purpose.

The first example of this type of transformation was reported in 1974 [76]. Three catalysts were investigated, namely $[Co_2(CO)_8]$, $[Co(CO)_8/PBu_3^n]$, and $[Rh_6(CO)_{16}]$. The $[Co(CO)_8/PBu_3^n]$ catalyst showed activity for reductive amination using ammonia and aromatic amines. The $[Rh_6(CO)_{16}]$ catalyst could be used for reductive amination using the more basic aliphatic amines that were found to poison the cobalt catalyst. This early report pointed out that the successful reductive amination of *iso*-butanal (Me_2CCHO) with piperidine involves selective enamine hydrogenation, that reductive amination of cyclohexanone with isopropylamine probably involves imine hydrogenation, and that reductive amination of benzaldehyde with piperidine would presumably involve the reduction of a carbinolamine.

Although this report establishes some of the principles of this class of reaction, no turnover numbers or SCRs were reported, and harsh reaction conditions (100–300 bar H_2/CO, 110–200 °C) were employed. Subsequently, this process received sporadic attention, except as a process combined with a hydroformylation stage. In 1997, Knifton found that amination of linear aldehydes using ammonia could be achieved, and showed that the related domino hydroformylation-amination process was also possible [77]. In 2000, Borner and coworkers released preliminary results describing a more practical catalyst system for these reactions [78]. Benzaldehyde and piperidine could be reductively aminated using $[Rh(dppb)(COD)]BF_4$ or [Rh(1,2-bis-diphenylphosphinitoethane)(COD)]BF$_4$ under mild conditions (50 bar H_2, room temperature). A total of 500 catalytic turnovers could be achieved within a few hours, with the reaction being hampered by only moderate selectivity towards the tertiary amine (Scheme 15.9).

Selectivities of about 2:1 are the best found for this type of hydrogenation and are highly dependent on the secondary amine used: they seem to correlate with the nucleophilicity of the amine. Reductive amination of PhCHO with benzylamine can proceed through an imine intermediate, and thus gave better selectivities (12:1) but was found to be sluggish using this catalyst system.

Beller and coworkers recently realized a more practical system for reductive amination of aromatic aldehydes using ammonia [79]. Their preferred conditions, which require the addition of an acidic additive, are shown in Scheme 15.10. Without extensive optimization, turnover numbers of 1700 could be

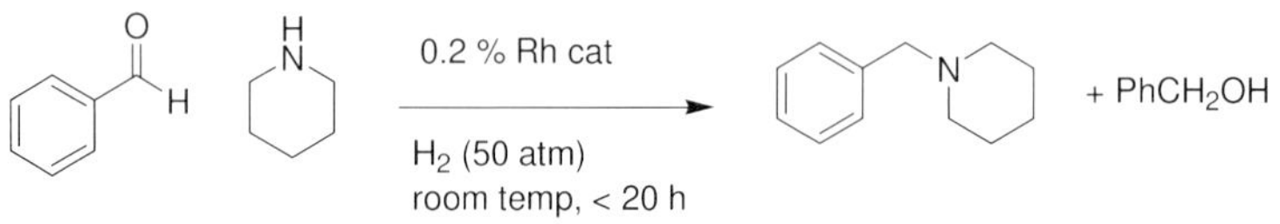

Scheme 15.9 Reductive amination of benzaldehyde.

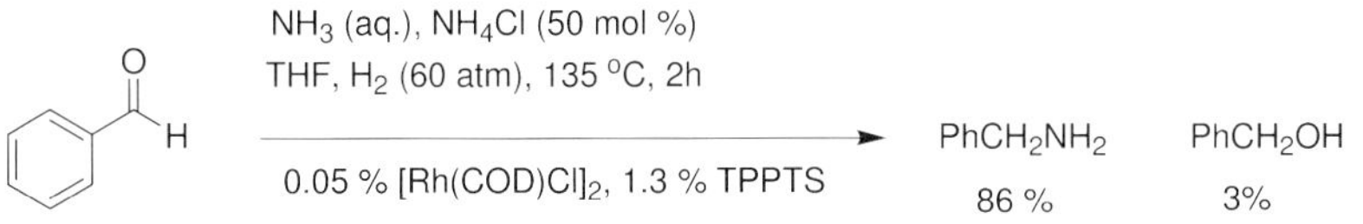

NH$_3$ (aq.), NH$_4$Cl (50 mol %)

THF, H$_2$ (60 atm), 135 °C, 2h

0.05 % [Rh(COD)Cl]$_2$, 1.3 % TPPTS

PhCH$_2$NH$_2$ PhCH$_2$OH

86 %   3%

Scheme 15.10 Reductive amination using ammonia.

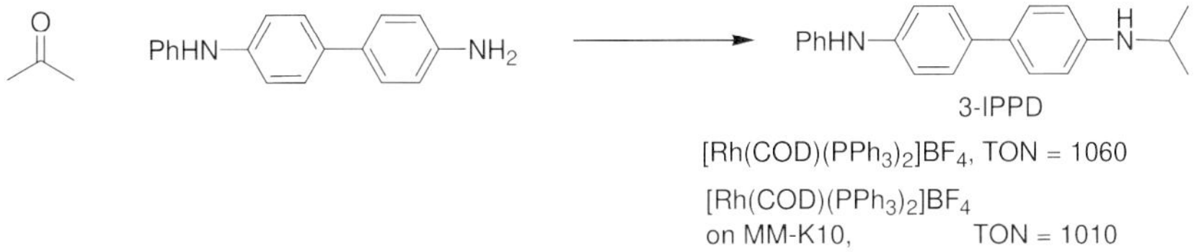

PhHN—⬡—⬡—NH$_2$ ⟶ PhHN—⬡—⬡—N(H)—⟨

3-IPPD

[Rh(COD)(PPh$_3$)$_2$]BF$_4$, TON = 1060

[Rh(COD)(PPh$_3$)$_2$]BF$_4$
on MM-K10,   TON = 1010

Scheme 15.11 Reductive amination of acetone.

achieved. A biphasic system is required in order to make use of aqueous ammonia. However, preliminary data show a second advantage in that the Rh-containing aqueous phase can be recovered by phase separation and re-used. Aliphatic aldehydes remain a problem for which further research is required.

[Rh(COD)(PPh$_3$)$_2$]BF$_4$ has been shown to be a good catalyst for reductive amination of acetone with 4-anilino-aniline to give the commercial product 3-IPPD. In laboratory-scale comparative experiments, this catalyst – both in homogeneous phase or immobilized on Montmorillonite K10 clay – was found to be superior to the commercially applied Pt/C catalyst (Scheme 15.11) [80].

In recent years there has been emerging interest in one-pot asymmetric amination of ketones, but this subject is beyond the scope of this chapter. However, an interesting observation by Borner and coworkers is that different catalysts seem to be required to carry out this process compared to those used for hydrogenation of the corresponding imines or enamines [81, 82].

15.6
Hydroaminomethylation of Alkenes
(Domino Hydroformylation-Reductive Amination)

Given the previous discussion on reductive amination, it is surprising that the potentially more complicated domino hydroformylation-reductive amination reactions have been more thoroughly developed. The first example of hydroaminomethylation was reported as early as 1943 [83]. The most synthetically useful procedures utilize rhodium [84–87], ruthenium [88], or dual-metal (Rh/Ir) catalysts [87, 89, 90]. This area was reviewed extensively by one of the leading research groups in 1999 [91], and so is only briefly outlined here as the second step in the domino process is reductive amination of aldehydes. Eilbracht's group have shown that linear selective hydroaminomethylation of 1,2-disubstituted alkenes

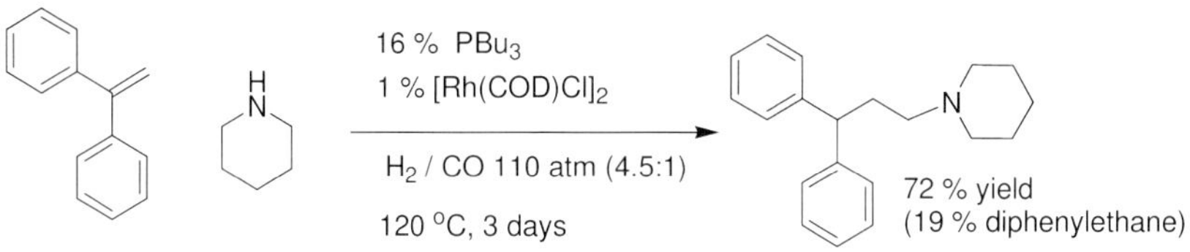

Scheme 15.12 Synthesis of fenpiprane using hydroaminomethylation of diphenylethene.

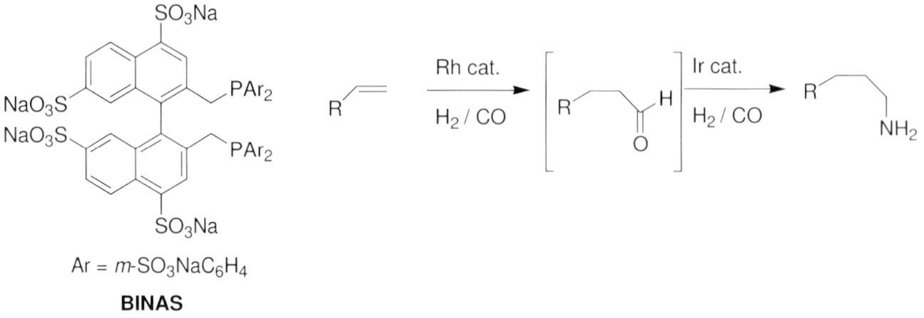

Scheme 15.13 Hydroaminomethylation of terminal alkenes to linear amines.

such as diphenylethene can give access to a series of compounds of pharmaceutical interest such as fenpiprane, diisopromine, tolpropamine, fendiline, prozapine [92], penfluridol [93], and fluspirilene [93]. An example of one of their procedures is shown in Scheme 15.12. The use of a relatively large amount of phosphine is required to suppress competing alkene hydrogenation reactions.

Eilbracht's group has done much to demonstrate the synthetic possibilities of using this reaction. However, the most recent developments in this field have also shown that the reaction could be applied as a practical method to prepare linear amines. Beller and coworkers have shown that linear selective hydroaminomethylation of propene, but-1-ene, and pent-1-ene with aqueous ammonia can be realized in a two-phase solvent system (water:methyl *tert*-butylether), using [Rh(COD)Cl]$_2$/[Ir(COD)Cl]$_2$ and water-soluble diphosphine ligand, BINAS as catalyst. If excess ammonia is used, primary amines can be produced with good primary:secondary selectivity and near-perfect linear-to-branched selectivity (Scheme 15.13, Table 15.12). Running the reaction with excess alkene allows for secondary amines to be synthesized with excellent chemo- and regioselectivity. The catalyst displays up to 4000 turnovers with respect to rhodium, although relatively high concentrations of phosphine ligand seem to be required [90].

This group subsequently invented a domino reaction consisting of isomerization of internal to terminal alkenes, followed by linear selective hydroformylation and reductive amination (Scheme 15.14) [89].

A more recent report thoroughly investigates hydroaminomethylation of terminal alkenes to give high yields of linear (linear:branched=99:1) tertiary amines from secondary amines and terminal alkenes or linear secondary

Table 15.12 Hydroaminomethylation of terminal alkenes to linear primary and secondary amines.[a]

Alkene	NH$_3$/alkene	Yield (amine)	n:iso	Primary:secondary
Propene	8:1	90	99:1	77:23
Propene	0.5:1	90	99:1	1:99
Pent-1-ene	8:1	75	99:1	87:13
Pent-1-ene	0.5:1	90	99:1	10:90

a) Conditions: temperature = 130 °C; 79 bar H$_2$/CO (5:1); time = 10 h; 0.026% Rh; 0.21% Ir; ligand:Rh ratio = 140.

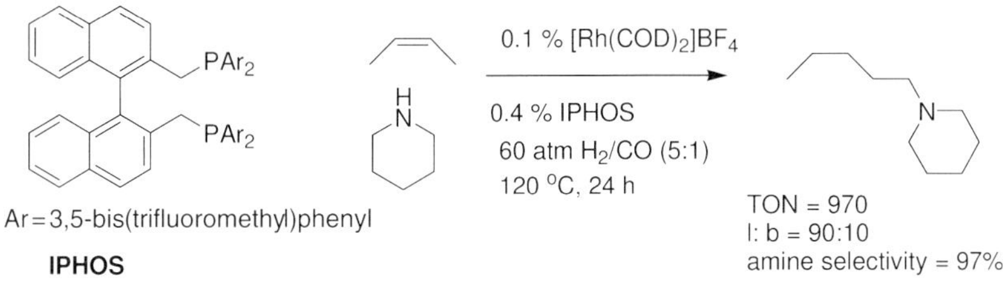

Ar = 3,5-bis(trifluoromethyl)phenyl

IPHOS

0.1 % [Rh(COD)$_2$]BF$_4$

0.4 % IPHOS
60 atm H$_2$/CO (5:1)
120 °C, 24 h

TON = 970
l:b = 90:10
amine selectivity = 97%

Scheme 15.14 Domino isomerization hydroaminomethylation.

amines from primary amines and alkenes. Reactions were conducted at 125 °C with TOF of ca. 160 h^{-1} [87].

The recent improvements described above suggest that hydroaminomethylation is approaching use as a practical process for preparing a range of amines with good linear selectivity, and good catalytic activity.

15.7
Hydrogenation of Carboxylic Acid Derivatives

The hydrogenation of acids, esters and anhydrides using molecular hydrogen is a neglected and difficult challenge. Lithium aluminum hydride (LiAlH$_4$) and certain boron hydrides are traditionally used for this reduction. However, a stoichiometric aluminum reagent is not atom-efficient and requires the separation and disposal of aluminum reagents at the end of the reaction. Catalytic hydrogenation using molecular hydrogen is potentially the ideal "green" alternative to any of the stoichiometric procedures, and would attract industrial attention if a catalyst were sufficiently active. Heterogeneous catalysts (especially copper-chromite) can carry out this process, albeit under severe conditions (200–250 °C; 14 000–20 000 kPa), which limits their application.

15.7.1
Hydrogenation of Acids and Anhydrides

The first examples of a homogeneous reduction of this type were reported in 1971. Cobalt carbonyl was found to reduce anhydrides such as acetic anhydride, succinic anhydride and propionic anhydride to mixtures of aldehydes and acids. However, scant experimental details were recorded [94]. In 1975, Lyons reported that [Ru(PPh$_3$)$_3$Cl$_2$] catalyzes the reduction of succinic and phthalic anhydrides to the lactones γ-butyrolactone and phthalide, respectively [95]. The proposed reaction sequence for phthalic anhydride is shown in Scheme 15.15. Conversion of phthalic anhydride was complete in 21 h at 90 °C, but yielded an equal mixture of the lactone, phthalide (TON = 100; TOF ~ 5) and *o*-phthalic acid, which is presumably formed by hydrolysis of the anhydride by water during lactonization. Neither acid or lactone were further hydrogenated to any extent using this catalyst system, under these conditions.

This catalyst was subsequently applied in the regioselective hydrogenation of 2,2-dimethylsuccinic anhydride [96]. An interesting reversal of regioselectivity towards the isomer B was found when switching from LiAlH$_4$ reduction to the catalytic method. Quite good isolated yields and selectivity were recorded, though no data on catalytic turnover were reported (Scheme 15.16).

Mitsubishi have reported several processes based on Ru-catalyzed hydrogenation of anhydrides and acids. Succinic anhydride can be converted into mixtures of 1,4-butane-diol and γ-butyrolactone using [Ru(acac)$_3$]/trioctylphosphine and an activator (often a phosphonic acid) [97]. Relatively high temperatures are required (~200 °C) for this reaction. The lactone can be prepared selectively under the appropriate reaction conditions, and a process has been developed for isolating the products and recycling the ruthenium catalyst [98–100].

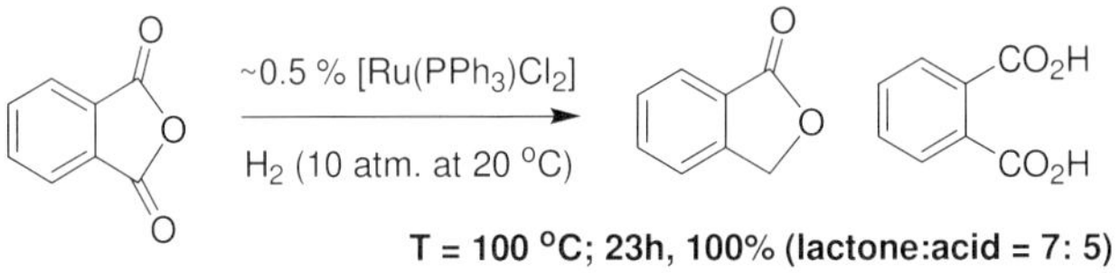

T = 100 °C; 23h, 100% (lactone:acid = 7: 5)

Scheme 15.15 Hydrogenation of *o*-phthalic acid.

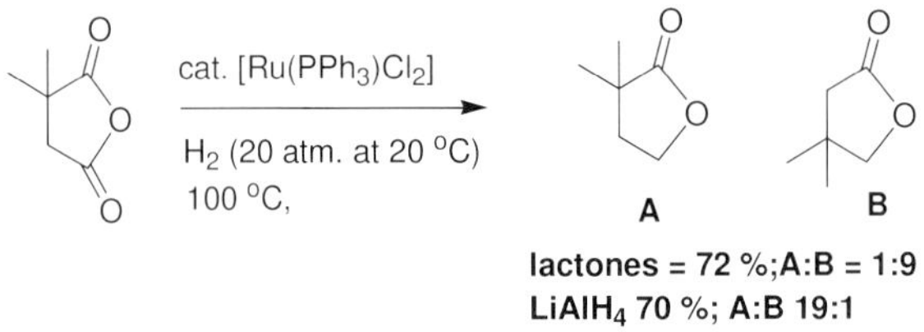

lactones = 72 %; A:B = 1:9
LiAlH$_4$ 70 %; A:B 19:1

Scheme 15.16 Regioselective hydrogenation of an unsymmetrical succinic acid derivative.

Table 15.13 Hydrogenation of succinic acid and anhydride using $[Ru_4H_4(CO)_8(P^nBu_3)_4]$. [a]

Substrate	Temperature [°C]	Time [h]	Conversion [%]	Yield of γ-butyro-lactone [%] [b]
Succinic acid	150	20	11	11
Succinic acid	180	22	83	83
Succinic acid	180	48	100	100
Succinic anhydride	100	22	40	16
Succinic anhydride	100	48	78	36
Succinic anhydride	170	40	100	100

a) TON were not reported but, based on the 100 mg of catalyst reported, are approximately 300.
b) The remaining mass is succinic acid.

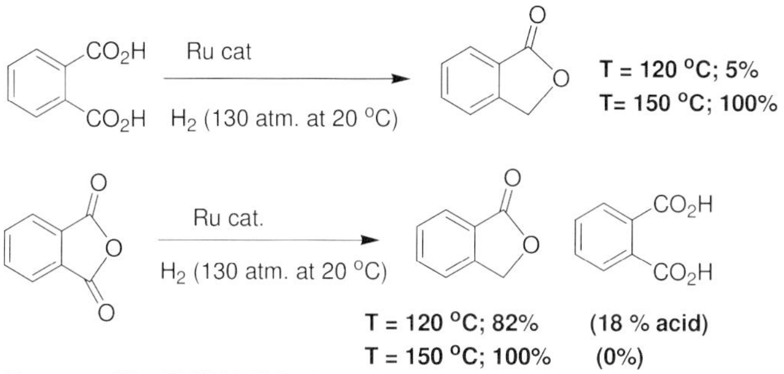

Ru cat. = $[Ru_4H_4(CO)_8(PBu_3)_4]$

Scheme 15.17 Hydrogenation of o-phthalic acid and anhydride.

The first example of carboxylic acid hydrogenation was reported as a side product in the hydrogenation of citraconnic acid using the chiral catalyst $[RuH_4(CO)_8\{(-)-DIOP)\}]$ [101]. This research team subsequently investigated acid, ester, and anhydride hydrogenation in some detail in studies which exclusively used Ru carbonyl clusters with monodentate trialkylphosphine ligands as catalysts. The reduction of succinic acid, $(CH_2CO_2H)_2$ with succinic anhydride, is compared in Table 15.13 [102].

Succinic anhydride is clearly hydrogenated more readily than the acid, as was the case with phthalic acid (Scheme 15.17), but faster absolute rates were observed in the hydrogenation of o-phthalic acid and phthalic anhydride to phthalide. In these reactions, the problem of anhydride hydrolysis is less significant as the acid can also be reduced to the same lactone product.

The effect of carboxylic acid structure was also investigated. Oxalic acid and malonic acids were found to decompose, while glutaric acid $HO_2C(CH_2)_3CO_2H$ was hydrogenated, though with poor selectivity. Although the glutaric acid results were not synthetically useful, the products included 1,5-pentane-diol and 2-hydroxy-tetrahydropyran, which showed that ester hydrogenation was a possibility. Adipic acid ($HO_2C(CH_2)_4CO_2H$) was only 25% converted to ε-caprolactone, while aze-laic acid ($HO_2C(CH_2)_7CO_2H$) was not reduced under similar conditions. The importance of a neighboring carboxylate group was therefore demonstrated, although it is not clear from these results whether the origin of this effect is the formation of stable lactones, secondary coordination to the ruthenium catalyst, or the presence of an electron-withdrawing substituent. Benzoic and phenyl acetic acids are not reduced under the conditions shown in Table 15.13, and are only slowly hydrogenated at 200 °C (9% in 48 h for benzoic acid). Although this study provides some important information regarding the feasibility of acid and anhydride hydrogenation, a number of questions remain unanswered. The effect of different ligands on ruthenium, and the importance of the cluster species on catalytic activity were not investigated. It would therefore be unwise to conclude that hydrogenation of a certain acid substrate is impractically difficult. In particular, a rough comparison of the results in Table 15.13 for succinic anhydride hydrogenation (Entry 4) with those previously described with [$Ru(PPh_3)_3Cl_2$] for succinic anhydride (90 °C, 100% conversion in 21 h, 50:50 mix of lactone:acid) does not suggest that the cluster catalysts used by this group are necessarily the most reactive catalysts possible. Davy Process and Technology have recently developed a useful catalyst for hydrogenation of acids, whereby unactivated propionic acid can be hydrogenated to propanol at 240 °C with good productivity and selectivity using a catalyst derived from a ruthenium (III) salt such as [$Ru(acac)_3$] and the tridentate phosphine, triphos (see also Table 15.17). The choice of ligand is essential for high catalytic activity [103].

An investigation of several transition-metal catalysts – including those that could be considered heterogeneous – were investigated in the hydrogenation of pentadecanoic acid. A strong promotional effect of metal carbonyls such as $Re_2(CO)_{10}$ and $Mo(CO)_6$ on catalysts such as $M(acac)_3$ (M = Ru, Rh), increasing yields of pentadecanol from 2% to 97% (TON = 97) at 160 °C and 100 bar H_2 pressure. A chemoselective reduction of pentadecanedioic acid monomethyl ester was also reported using these catalysts. The authors note that these reactions gave alcohols relatively cleanly, without ester side products [104].

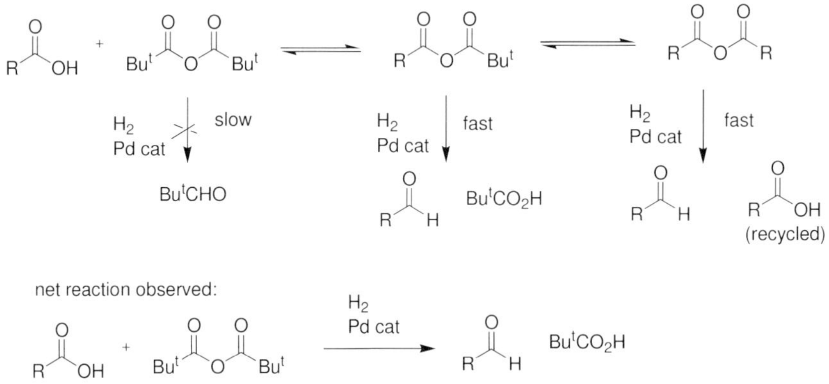

Scheme 15.18 Hydrogenation of acids *via* anhydride intermediates.

The hydrogenation of cyclic anhydrides using [Pd(PPh$_3$)$_4$] as catalyst was reported by Yamamoto and coworkers. The reaction proceeds by oxidative addition of the anhydride followed by hydrogenolysis, and proceeds well in THF at 80 °C (~100 turnovers, unoptimized). However, aldehyde productivity is limited to 50% by the reaction mechanism that involves hydrogenolysis of Pd-acyl and Pd-carboxylato groups in [Pd(PPh$_3$)$_2$(C(O)R)(O$_2$CR)] to give an equal mixture of aldehydes and acids [105]. Very bulky anhydrides were significantly more difficult to reduce, which led this group to design a process for converting carboxylic acids into aldehydes in the presence of bulky anhydrides [105–107]. Thus, heating a wide range of less sterically demanding acids in the presence of [Pd(PPh$_3$)$_4$] (~1%), (*t*BuCO)$_2$O (3 equiv.) and H$_2$ (~30 bar) delivers both *t*BuCO$_2$H and aldehyde in high yield. The reaction is proposed to occur via transesterification between acid, (RCO$_2$H) and (*t*BuCO)$_2$O to give mixed anhydride RC(O)OC(O)*t*Bu and new anhydride (RCO)$_2$O. These anhydrides are hydrogenated much more rapidly than (*t*BuCO)$_2$O and the oxidative addition of the mixed anhydride is regioselective, giving the acyl complexes of type [Pd(L)$_2$(C(O)R)(OC(O)*t*Bu], which hydrogenate to RCHO and *t*BuCO$_2$H.

The reaction tolerates ketone, chloride, internal C=C bonds, esters, nitriles, and ether functional groups. Given that the DIBAL-H reduction of acid derivatives often suffers from over-reduction to alcohols, these catalytic procedures are of synthetic value for laboratory-scale syntheses. However, it is likely that the requirement for excess (*t*BuCO)$_2$O will prevent this reaction from ever being used in commercial production.

15.7.2
Hydrogenation of Esters

The first examples of a clean hydrogenation of an ester to an alcohol was reported by Grey et al. [108]. A catalyst prepared by potassium naphthalide reduction of [RuH(PPh$_3$)$_2$Cl]$_2$, formulated as K$_2$[Ru$_2$(PPh$_3$)$_3$(PPh$_2$)H$_4$]$_2$. Diglyme hydrogenated methyl trifluoroacetate (MTFA) to trifluoroethanol and methanol at 90 °C (6 bar H$_2$). The 88% yield obtained corresponds to 290 turnovers. Trifluoroethyl trifluoroacetate (TFETFA) was hydrogenated more readily using the same catalyst system, while methyl acetate could be hydrogenated for the first time (TON = 35), but with considerable difficulty. Formate esters decompose with the liberation of carbon monoxide under these reaction conditions. The anionic catalysts used by this group were compared with [RuH(PPh$_3$)$_3$Cl], and found to be significantly more active (Table 15.14).

In addition to the successful hydrogenation of the two fluorinated esters, this report describes the hydrogenation of dimethyl oxalate. Using the reactive anionic ruthenium catalyst, a 70% conversion to methyl glycolate could be achieved (TON = 235; TOF ~12 h^{-1}) (Scheme 15.19, Table 15.14, final entry).

The results suggest a pronounced electronic effect on ester hydrogenations. This substrate effect has not been studied exhaustively by any means, but led to

Table 15.14 Hydrogenation of esters using ruthenium catalysts.[a]

Substrate	Catalyst	Conversion [%]	Remarks
MeOAc	[RuH(PPh$_3$)$_3$Cl]	0	–
MeOAc	K$_2$[Ru$_2$(PPh$_3$)$_3$(PPh$_2$)H$_4$]	22	Toluene, 13% EtOAc product (by transesterification) and EtOH (9%)
MeOAc	K$_2$[Ru$_2$(PPh$_3$)$_3$(PPh$_2$)H$_4$]	5	THF
MTFA	[RuH(PPh$_3$)$_3$Cl]	0	Toluene
MTFA	K$_2$[Ru$_2$(PPh$_3$)$_3$(PPh$_2$)H$_4$]	88	Toluene
TFETFA	[RuH(PPh$_3$)$_3$Cl]	20	Toluene
TFETFA	K$_2$[Ru$_2$(PPh$_3$)$_3$(PPh$_2$)H$_4$]	100	Toluene, 4 h
DMO	K$_2$[Ru$_2$(PPh$_3$)$_3$(PPh$_2$)H$_4$]	70	Toluene, 70% MG, 0% EG

a) Conditions: 5.7 mmol ester, 0.017 mmol K$_2$[Ru$_2$(PPh$_3$)$_3$(PPh$_2$)H$_4$]$_2$ diglyme, 0.045 mmol [RuH(PPh$_3$)$_3$Cl]; reaction time = 20 h; temperature = 90 °C; P = 620 kPa H$_2$.

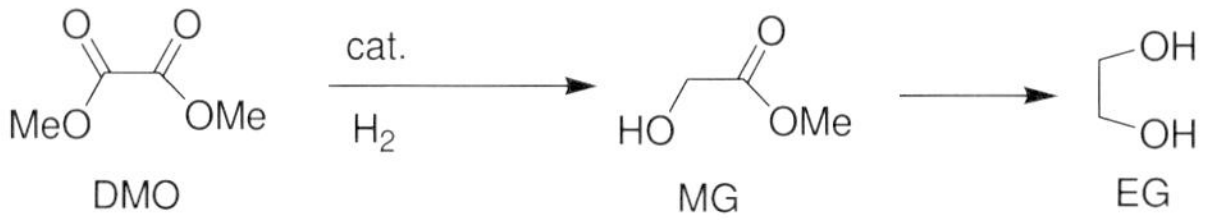

Scheme 15.19 Dimethyl oxalate hydrogenation.

a considerable research effort aimed at reducing dimethyl oxalate (DMO) to either methyl glycolate (MG) or ethylene glycol (EG).

By using the [Ru$_4$H$_4$(CO)$_8$(PnBu$_3$)$_4$] catalyst system reported for acid hydrogenation of acids [102], Matteoli and coworkers investigated the hydrogenation of dicarboxylic acid ester derivatives at 130 bar pressure and 180 °C [109]. Using relatively high catalyst loadings (maximum TON ∼ 150), DMO could be converted cleanly into the hydroxyl-ester, MG. The hydrogenations of various dicarboxylate esters under similar conditions are listed in Table 15.15. No TOF were reported, though these data do show the relative reactivity of several substrates. Consistent with Grey's observation regarding the activating effect of electron-withdrawing substituents, striking differences in hydrogenation rates were seen, depending on the proximity of the second carboxylate ester group in the substrate.

As can be seen from the data in Table 15.15, increasing the tether length results in significantly less hydrogenation. The results obtained with the C$_4$ esters, dimethyl-o-phthalate, dimethyl-cis-cyclohexane-1,2-carboxylate and dimethyl succinate are informative (Table 15.15, Entries 4–6, respectively). The close proximity of the second carboxylate ester in the substrates that are readily hydrogenated suggests two possibilities: an electronic effect, or a chelate effect. It can be envisaged that the electron-withdrawing effects of the ester group are more readily

Table 15.15 Hydrogenation of dicarboxylic esters under similar conditions. [a]

Entry	Substrate	Conversion [%]	Product(s) [b]
1	$(CO_2Me)_2$	51	Methylglycolate (MG) (51)
2	$CH_2(CO_2Me)_2$	38	$HOCH_2CH_2CO_2Et$ (17) $CH_3CH_2CO_2Et$ (10) Ethyl acetate and transesterification products (11)
3	$(CH_2CO_2Me)_2$	7	γ-Butyrolactone (7)
4	(benzene-1,2-diyl)$(CO_2Me)_2$	21	Phthalide (11) Methyl benzoate (10)
5	cyclohexane-1,2-diyl $(CO_2Me)_2$ (cis)	1	–
6	$CH_2(CH_2CO_2Me)_2$	0	–
7	(benzene-1,3-diyl)$(CO_2Me)_2$	0	–

a) Conditions: 144 h; 25 mg [Ru$_4$H$_4$(CO)$_8$(PnBu$_3$)$_4$]; 6 g substrate, 130 bar H$_2$; 180 °C.
b) Values in brackets are product yields (%).

transmitted through the aromatic system in dimethyl-o-phthalate than in dimethyl succinate. If chelate coordination of the substrate was primarily responsible for the high reaction rates, then dimethyl-cis-cyclohexane-1,2-carboxylate and dimethyl-o-phthalate should give similar yields. Since this is not the case it seems that, for ester hydrogenations – at least using this type of catalytic system – reactivity is primarily controlled by electronic effects within the ester substrate.

In 1986, the same research group reported an improved pre-catalyst, [Ru(CO)$_2$(CO$_2$CH$_3$)$_2$(PnBu$_3$)$_2$] [110]. Using this catalyst system in hydroxylated solvents, the hydrogenation of DMO produced ethylene glycol in addition to methyl glycolate, therefore inferring the hydrogenation of the less-activated ester methyl glycolate. When this system was studied in detail under standard conditions, the gradual conversion of DMO to MG then to EG is clear to see ($t=1$ h: DMO 48%, MG 52%, EG 0%, $t=2.5$ h: DMO 0%, MG 100%, EG 0%, $t=72$ h: DMO 0%; MG 78.4%; EG 21.6%). The DMO hydrogenation shows half-order reliance on DMO concentration, whereas the MG hydrogenation does not fit any steady-

Table 15.16 Hydrogenation of dimethyl oxalate using $[Ru(CO)_2(CO_2CH_3)_2(PnBu_3)_2]$.

Pressure [bar]	Catalyst concentration [mmol L^{-1}]	Temperature [°C]	Methyl glycolate [%]	Ethylene glycol [%]
10	4.9	180	96.2	3.8
90	4.9	180	81.2	18.8
130	4.9	180	78.4	21.6
130	2.45	180	90.8	9.2
130	9.70	180	69.2	30.8
130	4.9	120	51.4	0

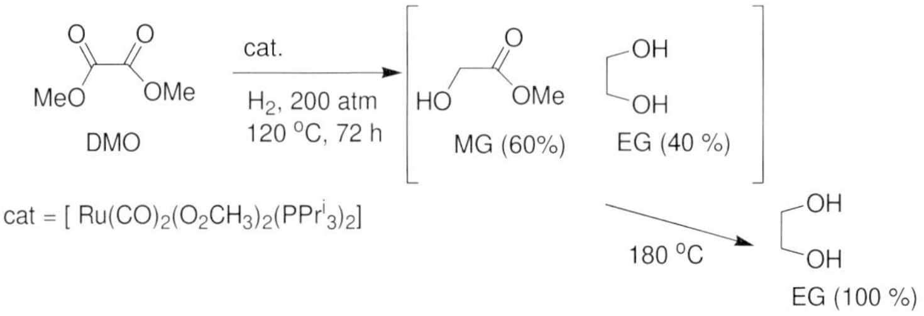

Scheme 15.20 Two-stage hydrogenation of esters giving ethylene glycol (EG), without decomposition products. MG = methyl glycolate.

rate laws. The conversion did not surpass 31%, inferring a decomposition pathway for the catalyst – not surprisingly, after many days at 180 °C. A careful set of optimization experiments were carried out focused on increasing the yields of EG from DMO. Increasing hydrogen pressures, catalyst loading, and temperature all have beneficial effects on the hydrogenation. Informative results from these experiments are in Table 15.16. Finally, a pronounced improvement on conversion was realized by the interesting – but not entirely satisfactory – addition of ∼ 1 equiv. of product in 0.5 mL benzene as additive. A 95% conversion to EG after 144 h at 180 °C (200 bar H$_2$) was observed.

In a subsequent report, the authors compared the more bulky triisopropylphosphine-based catalyst in DMO hydrogenation [111]. This initially appeared worse than the first system, as it produced considerable decomposition products (65%). However, the rates for hydrogenation of isolated MG using this system are superior to those with $[Ru(CO)_2(CO_2CH_3)_2(PnBu_3)_2]$, and do not produce decomposition products, which were proven to come only from DMO. A two-stage (two-temperature) procedure using the PiPr$_3$-based catalyst was therefore developed, which uses a lower initial temperature to suppress substrate decomposition (Scheme 15.20).

Table 15.17 Hydrogenation of dimethyl oxalate using bi-, tri-, and tetra-dentate ligands. [a]

Ligand	Catalyst [μmol]	L : Ru ratio	Conversion [%]	MG [%]	EG [%]	TON [h⁻¹]
dppe	16.1	3	18	11	0	6
PPh$_3$	9.6	5.9	73	36	0	18
PhP(C$_2$H$_4$PPh$_2$)$_2$	20.1	1.7	76	67	0	38
MeC(CH$_2$PPh$_2$)$_3$	21.1	1.4	100	1	95	160
(CH$_2$P(Ph)C$_2$H$_4$PPh$_2$)$_2$	22.8	1.0	91	85	0	36

a)   Conditions: MeOH solvent, 80 bar H$_2$; 120 °C; 0.3% Zn as additive.

There has been one further, recent development in this area. Elsevier and co-workers studied DMO hydrogenation using a broader range of catalysts, and under milder conditions than those used by Matteoli et al. [112, 113]. Elsevier and colleagues showed that tetra- and tri-dentate phosphines, when used in combination with [Ru(acac)$_3$], are very promising pre-catalysts for this class of reaction (Table 15.17). Although no direct comparisons to Matteoli's or Grey's system were reported, a comparison of the preceding discussion with the data in Table 15.16 suggests that, at present, this catalyst system is the most active one known.

The data in Table 15.17 clearly show the improved activity of all three multidentate ligands, and more strikingly, the selective formation of EG using the TRIPHOS ligand. The most significant difference between triphos (MeC(CH$_2$PPh$_2$)$_3$) and PhP(CH$_2$CH$_2$PPh$_2$)$_2$ is that triphos is a facially coordinating ligand to octahedral ruthenium complexes. This type of coordination chemistry could therefore prove a key to further improved ester hydrogenation catalysts.

The Elsevier system has since been shown to carry out several ester hydrogenations that were previously deemed impossible [114]. The hydrogenation of dimethyl phthalate to phthalide with ruthenium cluster catalysts has already been discussed (Table 15.15, Entry 4). The application of [Ru(acac)$_3$] and triphos – this time with a 20-fold excess of Et$_3$N as additive – delivers good yields of phthalide. However, the use of isopropanol (IPA) as solvent and 24% HBF$_4$ allows further hydrogenation to 1,2,-bis-hydroxylmethyl benzene for the first time. Both of these reactions were carried out under milder conditions (100 °C, 85 bar H$_2$, 16 h) than those reported previously.

A striking improvement in catalytic activity was observed when hydrogenating the esters benzyl benzoate (BZB) and methyl palmitate (MP; C$_{15}$H$_{31}$CO$_2$Me). An increase from the TON of ∼100 observed in IPA to ∼2000 (BZB hydrogenation) and 600 (MP hydrogenation) were found by using hexafluoroisopropanol as solvent with 9 mol% Et$_3$N as additive. Although this solvent is rather expensive, these are high turnover numbers for the hydrogenation of substrates that previously could not be hydrogenated at all using homogeneous catalysts. Hopefully, these two reports will lead the way towards developing practical ester hydrogenation in the not too distant future. Indeed, a recent patent from Davy Process and Technology has explored this type of catalyst system in the hydrogenation of unactivated esters such as methyl propionate and dimethylmaleate. In methyl propionate hydrogenation at ~190°C, good conversions to the propanol can be achieved, provided that water is present in the reaction vessel. The role of the water is to regenerate the catalyst which is deactivated during the reaction. This was proven by an experiment in which a catalyst that was no longer active for hydrogenation was reactivated by heating in the presence of water [103]. This catalyst system also hydrogenates anhydrides and acids. In these cases, the water produced by the hydrogenation is sufficient to allow the reaction to be run without any added water. Another patent on effective solutions for ester hydrogenation has also recently appeared [115].

The field of ester hydrogenation is significantly less developed in comparison with the hydrogenation of other double bonds. Many of the studies are limited to DMO hydrogenation, and the full scope of the reaction needs to be evaluated. At present, the research findings suggest that electron-withdrawing substituents activate substrates considerably, but the breakthrough by Elsevier's group suggest that a more broadly applicable procedure for ester hydrogenation might become reality.

Catalyst development has also been relatively unexplored. It is noteworthy that two of the most significant developments were made when the effect of different phosphine ligands were being investigated in more detail for the first time [111, 114]. At present, it is difficult to predict the future for ester hydrogenation, but if the "catalysis community" invests time into the development of the process it could prove to be an environmentally benign method for carrying out reductions in the fine chemical and pharmaceutical industries. Indeed, recent developments in industry suggest that the reaction could be viable for production-scale synthesis, and that the discovery of more active catalysts would be of considerable value.

15.8
Summary and Outlook

Since the first report of a homogeneously catalyzed reduction of a C=O bond, various research groups have endeavored to develop catalysts that show sufficient activity and high chemoselectivity to be used as a viable alternative to heterogeneous catalysts in the production of primary and racemic/achiral secondary alcohols. Much of this research effort has been conducted side-by-side with, and informed developments in, the diastereoselective and enantioselective hydrogenation of polar bonds, since activity and chemoselectivity are also key issues for these catalysts. This research effort has brought some catalysts to a level of development that suggests they might be applied in commercial production.

The ruthenium-phosphine-diamine catalysts exhibit high turnover numbers and frequencies, and near-perfect chemoselectivity for ketone/aldehyde over

C=C reduction. The achiral catalysts are relatively cheap, easy to prepare/handle, and sufficiently active to set SCRs near the threshold for Ru content in pharma products. This suggests that these catalysts could certainly be competitive, easy to operate in hydrogenations in which heterogeneous catalysts are less effective, and thereby also worthy of investigation in other cases.

The improvements made in hydroaminomethylation technology suggest that certain variants of this reaction are sufficiently developed for the potential production of amines. The synthesis of linear tertiary and secondary amines from terminal alkenes shows promise in this regard. Beller's recent contributions towards hydroaminomethylation using ammonia to produce linear primary amines, which are of industrial significance due to their abundance, suggest a bright future for this reaction. Branched selective hydroaminomethylation remains relatively underdeveloped and needs further study.

The hydrogenation of carboxylic acid derivatives using molecular hydrogen represents a major challenge, but has considerable importance from a "green chemistry" point of view. The heterogeneous catalysts capable of achieving this transformation function under energy-consuming and harsh conditions, and homogeneous catalysts for ester hydrogenation would clearly attract industrial interest if they were adequately efficient. Given that only a handful of reports have appeared on this subject, and that both substrate scope and catalyst structure–activity relationships have barely been defined, considerable further research is required in this area.

In summary, the research effort aimed towards active, chemoselective hydrogenations of certain C=O and C=N bonds have delivered several catalysts that approach the level of activity required for use in the synthesis of alcohols and amines. However, other classes of substrate require considerable additional investigations to be conducted before homogeneous catalysts may be considered for this purpose.

Abbreviations

DCM	dichloromethane
DMO	dimethyl oxalate
EG	ethylene glycol
EtOAc	ethyl acetate
MG	methyl glycolate
MTFA	methyl trifluoroacetate
SCR	substrate-catalyst-ratio
TFETFA	trifluoroethyl trifluoroacetate
THF	tetrahydrofuran
TOF	turnover frequency
TON	turnover number
TPPMS	triphenylphosphine, mono-sulfonated
TPPTS	3,3′,3″-phosphinidynetris-, trisodium salt

References

1 R. S. Coffey, *J. Chem. Soc. Chem. Commun.* **1967**, 923.

2 W. Strohmeier, H. Steigerwald, *J. Organomet. Chem.* **1977**, *129*, C43.

3 C. S. Chin, B. Lee, S. C. Park, *J. Organomet. Chem.* **1990**, *393*, 131.

4 M. Visintin, R. Spogliarich, J. Kaspar, M. Graziani, *J. Mol. Catal.* **1985**, *32*, 349.

5 E. Farnetti, M. Pesce, J. Kaspar, R. Spogliarich, M. Graziani, *J. Mol. Catal.* **1987**, *43*, 35.

6 R.-X. Li, X.-J. Li, N.-B. Wong, K.-C. Tin, Z.-Y. Zhou, T. C. W. Mak, *J. Mol. Catal. Sect. A* **2002**, *178*, 181.

7 R.-X. Li, N.-B. Wong, X.-J. Li, T. C. W. Mak, Q.-C. Yang, K.-C. Tin, *J. Organomet. Chem.* **1998**, *571*, 223.

8 J. A. Osborn, F. M. Jardine, J. F. Young, G. Wilkinson, *J. Chem. Soc. Sect. A* **1966**, 1711.

9 B. Heil, L. Marko, *Chem. Abstr.* **1969**, *71*, 286.

10 T. Mizoroki, K. Seki, S.-I. Meguro, A. Ozaki, *Bull. Chem. Soc. Jpn.* **1977**, *50*, 2148.

11 K. Kaneda, M. Yasumura, T. Imanaka, S. Teranishi, *J. Chem. Soc. Chem. Commun.* **1982**, 935.

12 H. Fujitsu, E. Matsumura, K. Takeshita, I. Mochida, *J. Org. Chem.* **1981**, *46*, 5353.

13 M. J. Burk, T. G. P. Harper, J. R. Lee, C. Kalberg, *Tetrahedron Lett.* **1994**, *35*, 4963.

14 M. J. Burk, A. Gerlach, D. Semmeril, *J. Org. Chem.* **2000**, *65*, 8933.

15 K. Kaneda, T. Mizugaki, *Organometallics* **1996**, *15*, 3247.

16 T. Mizugaki, Y. Kanayama, K. Ebitani, K. Kaneda, *J. Org. Chem.* **1998**, *63*, 2378.

17 A. N. Ajjou, J.-L. Pinet, *J. Mol. Catal. Sect. A.* **2004**, *214*, 203.

18 J. M. Grosselin, C. Mercier, G. Allmang, F. Grass, *Organometallics* **1991**, *10*, 2126.

19 R. A. Sanchez-Delgado, J. S. Bradley, G. Wilkinson, *J. Chem. Soc. Dalton Trans.* **1976**, 399.

20 J. Tsuji, H. Suzuki, *Chem. Lett.* **1977**, 1085.

21 J. F. Knifton, *Tetrahedron Lett.* **1975**, *16*, 2163.

22 W. Strohmeier, L. Weigelt, *J. Organomet. Chem.* **1978**, *145*, 189.

23 R. A. Sanchez-Delgado, O. L. De Ochoa, *J. Mol. Catal.* **1979**, *6*, 303.

24 R. A. Sanchez-Delgado, A. Andriollo, O. L. De Ochoa, T. Suarez, N. Valencia, *J. Organomet. Chem.* **1981**, *209*, 77.

25 R. A. Sanchez-Delgado, A. Andriollo, N. Valencia, *J. Mol. Catal.* **1984**, *24*, 217.

26 K. Hotta, *J. Mol. Catal.* **1985**, *29*, 105.

27 T. Suarez, B. Fontal, *J. Mol. Catal.* **1988**, *45*, 335.

28 R.-X. Li, K.-C. Tin, N.-B. Wong, T. C. W. Mak, Z.-Y. Zhang, X.-J. Li, *J. Organomet. Chem.* **1998**, *557*, 207.

29 T. Ohkuma, H. Ooka, T. Ikariya, R. Noyori, *J. Am. Chem. Soc.* **1995**, *117*, 10417.

30 R. Noyori, T. Ohkuma, *Angew. Chem. Int. Ed.* **2001**, *40*, 40.

31 M. Hernandez, P. Kalck, *J. Mol. Catal. Sect. A.* **1997**, *116*, 131.

32 A. Benyei, F. Joo, *J. Mol. Catal.* **1990**, *58*, 151.

33 F. Joo, A. Benyei, *J. Organomet. Chem.* **1989**, *363*, C19.

34 R. A. Sanchez-Delgado, M. Medina, F. Lopez-Linares, A. Fuentes, *J. Mol. Catal. Sect. A* **1997**, *116*, 167.

35 J.-X. Chen, J. F. Daeuble, D. M. Brestensky, J. M. Stryker, *Tetrahedron* **2000**, *56*, 2153.

36 C. P. Lau, C. Y. Ren, C. H. Yeung, M. T. Chu, *Inorg. Chim. Acta* **1992**, *191*, 21.

37 E. Farnetti, G. Nardin, M. Graziani, *J. Chem. Soc. Chem. Commun.* **1988**, 1264.

38 E. Farnetti, J. Kaspar, R. Spogliarich, M. Graziani, *J. Chem. Soc., Dalton Trans.* **1988**, 947.

39 C. Bianchini, E. Farnetti, M. Graziani, G. Nardin, A. Vacca, F. Zanobini, *J. Am. Chem. Soc.* **1990**, *112*, 9190.

40 J. R. Shapely, R. R. Schrock, J. A. Osborn, *J. Am. Chem. Soc.* **1969**, *91*, 2816.

41 R. R. Schrock, J. A. Osborn, *J. Chem. Soc. Chem. Commun.* **1970**, 567.

42 M. Gargano, P. Giannoccaro, M. Rossi, *J. Organomet. Chem.* **1977**, *129*, 239.

43 G. Mestroni, G. Zassinovich, A. Camus, *J. Organomet. Chem.* **1977**, *140*, 63.

44 G. Mestroni, R. Spogliarich, A. Camus, F. Martinelli, G. Zassinovich, *J. Organomet. Chem.* **1978**, *157*, 345.

45 B. Heil, S. Toros, J. Bakos, L. Marko, *J. Organomet. Chem.* **1979**, *175*, 229.

46 S. Toros, L. Kollar, B. Heil, L. Marko, *J. Organomet. Chem.* **1983**, *253*, 375.

47 P. Frediani, U. Matteoli, *J. Organomet. Chem.* **1978**, *150*, 273.

48 M. Bianchi, U. Matteoli, G. Menchi, P. Frediani, S. Pratesi, F. Piacenti, C. Botteghi, *J. Organomet. Chem.* **1980**, *198*, 73.

49 R. A. Sanchez-Delgado, O. L. Ochoa, *J. Organomet. Chem.* **1980**, *202*, 427.

50 W. Strohmeier, L. Weigelt, *J. Organomet. Chem.* **1979**, *171*, 121.

51 Y. Watanabe, T. Ohta, Y. Tsuji, *Bull. Chem. Soc. Jpn.* **1982**, *55*, 2441.

52 T. Ohkuma, H. Ooka, S. Hashiguchi, T. Ikariya, R. Noyori, *J. Am. Chem. Soc.* **1995**, *117*, 2675.

53 T. Ohkuma, M. Koizumi, H. Ikehira, T. Yokozawa, R. Noyori, *Org. Lett.* **2000**, *2*, 659.

54 M. L. Clarke, M. B. Diaz-Valenzuela, unpublished results.

55 V. Rautenstrauch, X. Hoang-Cong, R. Churlaud, K. Abdur-Rashid, R. H. Morris, *Chem. Eur. J.* **2003**, *9*, 4954.

56 L. Dahlenburg, C. Kuhnlein, *J. Organomet. Chem.* **2005**, *690*, 1.

57 C. P. Lau, L. Cheng, *Inorg. Chim. Acta* **1992**, *195*, 133.

58 K.-M. Sung, S. Huh, M.-J. Jun, *Polyhedron* **1999**, *18*, 469.

59 V. Cadierno, P. Crochet, J. Diez, S. E. Garcia-Garrido, J. Gimeno, *Organometallics* **2004**, *23*, 4836.

60 W. S. Mahoney, J. M. Stryker, *J. Am. Chem. Soc.* **1989**, *111*, 8818.

61 L. Marko, Z. Nagy-Magos, *J. Organomet. Chem.* **1985**, *285*, 193.

62 L. Marko, J. Palagyi, *Trans. Met. Chem.* **1983**, *8*, 207.

63 L. H. Slaugh, R. D. Mullineaux, *J. Organomet. Chem.* **1968**, *13*, 469.

64 J. L. van Winkle, S. Lorenzo, R. C. Morris, R. F. Mason, US3420898, **1969**.

65 C. Crause, L. Bennie, L. Damoense, C. L. Dwyer, C. Grove, N. Grimmer, W. J. van Rensburg, M. M. Kirk, K. M. Mokheseng, S. Otto, P. J. Steynberg, *J. Chem. Soc. Dalton Trans.* **2003**, 2036.

66 P. C. J. Kamer, P. W. N. van Leeuwen, J. N. H. Reek, *Acc. Chem. Res.* **2001**, *34*, 895.

67 E. Drent, European Patent 151822, **1985**.

68 M. J. Lawrenson, G. Foster, Ger. Offen 1901145, **1969**.

69 M. J. Lawrenson, UK Patent 1284615, **1972**.

70 B. Fell, A. Guerts, *Chem. Ing. Tech.* **1972**, *44*, 708.

71 L. D. Jurewicz, L. D. Rollman, D. D. Whitehurst, *Adv. Chem. Ser.* **1974**, *132*, 240.

72 P. Cheliatsidou, D. F. S. White, D. J. Cole-Hamilton, *J. Chem. Soc. Dalton Trans.* **2004**, 3425.

73 J. K. MacDougall, M. C. Simpson, M. J. Green, D. J. Cole-Hamilton, *J. Chem. Soc. Dalton Trans.* **1996**, 1161.

74 M. C. Simpson, K. Porteous, J. K. MacDougall, D. J. Cole-Hamilton, *Polyhedron* **1993**, *12*, 2883.

75 A. J. Sandee, J. N. H. Reek, P. C. J. Kamer, P. W. N. M. van Leeuwen, *J. Am. Chem. Soc.* **2001**, *123*, 8468.

76 L. Marko, J. Bakos, *J. Organomet. Chem.* **1974**, *81*, 411.

77 J. F. Knifton, *Catal. Today* **1997**, *36*, 305.

78 V. I. Tararov, R. Kadyrov, T. H. Riermeier, A. Borner, *J. Chem. Soc. Chem. Commun.* **2000**, 1867.

79 T. Gross, A. M. Seayad, M. Ahmad, M. Beller, *Org. Lett.* **2002**, *4*, 2055.

80 R. Margalef-Catala, C. Claver, P. Salagre, E. Fernandez, *Tetrahedron Lett.* **2000**, *41*, 6583.

81 V. I. Tararov, R. Kadyrov, T. H. Riermeier, A. Borner, *Adv. Synth. Catal.* **2002**, *344*, 200.

82 V. I. Tararov, R. Kadyrov, K. H. Riermeier, C. Fischer, A. Borner, *Adv. Synth. Catal.* **2004**, *346*, 561.

83 W. Reppe, *Experientia* **1949**, *5*, 93.

84 F. Jachimowicz, J. W. Raksis, *J. Org. Chem.* **1982**, *47*, 445.

85 T. Baig, P. Kalck, *J. Chem. Soc. Chem. Commun.* **1992**, 1373.

86 T. Baig, J. Molinier, P. Kalck, *J. Organomet. Chem.* **1993**, *455*, 219.

87 M. Ahmed, A. M. Seayad, R. Jackstell, M. Beller, *J. Am. Chem. Soc.* **2003**, *125*, 10311.

88 H. Schaffrath, W. Keim, *J. Mol. Catal. Sect. A* **1999**, *140*, 107.

89 A. Seayad, M. Ahmed, H. Klein, R. Jackstell, T. Gross, M. Beller, *Science* **2002**, *297*, 1676.

90 B. Zimmermann, J. Herwig, M. Beller, *Angew. Chem. Int. Ed.* **1999**, *38*, 2372.

91 P. Eilbracht, L. Barfacker, C. Buss, C. Hollmann, B. E. Kitsos-Rzychon, C. L. Kranemann, T. Rische, R. Roggenbuck, A. Schmidt, *Chem. Rev.* **1999**, *99*, 3329.

92 T. Rische, P. Eilbracht, *Tetrahedron* **1999**, *55*, 1915.

93 A. Schmidt, M. Marchetti, P. Eilbracht, *Tetrahedron* **2004**, *60*, 11487.

94 H. Wakamatsu, J. Furukawa, N. Yama-kami, *Bull. Chem. Soc. Jpn.* **1971**, *44*, 288.

95 J. E. Lyons, *J. Chem. Soc. Chem. Commun.* **1975**, 412.

96 P. Morand, M. Kayser, *J. Chem. Soc. Chem. Commun.* **1976**, 314.

97 H. Yoshinori, I. Hiroko, US5077442, **1991**.

98 W. Keisure, H. Yoshinori, K. Sasaki, US5079372, **1992**.

99 M. Chihiro, T. Kazunari, K. Hiroshi, I. Shinji, O. Masayuki, US5047561, **1991**.

100 S. Hitoshi, T. Kazunari, K. Haruhiko, US5580991, **1996**.

101 M. Pianchi, F. Piacenti, P. Frediani, U. Matteoli, C. Botteghi, S. Gladiali, E. Benedetti, *J. Organomet. Chem.* **1977**, *141*, 107.

102 M. Bianchi, G. Menchi, F. Francalanci, F. Piacenti, U. Matteoli, P. Frediani, C. Botteghi, *J. Organomet. Chem.* **1980**, *188*, 109.

103 D. V. Tyers, M. Kilner, S. P. Crabtree, M. A. Wood, WO0309308, **2003**.

104 D. He, N. Wakasa, T. Fuchikami, *Tetrahedron Lett.* **1995**, *36*, 1059.

105 A. Yamamoto, Y. Kayaki, K. Nagayama, I. Shimizu, *Synlett* **2000**, 925.

106 K. Nagayama, I. Shimizu, A. Yamamo-to, *Chem. Lett.* **1998**, 1143.

107 K. Nagayama, I. Shimizu, A. Yamamo-to, *Bull. Chem. Soc. Jpn.* **2001**, *74*, 1803.

108 R. A. Grey, G. P. Pez, A. Wallo, *J. Am. Chem. Soc.* **1981**, *103*, 7536.

109 U. Matteoli, M. Bianchi, G. Menchi, P. Frediani, F. Piacenti, *J. Mol. Catal.* **1984**, *22*, 353.

110 U. Matteoli, G. Menchi, M. Bianchi, F. Piacenti, *J. Organomet. Chem.* **1986**, *299*, 233.

111 U. Matteoli, G. Menchi, M. Bianchi, F. Piacenti, *J. Mol. Catal.* **1991**, *64*, 257.

112 H. T. Teunissen, C. J. Elsevier, *J. Chem. Soc. Chem. Commun.* **1997**, 667.

113 M. C. van Engelen, H. T. Teunissen, J. G. de Vries, C. J. Elsevier, *J. Mol. Catal. Sect. A*, **2003**, *206*, 185.

114 H. T. Teunissen, C. J. Elsevier, *J. Chem. Soc. Chem. Commun.* **1998**, 1367.

115 K. Yamamoto, T. Watanabe, T. Abe, JP2004300131, **2004**.

16
Hydrogenation of Arenes and Heteroaromatics

Claudio Bianchini, Andrea Meli, and Francesco Vizza

16.1
Introduction

The hydrogenation of arenes and heteroaromatics to partially or fully saturated cyclic hydrocarbons is a reaction of paramount industrial importance, typically catalyzed in heterogeneous phase by a number of transition metals [1]. Just to mention a few huge applications, each year a million tons of benzene are hydrogenated on Raney nickel to cyclohexane for the production of nylon *via* adipic acid [2], and much larger amounts of liquid fossil fuels are hydrotreated in refineries to remove sulfur, nitrogen, and oxygen from various hetero-aromatics – the hydrodesulfurization (HDS), hydrodenitrogenation (HDN), and hydrodeoxygenation (HDO) processes [3]. The hydrogenation of aromatics and heteroaromatics will become increasingly important if coal, which contains a huge amount of such compounds, continues to be used for the production of petrochemicals.

Aromatic hydrogenation reactions in homogeneous phase are much less numerous, and also much less efficient than in heterogeneous phase, especially in terms of turnover frequencies and catalyst stability [4]. On the other hand, soluble metal complexes still provide a better regio- and stereo-control in the reduction of heteroaromatics, although this supremacy over heterogeneous catalysis is being threatened by the development of increasingly efficient chiral phase-transfer reagents and chiral auxiliaries adsorbed onto the support materials [5]. There is little doubt, however, that organometallic compounds will always play an irreplaceable role as models systems to gain insight into the mechanisms of substrate binding and activation as well as hydrogen adsorption, activation, and transfer. The origin of the chemo-, regio-, and stereoselectivity is another issue that can be better addressed at the molecular level than using a supported metal particle.

The scarce number of metal complexes capable of catalyzing the hydrogenation of arenes is a direct consequence of the tendency of these substrates to use all the available π-electrons for coordination, and hence to occupy three contigu-

The Handbook of Homogeneous Hydrogenation.
Edited by J. G. de Vries and C. J. Elsevier
Copyright © 2007 WILEY-VCH Verlag GmbH & Co. KGaA, Weinheim
ISBN: 978-3-527-31161-3

ous coordination sites [6]. Indeed, the barrier to disruption of the aromaticity is generally very high and other bonding modes, such as the η^2 and the η^4, which would allow a more facile metal coordination and a lower barrier to reduction (H_2 activation and transfer), are extremely difficult to accomplish, even with highly energetic metal fragments [7]. Just the presence of the heteroatom, with suitable lone pairs for σ-bonding to the metal, is the main reason for the relatively large number of homogeneous catalysts capable of hydrogenating sulfur-, oxygen-, and nitrogen-heteroaromatics [8]. The most effective molecular catalysts for the hydrogenation of arenes and heteroaromatics are complexes consisting of a central metal ion (generally ruthenium, rhodium, or iridium), one or more ligands, and anions. The ensemble of these three components is responsible for the activation of hydrogen (either heterolytic or homolytic) and its selective transfer to an acceptor substrate. Experience has shown that low-valent metal complexes stabilized by ligands with phosphorus and/or nitrogen donor atoms constitute the most active and versatile catalysts [2, 8, 9].

The aim of this chapter is to provide the reader with a survey of the molecular catalysts that are able to hydrogenate aromatics, and to demonstrate the advantages and limits of the homogeneous approach. This review includes hydrogenation reactions performed in aqueous-biphasic systems, while the many structural and mechanistic analogies between molecular catalysts and heterogenized single-site metal catalysts induced us to comment also about aromatic hydrogenation by metal complexes tethered to both inorganic and organic support materials.

Several excellent reviews on the selective hydrogenation of arenes and heteroaromatics by single-site metal catalysts have been published over the past few years [8–10]. Consequently, the reader is advised to become acquainted with these accounts in order to obtain a deeper insight into the subject.

16.2
Hydrogenation of Arenes

16.2.1
Molecular Catalysts in Different Phase-Variation Systems

Very few metal complexes have been reported to generate effective catalysts for the hydrogenation of carbocyclic aromatic rings in homogeneous phase. Moreover, even the reported cases are not completely convincing, as black metal often precipitates during the catalysis. In fact, the reduction of arenes is the domain of heterogeneous catalysts, especially those based on noble metals among which rhodium, ruthenium, and platinum generate the most active systems [11]. The chemoselectivity is generally low, as most of the functional groups are hydrogenated prior to the aromatic ring. In contrast, several regioselective examples of *cis* hydrogen addition have been reported [12], while no example of asymmetric hydrogenation of prochiral arenes in homogeneous phase has been reported so far.

Table 16.1 Homogeneous catalysts, tethered single-site catalysts, and biphase catalysts for the hydrogenation of aromatic hydrocarbons.

Catalyst	Substrate	T [°C]	pH_2 [bar]	Reference(s)
$M(OAr)(H)_3L_2$ (M = Ta, Nb; L = PM_2Ph, $PMePh_2$)	benzenes, polyaromatics	80–100	3–100	30
$[Ru(\eta^6\text{-}C_{10}H_{14})(\eta^2\text{-triphos})Cl]PF_6$	benzenes [a]	90	60	19
$RuH_2(H_2)_2(PCy_3)_2$	benzenes, polyaromatics	80	3–20	18
$[Ru_3(\eta^6\text{-}C_6Me_6)_2(\eta^6\text{-}C_6H_6)(\mu_3\text{-}O)(\mu_2\text{-OH})(\mu_2\text{-H})_2]BF_4$	benzenes [a]	20	40	22
$RuCl_2(PTA)(\eta^6\text{-}C_{10}H_{14})$; $RuCl(PTA)_2(\eta^6\text{-}C_{10}H_{14})$	benzenes	90	60	21
$Co(\eta^3\text{-}C_3H_5)\{P(OR)_3\}_3$	benzenes, polyaromatics	25	1	13
$Ni(\eta^6\text{-}CH_3C_6H_5)$ $(C_6F_5)_2$	benzene		35	45
Metal alkoxides, acac, or carboxylates + AlR_3	benzenes, polyaromatics	150–210	70	34, 35
$Co(Cy_2PC_8H_{11})(\eta^5\text{-}C_8H_{13})$	benzene	25	1	46
$[Cp^*RhCl_2]_2$	benzenes, anthracene	50	50	25
$L_2RhH(\mu\text{-H})_3RhL_2$ (L = $P(O^iPr)_3$)	benzenes	25	1	14
$Rh(acac)\{P(OPh)_3\}_2$	benzenes	80	10	47
$RhH\{P(NC_4H_4)_3\}_4$; $RhH(CO)\{P(NC_4H_4)_3\}_4$	benzenes	25	5	48
$[Rh(diphos)(MeOH)_2]BF_4$	anthracenes	50–75	1	17
$[RhCl(diene)]_2^+[NR_4]X$	benzenes [b], naphthalene	25	1	49
$Rh(cod)(sulphos)/Pd^0/SiO_2$	benzenes [c]	40	30	42
Rh or Pt complexes on SiO_2-supported metals (Pd, Pt, Ru)	benzenes, naphthalene [c]	40	1	38, 40
$Ru(\eta^6\text{-}C_6Me_6)(\eta^4\text{-}C_6Me_6)$	benzenes	90	2–3	50
$[Ru(\eta^6\text{-}C_6Me_6)_2(\mu\text{-H})_2(\mu\text{-Cl})]Cl_2$	benzenes	50	50	25
$Ru(H)_3(PPh_3)_3$; $Ru(H)_2(H_2)(PPh_3)_3$	anthracenes	50–100	5	23
$Ru_4H_4(\eta^6\text{-}C_6Me_6)$; $Ru_2Cl_4(\eta^6\text{-}C_6Me_6)$	benzenes [a]	90	60	20, 52
Fe, Co, Mn, Rh, Ru, W, Mo, Cr carbonyls	polyaromatics [d, e]	180	25	24
$Fe(CO)_5$ + ammonium salt	anthracene	150	35	51
Early metal complexes on oxides (Th, U, Nb, Ta, Zr)	benzenes, polyaromatics [c]	100–120	70–90	44
$Co_2(CO)_8$	polyaromatics [e]	135–185	230–270	36

a) Liquid biphasic catalysis.
b) Phase transfer catalysis.
c) Supported metal complexes.
d) CO/H_2O as reducing agent.
e) Syngas as reducing agent.

As a general trend in both homogeneous and heterogeneous phase, the hydrogenation of arenes requires higher H_2 pressures and higher temperature as compared to the hydrogenation of olefins. Naphthalenes are extremely difficult to reduce, while higher polynuclear arenes are hydrogenated more easily than benzenes, especially at the outer rings, the resonance-stabilization of which is not as efficacious as that of the inner benzene ring.

A list of the metal complexes that have been claimed to generate catalysts for the hydrogenation of carbocyclic aromatic rings is provided in Table 16.1. This list includes homogeneous catalysts, biphase catalysts, and tethered single-site catalysts.

Ruthenium (Ru), rhodium (Rh), and cobalt (Co) form the most active and versatile catalysts, with a prevalence for Ru; effective catalysts have been reported also for other metals such as Ni, Pd, Pt, Cr, W, Mo, Mn, Nb, and Ta, some of which, however, are selective for the partial reduction of polynuclear aromatics.

The first well-documented case of homogeneous catalytic hydrogenation of a carbocyclic aromatic ring was reported by Muetterties and coworkers, who employed allyl cobalt complexes of the general formula $(\eta^3\text{-}C_3H_5)Co(PR_3)_3$ (PR_3=phosphite, phosphine) to hydrogenate benzene, alkylbenzenes, anisole, naphthalene, anthracene, and phenanthrene under mild conditions (25 °C, 1–3 bar H_2) [13]. The catalytic activity was found to increase with the size of the phosphite/phosphine ligand in the order $P(OMe)_3 < P(OEt)_3 < PMe_3 < P(OiPr)_3$. Remarkably, the hydrogenation of benzene to cyclohexane could be achieved already at 25 °C and 1 bar H_2, yet only 25 turnovers were observed prior to catalyst deactivation. Alkyl substituents on the benzene ring were also found to inhibit the reduction. In contrast to what is generally observed in both homogeneous and heterogeneous phase, the cobalt catalysts proved more active for benzene than for polyaromatics. The monohydride Co^I fragment $CoH(PR_3)_2$, generated by hydrogenation of the precursor, was proposed as the catalytically active species. This unsaturated 14-electron fragment reacts with further phosphite/phosphine, yielding $CoH(PR_3)_n$ (n=3, 4) and with H_2 yielding the trihydride $CoH_3(PR_3)_3$ (Scheme 16.1). As both these species are catalytically inactive, their unavoidable formation during the catalysis was suggested to be the main factor for catalyst deactivation.

Scheme 16.2 illustrates the catalytic mechanism proposed by Muetterties and coworkers [13]. Salient features of this mechanism are the coordination of benzene in the η^4-fashion, to give a transient $CoH(\eta^4\text{-}C_6H_6)(PR_3)_2$ complex, and the intramolecular hydride transfer to form the allylic intermediate $Co(\eta^3\text{-}C_3H_7)(PR_3)_2$. Hydrogen addition would give an η^4-1,3-cyclohexadiene complex that ultimately releases cyclohexane *via* H_2 addition/hydride migration steps. Complete *cis* stereoselectivity of hydrogen addition was demonstrated by replacing H_2 with D_2.

A mechanism similar to that shown in Scheme 16.2 has also been proposed to rationalize the arene hydrogenation activity of the triply-bridged dirhodium complex $L_2HRh(\mu\text{-}H_3)RhL_2$ ($L=P(OiPr)_3$) [14].

Some steps of the mechanism proposed by Muetterties have been proved experimentally by Bianchini and coworkers [15]. These authors synthesized the η^4-benzene Ir^I complex $[Ir(triphos)(\eta^4\text{-}C_6H_6)]^+$(triphos$=CH_3C(CH_2PPh_2)_3$), and studied

$$Co(\eta^3\text{-}C_3H_5)(PR_3)_3 \rightleftharpoons Co(\eta^3\text{-}C_3H_5)(PR_3)_2 + PR_3$$

PR_3 = phosphite, phosphine $\quad\downarrow H_2$

$$CoH(PR_3)_2 + C_3H_8$$

$$CoH(PR_3)_2 \xrightarrow{PR_3} CoH(PR_3)_3 \xrightarrow{PR_3} CoH(PR_3)_4$$

$$\Updownarrow H_2$$

$$CoH_3(PR_3)_3 \qquad\qquad \textbf{Scheme 16.1}$$

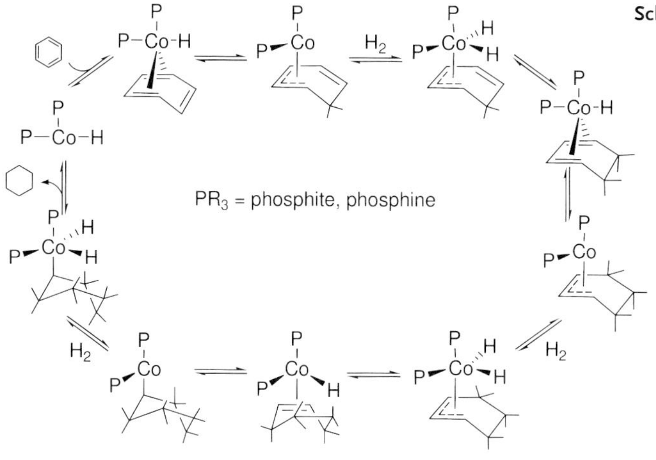

Scheme 16.2

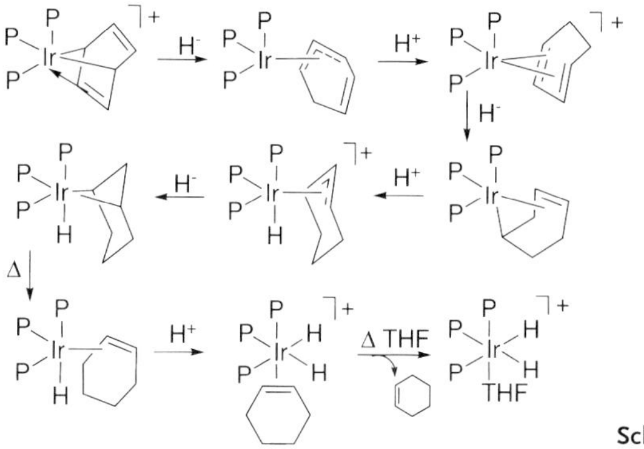

PR$_3$ = phosphite, phosphine

Scheme 16.3

in detail each reduction step down to the conversion of benzene to cyclohexene by sequential addition of H$^-$ and H$^+$. The overall reaction sequence is illustrated in Scheme 16.3. All the metal intermediates along the conversion of benzene to cyclohexene were unambiguously characterized and the region/stereochemistry of each "H" addition was determined.

It is noteworthy that the starting η^4-benzene complex was prepared by cyclotrimerization of acetylene by IrCl(C$_2$H$_4$)(triphos) [16]. All of the attempts to react the fragment [Ir(triphos)]$^+$ with benzene were unsuccessful, which reflects the difficulty met by a transition-metal fragment to overcome the energy barrier to η^4-benzene coordination.

The complex [Rh(MeOH)$_2$(diphos)]$^+$ (diphos = 1,2-bis(diphenylphosphino)ethane) has been reported to hydrogenate polynuclear aromatic hydrocarbons under mild conditions (60 °C, 1 bar H$_2$) [17]. A kinetic study of the hydrogenation of 9-CF$_3$CO-anthracene to the corresponding 1,2,3,4-tetrahydroanthracene was consistent with a rapid conversion of the precursor to [Rh(η^6-anthracene)(diphos)]$^+$ and a rate-determining step involving the reaction of the latter complex with H$_2$

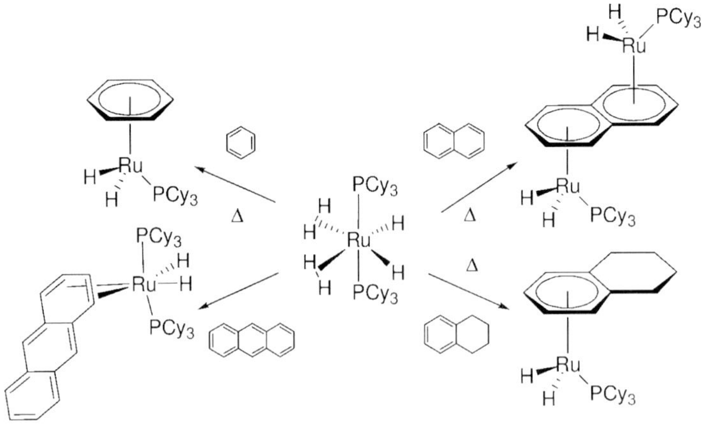

Scheme 16.4

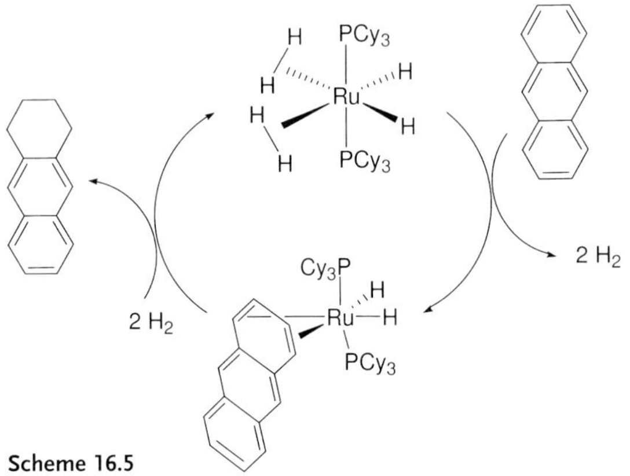

Scheme 16.5

to give 1,2-dihydroanthracene. A second-order rate kinetic law ($-d$[anthracene]/$dt = k$ [Rh] [H$_2$]) was determined with $k(59.7\,^\circ\text{C}) = (9.0 \pm 1.0) \times 10^{-2}$ M^{-1} s^{-1}.

Whilst the metals of the cobalt group have provided valuable mechanistic information on the mechanism of homogeneous hydrogenation of arenes, there is little doubt that ruthenium forms the most active and versatile catalysts.

Borowski and coworkers have reported that benzene, naphthalene, and anthracene are reduced to cyclohexane, tetralin and a mixture of 1,2,3,4-tetrahydroanthracene (4H-An) and 1,2,3,4,5,6,7,8-octahydroanthracene (8H-An), respectively, in the presence of the dihydride bishydrogen complex RuH$_2$(H$_2$)$_2$(PCy$_3$)$_2$ (80 °C, 3–30 bar H$_2$) [18]. Notably, the latter was found to react at 80 °C with neat benzene or with cyclohexane solutions of naphthalene or tetralin to form the corresponding η^6-adducts (Scheme 16.4). These products were also isolated from the final catalytic mixtures.

Unlike the previous arenes, anthracene reacted with RuH$_2$(H$_2$)$_2$(PCy$_3$)$_2$ already at room temperature to form an η^4-anthracene adduct which was found to be an effective catalyst for anthracene hydrogenation. It was suggested, therefore, that all these arene adducts may have an active role in the catalytic cycle. A simplified cycle for the hydrogenation of anthracene to 4H-An by $RuH_2(H_2)_2(PCy_3)_2$ is shown in Scheme 16.5. This involves the preliminary dissociation of two H_2 molecules to generate a coordination vacancy for the incoming molecule that ultimately binds the metal in η^4 fashion. The occurrence of this step was supported by evidence that the reaction rate decreased by increasing the H_2 pressure. The reduction of the second external ring of 4H-An to 8H-An would follow a similar mechanism as it appreciably started only when most – if not all – anthracene was consumed. 9,10-Dihydroanthracene – a typical product of catalysis proceeding through a radical mechanism – was not detected.

It is worth noting, however, that the real homogeneous character of the reactions catalyzed by the hexahydride $RuH_2(H_2)_2(PCy_3)_2$ remains questionable, as elemental mercury was found to inhibit the hydrogenation reaction, which may indicate the formation of catalytically active ruthenium metal. In contrast, a truly homogeneous ruthenium catalyst for the hydrogenation of benzenes seems to be generated by the precursor $[RuCl(\eta^2\text{-triphos})(\eta^6\text{-}p\text{-cymene})]PF_6$, recently described by Dyson and coworkers (Scheme 16.6) [19]. The catalytic activity of this complex was evaluated at 90 °C and 60 bar H_2 either in dichloromethane or in a biphasic system comprising the substrate and 1-butyl-3-methylimidazolium tetrafluoroborate. Due to its solubility in the ionic liquid (IL), the catalyst could be recovered and recycled after use. Interestingly, 1-alkenyl-substituted arenes, such as styrene and 1,3-divinylbenzene, were not hydrogenated, whereas allylbenzene was selectively converted to allylcyclohexane with a turnover frequency (TOF; mol. product mol^{-1} catalyst h^{-1}) of 329 and complete regioselectivity.

The catalytic hydrogenation of various benzene derivatives by the ruthenium tetrahydride clusters $[Ru_4H_4(\eta^6\text{-}C_6H_6)_4]^{2+}$ was investigated by Süss-Fink in both

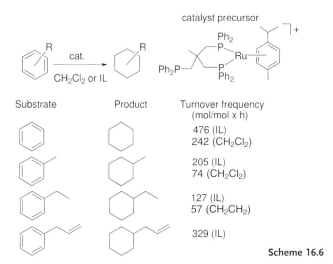

Scheme 16.6

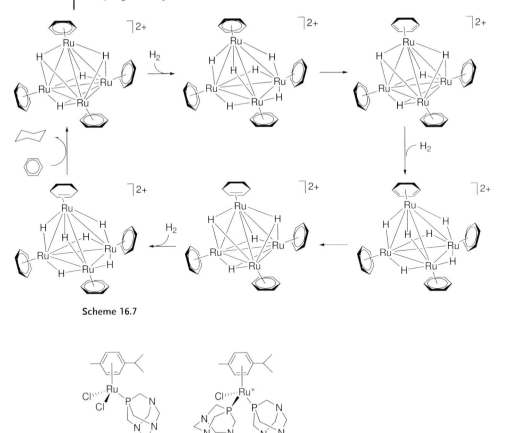

Scheme 16.7

Fig. 16.1 Sketches of Ru(PTA)Cl$_2$(η^6-C$_{10}$H$_{14}$) and [RuCl(PTA)$_2$(η^6-C$_{10}$H$_{14}$)]$^+$ (PTA = 1,3,5-triaza-7-phosphadamantane).

biphasic and aqueous systems [20]. Under aqueous biphasic conditions, cyclohexanes were produced with TOFs varying from 20 to 2000, depending on the substrate. On a quite speculative basis, a hydrogenation mechanism was proposed involving $\eta^6 \rightarrow \eta^4 \rightarrow \eta^2$ arene intermediates (Scheme 16.7). However, the only proven step was the hydrogenation of the starting [Ru$_4$H$_4$(η^6-C$_6$H$_6$)$_4$]$^{2+}$ cluster to [Ru$_4$H$_6$(η^6-C$_6$H$_6$)$_4$]$^{2+}$.

The ruthenium cluster [Ru$_4$H$_4$(η^6-C$_6$H$_6$)$_4$]$^{2+}$ was also employed for the hydrogenation of arenes in a biphasic water/1-butyl-3-methylimidazolium tetrafluoroborate biphasic system. At 90 °C and 60 bar H$_2$, benzene was reduced to cyclohexane with a TOF of 364.

The two water-soluble complexes Ru(PTA)Cl$_2$(η^6-C$_{10}$H$_{14}$) and [RuCl(PTA)$_2$(η^6-C$_{10}$H$_{14}$)]$^+$ (PTA = 1,3,5-triaza-7-phosphadamantane) (Fig. 16.1) have been tested as catalyst precursors for the hydrogenation of benzenes at 90 °C and 60 bar H$_2$ [21]. After catalysis, the former complex was converted to a triruthenium cluster

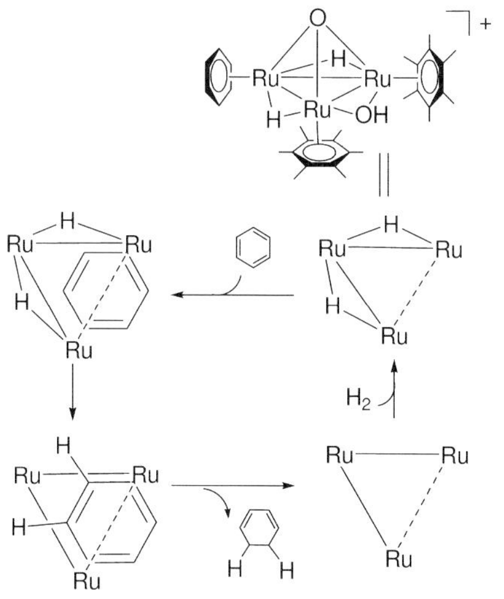

Scheme 16.8

with no coordinated PTA (NMR and electrospray mass spectrometry). In contrast, the starting complex with two PTAs gave a termination-metal product containing these ligands.

It is worth highlighting a very particular case of arene hydrogenation involving a triruthenium cluster [22]. In contrast to any other previous report, the hydrogenation of benzene was suggested by Süss-Fink to involve a direct H-transfer without substrate coordination to the metal. The proposed mechanism is shown in Scheme 16.8. The salient feature of this mechanism is adsorption of the arene in the hydrophobic pocket formed by the three arene ligands of the trimetallic precursor. The lack of substrate exchange with the originally coordinated arenes and the mass-spectrometry detection of a benzene adduct of the starting cluster were brought forward as substantial evidence for the proposed mechanism.

Many other mononuclear and binuclear Ru^{II} complexes stabilized by phosphine, cyclopentadienyl or arene ligands – for example $Ru(H)_2(H_2)(PPh_3)_3$ [23], $RuCl_2(CO)_2(PPh_3)_3$ [24], $[Ru(\mu-H_2)(\mu-Cl)(\eta^6-C_6H_6)_2]Cl_2$ [25], $[Rh(\eta^5-C_5Me_5)Cl_2]_2$ [26], and $Ru(\eta^6-C_6Me_6)(O_2CMe)_2$ [27] – have been claimed to catalyze the hydrogenation of polynuclear aromatic hydrocarbons in homogeneous fashion. Later evidence has suggested and, in some cases proved, that most of these systems are heterogeneous [28]. A paradigmatic case is the binuclear complex $[Rh(\eta^5-C_5Me_5)Cl_2]_2$ that was reported to hydrogenate benzene and substituted benzenes to cyclohexanes under relatively mild conditions (50 °C, 50 bar H_2) in the presence of a base that would promote the heterolytic splitting of H_2 as well as tie up the evolved HCl. Based on light-scattering experiments and on the good *cis* stereospecificity of hydrogen addition (e.g., *o*-xylene gave *cis*- and *trans*-1,2-dimethylcyclohexanes in a 62:1 ratio and *m*-xylene gave 1,3-dimethylcyclohexanes in *cis*:*trans* 38:1 ratio), this catalyst was thought to be homogeneous. However, later studies suggested that the true catalyst is likely heterogeneous [29].

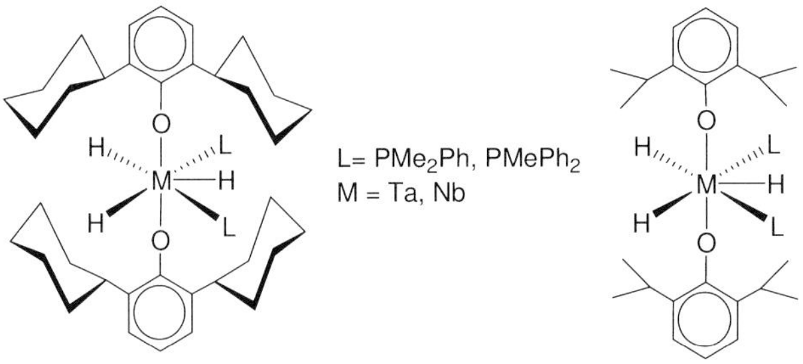

Fig. 16.2 The Nb^V and Ta^V hydride complexes containing bulky aryloxide ligands (as described by Rothwell).

Arene hydrogenation catalysts based on other metals than late transition ones are less numerous. Of particular relevance are the results reported by Rothwell, who found that Nb^V and Ta^V hydride complexes containing bulky aryloxide ligands (Fig. 16.2) are active for the homogeneous hydrogenation of arenes [30].

These catalytic systems demonstrated high regio- and stereoselectivity in the hydrogenation of benzene and of polynuclear aromatic hydrocarbons. For instance, the isolated tantalum trihydrides Ta{OC$_6$H$_3$(C$_6$H$_{11}$)$_2$-2,6}$_2$(H$_3$)(PMe$_2$Ph)$_2$ and Ta{OC$_6$H$_3$-Pr$^i_2$-2,6)}$_2$(H$_3$)(PMe$_2$Ph)$_2$ catalyzed the hydrogenation of naphthalene and anthracene at 80 °C and 3–100 bar H$_2$. The former substrate was converted to tetralin, while anthracene was reduced to 1,2,3,4,5,6,7,8-octahydroanthracene *via* 1,2,3,4-tetrahydroanthracene. Since no trace of 9,10-dihydroanthracene was observed, the occurrence of either a radical reaction [31] or a Birch-type reduction was ruled out [32].

As shown in Scheme 16.9, the intermolecular hydrogenation of [^{2}H$_8$]toluene, [^{2}H$_{10}$]acenaphthene, [^{2}H$_8$]naphthalene, and [^{2}H$_{10}$]anthracene produced single isotopomers. The ^{1}H- and ^{13}C{^{1}H}-NMR spectra confirmed a high selectivity: all-*cis* hydrogenation occurred without H/D scrambling between unreacted substrates and products. The all-*cis* nature of [^{2}H$_8$]tetralin was also proved using mass-spectrometry techniques. A unique characteristic of the niobium compound is its ability to rapidly hydrogenate arylphosphine ligands, thereby providing a new interesting procedure for the synthesis of cyclohexylphosphine ligands [33].

The only "homogeneous or substantially homogeneous" system which seems to offer a viable alternative to heterogeneous catalysis for the large-scale hydrogenation of arenes remains the IFP process [34]. This process utilizes Ziegler-type catalysts obtained by reacting at least two different metal salts (e.g., nickel and cobalt alkoxides, acetylacetonates or carboxylates), and a metal, selected among iron, zinc, and molybdenum, with trialkylaluminum as reducing agent. The hydrogenation of aromatic hydrocarbons is carried out under relatively mild conditions (155–180 °C, 10–30 bar H$_2$) in a solvent or in neat substrate. Bis-phenol A, phenol and benzene are hydrogenated to propane-dicyclohexanol, cyclohexanol and cyclohexane, respectively (Table 16.2).

cat = Ta or Nb trihydrides   Scheme 16.9

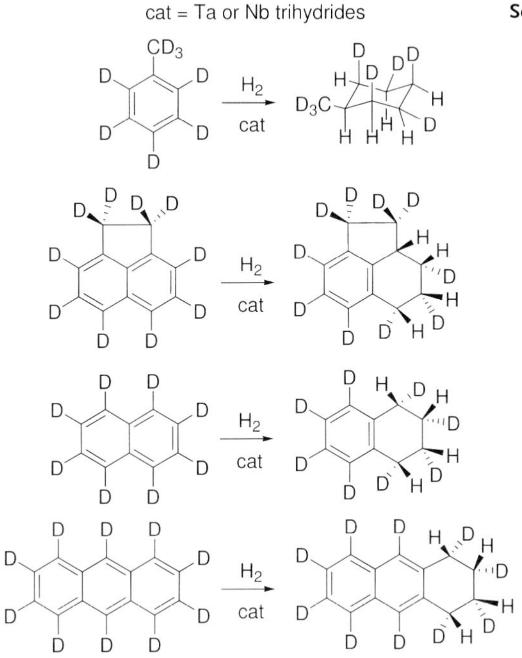

Table 16.2 Hydrogenation of aromatic hydrocarbons with the IFP process.[a]

Catalyst	Substrate	Substrate/M ratio	% Conversion (TOF)[b] product
1	bis-phenol A	250	99 (62) propane-dicyclohexanol
2	benzene	1829	99 (3621) cyclohexane
3	phenol	708	99 (2805) cyclohexanol

a) Experimental conditions: **1** (nickel octoate 0.35 mmol, iron octoate 0.35 mmol, triethylaluminum 5.6 mmol, solvent = 100 g cyclohexanol, 30 bar H_2, 4 h, 180 °C); **2** (cobalt stearate 2.2 mmol, iron stearate 0.2 mmol, triisobutylaluminum 2 mmol, 10 bar H_2, 30 min, 155 °C); **3** (nickel octoate 0.25 mmol, zinc octoate 0.25 mmol, triethylaluminum 2.1 mmol, 30 bar H_2, 15 min, 155 °C).
b) Mol of product (mol M × h)$^{-1}$.

Other examples of arene hydrogenation by Ziegler-type catalysts have been reported [35]. However, none of them is discussed at this point as they are generally poorly defined. Likewise, some hydrogenation catalytic systems in either oxo or water-gas-shift conditions are only reported in the list of references for sake of information [24, 36].

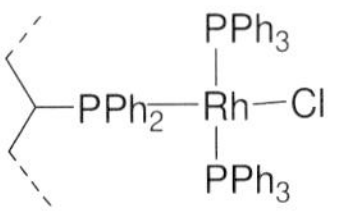

Fig. 16.3 Heterogenization of RhCl(PPh₃)₂ by grafting to a cross-linked phosphinated styrene/divinylbenzene resin.

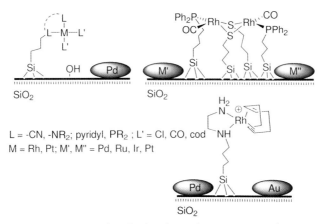

L = -CN, -NR₂; pyridyl, PR₂ ; L' = Cl, CO, cod
M = Rh, Pt; M', M" = Pd, Ru, Ir, Pt

Fig. 16.4 Some examples of tethered complexes on supported metals (as described by Angelici).

16.2.2
Molecular Catalysts Immobilized on Support Materials

One of the very first attempts to hydrogenate aromatic compounds by means of a single-site metal catalyst was reported by Fish and coworkers, who were able to tether "RhCl(PPh₃)₂" moieties to a cross-linked phosphinated styrene/divinylbenzene (DVB) resin (RhCl(PPh₃)₂/P) (Fig. 16.3).

The resulting catalyst proved active for the hydrogenation of pyrene, tetralin, p-cresol, and methylnaphthalene [37]. A rate-enhancement effect was observed which was attributed to the ability of some substrates (especially p-cresol) to stabilize unsaturated rhodium species formed during the course of the catalysis. Since then, no remarkable progress in arene hydrogenation by single-site metal catalysts has been made, until a new class of catalysts emerged from the combination on the same support of both molecular complexes and metal particles. These systems, known under the name of tethered complexes on supported metals (TCSM), were introduced by Angelici [38a] and developed independently by Angelici [38b,c] and Bianchini [39].

Angelici's approach to heterogenization involves the functionalization of a ligand, either monodentate or bidentate, with a tail bearing a reactive group capable of forming covalent bonds to silica (e.g., alkyl-Si(OR)₃). Three TCSM catalysts, among several Rh/Pd, Pt/Pd and Rh/Pd:Au examples reported by Angelici, are shown in Figure 16.4 [38].

It has been found that the complexed metal and the supported metal(s) act synergistically, to provide enhanced results, superior to those of the component catalysts, in various reactions that include the hydrogenation of arenes. Typical tethered complexes contain Rh^I, while the silica-grafted ligands can be either monodentate with N and P donors or chelating with P-N and N-N donors (diamines, pyridylphosphines). Benzenes bearing a variety of functional groups (ester, ether, hydroxy, acyl, vinyl) have been hydrogenated with TOFs much higher than those of the single components which in some cases are completely inactive. To report one such example, the hydrogenation of phenol to cyclohexanol occurs with a TOF of 3400 with a Rh(N-N)/Pd-SiO$_2$ catalyst, whereas both Pd-SiO$_2$ and unsupported Rh(N-N) are inactive (N-N=bipyridyl) [40].

Angelici has proposed that the enhanced activity might be due to a hydrogen-spillover process, promoted by the supported metallic phase, that would enhance specifically the hydrogenation activity of the molecular catalyst [40]. A later study of the hydrogenation of arenes with a catalyst obtained by silica sol-gel co-entrapment of metallic palladium and $[Rh(cod)(\mu\text{-}Cl)]_2$ (cod=cyclohexa-1,5-diene) disagreed with the hydrogen spillover hypothesis and suggested that the action of both metals is caused by a type of synergistic effect [41]; however, no clear-cut explanation was provided.

A synergistic effect operating at the level of the first H$_2$ addition (e.g., conversion of benzene to cyclohexadiene) was demonstrated by Bianchini and co-workers for the hydrogenation of various benzenes to cyclohexanes by means of a different class of TCSM catalysts [42]. These differ substantially from Angelici's catalysts for the bonding interaction of the molecular complexes to the support material. The ligands of the molecular complexes were functionalized with sulfonate tails capable of forming robust hydrogen bonds to the isolated silanols of silica (Fig. 16.5 a) [42, 43].

For this reason, these catalysts are also known under the name of supported hydrogen-bonded (SHB) catalysts and, in conjunction with Pd^0 particles on the same support material, have contributed to generate active heterogeneous systems for the hydrogenation of benzenes in aprotic solvents. Irrespective of the substrate, the combined single-site/dispersed-metal catalyst Rh^I-Pd^0/SiO$_2$ shown in Figure 16.5 a was from four- to six-fold more active than supported palladium

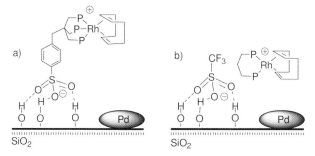

Fig. 16.5 Some examples of supported hydrogen-bonded catalysts (as described by Bianchini).

Table 16.3 Hydrogenation of benzenes with Pd^0/SiO_2, Rh^I/SiO_2 or Rh^I-Pd^0/SiO_2. [a]

Catalyst	Temp. [°C]	Substrate	Substrate/M ratio	% Conversion [b] (TOF, M) [c] product
Pd^0/SiO_2	40	benzene	525	4 (11) cyclohexane
Rh^I/SiO_2	40	benzene	9200	0
Rh^I-Pd^0/SiO_2	40	benzene	525/9200	15 (39, Pd) cyclohexane
Pd^0/SiO_2 [d]	40	toluene	426	4 (8) methylcyclohexane
Rh^I/SiO_2 [e]	40	toluene	7520	0
Rh^I-Pd^0/SiO_2 [f]	40	toluene	426/7520	16 (32, Pd) methylcyclohexane
Pd^0/SiO_2	60	styrene	400	97 (194) ethylbenzene; 3 (6) ethylcyclohexane
Rh^I/SiO_2	60	styrene	8750	98 (4287) ethylbenzene
Rh^I-Pd^0/SiO_2	60	styrene	400/8750	81 (162, Pd) ethylbenzene; 19 (38, Pd) ethylcyclohexane
Pd^0/SiO_2	60	ethylbenzene	400	3 (6) ethylcyclohexane
Rh^I/SiO_2	60	ethylbenzene	8750	0
Rh^I-Pd^0/SiO_2	60	ethylbenzene	400/8750	20 (40, Pd) ethylcyclohexane

a) Experimental conditions: Pd^0/SiO_2 (9.86 wt.% Pd), 0.044 mmol Pd; Rh^I/SiO_2 (0.56 wt.% Rh), 0.0025 mmol Rh; Rh^I-Pd^0/SiO_2 (0.56 wt.% Rh, 9.86 wt.% Pd), 0.044 mmol Pd, 0.0025 mmol Rh; 30 bar H_2; 30 mL n-pentane, 2 h, 1500 rpm.
b) Average values over at least three runs.
c) Mol product (mol M × h)$^{-1}$ (M = Pd, Rh).
d) 0.088 mmol Pd.
e) 0.005 mmol Rh.
f) 0.088 mmol Pd, 0.005 mmol Rh.

alone (Pd^0/SiO_2), while the tethered Rh^I complex alone (Rh^I/SiO_2) proved to be totally inactive (Table 16.3).

Separate experiments with cyclohexadienes and cyclohexenes showed that 1,3-cyclohexadienes are more rapidly reduced to cyclohexenes at rhodium, while the latter are predominantly reduced at palladium. It was also found that the 1,3-cyclohexadiene disproportionation, occurring on palladium, is inhibited by the grafted rhodium complex. Based on this information, as well as a number of experiments (including the isolation of relevant intermediates), the authors concluded that the enhanced activity of the Rh^I-Pd^0/SiO_2 catalyst is not due to hydrogen spillover, but to the fact that the rate-limiting hydrogenation of benzenes to cyclohexa-1,3-dienes is assisted by both palladium and rhodium, the concerted action of which, besides preventing the competitive diene disproportionation to benzene and cyclohexene, speeds up the reduction of the first double bond (Scheme 16.10).

The intimate mechanism by which the single rhodium sites and the neighboring palladium particles interact with benzene to accelerate its reduction to cyclohexadiene remains somewhat obscure.

A variation of the SHB technology to immobilize cationic molecular catalysts on silica is shown in Figure 16.5 b. This involves SHB immobilization of the counter-

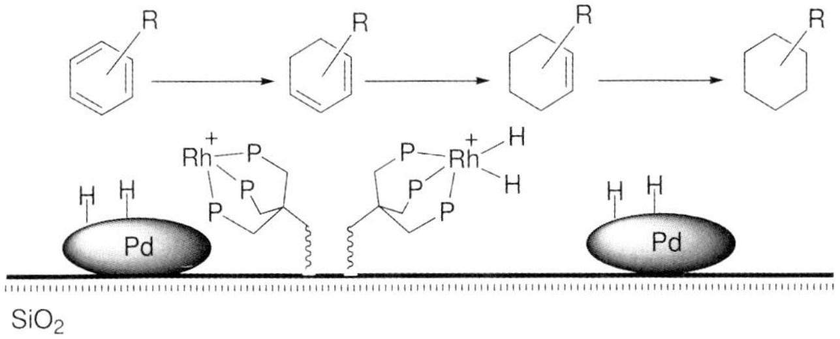

Scheme 16.10

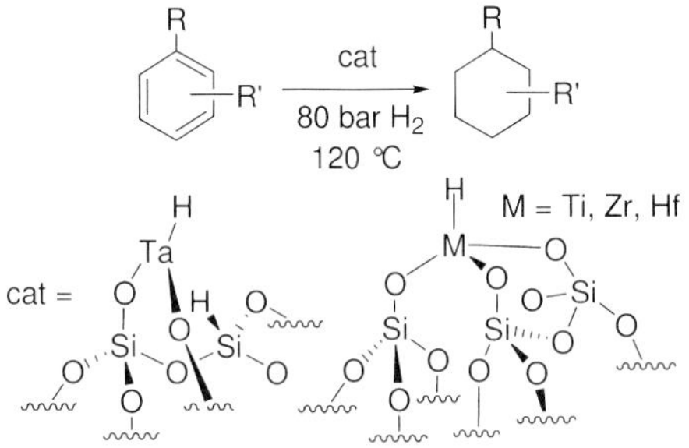

Scheme 16.11

anion, provided that the latter is capable of forming robust hydrogen bonds to the surface silanols, as is the case of the triflate counter-anion. Clearly, only aprotic solvents are viable for the successful use in catalysis of this SHB technique. It has been shown, using ^{31}P- and ^{19}F-NMR spectroscopy in CD_2Cl_2, that the metal cations reside close to the silica surface by electrostatic interaction with the SHB triflate. Therefore, only the counter-anions are truly immobilized on the support, whereas the cationic catalysts can interact freely with the substrate and H_2 as if they were in solution. Following this protocol, several chiral catalysts, for example [Rh((+)-DIOP)(nbd)](SO$_3$CF$_3$) and [Rh((S)-BINAP)(nbd)](SO$_3$CF$_3$) (nbd=norbornadiene), have been immobilized and successfully employed for the enantioselective hydrogenation of prochiral alkenes [39c]. Recently, this SHB approach was successfully extended to arene hydrogenation through the immobilization of cationic catalysts, such as [Rh(diphos)(cod)] (SO$_3$CF$_3$), on silica containing supported palladium particles [39c]. Enhanced conversions to saturated cyclic hydrocarbons, as compared to the single components, were observed for the hydrogenation of benzenes and anthracenes [43].

A distinct class of single-site metal catalysts for arene hydrogenation is known under the name of surface organometallics. The surface organometallic technique has been introduced and largely developed by Basset and Marks, and is currently utilized in a number of catalytic processes [44]. Silica- or alumina-supported Ta, Ti, Zr and Hf hydrides, generated *in situ* by hydrogenation of alkyl derivatives, have been found capable of catalyzing the reduction of benzene and alkyl-substituted benzenes with TOFs as high as 1000 (Scheme 16.11) [44b,c].

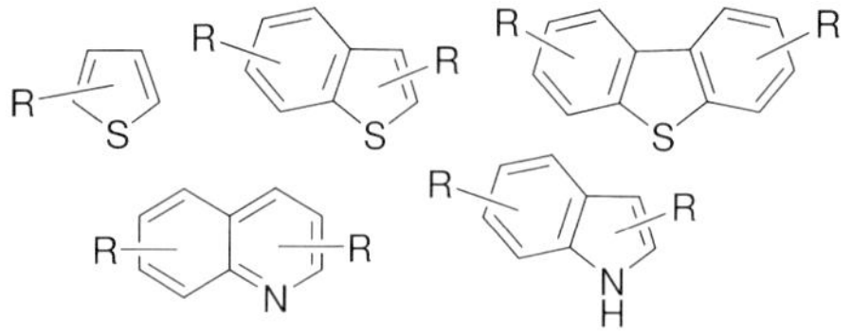

Fig. 16.6 Common *S*- and *N*-heterocycles contained in fossil fuels.

16.3
Hydrogenation of Heteroaromatics

16.3.1
Molecular Catalysts in Different Phase-Variation Systems

The number of homogeneous catalysts available for the hydrogenation of *N*-, *S*-and *O*-heteroaromatics is exceedingly greater than that of the catalysts for arenes. A crucial role in making the reduction of heteroaromatics easier than that of carbocyclic aromatic rings is just played by the heteroatom that possesses at least one σ-lone pair for occupying a coordination vacancy at the metal center. The heteroatom also has the effect of decreasing the overall aromatic character of the molecule, favoring the localization of electron density on the proximal X=C bond, hence allowing for the coordination of the substrate in the easily reducible olefin-like η^2-C-X mode (X=heteroatom) [8, 9].

A great impulse to design homogeneous catalysts for the hydrogenation heteroaromatics stems from the need to understand the mechanisms of the HDS, HDN, and HDO reactions [3]. These three processes form the heart of fossil fuels hydrotreatments, and have a vast commercial and environmental importance. It is not surprising, therefore, that most studies have been centered on the development of molecular catalysts for the hydrogenation of thiophenes, quinolines, and indoles (Fig. 16.6), as these substrates are largely abundant in crude oils and their degradation remains incomplete, even with the most efficient heterogeneous catalysts [3, 8, 9].

16.3.1.1 S-Heteroaromatics
The homogeneous hydrogenation of thiophenes and benzothiophenes to the corresponding cyclic thioethers is a reaction which is catalyzed, under relatively mild experimental conditions, by a number of metal complexes, generally comprising noble metals modified with phosphine ligands: $RuCl_2(PPh_3)_3$ [53], $RuHCl(CO)(PPh_3)_3$ [53], $RuH_2(\eta^2\text{-}H_2)(PCy_3)_2$ [54], $OsHCl(CO)(PPh_3)_3$ [53], $RhCl(PPh_3)_3$ [53], $[Rh(MeCN)_3(Cp^*)](BF_4)_2$ [55], $[Rh(PPh_3)_2(cod)]PF_6$ [53, 56], $[Ir(PPh_3)_2(cod)]PF_6$ [53, 57, 58], and $[Ru(MeCN)_3(triphos)](SO_3CF_3)_2$ [39b, 59, 60]. In contrast, no metal complex has been ever reported to hydrogenate diben-

zo[*b,d*]thiophene (DBT) to either tetrahydrodibenzothiophenes or hexahydrodi-benzothiophenes, which reflects the strong aromatic character of this substrate.

A common feature of all hydrogenation catalysts for thiophene (T) and ben-zo[*b*]thiophene (BT) is apparently a d^6 electronic configuration of the metal ion, which favors the η^2-C,C coordination of the thiophenic molecule over the alternative η^1-S bonding mode. The latter is more frequent for low-valent metal fragment and is precursor to C-S insertion, hence to hydrogenolysis rather than to hydrogenation [8, 9a]. As a general trend, the hydrogenation activity decreases in the order Ru^{II} > Rh^{III} > Os^{II} > Ir^{III} as well as with increasing nucleophilicity of the solvent that may compete with the substrate for coordination.

The highest activity for BT hydrogenation to dihydrobenzo[*b*]thiophene (DHBT) has been reported for the Ru^{II} catalyst [(triphos)RuH]$^+$ obtained by hydrogenation of the precursor [Ru(NCMe)$_3$(triphos)](SO$_3$CF$_3$)$_2$ in basic solvents capable of promoting the heterolytic splitting of H$_2$ (Scheme 16.12) [59]. Interestingly, the hydrogenation of [Ru(NCMe)$_3$(triphos)](SO$_3$CF$_3$)$_2$ in apolar or non-basic solvents (e.g., CH$_2$Cl$_2$) produced the 16-electron system [Ru(H)$_2$(triphos)]$^+$, which was slightly less active than the monohydride fragment (TOF = 1340) [39b, 60].

The hydrogenation mechanism of BT has been widely studied using a variety of techniques, including operando HP-NMR, kinetic studies, and deuterium labeling. A unique mechanism has been proposed, irrespective of the metal catalyst: η^2-C,C coordination of the substrate (eventually in equilibrium with η^1-S coordination), addition of H$_2$ in either oxidative [M(H)$_2$] or intact form [M(H$_2$)] (this step may also precede the previous one), hydride transfer to form dihydrobenzothienyl, and elimination of DHBT by hydride/dihydrobenzothienyl reductive coupling. Scheme 16.13 exemplifies this mechanism for a model catalyst bearing one hydride ligand, as is the case of the 14-electron fragment [RuH(triphos)]$^+$ [39b, 59, 60].

Kinetic studies of the hydrogenations of BT to DHBT catalyzed by [Rh(PPh$_3$)$_2$(cod)]PF$_6$ [56] and [Ir(PPh$_3$)$_2$(cod)]PF$_6$ [57] indicated the hydride migration yielding the dihydrobenzothienyl intermediate as the rate-determining step. In contrast, the rate-determining step of the reaction catalyzed by [RuH(triphos)]$^+$ was shown to be the reversible dissociation of DHBT from the metal center [59].

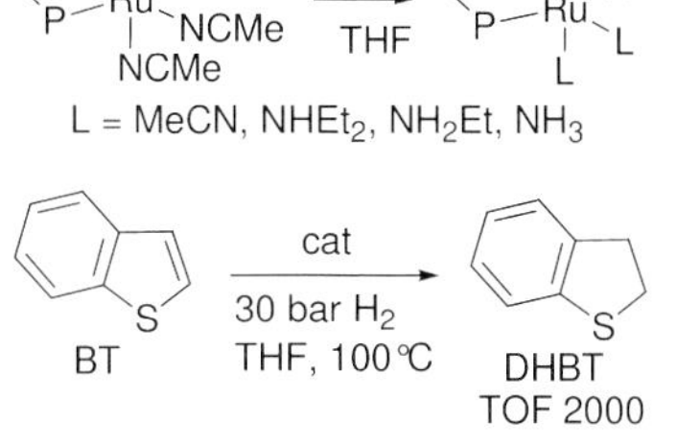

L = MeCN, NHEt$_2$, NH$_2$Et, NH$_3$

BT

DHBT
TOF 2000

Scheme 16.12

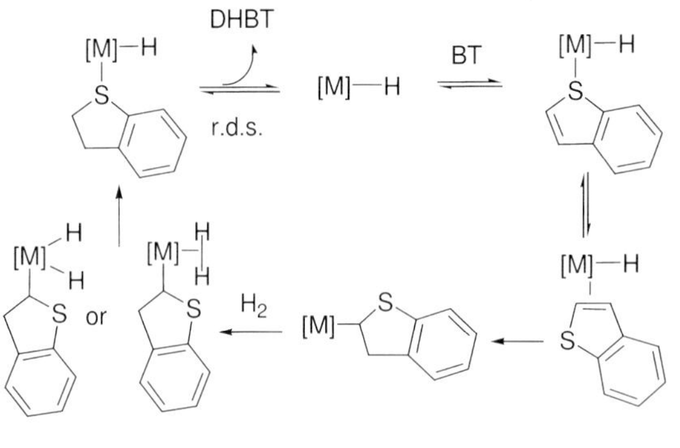

Scheme 16.13

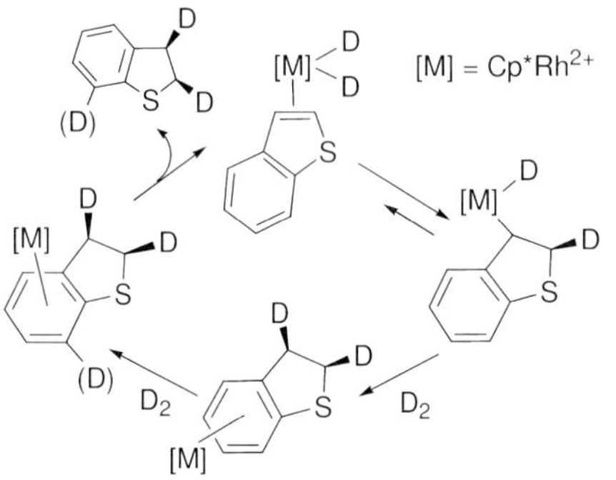

Scheme 16.14

Substituting deuterium for hydrogen gas in the reduction of BT to DHBT with the catalyst precursor $[Rh(NCMe)_3(Cp^*)](BF_4)_2$ has shown that the stereoselective *cis*-deuteration of the double bond is kinetically controlled by the η^2-*C,C* coordination of BT. The incorporation of deuterium in the 2- and 3-positions of unreacted substrate and in the 7-position of DHBT has been interpreted in terms of reversible double-bond reduction and arene-ring activation, respectively (Scheme 16.14) [55].

Overall, the hydrogenation of thiophene to tetrahydrothiophene (THT) is quite similar to that of BT, the only remarkable difference being the formation of a thioallyl complex via regio- and stereospecific hydride migration (*endo* migration). Scheme 16.15 shows the catalytic mechanism proposed for $IrH_2(\eta^1$-*S*-T)(PPh$_3$)$_2$]PF$_6$ [58]. Upon hydride addition, the thioallyl intermediate formed a 2,3-dihydrothiophene ligand which was then hydrogenated like any other alkene. The substitution of either 2,3- or 2,5-dihydrothiophene for thiophene showed that only the 2,3-isomer was hydrogenated to THT.

The use of water-soluble metal catalysts for the hydrogenation of thiophenes in aqueous biphasic systems has been primarily introduced by Sanchez-Delgado and coworkers at INTEVEP S.A. [61]. The precursors $RuHCl(TPPTS)_2(L_2)$ (TPPTS = triphenylphosphine trisulfonate; L = aniline, 1,2,3,4-tetrahydroquinoline) and $RuHCl(TPPMS)_2(L_2)$ (TPPMS = triphenylphosphine monosulfonate) were

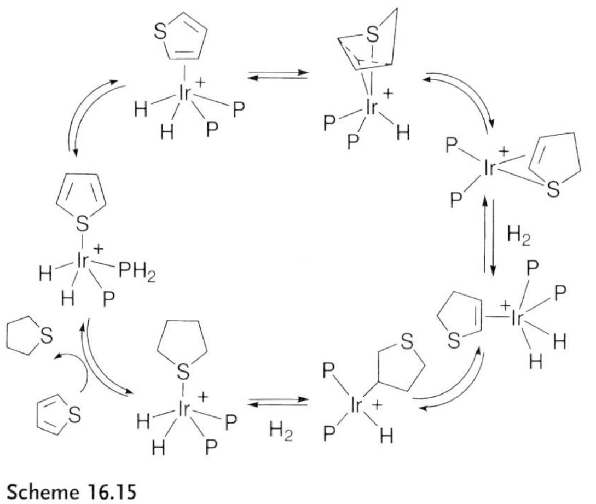

Scheme 16.15

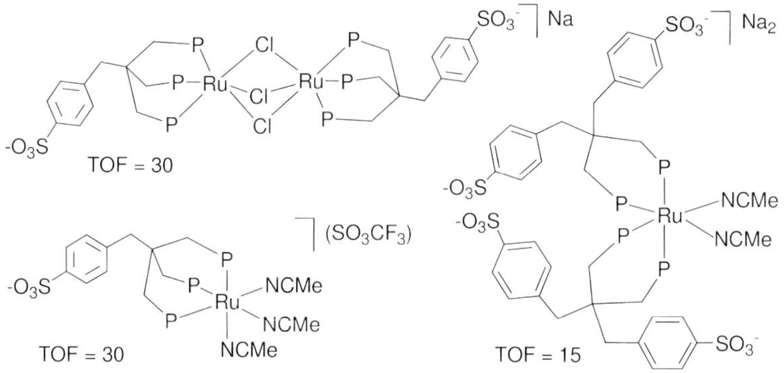

Fig. 16.7 Water-soluble polyphosphine metal catalysts used to hydrogenate thiophenes.

employed to hydrogenate BT to DHBT in water-decaline under relatively harsh experimental conditions (130–170 °C, 70–110 bar H_2). It was observed that nitrogen compounds did not inhibit the hydrogenation; on the contrary, a promoting effect was observed. Later, rhodium and ruthenium catalysts with the polydentate water-soluble ligands $(NaO_3S(C_6H_4)CH_2)_2C(CH_2PPh_2)_2$ ($Na_2DPPPDS$) [62] and $NaO_3S(C_6H_4)CH_2C(CH_2PPh_2)_3$ (Nasulphos) [63], which differ from traditional water-soluble phosphines for the presence of the hydrophilic groups in the ligand backbone were successfully employed to hydrogenate BT under biphasic conditions (Fig. 16.7) [39b, 59, 60, 64, 65].

The aqueous-biphase hydrogenation reactions of thiophenes to the corresponding cyclic thioethers have been shown to be mechanistically similar to those in truly homogenous phase.

16.3.1.2 *N*-Heteroaromatics

As a general trend, six-membered mononuclear *N*-heteroaromatics such as pyridine and derivatives are much less prone to undergo hydrogenation than bi- and trinuclear *N*-ring compounds (e.g., quinolines, benzoquinolines, acridines) due to their higher resonance stabilization energy.

The first examples of selective hydrogenation of pyridine to piperidine and of quinoline to 1,2,3,4-tetrahydroquinoline (THQ) by a homogeneous metal catalyst (Rh(PY)$_3$Cl$_3$/NaBH$_4$ in DMF under 1 bar H$_2$) were reported in 1970 by Jardine and McQuillin [66]. The first mechanistic studies appeared much later, when Fish employed the Rh^I and Ru^{II} precatalysts RhCl(PPh$_3$)$_3$ [67] and RuHCl(PPh$_3$)$_3$ [68] to hydrogenate various *N*-polyaromatics (85 °C, 20 bar H$_2$, benzene). The hydrogenation rates decreased in the order phenanthridine > acridine > quinoline >5,6-benzoquinoline (5,6-BQ) >7,8-BQ, which reflects the influence of both steric and electronic effects. All substrates were hydrogenated regioselectively at the heteroaromatic ring; only acridine was converted to a mixture of 9,10-dihydroacridine and 1,2,3,4-tetrahydroacridine. The hydrogenation of quinoline was found to be inhibited by the presence of pyridines and of THQ in the reaction mixture, due to competing coordination to the metal center, while all the other substrates had no appreciable effect on the hydrogenation rate.

The substitution of D$_2$ for H$_2$ in the reduction of quinoline catalyzed by either RhCl(PPh$_3$)$_3$ [67] or RuHCl(PPh$_3$)$_3$ [68] showed that: 1) hydrogenation of the C=N bond is reversible; 2) the C$_3$–C$_4$ double bond is irreversibly hydrogenated in stereoselective *cis* manner; and 3) the C$_8$–H bond in the carbocyclic ring is activated, likely *via* cyclometalation. Later, Fish studied the hydrogenation of 2-methylpyridine to 2-methylpiperidine catalyzed by [Rh(NCMe)$_3$Cp*]$^{2+}$, again by means of deuterium labeling experiments [55]. The rate-limiting step of the reaction was identified as being the initial C=N bond hydrogenation, which actually disrupts the aromaticity of the molecule. It was also proposed that the reversible reduction of the C=N and C=C bonds was promoted by the allylic nature of the reduction product, NH-CH$_2$-C=C, which is highly activated toward re-aromatization of the N-ring.

The reduction of 1,2,5,6-tetrahydropyridine (THPY) with D$_2$ in the presence of [Rh(NCMe)$_3$Cp*]$^{2+}$, yielding exclusive deuterium incorporation in the C$_3$ and C$_4$ carbon atoms, and the independent synthesis of [Rh(η^1(N)-THPY(NC-Me)$_2$Cp*]$^{2+}$ showed that: 1) η^1(N)-THPY complexes are not intermediate to piperidine production; and 2) partially hydrogenated N-heterocycles are easily dehydrogenated to their aromatic precursors [55].

Deuterium gas experiments, continuous NMR and GC/MS analysis, *in situ* high-pressure NMR spectra and the isolation of some intermediates provided Fish with sufficient information to propose the mechanism shown in Scheme 16.16 for the hydrogenation of quinoline to THQ, catalyzed by [Rh(NCMe)$_3$Cp*]$^{2+}$ (40 °C, 33 bar H$_2$, CH$_2$Cl$_2$) [55].

The salient features of this mechanism are:

- η^1(N) bonding of quinoline to rhodium with loss of complexed MeCN, followed by the formation of a hydride.

- Reversible 1,2-N=C bond hydrogenation, likely *via* η^2(N,C) coordination.
- Migration of Cp*Rh from nitrogen to the C_3–C_4 double bond.
- Reversible C_3–C_4 double bond reduction.
- Cp*Rh complexation to the carbocyclic ring, followed by C_6–H and C_8–H bond activation.
- η^6(πC) coordination of THQ, followed by ligand exchange with quinoline to continue the catalytic cycle.

In this mechanistic picture, the rhodium center goes through the catalysis with the unusual III → V → III oxidation/reduction cycle.

Various late transition-metal carbonyls, alone or modified by phosphine ligands, have been found to hydrogenate pyridine and polyaromatic heterocycles such as quinoline, 5,6-BQ, 7,8-BQ, acridine, and isoquinoline (IQ) using either H_2 obtained from water-gas-shift (WGS) or syngas (SG) [69, 70]. Selective hydrogenation of the heterocyclic ring has been achieved with $Fe(CO)_5$, $Mn_2(CO)_8$-$(PBu_3)_2$ and $Co_2(CO)_6(PPh_3)_2$ [24a]. The cobalt catalyst was the most active for the hydrogenation of both acridine to 9,10-dihydroacridine (TOF = 10) and quinoline to THQ (TOF = 14). The iron and manganese catalysts converted appreciably only acridine, with TOFs of 5 and 2 (or 10 under SG conditions), respectively. Under WGS conditions, $RuCl_2(CO)_2(PPh_3)_2$ and $Ru_4H_4(CO)_{12}$ proved to be inactive due to competitive coordination of CO, while efficient regioselective hydrogenation of the substrate was achieved using H_2 gas ($TOF_{acridine/9,10-dihydroacridine}$ = $TOF_{quinoline/THQ}$ = 5) [24a]. In all cases, however, high temperatures (180–200 °C) were required for appreciable conversions.

Using as catalyst precursors the clusters $Os_3H_2(CO)_{10}$ and $Os_3(CO)_{12}$ [71, 72], Laine and coworkers found a deuteration pattern of quinoline hydrogenation similar to that shown in Scheme 16.16, except for the presence of more deuterium in the 4-position and less in the 2-position, which has been interpreted in terms of the occurrence of oxidative addition of the osmium cluster to C–H bonds in quinoline, and also 1,4-hydrogenation (Scheme 16.17).

In an attempt to correlate the catalytic performance of comparable precursors with the nature of the metal center, Sánchez-Delgado and Gonzáles have investigated the selective hydrogenation of quinoline to THQ (150 °C, 30 bar H_2, toluene) by $RuCl_2(PPh_3)_3$ (TOF = 63), $RhCl(PPh_3)_3$ (TOF = 52), $RuHCl(CO)(PPh_3)_3$ (TOF = 29), $OsHCl(CO)(PPh_3)_3$ (TOF = 5), $[Rh(PPh_3)_2(cod)]^+$ (TOF = 199), and $[Ir(PPh_3)_2(cod)]^+$ (TOF = 17) [73]. The cationic rhodium complex was by far the most active.

Several ruthenium complexes have been found capable of hydrogenating N-heteroaromatics (acridine, quinoline, 5,6-BQ, 7,8-BQ, indole, IQ, for example: $[Ru(NCMe)_3(triphos)](SO_3CF_3)_2$ ($TOF_{indole/indoline}$ = 17) in conjunction with protic acids [59, 65a, 74–76], $[RuH(CO)(NCMe)_2(PPh_3)_2]BF_4$ ($TOF_{quinoline/THQ}$ = 16) [77,78], and $RuH_2(\eta^2\text{-}H_2)_2(PCy_3)_2$ ($TOF_{quinoline/5,6,7,8-THQ}$ = 2; $TOF_{indole/indoline}$ < 1) [79]. The latter complex also led to saturation of the aromatic ring, which has been proposed to involve the coordination of the substrate through the aromatic ring, in a manner similar to that reported for η^4-arene complexes (see Scheme

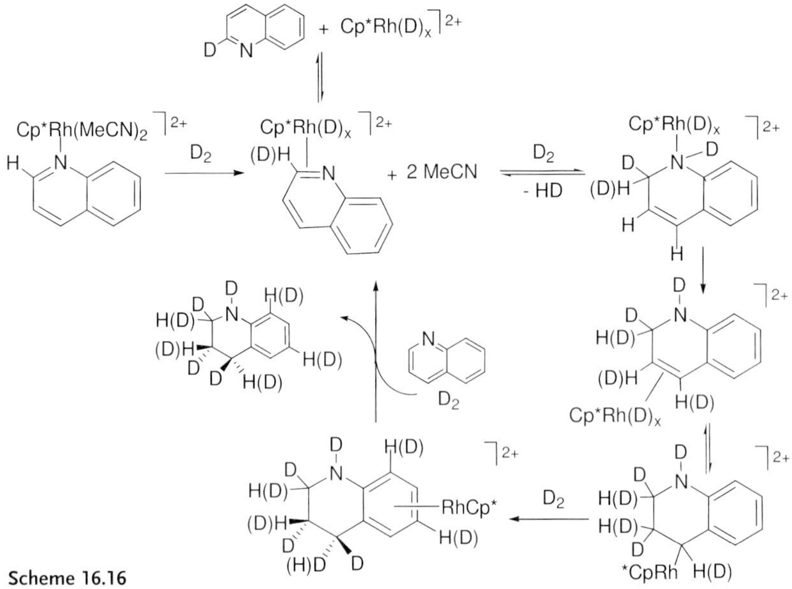

Scheme 16.16

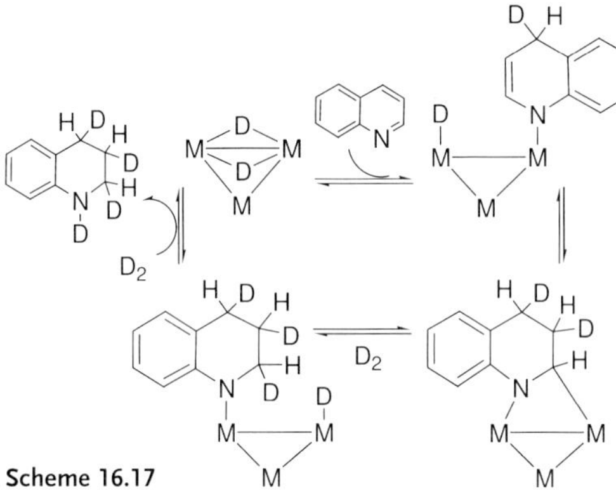

Scheme 16.17

16.4). However, it must be remembered that the true homogeneous nature of this system remains a matter of debate.

Kinetic studies of the hydrogenation of *N*-heteroaromatics have been reported wherein quinoline is the most studied substrate. Sánchez-Delgado and co-workers have identified the experimental rate law $r_i = k_{cat} [Rh][H_2]^2$, with $k_{cat} = 50 \pm 6 \ M^{-2} \ s^{-1}$ at 370 K for the hydrogenation of quinoline by $[Rh(PPh_3)_2(cod)]PF_6$ [80]. Kinetic studies for quinoline reduction to THQ have also been reported by Rosales for the reactions catalyzed by $[RuH(CO)(MeCN)(PPh_3)_2]BF_4$ [77]. At low hydrogen pressure, the experimental rate law was $r_i = k_{cat} [Ru_0][H_2]^2$ ($k_{cat} = 28.5 \ M^{-2} \ s^{-1}$ at 398 K), while a first-order dependence of the reaction rate with respect to the hydrogen concentration was observed at high H_2 pressure. The proposed mechanism involves a rapid and reversible partial hydrogenation of bonded quinoline, followed by a rate-determining second hydrogenation of dihydroquinoline.

A much more complex kinetic law has been reported by Bianchini and co-workers for the hydrogenation of quinoline catalyzed by the Rh^I complex $[Rh(DMAD)(triphos)]PF_6$ (DMAD = dimethyl acetylenedicarboxylate) [74, 75]. The rate was first order with respect to both H_2 in the pressure range from 4 to 30 bar, and in the catalyst concentration range from 36 to 110 mM, while the hydrogenation rate showed an inverse dependence with respect to quinoline concentration. The empiric rate law $r = k'' [Rh][H_2][Q]^2$, where $k'' = k (a + b[Q] + c[Q]^2)^{-1}$, was proposed to account for the inhibiting effect of quinoline (Q) concentration and the experimental observation that the rate tends to be second order for very low quinoline concentrations (< 30 mM) and zero-order for very high quinoline concentrations (> 70 mM). On the basis of the kinetic study, deuterium labeling and high-pressure NMR experiments under catalytic conditions, as well as the identification of catalytically relevant intermediates, a mechanism was proposed (Scheme 6.18) which essentially differs from that proposed by Sánchez-Delgado for the rate-limiting step (i.e., reversible reduction of the C=N bond instead of the irreversible one of the $C_3=C_4$ bond). The overall hydrogenation of the C=N bond, which actually disrupts the aromaticity of quinoline, was proposed as the rate-determining step, which was consistent with the fact that 2,3-dihydroquinoline was reduced faster than quinoline, while the lack of deuterium incorporation into the carbocyclic ring of both THQ and quinoline ruled out the formation of η^6-quinoline or η^6-THQ intermediates [74, 75].

The reduction of acridine to 9,10-dihydroacridine by the precursor $[RuH(CO)(NCMe)_2(PPh_3)_2]BF_4$ has been found to occur with the experimental rate law $r = k_{cat} [Ru][H_2]$ and the postulated mechanism involves, as the determining

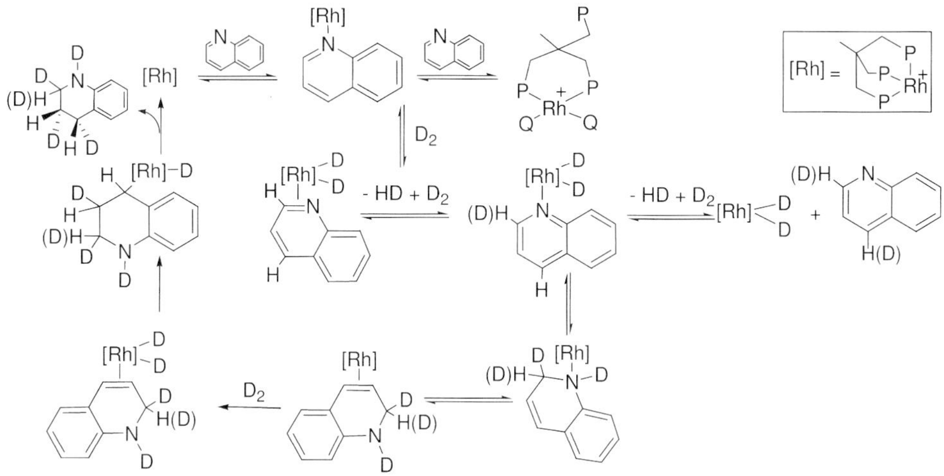

Scheme 16.18

step, the hydrogenation of coordinated acridine in $[RuH(CO)(\eta^1(N)-AC)$ $(NCMe)(PPh_3)_2]^+$ to yield 9,10-dihydroacridine and the coordinatively unsaturated complex $[RuH(CO)(NCMe)(PPh_3)_2]^+$ [78].

In homogeneous phase, indole is much more difficult to reduce than quinoline, as shown by the limited number of known catalysts (e.g., $RuHCl(PPh_3)_3$ [68] and $[RuH(CO)(NCMe)(PPh_3)_2]BF_4$ [78]) as well as their very scarce activity (TOFs ≤ 1). Indeed, the $\eta^1(N)$ coordination, which is critical for selective nitrogen ring reduction in quinoline, is virtually unknown for indole, which prefers to bind metal centers using the carbocyclic ring. In the latter coordination mode, the C=N bond is not activated and the many occupied coordination sites at the metal center make oxidative addition of H_2 very difficult to accomplish. Consistently, the hydrogenation of indole is generally inhibited when the reaction mixture contains basic substrates such as quinoline, THQ, and pyridine. The only catalysts that have proved able to regioselectively hydrogenate indole to indoline with an acceptable TOF are $[Rh(DMAD)(triphos)]PF_6$ and $[Ru(NCMe)_3(triphos)]$ $(SO_3CF_3)_2$, though on condition that a protic acid is added to the catalytic mixture [74, 76]. The rhodium catalyst was more efficient than the ruthenium catalyst, and allowed for hydrogenation of the substrate even at $60\,°C$ and 30 bar H_2, with TOFs as high as 100. It was shown experimentally that indoline was actually formed by reduction of the protonated form of indole, the 3H-indolium cation which possesses a localized C=N bond.

The selective hydrogenation of N-heterocycles has been achieved with the use of water-soluble Ru^{II} catalysts prepared *in situ* by reacting an excess of either triphenylphosphine trisulfonate (TPPTS) or triphenylphosphine monosulfonate (TPPMS) with $RuCl_3 \cdot 3H_2O$. The resulting solutions were added to a hydrocarbon solution containing various N-heteroaromatics such as quinoline, acridine, and IQ. The biphasic reactions were performed under relatively drastic experimental conditions ($130–170\,°C$, $70–110$ bar H_2) and led to selective reduction of the heterocyclic ring [61, 81].

The regioselective reduction of quinoline to THQ in water/hydrocarbon has also been achieved with bidentate and tridentate water-soluble ligands. The Rh^I complex $[Rh(H_2O)_2(DPPPDS)]Na$ was employed in water/n-octane to hydrogenate 1:1 mixtures of quinoline and BT at high temperature ($160\,°C$). Only the N-heterocycle was efficiently reduced (TOF = 50), with BT hydrogenation to DHBT being only marginal (TOF = 2) [8c]. A similar selectivity has been reported for the catalytic system $RuCl_3 \cdot H_2O/2Na_2DPPPDS$ prepared *in situ*. In contrast, the individual hydrogenation rates for quinoline and BT have been reported to be similar (TOF = 30 at $140\,°C$, 30 bar H_2, water/n-heptane) and independent of the presence of either substrate by using the binuclear precursor $Na[\{(sulphos)Ru\}_2(\mu\text{-Cl})_3]$ (sulphos = $(PPh_2CH_2)_3CCH_2(C_6H_4)SO_3^-$) [8c, 82].

Under biphasic conditions, the zwitterionic Rh^I complex $Rh(cod)(sulphos)$ proved to be very efficient for the hydrogenation of quinoline to THQ (TOF = 20 at $160\,°C$, 30 bar H_2, water/n-heptane) [8c].

16.3.1.3 **O-Heteroaromatics**

Very few examples of hydrogenation of O-heteroaromatics with molecular metal catalysts have appeared in the literature to date. Besides some cases of enantioselective catalysis (see Section 16.4), there is only one example reported by Fish dealing with the homogeneous hydrogenation of benzofuran to 2,3-dihydrobenzofuran using [Rh(NCMe)$_3$(Cp*)](BF$_4$)$_2$ as the catalyst precursor [55]. As for the hydrogenation of BT performed by the same catalyst, the hydrogenation of benzofuran has been proposed to involve coordination of the substrate through the 2,3 double bond to a Rh–H species, followed by hydrogen transfer to yield 2,3-dihydrobenzofuran.

16.3.2
Molecular Catalysts Immobilized on Support Materials

Rh(PPh$_3$)$_3$Cl tethered to 2% cross-linked phosphinated polystyrene-divinylbenzene was the first heterogenized single-site metal catalyst to be used in the hydrogenation of N- and S-heteroaromatics (see Fig. 16.3) [37]. This catalyst was able to hydrogenate quinoline, acridine, 5,6-BQ and 7,8-BQ in benzene solution (85 °C, 20 bar H$_2$) with an order of activity (acridine > quinoline > 5,6-BQ > 7,8-BQ) that is identical to that in homogeneous phase with the unsupported catalysts, except for the initial rates that were from 10- to 20-fold faster than in homogeneous phase [67]. This remarkable rate enhancement was attributed to steric requirements surrounding the active metal center in the tethered complex, which apparently would favor the coordination of the N-heterocycles by disfavoring that of PPh$_3$. The regioselectivity of hydrogenation was even higher than that in homogeneous phase as no formation of 1,2,3,4-tetrahydroacridine was observed. The deuteration pattern of the heteroaromatic ring after a catalytic reaction with D$_2$ was identical to that observed in homogeneous phase, except for the absence of deuterium incorporation at position 8 of the carbocyclic ring. The tethered catalyst proved able also to hydrogenate BT to DHBT (benzene, 85 °C, 20 bar H$_2$) with rates three-fold faster than those observed in homogeneous phase with the parent precursor RhCl(PPh$_3$)$_3$ [37].

The most active and fully recyclable single-site catalyst for the hydrogenation of thiophenes is still that generated by the SHB precursor [Ru(NCMe)$_3$(sulphos)](OSO$_2$CF$_3$)/SiO$_2$ (Ru^{II}/SiO$_2$), obtained by tethering [Ru(NCMe)$_3$(sulphos)](OSO$_2$CF$_3$) to silica (Scheme 16.19). In this case, immobilization of the molecular catalyst involves the formation of hydrogen-bonds to the surface silanols by SO$_3^-$ groups from both the sulphos ligand and the triflate counter-anion [39b]. Upon hydrogenation (30 bar H$_2$), Ru^{II}/SiO$_2$ has been found to generate a very active, recyclable and stable catalyst for the selective hydrogenation of BT to DHBT, with TOFs as high as 2000. The TOF with Ru^{II}/SiO$_2$ did not practically change even when a new feed containing 2000 equiv. BT in *n*-octane was injected into the reactor after 1 h reaction, which means that DHBT does not compete with BT for coordination to the Ru^{II} center.

All attempts to hydrogenate thiophenes by using TCSM catalysts of the types shown in Figures 16.4 and 16.5 have, so far, been unsuccessful. Rh^{I}-Pd^0/SiO$_2$

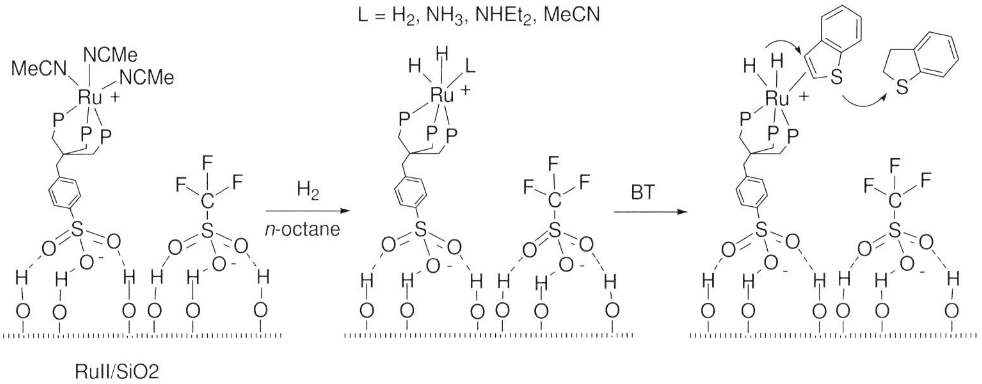

L = H₂, NH₃, NHEt₂, MeCN

Scheme 16.19

was tested in the hydrogenation (30 bar H₂) of BT in *n*-octane under 30 bar at 100–170 °C, but the production of a DHBT was the same as that obtained with silica-supported Pd⁰ nanoparticles alone (TOF = 8–10) [43]. Apparently, no synergistic effect between the isolated rhodium sites and the surface palladium atoms takes place for the hydrogenation of thiophenes. This was not totally unexpected, as neither silica-supported Rh(cod)(sulphos)/SiO₂ in *n*-octane [43] nor free Rh(cod)(sulphos) [83] in MeOH or [Rh(cod)(triphos)]PF₆ [83] in THF proved able to hydrogenate appreciably BT and thiophene below 150–170 °C. In fact, at these high temperatures hydrogenolysis to the corresponding thiol occurred [83 b, c]. In contrast, the SHB rhodium complexes Rh(cod)(sulphos)/SiO₂ and [Ru(NCMe)₃(sulphos)](SO₃CF₃)/SiO₂ have been used successfully to hydrogenate quinoline in *n*-octane (100 °C, 30 bar H₂), yielding selectively THQ with TOFs as high as 100 [43]. In line with the behavior of the Fish catalyst RhCl(PPh₃)₃/P [37], both Rh(cod)(sulphos)/SiO₂ and [(sulphos)Ru(NCMe)₃] (SO₃CF₃)/SiO₂ have been found to be more efficient catalysts than the homogeneous and aqueous-biphasic counterparts with triphos or sulphos ligands. The rate enhancement observed for the heterogeneous reactions has been attributed to the fact that, unlike in fluid solution systems, the heterogenized complexes do not undergo dimerization to give catalytically inactive species.

The supported complex [Rh(cod)(POLYDIPHOS)]PF₆, obtained by stirring a CH₂Cl₂ solution of [RhCl(cod)]₂ and Bu₄NPF₆ in the presence of a diphenyl-phosphinopropane-like ligand tethered to a cross-linked styrene/divinylbenzene matrix (POLYDIPHOS), forms an effective catalyst for the hydrogenation of quinoline (Fig. 16.8) [84]. Under relatively mild experimental conditions (80 °C, 30 bar H₂), quinoline was mainly converted to THQ, though appreciable formation of both 5,6,7,8-THQ and decahydroquinoline also occurred (Scheme 16.20).

An effective catalyst recycling with no loss of catalytic activity was accomplished by removing the liquid phase via the liquid sampling valve and re-charging the autoclave with a solution containing the substrate. In all cases, no rhodium leaching occurred. Remarkably, the hydrogenation activity of the 1,3-bis-

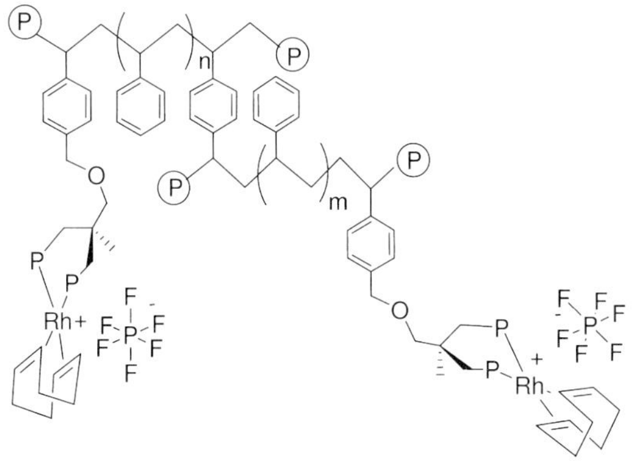

Fig. 16.8 Schematic of a diphosphine rhodium complex covalently tethered to a cross-linked styrene/divinylbenzene matrix, used for the hydrogenation of quinoline.

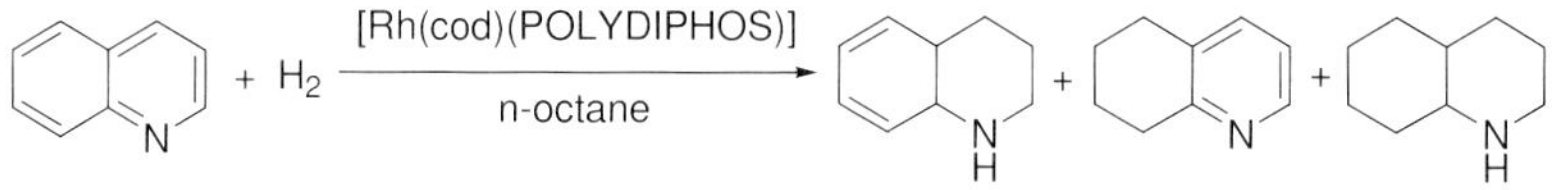

Scheme 16.20

diphenylphosphinopropane complex $[Rh(dppp)(cod)]PF_6$ in THF was much lower, as well as being selective, for THQ.

16.4
Stereoselective Hydrogenation of Prochiral Heteroaromatics

16.4.1
Molecular Catalysts in Homogeneous Phase

The enantioselective hydrogenation of prochiral heteroaromatics is of major relevance for the synthesis of biologically active compounds, some of which are difficult to access via stereoselective organic synthesis [4]. This is the case for substituted N-heterocycles such as piperazines, pyridines, indoles, and quinoxalines. The hydrogenation of these substrates by supported metal particles generally leads to diastereoselective products [4], while molecular catalysts turn out to be more efficient in enantioselective processes. Rhodium and chiral chelating diphosphines constitute the ingredients of the vast majority of the known molecular catalysts.

Relevant examples of enantioselective hydrogenation of aromatic N-heterocycles are given below. Scheme 16.21 shows the hydrogenation of a 2-ester substituted piperazine to the corresponding 2-substituted pyrazine with a catalyst

[Rh(nbd)₂]BF₄ / Fe

50 bar H₂, MeOH
70 °C

COOtBu

COOtBu

yield 41%, ee 78%

Scheme 16.21

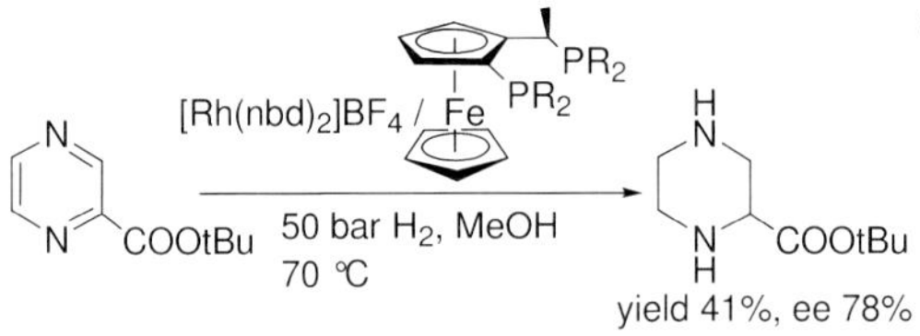

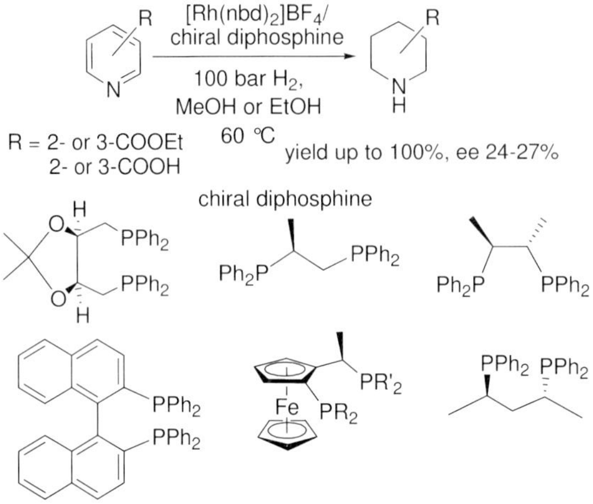

R

[Rh(nbd)₂]BF₄/
chiral diphosphine

100 bar H₂,
MeOH or EtOH
60 °C

R

yield up to 100%, ee 24-27%

R = 2- or 3-COOEt
2- or 3-COOH

chiral diphosphine

PPh₂
PPh₂

Ph₂P
PPh₂

Ph₂P
PPh₂

PPh₂
PPh₂

Fe
PR'₂
PR₂

PPh₂ PPh₂

Scheme 16.22

prepared *in situ* by mixing [Rh(nbd)₂Cl]₂ with a *Josiphos*-type ferrocenyldiphosphine, preferentially 1-[1(R)-(dicyclohexylphosphino)ethyl]-2(S)-(diphenylphosphino)ferrocene [85]. Under relatively mild conditions, the conversions were low, but the ee-values were quite satisfactory.

Josiphos-rhodium systems have been also used to hydrogenate 2- or 3-substituted pyridines and furans, yet both the activities (TOF = 1–2) and enantioselectivities were rather low (Scheme 16.22) [86, 87]. Comparable results were obtained with a number of chiral chelating diphosphines of various symmetries.

The diphosphines leading to the formation of six- or seven-membered metallarings have been found to give higher ee-values as compared to 1,2-diphosphines. With most ligands, the 2- or 3-substituted furans were hydrogenated with much lower enantioselectivity (ee 1–7%). Only the *Josiphos* ligand with R = t-Bu gave a significant ee (24%) for the reduction of substituted furans, yet the activity was almost negligible (3%) [85]. It is worth noting that black precipitates were observed in some experiments, which may indicate catalyst decomposition.

Excellent ee-values (up to 94%) have been obtained for the hydrogenation of various 2-substituted N-acetyl indoles with an *in-situ* prepared rhodium catalyst modified with the *trans*-chelating diphosphine (S,S)-(R,R)-2,2''-bis[1-(diphenylphosphino)ethyl]-1,1''-biferrocene (Scheme 16.23) [88]. A strong base was required as co-reagent to observe both high conversion (TOFs of 50–100) and en-

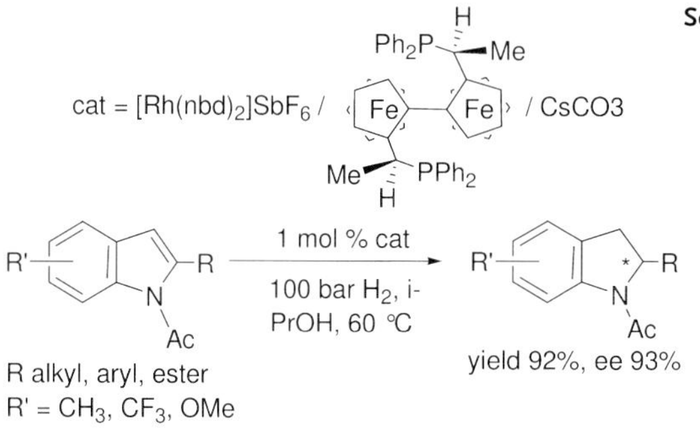

Scheme 16.23

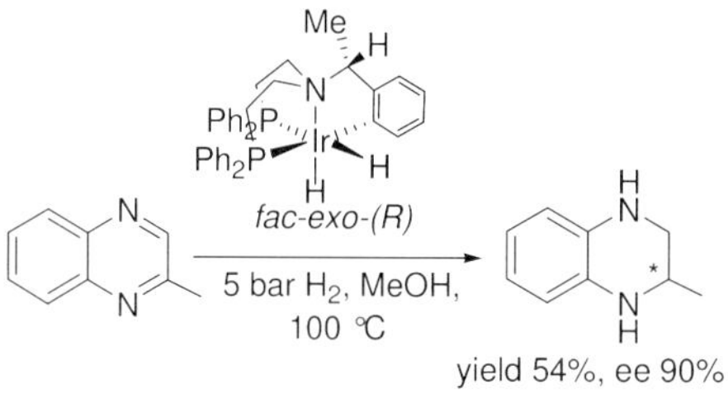

Scheme 16.24

antioselectivity. Best results were achieved with catalytic systems comprising $CsCO_3$ and $[Rh(nbd)_2]SbF_6$.

A quite different ligand system has been found to generate a selective iridium catalyst for the hydrogenation of 2-methylquinoxaline to (–)-(2S)-2-methyl-1,2,3,4-tetrahydroquinoxaline (Scheme 16.24) [89]. Unlike all previous examples of enantiomeric hydrogenation, an isolated catalyst precursor, namely the Ir^{III} *o*-metalated dihydride *fac-exo-(R)*-[IrH_2{C_6H_4C*H(Me)N(CH_2CH_2PPh_2)_2}], was employed. Under quite mild experimental conditions, ee-values of up to 90% were obtained at 50% conversion, while at 100% conversion the ee decreased to 75%. An *operando* high-pressure NMR study showed that the catalytically active species was generated by de-orthometalation rather than by H_2-reductive elimination. It was also shown that the two C=N moieties of 2-methylquinoxaline were reduced at comparable rates. Notably, the use of the *fac-exo-(S)* dihydride precursor gave the product with opposite configuration, that is (+)-(2R)-2-methyl-1,2,3,4-tetrahydroquinoxaline [89].

The only other example of enantioselective hydrogenation of 2-methylquinoxaline has been reported by Murata and coworkers, who used a [(+)-(DIOP)RhH] catalyst to produce 2-methyl-1,2,3,4-tetrahydroquinoxaline, albeit in 3% ee [90].

Scheme 16.25

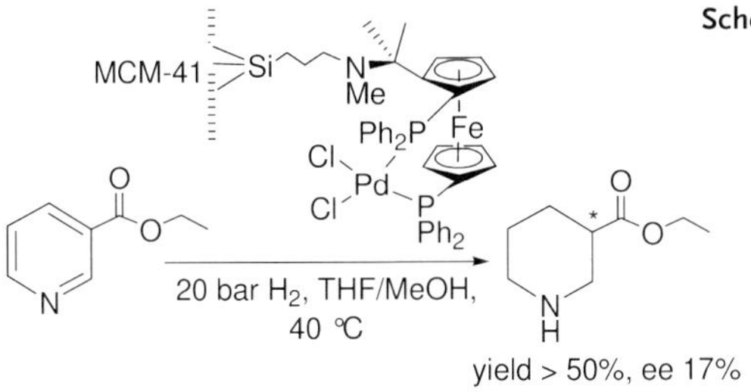

yield > 50%, ee 17%

16.4.2
Molecular Catalysts Immobilized on Support Materials

The enantioselective hydrogenation of ethyl nicotinate to ethyl nipecotinate is a difficult process of which only a few heterogeneous examples are known, generally catalyzed by Pd/C modified with supported chiral auxiliaries [91]. No example in homogeneous phase has been reported to date. Palladium(II) complexes with the chelating ligand (S)-1-[(R)-1,2′-bis(diphenylphosphino)ferrocene are equally inactive, but their immobilization onto the inner walls of MCM-41 has surprisingly generated an effective catalyst, albeit with a modest ee (Scheme 16.25) [92]. Nonetheless, this reaction deserves to be highlighted for the elegant approach to heterogenization as well as for developing the concept of catalyst confinement as an innovative method to magnify both the catalytic activity and the asymmetric induction [92, 93].

Abbreviations

4H-An	1,2,3,4-tetrahydroanthracene
8H-An	1,2,3,4,5,6,7,8-octahydroanthracene
BQ	benzoquinoline
BT	benzo[b]thiophene
DBT	dibenzo[b,d]thiophene
DHBT	dihydrobenzo[b]thiophene
DVB	divinylbenzene
ee	enantiomeric excess
HDN	hydrodenitrogenation
HDO	hydrodeoxygenation
HDS	hydrodesulfurization
IQ	isoquinoline
PTA	1,3,5-triaza-7-phosphadamantane
SG	syngas
SHB	supported hydrogen-bonded
TCSM	tethered complexes on supported metals
THPY	1,2,5,6-tetrahydropyridine

THQ   1,2,3,4-tetrahydroquinoline
THT   tetrahydrothiophene
TOF   turnover frequency
TPPMS   triphenylphosphine monosulfonate
TPPTS   triphenylphosphine trisulfonate
WGS   water-gas-shift

References

1 S. Siegel. In: B. M. Trost, I. Fleming (Eds.), *Comprehensive Organic Synthesis*. Volume 9. Pergamon Press, New York, **1991**.

2 A. F. Noels, A. J. Hubert. In: A. Montreux, F. Petit (Eds.), *Industrial Applications of Homogeneous Catalysis*. Riedel Publishing Company, Dordrecht, Netherlands, **1988**.

3 (a) T. Kabe, A. Ishihara, W. Qian, *Hydrodesulfurization and Hydrodenitrogenation*, Wiley-VCH, Tokyo, Japan, **1999**; (b) H. Topsøe, B. S. Clausen, F. E. Massoth, *Hydrotreating Catalysis*, Springer, Heidelberg, Germany, **1996**.

4 T. J. Donohoe, R. Garg, C. A. Stevenson, *Tetrahedron: Asymmetry* **1996**, *7*, 317.

5 H.-U. Blaser, C. Malan, B. Pugin, F. Spindler, H. Steiner, M. Studer, *Adv. Synth. Catal.* **2003**, *345*, 103.

6 M. F. Semmelhack. In: E. W. Abel, F. G. A. Stone, G. Wilkinson (Eds.), *Comprehensive Organometallics Chemistry II*. Pergamon, New York, **1995**, Volume 12, p. 979.

7 P. A. Chaloner, M. A. Esteruelas, F. Joo, L. A. Oro, *Homogeneous Hydrogenation*, Kluwer Academic Publishers, Dordrecht, Netherlands, **1993**.

8 (a) C. Bianchini, A. Meli, F. Vizza, *J. Organomet. Chem.* **2004**, *689*, 4277; (b) C. Bianchini, A. Meli, F. Vizza. In: B. Cornils, W. A. Herrmann (Eds.), *Applied Homogeneous Catalysis with Organometallic Compounds*. Wiley-VCH, New York, **2002**, Volume 3, p. 1099; (c) C. Bianchini, A. Meli, F. Vizza, *Eur. J. Inorg. Chem.* **2001**, 43.

9 (a) R. A. Sánchez-Delgado, *Organometallic Modeling of the Hydrodesulfurization and Hydrodenitrogenation Reactions*, Kluwer Academic, Dordrecht, Netherlands, **2002**;

(b) R. H. Fish. In: R. Ugo (Ed.), *Aspects of Homogeneous Catalysis*. Kluwer Academic Publishers, Dordrecht, Netherlands, **1990**, Volume 7, p. 65.

10 (a) B. R. James, *Homogeneous Hydrogenation*, Wiley, New York **1973**; (b) B. R. James. In: G. Wilkinson, F. G. A. Stone, E. Abel (Eds.), *Comprehensive Organometallic Chemistry*. Pergamon Press, Oxford, **1982**, Volume 8, Chapter 51.

11 (a) P. N. Rylander, *Hydrogenation Methods*, Academic Press, New York, **1990**; (b) S. Nishimur, *Hand-book of Heterogeneous Catalytic Hydrogenation for Organic Synthesis*, Wiley, New York, **2001**; (c) J. G. Donkervoort, E. G. M. Kuijpers. In: R. A. Sheldon, H. van Bekkum (Eds.), *Fine Chemicals through Heterogeneous Catalysis*. Wiley-VCH, Weinheim, **2001**.

12 (a) R. Burmeister, A. Freund, P. Panster, T. Tacke, S. Wieland, *Stud. Surf. Sci. Catal.* **1995**, *92*, 343; (b) T. Q. Hu, B. R. James, J. S. Retting, C.-L. Lee, *Can. J. Chem.* **1977**, *75*, 1234.

13 (a) E. L. Muetterties, F. J. Hirsekorn, *J. Am. Chem. Soc.* **1974**, *96*, 4063; (b) E. L. Muetterties, F. J. Hirsekorn, *J. Am. Chem. Soc.* **1974**, *96*, 7920; (c) E. L. Muetterties, M. C. Rakowski, F. J. Hirsekorn, W. D. Larson, V. J. Bauss, F. A. L. Anet, *J. Am. Chem. Soc.* **1975**, *97*, 1266; (d) M. C. Rakowski, F. J. Hirsekorn, L. S. Stuhl, E. L. Muetterties, *Inorg. Chem.* **1976**, *15*, 2379; (e) L. S. Stuhl, M. C. Rakowski, A. Du Bois, F. J. Hirsekorn, J. R. Bleeke, A. E. Stevens, E. L. Muetterties, *J. Am. Chem. Soc.* **1978**, *100*, 2405; (f) J. R. Bleeke, E. L. Muetterties, *J. Am. Chem. Soc.* **1981**, *103*, 556.

14 A. J. Sivak, E. L. Muetterties, *J. Am. Chem. Soc.* **1979**, *101*, 4878.

15 C. Bianchini, K.G. Caulton, K. Folting, A. Meli, M. Peruzzini, A. Polo, F. Vizza, *J. Am. Chem. Soc.* **1992**, *114*, 7290.

16 C. Bianchini, K.G. Caulton, C. Chardon, M.L. Doublet, O. Eisenstein, S.A. Jackson, T.J. Johonson, A. Meli, M. Peruzzini, W.E. Streib, A. Vacca, F. Vizza, *Organometallics* **1994**, *13*, 2010.

17 C.R. Landis, J. Halpern, *Organometallics* **1983**, *2*, 840.

18 A.F. Borowski, S. Sabo-Etienne, B. Chaudret, *J. Mol. Catal. A: Chem.* **2001**, *174*, 69.

19 C.J. Boxwell, P.J. Dyson, D.J. Ellis, T. Welton, *J. Am. Chem. Soc.* **2002**, *124*, 9334.

20 (a) L. Plasseraud, G. Süss-Fink, *J. Organomet. Chem.* **1997**, *539*, 163; (b) E.G. Fidalgo, L. Plasseraud, G. Süss-Fink, *J. Mol. Cat. A: Chem.* **1998**, 132, 5; (c) M. Faure, A.T. Vallina, H. Stoeckli-Evans, G. Süss-Fink, *J. Organomet. Chem.* **2001**, *621*, 103.

21 P.J. Dyson, D.J. Ellis, G. Laurenczy, *Adv. Synth. Catal.* **2003**, 345.

22 G. Süss-Fink, M. Faure, T.R. Ward, *Angew. Chem. Int. Ed.* **2002**, *41*, 99.

23 (a) R.A. Grey, G.P. Pez, A. Wallo, *J. Am. Chem. Soc.* **1980**, *102*, 5949; (b) R. Wilczynski, W.A. Fordyce, J. Halpern, *J. Am. Chem. Soc.* **1983**, *105*, 2066; (c) D.E. Linn, J. Halpern, *J. Am. Chem. Soc.* **1987**, *109*, 2969.

24 (a) R.H. Fish, A.D. Thormodsen, G.A. Cremer, *J. Am. Chem. Soc.* **1982**, *104*, 5234; (b) R.H. Fish, *Ann. N. Y. Acad. Sci.* **1983**, *415*, 292.

25 M.A. Bennet, T.-N. Huang, T.W. Turney, *J. Chem. Soc. Chem. Commun.* **1979**, 312.

26 M.J. Russell, C. White, P.M. Maitlis, *J. Chem. Soc. Chem. Commun.* **1977**, 427.

27 D.A. Tocker, R.O. Gould, T.A. Stephenson, M.A. Bennett, J.P. Ennett, T.W. Matheson, L. Sawyer, V.K. Shah, *J. Chem. Soc., Dalton Trans.* **1983**, 1571.

28 A. Widegren, R.G. Finke, *J. Mol. Catal. A: Chem.* **2003**, *198*, 317.

29 J.P. Collman, K.M. Kosydar, M. Bressan, W. Lamanna, T. Garrett, *J. Am. Chem. Soc.* **1984**, *106*, 7228.

30 (a) J.S. Yu, B.C. Ankianiec, M.T. Nguyen, I.P. Rothwell, *J. Am. Chem. Soc.* **1992**, *114*, 1927; (b) B.C. Ankianiec, P.E. Fanwick, I.P. Rothwell, *J. Am. Chem. Soc.* **1991**, *113*,

4710; (c) V.M. Visciglio, J.R. Clark, M.T. Nguyen, D.R. Mulford, P.E. Fanwick, I.P. Rothwell, *J. Am. Chem. Soc.* **1997**, *119*, 3490; (d) I.P. Rothwell, *J. Chem. Soc. Chem. Commun.* **1997**, 1331; (e) J.R. Clark, P.E. Fanwick, I.P. Rothwell, *J. Chem. Soc. Chem. Commun.* **1995**, 553.

31 M. Feder, J. Halpern, *J. Am. Chem. Soc.* **1975**, *97*, 7186.

32 J. Birch, *Quart. Rev.* **1950**, *4*, 69.

33 J.S. Yu, I.P. Rothwell *J. Chem. Soc. Chem. Commun.* **1992**, 632.

34 D. Durand, G. Hillion, C. Lassau, C. Sajus, US Patent 4.271.323, **1981**.

35 (a) A. Alvanipour, L.D. Kispert, *J. Mol. Cat.* **1998**, *48*, 277; (b) M.F. Sloan, A.S. Matlack, D.S. Breslow, *J. Am. Chem. Soc.* **1963**, *85*, 4014; (c) S.J. Lapporte, W.R. Schuett *J. Org. Chem.* **1963**, *28*, 1947; (d) S.J. Lapporte, *Ann. N.Y. Acad. Sci.* **1969**, *158*, 510.

36 I. Wender, R. Levine, M. Orchin, *J. Am. Chem. Soc.* **1950**, *72*, 4375.

37 R.H. Fish, A.D. Thormodsen, H. Heinemann, *J. Mol. Catal.* **1985**, *31*, 191.

38 (a) H. Gao, R.J. Angelici, *J. Am. Chem. Soc.* **1997**, *119*, 6937; (b) H. Gao, R.J. Angelici, *Organometallics* **1999**, *18*, 989; (c) H. Gao, R.J. Angelici, *J. Mol. Catal. A: Chem.* **1999**, *145*, 83.

39 (a) C. Bianchini, D.G. Burnaby, J. Evans, P. Frediani, A. Meli, W. Oberhauser, R. Psaro, L. Sordelli, F. Vizza, *J. Am. Chem. Soc.* **1999**, *121*, 5961; (b) C. Bianchini, V. Dal Santo, A. Meli, W. Oberhauser, R. Psaro, F. Vizza, *Organometallics* **2000**, *19*, 2433; (c) C. Bianchini, P. Barbaro, V. Dal Santo, R. Gobetto, A. Meli, W. Oberhauser, R. Psaro, F. Vizza, *Adv. Synth. Catal.* **2001**, *343*, 41.

40 H. Yang, H. Gao, R.J. Angelici, *Organometallics* **2000**, *19*, 622.

41 R. Abu-Reziq, D. Avnir, I. Miloslavski, H. Schumann, J. Blum, *J. Mol. Catal. A* **2002**, *185*, 179.

42 C. Bianchini, V. Dal Santo, A. Meli, S. Moneti, M. Moreno, W. Oberhauser, R. Psaro, L. Sordelli, F. Vizza, *Angew. Chem. Int. Ed.* **2003**, *42*, 2636.

43 C. Bianchini, F. Vizza (manuscript in preparation).

44 (a) M.S. Eisen, T.J. Marks, *J. Am. Chem. Soc.* **1992**, *114*, 10358; (b) H. Ahn,

C. P. Nicholas, T. J. Marks, *J. Am. Chem. Soc.* **2003**, *125*, 4325; (c) C. Coperet, M. Chabanas, R. P. Saint-Arroman, J. M. Basset, *Angew. Chem. Int. Ed.* **2003**, *42*, 157.

45 K. J. Klabunde, B. B. Anderson, M. Bader, L. J. Radonovich, *J. Am. Chem. Soc.* **1978**, *100*, 1313.

46 K. Jonas, *Angew. Chem. Int. Ed. Engl.* **1985**, *24*, 295.

47 D. Pieta, A. M. Trzeciak, J. J. Ziolkowiski, *J. Mol. Cat.* **1983**, *18*, 193.

48 M. Trzeciak, T. Glowiak J. Ziolkowiski, *J. Organomet. Chem.* **1998**, *552*, 159.

49 K. R. Januszklewicz, H. Alper, *Organometallics,* **1983**, *2*, 1055.

50 (a) J. W. Johnson, E. L. Muetterties, *J. Am. Chem. Soc.* **1977**, *99*, 7395; (b) M. Y. Darensburg, E. L. Muetterties *J. Am. Chem. Soc.* **1978**, *100*, 7425.

51 T. J. Lynch, M. Banah, H. D. Kaesz, C. R. Porter, *J. Org. Chem.* **1984**, *49*, 1266.

52 P. J. Dyson, D. J. Ellis, D. G. Parker, T. Welton *J. Chem. Soc. Chem. Commun.* **1999**, 25.

53 R. A. Sánchez-Delgado, E. González, *Polyhedron* **1989**, *8*, 1431.

54 A. F. Borowski, S. Sabo-Etienne, B. Domadieu, B. Chaudret, *Organometallics* **2003**, *22*, 4803.

55 E. Baralt, S. J. Smith, I. Hurwitz, I. T. Horváth, R. H. Fish, *J. Am. Chem. Soc.* **1992**, *114*, 5187.

56 R. A. Sánchez-Delgado, V. Herrera, L. Rincón, A. Andriollo, G. Martín, *Organometallics* **1994**, *13*, 553.

57 V. Herrera, A. Fuentes, M. Rosales, R. A. Sánchez-Delgado, C. Bianchini, A. Meli, F. Vizza, *Organometallics* **1997**, *163*, 2465.

58 C. Bianchini, A. Meli, M. Peruzzini, F. Vizza, V. Herrera, R. A. Sánchez-Delgado, *Organometallics* **1994**, *13*, 721.

59 C. Bianchini, A. Meli, S. Moneti, W. Oberhauser, F. Vizza, V. Herrera, A. Fuentes, R. A. Sánchez-Delgado, *J. Am. Chem. Soc.* **1999**, *121*, 7071.

60 C. Bianchini, V. Dal Santo, A. Meli, S. Moneti, M. Moreno, W. Oberhauser, R. Psaro, L. Sordelli, F. Vizza, *J. Catal.* **2003**, *213*, 47.

61 (a) INTEVEP S. A. (D. E. Páez, A. Andriollo, R. A. Sánchez-Delgado, N. Valencia, R. E. Galiasso, F. A. López), US Patent 5,753,584, **1998**; (b) INTEVEP S. A. (D. E. Páez, A. Andriollo, R. A. Sánchez-Delgado, N. Valencia, R. E. Galiasso, F. A. López), **1999**, US Patent 5,958,223; (c) INTEVEP S. A. (D. E. Páez, A. Andriollo, R. A. Sánchez-Delgado, N. Valencia, R. E. Galiasso, F. A. López), US Patent 5,981,421, **1999**; (d) M. A. Busolo, F. Lopez-Linares, A. Andriollo, D. E. Páez, *J. Mol. Catal. A.* **2002**, *189*, 211.

62 (a) CNR (C. Bianchini, A. Meli, F. Vizza), **1999**, PCT/EP97/06493; (b) CNR (C. Bianchini, A. Meli, F. Vizza), **1996**, IT FI96A000272.

63 C. Bianchini, P. Frediani, V. Sernau, *Organometallics* **1995**, *14*, 5458.

64 I. Rojas, F. López-Linares, N. Valencia, C. Bianchini, *J. Mol. Catal. A* **1999**, *144*, 1.

65 (a) C. Bianchini, A. Meli, S. Moneti, F. Vizza, , *Organometallics* **1998**, *17*, 2636; (b) C. Bianchini, D. Masi, A. Meli, M. Peruzzini, F. Vizza, F. Zanobini, *Organometallics* **1998**, *17*, 2495.

66 I. Jardine, F. J. McQuillin, *J. Chem. Soc. D* **1970**, 626.

67 R. H. Fish, J. L. Tan, A. Thormodsen, *J. Org. Chem.* **1984**, *49*, 4500.

68 R. H. Fish, J. L. Tan, A. Thormodsen, *Organometallics* **1985**, *4*, 1743.

69 R. M. Laine, D. W. Thomas, L. W. Cary, *J. Org. Chem.* **1979**, *44*, 4964.

70 K. Kindler, D. Mathies, *Chem. Abstr.* **1960**, *54*, 19731b.

71 R. M. Laine, *New J. Chem.* **1987**, *11*, 543.

72 A. Eisenstadt, C. M. Giandomenico, M. F. Frederick, R. M. Laine, *Organometallics* **1985**, *4*, 2033.

73 R. A. Sánchez-Delgado, E. Gonzalez, *Polyhedron* **1989**, *8*, 1431.

74 M. Macchi, PhD Dissertation, Università di Trieste (Italy), **1999**.

75 C. Bianchini, P. Barbaro, M. Macchi, A. Meli, F. Vizza, *Helv. Chim. Acta* **2001**, *84*, 2895.

76 P. Barbaro, C. Bianchini, A. Meli, M. Moreno, F. Vizza, *Organometallics* **2002**, *21*, 1430.

77 M. Rosales, Y. Alvarado, M. Boves, R. Rubio, H. Soscun, R. Sánchez-Delgado, *Trans. Met. Chem.* **1995**, *20*, 246.

78 M. Rosales, J. Navarro, L. Sanchez, A. Gonzales, Y. Alvarado, R. Rubio,

C. De la Cruz, T. Rajmankina, *Trans. Met. Chem.* **1996**, *21*, 11.

79 A. F. Borowski, S. Sabo-Etienne, B. Domadieu, B. Chaudret, *Organometallics* **2003**, *22*, 1630.

80 R. A. Sánchez-Delgado, D. Rondón, A. Andriollo, V. Herrera, G. Martin, B. Chaudret, *Organometallics* **1993**, *12*, 4291.

81 D. E. Páez, A. Andriollo, F. López-Linares, R. E. Galiasso, J. A. Revete, R. A. Sánchez-Delgado, A. Fuentes, *Am. Chem. Soc. Div. Fuel Chem. Symp. Prepr.* **1998**, *43*, 563.

82 I. Rojas, F. López-Linares, N. Valencia, C. Bianchini, *J. Mol. Cat. A* **1999**, *144*, 1.

83 (a) C. Bianchini, A. Meli, V. Patinec, V. Sernau, F. Vizza, *J. Am. Chem. Soc.* **1997**, *119*, 4945; (b) C. Bianchini, J. Casares, A. Meli, V. Sernau, F. Vizza, R. A. Sánchez-Delgado, *Polyhedron* **1997**, *16*, 3099; (c) C. Bianchini, V. Herrera, M. V. Jiménez, A. Meli, R. A. Sánchez-Delgado, F. Vizza, *J. Am. Chem. Soc.* **1995**, *117*, 8567.

84 C. Bianchini, M. Frediani, G. Mantovani, F. Vizza, *Organometallics* **2001**, *20*, 2660.

85 Lonza AG (R. Fuchs) EP 0803502 A2, **1997**.

86 C. Bianchini, P. Barbaro, G. Giambastiani, S. Parisel, *Coord. Chem. Rev.* **2004**, *248*, 2131.

87 M. Studer, C. Wedemeyer-Exl, F. Spindler, H.-U. Blaser, *Monatsh. Chem.* **2000**, *131*, 1335.

88 R. Kuwano, K. Sato, T. Kurosawa, D. Karube, Y. Ito, *J. Am. Chem. Soc.* **2000**, *122*, 7614.

89 C. Bianchini, P. Barbaro, G. Scapacci, E. Farnetti, M. Graziani, *Organometallics* **1998**, *17*, 3308.

90 S. Murata, T. Sugimoto, S. Matsuura, *Heterocycles* **1987**, *26*, 883.

91 H.-U. Blaser, H. Konig, M. Studer, C. Wedemeyer-Exl, *J. Mol. Catal. A: Chem.* **1999**, *139*, 253.

92 S. A. Raynor, J. M. Thomas, R. Raja, B. F. G. Johnson, R. G. Bell, M. D. Mantle, *Chem. Commun.* **2000**, 1925.

93 B. F. G. Johnson, S. A. Raynor, D. S. Shephard, T. Mashmayer, J. M. Thomas, G. Sankar, S. Bromley, R. Oldroyd, L. Gladden, M. D. Mantle, *Chem. Commun.* **1999**, 1167.

17
Homogeneous Hydrogenation of Carbon Dioxide

Philip G. Jessop

17.1
Introduction

Carbon dioxide (CO_2) fixation for synthetic purposes requires either reduction or coupling reactions. Reduction of CO_2 by catalytic hydrogenation has been extensively studied. Industrial application at present is minor; CO_2 is included with CO in the feed for the currently-practiced heterogeneous hydrogenation to methanol, while Lurgi Öl-Gas-Chemie has explored the hydrogenation of CO_2 without CO as an alternative process [1, 2]. Homogeneously catalyzed hydrogenation of CO_2 is not in production at this time, although it has been 35 years since the first catalysts for the production of formamides were discovered by Haynes and coworkers [3, 4]. During the past 15 years, much more efficient catalysts have been developed; before 1990, the only system to give high yields of formic acid (>1 000 TON) operated at 160 °C [5], but now a catalyst giving 32 000 TON at only 50 °C is known [6]. Even greater improvements have been observed for the synthesis of formamides by CO_2 reduction. Unfortunately, the same progress has not been observed for homogeneously catalyzed hydrogenation of CO_2 to other products such as methanol or oxalic acid.

This chapter presents an overview of the primary findings from 1970 to the present. Hydrogenation using H_2 as the reductant will be described, although there are examples of the electrocatalytic reduction of CO_2 [7–10] and the use of other reductants [11]. More detailed reviews on homogeneous hydrogenation of CO_2 have been published, covering the years up to 1994 [12, 13] and from 1995 to 2003 [14].

The motivation for studying or contemplating industrial applications of CO_2 hydrogenation does not stem from a desire to use up excess CO_2 in order to lessen global warming. If the H_2 that is used in the reduction is derived from either the water-gas shift reaction (WGSR) or from steam reforming of methane, then 1 or 0.25 equivalents of CO_2, respectively, were produced per molecule of H_2. Thus, the hydrogenation of CO_2 to formic acid, using H_2 from WGSR, results in neither a net consumption nor a net production of CO_2. Hy-

The Handbook of Homogeneous Hydrogenation.
Edited by J.G. de Vries and C.J. Elsevier
Copyright © 2007 WILEY-VCH Verlag GmbH & Co. KGaA, Weinheim
ISBN: 978-3-527-31161-3

	$\Delta G°$ (kJ/mol)	$\Delta H°$ (kJ/mol)	$\Delta S°$ J/(mol•K)
$CO_2(g) + H_2(g) \rightleftharpoons HCO_2H(l)$	32.8	-31.5	-216
$CO_2(g) + H_2(g) + NH_3(aq) \rightleftharpoons HCO_2^-(aq) + NH_4^+(aq)$	-9.5	-84.3	-250
$CO_2(g) + H_2(g) + NHMe_2(liq) \rightleftharpoons HC(O)NMe_2(liq) + H_2O(l)$	na	-239	na
$CO_2(g) + 3H_2(g) \rightleftharpoons CH_3OH(l) + H_2O(l)$	-9.5	-131	-409

Scheme 17.1 The thermodynamics of CO_2 reductions.

drogenation of CO_2 to formic acid using H_2 from methane does result in the net consumption of CO_2, but in order to consume a significant portion of the excess CO_2 one would have to synthesize a ridiculously large amount of formic acid!

Multiple products are possible from CO_2 hydrogenation, but all of the products are entropically disfavored compared to CO_2 and H_2 (Scheme 17.1). As a result, the reactions must be driven by enthalpy, which explains why formic acid is usually prepared in the presence of a base or another reagent with which formic acid has an exothermic reaction. Of the many reduction products that are theoretically possible, including formic acid, formates, formamides, oxalic acid, methanol, CO, and methane, only formic acid and its derivatives are readily prepared by homogeneous catalysis.

17.2
Reduction to Formic Acid

Formic acid, a synthetic precursor and a commercial product for use in the leather, agriculture and dye industries, is traditionally prepared by carbonylation of NaOH at elevated pressure and temperature [15, 16]. The use of toxic CO begs the question; can a competitive route be found using CO_2 instead? Given that CO and H_2 are roughly interchangeable thanks to the WGSR and that CO_2 in the post-Kyoto world is essentially free, CO_2/H_2 should be competitive with CO if sufficiently active and inexpensive catalysts can be identified. Of course, the economics are considerably more complicated than that simple analysis: complicating factors include the costs of water addition or removal and the costs of the bases used. Nevertheless, the homogeneous hydrogenation of CO_2 as an alternative route to formic acid has been the subject of a large number of studies, starting with the first reports in the mid-1970s [17, 18].

Hydrogenation of CO_2 to formic acid (Eq. (1)) is thermodynamically unfavorable unless a base is present (see Scheme 17.1); the proton transfer to the base drives the reaction. In the absence of a base, the reaction usually fails completely or provides only small yields [19, 20], even though the initial rate of reaction was high in one case [21]. For the reaction in water, NaOH, bicarbonates, carbonates and even dialkylamines [22] are used as the base. For the reaction in organic solvents, amines are most commonly used, but some amines are decidedly superior to others in this respect. Whilst NEt$_3$ is most often used because it is inexpensive, far greater rates are observed (at least for the RuCl(OAc) (PMe$_3$)$_4$-catalyzed hydrogenation) if an amidine or guanidine base such as 1,8-diazabicyclo-[5.4.0]-undec-7-ene (DBU) is used instead [6]. The increase in rate with some bases has never been satisfactorily explained, but could be related to a base-assisted step in the mechanism (e.g., a hydrogen-bonding interaction between metal-bound H$_2$ and base during hydrogenolysis of a metal formate; see Section 17.2.1) or possibly to the ability of amidine bases to solubilize CO$_2$ in the form of bicarbonate in wet organic solvents [23]. Whichever base is used, there remains the question of how to separate the base from the formic acid product. The addition of an acid to the raw amine/formic acid mixture would liberate the formic acid and leave the amine in the form of a salt [24]. A base-exchange [25, 26] with a high-boiling amine such as imidazole would allow for thermal decomposition of the salt to give free formic acid and recyclable amine [27, 28].

$$CO_2 + H_2 \rightleftharpoons HCO_2H \tag{1}$$

The equilibrium yield of formic acid is limited by the amount of base. Many reports describing the hydrogenation in aqueous solution indicate yields of up to 1 mol formic acid per mol base, although Karakhanov was able to obtain 1.34 mol per mol NHEt$_2$ in water at 155 °C [29]. Higher yields of 1.6 to 1.9 are found for the reaction in organic solvents [18] and in supercritical CO$_2$ (scCO$_2$) [30, 31]. Because the reaction in the absence of base is thermodynamically unfavorable, added base must still stabilize the formic acid product, even after 1 equiv. acid has been produced and essentially all of the base is protonated. In fact, HCO$_2$H and NEt$_3$ are known [32–34] to exist in relatively stable adducts of various ratios, including 1:1, 2:1, and 3:1, plus unprotonated amine (even at 3:1 acid:amine ratios) and oligomeric chains of formic acid molecules ending with an amine. It is not clear, given that these higher ratios are possible, why the yield in practice is limited to a ratio of 1.9:1, or lower.

Solvent choice is important for optimizing the rate of this reaction. One of the most active catalysts to date, RhCl(TPPTS)$_3$, is used in water, despite the very poor solubility of H$_2$ in that medium. For the hydrophobic catalysts, the best solvents so far have been scCO$_2$ and polar aprotic solvents. The supercritical medium is highly effective because of the enormous concentrations of CO$_2$ and H$_2$ in the same phase as the catalyst, and also the elimination of mass-transfer limitations on the transfer of these reagent gases into traditional liquid solvents. These advantages are particularly important for the RuH$_2$(PMe$_3$)$_4$ and RuCl(OAc)(PMe$_3$)$_4$ catalyst precursors [30, 31, 35] because the hydrogenation rate is first order with respect to both H$_2$ and CO$_2$ [36]. For the same catalysts in conventional liquid solvents, the rate was found to decrease in the order DMSO > MeOH > MeCN > THF > H$_2$O. Performing the reaction in MeOH/NEt$_3$ and diluting the solvent with hexane causes the rate to decrease, a result which was attributed to the decrease in the dielectric constant of the medium [36]. Similarly, dissolving large amounts of ethane or fluoroform (CHF_3) into the MeOH/NEt$_3$ mixture causes the rate to drop or rise, respectively [36]. Rhodium(I) catalyst precursors [20, 37] have similar dependence of rate on solvent polarity, with DMSO always being excellent and less-polar aprotic solvents such as acetone, THF, or benzene being inferior. However, methanol was found to be better than DMSO for RhCl(PPh$_3$)$_3$ [37] and decidedly inferior for [RhCl(COD)]$_2$/dppb [20]. The overall need for a highly polar solvent is not related to gas solubility (H$_2$ is more soluble in the nonpolar solvents), but may be related to the ability of polar solvents such as DMSO to increase the entropy of the formic acid product by disrupting hydrogen bonding [20].

Even in aprotic solvents, the hydrogenation with most catalysts requires at least a small quantity of water or an alcohol as a co-catalyst or promoter; using thoroughly dried solvents leads to poor rates of reaction. This phenomenon was first observed by Inoue et al. [18] for Pd(dppe)$_2$, but has since been found to exist also for many of the Rh- and Ru-based catalysts. However, for the catalyst precursor [RhCl(COD)]$_2$/dppb, added water was found to inhibit the hydrogenation, and added molecular sieves to remove trace water were found to help the rate [20]. For those systems which require water, the amount needed is typically very small; on the order of a few equivalents per catalyst [6, 18]. There are so many possible roles for water in a hydrogenation mechanism that it is very difficult to demonstrate conclusively how the water may be helping. Possibilities include hydrogen bonding during CO_2 insertion, hydrolysis of formate ligands, capturing CO_2 in the form of bicarbonate, or even supplying protons in an ionic hydrogenation. Some alcohols can be more effective than water at promoting the reaction. Munshi et al. [6] surveyed a large number of alcohols as co-catalysts for the RuCl(OAc)(PMe$_3$)$_4$ and found the most effective alcohols to be those that have an aqueous-scale pK_a below that of the protonated amine (10.7 for HNEt$_3^+$) and that have a potentially coordinating conjugate base. Thus, 3,5-(CF$_3$)$_2$C$_6$H$_3$OH and triflic acid, which fulfill both requirements, are particularly effective, whereas MeOH and water are only moderately active because they are insufficiently acidic, and 2,6-tBu$_2$C$_6$H$_3$OH and HBF$_4$ have no activity because their conjugate bases are too sterically encumbered or insufficiently nucleophilic to have any coordinating ability.

A large number of catalysts or catalyst precursors for the reaction in Eq. (1) have been identified (Table 17.1). Almost all of these are complexes of Rh and Ru, although there are a few, less active, examples of Ir, Pd, Ni, Fe, Ti, and Mo. The most active catalysts are RhCl(TPPTS)$_3$, Rh(hfacac)(dcpb) and RuCl(OAc)(PMe$_3$)$_4$, with little difference between them after rough correction of the turnover frequencies (TOF) for pressure and temperature differences [38]. Musashi and Sakaki [39] argue that the barrier to CO_2 insertion into M–H bonds increases in the order Rh(I) < Ru(II) < Rh(III) because Rh(III)–formate bonds are too weak and Ru(II)–H bonds are too strong compared to the case of Rh(I).

Table 17.1 Homogeneous catalyst precursors for the hydrogenation of CO_2 to formic acid.

Catalyst precursor	Base	$P_{H2, CO2}$[a] [atm]	Temperature [°C]	Time [h]	TON	TOF[b] [h^{-1}]	Reference(s)
Ruthenium							
$Ru_2(CO)_5(dppm)_2$	NEt_3	38, 38	RT	21	2160	103	40
$RuCl_3$, PPh_3	NEt_3	60, 60	60	5	200	40	41
$RuH_2(PPh_3)_4$	NEt_3	25, 25	RT	20	87	4	18
$RuH_2(PPh_3)_4$	Na_2CO_3	25, 25	100	4	169	42	42
$RuH_2(PMe_3)_4$	NEt_3	85, 120	50	1	1400	1400	30
$RuCl_2(PMe_3)_4$	NEt_3	80, 140	50	47	7200	153	31
$RuCl(OAc)(PMe_3)_4$	NEt_3	70, 120	50	0.3	32 000	95 000	6
$TpRuH(PPh_3)(CH_3CN)$	NEt_3	25, 25	100	16	1815	113	43, 44
$[Ru(Cl_2bpy)_2(H_2O)_2][O_3SCF_3]_2$	NEt_3	30, 30	150	8	5000	625	45
$[(C_5H_4(CH_2)_3NMe_2)Ru(dppm)]BF_4$	None	40, 40	80	16	8	0.5	46
$[RuCl_2(CO)_2]_n$	NEt_3	81, 27	80	0.3	400	1300	24
$K[RuCl(EDTA-H)]$	None	3, 17	40	0.5	na	250	21
$[RuCl_2(TPPMS)_2]_2$	$NaHCO_3$	60, 35	80	0.03	320	9600	47
$[RuCl(C_6Me_6)(DHphen)]Cl$	KOH	30, 30	120	24	15 400	642	48
$CpRu(CO)(\mu\text{-dppm})Mo(CO)_2Cp$	NEt_3	30, 30	120	45	43	1	49
Rhodium							
$[RhCl(COD)]_2$+dppb	NEt_3	20, 20	RT	22	1150	52	25
$[RhCl(COD)]_2$+dippe	NEt_3	40 total	24	18	205	11	50
$[RhH(COD)]_4$+dppb	NEt_3	40 total	RT	18	2200	122	20
$RhCl(PPh_3)_3$	Na_2CO_3	60, 55	100	3	173	58	51
$RhCl(PPh_3)_3$	NEt_3	20, 40	25	20	2700	125	37
$RhCl(TPPTS)_3$	$NHMe_2$	20, 20	81	0.5		7260	22, 52
$[RhCl(\eta^2\text{-CYPO})_2]BPh_4$	NEt_3	25, 25	55	4.2	420	100	53
$Rh(hfacac)(dcpb)$	NEt_3	20, 20	25	–	–	1335	54
$[Rh(NBD)(PMe_2Ph)_3]BF_4$	None	48, 48	40	48	128	3	19
$RhCl_3$+PPh_3	$NHMe_2$	10, 10	50	10	2150	215	29
Palladium							
$Pd(dppe)_2$	NEt_3	25, 25	110	20	62	3	18
$Pd(dppe)_2$	NaOH	24, 24	RT	20	17	0.9	55
$PdCl_2$	KOH	110, na	160	3	1580	530	5
$PdCl_2(PPh_3)_2$	NEt_3	50, 50	RT	na	15	na	26

Table 17.1 (continued)

Catalyst precursor	Base	$P_{H2, CO2}$[a] [atm]	Tempera- ture [°C]	Time [h]	TON	TOF[b] [h^{-1}]	Reference(s)
Other metals							
TiCl$_4$/Mg	None	1, 1	RT	na	15	na	17
Ni(dppe)$_2$	NEt$_3$	25, 25	RT	20	7	0.4	18
NiCl$_2$(dcpe)	DBU	40, 160	50	216	4400	20	56
FeCl$_3$/dcpe	DBU	40, 60	50	7.5	113	15	56
MoCl$_3$/dcpe	DBU	40, 60	50	7.5	63	8	56
[Cp*IrCl(DHphen)]Cl	KOH	30, 30	120	10	21 000	2100	48

a) In some cases, the pressure of CO$_2$ was not given and was calculated from the
total stated pressure minus the pressure of H$_2$.
b) The TOF values are not directly comparable to each other because some are at
complete conversion and some are at partial conversion. They can, however, give
an order of magnitude indication. Initial TOF values will be even higher.
na = data not available; RT = room temperature.

17.2.1
Insertion Mechanisms

Carbon dioxide is known to readily insert into a metal–hydride bond to give a
metal formate [57, 58]; this forms the first step in insertion mechanisms of CO$_2$
hydrogenation (Scheme 17.2). Both this insertion step and the return path from
the formate complex to the hydride, generating formic acid, have a number of
possible variations.

Insertion of CO$_2$ into a metal–hydride bond normally requires the prior dissociation of an ancillary ligand to generate a coordinatively unsaturated complex, because CO$_2$ coordination to the metal usually precedes the formal insertion
(Scheme 17.3, lower pathway). *Ab initio* calculations [59] support this mechanism
for the insertion of CO$_2$ into the Ru–H bond of RuH$_2$(PH$_3$)$_4$, a model for the catalyst RuH$_2$(PMe$_3$)$_4$. However, it is theoretically possible for CO$_2$ insertion to take
place without prior CO$_2$ coordination (Scheme 17.3, upper pathway) [60, 61]. The

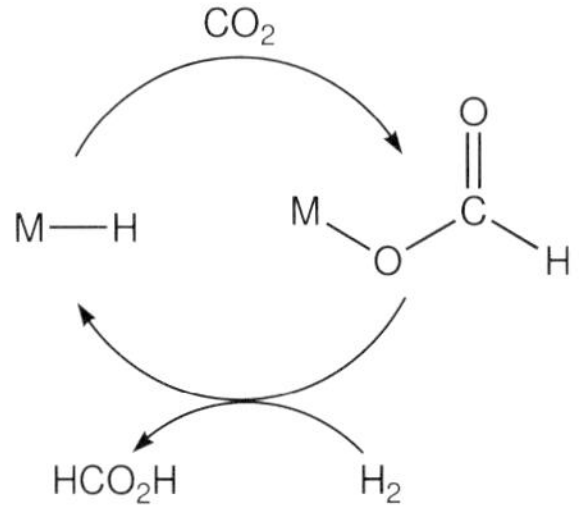

Scheme 17.2 Simplified insertion mechanism.

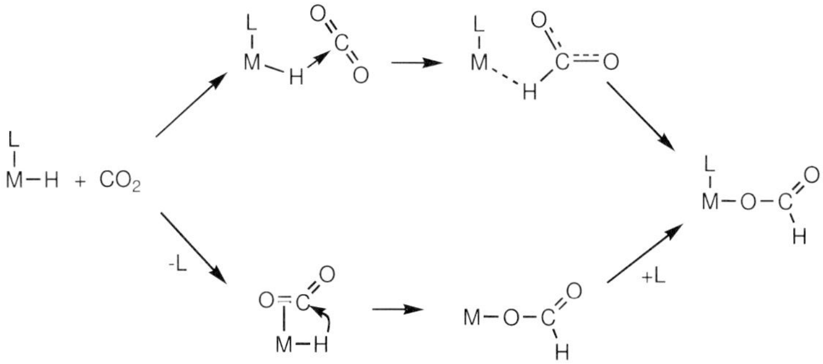

Scheme 17.3 Mechanisms of CO_2 insertion into a metal-hydrogen bond. "L" represents a potentially dissociable ligand. Ancillary ligands are not shown.

upper pathway of Scheme 17.3 was supported for the CO_2 insertion step of the catalyst in Eq. (2) by hybrid density functional calculations [62, 63].

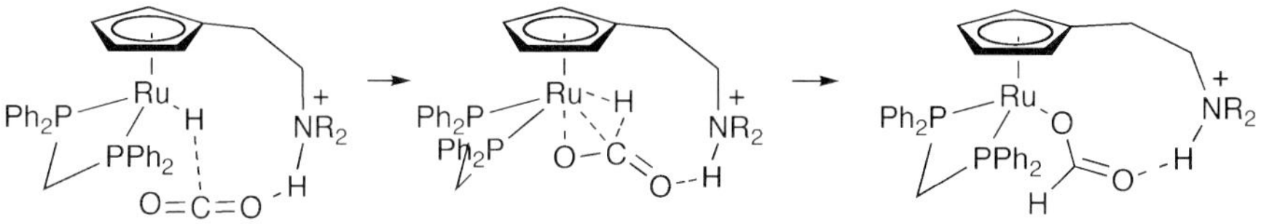

(2)

Hydrogen-bonding during the CO_2 insertion step may bring down the kinetic barrier and thereby explain the promoting effect of water and alcohols. Tsai and Nicholas [19] first proposed this for their system based on the catalyst precursor $[Rh(nbd)(PMe_2Ph)_3]BF_4$ (Scheme 17.4a). However, for the insertion of CO_2 into the Rh–H bonds of $[RhH_2(PH_3)_2(H_2O)]^+$ and $[RhH_2(PH_3)_3]^+$, proposed to be models of the actual catalyst in Tsai and Nicholas' system, Musashi and Sakaki's calculations [39] show that for this particular system the barrier-lowering effect of water during the attack of CO_2 is due not to hydrogen-bonding but to differing trans effects present in the complexes with and without water. In contrast, for the $TpRuH(PPh_3)(MeCN)$ system, hydrogen-bonding to external water was proposed (Scheme 17.4b). Density functional theory (DFT) calculations suggest that this interaction reduces the transition state energy by only 23 kJ mol^{-1} [43].

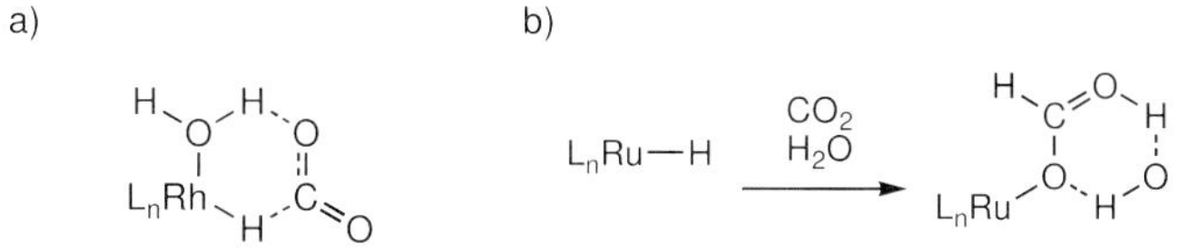

a)   b)

Scheme 17.4 Possible hydrogen-bonding interactions assisting in the insertion of CO_2 for the (a) $[Rh(nbd)(PMe_2Ph)_3]BF_4$ [19] and (b) $TpRuH(PPh_3)(MeCN)$ systems.

The hydrogen-bonding interaction in Eq. (2) greatly decreases the activation energy for CO_2 insertion [62, 63].

The formate complex that results from CO_2 insertion can liberate formic acid and return to the starting hydride species by one of three return pathways, the most obvious of which is hydrogenolysis (path A of Scheme 17.5). Hydrogenolysis involves either oxidative addition of H_2 or coordination of H_2 as a molecular hydrogen ligand, followed by elimination of formic acid, although technically it is possible for hydrogenolysis via a "sigma-bond metathesis" mechanism without prior H_2 coordination. Evidence for the hydrogenolysis mechanism for the intermediate $[M(O_2CH)(CO)_5]^-$ (M = Cr, W) was the fact that the related carboxylate species $[M(O_2CMe)(CO)_5]^-$ reacts with MeOH to form $MeCO_2Me$ only when H_2 is present [64]. In that system, H_2 was believed to bind to a site made available by prior dissociation of a carbonyl ligand. The hydrogenolysis pathway was also proposed for the catalytic intermediate $Rh(O_2CH)(dppb)$; the dihydride formed by H_2 oxidative addition was calculated to be lower in energy than the molecular hydrogen complex, but the latter was more reactive and therefore was the dominant pathway [65, 66]. The reaction between the bound H_2 ligand and formate was facilitated by a hydrogen-bonding interaction with external base (Scheme 17.6b). For some systems, the H atom from the H_2 ligand might transfer to the free carbonyl oxygen atom by a six-center transition state (Scheme 17.6c) rather than to the metal-bound oxygen via a four-center transition state (Scheme 17.6a). The six-membered transition state was shown, by calculation, to be preferred for the Ru intermediate $[RuH(O_2CH)(H_2)(PR_3)_3]^+$ [59].

Path B in Scheme 17.5 involves hydrolysis or alcoholysis of the formate complex, yielding formic acid and a hydroxide or alkoxide complex, which then undergoes hydrogenolysis. This pathway would explain the observations that water

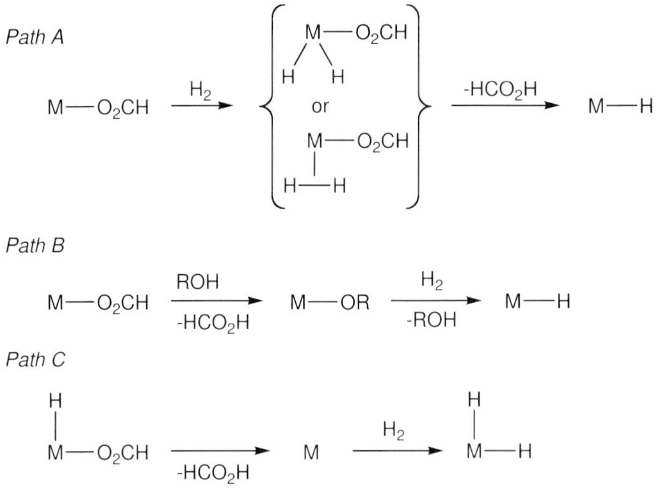

Scheme 17.5 Return pathways from the formate intermediate to the starting hydride.

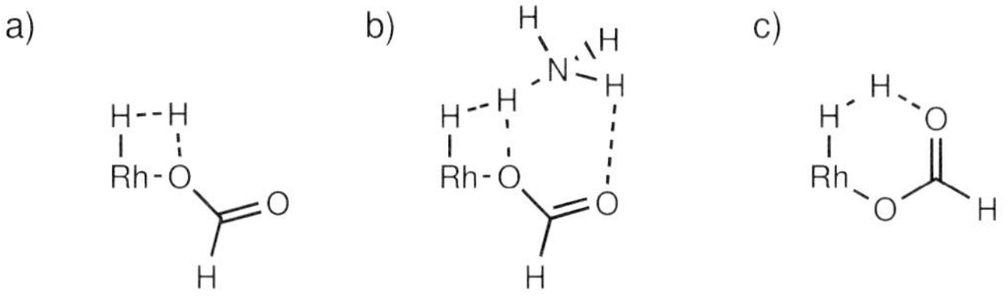

Scheme 17.6 Transition states for the transfer of a hydrogen atom from a molecular hydrogen ligand to a formate ligand: a) four-centered; b) four-centered and base-assisted [66]; and c) six-centered [59].

and alcohols increase the rate of reaction for many catalysts. Inoue et al. [18] proposed this mechanism for the reaction catalyzed by $Pd(dppe)_2$ for precisely this reason. Further, it has been observed for the $RuCl(OAc)(PMe_3)_4$ catalyst precursor that the most effective alcohols are those for which the conjugate bases are not sterically encumbered (and therefore are potentially coordinating) [6], although none of these arguments can be considered to be more than circumstantial.

Path C (Scheme 17.5), which involves reductive elimination of formic acid before reaction with H_2, is only possible for dihydride and polyhydride catalysts. This pathway was proposed for the catalyst precursor $[Rh(nbd)(PMe_2Ph)_3]^+$ by Tsai and Nicholas [19], who observed the intermediates $[RhH_2(PMe_2Ph)_3L]^+$ and $[RhH(\eta^2\text{-}O_2CH)(PMe_2Ph)_2L]^+$ ($L=H_2O$ or solvent).

17.2.2
Ionic Hydrogenation

Ionic hydrogenation mechanisms involve the sequential transfer of hydride and proton to the substrate [67]. This was suggested by the Leitner group for the hydrogenation of CO_2 with the catalyst precursor $RhH(dppp)_2$ (Scheme 17.7) [50]. Spectroscopic evidence for each of the three intermediates was obtained by studying the steps as stoichiometric reactions. However, catalyst precursors that generate the highly active RhH(diphosphine) species in solution were subsequently found to operate by a more conventional insertion mechanism [20].

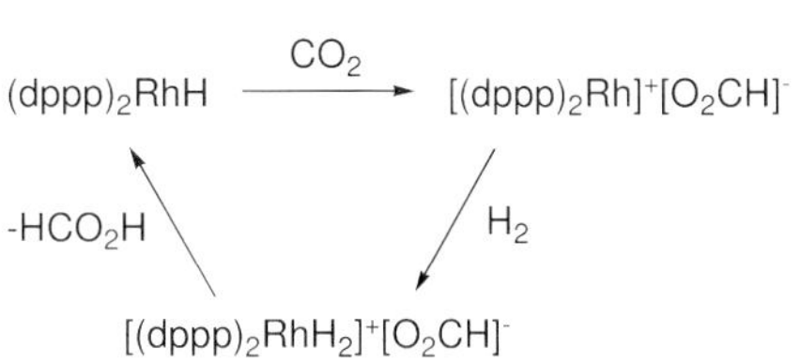

Scheme 17.7 An ionic hydrogenation mechanism for CO_2 hydrogenation [50].

17.2.3
Concerted Ionic Hydrogenation

The Noyori mechanism for the concerted ionic hydrogenation of ketones involves the simultaneous donation of a hydride ion from a metal hydride complex and a proton from an acidic ligand bound to the same metal (Scheme 17.8a). There is no reason why this mechanism should not also be possible for CO_2 hydrogenation (Scheme 17.8b) [6, 43, 68], although it has not yet been possible to demonstrate that this mechanism operates for any of the catalysts shown in Table 17.1. The mechanism does not require the ketone or CO_2 to bind to the metal or to insert into the M–H bond, as is required in more conventional mechanisms. Pomelli et al. [68] evaluated the possibility that bound formic acid could serve as the proton source in such a mechanism for RhH(diphosphine) catalyst, but found that the energy of the intermediates in the conventional insertion pathway were lower. The concerted mechanism for CO_2 hydrogenation was more recently suggested by Lau's group [43, 44] to explain the catalytic activity of the complex TpRuH(-MeCN)(PPh$_3$) in the presence of water, methanol and the acidic alcohol CF_3CH_2OH, the last of which offers the greatest rates of reaction. NMR spectroscopy detected TpRu(OCH$_2$CF$_3$)(PPh$_3$)(MeCN) and possibly TpRu(O$_2$-COCH$_2$CF$_3$)(PPh$_3$)(MeCN). The ionic hydrogenation mechanism does not necessarily require an inner-sphere alcohol ligand. The alcohol could merely be outer-sphere, bound by an unconventional hydrogen bond to the metal hydride (i.e., M–H$\cdots$HOR). Hybrid density functional calculations have shown that the transfer of H_2 from an unconventional hydrogen bond to CO_2 is energetically feasible [63].

17.2.4
Bicarbonate Hydrogenation

In basic aqueous solutions, CO_2 exists primarily in the bicarbonate or carbonate forms. Thus, the hydrogenation of either CO_2 or bicarbonate salts in such solutions could conceivably proceed by hydrogenation of the bicarbonate anion. Pd

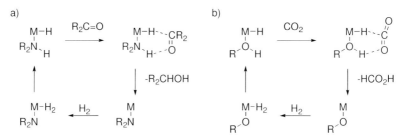

Scheme 17.8 Concerted ionic hydrogenation mechanisms for the hydrogenation of (a) ketones and (b) CO_2. The acidic ligand is shown as an alcohol in Scheme 17.7b, but could equally well be water, a secondary amine, or a carboxylic acid.

complexes are known [5, 18] to catalyze the hydrogenation of bicarbonates and carbonates, although the mechanism is unknown. This mechanism has been proposed for the catalyst precursors $PdCl_2/KOH$ [5, 69], $RuCl_2(PTA)_4$ [70], and $[RuCl_2(TPPMS)_2]_2$ [47]. The complex $RuCl_2(PTA)_4$ hydrogenates CO_2 rapidly in basic aqueous solutions but not in the absence of base, although this may be due to thermodynamics rather than evidence of a bicarbonate hydrogenation mechanism. Reaction of $RuCl_2(TPPMS)_2$ with $NaHCO_3$ gives $Ru(O_3CH)_2(TPPMS)_2$, which reacts stoichiometrically with H_2 to produce HCO_2^-.

It must keep be borne in mind that insertion and other pathways for CO_2 hydrogenation are not only possible but are believed to operate for some catalyst precursors, such as $RhCl(TPPTS)_3$, even in aqueous solutions [52].

17.2.5
Other Mechanisms

Hydrogenation of CO_2 to formic acid could potentially proceed first by reduction to CO, followed by a reaction between CO and water to give formic acid, a reaction which is known (Eq. (3)). It is unlikely that this pathway to formic acid is common because very few homogeneous catalysts (primarily homoleptic carbonyl complexes) [71–73] have been reported for the hydrogenation of CO_2 to CO, and because the few CO_2 hydrogenation catalysts that have deliberately been exposed to CO, in order to check whether this pathway is operating, have been poisoned as a result [18, 19, 31, 74].

$$CO_2 + H_2 \rightleftharpoons CO + H_2O \rightleftharpoons HCO_2H \qquad (3)$$

17.3
Reduction to Oxalic Acid

Hydrogenation of CO_2 could also generate oxalic acid (Eq. (4)); the electrochemical reduction of CO_2 to oxalate dianion has been well studied [75–77]. However, there have been almost no reports of oxalic acid or its esters being detected among the products of homogeneous CO_2 hydrogenation. Denise and Sneeden [78] reported the detection of traces of diethyloxalate (0.002 TON) in a similar reaction in ethanol solvent at 120 °C. The lack of other reports may not necessarily be evidence that oxalic acid is not formed; the conventional methods for detecting formic acid (especially ^{1}H-NMR spectroscopy) are not effective for oxalic acid. Therefore, oxalic acid – were it formed – would go undetected by many researchers studying the hydrogenation of CO_2 to formic acid. In order to determine whether oxalic acid is a co-product of formic acid production, a method was developed in the Jessop group which used dimethylsulfate to methylate the oxalate anion. The dimethyloxalate was then detectable by gas chromatography. By using this assay method, it was possible to show that catalysts having high activity for the hydrogenation of CO_2 to formate anion do not simultaneously produce detectable quantities of oxalate anion [79].

$$2CO_2 + H_2 \rightleftharpoons HO_2CCO_2H \qquad (4)$$

It may not be that surprising that an effective homogeneous catalyst for the reaction shown by Eq. (4) has not been found; it is difficult to imagine a facile mechanism by which oxalate anion or oxalic acid could be generated at a metal center.

17.4
Reduction to Formate Esters

17.4.1
In the Presence of Alcohols

Formate esters can be synthesized by the hydrogenation of CO_2 in the presence of alcohols in addition to the usual base (Eq. (5) and Table 17.2). The alcohol used is typically methanol or ethanol, giving methyl- or ethylformate; the yield decreases with increasing length of the alkyl chain [80, 81]. Most typically, the pathway is catalytic hydrogenation of CO_2 to formic acid followed by an uncatalyzed thermal esterification (Eq. (6)). The formic acid intermediate, which has been observed spectroscopically for the $RhCl(PPh_3)_3$ [80] and $RuCl_2(PMe_3)_4$ [82] systems, builds up to a concentration somewhat lower than 1:1 with the base. The thermal esterification is slow because the reaction is performed in the presence of excess base. Tests of thermal esterification under basic conditions have been performed separately; acetic acid is esterified, in 25% yield after 44 h at 100 °C, by an excess of methanol in the presence of 1 equiv. NEt_3 [31]. The yield of methyl formate from CO_2 is optimum at a temperature which is sufficiently high to allow substantial esterification, but not so high that the hydrogenation catalyst is deactivated. This optimum temperature is ~80 °C for the $RuCl_2(PMe_3)_4$ catalyst precursor [31, 82] and 100–125 °C for the $RhCl(PPh_3)_3$ [80]. Selectivity, on the other hand, is poor at the optimum temperature but greater at higher temperatures, where the yield is poor instead [31, 82]. Complete selectivity is not obtained in the presence of base, although Lodge and Smith [83] showed that the liquid products obtained after the reaction in the presence of basic Al_2O_3 contained only the ester and no formic acid; presumably the acid was captured by the solid base.

$$CO_2 + H_2 + ROH \rightleftharpoons HCO_2R + H_2O \qquad (5)$$

$$CO_2 + H_2 \xrightarrow{\text{catalyst}} HCO_2H \xrightarrow[\Delta]{ROH} HCO_2R + H_2O \qquad (6)$$

Formate ester synthesis by CO_2 hydrogenation in the absence of base has also been observed [45, 64, 74, 84]. As one might expect, the selectivity is much

Table 17.2 Homogeneous catalyst precursors for the hydrogenation of CO_2, in the presence of methanol or ethanol, to the corresponding formate esters.

Catalyst precursor	Base	$P_{H2, CO2}$ [a) [atm]	Temperature [°C]	Time [h]	TON	TOF [b) [h^{-1}]	Reference(s)
Ruthenium							
$Ru_4H_3(CO)_{12}^-$	None	17, 17	125	24	7	0.3	74
$RuHCl(PPh_3)_3/BF_3$	None	30, 30	100	na	17	na	86
$RuCl_2(PPh_3)_3$	Al_2O_3	60, 20	100	64	470	7.3	83
$RuCl_2(PMe_3)_4$	NEt_3	80, 125	80	64	3500	55	82
$[Ru(Cl_2bpy)_2(H_2O)_2][O_3SCF_3]_2$	NEt_3	30, 30	100	8	160	20	45
$^*RuCl_2(dppe)_2$	NEt_3	85, 130	100	16	12900	830	87
$TpRuH(PPh_3)(MeCN)$	NEt_3	25, 25	100	16	75	4.7	44
Rhodium							
$RhCl(PPh_3)_3$	NEt_3	25, 25	140	21	30	1.4	81
$RhCl(PPh_3)_3$	NaOMe	68, 29	140	21	27	1.3	55
$RhCl(PPh_3)_3$	TED	>200, 48	100	5	120	24	80
Palladium							
$Pd(dppe)_2$	NEt_3	25, 25	140	21	24	1.1	81
$Pd(dppm)_2$	NEt_3	70, 30	160	21	58	2.8	88
$Pd(dppm)_2$	NEt_3	15, 15	120	24	9	0.4	89, 90
$MnPdBr(CO_3)(dppm)_2$	NEt_3	6, 6	130	na	na	7	91
Other metals							
$Fe_3H(CO)_{11}^-$	None	20, 20	175	96	6	0.06	84
$W(O_2CH)(CO)_5^-$	None	17, 17	125	24	16	0.7	64
$IrH_3(PPh_3)_3/BF_3$	None	30, 30	100	na	38	na	86

a) In some cases, the pressure of CO_2 was not given and was calculated from the total stated pressure minus the pressure of H_2.

b) The TOF values are not directly comparable to each other because some are at complete conversion and some are at partial conversion. They can, however, give an order of magnitude indication. Initial TOF values will be even higher.

na = data not available.

greater by this method because there is no base to interfere with the esterification or to stabilize the formic acid intermediate. Although there have not yet been reports of very high TON by this method, the value of 160 obtained by Lau and Chen using the $[Ru(Cl_2bpy)_2(H_2O)_2][O_3SCF_3]_2$ precursor is the most impressive [45].

Other mechanisms for the synthesis of alkylformates, not via formic acid esterification, are possible. Hydrogenation of CO_2 to CO, followed by catalytic carbonylation of alcohol, would produce alkyl formate. This mechanism seems more likely for the anionic metal carbonyls because they are known catalysts for alcohol carbonylation. However, Darensbourg and colleagues [64, 74, 85] showed that $[W(O_2CH)(^{13}CO)_5]^-$ gives $H^{12}CO_2H$ as a product, and they proposed that formic acid was again the intermediate for the $[MH(CO)_5]^-$ (M=Cr, W) system, even though formic acid was not directly observed in the presence of alcohol. Another possible mechanism is the hydrogenation of alkyl carbonate anion formed by the reaction of alcohol, amine, and CO_2, although to date there are no known examples of this mechanism. Finally, alcoholysis of a metal formate complex to give a metal hydroxide and alkyl formate (Eq. (7), in contrast to path B in Scheme 17.5), was proposed by Kolomnikov et al. [86]. However, Darensbourg showed that this mechanism could be ruled out for $[M(O_2CH)(CO)_5]^-$ because that complex does not react with alcohol [64].

$$MO_2CH + ROH \rightleftharpoons MOH + RO_2CH \qquad (7)$$

17.4.2
In the Presence of Alkyl Halides

Given the lack of reactivity of higher alcohols in this reaction, alternative means of producing alkylformates have been explored using more reactive reagents instead of alcohols. Alkyl halides have been known, since Kolomnikov's 1972 study [86], to produce alkyl formates when they are present during CO_2 hydrogenation (Eq. (8)). That study described the production of methyl formate from methyl iodide in very low yield and only 5 TON using Ru, Ir, and Os phosphine complexes. Better yields (up to 64% but only 15 TON) were obtained by Darensbourg and Ovalles [92] using $[Cr_2H(CO)_{10}]^-$ or $[WCl(CO)_5]^-$ catalyst precursors and NaO_2COH or NaOMe as base. The base was necessary to trap the HX acid. The proposed mechanism is shown in Scheme 17.9. The last step of the mechanism, the reaction of bound RCl with a formate ligand, was investigated kinetically and is proposed to proceed by oxidative addition of the RCl followed by reductive elimination of the formate ester. Alkyl chlorides were found to be more reactive than the bromides or iodides because the heavier halides led to excessive stability of the catalytic intermediate $[MX(CO)_5]^-$. Bulky alkyl halides react

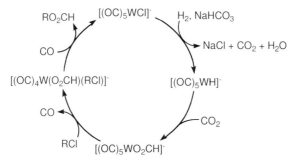

Scheme 17.9 Proposed mechanism for the preparation of formate esters using alkyl halides [92].

slowly, or not at all. Yields would have been better had there not been production of alcohol by formate ester hydrolysis and production of alkane by a side reaction (Eq. (9)).

$$H_2 + CO_2 + RX \rightleftharpoons RO_2CH + HX \tag{8}$$

$$[MH(CO)_5]^- + RX \rightleftharpoons [MX(CO)_5]^- + RH \tag{9}$$

17.4.3
In the Presence of Epoxides

Hydrogenation of carbon dioxide in the presence of an epoxide generates a mixture of the diol, its formate esters, and the cyclic carbonate. While the reaction has been shown to operate in high yield (1300 TON for the cyclic carbonate; Eq. (10)) [93], the fact that a mixture is generated and that the cyclic carbonate could be made more cleanly in the absence of H_2 makes the reaction uninteresting for synthesis. Sasaki's group showed that this reaction in the presence of an amine base gives CO rather than cyclic carbonate (Eq. (11)) [94]. The epoxide then serves as a trap for the water.

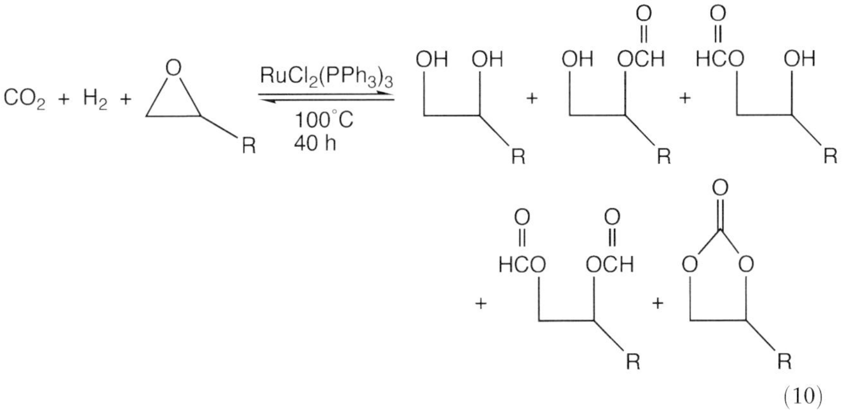

(10)

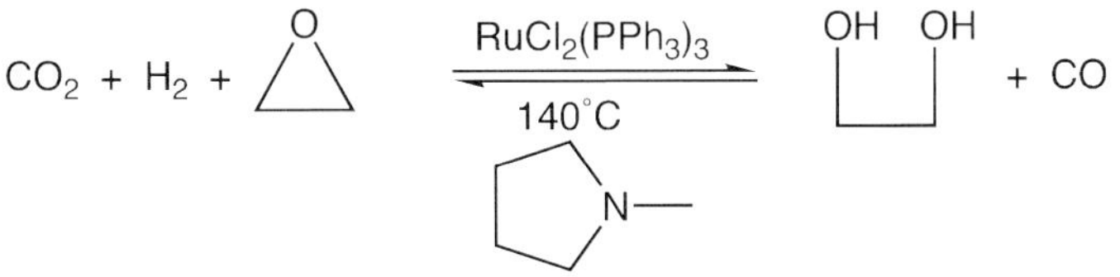

(11)

17.5
Reduction to Formamides

Formamides have applications as solvents (particularly dimethylformamide (DMF) and formamide itself), as intermediates for insecticides and pharmaceuticals, and (for formanilide, PhNHC(O)H) as an antioxidant [95]. Formamides are prepared either from CO or from methylformate or formamide, which are themselves prepared from CO.

Preparation of formamides from CO_2 and a non-tertiary amine by homogeneous hydrogenation has been well studied and is extremely efficient (Eq. (12)). Essentially complete conversions and complete selectivity can be obtained (Table 17.3). This process seems more likely to be industrialized than the syntheses of formic acid or formate esters by CO_2 hydrogenation. The selectivity is excellent, in contrast to the case for alkyl formates, because the amine base which would stabilize the formic acid is used up in the synthesis of the formamide; consequently little or no formic acid contaminates the product. The only byproducts that are likely to crop up in industrial application are the methylamines by overreduction of the formamide. This has been observed [96], but not with such high conversion that it could constitute a synthetic route to methylamines.

$$CO_2 + H_2 + NHR_2 \rightleftharpoons HC(O)NR_2 + H_2O \tag{12}$$

The formamide that has most often been synthesized in studies of this reaction is DMF. The first report of the reaction was by Haynes et al. in 1970 [3]. By far the most active catalyst precursors are the Ru(II) complexes $RuCl_2(PMe_3)_4$ [31, 97] and $RuCl_2(dppe)_2$ [87], although $RuCl_3$ and dppe or PPh_3 can be combined to make an active catalyst [98]. The mechanism, at least for $RuCl_2(PMe_3)_4$, is believed to be hydrogenation of CO_2 to formic acid by the usual insertion mechanism followed by uncatalyzed thermal condensation between the formic acid and the dimethylamine (Eq. (13)). The formic acid/formate salt intermediate is observed spectroscopically [31, 97]. Fortunately, the condensation step seems to be essentially irreversible.

$$CO_2 + H_2 + NHR_2 \xrightleftharpoons{\text{catalyst}} [NH_2R_2][O_2CH] \longrightarrow HC(O)NR_2 + H_2O \tag{13}$$

Three alternative mechanisms have been mentioned in the literature. Reduction of CO_2 to CO followed by carbonylation of dimethylamine was ruled out by Haynes et al. [3] for $RhCl(PPh_3)_3$ because no carbonyl complexes were detected. Aminolysis of formate complexes (Eq. (14)) was proposed by Kudo et al. [69], but strong evidence has not been obtained. Finally, CO_2 is known to react with the amine to produce a carbamate salt (Eq. (15)), and it is possible that the pathway to the formamide is by hydrogenation of the carbamate rather than of the CO_2.

$$MO_2CH + NHR_2 \rightarrow MOH + HC(O)NR_2 \qquad (14)$$

$$CO_2 + 2NHR_2 \rightleftharpoons [NH_2R_2][O_2CNR_2] \qquad (15)$$

For longer-chain or more bulky amines, yields and conversions are often significantly lower, possibly because the carbamate formed (Eq. (15)) is solid [99], but some successful syntheses have been reported. Süss-Fink et al. were able to prepare the formamides of piperidine and pyrrolidine at 140 °C using

Table 17.3 Homogeneous catalyst precursors for the hydrogenation of CO_2, in the presence of dimethylamine, to dimethylformamide.

Catalyst precursor	$P_{H2, CO2}$[a] [atm]	Tempera- ture [°C]	Time [h]	TON	TOF[b] [h^{-1}]	Yield[c] [%]	Reference(s)
Ruthenium							
RuCl$_3$/dppe/AlEt$_3$	29, 29	130	6	3400	570	73	105
RuCl$_2$(PPh$_3$)$_3$	29, 29	130	6	2650	440	51	105
RuCl$_2$(PMe$_3$)$_4$	80, 130	100	70	420 000	6000	71	31, 97
RuCl$_2$(PMe$_3$)$_4$	80, 130	100	19	62 000	3000	99	31, 97
RuCl$_2$(dppe)$_2$	85, 130	100	2	740 000	360 000	18	87
RuCl$_2$PH(CH$_2$OH)$_{22}$P(CH$_2$OH)$_{32}$	85, 133	100	48	10 000	210	100	106
Rhodium							
RhCl(PPh$_3$)$_3$	28, 28	100	17	43	2.5	11	3
RhCl(PPh$_3$)$_3$	80, 40	150	5	36	7	65	107
Palladium							
Pd(CO$_3$)(PPh$_3$)$_2$	28, 28	100	17	120	7	60	3
PdCl$_2$/KHCO$_3$	80, 40	170	1.5	34	23	99	69
Other metals							
CdCl$_2$(PPh$_3$)$_2$	28, 28	125	17	10.5	0.6	4	4
Pt(CO$_3$)(PPh$_3$)$_2$	28, 28	100	17	104	6	47	3
Pt$_2$(dppm)$_3$	67–94, 10–12	75	24	1460	61	na	96
Pt$_2$(dppm)$_3$	114 total	100	24	1375	57	na	108
IrCl(CO)(PPh$_3$)$_2$[d]	50–68, 13–17	125	240	1145	5	na	109

a) In some cases, the pressure of CO_2 was not given and was calculated from the total stated pressure minus the pressure of H_2.
b) The TOF values are not directly comparable to each other because some are at complete conversion and some are at partial conversion. They can, however, give an order of magnitude indication. Initial TOF values will be even higher.
c) Data in this column may be either isolated yield or % conversion, depending on data available.
d) Using NH$_3$ to produce formamide.
na = data not available.

Na[Ru₃H(CO)₁₁] catalyst precursor [100]. Tumas' group [101] obtained 100% conversion to di(n-propyl)formamide from di(n-propyl)amine by performing the reaction in an ionic liquid at 80 °C. Jessop's group [102] found that adding C_6F_5OH to the $RuCl_2(PMe_3)_4$ catalyst precursor allows one to obtain 96–100% conversion to a variety of bulky formamides, including those from heptylamine, piperidine, and benzylamine, at 100 °C (3 μmol Ru, 100 μmol C_6F_5OH, 5 mmol amine). For less basic amines such as aniline [103], no formamide is obtained unless an additional base is added such as DBU instead of, and not in addition to, the C_6F_5OH. Baiker's group [99, 104] used $RuCl_2(dppe)_2$ catalyst precursor and water as a promoter to prepare a range of formamides from aliphatic amines in very high TON at 100 °C.

17.6
Reduction to Other Products

Reduction of CO_2 past formic acid generates formaldehyde, methanol or methane (Eqs. (16–18)), and ethanol can be produced by homologation of the methanol. The liberation of water makes these reactions thermodynamically favorable but economically less favorable. The reductions typically require much higher temperatures than does the reduction to formic acid, and consequently few homogeneous catalysts are both kinetically capable and able to withstand the operating conditions.

$$CO_2 + H_2 \rightarrow HC(O)H + H_2O \tag{16}$$

$$CO_2 + 3H_2 \rightarrow CH_3OH + H_2O \tag{17}$$

$$CO_2 + 4H_2 \rightarrow CH_4 + 2H_2O \tag{18}$$

The hydrogenation of CO_2 to CO is the reverse water-gas shift reaction, which has been reviewed elsewhere [110, 111].

Hydrogenation to both formaldehyde and formic acid is catalyzed by K[RuCl(EDTA-H)] [21, 112, 113], but the proposed mechanism involved a highly unlikely reverse insertion of CO_2 (Eq. (19)).

$$MH + CO_2 \rightarrow MCO_2H \tag{19}$$

Methane production by CO_2 hydrogenation has been catalyzed by [PdCl(dppm)]₂ (1.5 TON at 120 °C) [78, 89] and $Ru_3(CO)_{12}/I_2$ (76 TON at 240 °C) [72, 73, 114]. The inclusion of KI in the latter system shifted the selectivity over to methanol production (95 TON) by preventing deposition of Ru metal which would have catalyzed methane production. The mechanism was believed [114] to be reduction of CO_2 to CO followed by hydrogenation of the CO to CH_3OH via Dombek's [115] mechanism.

Ethanol production by homologation of methanol can be achieved by hydrogenation of CO_2 (Eq. (20)) [116]. Sasaki's group showed that both Ru and Co carbonyl catalysts are required, as is an iodide salt; in the absence of iodide, methyl formate is obtained. The high-boiling solvent 1,3-dimethyl-2-imidazolidinone (DMI) was required for the reaction, which was typically performed at 180 °C. In a separate study [71], the same group showed that hydrogenation of CO_2 to methanol by these same catalysts also produces ethanol (up to TON 17, TOF $1 h^{-1}$) because of the *in-situ* homologation of the methanol product.

$$CH_3OH + CO_2 + 3H_2 \xrightarrow[\text{LiI, DMI}]{\substack{Ru_3(CO)_{12} \\ Co_2(CO)_8}} CH_3CH_2OH + 2H_2O \tag{20}$$

Carboxylic acids have been prepared by CO_2 hydrogenation in the presence of alkyl iodides and catalyzed by the $Ru_3(CO)_{12}/Co_2(CO)_8$ combination (Eq. (21)). Although the yield was not high (2.5 TON, mol acetic acid per mol Co), the reaction is an interesting variation [117].

$$CH_3I + CO_2 + H_2 \xrightarrow[\text{150 pC}]{\substack{Ru_3(CO)_{12} \\ Co_2(CO)_8}} CH_3CO_2H + HI \tag{21}$$

17.7
Concluding Remarks

Highly efficient catalysts have been developed for the hydrogenation of CO_2 to formic acid and formamides, to the point where industrialization could be considered. Researchers have been far less successful in developing efficient homogeneous catalysts and optimum conditions for the hydrogenation of CO_2 to alkyl formates, methanol, methane, and especially oxalic acid. These are the areas in which research efforts are most needed.

Great progress has been made in the understanding of the insertion mechanism of CO_2 hydrogenation to formic acid. However, the evidence for other mechanisms remains circumstantial. Strong evidence for exciting new mechanisms, such as concerted ionic hydrogenation, is unavailable. Because research in this area has progressed far more rapidly during the past decade than in earlier decades, one can anticipate many more new results in the near future and look forward to a greater clarification on the subject of mechanism.

Acknowledgments

The author acknowledges support from the Division of Chemical Sciences, Office of Basic Energy Sciences, Office of Science, U.S. Department of Energy (grant number DE-FG03-99ER14986) and the support of the Canada Research Chair program (http://www.chairs.gc.ca).

Abbreviations

Cl_2bpy	6,6'-dichloro-2,2'-bipyridine
COD	1,5-cyclooctadiene
CYPO	$Cy_2PCH_2CH_2OCH_3$
DBU	1,8-diazabicyclo[5.4.0]undec-7-ene
dcpb	1,4-bis(dicyclohexylphosphino)butane
dcpe	1,2-bis(dicyclohexylphosphino)ethane
DFT	density functional theory
DHphen	4,7-dihydroxy-1,10-phenanthroline
dippe	1,2-bis(diisopropylphosphino)ethane
DMF	dimethylformamide
DMI	1,3-dimethyl-2-imidazolidinone
DMSO	dimethylsulfoxide
dppb	1,4-bis(diphenylphosphino)butane
dppe	1,2-bis(diphenylphosphino)ethane
dppp	1,3-bis(diphenylphosphino)propane
dppm	1,1-bis(diphenylphosphino)methane
EDTA-H	protonated ethylenediaminetetraacetic acid
hfacac	1,1,1,5,5,5-hexafluoroacetylacetonate
NBD	norbornadiene
PTA	1,3,5-triaza-7-phosphaadamantane
$scCO_2$	supercritical CO_2
TED	triethylenediamine
THF	tetrahydrofuran
TOF	turnover frequency (TON per h)
TON	turnover number (mol product per mol catalyst)
Tp	tris(pyrazolyl)borate
TPPMS	sodium triphenylphosphine monosulfonate
TPPTS	sodium triphenylphosphine trisulfonate
WGSR	water-gas shift reaction

References

1 J. Haggin, *Chem. Eng. News* **1994**, 29.

2 H. Goehna, P. Koenig, *ChemTech* **1994**, 36.

3 P. Haynes, L. H. Slaugh, J. F. Kohnle, *Tetrahedron Lett.* **1970**, 365.

4 P. Haynes, J. F. Kohnle, L. H. Slaugh, U.S. Patent 3 530 182, **1970**.

5 K. Kudo, N. Sugita, Y. Takezaki, *Nihon Kagaku Kaishi* **1977**, 302.

6 P. Munshi, A. D. Main, J. Linehan, C. C. Tai, P. G. Jessop, *J. Am. Chem. Soc.* **2002**, *124*, 7963.

7 B. P. Sullivan, K. Krist, H. E. Guard (Eds.), *Electrochemical and Electrocatalytic Reactions of Carbon Dioxide*. Elsevier, Amsterdam, **1993**.

8 A. Miedaner, C. J. Curtis, R. M. Barkley, D. L. DuBois, *Inorg. Chem.* **1994**, *33*, 5482.

9 I. Bhugun, D. Lexa, J.-M. Savéant, *J. Am. Chem. Soc.* **1994**, *116*, 5015.

10 I. Bhugun, D. Lexa, J.-M. Savéant, *J. Am. Chem. Soc.* **1996**, *118*, 1769.

11 H. Hayashi, S. Ogo, T. Abura, S. Fuku-
zumi, *J. Am. Chem. Soc.* **2003**, *125*,
14266.

12 P. G. Jessop, T. Ikariya, R. Noyori, *Chem.
Rev.* **1995**, *95*, 259.

13 W. Leitner, *Angew. Chem., Int. Ed. Engl.*
1995, *34*, 2207.

14 P. G. Jessop, F. Joo, C.-C. Tai, *Coord.
Chem. Rev.* **2004**, *248*, 2425.

15 E. B. Reid. In: *McGraw-Hill Encyclopedia of
Science and Technology*, Volume 5.
McGraw-Hill Book Co., New York, **1982**,
p. 670.

16 W. Reutemann, H. Kieczka. In: B. Elvers, S. Hawkins, M. Ravenscroft, J. F.
Rounsaville, G. Schulz (Eds.), *Ullmann's
Encyclopedia of Industrial Chemistry*, Volume A12, 5th edn. VCH, Weinheim,
1989, p. 13.

17 B. Jezowska-Trzebiatowska, P. Sobota,
J. Organomet. Chem. **1974**, *80*, C27.

18 Y. Inoue, H. Izumida, Y. Sasaki, H. Hashimoto, *Chem. Lett.* **1976**, 863.

19 J.-C. Tsai, K. M. Nicholas, *J. Am. Chem.
Soc.* **1992**, *114*, 5117.

20 W. Leitner, E. Dinjus, F. Gaßner, *J. Organomet. Chem.* **1994**, *475*, 257.

21 M. M. T. Khan, S. B. Halligudi, S. Shukla,
J. Mol. Catal. **1989**, *57*, 47.

22 F. Gassner, W. Leitner, *J. Chem. Soc.,
Chem. Commun.* **1993**, 1465.

23 D. J. Heldebrant, P. G. Jessop, C. A. Thomas, C. A. Eckert, C. L. Liotta, *J. Org.
Chem.* **2005**, *70*, 5335.

24 D. J. Drury, J. E. Hamlin, European Patent Application 0 095 321, **1983**.

25 E. Graf, W. Leitner, *J. Chem. Soc., Chem.
Commun.* **1992**, 623.

26 M. Sakamoto, I. Shimizu, A. Yamamoto,
Organometallics **1994**, *13*, 407.

27 J. J. Anderson, J. E. Hamlin, European
Patent Appl. 0 126 524, **1984**.

28 J. J. Anderson, D. J. Drury, J. E. Hamlin,
A. G. Kent, European Patent Application
0 181 078, **1985**.

29 E. A. Karakhanov, S. V. Egazar'yants, S. V.
Kardashev, A. L. Maksimov, S. S. Minos'-
yants, A. D. Sedykh, *Petroleum Chem.*
2001, *41*, 268.

30 P. G. Jessop, T. Ikariya, R. Noyori, *Nature*
1994, *368*, 231.

31 P. G. Jessop, Y. Hsiao, T. Ikariya, R.
Noyori, *J. Am. Chem. Soc.* **1996**, *118*, 344.

32 F. Kohler, H. Atrops, H. Kalali, E. Liebermann, E. Wilhelm, F. Ratkovics, T. Sala-
mon, *J. Phys. Chem.* **1981**, *85*, 2520.

33 K. Wagner, *Angew. Chem., Int. Ed. Engl.*
1970, *9*, 50.

34 F. Kohler, R. Gopal, G. Goetze, H.
Atrops, M. A. Demeriz, E. Liebermann,
E. Wilhelm, F. Ratkovics, B. Palagyi,
J. Phys. Chem. **1981**, *85*, 2524.

35 T. Ikariya, P. G. Jessop, R. Noyori, Japan
Tokkai 5-274721, **1993**.

36 C. A. Thomas, R. J. Bonilla, Y. Huang,
P. G. Jessop, *Can. J. Chem.* **2001**, *79*, 719.

37 N. N. Ezhova, N. V. Kolesnichenko, A. V.
Bulygin, E. V. Slivinskii, S. Han, *Russ.
Chem. Bull., Int. Ed.* **2002**, *51*, 2165.

38 The TOF data from the table were corrected to 20 °C by assuming that every
10° rise doubled the rate. The TOF data
were corrected to 1 atm each of H_2 and
CO_2 by assuming that the rates were
first order with respect to both H_2 and
CO_2, but switched to 0 order above
80 atm due to saturation kinetics. The
same sequence was obtained if saturation kinetics were not assumed.

39 Y. Musashi, S. Sakaki, *J. Am. Chem. Soc.*
2002, *124*, 7588.

40 Y. Gao, J. K. Kuncheria, H. A. Jenkins,
R. J. Puddephatt, G. P. A. Yap, *J. Chem.
Soc., Dalton Trans.* **2000**, 3212.

41 J. Z. Zhang, Z. Li, H. Wang, C. Y. Wang,
J. Mol. Catal. A: Chem. **1996**, *112*, 9.

42 T. Yamaji, Japan Kokai Tokkyo Koho
140 948, **1981**.

43 C. Yin, Z. Xu, S.-Y. Yang, S. M. Ng, K. Y.
Wong, Z. Lin, C. P. Lau, *Organometallics*
2001, *20*, 1216.

44 S. M. Ng, C. Yin, C. H. Yeung, T. C. Chan,
C. P. Lau, *Eur. J. Inorg. Chem.* **2004**, 1788.

45 C. P. Lau, Y. Z. Chen, *J. Mol. Catal. A:
Chem.* **1995**, *101*, 33.

46 H. S. Chu, C. P. Lau, K. Y. Wong, *Organometallics* **1998**, *17*, 2768.

47 J. Elek, L. Nádasdi, G. Papp, G. Laurenczy, F. Joó, *Appl. Catal. A: General* **2003**,
255, 59.

48 Y. Himeda, N. Onozawa-Komatsuzaki,
H. Sugihara, H. Arakawa, K. Kasuga. In:
*50th Symposium on Organometallic Chem-
istry*, Osaka, Japan, **2003**.

49 M. L. Man, Z. Zhou, S. M. Ng, C. P. Lau,
J. Chem. Soc., Dalton Trans. **2003**, 3727.

50 T. Burgemeister, F. Kastner, W. Leitner, *Angew. Chem., Int. Ed. Engl.* **1993**, *32*, 739.

51 T. Yamaji, Teijin Ltd., Japan Kokai Tokkyo Koho, 166.146, **1981**.

52 W. Leitner, E. Dinjus, F. Gassner. In: B. Cornils, W.A. Herrmann (Eds.), *Aqueous-Phase Organometallic Catalysis.* Wiley-VCH, Weinheim, **1998**, p. 486.

53 E. Lindner, B. Keppeler, P. Wegner, *Inorg. Chim. Acta* **1997**, *258*, 97.

54 R. Fornika, H. Görls, B. Seemann, W. Leitner, *J. Chem. Soc., Chem. Commun.* **1995**, 1479.

55 H. Hashimoto, S. Inoue, Japanese Patent 07.612, **1978**.

56 C.C. Tai, T. Chang, B. Roller, P.G. Jessop, *Inorg. Chem.* **2003**, *42*, 7340.

57 T. Ito, A. Yamamoto. In: S. Inoue, N. Yamazaki (Eds.), *Organic and Bioorganic Chemistry of Carbon Dioxide.* Kodansha Ltd., Tokyo, **1982**, p. 79.

58 A. Behr. In: W. Keim (Ed.), *Catalysis in C1 Chemistry.* D. Reidel Publishing Co., Dordrecht, **1983**, p. 169.

59 Y. Musashi, S. Sakaki, *J. Am. Chem. Soc.* **2000**, *122*, 3867.

60 C. Bo, A. Dedieu, *Inorg. Chem.* **1989**, *28*, 304.

61 S. Sakaki, K. Ohkubo, *Inorg. Chem.* **1989**, *28*, 2583.

62 T. Matsubara, K. Hirao, *Organometallics* **2001**, *20*, 5759.

63 T. Matsubara, *Organometallics* **2001**, *20*, 19.

64 D.J. Darensbourg, C. Ovalles, *J. Am. Chem. Soc.* **1984**, *106*, 3750.

65 F. Hutschka, A. Dedieu, W. Leitner, *Angew. Chem., Int. Ed. Engl.* **1995**, *34*, 1742.

66 F. Hutschka, A. Dedieu, M. Eichberger, R. Fornika, W. Leitner, *J. Am. Chem. Soc.* **1997**, *119*, 4432.

67 D.N. Kursanov, Z.N. Parnes, M.I. Kalinkin, N. Loim, *Ionic Hydrogenation and Related Reactions*, Harwood Academic Publishers, New York, **1985**.

68 C.S. Pomelli, J. Tomasi, M. Sola, *Organometallics* **1998**, *17*, 3164.

69 K. Kudo, H. Phala, N. Sugita, Y. Takezaki, *Chem. Lett.* **1977**, 1495.

70 G. Laurenczy, F. Joó, L. Nádasdi, *Inorg. Chem.* **2000**, *39*, 5083.

71 K.-I. Tominaga, Y. Sasaki, M. Saito, K. Hagihara, T. Watanabe, *J. Mol. Catal.* **1994**, *89*, 51.

72 K. Tominaga, Y. Sasaki, M. Kawai, T. Watanabe, M. Saito, *J. Chem. Soc., Chem. Commun.* **1993**, 629.

73 K. Tominaga, Y. Sasaki, K. Hagihara, T. Watanabe, M. Saito, *Chem. Lett.* **1994**, 1391.

74 D.J. Darensbourg, C. Ovalles, M. Pala, *J. Am. Chem. Soc.* **1983**, *105*, 5937.

75 L. Skarlos, U.S. Patent 3720591, **1973**.

76 F.R. Keene, B.P. Sullivan. In: B.P. Sullivan (Ed.), *Electrochemical and Electrocatalytic Reactions of Carbon Dioxide.* Elsevier, Amsterdam, **1993**, p. 118.

77 M.M. Ali, H. Sato, T. Mizukawa, K. Tsuge, M. Haga, K. Tanaka, *Chem. Commun.* **1998**, 249.

78 B. Denise, R.P.A. Sneeden, *J. Organomet. Chem.* **1981**, *221*, 111.

79 C.-C. Tai, P.G. Jessop, unpublished material, **2002**.

80 H. Phala, K. Kudo, S. Mori, N. Sugita, *Bull. Inst. Chem. Res., Kyoto Univ.* **1985**, *63*, 63.

81 Y. Inoue, Y. Sasaki, H. Hashimoto, *J. Chem. Soc., Chem. Commun.* **1975**, 718.

82 P.G. Jessop, Y. Hsiao, T. Ikariya, R. Noyori, *J. Chem. Soc., Chem. Commun.* **1995**, 707.

83 P.G. Lodge, D.J.H. Smith, European Patent Application 0 094 785, **1983**.

84 G.O. Evans, C.J. Newell, *Inorg. Chim. Acta* **1978**, *31*, L387.

85 D.J. Darensbourg, C. Ovalles, *CHEMTECH* **1985**, *15*, 636.

86 I.S. Kolomnikov, T.S. Lobeeva, M.E. Voľpin, *Izv. Akad. Nauk SSSR, Ser. Khim.* **1972**, 2329.

87 O. Kröcher, R.A. Köppel, A. Baiker, *Chem. Commun.* **1997**, 453.

88 H. Hashimoto, S. Inoue, Japanese Patent 138614, **1976**.

89 B. Denise, R.P.A. Sneeden, *CHEMTECH* **1982**, February, 108.

90 B. Beguin, B. Denise, R.P.A. Sneeden, *J. Organomet. Chem.* **1981**, *208*, C18.

91 B.F. Hoskins, R.J. Steen, T.W. Turney, *Inorg. Chim. Acta* **1983**, *77*, L69.

92 D.J. Darensbourg, C. Ovalles, *J. Am. Chem. Soc.* **1987**, *109*, 3330.

93 H. Koinuma, H. Kato, H. Hirai, *Chem. Lett.* **1977**, 517.

94 K.-I. Tominaga, Y. Sasaki, T. Watanabe, M. Saito, *J. Chem. Soc., Chem. Commun.* **1995**, 1489.

95 H. Bipp, H. Kieczka. In: B. Elvers, S. Hawkins, M. Ravenscroft, J.F. Rounsaville, G. Schulz (Eds.), *Ullmann's Encyclopedia of Industrial Chemistry*, Volume A12, 5th edn. VCH, Weinheim, **1989**, p. 1.

96 S. Schreiner, J.Y. Yu, L. Vaska, *Inorg. Chim. Acta* **1988**, *147*, 139.

97 P.G. Jessop, Y. Hsiao, T. Ikariya, R. Noyori, *J. Am. Chem. Soc.* **1994**, *116*, 8851.

98 M. Rohr, J.-D. Grunwaldt, A. Baiker, *J. Mol. Catal. A: Chem.* **2005**, *226*, 253.

99 L. Schmid, M.S. Schneider, D. Engel, A. Baiker, *Catal. Lett.* **2003**, *88*, 105.

100 G. Süss-Fink, M. Langenbahn, T. Jenke, *J. Organomet. Chem.* **1989**, *368*, 103.

101 F.C. Liu, M.B. Abrams, R.T. Baker, W. Tumas, *Chem. Commun.* **2001**, 433.

102 P. Munshi, P.G. Jessop, E. McKoon, unpublished material, **2002**.

103 P. Munshi, D. Heldebrant, E. McKoon, P.A. Kelly, C.-C. Tai, P.G. Jessop, *Tetrahedron Lett.* **2003**, *44*, 2725.

104 L. Schmid, A. Canonica, A. Baiker, *Applied Catalysis A: General* **2003**, *255*, 23.

105 Y. Kiso, K. Saeki, Vol. 22 March, Japan Kokai 36617, **1977**, p. 9.

106 Y. Kayaki, T. Suzuki, T. Ikariya, *Chem. Lett.* **2001**, 1016.

107 H. Phala, K. Kudo, N. Sugita, *Bull. Inst. Chem. Res., Kyoto Univ.* **1981**, *59*, 88.

108 S. Schreiner, J.Y. Yu, L. Vaska, *J. Chem. Soc., Chem. Commun.* **1988**, 602.

109 L. Vaska, S. Schreiner, R.A. Felty, J.Y. Yu, *J. Mol. Catal.* **1989**, *52*, L11.

110 P. Ford. In: B.P. Sullivan, K. Krist, H.E. Guard (Eds.), *Electrochemical and electrocatalytic reactions of carbon dioxide.* Elsevier, Amsterdam, **1993**, Chapter 3.

111 M. Torrent, M. Sola, G. Frenking, *Chem. Rev.* **2000**, *100*, 439.

112 M.M.T. Khan, S.B. Halligudi, S. Shukla, *J. Mol. Catal.* **1989**, *53*, 305.

113 M.M.T. Khan, S.B. Halligudi, N.N. Rao, S. Shukla, *J. Mol. Catal.* **1989**, *51*, 161.

114 K. Tominaga, Y. Sasaki, T. Watanabe, M. Saito, *Bull. Chem. Soc. Jpn.* **1995**, *68*, 2837.

115 B.D. Dombek, *J. Organometal. Chem.* **1983**, *250*, 467.

116 K.-I. Tominaga, Y. Sasaki, T. Watanabe, M. Saito. In: T. Inui, M. Anpo, K. Izui, S. Yanagida, T. Yamaguchi (Eds.), *Advances in Chemical Conversions for Mitigating Carbon Dioxide.* Elsevier, Amsterdam, **1998**, p. 495.

117 A. Fukuoka, N. Gotoh, N. Kobayashi, M. Hirano, S. Komiya, *Chem. Lett.* **1995**, 567.